STRESS: CONCEPTS, COGNITION, EMOTION, AND BEHAVIOR

STRESS: CONCEPTS, COGNITION, EMOTION, AND BEHAVIOR

Handbook of Stress, Volume 1

Edited by

George Fink
Florey Institute of Neuroscience and Mental Health,
University of Melbourne,
Parkville, Victoria, Australia

AMSTERDAM • BOSTON • HEIDELBERG • LONDON
NEW YORK • OXFORD • PARIS • SAN DIEGO
SAN FRANCISCO • SINGAPORE • SYDNEY • TOKYO
Academic Press is an imprint of Elsevier

Academic Press is an imprint of Elsevier
125 London Wall, London, EC2Y 5AS, UK
525 B Street, Suite 1800, San Diego, CA 92101-4495, USA
50 Hampshire Street, 5th Floor, Cambridge, MA 02139, USA
The Boulevard, Langford Lane, Kidlington, Oxford OX5 1GB, UK

Cover figure
Figure 9a from: Lucassen PJ, Pruessner J, Sousa N, Almeida OF, Van Dam AM, Rajkowska G, Swaab DF, Czéh B. Neuropathology of stress. Acta Neuropathol. 2014 Jan;127(1):109–35. doi: 10.1007/s00401-013-1223-5. Open access.

Notices
Knowledge and best practice in this field are constantly changing. As new research and experience broaden our understanding, changes in research methods, professional practices, or medical treatment may become necessary.

Practitioners and researchers must always rely on their own experience and knowledge in evaluating and using any information, methods, compounds, or experiments described herein. In using such information or methods they should be mindful of their own safety and the safety of others, including parties for whom they have a professional responsibility.

To the fullest extent of the law, neither the Publisher nor the authors, contributors, or editors, assume any liability for any injury and/or damage to persons or property as a matter of products liability, negligence or otherwise, or from any use or operation of any methods, products, instructions, or ideas contained in the material herein.

ISBN: 978-0-12-800951-2

British Library Cataloguing in Publication Data
A catalogue record for this book is available from the British Library

Library of Congress Cataloging-in-Publication Data
A catalog record for this book is available from the Library of Congress

Printed in the United States of America.

Last digit is the print number: 10 9 8 7 6 5 4 3 2

Contents

List of Contributors

N.B. Allen The University of Melbourne, Melbourne, VIC, Australia

O. Almkvist Stockholm University; Karolinska Institutet, Stockholm, Sweden

W.R. Avison The University of Western Ontario; Children's Health Research Institute, Lawson Health Research Institute, London, ON, Canada

S. Ayers City University London, London, UK

S. Bhatnagar Children's Hospital of Philadelphia and the University of Pennsylvania School of Medicine, Philadelphia, PA, USA

A.H. Brooke University of Sussex, Brighton, UK

B. Buwalda University of Groningen, Groningen, The Netherlands

M.G. Calvo University of La Laguna, Tenerife, Spain

D. Carroll University of Birmingham, Birmingham, UK

D.J. Castle University of Melbourne; St. Vincent's Mental Health, Melbourne, VIC, Australia; The University of Cape Town, Cape Town, South Africa; Australian Catholic University, Banyo, Australia

C.L. Chollak NYU School of Medicine, New York, NY, USA

H.M. Cochran Ann Arbor Veterans Healthcare Administration; University of Michigan, Ann Arbor, MI, USA

S.M. Conn University of British Columbia, Vancouver, BC, Canada

B. Czéh MTA—PTE, Neurobiology of Stress Research Group, Szentágothai Research Center, University of Pécs, Pécs, Hungary

R. Dantzer MD Anderson Cancer Center, Houston, TX, USA

S.F. de Boer University of Groningen, Groningen, The Netherlands

R.M. Deighton The Cairnmillar Institute, Melbourne, VIC, Australia

E.R. de Kloet Leiden University Medical Center, Leiden, The Netherlands

A. DeLongis University of British Columbia, Vancouver, BC, Canada

V.A. Diwadkar Wayne State University School of Medicine, Detroit, MI, USA

W.W. Dressler The University of Alabama, Tuscaloosa, AL, USA

L.M. Dunn-Jensen San Jose State University, San Jose, CA, USA

D.B. Dwyer The University of Melbourne and Melbourne Health, Melbourne, VIC, Australia

K.P. Ebmeier University of Oxford Department of Psychiatry, Warneford Hospital, Oxford, UK

B. Ellis University of Arizona Norton School of Family and Consumer Sciences, Tucson, AZ, USA

G.A. Fava University of Bologna, Bologna, Italy; State University of New York at Buffalo, Buffalo, NY, USA

G. Fábián Semmelweis University, Budapest, Hungary

T.K. Fábián Private Practitioner, Faaborg, Denmark

P. Fejérdy Semmelweis University, Budapest, Hungary

G. Fink Florey Institute of Neuroscience and Mental Health, University of Melbourne, Parkville, VIC, Australia

A. Fornito Monash University, Clayton, VIC, Australia

E. Fuchs German Primate Center, Göttingen, Germany

J. Gerhart Rush University Medical Center, Chicago, IL, USA

E.L. Gibson University of Roehampton, London, UK

A.T. Ginty University of Pittsburgh, Pittsburgh, PA, USA

G.M. Goodwin University Department of Psychiatry, University of Oxford, Warneford Hospital, Oxford, UK

J.F. Gunn III Montclair State University, Montclair, NJ, USA

A. Gutiérrez-García University of La Laguna, Tenerife, Spain

K.L. Harkness Queen's University, Kingston, ON, Canada

B.J. Harrison The University of Melbourne and Melbourne Health, Melbourne, VIC, Australia

N.A. Harrison University of Sussex, Brighton, UK

P. Hermann Semmelweis University, Budapest, Hungary

S.E. Hobfoll Rush University Medical Center, Chicago, IL, USA

L. Holmgreen Rush University Medical Center, Chicago, IL, USA

A. Horsch University Hospital Lausanne, Lausanne, Switzerland

C. Hundert VA Boston Healthcare System, Boston, MA, USA

B. Iffland Bielefeld University, Bielefeld, Germany

A. Keil University of Florida, Gainesville, FL, USA

H. Kessler The Ruhr University Bochum, Bochum, Germany

C.D. King Cincinnati Children's Hospital Medical Center, Cincinnati, OH, USA

D.B. King University of British Columbia, Vancouver, BC, Canada

J.D. Kinzie Department of Psychiatry, Oregon Health and Science University, Portland, OR, USA

J.M. Koolhaas University of Groningen, Groningen, The Netherlands

W.J. Kop Tilburg University, Tilburg, the Netherlands

H.M. Kupper Tilburg University, Tilburg, the Netherlands

M.P. Leiter Acadia University, Wolfville, NS, Canada

D. Lester The Richard Stockton College of New Jersey, Galloway, NJ, USA

M. Lindau Stockholm University, Stockholm, Sweden

B. Litz VA Boston Healthcare System; Boston University School of Medicine, Boston, MA, USA

S. Maguen UCSF Medical School; San Francisco VA Medical Center, San Francisco, CA, USA

C. Maslach University of California, Berkeley, CA, USA

G. Matthews University of Central Florida, Orlando, FL, USA

R. McCarty Vanderbilt University, Nashville, TN, USA

B.S. McEwen The Rockefeller University, New York, NY, USA

J.T. Mitchell University of Maryland Baltimore County, Baltimore; International Critical Incident Stress Foundation, Ellicott City, MD, USA

A.H. Mohammed Karolinska Institutet, Stockholm; Linnaeus University, Växjö, Sweden

S.M. Monroe University of Notre Dame, Notre Dame, IN, USA

S. Musić Swinburne University of Technology, Melbourne, VIC, Australia

R.M. Nesse Arizona State University School of Life Sciences, Tempe, AZ, USA

F. Neuner Bielefeld University, Bielefeld, Germany

R. Norbury University Department of Psychiatry, University of Oxford, Warneford Hospital, Oxford; Department of Psychology, University of Roehampton, London, UK

R.W. Novaco University of California, Irvine, CA, USA

J. Nursey Phoenix Australia, Centre for Posttraumatic Mental Health, Carlton, VIC, Australia

S. Packer Icahn School of Medicine at Mt. Sinai, New York, NY, USA

C. Pantelis The University of Melbourne and Melbourne Health, Melbourne, VIC, Australia

A.J. Phelps Phoenix Australia, Centre for Posttraumatic Mental Health, Carlton, VIC, Australia

A.C. Phillips University of Birmingham, Birmingham, UK

L. Poole University College London, London, UK

K.E. Porter Ann Arbor Veterans Healthcare Administration; University of Michigan, Ann Arbor, MI, USA

B.J. Ragen New York University; NYU School of Medicine, New York, NY, USA

S.K.H. Richards Ann Arbor Veterans Healthcare Administration, Ann Arbor, MI, USA

A.E. Roach University of South Carolina-Aiken, Aiken, SC, USA

D. Roger University of Canterbury, Christchurch, New Zealand

S.L. Rossell Swinburne University of Technology; The Alfred Hospital and Monash University Central Clinical School; St. Vincent's Mental Health, Melbourne, VIC, Australia

K.C. Ryan Indiana University, Bloomington, IN, USA

J. Savla Virginia Tech University, Blacksburg, VA, USA

B. Serwinski University College London, London, UK

M.B. Sexton Ann Arbor Veterans Healthcare Administration; University of Michigan, Ann Arbor, MI, USA

K.T. Sibille University of Florida, Gainesville, FL, USA

J. Siegrist University of Duesseldorf, Duesseldorf, Germany

E.E.A. Simpson Ulster University, Coleraine, UK

G.M. Slavich University of California, Los Angeles, CA, USA

L. Solberg Nes Oslo University Hospital, Oslo, Norway

V. Starcevic University of Sydney, Sydney, NSW; University of Melbourne; St. Vincent's Mental Health, Melbourne, VIC, Australia

M. Steinberg Independent Practice, Naples, FL, USA

E. Stephenson University of British Columbia, Vancouver, BC, Canada

A. Steptoe University College London, London, UK

A. Tankersley VA Boston Healthcare System, Boston, MA, USA

V. Tirone Rush University Medical Center, Chicago, IL, USA

H.C. Traue The University of Ulm, Ulm, Germany

P. Valent Melbourne, VIC, Australia

J. Wardle University College London, London, UK

D. Washburn Queen's University, Kingston, ON, Canada

A. Weinberg University of Salford, Salford, Lancashire, UK

E. Wethington Cornell University, Ithaca, NY, USA

S. Whittle The University of Melbourne and Melbourne Health, Melbourne, VIC, Australia

M. Yücel The University of Melbourne and Melbourne Health, Melbourne; Monash University, Clayton, VIC, Australia

A. Zalesky The University of Melbourne and Melbourne Health; The University of Melbourne, Melbourne, VIC, Australia

S.H. Zarit Penn State University, State College, PA, USA

E. Zsoldos University of Oxford Department of Psychiatry, Warneford Hospital, Oxford, UK

Preface

"Stress" has been dubbed the "Health Epidemic of the 21st Century" by the World Health Organization and is estimated to cost American businesses up to $300 billion a year. The effect of stress on our emotional and physical health can be devastating. In a recent US study, over 50% of individuals felt that stress negatively impacted work productivity. Between 1983 and 2009, stress levels increased by 10-30% among all demographic groups in the United States.

Numerous studies show that job stress is by far the major source of stress for American adults and that it has escalated progressively over the past few decades. Increased levels of job stress as assessed by the perception of having little control but many demands have been demonstrated to be associated with increased rates of heart attack, hypertension, obesity, addiction, anxiety, depression, and other disorders. In New York, Los Angeles, and other municipalities, the relationship between job stress and heart attacks is recognized, so that any police officer who suffers a coronary event on or off the job is assumed to have a work related injury and is compensated accordingly.

Stress is a highly personalized phenomenon that varies between people depending on individual vulnerability and resilience, and between different types of tasks. Thus one survey showed that having to complete paper work was more stressful for many police officers than the dangers associated with pursuing criminals. The severity of job stress depends on the magnitude of the demands that are being made and the individual's sense of control or decision-making latitude for dealing with the stress.

There is a vast literature on the possible role of stress in the causation and/or exacerbation of disease in most organ systems of the body. Inextricably linked to anxiety, stress plays a pivotal role in mental disorders including phobias, major depression, bipolar disorder, and schizophrenia.

The intense public, research, and clinical interest in stress coupled with our experience with the *Encyclopedia of Stress* (2000 and 2007) prompted the publication of a multiauthor *Handbook of Stress* comprised of self-contained volumes. The present volume on concepts, cognition, emotion and behavior, is the first in this new *Handbook* series.

George Fink
Florey Institute of Neuroscience and Mental Health, University of Melbourne, Parkville, Victoria, Australia
2015

PART 1

GENERAL CONCEPTS

CHAPTER

1

Stress, Definitions, Mechanisms, and Effects Outlined: Lessons from Anxiety

G. *Fink*

Florey Institute of Neuroscience and Mental Health, University of Melbourne, Parkville, VIC, Australia

OUTLINE

Abstract

The present volume on concepts, cognition, emotion, and behavior, is the first in this new *Handbook* series. The purpose of this first chapter is to provide an outline of stress, stress definitions, the response to stress and neuroendocrine mechanisms involved, and stress consequences such as anxiety and posttraumatic stress disorder. Study of the neurobiology of anxiety and related disorders has facilitated our understanding of the neural mechanisms that subserve stress and will therefore be underscored.

Now is the age of anxiety **W.H. Auden**

INTRODUCTION

"Stress" has been dubbed the "Health Epidemic of the 21st Century" by the World Health Organization and is estimated to cost American businesses up to $300 billion a year. The effect of stress on our emotional and physical health can be devastating. In a recent US study, over 50% of individuals felt stress negatively impacted work productivity. Between 1983 and 2009, stress levels increased by 10-30% among all demographic groups in the United States.

Numerous studies show that job stress is by far the major source of stress for American adults and that it has escalated progressively over the past few decades. Increased levels of job stress as assessed by the perception of having little control but many demands have been demonstrated to be associated with increased rates of heart attack, hypertension, obesity, addiction, anxiety, depression, and other disorders. In New York, Los Angeles, and other municipalities, the relationship between job stress and heart attacks is recognized, so that any police officer who suffers a coronary event on or off the job is assumed to have a work-related injury and is compensated accordingly.

Stress is a highly personalized phenomenon that varies between people depending on individual vulnerability and resilience, and between different types of tasks. Thus one survey showed that having to complete paper work was more stressful for many police officers than the dangers associated with pursuing criminals. The severity of job stress depends on the magnitude of the demands that are being made and the individual's sense of control or decision-making latitude for dealing with the stress.

Stress: Concepts, Cognition, Emotion, and Behavior
http://dx.doi.org/10.1016/B978-0-12-800951-2.00001-7

Stress is, of course, not limited to the workplace. There is vast literature on the possible role of stress in the causation and/or exacerbation of disease in most organ systems of the body. Inextricably linked to anxiety, stress plays a pivotal role in mental disorders including phobias, major depression, and bipolar disorder.[2–9] Stress and anxiety aggravate schizophrenia and people with schizophrenia often experience difficulties in coping with stress. Consequently, stress-inducing changes in lifestyle patterns place a substantial burden on mental health.

Posttraumatic stress disorder (PTSD) is a special form of stress that affects more than 7 million people in the United States. In 1980, largely as a consequence of the psychological trauma experienced by Vietnam War veterans, PTSD was recognized as a disorder with specific symptoms that could be reliably diagnosed and was, therefore, added to the American Psychiatric Association's *Diagnostic and Statistical Manual of Mental Disorders* (*DSM*). PTSD is recognized as a psychobiological mental disorder that can affect survivors of not only combat experience in war and conflict, but also terrorist attacks, natural disasters, serious accidents, assault, rape trauma syndrome, battered woman syndrome, child abuse syndrome, or sudden and major emotional losses. PTSD is associated with epigenetic changes in the brain as well as changes in brain function and structure. These changes provide clues to the origins and possible treatment, and prevention of PTSD. Stress, PTSD, and anxiety are linked to fear, fear memory and extinction, phenomena that together with their neural circuitry and neurochemistry remain the subject of intense research. Notwithstanding its links with anxiety, PTSD in the latest DSM-5 is now included in a new section/classification of trauma- and stressor-related disorders. This move of PTSD from its earlier classification in the DSM-IV as an anxiety disorder is among several changes approved for PTSD that heighten its profile as a disease entity that is increasingly at the center of public as well as professional attention.

Physiological and neurochemical approaches have elucidated the way in which stress is controlled by two major neuroendocrine systems, the hypothalamic-pituitary-adrenal (HPA) axis and the sympathetic-adrenomedullary (SAM) limb of the autonomic nervous system (ANS). Our understanding of stress mechanisms in man and animals has benefited significantly from several recent quantum leaps in technology and knowledge. First, advances in molecular genetics (including optogenetics and chemogenetics), sequencing of the human genome and genomics[10] have increased the rigor and precision of our understanding of the molecular neurobiology of stress and its effects on mental state, behavior, and somatic systems. Secondly, through genomics we are also beginning to understand the genetic and epigenetic factors that play a role in susceptibility, vulnerability, and resilience to stress and the various components of the stress response. Thirdly, functional brain imaging has enhanced our understanding of the neurobiology of stress in the human.

Here we provide an outline of stress with the focus on stress definitions, the response to stress, and neuroendocrine and central neurobiological mechanisms involved, and stress consequences, such as anxiety and PTSD. The neurobiology of anxiety is underscored because it has heuristically facilitated our understanding of the neural mechanisms that subserve stress.

KEY POINTS

- Stress is the (nonspecific) response of the body to any demand.[1]
- Stress consequences include anxiety, fear, depression, and PTSD. In addition, stress has adverse effects on other major mental disorders such as bipolar disorder and schizophrenia.
- Stress also has adverse effects on the cardiovascular, including the cerebrovascular system and other organ systems of the body.
- Stressors are perceived and processed by the brain which triggers the release of glucocorticoids (by way of the hypothalamic-pituitary-adrenocortical axis) and catecholamines (adrenaline and noradrenaline) by way of the sympathetic-adrenomedullary system (SAM).
- The glucocorticoids and the catecholamines act synergistically to raise blood glucose levels (by triggering the release of glucose from the liver) which facilitates the "flight or fight" response to stress, as does the ramp-up of cardiovascular output by the catecholamines. The rapid stress-induced release of the catecholamines also shunts blood from the skin and gut to the skeletal muscles.
- Central awareness of and response to stress, anxiety, and fear depends on extensive neural circuits that involve, for example, the amygdala, thalamus, hypothalamus, brain stem nuclei such as the locus coeruleus and the neocortex and limbic cortex.
- Our understanding of the neurobiology of stress has been enhanced by experimental, clinical, and human brain imaging studies of anxiety and other stress-related conditions.

STRESS DEFINITIONS

Stress has a different meaning for different people under different conditions. A working definition of stress that fits many human situations is a condition in which an individual is aroused and made anxious by an uncontrollable aversive challenge—for example, stuck in heavy traffic on a motorway, a hostile employer, unpaid bills, or a predator. Stress leads to a feeling of fear and anxiety. Depending on the circumstances, the fear response can lead to either *fight or flight*. The magnitude of the stress and its physiological consequences are influenced by the individual's perception of their ability to cope with the stressor.

Stress is difficult to define. As Hans Selye (the oft-called "Father of stress") opined, "Everyone knows what stress is, but nobody really knows." Selye's definition, "Stress is the nonspecific response of the body to any demand,"[1] is the most generic. This definition and Selye's stress-related concepts had several detractors which he systematically rebutted.[11] Other definitions are detailed by Fink.[3] Briefly, they include the following:

- "Perception of threat, with resulting anxiety discomfort, emotional tension, and difficulty in adjustment."
- "Stress occurs when environmental demands exceed one's perception of the ability to cope."
- In the group situation, lack of structure or loss of anchor "makes it difficult or impossible for the group to cope with the requirements of the situation. Leadership is missing and required for coping with the demands of the situation."
- For the sociologist, it is social disequilibrium, that is, disturbances in the social structure within which people live.
- A purely biological definition is that stress is any stimulus that will activate (i) the HPA system, thereby triggering the release of pituitary adrenocorticotropin (ACTH) and adrenal glucocorticoids and (ii) the SAM system with the consequent release of adrenaline and noradrenaline.
- In their seminal review "The Stressed Hippocampus, synaptic plasticity and lost memories," Kim and Diamond[12] suggest a three-component definition of stress that can be applied broadly across species and paradigms. First, stress requires heightened excitability or arousal, which can be operationally measured using electroencephalography, behavioral (motor) activity, or neurochemical (adrenaline, glucocorticoid) levels. Second, the experience must also be perceived as aversive. Third, there is lack of control. Having control over an aversive experience has a profound mitigating influence on how stressful the experience feels. The element of control (and "predictability") is the variable that ultimately determines the magnitude of the stress experience and the susceptibility of the individual to develop stress-induced behavioral and physiological sequelae.

Thus, the magnitude of neurocognitive stress (S) approximates to the product of:

- Excitability/arousal (E)
- Perceived aversiveness (A)
- Uncontrollability (U)

$$(S) = E \times A \times U$$

But Is the Stress Response "Nonspecific" as Proposed by Hans Selye?

In a seminal paper, Pacak and Palkovits[13] challenged Selye's doctrine of nonspecificity of the stress response. They studied the similarities and differences between the neuroendocrine responses ("especially the sympathoadrenal and the sympathoneuronal systems and the hypothalamo-pituitary-adrenocortical axis") among five different stressors: immobilization, hemorrhage, cold exposure, pain, or hypoglycemia. With the exception of immobilization stress, these stressors also differed in their intensities. Pacak and Palkovits found heterogeneity of neuroendocrine responses to these stressors: each stressor had its own specific neurochemical "signature." By examining changes of Fos immunoreactivity in various brain regions upon exposure to different stressors, Pacak and Palkovits also investigated the central stressor-specific neuroendocrine pathways. In a separate study on the aortic response to stress, Navarro-Oliveira et al.[14] showed that the SAM, but not the hypothalamic-pituitary-adrenal axis, participates in the adaptive responses of the aorta to stress.

There is now substantial literature on the specificity of stressors. Selye's definition of stress holds but the term "nonspecific" might be redundant. That is, Selye's definition of stress might now read "Stress is the response of the body to any demand."

FEAR VERSUS ANXIETY....WHAT ARE THE DIFFERENCES?

Stress is inextricably linked with fear and anxiety. Definitions of fear and anxiety vary greatly, and to an extent depend on subjective assessment. Nonetheless, in their seminal review, "What is anxiety disorder?", Craske and associates,[15] using Barlow's concepts, state; "anxiety is a future-oriented mood state associated with preparation for possible, upcoming negative events; and fear is an alarm response to present or imminent danger (real or perceived)." This view of human fear and anxiety is comparable to that in animals. "That is, anxiety corresponds to an animal's state during a potential predatory attack and fear corresponds to an animal's state during predator contact or imminent contact."

Table 1 shows the prototypes of self-report symptoms of fear, anxiety, and depression. The symptoms that represent prototypes of fear and anxiety lie at different places upon a continuum of responding. "Along such a continuum, symptoms of fear versus anxiety are likely

TABLE 1 Prototype of Self-Report Symptoms of Fear, Anxiety, and Depression

	Clusters[a]		
	Fear	**Anxiety**	**Depression**
Response-systems			
Verbal-subjective	Thoughts of imminent threat	Thoughts of future threat	Thoughts of loss, failure[b]
Somato-visceral	Sympathetic arousal	Muscle tension	Energy loss[b]
Overt motor	Escape	Avoidance	Withdrawal[b]

[a] *While represented as prototypes, fear and anxiety may be better represented as points along a continuum, with varying degrees of symptom overlap.*
[b] *More specifically, these features represent lack of positive affect, as represented by the absence of thoughts of success, the absence of energy, and the absence of desire to be with other people.*
Reproduced with permission from Craske MG, Rauch SL, Ursano R, Prenoveau J, Pine DS, Zinbarg RE. Depression and Anxiety. *John Wiley and Sons.*

to diverge and converge to varying degrees." For further details, the reader is referred to Craske et al.[15]

BIOLOGICAL RESPONSE TO STRESS

The biological response to stress involves activation of three major interrelated systems. First, the stressor is perceived by sensory systems of the brain, which evaluate and compare the stressful challenge with the existing state and previous stress experience of the organism. Second, on detection of a stressful challenge to homeostasis, the brain activates the ANS which through the SAM system triggers a rapid release of the catecholamines, noradrenaline, and adrenaline. The catecholamines increase cardiac output and blood pressure, shunt blood from the skin and gut to skeletal muscle, and trigger the release of glucose from the liver into the blood stream. Third, the brain simultaneously activates the HPA axis which results in the release of adrenal glucocorticoids, cortisol in man and fish, and corticosterone in rodents.

Increased glucocorticoid levels enhance the organism's resistance and adaptation to stress. However, the precise mechanisms of this defensive action of glucocorticoids remain to be elucidated. Glucocorticoids act synergistically with adrenaline to increase blood glucose, thus ensuring energy supplies often needed to overcome the stress by facilitating *fight or flight*. Glucocorticoids are also potent inhibitors of the immune response and inflammation, moderating the production of prostaglandins and inflammatory cytokines. In a seminal review, Munck and associates[16] proposed "that stress-induced increases in glucocorticoid levels protect not against the source of stress itself, but rather against the body's normal reactions to stress, preventing those reactions from overshooting and themselves threatening homeostasis." Munck's hypothesis has retained its currency. Thus, for example, Zhang et al.[17] have reported that glucocorticoids inhibit lipopolysaccharide-induced myocardial inflammation. Munck's theory does not necessarily conflict with the fact that in the uninjured brain, basal or acutely elevated glucocorticoid levels increase synaptic plasticity and facilitate hippocampal dependent cognition whereas chronically elevated glucocorticoid levels impair synaptic plasticity and cognition, decrease neurogenesis and spine density, and cause dendritic atrophy.[7,18,19]

Glucocorticoid actions are mediated by two biochemically distinct receptors which bind the same ligand (cortisol in humans, corticosterone in rodents), albeit with differing affinities. While glucocorticoid receptors (GRs) are ubiquitously distributed, the location of mineralocorticoid receptors (MRs) is more discrete. However, both receptors are expressed at particularly high levels in limbic areas that are responsible for the modulation of the stress response.[20] As compared with GR, MR have a much greater affinity for cortisol/corticosterone and are, therefore, highly occupied even under basal (stress-free) conditions.[20] In contrast, GR become increasingly occupied as circulating glucocorticoid levels rise in response to stress. MR have been implicated in the appraisal process and onset of the stress response, while GR are involved in the mobilization of energy substrates and most stress-induced changes in behavior.

The latter includes anxiety-like behavior and facilitated learning and memory (in particular, consolidation of memories). Long-term GR activation is associated with deleterious effects on several cognitive functions.[6,21–23] These deleterious effects have been correlated with neuroarchitectural changes in several brain regions, including the hippocampus, prefrontal cortex, and amygdala that are also implicated in modulating the negative feedback control within the HPA.[23,24]

The amount of glucococorticoid available for cells is "micromanaged" by 11β-hydroxysteroid dehydrogenase (HSD-11β) enzymes of which there are two isoforms. First, HSD11B1 which reduces cortisone to the active hormone cortisol that activates GRs. Second, HSD11B2 which oxidizes cortisol to cortisone and prevents illicit activation of the MR.[25–29]

While elevated glucocorticoid levels characterize stress, high levels of glucocorticoids per se do not mimic stress. Of the several central neurochemical neurotransmitters involved in the stress response, attention has focused on the corticotropin releasing factor (CRF) peptide family (and especially the urocortins) as possible orchestrators of the stress response. For details of the CRF peptide family and their cognate receptors, readers are referred to several reviews.[4,30–34]

Stress Neuroendocrinology Outlined

As mentioned above, stressors are perceived and processed by the sensory cortex which drives the hypothalamus by several pathways that include the thalamus and the limbic forebrain and hindbrain systems.[35,36] The hypothalamus triggers the release of glucocorticoids and the catecholamines, the primary stress hormones, by way of the paraventricular nuclei (PVN) in the case of the HPA and the PVN, lateral hypothalamus, arcuate, and brainstem nuclei in the case of the SAM system.[3,31,37–40] The amygdala, a prominent component of the limbic system that plays a key role in the evaluation of emotional events and formation of fearful memories, is a prime target of the neurochemical and hormonal mediators of stress. Clinical and experimental data have correlated changes in the structure/function of the amygdala with emotional disorders such as anxiety.[3,8,31,33,38]

The PVN is subject to differential activation by distinct neuronal pathways, depending on the quality and/or immediacy of the demand for an appropriate response.[41,42] Stressors such as hemorrhage, respiratory distress, or systemic inflammation, which represent an immediate threat to homeostasis, directly activate the PVN, bypassing cortical and limbic areas, by activation of somatic, visceral, or circumventricular sensory pathways.[43,44] Excitatory ascending pathways originating in the brainstem nuclei that convey noradrenergic inputs from the nucleus of tractus solitarius,[45–48] serotonergic inputs from the raphe nuclei,[49,50] or inputs from adjacent hypothalamic nuclei[42] are well positioned to receive visceral and autonomic inputs so as to evoke rapid neuroendocrine responses.

The hypothalamic control of the release of pituitary ACTH is mediated by the 41-amino acid residue neuropeptide CRF transported from PVN nerve terminals to the anterior pituitary gland by way of the hypophysial portal vessels.[51,52] The action of CRF is potentiated by the synergistic action of the nonapeptide, arginine vasopressin (AVP), which, like CRF, is synthesized in the PVN.[53] ACTH stimulates the secretion of adrenal glucocorticoids which have powerful metabolic effects that promote the stress response. Homeostasis within the HPA is maintained by a negative feedback system by which the adrenal glucocorticoids (the afferent limb) moderate ACTH synthesis and release (the efferent limb). Allostasis, that is, maintenance of constancy through change in HPA activity to cope with increased stress load is brought about by change in feedback set point. It must be stressed that in biology, "set point" is a conceptual construct rather than a precise structural entity.[31]

The major sites of glucocorticoid negative feedback are the PVN, where glucocorticoids inhibit CRF and AVP synthesis and release, and the pituitary gland, where they block the ACTH response to CRF and inhibit the synthesis of ACTH and its precursor, proopiomelanocortin. The limbic system of the brain, especially the hippocampus and amygdala, also plays a role in glucocorticoid negative feedback.[4,7,8,31,38,39,54]

Central control of the ANS involves the hypothalamic PVN together with various brainstem and limbic nuclei (caudal raphe, ventromedial and rostral ventrolateral medulla, the ventrolateral pontine tegmentum). The ANS plays the pivotal role in the early (immediate) response to stress. ANS action is mediated mainly by way of the release of noradenaline from nerve terminals and adrenaline from the chromaffin cells of the adrenal medulla.[40,55] Adrenaline and noradrenaline facilitate the stress response by triggering the synthesis and release of glucose from the liver into the blood stream, increasing the rate and force of cardiac contraction and shunting blood from the skin and the gastrointestinal system to the skeletal muscles.

Central Neural Stress-Response Mechanisms

In addition to the two canonical neural outflow systems (ANS and HPA), the stress response involves central nuclei, such as the locus coeruleus (LC), the principle brain nucleus for the production of noradrenaline. Located in the pons, the LC and its noradrenergic projections to the forebrain play a key role in the central control of arousal, attention, and the response to stress. The LC receives afferents from the medial prefrontal cortex, cingulate gyrus, amygdala, hypothalamus, and raphe nuclei. In turn LC noradrenergic projections innervate the spinal cord, the brain stem, cerebellum, hypothalamus, the thalamic relay nuclei, the amygdala, hippocampus, and the neocortex. Noradrenaline released from the LC neuronal projections has an excitatory effect on most of the brain, inducing arousal and priming central neurons to stimulus-activation. Stress shifts LC noradrenergic cell firing, normally moderated by glutamatergic input, to a high tonic firing. This shift is mediated by CRF projections from the central amygdala and mediated by CRF-R1 receptors in the LC.[39] In turn, LC noradrenergic cells project to the basolateral amygdala (BLA), hippocampal CA1, and the dentate gyrus (DG) where noradrenaline released shortly after stress exposure, enhances excitability, promoting the encoding of stress-related information. Glutamatergic output from the BLA to the hippocampal DG is thought to provide a means to "emotionally tag" information processed in the hippocampus.[39] After cessation of the stress, the stress-induced enhancement in activity of the LC, the BLA, the DG, and CA1 is gradually reversed, resulting in a return to the pre-stress activity level. In the LC, the frequency of tonic firing is reduced by opiates that bind to κ- and μ-opioid receptors. In the BLA, the DG, and CA1, these gradual normalizing effects are produced by glucocorticoids, presumably through GR-mediated gene-dependent cascades.[39]

Amygdala: Pivotal Role in Fear, Memory, Attention, and Anxiety

Animal studies have shown that the amygdala receives sensory information rapidly through the sensory thalamus and more slowly and precisely (in terms of topography) through the sensory cortex.[56–58] The thalamic or cortical pathway can be used for simple sensory stimuli such as those typically used in animal conditioning. Brain imaging findings on the role of the human amygdala in fear learning are consistent with those in animal models. As assessed by functional magnetic resonance imaging, fear conditioning in humans results in an increased blood-oxygen-level-dependent (BOLD) signal in the amygdala.[59,60] The magnitude of this BOLD response is predictive of the strength of the conditioned response.[60,61] In addition, a subliminally presented conditioned stimulus (CS)—one presented so quickly that subjects are unaware of its presentation—leads to coactivation of the amygdala and the superior colliculus and pulvinar.[62]

The pivotal role of the amygdala in the response to fear is underscored by the effects of brain lesions, in that patients in which the amygdala has been lesioned show no conditioned fear. However, providing the hippocampus is intact, these patients are able explicitly to recollect and report the events of fear conditioning procedures.[63–65] In contrast, bilateral lesions of the hippocampus that spare the amygdala, impair the ability consciously to report the events of fear conditioning, although there is normal expression of conditioned fear as assessed physiologically by skin conduction responses.[63] This dissociation following amygdala or hippocampal damage between indirect physiological assessments of the conditioned fear response (amygdala dependent) and awareness of the aversive properties of the CS (hippocampal dependent) supports the proposition that there are multiple systems for the encoding and expression of emotional learning in the human.

The amygdala, in addition to modulating memory systems, also alters processing in cortical systems involved in attention and perception and thereby potentially influences downstream cognitive functions both by direct projections and possibly also by way of the nucleus basalis of Meynert (NBM) that receives afferents from the central nucleus of the amygdala.[66] The NMB projects widely to the cortical sensory-processing regions. The NBM projections release acetylcholine, which has been shown to facilitate neuronal responsivity.[67,68] Transitory modulation of cortical regions by the amygdala might increase cortical attention and vigilance in situations of danger.[69–71] This view receives support from brain imaging that showed amygdala activation to fearful (versus neutral) faces does not depend on subjects' awareness of the presentation of the faces,[72] or whether or not the faces are the focus of attention.[73–75] These studies indicate that the amygdala responds to a fear stimulus automatically and prior to awareness.

The bed nucleus of the stria terminalis (BNST), considered to be an extension of the amygdala,[76] receives dense projections from the BLA, and projects in turn to hypothalamic and brainstem target areas that mediate autonomic and behavioral responses to aversive or threatening stimuli. The BNST participates in certain types of anxiety and stress responses and seems to mediate slower-onset, longer-lasting responses that frequently accompany sustained threats, and that may persist even after threat termination.[76,77]

In summary, a combination of lesion and imaging studies have shown that transitory feedback from the human amygdala to sensory cortical regions can facilitate attention and perception. The amygdala's influence on cortical sensory plasticity may also result in enhanced perception for stimuli that have acquired emotional properties through learning. By influencing attention and perception, the amygdala modulates the gateway of information processing. The amygdala enables preferential processing of stimuli that are emotional and potentially threatening, thus assuring that information of importance to the organism is more likely to influence behavior.

CONCLUSIONS AND RELEVANCE FOR STRESS AND ANXIETY MANAGEMENT

The neurobiology of stress and anxiety has highlighted new potential therapeutic targets for the management of anxiety. There has been considerable investment, for example, into new strategies such as the design and development of central CRF receptors antagonists. Furthermore, much ongoing research is focused on cognitive-behavioral (e.g., exposure) therapy, as well as possible pharmacological fear extinction. Care obviously needs to be taken to avoid "conditional reinstatement." Reinstatement of extinguished fear can be triggered by exposure to conditional as well as unconditional aversive stimuli, and this may help to explain why relapse is common following clinical extinction therapy in humans.[78] Neuropharmacologically we know that noradrenergic augmentation in the amygdala following retrieval of a traumatic memory enhances memory reconsolidation and makes the memory less susceptible to fear extinction. Elevated noradrenergic activity is associated with persistence and severity of PTSD symptoms. That is, noradrenergic-modulated reconsolidation processes contribute to the maintenance and exacerbation of trauma-related memories in PTSD.[79] These and other factors that need to be considered in devising management strategies for stress and its consequences, especially anxiety, PTSD and depression, will be covered in detail in specialist chapters in this and subsequent volumes of the *Handbook of Stress* series.

Glossary

Allostasis Homeostasis, "stability through constancy" is maintained by a self-limiting process involving negative feedback control by the output variable, which in the case of the HPA, is the secretion of adrenal glucocorticoids. The limits of feedback control are set by a notional regulatory "set point."[31] Sterling and Eyer[81] introduced the term Allostasis, "stability through change" brought about by central nervous regulation of the set points that adjust physiological parameters in anticipation to meet the stress/challenge. McEwen[82] integrated the concept of *allostasis* to describe the adaptation process of the organism in the face of different stressors and different circumstances. That is, allostasis incorporates circadian, circannual, and other life-history changes that might affect the animal's "internal balance."

Allostatic load Allostatic load represents the cumulative impact of stressors on the body's physiological systems over the life course.[83] That is, allostatic load can be defined as the "long-term cost of allostasis that accumulates over time and reflects the accumulation of damage that can lead to pathological states." Allostatic load has been shown to predict various health outcomes in longitudinal studies, such as declines in physical and cognitive functioning, and cardiovascular morbidity and mortality. Allostatic load is measured using a point scale that combines a series of stress biomarkers of cardiovascular, immune, and metabolic function.[84,85]

Homeostasis Aristotle, Hippocrates, and the other Ancients were aware of stress and its adverse effects. However, Claude Bernard was the first to formally explain how cells and tissues in multicelled organisms might be protected from stress. Bernard, working in Paris during the second half of the nineteenth century, first pointed out (1859) that the internal medium of the living organism is not merely a vehicle for carrying nourishment to cells. Rather, "it is the fixity of the milieu intérieur which is the condition of free and independent life." That is, cells are surrounded by an internal medium that buffers changes in acid-base, gaseous (O_2 and CO_2) and ion concentrations and other biochemical modalities to minimize changes around biologically determined set points, thereby providing a steady state. Fifty years later, Walter Bradford Cannon, working at Harvard, suggested the designation *homeostasis* (from the Greek homoios, or similar, and stasis, or position) for the coordinated physiological processes that maintain most of the steady states in the organism. Cannon popularized the concept of "homeostasis" in his 1932 book, *Wisdom of the Body*.[80]

Fight or flight Walter Cannon also coined the term *fight or flight* to describe an animal's response to threat. This concept proposed that animals react to threats with a general discharge of the sympathetic nervous system, priming the animal for fighting or fleeing.

Acknowledgments

The author acknowledges with gratitude the support and facilities provided by the Florey Institute of Neuroscience and Mental Health, University of Melbourne, 30 Royal Parade, Parkville, Victoria 3010, Australia.

References

1. Selye H. A syndrome produced by diverse nocuous agents. *Nature*. 1936;138:32.
2. Charney DS. Psychobiological mechanisms of resilience and vulnerability: implications for successful adaptation to extreme stress. *Am J Psychiatry*. 2004;161:195–216.
3. Fink G. Stress: definition and history. In: Squire L, ed-in-chief. *Encyclopedia of Neuroscience*. Oxford: Elsevier Ltd; 2009:549–555.
4. Fink G. Neural control of the anterior lobe of the pituitary gland (pars distalis). In: Fink G, Pfaff DW, Levine JE, eds. *Handbook of Neuroendocrinology*. London, Waltham, San Diego: Academic Press, Elsevier; 2012:97–138.
5. Hammen C. Stress and depression. *Annu Rev Clin Psychol*. 2005;1:293–319.
6. McEwen BS. Glucocorticoids, depression, and mood disorders: structural remodeling in the brain. *Metabolism*. 2005;54(5 Suppl 1): 20–23.
7. McEwen BS, Eiland L, Hunter RG, Miller MM. Stress and anxiety: structural plasticity and epigenetic regulation as a consequence of stress. *Neuropharmacology*. 2012;62:3–12.
8. Pêgo JM, Sousa JC, Almeida OF, Sousa N. Stress and the neuroendocrinology of anxiety disorders. *Curr Top Behav Neurosci*. 2010;2:97–117.
9. Risbrough VB, Stein MB. Role of corticotropin releasing factor in anxiety disorders: a translational research perspective. *Horm Behav*. 2006;50:550–561.
10. Geschwind DH, Flint J. Genetics and genomics of psychiatric disease. *Science*. 2015;349:1489–1494.
11. Selye H. Confusion and controversy in the stress field. *J Hum Stress*. 1975;1:37–44.
12. Kim JJ, Diamond DM. The stressed hippocampus, synaptic plasticity and lost memories. *Nat Rev Neurosci*. 2002;3:453–462.
13. Pacak K, Palkovits M. Stressor specificity of central neuroendocrine responses: implications for stress-related disorders. *Endocr Rev*. 2001;22:502–548.
14. Navarro-Oliveira CM, Vassilieff VS, Cordellini S. The sympathetic adrenomedullary system, but not the hypothalamic-pituitary-adrenal axis, participates in aortaadaptive response to stress: nitric oxide involvement. *Auton Neurosci*. 2000;83:140–147.
15. Craske MG, Rauch SL, Ursano R, Prenoveau J, Pine DS, Zinbarg RE. What is an anxiety disorder? *Depress Anxiety*. 2009;26:1066–1085. http://dx.doi.org/10.1002/da.20633.
16. Munck A, Guyre PM, Holbrook NJ. Physiological functions of glucocorticoids in stress and their relation to pharmacological actions. *Endocr Rev*. 1984;5:25–44.
17. Zhang HN, He YH, Zhang GS, et al. Endogenous glucocorticoids inhibit myocardial inflammation induced by lipopolysaccharide: involvement of regulation of histone deacetylation. *J Cardiovasc Pharmacol*. 2012;60:33–41. http://dx.doi.org/10.1097/FJC.0b013e3182567fef.
18. McEwen BS, Magarinos AM. Stress and hippocampal plasticity: implications for the pathophysiology of affective disorders. *Hum Psychopharmacol*. 2001;16:S7–S19.
19. Sorrells SF, Caso JR, Munhoz CD, Sapolsky RM. The stressed CNS: when glucocorticoids aggravate inflammation. *Neuron*. 2009; 64:33–39.
20. Reul JM, de Kloet ER. Two receptor systems for corticosterone in rat brain: microdistribution and differential occupation. *Endocrinology*. 1985;117:2505–2511.
21. Cerqueira JJ, Almeida OF, Sousa N. The stressed prefrontal cortex. Left? Right!. *Brain Behav Immun*. 2008;22:630–638.
22. Sapolsky RM, Krey LC, McEwen BS. The neuroendocrinology of stress and aging: the glucocorticoid cascade hypothesis. *Endocr Rev*. 1986;7:284–301.

23. Sousa N, Cerqueira JJ, Almeida OF. Corticosteroid receptors and neuroplasticity. *Brain Res Rev*. 2008;57:561–570.
24. McEwen BS. Physiology and neurobiology of stress and adaptation: central role of the brain. *Physiol Rev*. 2007;87:873–904.
25. Yau JL, Noble J, Seckl JR. 11 Beta-hydroxysteroid dehydrogenase type 1 deficiency prevents memory deficits with aging by switching from glucocorticoid receptor to mineralocorticoid receptor-mediated cognitive control. *J Neurosci*. 2011;31:4188–4193. http://dx.doi.org/10.1523/jneurosci.6145-10.2011.
26. Monder C, White PC. 11 Beta-hydroxysteroid dehydrogenase. *Vitam Horm*. 1993;47:187–271.
27. Seckl JR. 11 Beta-hydroxysteroid dehydrogenase in the brain: a novel regulator of glucocorticoid action? *Front Neuroendocrinol*. 1997;18:49–99. http://dx.doi.org/10.1006/frne.1996.0143.
28. Seckl JR, Walker BR. Minireview: 11 beta-hydroxysteroid dehydrogenase type 1—a tissue-specific amplifier of glucocorticoid action. *Endocrinology*. 2001;142:1371–1376. http://dx.doi.org/10.1210/en.142.4.1371.
29. White PC, Mune T, Agarwal AK. 11 Beta-hydroxysteroid dehydrogenase and the syndrome of apparent mineralocorticoid excess. *Endocr Rev*. 1997;18:135–156. http://dx.doi.org/10.1210/er.18.1.135. PMID 9034789.
30. Bale TL, Vale WW. CRF and CRF receptors: role in stress responsivity and other behaviors. *Annu Rev Pharmacol Toxicol*. 2004;44:525–557.
31. Fink G. Neuroendocrine feedback control systems: an introduction. In: Fink G, Pfaff DW, Levine JE, eds. *Handbook of Neuroendocrinology*. London, Waltham, San Diego: Academic Press, Elsevier; 2012:55–72.
32. Reul JM, Holsboer F. On the role of corticotropin-releasing hormone receptors in anxiety and depression. *Dialogues Clin Neurosci*. 2002;4:31–46.
33. Sztainberg Y, Chen A. Neuropeptide regulation of stress-induced behavior: insights from the CRF/urocortin family. In: Fink G, Pfaff DW, Levine JE, eds. *Handbook of Neuroendocrinology*. London, Waltham, San Diego: Academic Press, Elsevier; 2012:355–376.
34. Zorrilla EP, Heilig M, de Wit H, Shaham Y. Behavioral, biological, and chemical perspectives on targeting CRF(1) receptor antagonists to treat alcoholism. *Drug Alcohol Depend*. 2013;128:175–186. http://dx.doi.org/10.1016/j.drugalcdep.2012.12.017.
35. Maclean PD. The limbic system with respect to self-preservation and the preservation of the species. *J Nerv Ment Dis*. 1958;127:1–11.
36. Nauta WJ. Limbic system and hypothalamus: anatomical aspects. *Physiol Rev Suppl*. 1960;4:102–104.
37. Aguilera G. The hypothalamic–pituitary–adrenal axis and neuroendocrine responses to stress. In: Fink G, Pfaff DW, Levine JE, eds. *Handbook of Neuroendocrinology*. London, Waltham, San Diego: Academic Press, Elsevier; 2012:175–196.
38. Fink G. Stress controversies: posttraumatic stress disorder, hippocampal volume, gastro-duodenal ulceration. *J Neuroendocrinol*. 2011;23:107–117.
39. Joels M, Baram TZ. The neuro-symphony of stress. *Nat Rev Neurosci*. 2009;10:459–466.
40. Palkovits M. Sympathoadrenal system: neural arm of the stress response. In: Squire L, ed-in-chief. *Encyclopedia of Neuroscience*. Oxford: Elsevier Ltd; 2009:679–684.
41. Herman JP, Cullinan WE. Neurocircuitry of stress: central control of the hypothalamo-pituitary-adrenocortical axis. *Trends Neurosci*. 1997;20:78–84.
42. Herman JP, Figueiredo H, Mueller NK, et al. Central mechanisms of stress integration: hierarchical circuitry controlling hypothalamo-pituitary-adrenocortical responsiveness. *Front Neuroendocrinol*. 2003;24:151–180.
43. Chan RK, Brown ER, Ericsson A, Kovács KJ, Sawchenko PE. A comparison of two immediate-early genes, c-fos and NGFI-B, as markers for functional activation in stress-related neuroendocrine circuitry. *J Neurosci*. 1993;13:5126–5138.
44. Cole RL, Sawchenko PE. Neurotransmitter regulation of cellular activation and neuropeptide gene expression in the paraventricular nucleus of the hypothalamus. *J Neurosci*. 2002;22:959–969.
45. Abercrombie ED, Jacobs BL. Single-unit response of noradrenergic neurons in the locus coeruleus of freely moving cats. II. Adaptation to chronically presented stressful stimuli. *J Neurosci*. 1987;7:2844–2848.
46. Cullinan WE, Herman JP, Battaglia DF, Akil H, Watson SJ. Pattern and time course of immediate early gene expression in rat brain following acute stress. *Neuroscience*. 1995;64:477–505.
47. Gann DS, Ward DG, Baertschi AJ, Carlson DE, Maran JW. Neural control of ACTH release in response to hemorrhage. *Ann N Y Acad Sci*. 1977;297:477–497.
48. Smith MA, Brady LS, Glowa J, Gold PW, Herkenham M. Effects of stress and adrenalectomy on tyrosine hydroxylase mRNA levels in the locus ceruleus by in situ hybridization. *Brain Res*. 1991;544:26–32.
49. Feldman S, Conforti N, Melamed E. Paraventricular nucleus serotonin mediates neurally stimulated adrenocortical secretion. *Brain Res Bull*. 1987;18:165–168.
50. Sawchenko PE, Swanson LW, Steinbusch HW, Verhofstad AA. The distribution and cells of origin of serotonergic inputs to the paraventricular and supraoptic nuclei of the rat. *Brain Res*. 1983;277:355–360.
51. Fink G, Smith JR, Tibballs J. Corticotrophin releasing factor in hypophysial portal blood of rats. *Nature*. 1971;203:467–468.
52. Vale W, Spiess J, Rivier C, Rivier J. Characterization of a 41 residue ovine hypothalamic peptide that stimulates the secretion of corticotropin and beta-endorphin. *Science*. 1981;213:1394–1397.
53. Fink G, Robinson IC, Tannahill LA. Effects of adrenalectomy and glucocorticoids on the peptides CRF-41, AVP and oxytocin in rat hypophysial portal blood. *J Physiol*. 1988;401:329–345.
54. de Kloet ER, Joëls M, Holsboer F. Stress and the brain: from adaptation to disease. *Nat Rev Neurosci*. 2005;6:463–475.
55. Dahlstrom A. Sympathetic nervous system. In: Squire L, ed-in-chief. *Encyclopedia of Neuroscience*. Oxford: Elsevier Ltd; 2009:663–671.
56. LeDoux JE. Emotion, memory and the brain. *Sci Am*. 1994;270:50–57.
57. LeDoux JE, Sakaguchi A, Reis DJ. Subcortical efferent projections of the medial geniculate nucleus mediate emotional responses conditioned to acoustic stimuli. *J Neurosci*. 1984;4:683–698.
58. Schiller D, Levy I, Niv Y, LeDoux JE, Phelps EA. From fear to safety and back: reversal of fear in the human brain. *J Neurosci*. 2008;28:11517–11525.
59. Büchel C, Morris J, Dolan RJ, Friston KJ. Brain systems mediating aversive conditioning: an event-related fMRI study. *Neuron*. 1998;20:947–957.
60. LaBar KS, Gatenby JC, Gore JC, LeDoux JE, Phelps EA. Human amygdala activation during conditioned fear acquisition and extinction: a mixed-trial fMRI study. *Neuron*. 1998;20:937–945.
61. Phelps EA, Delgado MR, Nearing KI, LeDoux JE. Extinction learning in humans: role of the amygdala and vmPFC. *Neuron*. 2004;43:897–905.
62. Morris JS, Ohman A, Dolan RJ. Conscious and unconscious emotional learning in the human amygdala. *Nature*. 1998;393:467–470.
63. Bechara A, Tranel D, Damasio H, Adolphs R, Rockland C, Damasio AR. Double dissociation of conditioning and declarative knowledge relative to the amygdala and hippocampus in humans. *Science*. 1995;269:1115–1118.
64. LaBar KS, LeDoux JE, Spencer DD, Phelps EA. Impaired fear conditioning following unilateral temporal lobectomy in humans. *J Neurosci*. 1995;15:6846–6855.
65. LaBar KS, LeDoux JE. Partial disruption of fear conditioning in rats with unilateral amygdala damage: correspondence with unilateral temporal lobectomy in humans. *Behav Neurosci*. 1996;110:991–997.
66. Amaral DG, Behniea H, Kelly JL. Topographic organization of projections from the amygdala to the visual cortex in the macaque monkey. *Neuroscience*. 2003;118:1099–1120.

67. Chiba AA, Bucci DJ, Holland PC, Gallagher M. Basal forebrain cholinergic lesions disrupt increments but not decrements in conditioned stimulus processing. *J Neurosci*. 1995;15:7315–7322.
68. Edeline JM. Learning-induced physiological plasticity in the thalamo-cortical sensory systems: a critical evaluation of receptive field plasticity, map changes and their potential mechanisms. *Prog Neurobiol*. 1999;57:165–224.
69. Armony JL, LeDoux JE. How the brain processes emotional information. *Ann N Y Acad Sci*. 1997;821:259–270.
70. Armony JL, Servan-Schreiber D, Cohen JD, Ledoux JE. Computational modeling of emotion: explorations through the anatomy and physiology of fear conditioning. *Trends Cogn Sci*. 1997;1:28–34.
71. Davis M, Whalen PJ. The amygdala: vigilance and emotion. *Mol Psychiatry*. 2001;6:13–34.
72. Whalen PJ, Rauch SL, Etcoff NL, McInerney SC, Lee MB, Jenike MA. Masked presentations of emotional facial expressions modulate amygdala activity without explicit knowledge. *J Neurosci*. 1998;18:411–418.
73. Anderson AK, Christoff K, Panitz D, De Rosa E, Gabrieli JD. Neural correlates of the automatic processing of threat facial signals. *J Neurosci*. 2003;23:5627–5633.
74. Vuilleumier P, Armony JL, Driver J, Dolan RJ. Effects of attention and emotion on face processing in the human brain: an event-related fMRI study. *Neuron*. 2001;30:829–841.
75. Williams LM, Das P, Liddell B, et al. BOLD, sweat and fears: fMRI and skin conductance distinguish facial fear signals. *Neuroreport*. 2005;16:49–52.
76. Walker DL, Davis M. Role of the extended amygdala in short-duration versus sustained fear: a tribute to Dr. Lennart Heimer. *Brain Struct Funct*. 2008;213:29–42.
77. Walker DL, Toufexis DJ, Davis M. Role of the bed nucleus of the stria terminalis versus the amygdala in fear, stress, and anxiety. *Eur J Pharmacol*. 2003;463:199–216.
78. Halladay LR, Zelikowsky M, Blair HT, Fanselow MS. Reinstatement of extinguished fear by an unextinguished conditional stimulus. *Front Behav Neurosci*. 2012;6:18. http://dx.doi.org/10.3389/fnbeh.2012.00018.
79. Dębiec J, Bush DE, LeDoux JE. Noradrenergic enhancement of reconsolidation in the amygdala impairs extinction of conditioned fear in rats—a possible mechanism for the persistence of traumatic memories in PTSD. *Depress Anxiety*. 2011;28:186–193.
80. Cannon WB. *The Wisdom of the Body*. New York: Norton; 1932.
81. Sterling P, Eyer J. Allostasis: a new paradigm to explain arousal pathology. In: Fisher S, Reason J, eds. *Handbook of Life Stress, Cognition, and Health*. New York: John Wiley and Sons; 1988: 629–649.
82. McEwen BS. Stress, adaptation, and disease. Allostasis and allostatic load. *Ann N Y Acad Sci*. 1998;840:33–44.
83. McEwen BS, Stellar E. Stress and the individual. Mechanisms leading to disease. *Arch Intern Med*. 1993;153:2093–2101.
84. McEwen BS. Protective and damaging effects of stress mediators: central role of the brain. *Dialogues Clin Neurosci*. 2006;8:367–381.
85. Peters A, McEwen BS. Introduction for the allostatic load special issue. *Physiol Behav*. 2012;106:1–4.

CHAPTER

2

The Alarm Phase and the General Adaptation Syndrome: Two Aspects of Selye's Inconsistent Legacy

R. McCarty
Vanderbilt University, Nashville, TN, USA

OUTLINE

Abstract

The general adaptation syndrome (GAS) was first proposed by Hans Selye in his classic 1936 letter to the editor of *Nature*. The GAS consisted of three phases: (i) the alarm phase, (ii) the phase of adaptation, and (iii) the phase of exhaustion. Selye held that the stress syndrome was always a nonspecific response of the body to any demand and included a triad of responses: enlargement of the adrenal cortex, decrease in size of the thymus and lymphatic tissue, and ulceration of the stomach and duodenum. Selye also promoted the concept of diseases of adaptation that were connected to stressful stimulation. Much of Selye's work has been discounted as knowledge of neural and endocrine systems expanded and new analytical techniques were introduced. In particular, the doctrine of nonspecificity has been rejected and replaced with the indication that a given stressor stimulates a unique neuroendocrine signature in test subjects. In addition, many studies have demonstrated that prior stress history affects future stress responses across several neural and endocrine systems. Stress remains a key component of the etiology of many diseases and that is an enduring part of Selye's legacy.

INTRODUCTION

On July 4, 1936, Hans Selye published a brief letter to the editor of the journal, *Nature*, in which he summarized a series of experiments on laboratory rats exposed to a variety of "nocuous agents."[1] The nocuous agents included cold, surgical injury, spinal shock, muscular exercise, or sublethal doses of a variety of drugs or tissue extracts. He concluded that the rats developed a consistent constellation of symptoms in three stages that were independent of the nocuous agent employed. The three phases and an abbreviated selection of the symptoms that were observed included:

- *General alarm reaction*: occurred 6-48 h following exposure to a nocuous stimulus and was attended by decreases in the weights of the thymus, spleen, lymph glands, and liver; reduced fatty tissue, loss of muscle tone, decrease in body temperature, gastrointestinal erosions.
- *Adaptation*: from 48 h until 1-3 months after the beginning of repetitive exposure to a nocuous stimulus, the adrenals were greatly enlarged, body growth ceased, there was atrophy of the gonads, cessation of lactation, and enhanced production of thyrotropic and adrenotropic factors from the pituitary. Animals adapted to the deleterious effects of

Stress: Concepts, Cognition, Emotion, and Behavior
http://dx.doi.org/10.1016/B978-0-12-800951-2.00002-9

the nocuous stimulus but were more susceptible than controls to the deleterious effects of another stressor.
- *Exhaustion*: depending on the severity of the nocuous agent, animals died at some unspecified point (usually within 3 months) with symptoms similar to those observed during the general alarm reaction. Selye hypothesized that animals died because they had exhausted their stores of "adaptation energy," though he was never able to measure it.

As recounted by Selye in *The Stress of Life* (1978), his initial experiments with laboratory rats involved injections of ovarian extracts and he hoped to identify a new hormone based upon the stimulation by the extract of a triad of responses, including: (i) adrenocortical hypertrophy, (ii) atrophy of the thymus and spleen, and (iii) gastric ulceration. Much to his dismay, Selye also observed this same triad of responses following injections of other tissue extracts, and also formalin, a painful irritant to tissue. Exposure of rats to cold also resulted in the same triad of responses, and that finding prompted Selye to dismiss his disappointment in not discovering a new hormone and embrace the new opportunity to explore the "stress syndrome" and its potential relevance to diseases in humans.

Remarkably, Selye's 74-line letter to the editor contained no experimental details, no quantitative data, no photomicrographs, and no references.[2] Yet, it has been cited more than 3000 times according to Google Scholar (as of February 2015) and has stimulated an explosion of research on stress and disease since its appearance. How has the field of stress research been affected by this initial report and the subsequent voluminous publication record of Selye over the next 46 years until his death in 1982?

Being the first to propose a new theory doesn't guarantee that one's views will hold up over time. Such may the case with Selye's initial report in 1936 and many of his publications that followed. He very quickly became convinced that the general adaptation syndrome (GAS), or stress syndrome, represented a major new approach to human disease for clinical medicine and he was tireless in promoting his views. A first major step in the promotion of his theory of stress came in 1946 when he gave a series of lectures in Paris at the Collège de France. The significance was not lost on Selye that he would lecture at the same institution where, 100 years earlier, Claude Bernard had lectured on his theory of the *milieu intérieur*. Because there was no readily available translation for "stress" in the French language, those present for his first lecture agreed that *le stress* should be the French equivalent. Stress has now been added as a new word in many other languages, due in large part to the tireless efforts and extensive travel schedule of Hans Selye.[3]

In the prime years of Selye's research career (1940-1980), the exchange of ideas between scientists was limited in a large part to books and articles in print journals, exchange of correspondence through the mail, and discussions at scientific meetings. Over the course of his career, Selye published approximately 1700 articles and 40 books, including some for the general public.[4,5] In particular, he contributed a host of articles for specialty medical and nursing journals and scientific journals to ensure that a broad spectrum of physicians, nurses, and scientists would be aware of his work on the GAS and stress. In many cases, the titles of the journal articles were the same or very similar. He also lectured throughout the world and was fluent in five languages.[6] Imagine what he could have accomplished in promoting his theory of stress if the Internet was available!

KEY POINTS

- Hans Selye first proposed in a brief article in *Nature* a triad of biological responses to stress in laboratory rats that included enlargement of the adrenal cortex, reduction in the weight of the thymus, and gastric ulcerations.
- Selye's general adaptation syndrome (GAS) delineated the time-course of an organism's response to stress. The three phases of the GAS were (i) the alarm phase, (ii) the phase of adaptation, and (iii) the phase of exhaustion.
- Selye defined stress as the nonspecific response of the body to any demand placed upon it. This definition was so broad as to be of little use in designing experiments and making predictions about experimental outcomes.
- Selye's enduring contribution was to establish a link between stress and diseases of adaptation. He was vigorous in conveying his theory of diseases of adaptation to scientific and popular audiences through hundreds of articles, many books, and frequent lectures throughout the world.
- One approach to investigating the effects of prior experiences with stressful stimuli on subsequent responses of an organism to stressful stimulation is to employ experimental designs that include habituation, sensitization, and dishabituation of neural and endocrine systems to stressors.
- A formal and quite exhaustive series of tests of Selye's doctrine of nonspecificity revealed that animals respond with stressor-specific neuroendocrine signatures. Such a finding supports the notion that neural and endocrine systems are exquisitely tuned to respond to the different patterns of homeostatic challenges across a variety of stressors.

CRITICISMS OF THE GAS

From the time he was a medical student in Prague, Selye was subjected to criticisms of his theory of a syndrome of being sick, initially from his professors, and later from his colleagues and senior scientists whom he admired. At the beginning of his clinical training in medical school, Selye was impressed with the constellation of symptoms that was common to many diseases (e.g., coated tongues, swollen glands, generalized body aches, reduced appetite) while his medical school professors were more interested in symptoms that distinguished one disease from another, permitting a differential diagnosis.[3]

When Selye began his studies on endocrine responses to injections of tissue extracts or formalin, he made the connection between those early observations of sick patients as a medical student and a possible physiological basis for the symptoms that were common across several diseases. These nonspecific responses of the GAS could provide a new approach for medical science in the prevention of disease and he was single-minded in pushing for that new approach for his entire career.

Let's begin with Selye's definition of stress—"the nonspecific response of the body to any demand." Given this broad definition, virtually anything could be viewed as a stressor, from getting out of bed in the morning to confronting a robber in a dark alley, to surviving the horrors of an extended tour of duty in a combat zone. Further refinements made by Selye in his theory of stress included the following:

- With exposure to stress, the demand could be pleasant or unpleasant and could result in happiness or sadness. The pleasant demands were defined as *eustress* and the unpleasant demands were defined as *distress*. The biological changes were similar between eustress and distress but the damage to bodily systems was much greater with distress.
- Stress always expresses itself in a nonspecific syndrome and the whole body must be involved.
- Conditioning factors were introduced to explain why individuals differed in their responses to the same stressor. These included genetic differences, differences in prior experiences, and dietary differences. However, if one stripped away these conditioning factors, the constellation of nonspecific responses remained.[7]

Walter B. Cannon was one of the first senior scientists to criticize Selye's formulation of stress during a visit to McGill University. According to Selye's account of their meeting (1978), Cannon expressed reservations about the nonspecific nature of stress and how a nonspecific response pattern could be adaptive to survival of the organism given the plethora of stressors experienced over the lifespan. Ader[8] and Elliott and Eisdorfer[9] pointed out the lack of crisply defined concepts related to stress, such that no definition has yet been advanced that is embraced by a majority of researchers. Munck et al.[10] criticized Selye's concept of diseases of adaptation, noting that this concept was largely dismissed after the 1960s. The circular reasoning undergirding the GAS has also been a source of concern. For example, the only way one could distinguish between distress and eustress was to assess tissue damage and morbidity. If tissue damage occurred, it was a distress response, but if no damage occurred, then it was a eustress response.[11] Mason[12,13,33] argued that all of the stressors originally employed by Selye caused fear and anxiety in the laboratory rats and it was not surprising that the pattern of GAS responses was similar across treatment groups. Indeed, Selye did not incorporate central nervous system responses to emotional stimuli in his studies and yet emotions are especially salient in studies of human stress and disease. Finally, Selye's singular focus for the GAS was the hypothalamic-pituitary-adrenocortical system, to the general exclusion of all other stress-responsive neural and endocrine systems.[14–17,33]

BRIDGING THE CHASM BETWEEN CANNON AND SELYE

Two powerful figures who exerted profound influences on the development of stress research in the twentieth century were Walter B. Cannon and Hans Selye. Most discussions relating to the emergence of the field of stress research emphasize the pronounced differences in approach between these two distinguished scientists (refer to Table 1). In spite of their differences in the study of stress, there is also much to unite them. They could be represented by different points on the same continuum

TABLE 1 A Comparison of the Research Profiles of Walter B. Cannon and Hans Selye

Variable	Cannon	Selye
Underlying motivation	Basic science—mechanisms of homeostasis	Clinical medicine—mechanisms of human disease
Key concepts developed	Homeostasis Fight-or-flight Cannon-Bard theory of emotions	General adaptation syndrome Disease of adaptation Stress concept
Focus of research	Adrenal medulla (epinephrine)	Adrenal cortex (cortisol)
Time scale of experiments	Acute responses to stress	Chronic responses to stress

and their studies appear to me to be more complementary than antagonistic. When their findings are combined across systems, time scales, and stimuli, their combined impact informs basic and clinical approaches to the study of stress and enhances the relevance of both investigators to contemporary stress researchers.

LEARNING ABOUT STRESS

Over the past 30 years, research from many laboratories has demonstrated that prior stress history affects future stress responses. To place these studies into a well-accepted theoretical context, I have drawn upon three types of nonassociative learning in interpreting the results of such studies.[18] The three types of learning are (1) *habituation*, a decrease in the amplitude of the variable being measured after repeated exposure of an individual to the same low- or moderate-intensity stressor; (2) *sensitization*, the enhancement of a nonhabituated stress response by repeated exposure to a high-intensity stressor; and (3) *dishabituation*, the facilitation of an habituated stress response following exposure to a novel stressor.

In the following sections, I will summarize studies from my laboratory on plasma catecholamine responses of laboratory rats to various chronic intermittent stress paradigms as an illustration of the utility of the nonassociative learning framework.[19] An overview of these experiments is presented in Figure 1.[20,21]

Habituation

Chronic intermittent exposure of laboratory rats to a stressful stimulus that is of low to moderate intensity results in a significant reduction in plasma levels of norepinephrine (NE) and epinephrine (EPI) compared to controls stressed for the first time (Figure 1). Several stressors have been tested, including forced swimming, restraint, immobilization, footshock, and exercise. These chronic intermittent stress paradigms provide animals with a high degree of predictability regarding such parameters as the type of stressor, intensity of stressor, duration of each daily stress session, and the time of onset of the stressor session each day. When provided with this information after several daily stress sessions, animals are able to activate stress-responsive neural and hormonal systems to the minimum extent necessary to maintain cardiovascular and metabolic homeostasis during each subsequent exposure to the same stressor. This progressive dampening of plasma catecholamine responses to a familiar and highly predictable stressor affords significant conservation of energy expenditure.

Sensitization

When laboratory rats are exposed to a high-intensity stressor, there are recurring alterations in cardiovascular and metabolic homeostasis. One such stressor that has been studied is immersion of laboratory rats in water maintained at 18 or 24 °C for 15 min per day for 27

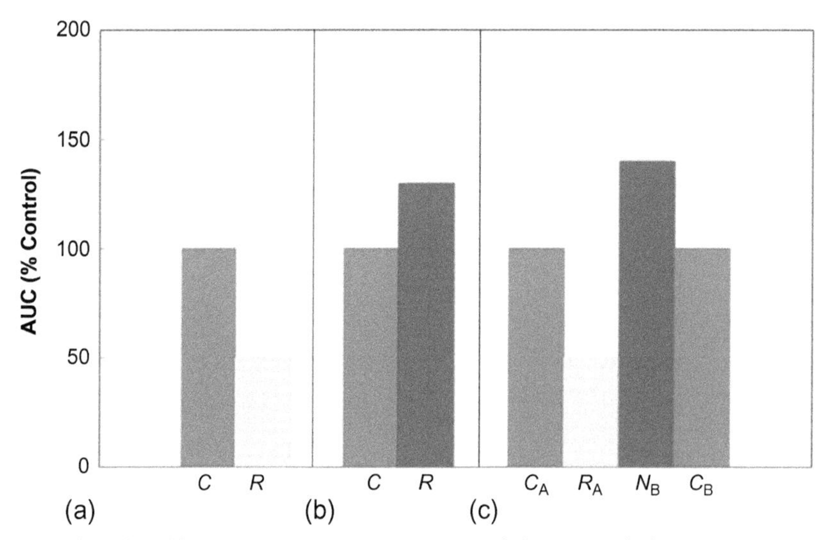

FIGURE 1 Typical experimental results addressing nonassociative properties of plasma catecholamine responses (expressed as a percentage of controls) of rats exposed to chronic intermittent stress. (a) Habituation of plasma catecholamine responses of repeatedly stressed rats to a mild stressor (*R*) compared to first-time stressed controls; (b) sensitization of plasma catecholamine responses of repeatedly stressed rats to an intense stressor (*R*) compared to first-time stressed controls (*C*); and (c) dishabituation of plasma catecholamine responses of rats repeatedly exposed to stressor A and then presented with a novel stressor B. The plasma catecholamine responses of the A and B controls are presented for comparison. Note that in the same group of chronically stressed animals, the response to a familiar stressor (R_A) is diminished compared to the appropriate control (C_A), whereas the response to a novel stressor (N_B) is enhanced compared to the appropriate control (C_B). AUC, integrated area under the curve for plasma catecholamine responses.

consecutive days. When compared to controls exposed to swim stress for the first time, animals exposed to chronic intermittent swim stress had significantly greater elevations in plasma levels of NE and EPI. This sensitized response of the sympathetic-adrenal medullary system to chronic intermittent swim stress is dependent on stressor intensity (i.e., water temperature). In contrast, chronic intermittent swim stress in water at 30 °C results in habituated plasma catecholamine responses.

Dishabituation

Dishabituation involves an enhancement of plasma catecholamine responses to a novel stressor in animals previously exposed each day to an unrelated stressor. For example, if laboratory rats are exposed to a brief period of footshock stress each day for several weeks, the plasma catecholamine response to footshock gradually decreases over time compared to first-time stressed controls (e.g., habituation). However, if these same rats are then exposed to a novel stressor such as restraint stress, their plasma catecholamine responses are amplified compared to the response to restraint stress of first-time stressed controls (Figure 1). Data from our laboratory and the work of others suggest that exposing chronically stressed animals to a novel stressor presents a much greater challenge to behavioral and physiological homeostasis than presentation of that same stressor to a naïve control. In this case, the abrupt departure from the expected appears to be a critical factor in eliciting an enhanced plasma catecholamine response to the novel stressor. These chronically stressed animals live in a moderately challenging, but highly predictable, environment. The experimenter arrives at approximately the same time each day, the stress session begins soon thereafter, the intensity of the stressor and its duration are constant, and most of the remainder of the day is free from disruption. Against this background of consistency, a novel stressor may be viewed as an abrupt, and at times, dramatic departure from what is expected.

Related Studies

Other researchers have employed the experimental paradigms described above and some have expanded the generalizability of these findings by focusing on other neural and endocrine systems. For example, Fernandes et al.[22] reported decreases in plasma corticosterone and decreased expression of CRH mRNA in the paraventricular nucleus of laboratory rats exposed to chronic intermittent restraint stress (30 min per day) for 15 consecutive days. More recently, Babb et al.[23] reported that male and female laboratory rats had reduced plasma adrenocorticotropin hormone (ACTH) and corticosterone responses to chronic intermittent audiogenic stress or restraint stress for 10 days. Other groups have reported evidence of sensitization of neuroendocrine responses to novel stressors in chronically stressed laboratory rats.[24,25] Limitations of space prevent me from a more exhaustive review of the literature. However, nonassociative experimental designs in stress research may have particular relevance for the development of animal models of posttraumatic stress disorder or chronic drug use.

EVIDENCE FOR STRESSOR-SPECIFIC NEUROENDOCRINE SIGNATURES

A consistent hallmark of Selye's theories of stress was the nonspecific nature of the stress response. He emphasized the triad of enlargement of the adrenal cortex, involution of the thymus, and ulcerations of the stomach and duodenum to the exclusion of other neuroendocrine systems and target organs.[26–30] Selye conceded that other neural and endocrine changes could occur during exposure to various stressors; however, there would remain a nonspecific component of stress after deletion of the stressor-specific elements from the total response.

In an elegant series of experiments, Pacak et al.[31] and Pacak and Palkovits[32] subjected the "doctrine of nonspecificity" to an exhaustive test by comparing neuroendocrine response profiles of laboratory rats across a range of different stressors and stressor intensities. In their initial paper, the authors[31] argued that Selye's theory of nonspecific responses is impossible to disprove without two simplifying assumptions. The first is that, regardless of the stressor, the ratio of the intensity-related increment in response for neurohormone X to the intensity-related increment in response for neurohormone Y is a constant. The second assumption is that the magnitudes of both the specific and nonspecific components of the various neurohormones vary directly across the whole range of stressor intensities.[31]

The stressors employed in these experiments included the following:

- Handling and subcutaneous (s.c.) or intravenous (i.v.) injection of 0.9% saline
- Insulin-induced hypoglycemia: i.v. injection of insulin in doses of 0.3, 1.0, or 3.0 IU/kg body weight
- Formalin-induced pain: 1.0% or 4.0% formaldehyde solution injected s.c. in right hind leg
- Hypotensive hemorrhage: removal of arterial blood from an indwelling catheter equal to 10% or 25% of estimated blood volume
- Cold exposure for 3 h at 4 or −3 °C
- Immobilization for 2 h

Blood samples were collected before and at timed intervals during and after exposure of rats to one of the stressors. Plasma levels of NE, EPI, and ACTH were quantified and expressed as an integrated area under the curve (AUC) measure for each animal to reflect the amplitude of the response to a given stressor.

The data for ACTH and EPI responses to hemorrhage and to formalin injection appeared to fulfill the two assumptions required to test the doctrine of nonspecificity. The findings indicated that the ratio of the EPI responses to the more severe versus the less severe hemorrhage was smaller than the ratio of the EPI responses to the more severe versus the less severe formalin injection. In contrast, the ratio of the plasma ACTH response for the more severe versus the less severe hemorrhage was much greater than the ratio of the plasma ACTH response to the more severe versus the less severe formalin injection (Figure 2). The results of this study clearly did not support Selye's doctrine of nonspecificity, and instead were consistent with each stressor evoking a distinct neuroendocrine signature based upon the three hormonal measures employed in this study (NE, EPI, and ACTH).

Pacak and Palkovits[32] expanded upon these initial findings by collecting data on additional neural and endocrine responses to the five stressors described above. The other measures included: plasma corticosterone, NE release in the paraventricular nucleus of the hypothalamus measured by microdialysis, and activation of the immediate early gene, *c-fos*, by quantifying Fos immunoreactivity in 34 brain areas. In combining the results of these two exhaustive studies (refer to Table 2), there is strong support for the existence of distinct stressor-specific neural and endocrine signatures and, based upon their findings, Selye's doctrine of nonspecificity was clearly rejected.

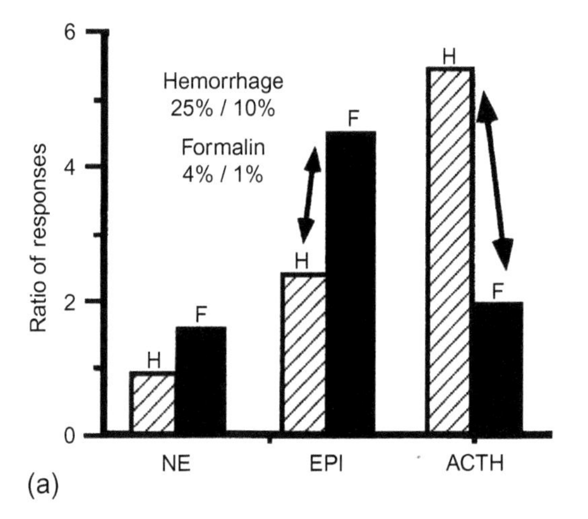

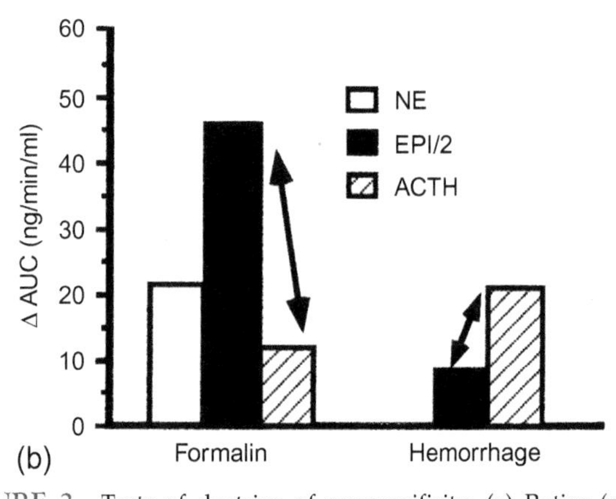

FIGURE 2 Tests of doctrine of nonspecificity. (a) Ratios (greater stress/less stress) of responses of plasma levels of NE, EPI, and ACTH for hemorrhage (H) and formalin (F). (b) Increments in area under curve (ΔAUC) for plasma NE, EPI, and ACTH for formalin and hemorrhage. Doctrine of nonspecificity would predict that arrows should be parallel to each other and the same length. Taken from Pacak et al.[31] with minor modifications.

TABLE 2 Overview of the Effects of Various Stressors on Fos Immunoreactivity in Selected Brain Areas, NE Release in the Paraventricular Nucleus, and Plasma Levels of Various Hormones in Laboratory Rats

Variable	IMMO	COLD	HYPO	HEMO	PAIN
Fos in hypothalamus					
Paraventricular nucleus	+++	+	+	++	+++
Supraoptic nucleus	++	−	±	++	++
Medial preoptic nucleus	+	+++	−−	+	++
Fos in limbic system					
Central amygdala	++	−−	+	−−	++
Hippocampus	±	−−	−−	−−	±
Fos in midbrain					
Substantia nigra	−−	−−	−−	−−	−−
Dorsal raphe	+	−−	−−	−−	+
Fos in pons					
Raphe nuclei	++	+	−−	−−	+
Fos in medulla oblongata					
NTS	+	++	+	+	++
Fos in CA cell groups					
A1	+++	+	−−	+	++
A2	++	±	+	+	+
A6	+++	±	−−	+	+++
A12	−−	−−	−−	−−	−−
NE in paraventricular nucleus	+++	+	+	±	++
Plasma hormone levels					
NE	+++	+++	++	±	+++
EPI	++	±	+++	±	+
ACTH	+++	±	++	+	++
Corticosterone	+++	++	++	+++	+++

+++, *high response*; ++, *moderate response*; +, *low response*; ±, *barely detectable response*; −−, *no response detected*.
Abbreviations: *IMMO, immobilization; HYPO, hypoglycemia; HEMO, hemorrhage; CA, catecholamine.*
Data are summarized from Pacak and Palkovits[32] and Pacak et al.[31]

CONCLUSIONS

It has been almost 80 years since Hans Selye published his classic paper on the GAS in the journal, *Nature*. With Selye as its main spokesperson and cheerleader, the field of stress research attracted many dedicated researchers and the number of published articles is staggering. Thanks in large measure to Selye's efforts, the word *stress* has been added to many languages and it is a frequent topic of discussion in the popular press. The strong link between stress and many diseases is all but taken for granted. These are important aspects of Selye's rich legacy as a thought leader in medical science in the middle part of the twentieth century.

In contrast, Selye's narrow focus on the hypothalamic-pituitary-adrenocortical axis and his insistence on the nonspecificity of the stress response have not stood the test of time. The evidence in favor of stressor-specific neuroendocrine signatures is compelling and underscores the necessity of examining multiple neural and endocrine systems in studies of stressful stimulation. Future breakthroughs in understanding stress-related diseases in humans through the study of animal models will depend on such an approach.[16]

References

1. Neylan TC. Hans Selye and the field of stress research. *J Neuropsychiatry*. 1998;10:230–231.
2. Selye H. A syndrome produced by diverse nocuous agents. *Nature*. 1936;138:32.
3. Selye H. *The Stress of Life*. revised ed. New York: McGraw-Hill; 1978.
4. Szabo S. The creative and productive life of Hans Selye: a review of his major scientific discoveries. *Experientia*. 1985;41:564–567.
5. Taché Y, Brunnhuber S. From Hans Selye's discovery of biological stress to the identification of corticotropin-releasing factor signaling pathways: implications in stress-related functional bowel diseases. *Ann N Y Acad Sci*. 2008;148:29–41.
6. Guillemin R. A personal reminiscence of Hans Selye. *Experientia*. 1985;41:560–561.
7. Taché J, Selye H. On stress and coping mechanisms. *Issues Ment Health Nurs*. 1985;7:3–24.
8. Ader R. Psychosomatic and psychoimmunologic research. *Psychosom Med*. 1980;42:307–321.
9. Elliott GR, Eisdorfer C. *Stress and Human Health*. New York: Springer; 1982.
10. Munck A, Guyre PM, Holbrook NJ. Physiological functions of glucocorticoids in stress and their relation to pharmacological action. *Endocr Rev*. 1984;5:25–44.
11. Goldstein DS. *Adrenaline and the Inner World*. Baltimore, MD: The Johns Hopkins University Press; 2006.
12. Mason JW. A historical view of the stress field: part I. *J Hum Stress*. 1975;1:6–12.
13. Mason JW. A historical view of the stress field: part II. *J Hum Stress*. 1975;1:22–36.
14. Chrousos GP. Stress and disorders of the stress system. *Nat Rev Endocrinol*. 2009;5:374–381.
15. McEwen BS. *The End of Stress as We Know It*. Washington, DC: Joseph Henry Press; 2002.
16. McEwen BS. The neurobiology and neuroendocrinology of stress: implications for post-traumatics stress disorder from a basic science perspective. *Psychiatr Clin N Am*. 2002;25:469–494.
17. McEwen BS. Physiology and neurobiology of stress and adaptation: central role of the brain. *Physiol Rev*. 2007;87:873–904.
18. McCarty R. Stress research: principles, problems and prospects. In: Van Loon GR, Kvetnansky R, McCarty R, Axelrod J, eds. *Stress: Neurochemical and Humoral Mechanisms*. New York: Gordon and Breach; 1989:3–13.
19. Groves PH, Thompson RF. Habituation: a dual-process theory. *Psychol Rev*. 1970;77:419–450.
20. McCarty R, Gold PE. Catecholamines, stress, and disease: a psychobiological perspective. *Psychosom Med*. 1996;58:590–597.
21. McCarty R, Horwatt K, Konarska M. Chronic stress and sympathetic-adrenal medullary responsiveness. *Soc Sci Med*. 1988;26:333–341.
22. Fernandes GA, Perks P, Cox NK, et al. Habituation and cross-sensitization of stress-induced hypothalamic-pituitary-adrenal activity: effect of lesions in the paraventricular nucleus of the thalamus or bed nuclei of the stria terminalis. *J Neuroendocrinol*. 2002;14:593–602.
23. Babb JA, Masini CV, Day HE, et al. Habituation of hypothalamic-pituitary-adrenocortical axis hormones to repeated homotypic stress and subsequent heterotypic stressor exposure in male and female rats. *Stress*. 2014;17:224–234.
24. Belda X, Daviu N, Nadel R, et al. Acute stress-induced sensitization of the pituitary-adrenal response to heterotypic stressors: independence of glucocorticoid release and activation of CRH1 receptors. *Horm Behav*. 2012;62:515–524.
25. Uschold-Schmidt N, Nyuyki KD, Fuchsl AM, et al. Chronic psychosocial stress results in sensitization of the HPA axis to acute heterotypic stressors despite a reduction of adrenal in vitro ACTH responsiveness. *Psychoneuroendocrinology*. 2012;37:1676–1687.
26. Selye H. Stress and disease. *Science*. 1955;122:625–631.
27. Selye H. What is stress? *Metabolism*. 1956;5:525–530.
28. Selye H. The evolution of the stress concept. *Am Sci*. 1973;61:692–699.
29. Selye H. Implications of stress concept. *N Y State J Med*. 1975;75:2139–2145.
30. Selye H. Confusion and controversy in the stress field. *J Hum Stress*. 1975;1:37–44.
31. Pacak K, Palkovits M, Yadid G, et al. Heterogeneous neurochemical responses to different stressors: a test of Selye's doctrine of nonspecificity. *Am J Physiol*. 1998;275:R1247–R1255.
32. Pacak K, Palkovits M. Stressor specificity of central neuroendocrine responses: implications for stress-related disorders. *Endocr Rev*. 2001;22:502–548.
33. Mason JW. A re-evaluation of the concept of nonspecificity in stress theory. *J Psychiatr Res*. 1971;8:323–333.

CHAPTER

3

Corticosteroid Receptor Balance Hypothesis: Implications for Stress-Adaptation

E.R. de Kloet

Leiden University Medical Center, Leiden, The Netherlands

OUTLINE

Abstract

The corticosteroid receptor balance hypothesis refers to the central action of cortisol and corticosterone (CORT) on stress adaptation, which is mediated by mineralocorticoid receptors (MR) and glucocorticoid receptors (GR). Upon imbalance of MR:GR-regulated limbic-cortical signaling pathways, the initiation and/or management of the neuroendocrine stress response is compromised. At a certain threshold this may lead to a condition of hypothalamus-pituitary-adrenal axis dysregulation and impaired behavioral adaptation, which can enhance susceptibility to stress-related neurodegeneration and mental disorders. Here the progress is presented to test this hypothesis from the perspective of CORT coordinating three complementary phases of stressful information processing. First is the onset of the stress reaction when MR mediates CORT action on appraisal of novel information and emotional reactivity. Second is the termination characterized by CORT promoting behavioral adaptation and memory storage. Third is the basal phase when ultradian and circadian oscillations of CORT permit recovery and growth, while maintaining responsivity to stress.

INTRODUCTION

The stress concept has been a source of inspiration for almost a century for numerous investigators to explain diseases of adaptation.[1] Central to this concept of stress and adaptation are the sympathetic nervous system and the hypothalamus-pituitary-adrenal (HPA)-axis, the latter with its end products cortisol and corticosterone (collectively abbreviated as CORT). In recent years, extensive overviews of the role of HPA-axis hormones in the pathophysiology of stress-related mental disorders have appeared.[2–5] This chapter is about CORT and its receptors in the brain that act in concert with neurotransmitters, neuropeptides, and other factors in the onset, termination, and recovery of the stress response.

Stress: Concepts, Cognition, Emotion, and Behavior
http://dx.doi.org/10.1016/B978-0-12-800951-2.00003-0

KEY POINTS

- The failure to cope with a stressful experience enhances vulnerability to disease for which the individual is predisposed.
- The nature and context of this experience determines circuit connectivity involved in processing of stressful information.
- Circuit connectivity is biased by stress hormones such as corticosteroids.
- Corticosteroids act as a double-edged sword by affecting in a complementary manner first appraisal of and then adaptation to stressful information via brain MR and GR, respectively.

The CORT receptor balance hypothesis, or corticosteroid receptor (CoRe) Balance hypothesis, is fundamental for understanding the role of CORT in stress coping, adaptation, and recovery. The hypothesis is based on the action of CORT, which is mediated by two receptor systems: the mineralocorticoid receptors (NR3C2, MR) and the glucocorticoid receptors (NR3C1, GR). MR and GR act as nuclear receptors in gene transcription and are as membrane variants engaged in rapid nongenomic membrane actions. These actions exerted by CORT via MR and GR coordinate and integrate—as hormones do—the function of systems, tissues, organs, circuits, and cells in body and brain. The focus here is on brain because of its crucial role in anticipation and processing of stressful information, which upon dysregulation may result in pathogenesis of stress-related diseases. Hence the CoRe Balance hypothesis formulated almost 30 years ago,[6–10] reads as follows (Box 1).

The CoRe Balance hypothesis has its roots in Selye's original *pendulum hypothesis* where mineralocorticoids and glucocorticoids were considered antagonistic adaptive hormones. In Selye's hypothesis, a relative excess of mineralocorticoid was thought to predispose the individual to inflammation and hypertension, while excess glucocorticoids increased the risk for infection.[1] Importantly in the current CoRe Balance hypothesis, Selye's opposing hormone actions are now represented by MR and GR that can mediate in a complementary fashion the action of one single adaptive hormone: CORT. MR overactivity relative to GR seems linked to anxiety and aggression-related disorders, while MR underactivity appears linked to depression (Figure 1).

Here the evolution of the stress concept (Box 2) is highlighted from the perspective of CORT action aimed to promote stress adaptation. The MR and GR that mediate CORT action differ in localization and properties that enable activation of signaling pathways in three distinct domains of stressful information processing: onset, termination, and recovery. The importance of the MR:GR balance is tested by either manipulation of the MR- and GR-multimeric complexes or the bioavailability of their ligands. The chapter concludes with a translational perspective of the current knowledge of diagnosis and treatment of stress-adaptation diseases.

> **BOX 1**
>
> **CORTICOSTEROID RECEPTOR BALANCE HYPOTHESIS**
>
> Upon imbalance in MR:GR-regulated limbic-cortical signaling pathways, the initiation and/or management of the stress response is compromised. At a certain threshold, this may lead to a condition of HPA-axis dysregulation and impaired behavioral adaptation, which can enhance susceptibility to stress-related neurodegeneration and mental disorders.

MR, GR, AND CELLULAR HOMEOSTASIS

GR is expressed in all neurons and glial cells, but expression is particularly high in cells engaged in the stress response, (e.g., ascending serotonergic, (nor)adrenergic and dopaminergic neurons, paraventricular (PVN), hippocampus, amygdala, and prefrontal cortex). MR expression is more restricted and occurs in abundance in all hippocampus and dentate gyrus neurons, lateral septum, and other areas of the limbic network involved in processing salient information. While thus MR and GR are coexpressed abundantly in limbic neurons, the most profound difference is, however, the 10-fold higher affinity of MR for CORT than displayed by GR (Box 3).[11]

Genomic effects mediated by MR and GR have a slow onset and are long-lasting. The effects occur with a delay of at least 30 min and may last for several hours. Marian Joëls demonstrated with intracellular recording of CA1 neurons in in vitro hippocampal slices that cellular excitability displayed an U-shaped curve with rising concentrations of CORT.[14] Thus ion conductances and transmitter responses were large in the tissue of adrenalectomized animals lacking steroid as well as during very high CORT concentrations occupying both MR + GR. The cellular responses were small in the presence of low CORT concentrations occupying mainly MR.

Three MR-mediated responses seem to be most important in hippocampal CA1 neurons: the reduction of 5HT1A hyperpolarization, the stability of amino acid transmission, and the decrease of Ca and Ca-dependent conductances. The small magnitude of these responses during MR activation results in stabilization of ongoing activity,

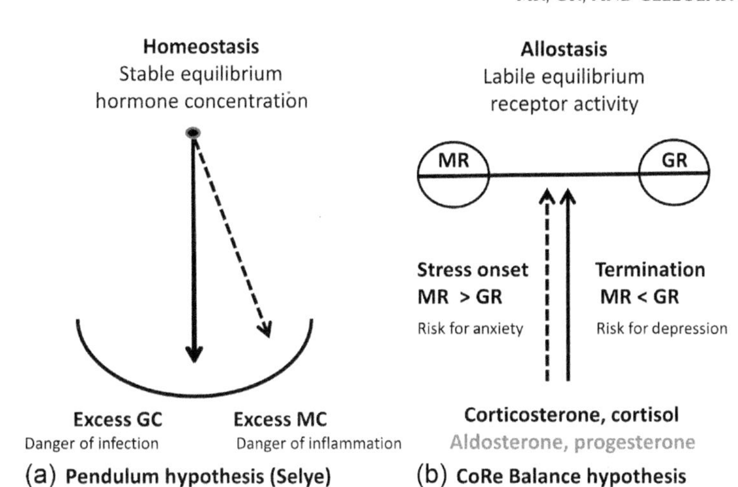

FIGURE 1 *Hypotheses.* (a) Pendulum hypothesis as formulated by Selye[1] more than 60 years ago showing the balance in action between mineralocorticoid (MC) and glucocorticoid hormones (GC) maintaining. Excess MC was thought to enhance the danger of inflammation; excess GC the danger of infection. This hypothesis was focused on maintaining a stable equilibrium or homeostasis. (b) Corticosteroid receptor balance hypothesis (CoRe Balance) showing the balance of MR:GR-mediated actions maintaining a labile equilibrium.[6–9] The extent of MR and GR activity can slide over the two arms as a function of hormone concentration and receptor properties. MR is concerned with emotional reactivity and excess MR over GR seems linked with anxiety and aggression-driven disorders. Inadequate MR vs. GR is a characteristic feature of major depressive disorder. Dashed line represents a different hormone concentration.

BOX 2

THE STRESS CONCEPT

Stress: Hans Selye coined the term "stress"[7] as "*a state of nonspecific tension in living matter, which manifests itself by tangible morphologic changes in various organs and particularly in the endocrine glands which are under anterior pituitary control.*" This "state of stress" is evoked by a stressor, which is defined as any stimulus that disrupts cellular "homeostasis" or, on the organismic level, as "a real or interpreted threat to the physiological and psychological integrity."[2,61] The stressor activates the organism's defense reactions, which are coordinated by the sympathetic nervous system and neuroendocrine HPA-axis.

CORT and stress: Alan Munck[4,62] reasoned that the action of the stress mediators are integrated over time. Hence excitatory transmitters, adrenaline and the neuropeptides of the HPA-axis are secreted rapidly in response to a stressor; part of their action is also fast, lasting over seconds to minutes. Because adrenal CORT secretion rises only minutes after the stressor, the hormone has actually a special role in the timing of this cascade of rapid initial stress reaction. As stated by Munck: "*CORT prevents initial stress reactions* (e.g., *autonomic, immune, inflammatory, metabolic, brain) from overshooting and becoming damaging to themselves.*"

Coping: Since all physical stressors have a psychological component, some researchers rather restrict the definition of stress to "*conditions where an environmental demand exceeds the regulatory and adaptive capacity of an organism, in particular in case of unpredictability and uncontrollability.*"[12] The most stressful condition is: no information, no control, and no prediction of upcoming events with an uncertain feeling of threat. A safe place, social context, and self-esteem help to cope.[63] It is not so much what happens but rather how the individuals takes it and copes. It implies that anticipation and appraisal of a stressful condition is extremely important, and this is governed by MR.

Allostasis: To capture this "state of readiness" in the face of a presumed threat, the concept of "allostasis" was introduced, when structure and function of brain networks already adjust in order to be prepared. Allostasis describes a labile equilibrium characterized by variable setpoints (cf a juggler who keeps dinner plates in delicate balance on a pointer) as opposed to homeostasis where the stability of pH, electrolyte concentration or body temperature is a *conditio sine qua non*. The cost of allostasis through energy consuming anticipatory adaptations is called allostatic load.[3,64] (See Chapter 5 by McEwen.)

which is favorable for neuronal integrity and survival. This condition can be achieved after substitution of adrenalectomized animals with small amounts of CORT or aldosterone in sufficient amounts to occupy MR, suggesting that the changes are due to steroids and not due to disturbances in catecholamines released by the adrenal medulla.[14,15] Brief GR activation is required to reverse the MR-mediated increase in excitability. However, the CA1 is unique for such U-shaped responses, since different patterns are observed elsewhere in the brain.[15]

Karst and Joëls demonstrated that MR promotes via a rapid membrane effect the release of glutamate and simultaneous decreases postsynaptically K^+ to increase excitability.[16] The enhanced glutamate release

BOX 3

CORT RECEPTORS AND BLOOD-BRAIN-BARRIER

Bruce McEwen[65] discovered that ^{3}H-corticosterone administered to adrenalectomized rats was retained in cell nuclei of the hippocampus and other limbic brain regions through binding to nuclear receptors. Yet, if the potent synthetic glucocorticoid dexamethasone was administered, it did not compete for CORT binding and neither was it retained very well in brain cell nuclei.[66] First it was found that the receptors with very high affinity for CORT actually are MR.[11,67,68] Then, 30 years later we discovered that dexamethasone is recognized by the mdr-Pgp in the blood-brain-barrier, which severely hampers its central penetration. The access of cortisol, but not of corticosterone, to the brain, is also hampered.[44,45]

downregulates the presynaptic mGLU2 receptor.[17] This rapid MR-mediated CORT action was not inhibited by protein synthesis inhibitors and occurred only in the presence of hormone, also when CORT was conjugated to BSA to prevent it from acting intracellularly. Trafficking of AMPA receptors at the postsynaptic membrane was enhanced by CORT with a short onset and long-lasting.[18,19] Rapid actions were also observed for GR involving endocannabinoid release causing presynaptically an immediate suppression of the release of excitatory transmitters.[20] Interestingly, membrane MR has a 10-fold lower affinity than its nuclear variant, allowing this membrane MR version to respond to rising CORT levels.[16]

PROCESSING OF STRESSFUL INFORMATION

After having attended numerous panel discussions centered around the seminal question: "what is stress?" one of the pioneers, Seymour Levine, used an operational definition: *"Stress is defined as a composite multidimensional construct in which three components interact:* (i) *the input, when a stimulus, the stressor, is perceived and appraised,* (ii) *the processing of stressful information, and* (iii) *the output, or stress response. The three components interact* via *complex self-regulating feedback loops with the goal to restore homeostasis through behavioral and physiological adaptations."*[21]

With this operational definition in mind, in this section the autonomous, neuroendocrine, and behavioral features of stress adaptation will be integrated with the action of CORT via nongenomic and genomic MR and GR. Regarding the neuroendocrine feature the focus is on the well-known cascade of HPA-axis hormones starting with hypothalamic CRH and vasopressin that stimulate after transport via the pituitary portal vessels the synthesis of pro-opiomelanocortin in the anterior pituitary corticotrophs and the release in the circulation of corticotropin (ACTH) which stimulates melanocortin-2 receptors on the adrenocortical zona reticularis to secrete CORT.

In the next sections, the action of CORT is discussed on the three phases of stressful information processing (Box 4). These include the initial phase of the stress reaction followed by stress adaptation which terminates the stress response while reinstating the basal function of CORT. The latter phase is characterized by oscillating CORT levels which act permissively to promote growth and the storage of energy resources important for the state of readiness and stress responsivity. Anticipation and accompanying arousal are important determinants for readiness.

Anticipation

Anticipation of upcoming stressful events is affected by CORT as is demonstrated by schedule-induced behavior. This occurs when rats are presented restricted feeding at a fixed point of the day. For instance, if rats receive at 16.00 h only 80% of their daily need of food, the animals

BOX 4

THREE PHASES IN CORT ACTION

Onset stress reaction: MR localized in limbic regions mediates rapid CORT action in anticipation and appraisal of novel information, retrieval of previous experience, response selection, and emotional reactivity expressed as fear and aggression. MR is a determinant in the initiation of the integrated autonomic, neuroendocrine, and behavioral stress reaction.

Stress adaptation: When the experience is perceived as a stressor, CORT levels rise, GR activation reallocates energy resources to limbic-frontocortical circuits engaged in executive function and memory storage, and behavioral adaptation occurs, terminating the stress reaction and reinstating a basal state.

Recovery: During this basal phase CORT secretion shows ultradian and circadian oscillations and acts permissively to support neuronal plasticity, growth, and storage of energy resources to maintain stress responsivity for upcoming challenges. This state of readiness can vary depending on arousal triggered by anticipation of all kinds.

show increased arousal and strongly enhanced HPA-axis activation in anticipation of food. Arousal and HPA-axis activity drop precipitously when the food is presented. The animals also show displacement behavior by excessive wheel running or polydipsia in anticipation of the food. Adrenalectomy prevents normal acquisition of this schedule-induced behavior, while it can be reinstated by CORT but not dexamethasone replacement. Since dexamethasone alone is not active, CORT activation of MR seems a prerequisite.[22]

Appraisal and Selection of Behavioral Response

The perception (or anticipation) of a salient situation instantaneously triggers an alarm reaction expressed as generalized arousal caused by enhanced activity of ascending excitatory pathways stemming from the n. gigantocellularis.[23] At the same time the novel information is processed in the limbic-cortical circuitry to make decisions, such as: is this individual a friend or a foe? Was a similar situation encountered before and how was it dealt with? How did I cope? These are questions that require retrieval of previous experiences to support appraisal and decision-making. If the outcome of this appraisal process is interpreted as a threat to integrity, defensive reactions are activated. Central norepinephrine and epinephrine pathways from the n. tractus solitarii (NTS)/A2 as well as the locus coeruleus/A6 are activated resulting in increased attention, alertness and vigilance. The CRH neurons in the PVN and amygdala orchestrate the sympathetic, neuroendocrine and behavioral stress response.[24]

The amygdala generates emotionally loaded information, which is labeled in time, place, and context when processed in the hippocampus. In mutual feedback and feedforward loops, the amygdala and hippocampus communicate with frontal brain regions. Over time, energy resources are allocated, notably to the mesolimbic-cortical DA pathways of reward and adversity, and the prefrontal cortex in which subregions are involved in specific higher cognitive functions. The appraisal process drives coping behavior to fulfill an expectancy. If coping behavior is adequate the stress response is extinguished. If inadequate, then feelings of uncertainty and anxiety, as well as the stress response are reinforced.

Anxiolytic and Antiaggressive Activity of MR Blockade

Mary Dallman[25] identified a fast rate-sensitive feedback mechanism of CORT which operates within minutes to attenuate the initial activation of the neuroendocrine stress response. Subsequent research demonstrated that this fast rate-sensitive feedback action probably proceeds via an inhibitory transmitter network around the PVN (Box 5). One candidate for fast feedback is the GABA-ergic network that receives projections from limbic structures, another candidate the endocannabinoids.[26]

MR functioning is tightly linked to this initial phase of the stress response and mediates the effect of CORT on the appraisal process, risk assessment, response selection, and emotional expressions of fear and aggression. Thus, blockade of MR by administering MR-antagonists interferes with acquisition and retrieval of fear-motivated behavior.[5] Blockade of MR also interferes with the propensity to aggressive behavior, provided this blockade is present prior to the first violent encounter.[27] Blockade of MR attenuates sympathetic outflow as measured from the attenuated stress-induced blood pressure response.[28] MR antagonism also rapidly blocks excitatory outflow from the hippocampus toward the inhibitory GABA-ergic PVN network governed by excitatory transmission, and thus can activate the HPA-axis resulting in higher basal and stress-induced ACTH and CORT levels.[29] Interestingly, the opposite effect is observed after administration of antiglucocorticoids in the hippocampus, causing enhanced inhibition of stress-induced ACTH.[7] The blockade of hippocampal GR apparently enhances MR-mediated hippocampal inhibition of the HPA-axis.

BOX 5

THE DIFFERENT MODES OF NEUROENDOCRINE CORT FEEDBACK[9,69]

Fast feedback: Operating within minutes via a GABA-ergic network at the PVN level that attenuates stress-induced HPA-axis activation and involves a rapid action of CORT via membrane MR and GR, the latter mediated by endocannabinoid.

Intermediate feedback: From 30 min to 2 h, that is regulated by the action of CORT in higher brain regions with the goal to promote stress adaptation and as a consequence the drive toward the PVN subsides.

Emergency brake: Exerted by extreme levels of CORT that exceed CBG capacity in blood and pituitary corticotrophs. Dexamethasone bypasses the CBG barrier and targets the pituitary, but the synthetic glucocorticoid poorly penetrates the brain.

Variable setpoint regulation[69]: Epigenetic factors are recruited by GR to clamp CRH gene transcription reciprocally with the concentration of bioavailable CORT and the drive from higher brain centers. The variable setpoint regulation requires further proof.

Habit Versus Spatial Learning

If individuals are exposed daily to homotypic stressors, they habituate and stress-induced ACTH and CORT responses gradually diminish. If prior to the daily stressor exposure, the animals are given MR-antagonists, this habituation is prevented and thus CORT responses remain the same.[30] In the behavioral realm, hippocampal MR appeared to have a crucial role in the stress-induced switch from spatial declarative performance toward caudate stimulus-response (habit). For instance, in the circular hole board test, naive male mice locate with a hippocampal-associated spatial strategy an exit hole at a fixed location flagged by a proximal stimulus. However, if exposed to a stressful context, close to 50% of the mice perseverated in the previously learned response.[31] This habit response was associated with hypertrophy of the caudate and atrophy of the hippocampus under chronic stress conditions.[3,32]

The switch toward the caudate stimulus-response or habit strategy was accompanied by a rescue of performance. However, the performance declined of the stressed mice that kept using the hippocampal spatial strategy as could be predicted from their atrophied hippocampal circuitry. Pretreatment with an MR antagonist did prevent the switch toward the stimulus-response strategy, but did not improve the deterioration of hippocampus-dependent performance in these stressed mice. These findings show that MR-mediated CORT action is linked to flexibility in the transition from spatial to stimulus-response memory systems. Also in humans, stress prior to learning facilitated simple stimulus-response behavior at the expense of a more cognitive learning strategy.[31]

The sexes differ strikingly in their MR-dependent cognitive flexibility. In spatial tasks, male mice performed superior to females. As mentioned, under stress spatial memory of males was impaired, while females actually improved their spatial abilities, depending on the task and type of stressor. Moreover, we found in the females that the performance of a spatial strategy depends on the phase in the estrous cycle, with females in estrus being most resistant to cognitive deterioration during stressor exposure.[33] These findings further support the relevance of MR for behavioral flexibility.

Behavioral Adaptation

For study of the complementary role of MR and GR in behavioral performance, we selected the Morris water maze, because this is a test of the function of the hippocampus. Melly Oitzl observed that removal of the adrenal medulla did not affect maze performance, while complete adrenalectomy did.[34] This finding shows that physiological levels of peripheral catecholamines are not implicated in learning, but synergism with noradrenaline in the amygdala is required for adequate learning of the task.[5]

Distinctly different effects were observed with antagonists for MR and GR. intracerebroventricular (ICV) infusion of the MR antagonist—but not of the GR antagonist—impaired retrieval of learned information if given 24 h later at 15 min before the retrieval test. The rats used a different strategy to search for an escape route, than the intact animals. The GR antagonist mifepristone impaired learning of the maze when administered immediately after the learning trial, and thus interferes with consolidation of the learned information. This observation supports the evidence that CORT released after stressful learning tasks facilitates memory storage via GR. In addition, GR activation promotes extinction of fear-motivated behavior when given after retest in the absence of the unconditioned stimulus.

CORT acts on higher brain circuits to facilitate behavioral adaptation causing dissipation of the drive toward the PVN to release CRH. This so-called intermediate feedback mechanism takes 30 min to a few hours depending on the ability to cope and adapt. If a GR antagonist is given behavioral adaptation is impaired delaying the intermediate feedback.[29] A recent study using ChIP failed to show binding of GR near the *Crf*-promotor, while p-CREB stimulated by afferent projections did bind indicating that CRH cannot be a primary feedback site for CORT.[35] The pituitary also was ruled out as a primary feedback site because selective GR knockout in pituitary corticotrophs did not alter HPA-axis activity in the mutants.[36] This lack of effect of CORT in adult animals probably is due to the presence of tissue-like corticosteroid binding globulin (CBG) in pituitary corticotrophs that prevents CORT from binding to the GR.[9] Hence only very high CORT levels exceeding CBG capacity during extreme stress would serve as some kind of emergency brake (Box 5).

Setpoint in CORT Oscillations

Termination of stress-induced HPA-axis activation as a result of behavioral adaptation raises the question of what the CORT feedback on CRH transcription in the PVN means? One attractive viewpoint comes from the studies of Elliott and Chen[37] showing that elevated CORT activates GR to promote an epigenetic mechanism meant to keep a sustained level of CRH expression for a prolonged period of time. This epigenetic action implies a mechanism that seems more concerned with regulation of the HPA-axis setpoint involving recruitment of methyltransferases and histone (de)acetylases by CORT in the PVN[37,38] and possibly also elsewhere in the brain. This would explain why GR knockout from the mouse PVN parvocellular neuron increased circulating ACTH and CORT levels[39]; it would represent a rise of basal ACTH levels as is the case after adrenalectomy.

Also MR is involved in setpoint regulation, since MR-antagonists infused systemically, ICV or in the hippocampus elevate basal circulating CORT in animals and man for a prolonged period of time, both at the peak and the trough. This suggests that the nuclear MR which is largely occupied by circulating hormone seems to participate in regulation of the tone of the HPA-axis.[6] Chronic GR blockade with moderate doses antagonist also causes a pronounced adrenal hypertrophy with deeper troughs and higher peaks in circadian CORT patterns.[40]

The circadian pattern entrained by the day-night rhythm is based on the ultradian rhythm in CORT secretion of about 1 pulse per hour. The ultradian rhythm is—as a computational analysis suggests—intrinsic to the HPA-axis and entirely caused by the delay between central drive toward the HPA-axis and the negative feedback action of CORT.[41] The amplitude of the pulse is largest at the circadian peak in anticipation of the energy expenditure for the upcoming activity period. It appears that the affinity of MR is high enough to stay occupied over the interpulse interval[42]; in contrast, GR occupation follows the ultradian changes of cortisol levels.

The ultradian rhythm was shown to support responsiveness to stressors.[42] The pattern can change in frequency and amplitude, and becomes disorganized at old age, implying that the organism is less well prepared to deal with stressors. CORT oscillations under basal conditions are very important for recovery. First, under these conditions CORT is permissive for growth, recovery, and storage of energy resources in cooperation with the parasympathetic nervous system and trophic factors.[4] Second, because of the storage of resources, the individual is brought into a state of readiness to deal with upcoming challenges as appears from the CORT awakening response which occurs the first hour after awakening and prepares the individual and its energy resources in anticipation of the day to come.

TESTING THE BALANCE HYPOTHESIS

In this section, two categories of factors are briefly discussed that potentially can affect the outcome of the coordinate MR:GR-mediated actions: the bioavailability of CORT and changes in its receptors.

Changing Bioavailability of CORT

Bioavailability depends on binding to CBG, multidrug resistance P-glycoprotein (mdr-Pgp) in the blood-brain-barrier and intracellular enzymes such as 11βHSD-1 (11β-hydroxysteroid-dehydrogenease type 1) and -2. For instance, blockade of 11βHSD-1 by glycerrhetinic acid or its analogs limits regeneration of CORT and therefore may prevent unwanted effects. Alternatively, also gene delivery of additional 11βHSD-2 inactivating excess CORT in the hippocampal dentate gyrus can reverse its damaging effects.[43]

Mdr-Pgp recognizes in the rodent exogenous cortisol and synthetic glucocorticoids as substrate, but not endogenous corticosterone. Dexamethasone and cortisol therefore poorly penetrate the blood-brain-barrier.[44,45] If dexamethasone is given in low doses the HPA-axis is suppressed and CORT levels are decreased, but the synthetic steroid hardly substitutes in brain. Moreover, since dexamethasone does not have an affinity for MR, in particular, the MR becomes depleted of endogenous hormone and as a consequence the MR:GR balance is severely disturbed even though MR levels increase in compensation to lack of ligand.[46] One way to ameliorate this deficit is by giving CORT as add-on to dexamethasone treatment. Indeed in animal experiments, the need for additional intermittent administration of CORT to promote learning-dependent synaptic plasticity was demonstrated.[47] This finding provides proof-of-principle for a combination therapy with CORT to restore the MR:GR balance as an approach to limit the severe adverse effects of prolonged therapy with synthetic glucocorticoids.

Genetic and Epigenetic Changes in MR:GR Balance

The Lewis rat shows in comparison with Wistars enhanced inflammatory and immune responses in the face of a reduced HPA-axis response. The Lewis rat shows increased expression of hippocampal MR. Hence, an augmented hippocampal MR-mediated effect of CORT underlies likely its hyporeactive HPA-axis. The lower hypothalamic CRH mRNA and circulating ACTH and CORT in response to various stimuli are apparently the consequences of a life-long suppressive action of CORT via central MR.[48]

Using mice with forebrain MR over-expression (MR-high) and/or simultaneous global GR under-expression (GR-low), a significant interaction was found between MR and GR in control of the HPA-axis. With reduced overall GR levels, HPA-axis activity in response to restraint stress was enhanced. However, in combination with high limbic MR, this excessive stress-induced HPA-axis activation was very much reduced.[49] MR:GR balance also played a role in determining the behavioral strategy during Morris watermaze performance. In combination with GR-low, MR-high showed enhanced perseveration in the probe trial in the retrieval phase 24 h after maze learning, suggesting enhanced spatial memory recall or reduced exploratory flexibility. Other alterations in cognitive functions were specific to a single receptor without interaction, with both MR-high and GR-low manipulations independently impairing reversal learning in spatial and fear memory tasks.[49]

Genetic variants of MR, GR, and their regulatory proteins such as FKBP5 have been identified that appeared to be associated with risk for depression and efficacy of antidepressant therapy. In vitro the MR gene has three levels of control (promoter, translation, primary structure) that can be modified by genetic variability. In vivo these different levels of genetic control merge into three haplotypes in the MR promotor region with frequencies of 50%, 35%, and 12%, respectively.[50] In a series of human studies, MR haplotype 1 and 3 associate with basal and stress levels of cortisol, psychological measures, signs of depression, suicidal ideation, and life history. MR-haplotype 2 associates in women with dispositional optimism and protection against depression.[50]

The GR variant N363S was found hypersensitive to cortisol and associated with an unhealthy metabolic profile while ER22/23EK is linked to steroid resistance and enhanced risk of depression. The Bcl-1 polymorphism predicts cardiovascular risk and contributes to individual differences in emotional and traumatic memories as well as PTSD symptoms after intensive care treatment. Currently, trials are underway to exploit this knowledge on MR and GR variants for its potential as a biomarker to support diagnosis of anxiety disorders and depression.[51]

MR and GR expression shows life-long changes in parallel with the amount of maternal care the rodents experience,[52] probably because of stable changes in histone acetylation and DNA methylation. This epigenetic signature imposed by early life experience is a determinant of plasticity in neuronal networks underlying later emotional expression and cognitive performance, and seems a significant factor in the precipitation of stress-related mental disorders. In particular, the epigenetic change in GR was found to be associated with a programmed HPA-axis and behavioral response patterns. This is not restricted to rodents. Also in man, decreased expression of GR due to DNA methylation in the hippocampus induced by emotional neglect in early life was found to be associated with adult neuro-behavioral endpoints.[53]

Effect of Chronic Stress

According to Selye, *"the imperfections of the adaptation syndrome"*[7] coincide with an altered balance in adaptive hormones and are important in the pathogenesis of most stress-related diseases. Selye referred in this context to the pendulum hypothesis, where excess mineralocorticoid over glucocorticoid enhanced vulnerability to inflammation whereas the reverse enhanced risk of infection.[1] Chronically stressed animals show profound changes in neuroendocrine regulations due to an altered phenotype of the CRH neurons expressing much more vasopressin as co-secretagog[54] and profound changes in brain plasticity.[3]

In animal experiments using dentate gyrus (where neurogenesis occurs) of controls 26 different GO terms could be assigned in pathway analysis, but the diversity in the CORT responsive pathways was in the stressed group reduced to only 7. After chronic stress, CORT or acute stress induced particularly genes involved in chromatin modification, epigenetics, and the cytokine/NFκB pathway.[55] One highly responsive gene network revealed by this procedure is the mammalian target of rapamycin (mTOR) signaling pathway which is critical for different forms of synaptic plasticity and appears to be associated with depression.

Since CORT challenge was used to identify dysregulated pathways in limbic regions of the chronically stressed animals, it may also represent a target for treatment. Indeed, enhanced expression of MR locally in the hippocampus or amygdala was protective to the effect of stress. Reduced MR expression is observed during the aging process[56] and depression.[57] Furthermore, in such stressed animals blocking GR with an antagonist improved cognitive performance,[7,58] reversed suppression of neurogenesis, Ca current and long term potentiation (LTP),[5] and rescued the CREB-signaling pathway.[59] Antiglucocorticoid treatment or genetic deletion of GR after chronic stress restored the hyperactive dopaminergic mesolimbic/cortical-amygdala loop and social behavior.[60]

CONCLUDING REMARKS

According to the CoRe Balance hypothesis, imbalance in MR:GR-regulated limbic-cortical signaling pathways causes HPA-axis dysregulation which would after passing a certain threshold impair behavioral adaptation resulting in enhanced susceptibility to stress-related mental disorders. Studies over the past 25 years revealed that the strength of this hypothesis is in the integration and coordination of MR- and GR-mediated actions over time in neuroendocrine regulation associated with behavioral performance. Thus, reduced MR expression in limbic regions associates with an increased initial HPA-axis response to stress, while a reduced GR causes at a later time point in a more prolonged stress response.

These altered neuroendocrine responses over time actually reflect temporal changes in processing of stressful information (Figure 2). Thus, an impaired MR function frustrates proper appraisal and decision-making which impairs subsequent coping, leading to

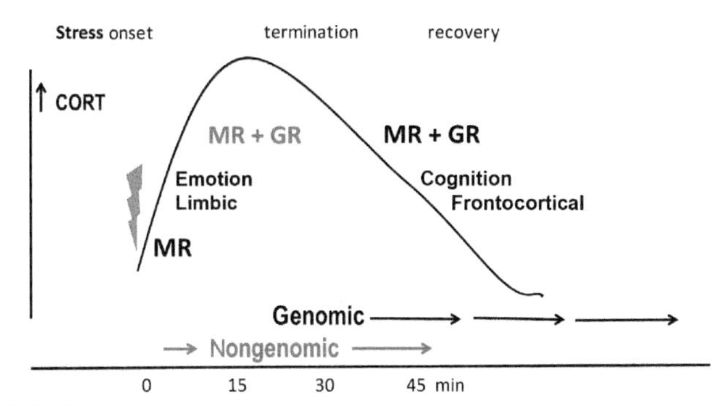

FIGURE 2 Time-dependent CORT action on processing of stressful information. CORT action on stress onset, termination, and recovery. Increasing CORT concentration initially affects emotional reactivity via nuclear MR and then progressively activates also nuclear GR to reallocate energy to circuits underlying behavioral adaptation and memory storage of the experience. CORT returns to baseline oscillations allowing recovery. In red: membrane MR and GR have a lower affinity and rapidly respond to rising CORT concentrations, while the nuclear receptors mediate CORT action on gene transcription with a slow onset producing primary, secondary, etc. waves of gene transcripts. Adapted from de Kloet.[8]

an altered autonomic and neuroendocrine response. An impaired brain GR function results in an ineffective adaptation as is apparent from prolonged CORT secretion. Hence, impaired behavioral adaptation is stored in the memory, which is retrieved again via MR when a similar salient event is anticipated (either real or imagined) or experienced. Accordingly, impaired MR and GR functions leave their mark on the third, restorative, phase changing the pattern of ultradian CORT oscillations permissive to the extent of growth and plasticity and altering the mechanism underlying stress responsivity. The three phases therefore interact and are mutually dependent. As a working hypothesis, we postulate that MR>GR increases the risk for anxiety- and aggression-driven disorders and MR<GR would enhance the risk for major depressive disorder (Figure 1).

The rationale for this thesis finds support from fMRI studies where acute stressors initially activate the MR-containing limbic salience network, while subsequently over time, energy resources are reallocated to frontocortical executive functions.[70] The mechanistic underpinning of this dynamic adaptation of brain networks is reflected to some extent in cellular homeostasis and circuit responses. Ion conductances, transmitter responses, and LTP show U-shaped CORT dose responses in hippocampal CA1,[14,15] which is part of the limbic-cortical network involved in spatial learning.

The distinction in the three mutually interactive phases in the stress response that are under differential control of CORT via genomic and nongenomic actions have heuristic value for endophenotypes featuring characteristics of each of these phases. One approach is the attempt to identify biomarkers of allostatic load, neuroplasticity, and stress-related pathophysiology.[64] Another approach is Hellhammer's Neuropattern™ tool that adopts 13 distinct constructs based on regulations in the HPA-axis, sympathetic and parasympathetic systems each characterized by an endophenotype of biological, psychological, and symptomatic measures.[71] This functional phenotype, if combined with genotyping of stress markers linked to clinical symptoms, is a possible answer to the ambition expressed in the NIMH Research Domain Criteria (RDoC http://www.nimh.nih.gov/research-priorities/rdoc/index.shtml).and the Roadmap for Mental Health Research in Europe[72].

Acknowledgments

The support by the Royal Netherlands Academy of Arts and Sciences, COST Action ADMIRE BM1301 and STW Take-off 14095 is gratefully acknowledged.

ERdk is scientific advisor to Pharmaseed Ltd, Dynacorts Therapeutics BV and Corcept Therapeutics Inc and owns stock of Corcept.

References

1. Selye H. *The Story of the Adaptation Syndrome.* Montreal: Acta Inc; 1952.
2. Chrousos GP, Gold PW. The concepts of stress and stress system disorders. Overview of physical and behavioral homeostasis. *JAMA.* 1992;267:1244–1252.
3. McEwen BS. Protective and damaging effects of stress mediators. *New Engl J Med.* 1998;338:171–179.
4. Sapolsky RM, Romero LM, Munck AU. How do glucocorticoids influence stress responses? Integrating permissive, suppressive, stimulatory, and preparative actions. *Endocr Rev.* 2000;21:55–89.
5. Joëls M, Sarabdjitsingh RA, Karst H. Unraveling the time domains of corticosteroid hormone influences on brain activity: rapid, slow, and chronic modes. *Pharmacol Rev.* 2012;64:901–938.
6. De Kloet ER. Brain corticosteroid receptor balance and homeostatic control. *Front Neuroendocrinol.* 1991;12:95–164.
7. De Kloet ER, Vreugdenhil E, Oitzl MS, Joëls M. Brain corticosteroid receptor balance in health and disease. *Endocr Rev.* 1998;19:269–301.
8. De Kloet ER, Joëls M, Holsboer F. Stress and the brain: from adaptation to disease. *Nat Rev Neurosci.* 2005;6:463–475.
9. De Kloet ER. From receptor balance to rational glucocorticoid therapy. *Endocrinology.* 2014;145:2754–2769.
10. Holsboer F. The corticosteroid receptor hypothesis of depression. *Neuropsychopharmacology.* 2000;23:477–501.
11. Reul JMHM, De Kloet ER. Two receptor systems for corticosterone in rat brain: microdistribution and differential occupation. *Endocrinology.* 1985;117:2505–2511.
12. Koolhaas JM, Bartolomucci A, Buwalda B, et al. Stress revisited: a critical evaluation of the stress concept. *Neurosci Biobehav Rev.* 2011;35:1291–1301.
13. de Kloet ER, Joëls M. Stress research: past, present and future. In: Pfaff DW, ed. *Neuroscience in the 21st Century.* New York: Springer; 2013:1979–2007.
14. Joëls M, de Kloet ER. Mineralocorticoid and glucocorticoid receptors in the brain. Implications for ion permeability and transmitter systems. *Prog Neurobiol.* 1994;43:1–36.
15. Joëls M. Corticosteroid effects in the brain: U-shape it. *Trends Pharmacol Sci.* 2006;27:244–250.
16. Karst H, Berger S, Erdmann G, Schütz G, Joëls M. Metaplasticity of amygdalar responses to the stress hormone corticosterone. *Proc Natl Acad Sci U S A.* 2010;107:14449–14454.

17. Nasca C, Bigio B, Zelli D, Nicoletti F, McEwen BS. Mind the gap: glucocorticoids modulate hippocampal glutamate tone underlying individual differences in stress susceptibility. *Mol Psychiatry*. 2014;20:755–763 (epub ahead of print).
18. Krugers HJ, Hoogenraad CC, Groc L. Stress hormones and AMPA receptor trafficking in synaptic plasticity and memory. *Nat Rev Neurosci*. 2010;11:675–681.
19. Sandi C. Glucocorticoids act on glutamatergic pathways to affect memory processes. *Trends Neurosci*. 2011;34:165–176.
20. Di S, Malcher-Lopes R, Halmos KC, Tasker JG. Nongenomic glucocorticoid inhibition via endocannabinoid release in the hypothalamus: a fast feedback mechanism. *J Neurosci*. 2003;23:4850–4857.
21. Levine S. Developmental determinants of sensitivity and resistance to stress. *Psychoneuroendocrinology*. 2005;30:939–946.
22. Cirulli F, van Oers H, De Kloet ER, Levine S. Differential influence of corticosterone and dexamethasone on schedule-induced polydipsia in adrenalectomized rats. *Behav Brain Res*. 1994;65:33–39.
23. Pfaff DW, Martin EM, Ribeiro AC. Relations between mechanisms of CNS arousal and mechanisms of stress. *Stress*. 2007;10:316–325.
24. Herman JP, Figueiredo H, Mueller NK, et al. Central mechanisms of stress integration: hierarchical circuitry controlling hypothalamo-pituitary-adrenocortical responsiveness. *Front Neuroendocrinol*. 2003;24:151–180.
25. Dallman MF. Fast glucocorticoid actions on brain: back to the future. *Front Neuroendocrinol*. 2005;26:103–108.
26. Inoue W, Bains JS. Beyond inhibition: GABA synapses tune the neuroendocrine stress axis. *Bioessays*. 2014;36(6):561–569.
27. Kruk MR, Haller J, Meelis W, de Kloet ER. Mineralocorticoid receptor blockade during a rat's first violent encounter inhibits its subsequent propensity for violence. *Behav Neurosci*. 2013;127:505–514.
28. Van den Berg DT, de Kloet ER, van Dijken HH, de Jong W. Differential central effects of mineralocorticoid and glucocorticoid agonists and antagonists on blood pressure. *Endocrinology*. 1990;126:118–124.
29. Ratka A, Sutanto W, Bloemers M, De Kloet ER. On the role of brain mineralocorticoid (type I) and glucocorticoid (type II) receptors in neuroendocrine regulation. *Neuroendocrinology*. 1989;50:117–123.
30. Cole MA, Kalman BA, Pace TW, Topczewski F, Lowrey MJ, Spencer RL. Selective blockade of the mineralocorticoid receptor impairs hypothalamic-pituitary-adrenal axis expression of habituation. *J Neuroendocrinol*. 2000;12:1034–1042.
31. Schwabe L, Schächinger H, de Kloet ER, Oitzl MS. Corticosteroids operate as switch between memory systems. *J Cogn Neurosci*. 2009;22:1362–1372.
32. Dias-Ferreira E1, Sousa JC, Melo I, et al. Chronic stress causes frontostriatal reorganization and affects decision-making. *Science*. 2009;325:621–625.
33. Ter Horst JP, Kentrop J, de Kloet ER, Oitzl MS. Stress and estrous cycle affect strategy but not performance of female C57BL/6J mice. *Behav Brain Res*. 2013;241:92–95.
34. Oitzl MS, De Kloet ER. Selective corticosteroid antagonists modulate specific aspects of spatial orientation learning. *Behav Neurosci*. 1992;106:62–71.
35. Evans AN, Liu Y, Macgregor R, Huang V, Aguilera G. Regulation of hypothalamic corticotropin-releasing hormone transcription by elevated glucocorticoids. *Mol Endocrinol*. 2013;27:1796–1807.
36. Schmidt MV, Sterlemann V, Wagner K, et al. Postnatal glucocorticoid excess due to pituitary glucocorticoid receptor deficiency: differential short- and long-term consequences. *Endocrinology*. 2009;150:2709–2716.
37. Elliott E, Ezra-Nevo G, Regev L, Neufeld-Cohen A, Chen A. Resilience to social stress coincides with functional DNA methylation of the Crf gene in adult mice. *Nat Neurosci*. 2010;13:1351–1353.
38. Sharma D, Bhave S, Gregg E, Uht R. Dexamethasone induces a putative repressor complex and chromatin modifications in the CRH promoter. *Mol Endocrinol*. 2013;27:1142–1152.
39. Laryea G, Schütz G, Muglia LJ. Disrupting hypothalamic glucocorticoid receptors causes HPA axis hyperactivity and excess adiposity. *Mol Endocrinol*. 2013;27:1655–1665.
40. Van Haarst AD, Oitzl MS, Workel JO, de Kloet ER. Chronic brain glucocorticoid receptor blockade enhances the rise in circadian and stress-induced pituitary-adrenal activity. *Endocrinology*. 1996;137:4935–4943.
41. Walker JJ, Spiga F, Waite E, et al. The origin of glucocorticoid hormone oscillations. *PLoS Biol*. 2012;10:6. e1001341.
42. Sarabdjitsingh RA, Joëls M, de Kloet ER. Glucocorticoid pulsatility and rapid corticosteroid actions in the central stress response. *Physiol Behav*. 2012;106:73–80.
43. Wyrwoll CS, Holmes MC, Seckl JR. 11β-Hydroxysteroid dehydrogenases and the brain: from zero to hero, a decade of progress. *Front Neuroendocrinol*. 2011;32:265–286.
44. De Kloet ER. Why dexamethasone poorly penetrates in brain. *Stress*. 1997;2(1):13–20.
45. Karssen AM, Meijer OC, Berry A, Sanjuan Piñol R, de Kloet ER. Low doses of dexamethasone can produce a hypocorticosteroid state in the brain. *Endocrinology*. 2005;146:5587–5595.
46. Reul JM, van den Bosch FR, de Kloet ER. Relative occupation of type-I and type-II corticosteroid receptors in rat brain following stress and dexamethasone treatment: functional implications. *J Endocrinol*. 1987;115:459–467.
47. Liston C, Cichon JM, Jeanneteau F, Jia Z, Chao MV, Gan WB. Circadian glucocorticoid oscillations promote learning-dependent synapse formation and maintenance. *Nat Neurosci*. 2013;16:698–705.
48. Oitzl MS, van Haarst AD, Sutanto W, de Kloet ER. Corticosterone, brain mineralocorticoid receptors (MRs) and the activity of the hypothalamic-pituitary-adrenal (HPA) axis: the Lewis rat as an example of increased central MR capacity and a hyporesponsive HPA axis. *Psychoneuroendocrinology*. 1995;20:655–675.
49. Harris AP, Holmes MC, de Kloet ER, Chapman KE, Seckl JR. Mineralocorticoid and glucocorticoid receptor balance in control of HPA axis and behaviour. *Psychoneuroendocrinology*. 2013;38:648–658.
50. Klok MD, Giltay EJ, van der Does AJ, et al. A common and functional mineralocorticoid receptor haplotype enhances optimism and protects against depression in females. *Transl Psychiatry*. 2011;13:e62.
51. Quax RA, Manenschijn L, Koper JW, et al. Glucocorticoid sensitivity in health and disease. *Nat Rev Endocrinol*. 2013;9:670–686.
52. Champagne DL, Bagot RC, van Hasselt F, et al. Maternal care and hippocampal plasticity: evidence for experience-dependent structural plasticity, altered synaptic functioning, and differential responsiveness to glucocorticoids and stress. *J Neurosci*. 2008;28:6037–6045.
53. Turecki G, Meaney MJ. Effects of the social environment and stress on glucocorticoid receptor gene methylation: a systematic review. *Biol Psychiatr*. 2014. pii: S0006-3223(14)00967-6 http://dx.doi.org/101016/j.biopsych.201411022 [Epub ahead of print].
54. Herman JP. Neural control of chronic stress adaptation. *Front Behav Neurosci*. 2013;7:61.
55. Datson NA, van den Oever JM, Korobko OB, Magarinos AM, de Kloet ER, McEwen BS. Previous history of chronic stress changes the transcriptional response to glucocorticoid challenge in the dentate gyrus region of the male rat hippocampus. *Endocrinology*. 2013;154:3261–3272.
56. Van Eekelen JA, Oitzl MS, De Kloet ER. Adrenocortical hyporesponsiveness and glucocorticoid feedback resistance in old male brown Norway rats. *J Gerontol A Biol Sci Med Sci*. 1995;50:B83–B89.
57. Klok MD, Alt SR, Irurzun Lafitte AJ, et al. Decreased expression of mineralocorticoid receptor mRNA and its splice variants in postmortem brain regions of patients with major depressive disorder. *J Psychiatr Res*. 2011;45:871–878.

58. Dumas TC, Gillette T, Ferguson D, Hamilton K, Sapolsky RM. Anti-glucocorticoid gene therapy reverses the impairing effects of elevated corticosterone on spatial memory, hippocampal neuronal excitability, and synaptic plasticity. *J Neurosci.* 2010;30: 1712–1720.
59. Datson NA, Speksnijder N, Mayer JL, et al. The transcriptional response to chronic stress and glucocorticoid receptor blockade in the hippocampal dentate gyrus. *Hippocampus.* 2012;22:359–371.
60. Barik J, Marti F, Morel C, et al. Chronic stress triggers social aversion via glucocorticoid receptor in dopaminoceptive neurons. *Science.* 2013;339:332–335.
61. McEwen BS. Definitions and concepts of stress. In: In: Fink G, ed. *Encyclopedia of Stress*; vol. 3: San Diego: Academic Press; 2000: 508–515.
62. Munck A, Guyre PM, Holbrook NJ. Physiological functions of glucocorticoids in stress and their relation to pharmacological actions. *Endocr Rev.* 1984;5:25–44.
63. Lazarus RS. Emotions and interpersonal relationships: toward a person-centered conceptualization of emotions and coping. *J Pers.* 2006;74:9–46.
64. McEwen BS, Gianaros PJ. Stress- and allostasis-induced brain plasticity. *Annu Rev Med.* 2011;62:431–445.
65. McEwen BS, Weiss JM, Schwartz LS. Selective retention of corticosterone by limbic structures in rat brain. *Nature.* 1968;220: 911–912.
66. De Kloet R, Wallach G, McEwen BS. Differences in corticosterone and dexamethasone binding to rat brain and pituitary. *Endocrinology.* 1975;96:598–609.
67. Moguilewsky M, Raynaud JP. Evidence for a specific mineralocorticoid receptor in rat pituitary and brain. *J Steroid Biochem.* 1980;12:309–314.
68. Evans RM, Arriza JL. A molecular framework for the actions of glucocorticoid hormones in the nervous system. *Neuron.* 1989;2:1105–1112.
69. Dallman MF, Akana SF, Cascio CS, Darlington DN, Jacobson L, Levin N. Regulation of ACTH secretion: variations on a theme of B. *Recent Prog Horm Res.* 1987;43:113–173.
70. Hermans EJ, Henckens MJ, Joëls M, Fernández G. Dynamic adaptation of large-scale brain networks in response to acute stressors. *Trends Neurosci.* 2014;37:304–314.
71. Hellhammer D, Hero T, Gerhards F, Hellhammer J. Neuropattern: a new translational tool to detect and treat stress pathology I. Strategical consideration. *Stress.* 2012;15:479–487.
72. Schumann G, Binder EB, Holte A, et al. Stratified medicine for mental disorders. *Eur Neuropsychopharmacol.* 2014;24:5–50.

CHAPTER

4

The Fight-or-Flight Response: A Cornerstone of Stress Research

R. McCarty

Vanderbilt University, Nashville, TN, USA

OUTLINE

Abstract

The fight-or-flight response was a concept developed by Walter B. Cannon in the course of his studies on the secretion of epinephrine from the adrenal medulla of laboratory animals. This concept was an outgrowth of his studies of homeostatic mechanisms, particularly as they related to the sympathetic-adrenal medulla system. Cannon's research on homeostasis and the fight-or-flight response led him to delve into mechanisms of "voodoo death" and to propose a new theory of emotions, known as the Cannon-Bard theory. Cannon thought that the sympathetic nervous system and the adrenal medulla operated as a functional unit, with epinephrine as the chemical messenger. He did not understand that the postganglionic sympathetic nerves utilized norepinephrine as a chemical transmitter. Cannon's research legacy is a rich one and his work is still cited frequently by contemporary researchers in the field of stress.

INTRODUCTION

Walter B. Cannon was one of the most influential physiologists of the first half of the twentieth century.[1] Following the lead of his advisor and mentor at Harvard Medical School, Professor Henry P. Bowditch, Cannon devoted much of his early research career to studies of the digestive system, including the use of newly discovered X-rays to visualize digestive processes.[2] Beginning in 1910, Cannon moved into a new area of investigation that included studies of the adrenal medulla and the sympathetic nervous system in laboratory animals. In addition to studying the physiology of the adrenal medulla, he broadened his focus to include psychological stimuli and emotional responses that were often associated with sympathetic nervous system discharge and adrenal medullary secretion.[3,4]

THE CONCEPT OF HOMEOSTASIS

To place Cannon's research on the fight-or-flight response in context, it is important to first consider his path-breaking work on homeostasis, which was an outgrowth of earlier studies in France. Beginning in the mid-nineteenth century and continuing until his death in 1878, the renowned French physiologist, Claude Bernard, advanced the theory that bodily systems operate in concert to maintain a relatively constant internal environment, or *milieu intérieur*.[5] These bodily systems would also join together to effect a return to a constancy of the *milieu intérieur* even after major disruptions to an organism. Thus, Bernard's view at the end of his career was that higher animals are in a close and informed relationship with the external world, and the relative constancy of the *milieu intérieur* results from moment-to-moment adjustments in various physiological systems.[6]

Stress: Concepts, Cognition, Emotion, and Behavior
http://dx.doi.org/10.1016/B978-0-12-800951-2.00004-2

KEY POINTS

- Walter B. Cannon expanded upon Claude Bernard's concept of the *milieu intérieur* and introduced the concept of homeostasis. He recognized that stressful stimuli could disrupt the constancy of the internal environment and he demonstrated that central control of epinephrine secretion from the adrenal medulla was important in reestablishing homeostatic balance.
- The fight-or-flight response was a term coined by Cannon to describe the activation of an organism when exposed to a conspecific or a predator. The physiological changes in these situations, including epinephrine release into the circulation, enhance survival by increasing the delivery of oxygen and glucose to skeletal muscles and brain at the expense of the viscera and skin.
- Cannon's investigations into "voodoo death" in primitive societies revealed his broad interests in behavioral sciences and the scientific rigor of his approach. He hypothesized that voodoo death resulted in hyperactivity of the adrenal medulla, which resulted in life-threatening changes in cardiac function. Later studies by Curt P. Richter using an animal model of sudden death suggested that the mechanism was more likely associated with increases in vagal drive to the heart.
- Cannon challenged the prevailing theory of emotions, the James-Lange theory, in a paper in the *American Journal of Psychology* in 1927. The Cannon-Bard theory of emotions, which drew on research by Cannon and his student, Philip Bard, focused attention on the hypothalamus and thalamus as critical brain areas for generating emotions and their associated peripheral physiological changes.
- Cannon's publications continue to serve as a strong foundation for contemporary studies of stress. In particular, his work on homeostasis, the fight-or-flight response, and the emergency functions of the adrenal medulla are still widely referenced by researchers.

As Cannon came into the picture, these major disruptions to an organism were expanded to include stressful stimuli. Cannon built directly upon Bernard's theory by introducing the concept of *homeostasis*, with the central nervous system playing a critical role in maintaining the constancy of the internal environment. A key component of this homeostatic balance was the central control of the adrenal medulla and the sympathetic nerves and the secretion of epinephrine. As Goldstein[7] has noted, Cannon was mistaken on several critical details relating to the sympathetic nervous system and the adrenal medulla but was largely correct on the bigger picture. For example, Cannon[8] argued that the sympathetic nervous system and the adrenal medulla were a single functional unit that employed the same chemical messenger, epinephrine. The chemical arsenal of the sympathetic nerves was expanded, Cannon thought, through the conversion of epinephrine into two other substances, sympathin E (excitation) and sympathin I (inhibition).[9,10] It was not until immediately after World War II that von Euler[11] provided definitive evidence that norepinephrine was the neurotransmitter released from sympathetic nerve endings, not epinephrine. It was later still that a host of investigators revealed the two broad classes of adrenergic receptors, alpha- and beta- and their subclasses.[12]

THE FIGHT-OR-FLIGHT RESPONSE

The fight-or-flight response presents a special challenge to the maintenance of homeostasis in animals and humans. Cannon recognized that the sympathetic-adrenal medullary system would at critical times, such as aggressive encounters with conspecifics or exposure to a predator, drive the individual out of homeostatic balance. On such occasions, increased blood flow to the skeletal muscles, release of glucose from the liver, dilation of bronchi to increase availability of oxygen, reduced blood flow to the skin and digestive system, and a host of other changes were all directed toward enhancing the immediate survival of the organism (see Table 1). Homeostasis could be reestablished at a later time, or even at different set points, when the threat was eliminated and the survival of the organism was certain.[13–15]

The linkage between epinephrine secretion from the adrenal medulla and increases in glucose in the circulation prompted Cannon to write the following entry in his journal on January 20, 1911—"Got idea that adrenals in excitement serve to affect muscular power and mobilize sugar for muscular use—thus in wild state readiness for fight or run!" In due course, *fight or run* was modified into the more familiar *fight or flight*.

It is important to note that Cannon, a classically trained physiologist, did not hesitate to weave psychological concepts into his work on the adrenal medulla and the control of epinephrine release. In addition, he also published important papers in behavioral science journals (e.g., Cannon[16,17]). His surprisingly broad embrace of scientific domains was captured in a letter from Cannon to his friend and colleague, Dr. Carl Binger, a Harvard psychiatrist, in October 1934[27]:

TABLE 1 Sympathetic-Adrenal Medullary Components of the Flight-or-Flight Response

System	Physiological effect	Physiological consequences
Heart	• Increased rate • Dilation of coronary vessels	• Increase in blood flow • Increased availability of O_2 and energy to cardiac myocytes
Circulation	• Dilation of vessels serving skeletal muscle cells • Vasoconstriction of vessels serving digestive organs and skin • Contraction of spleen	• Increased availability of O_2 to skeletal muscle cells • Facilitates shunting of blood to skeletal muscles and brain • Increased delivery of O_2 to metabolically active cells
Lungs	• Dilation of bronchi • Increased respiratory rate	• Increased availability of O_2 in blood • Increased availability of O_2 in blood
Liver	• Increased conversion of glycogen to glucose	• Increased availability of glucose in skeletal muscle and brain cells

> I personally conceive of the well-grounded work of the psychologist and the psychiatrist as being related to one aspect, while the work of the physiologist is related to another aspect of the same unit. Therefore, I do not hesitate to use psychological terms along with physiological terms in descriptions. If the physiologist has observations which support or yield interpretation of the views of the psychologist or psychiatrist, why should they not be accepted and incorporated into the general scheme of things?

Cannon's openness to combining psychology and evolutionary biology with physiology through his experiments in homeostasis, the fight-or-flight response, stress, and emotions has ensured his continuing influence on contemporary researchers. This is in spite of the fact that many of his fundamental findings related to the sympathetic nervous system and the adrenal medulla have not stood the test of time.

FIGHT-OR-FLIGHT TAKEN TO AN EXTREME

Toward the end of his career, Cannon[17] became interested in case studies, mostly anecdotal, of "voodoo death" from the anthropological and medical literature. These reports described instances from primitive peoples in Brazil, Africa, Australia, New Zealand, Haiti, and Hawaii that included curses placed on individuals through bone pointing or other forms of magic. The person on the receiving end of the curse moved very quickly into a persistent and catastrophic state of fear, and death often ensued within 48 h of the placement of the curse. Cannon focused on those cases where there were no apparent injuries to the individual and poisoning could be eliminated as a cause of death. He then proposed a possible explanation for "voodoo death" that encompassed the sympathetic-adrenal medullary system and its deleterious impact on the cardiovascular system.

As Sternberg and Walter[18] has pointed out, Cannon's hypotheses obviously could not include a role for neuroendocrine systems that had not yet been discovered; prominent among these is the hypothalamic-pituitary-adrenocortical system. However, within the limits of his knowledge and his own specialized research interests, Cannon did succeed in tackling a phenomenon from the anthropological literature that on the surface defied logic. His success was tied to his care in sorting through case studies, his attempts to eliminate alternative explanations (e.g., poisoning, the supernatural), and his vast knowledge of cardiovascular physiology. He is given high marks for linking "voodoo death" with the physiology of emotions and his work has, at least in part, stood the test of time (Sternberg and Walter[18]).

It was left to Curt P. Richter of The Johns Hopkins University to suggest a possible mechanism of "voodoo death" through development of an animal model. Richter employed wild-trapped and domesticated laboratory rats in his studies. Wild rats were initially restrained in a black bag to facilitate handling and experimental manipulations and they were then placed individually into cylinders of water. Many had their vibrissae shaved off prior to immersion in the water. Heart rates were recorded from implanted electrodes and revealed an initial significant increase in heart rate followed by a pronounced bradycardia, especially in those animals that died.

Richter marshaled evidence that this animal model of "voodoo" or sudden death was explained by hyperactivation of the vagus nerve and not over-activity of the sympathetic nerves and the adrenal medulla. Behaviorally, those wild rats that died quickly in the swim cylinders showed signs of hopelessness given that their avenues of fight-or-flight had been eliminated. Under normal conditions, these same rats were extremely aggressive and reactive and to manage them in the laboratory required great care. In contrast, domesticated laboratory rats that were more familiar with restraint and handling and the general laboratory environment survived much longer in the swim cylinders and usually died of exhaustion.[19]

THEORIES OF EMOTIONS

As Cannon extended his research into central control of sympathetic-adrenal medullary outflow, he also proposed a new theory of emotions.[16] The prevailing view of emotions was one developed more than 40 years earlier by the noted American psychologist William James and the Danish physiologist G.C. Lange. The James-Lange theory of emotions included the following important elements:

- An emotionally charged event occurs and stimulates sensory receptors.
- Afferent nerve signals reach the cerebral cortex and the event is perceived.
- Efferent nerve impulses are sent to the skeletal muscles and visceral tissues and alter their levels of activity in preparation for the event.
- Afferent nerve signals once again go to the cerebral cortex and the emotionally relevant details of the event are perceived.

James[20] emphasized the importance of visceral afferent signals and the stimulation of emotions, whereas Lange[21] focused in a more limited way upon the relationships between vascular afferent activity and emotions. The James-Lange theory dominated the field until Cannon's seminal paper was published in the *American Journal of Psychology* in 1927.

Cannon argued that a critical test of the James-Lange theory would involve an experiment on emotional expression in animals lacking visceral afferent feedback. In such an approach, the critical link between visceral afferent changes and cerebral expression of emotions would be decoupled. Cannon advanced five prevailing lines of evidence to discount the James-Lange theory:

- Sherrington[22] reported that dogs exhibited intense emotional expressions following complete destruction of the sympathetic and spinal sensory roots.
- Cannon observed that cats remained emotionally responsive following surgical destruction of the sympathetic nervous system.
- The timeframe for afferent impulses to signal an emotional expression in the brain was simply too slow to account for experimental observations.
- Artificial production of visceral changes (e.g., injection of epinephrine) did not in and of itself produce emotions.
- Clinical case studies demonstrated that patients with transections of the spinal cord remained emotionally responsive despite being completely paralyzed below the cervical area.

Cannon was joined in this line of research by one of his most notable students, Philip Bard, who spent most of his illustrious career as a professor at The Johns Hopkins University. Their combined efforts, beginning in the 1920s, changed the focus of research on emotions to include the central nervous system (hypothalamus and thalamus) as the primary site for receiving sensory information; directing peripheral nervous system, visceral and vascular changes; and generating the psychological manifestations of the emotional experience.[16,23]

What has become known as the Cannon-Bard theory of emotions was incomplete in that it did not encompass limbic areas of the brain, especially the amygdala, in the expression of emotions. Later work by Papez[24] and MacLean[25] expanded the brain areas involved in the control of emotions to include limbic areas through the Papez-MacLean theory of emotions.[26]

THE LEGACY OF WALTER B. CANNON

Articles, chapters, and books written by Walter B. Cannon are frequently referenced today by researchers in physiology, neuroscience, psychology, the medical sciences, and related fields more than 100 years after the start of his scientific career in 1900. Several aspects of his work and the approaches he took have clearly stood the test of time:

- He introduced the concept of homeostasis and emphasized the importance of the sympathetic nervous system and the adrenal medulla in maintaining a stable internal environment.
- He first described the fight-or-flight response and the importance of epinephrine secretion from the adrenal medulla in directing the body's response to potentially life-threatening stimuli.
- He provided a foundation for the continuing study of central mechanisms in controlling emotional expression through his contributions to the Cannon-Bard theory.
- He exhibited an unusually broad view of physiological experimentation that extended over to the behavioral sciences and psychiatry. His paper on "voodoo death" is but one example of his interdisciplinary approach to research.
- Not discussed in this article, but in 1917-1918, Cannon worked during World War I on traumatic shock in severely injured soldiers as a member of the Harvard University Hospital Unit in England and France. His studies emphasized the importance of restoring blood volume in severely injured soldiers to enhance their survival.
- Also not discussed in this article was Cannon's strong public support for the use of animals in medical experimentation. This was but one aspect of his distinguished service in the public arena as one of the nation's most distinguished researchers.

References

1. Benison S, Barger AC. *Walter B. Cannon: The Life and Times of a Young Scientist.* Cambridge, MA: Belknap Press; 1987.
2. Cannon WB. Movements of the intestines studied by means of the Röentgen rays. *J Med Res.* 1902;7:72–75.
3. Cannon WB. Studies on the conditions of activity of adrenal glands. V. The isolated heart as an indicator of adrenal secretion induced by pain, asphyxia and excitement. *Am J Physiol.* 1919;50: 399–432.
4. Cannon WB, de la Paz D. Emotional stimulation of adrenal secretion. *Am J Physiol.* 1911;28:64–70.
5. Bernard C; Hoff HE, Guillemin R, Guillemin L (Translators). Lectures on the Phenomena of Life Common to Animals and Plants. Springfield, IL: Charles C. Thomas; 1974.
6. Holmes FL. Claude Bernard, the *milieu intérieur* and regulatory physiology. *Hist Philos Life Sci.* 1986;8:3–25.
7. Goldstein DS. *Adrenaline and the Inner World.* Baltimore, MD: The Johns Hopkins University Press; 2006.
8. Cannon WB. The adrenal medulla. *Bull N Y Acad Med.* 1940;16: 3–13.
9. Cannon WB, Rosenblueth A. Studies on conditions of activity in endocrine organs. XXIX. Sympathin E and sympathin I. *Am J Physiol.* 1933;104:557–574.
10. Rosenblueth A, Cannon WB. Studies on conditions of activity in endocrine organs. XXVIII. Some effects of sympathin on the nictitating membrane. *Am J Physiol.* 1932;99:398–407.
11. von Euler US. Identification of the sympathomimetic ergone in adrenergic nerves of cattle (Sympathin N) with laevo-noradrenaline. *Acta Physiol Scand.* 1948;16:63–74.
12. Ahles A, Engelhardt S. Polymorphic variants of adrenoceptors: pharmacology, physiology and role in disease. *Pharmacol Rev.* 2014;66:598–637.
13. Cannon WB. The emergency function of the adrenal medulla in pain and the major emotions. *Am J Physiol.* 1914;33:356–372.
14. Cannon WB. *Bodily Changes in Pain, Hunger, Fear and Rage.* New York: Appleton-Century; 1915.
15. Cannon WB. *The Wisdom of the Body.* New York: W.W. Norton; 1932.
16. Cannon WB. The James-Lange theory of emotions: a critical examination and an alternative theory. *Am J Psychol.* 1927;39: 106–124.
17. Cannon WB. Voodoo death. *Am Anthropol.* 1942;44:169–181.
18. Sternberg EM, Walter B. Cannon and "Voodoo Death": a perspective from 60 years on. *Am J Public Health.* 2002;92:1564–1566.
19. Richter CP. On the phenomenon of sudden death in animals and man. *Psychosom Med.* 1957;19:191–198.
20. James W. What is an emotion? *Mind.* 1884;os-IX:188–205.
21. Lange CG. The mechanism of the emotions. In: Dunlap D, ed. *The Emotions.* Baltimore, MD: Williams and Wilkins; 1885:33–92.
22. Sherrington CS. *The Integrative Action of the Nervous System.* New Haven, CT: Yale University Press; 1906.
23. Bard P. A diencephalic mechanism for the expression of rage with special reference to the sympathetic nervous system. *Am J Physiol.* 1928;84:490–515.
24. Papez J. A proposed mechanism of emotion. *Arch Neur Psychiat.* 1937;38:725–743.
25. MacLean P. Psychosomatic disease and the 'visceral brain', recent developments bearing on the Papez theory of emotion. *Psychosom Med.* 1950;11:338–353.
26. Weisfeld GE, Goetz SMM. Applying evolutionary thinking to the study of emotion. *Behav Sci.* 2013;3:388–407.
27. Psychosom. Med. 1957;19:180.

CHAPTER

5

Central Role of the Brain in Stress and Adaptation: Allostasis, Biological Embedding, and Cumulative Change

B.S. McEwen

The Rockefeller University, New York, NY, USA

OUTLINE

Abstract

The brain is the central organ of stress and adaptation because it perceives what is threatening and determines behavioral and physiological responses. Brain circuits are remodeled by stress–which changes the ability to self-regulate anxiety and mood–to perform working and episodic memory, as well as executive function and decision making. The brain regulates the body via the neuroendocrine, autonomic, immune, and metabolic systems, and the mediators of these systems and those within the brain and other organs activate epigenetic programs that alter expression of genetic information so as to alter cellular and organ function. While the initial active response to stressors promotes adaptation ("allostasis"), there can be cumulative change (e.g., body fat, hypertension) from chronic stress and resulting unhealthy behaviors ("allostatic load") that can lead to disease, e.g., diabetes, cardiovascular disease ("allostatic overload"). Besides the embedding of early life experiences, the most potent of stressors are those arising from the social and physical environment that affects both brain and body. Gradients of socioeconomic status reflect the cumulative burden of coping with limited resources, toxic environments, and negative life events, as well as health-damaging behaviors that result in chronic activation of physiological systems that lead to allostatic load and overload. Can we intervene to change this progression? After describing the new view of epigenetics that negates the old notion that "biology is destiny," this chapter summarizes some of the underlying cellular, molecular, and neuroendocrine mechanisms of stress effects upon brain and body. It then discusses

Stress: Concepts, Cognition, Emotion, and Behavior
http://dx.doi.org/10.1016/B978-0-12-800951-2.00005-4

integrative or "top down" approaches involving behavioral interventions that take advantage of the increasing ability to reactivate plasticity in the brain. At the societal level, policies of government and the private sector affect health directly or indirectly and must be redirected, to allow people to make choices that improve their chances for a healthy life.

INTRODUCTION

"Stress" is a word that is with us in almost everything we do. Yet the word "stress" is ambiguous and we forget that the major "stress hormone," cortisol, so often associated with bad outcomes, is actually an important mediator of the ability of the body and brain to adapt to the diurnal cycle, as well as to experiences that we call stressors. This review presents a conceptual framework for understanding the protective, as well as damaging, aspects of mediators of adaptation like cortisol within a life course perspective that emphasizes biological embedding of early-life experiences along with cumulative change.

This perspective also encourages the possibility of interventions to prevent or ameliorate negative effects of stressors. Under this view, the brain is the central organ of stress and adaptation because it perceives what is threatening and determines behavioral and physiological responses (Figure 1). Brain circuits are remodeled by stress so as to change the ability to self-regulate anxiety and mood, to perform working and episodic memory, as well as executive function and decision making. The brain regulates the body via the neuroendocrine, autonomic, immune, and metabolic systems, and the mediators of these systems and those within the brain and other organs activate epigenetic programs that alter expression of genetic information so as to change cellular and organ function.

Besides developmental influences associated with parent-infant interactions and the quality of early life experiences that result in long-term effects ("biological embedding"), the most potent of influences as one proceeds through adult life are those arising from the family, neighborhood, workplace, and exposure to local, national, and international events in the media that can affect both brain and body health and progression toward a variety of diseases.

Social ordering in human society is associated with gradients of disease, with an increasing frequency of mortality and morbidity along a gradient of decreasing income and education (socioeconomic status, SES; http://www.macses.ucsf.edu/). Although the causes of these gradients of health are very complex, they likely reflect, with increasing frequency going down the SES ladder, the cumulative burden of coping with limited resources, toxic and otherwise stressful living environments and negative life events, as well as differences in health-related behaviors (aka "lifestyle"), and resulting chronic activation of physiological systems involved in adaptation (http://www.macses.ucsf.edu/) leading to allostatic overload,[2] as will be discussed below.

The brain mediates adaptation to changes in the physical and social environment through the autonomic, neuroendocrine, and immune systems, as well as through behavioral responses that include fighting or fleeing, as well as health-promoting or health-damaging responses. Adaptation to stressful events or environmental changes is an active process that involves the output of mediators such as neurotransmitters and modulators, as well as

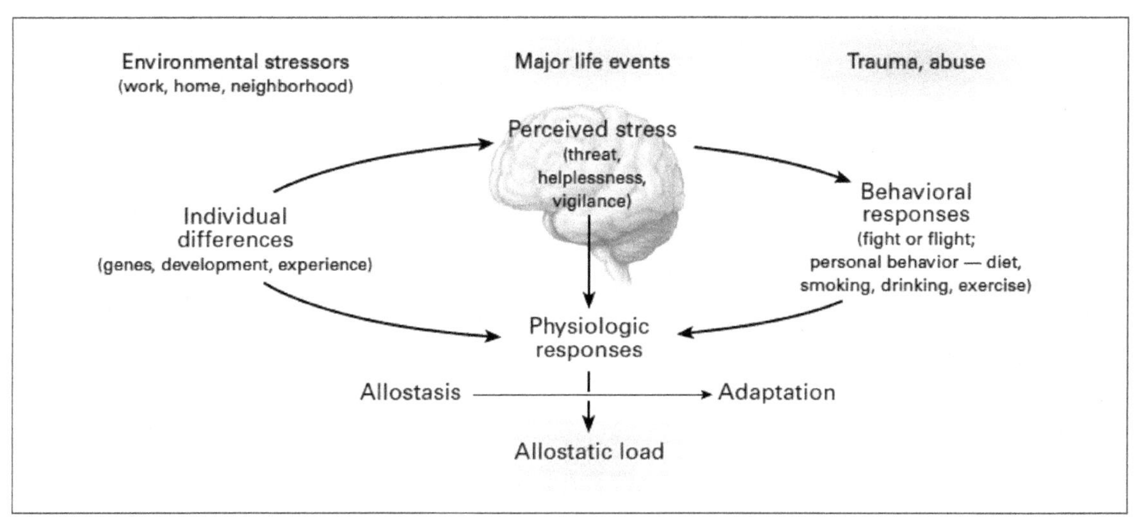

FIGURE 1 The stress response and development of allostatic load. The perception of stress is influenced by one's experiences, genetics, and behavior, and the social and physical environment provides the stressors that require adaptation via epigenetic processes. When the brain perceives an experience as stressful, physiologic and behavioral responses are initiated, leading to allostasis and adaptation. Over time, allostatic load can accumulate, and the overexposure to mediators of neural, endocrine, and immune stress can have adverse effects on various organ systems, including the brain, leading to disease. *From Ref. 1 by permission.*

many hormones and cytokines and chemokines of the immune system. The goal of this adaptation is to maintain homeostasis and promote survival of the organism. However, the process of adaptation also produces an almost inevitable wear and tear on the body and brain, and this wear and tear is exacerbated if there are many stressful events and/or if the mediators that normally promote adaptation are dysregulated, that is, not turned on when needed or not turned off efficiently when no longer needed. The description and analysis of this scenario has entailed the addition of some new terminology, such as allostasis, and the refinement of some classical terminology, such as homeostasis and stress, which we shall now discuss before describing the central role of the brain and its plasticity and ability to change. We begin this discussion by summarizing concepts related to homeostasis.

KEY POINTS

- Allostasis and allostatic load/overload are more precise biological concepts than "stress" to describe adaptation and maladaptation to "stressors" and they include the physiological effects of health-promoting and health-damaging behaviors as well as stressful experiences.
- "Good stress," "tolerable stress," and "toxic stress" are terms that better characterize use of the word "stress." Allostatic overload is the consequences of toxic stress.
- There are multiple interacting mediators of allostasis that operate nonlinearly.
- The brain is a target of mediators of stress and allostasis and it responds epigenetically, affecting brain architecture and neurochemical functions by both genomic and nongenomic mechanisms.
- Epigenetically regulated gene expression changes continuously with experience.
- Adverse early life events lead to long-lasting changes in the brain and body and increase vulnerability to later experiences that increase allostatic overload and contribute disproportionately to the health care burden.
- The healthy brain is resilient after stress; lack of resilience characterizes anxiety and depressive disorders and the aging brain, but these are treatable.
- Besides the urgent need for prevention, interventions to ameliorate allostatic load/overload include pharmaceutical agents or behaviors (e.g., physical activity) which open "windows of plasticity" that allow targeted behavioral interventions to change brain architecture and function.
- Policies of government and the private sector, including targeted community programs, can create the conditions to promote resilience and to prevent or reduce effects of early life adversity.

HOMEOSTASIS, ALLOSTASIS AND ALLOSTATIC LOAD AND OVERLOAD

Characteristics of Homeostatic Systems

Homeostasis refers to the ability of an organism to maintain the internal environment of the body within limits that allow it to survive. Homeostasis also refers to self-regulating processes that return critical systems of the body to a set point within a narrow range of operation, consistent with survival of the organism. Homeostasis is highly developed in warm-blooded animals living on land, which must maintain body temperature, fluid balance, blood pH, and oxygen tension within rather narrow limits, while at the same time obtaining nutrition to provide the energy to maintain homeostasis. This is because maintaining homeostasis requires the expenditure of energy. Energy is used for locomotion, as the animal seeks and consumes food and water, for maintaining body temperature via the controlled release of calories from metabolism of food or fat stores, and for sustaining cell membrane function as it resorbs electrolytes in the kidney and intestine and maintains neutral blood pH. Homeostasis also refers to the body's defensive mechanisms. These include protective reflexes against such things as inhaling matter into the lungs, the vomiting reflex as a protection to expel toxic materials from the esophagus or stomach, the eye blink reflex, and the withdrawal response to hot or otherwise painful skin sensations. There is also the defense against pathogens through innate and acquired immunity, the latter of which is stimulated by acute stress via cortisol and adrenalin and inhibited by chronic stress and high levels of cortisol.[3,4]

Need for Refocusing

Yet, there are paradoxes and problems with the notion of a set point that is always maintained, in the case of obesity, for example. Another problem with "homeostasis" is that, taken literally, it is a rather static concept, akin to the notion of equilibrium in classical thermodynamics, whereas in real life, an organism experiences changing set points around which it maintains stability over a finite period of time. Moreover, unlike the closed systems of classical thermodynamics, a living organism is an open system in which energy and matter flow in and out, with equilibrium being replaced by steady state.

Cannon recognized this (http://en.wikipedia.org/wiki/Walter_Bradford_Cannon) and used the term steady state and even suggested that "homeodynamics" might be a better term than "homeostasis." The same problem was also recognized by a number of other authors. Nicolaides introduced the term "homeorheusis," in which "stasis" is replaced by "rheusis," meaning "something flowing." Mrosovsky[5] introduced "rheostasis" in which, at any one instant, homeostatic defenses are still present but over a span of time there is a change in the regulated level, or set point of a system. Mrosovsky considers this concept advantageous in explaining how organisms adjust to changing environments such as seasons of the year by storing body fat or changing reproductive physiology, and he believes that "rheostasis" allows for the kinds of physiological plasticity that is involved in evolution.

What is Stress?

The word and concept of "stress" was introduced by Hans Selye[6,7] with emphasis on physical stressors such as physical injury, heat, and cold and later modified to include psychological stressors by the work of John Mason.[8] "Stress" is a word often used in daily life and has a number of meanings. In biomedicine, stress often refers to situations in which the adrenal glucocorticoids and catecholamines are elevated because of an experience—hence, the frequent negative associates of these mediators with "bad stress." Stress is also a subjective experience that may or may not correspond to physiological responses, and the word stress is widely used in many languages as part of daily discourse. There is "good stress" and "bad stress," and people talk about bad stress as "being stressed out."

Stress may be defined as a real or interpreted threat to the physiological or psychological integrity of an individual that results in physiological and/or behavioral responses. Yet this results in three different meanings (http://developingchild.harvard.edu/index.php/activities/council/): "Good stress" refers to rising to a challenge, like having to give a speech or taking an exam, and feeling rewarded by a successful outcome and "bad stress" takes two forms. "Tolerable stress" refers to a life experience that one can cope with and adapt to because one has sufficient internal and external support. "Toxic stress" means life experiences for which one does not have adequate internal and external resources and support and, as a result, the individual experiences adverse physical and mental consequences. Indeed, stress is a condition of the mind and a factor in the expression of disease that differs among individuals and reflects not only major life events, but also the conflicts and pressures of daily life that elevate physiological systems so as to cause a cumulative chronic stress burden on brain and body that can be referred to as "allostatic load and overload" (see below). This view is in agreement with a recent commentary on use of the word 'stress' and what it should mean: "We propose that the term 'stress' should be restricted to conditions where an environmental demand exceeds the natural regulatory capacity of an organism, in particular situations that include unpredictability and uncontrollability."[9]

This burden, which is greater in "toxic" than in "tolerable" stress, reflects not only the impact of life experiences but also of genetic variations; individual health-related behaviors such as diet, exercise, sleep and substance abuse, and epigenetic modifications in development and throughout life that set lifelong patterns of behavior and physiological reactivity through both biological embedding and cumulative change.[1] Epigenetics is the now popular way to describe gene × environment interactions via molecular mechanisms that do not change the genetic code but rather activate, repress, and modulate expression of the code.[10] Indeed, epigenetics denies the notion that "biology is destiny" and opens new opportunities for collaboration between the biological, behavioral, and social sciences and preventative and palliative care.

Acting epigenetically, hormones associated with stress protect the body in the short run and promote adaptation (allostasis), but, in the long run, the burden of chronic stress causes changes in the brain and body that lead to cumulative change such as accumulation of body fat (allostatic load) or disease such as diabetes or cardiovascular disease (allostatic overload) (Figure 1). As will be discussed below, brain circuits are plastic and appear to be continuously remodeled by stress, as well as by other experiences, so as to change the balance between anxiety, self-regulatory behaviors including mood control and impulsivity, memory, and decision making. Such changes may have adaptive value in danger but their persistence and lack of reversibility in brains that are not resilient can be maladaptive.

Stressors and Stress Responses

Stress involves a stressor and a stress response. A stressor may be a physical insult, such as trauma or injury, or physical exertion, particularly when the body is being forced to operate beyond its capacity. Other physical stressors include noise, overcrowding, excessive heat or cold. Stressors also include primarily psychological experiences such as time-pressured tasks, interpersonal conflict, unexpected events, frustration, isolation and loneliness, and traumatic life events, and all of these types of stressors may produce behavioral responses and evoke physiological consequences such as increased blood pressure, elevated heart rate, increased cortisol levels, impaired cognitive function, and altered metabolism, as well as anxiety and depression.

Behavioral responses to stressors may decrease risk and get the individual out of trouble or involve health-promoting activities such as a good diet and regular exercise, but they may also include responses that exacerbate the physiological consequences of stress, for example, self-damaging behaviors like smoking, drinking, overeating, or consuming a rich diet, or risk-taking behaviors like driving an automobile recklessly. The physiological stress responses include primarily the activation of the autonomic nervous system and the hypothalamo-pituitary-adrenal (HPA) axis, leading to increased blood and tissue levels of catecholamines and glucocorticoids. It is these physiological responses that have both protective and damaging effects (see sections on Allostasis and Allostatic Load and Overload).

There are two important features of the physiological stress response[1]: the first involves turning it on in amounts that are adequate to the challenge. The second is turning off the response when it is no longer needed (Figure 2). The physiological mediators of the stress response, namely, the catecholamines of the sympathetic nervous system and the glucocorticoids from the adrenal cortex, initiate cellular events that promote adaptive changes in cells and tissues throughout the body, which in turn protect the organism and promote survival. However, too much stress, or inefficient operation of the acute responses to stress, can cause wear and tear and exacerbate disease processes.

Individual Differences

There are enormous individual differences in interpreting and responding to what is stressful, as well as individual differences in the susceptibility to diseases, in which stress may play a role.[1] Genetic predispositions exist which increase the risk of certain disorders. In addition, developmental processes, such as prenatal stress or nurturing postnatal experiences, contribute to the lifelong responsiveness of the behavioral and physiological responses to stressors. Furthermore, experiences throughout the life course resulting in memories of particularly unpleasant or pleasant situations combine with the genetic and developmental influences to produce large differences among individuals in how they react to stress and what the long-term consequences may be. We now consider two terms, allostasis and allostatic load/overload, that are intended to eliminate ambiguity in using the word stress in so many ways and at the same time, highlight the biphasic role of the physiological mediators of adaptation that can also contribute to pathophysiology.

Allostasis

Allostasis means "achieving stability through change"; it was introduced by P. Sterling and J. Eyer in 1988.[11] Allostasis refers to the process that maintains homeostasis, as defined above, and it recognizes that "set points" and other boundaries of control may change with environmental conditions. There are primary mediators of allostasis such as, but not confined to, hormones of the HPA axis, catecholamines, the parasympathetic nervous system and pro- and anti-inflammatory cytokines. These operate as a nonlinear network in which changes in the output of each mediator influences the output of other mediators (Figure 3). Allostasis also clarifies the inherent ambiguity in the term homeostasis and distinguishes between the systems that are essential for life (homeostasis) and those that maintain these systems in balance (allostasis). Allostatic systems enable an organism to respond to its physical state (e.g., awake, asleep, supine, standing, exercising) and to cope with noise, crowding, isolation, hunger, extremes of temperature, physical danger, psychosocial stress, and to microbial or parasitic infections.

Allostatic States

An allostatic state[13] refers to the altered and sustained activity levels of the primary mediators, for example, glucocorticoids, that integrate energetic and associated behaviors in response to changing environments, challenges such as social interactions, weather, disease, predators, pollution, etc. An allostatic state results in an imbalance of the primary mediators reflecting excessive production of some and inadequate production of others. Examples are hypertension, a perturbed cortisol rhythm in major depression or after chronic sleep deprivation, chronic elevation of inflammatory cytokines and low cortisol in chronic fatigue syndrome, imbalance of cortisol, corticosteroid-releasing factor (CRF), and cytokines in the Lewis rat that increases risk for autoimmune and inflammatory disorders.[14] Allostatic states can be sustained for limited periods if food intake and/or stored energy, such as fat, can fuel homeostatic mechanisms (e.g., bears and other hibernating animals preparing for the winter). If imbalance continues for longer periods and becomes independent of maintaining adequate energy reserves, then symptoms of allostatic overload appear. Abdominal obesity is an example of this condition.[2]

Allostatic Load and Overload

Allostatic load and allostatic overload refers to the cumulative result of an allostatic state. For example, as noted, fat deposition in a bear preparing for the winter,

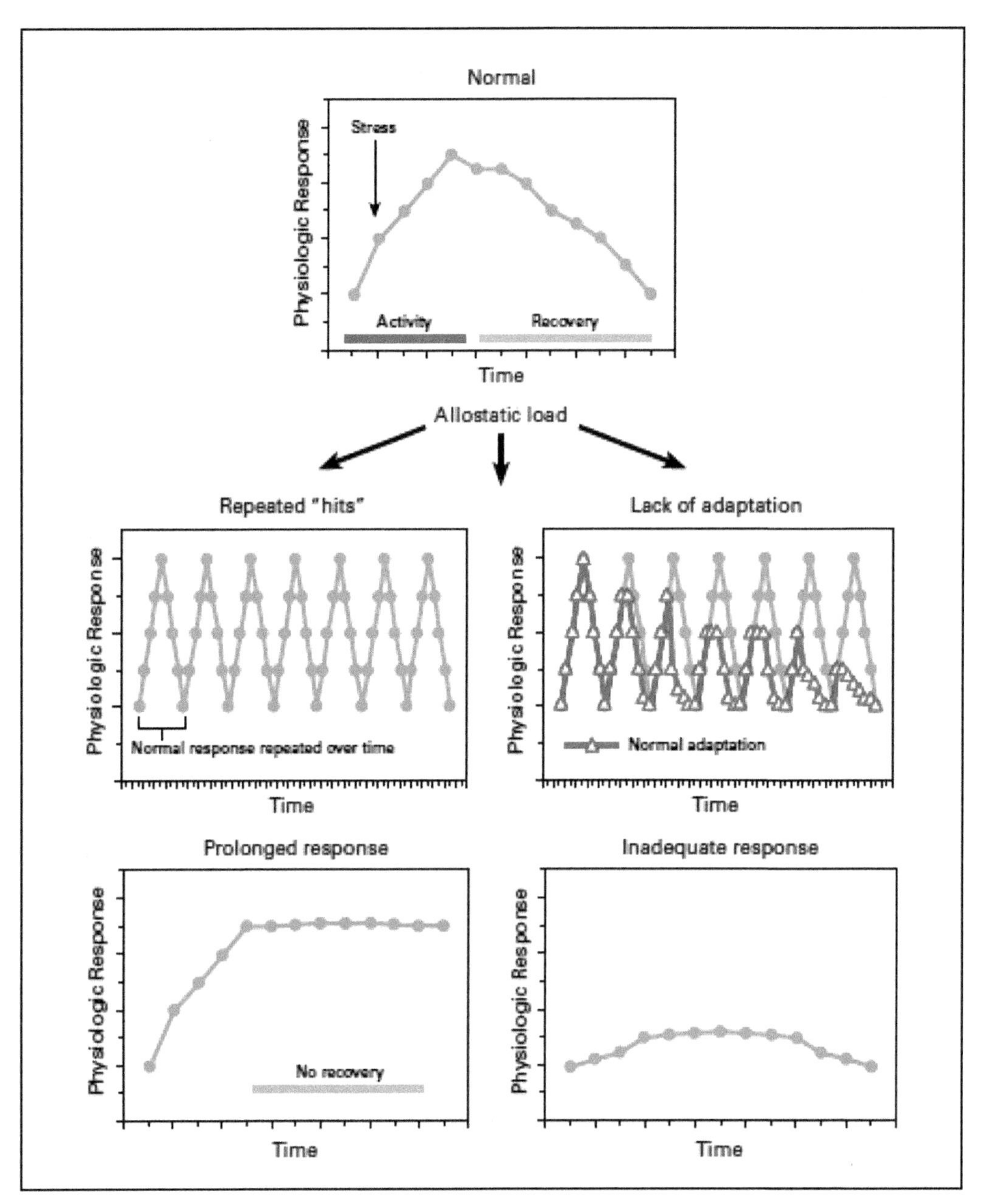

FIGURE 2 Three types of allostatic load. The top panel illustrates the normal allostatic response, in which a response is initiated by a stressor, sustained for an appropriate interval, and then turned off. The remaining panels illustrate four conditions that lead to allostatic load: repeated "hits" from multiple stressors; lack of adaptation; prolonged response due to delayed shutdown; and inadequate response that leads to compensatory hyperactivity of other mediators (e.g., inadequate secretion of glucocorticoids, resulting in increased concentrations of cytokines that are normally counterregulated by glucocorticoids). *From Ref. 1 by permission.*

a bird preparing to migrate or a fish preparing to spawn are examples of animals experiencing an allostatic load.[2] This is a largely beneficial and adaptive condition and can be considered the result of the daily and seasonal routines which organisms use to obtain food and survive, and obtain the extra energy needed to migrate, molt, and breed. Within limits, they are adaptive responses to seasonal and other demands. However, if one superimposes on this additional load of unpredictable events in the environment, disease, human disturbance (for animals in the wild), and social interactions, then allostatic load can increase dramatically and become allostatic overload, which can become a pathophysiological condition.

There are two distinctly different outcomes of an allostatic state in terms of allostatic load or overload.[2] First, if energy demands exceed energy intake, and also exceeds what can be mobilized from stores, then type 1 allostatic overload occurs. For example, breeding birds use increasing food abundance in spring to reproduce and raise their young. If inclement weather then increases costs of

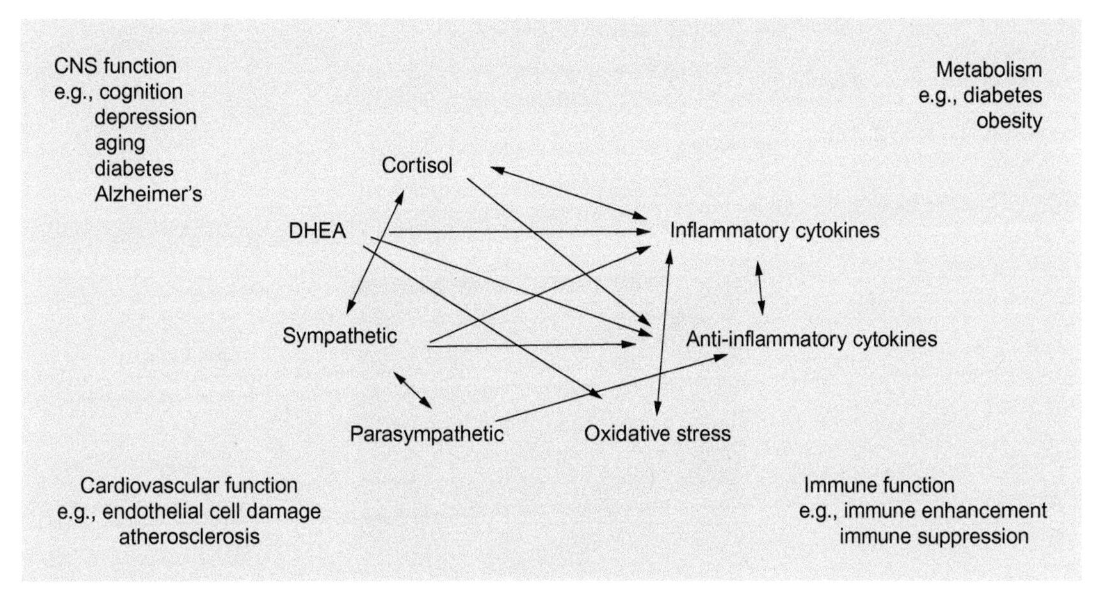

FIGURE 3 Nonlinear network of mediators of allostasis involved in the stress response. Arrows indicate that each system regulates the others in a reciprocal manner, creating a nonlinear network. Moreover, there are multiple pathways for regulation—e.g., inflammatory cytokine production is negatively regulated via anti-inflammatory cytokines as well as via parasympathetic and glucocorticoid pathways, whereas sympathetic activity increases inflammatory cytokine production. Parasympathetic activity, in turn, restrains sympathetic activity. *From Ref. 12 by permission.*

maintaining homeostasis and the allostatic load of breeding, and at the same time reduces food available to fuel that allostatic load, then negative energy balance results in loss of body mass and suppression of reproduction, in part via cortisol secretion. Second, if energy demands are not exceeded and the organism continues to take in or store as much or even more energy than it needs, perhaps as a result of diet, or metabolic imbalances (prediabetic state) that favor fat deposition, then type 2 allostatic overload occurs. Besides fat deposition, there are other cumulative changes in other systems that can result from repeated stressors, for example, neuronal remodeling or loss in hippocampus, atherosclerotic plaques, left ventricular hypertrophy of the heart, glycosylated hemoglobin and other proteins by advanced glycosylation end products as a measure of sustained hyperglycemia, high cholesterol with low high-density lipoprotein (HDL), increased oxidative stress, elevated proinflammatory mediators, and chronic pain and fatigue, for example, in arthritis or psoriasis, associated with imbalance of immune mediators.

Role of Behavior in Allostatic Overload

Anticipation and worry can also contribute to allostatic overload.[15] Anticipation is involved in the reflex that prevents us from blacking out when we get out of bed in the morning and is also a component of worry, anxiety, and cognitive preparation for a threat. Anticipatory anxiety can drive the output of mediators like ACTH, cortisol, and adrenalin; thus, prolonged anxiety and anticipation is likely to result in allostatic overload. For example, salivary cortisol levels increase within 30 min after waking in individuals who are under considerable psychological stress due to work or family matters. Moreover, intrusive memories from a traumatic event (e.g., in post-traumatic stress disorder (PTSD)) can produce a form of chronic, internal stress and can drive physiological responses.

Allostasis, allostatic states, and allostatic overload are also affected by health-damaging behaviors, such as smoking, drinking, excess calorie intake, and poor or limited sleep, or, on the opposite side, health-promoting behaviors, such as a healthy diet and regular, moderate exercise.[1] These behaviors are integral to the overall notion of allostasis—how individuals cope with a challenge—and also contribute to increasing or decreasing allostatic overload by known pathways. For example, a rich diet accelerates atherosclerosis and progression to noninsulin-dependent diabetes by increasing cortisol, leading to fat deposition and insulin resistance; smoking elevates blood pressure and atherogenesis. Yet, regular, moderate exercise protects one against cardiovascular disease, prevents diabetes, counteracts depression and increases decision making and memory.[1]

PLASTICITY AND VULNERABILITY OF THE BRAIN

Glucocorticoids, Stress, and the Hippocampus

The hippocampus was the first higher brain center that was recognized as a target of adrenal steroids[16] and it has

Medial prefrontal cortex

Decision making, working memory, self regulatory behaviors: mood, impulses

Helps shut off the stress response

Shrinkage of dendrites; loss of synapses

Hippocampus

Memory of daily events; spatial memory; mood regulation

Helps shut off stress response

Shrinkage of neurons; synapse loss

Reduced neurogenesis

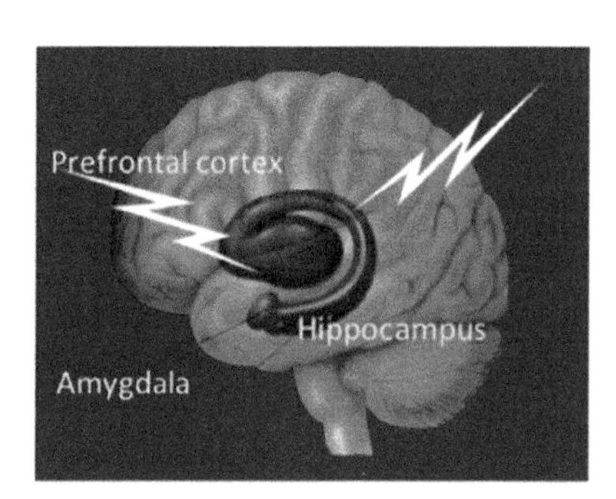

Amygdala

Anxiety, fear; aggression

Turns on stress hormones and increases heart rate

Increased volume and activity

Hypertrophy of neurons; increased synapses

FIGURE 4 Three brain regions that undergo remodeling of dendrites and synapses. The hippocampus, amygdala, and prefrontal cortex communicate with each other and mediate cognitive function, fear, aggression and self-regulation, as well as turning on and off the autonomic and HPA response to stressors.

figured prominently as a gateway to our understanding of how stress impacts neural architecture and behavior (Figure 4). The hippocampus expresses both Type I mineralocorticoid receptor (MR) and Type II glucocorticoid receptor (GR),[17] and these receptors mediate a biphasic response to adrenal steroids in the CA1 region, although only facilitation in the dentate gyrus,[18] nevertheless, shows a diminished excitability in the absence of adrenal steroids.[19] Other brain regions, such as the paraventricular nucleus, lacking in MR but having GR, show a monophasic negative response to increasing glucocorticoid levels.[18] Adrenal steroids exert biphasic effects on excitability of hippocampal neurons in terms of long-term potentiation and primed burst potentiation[20–23] and show parallel biphasic effects on memory.[24,25]

A form of structural plasticity is the remodeling of dendrites in the hippocampus, as well as in the amygdala and prefrontal cortex.[26] In hippocampus, chronic restraint stress (CRS), daily for 21 days, causes retraction and simplification of dendrites in the CA3 region of the hippocampus.[27,28] Such dendritic reorganization is found in both dominant and subordinate rats undergoing adaptation to psychosocial stress in the visible burrow system and it is independent of adrenal size.[29] It also occurs in psychosocial stress in intruder tree shrews in a resident-intruder paradigm, with a time course of 28 days,[30] a procedure that does not cause a loss of pyramidal neurons in the hippocampus.[31]

The mossy fiber input to the CA3 region in the stratum lucidum appears to drive the dendritic remodeling leading to the retraction of the apical dendrites above this input.[27] Moreover, the thorny excrescences, giant spines on which the mossy fiber terminals form their synapses, show stress-induced modifications.[32] And the number of active synaptic zones between thorny excrescences and mossy fiber terminals is rapidly modulated during hibernation and recovery from the hibernating state.[33] The thorny excrescences are not the only spines affected by CRS. Dendritic spines also show remodeling, with increased spine density reported after CRS on apical dendrites of CA3 neurons[34] and decreased spine density reported for CA1 pyramidal neurons.[35]

Exploration of the underlying mechanism for this remodeling of dendrites and synapses reveals that it is not adrenal size or presumed amount of physiological stress per se that determine dendritic remodeling, but rather a complex set of other factors that modulate neuronal structure.[27] Indeed, after repeated stress, dendritic remodeling is reversible,[36] and in species of mammals that hibernate, dendritic remodeling is a reversible process and occurs within hours of the onset of hibernation in European hamsters and ground squirrels, and it is also reversible within hours of wakening of the animals from torpor.[33,37–39] Along with data on post-translational modification of cytoskeletal proteins, this implies that reorganization of the cytoskeleton is taking place rapidly and reversibly[39] and that changes in dendrite length and branching are not damage, but a form of adaptive structural plasticity.

Cellular and molecular mechanisms involving steroids contribute to structural remodeling. Specifically, adrenal steroids are important mediators of remodeling of hippocampal neurons during repeated stress, and exogenous adrenal steroids can also cause remodeling in the absence of an external stressor.[40,41] The role of adrenal steroids in the hippocampus involves many interactions with neurochemical systems including serotonin, endogenous opioids, calcium currents, gamma amino butyric acid

(GABA)-benzodiazepine receptors, and excitatory amino acids.[42] Central to all of these interactions is the role of excitatory amino acids, such as glutamate. Excitatory amino acids released by the mossy fiber pathway play a key role in the remodeling of the CA3 region of the hippocampus, and regulation of glutamate release by adrenal steroids may play an important role.[27]

Among the consequences of restraint stress is the elevation of extracellular glutamate levels, leading to induction of glial glutamate transporters, as well as increased activation of the nuclear transcription factor, phospho-CREB.[43] Moreover, 21 days of CRS leads to depletion of clear vesicles from mossy fiber terminals and increased expression of presynaptic proteins involved in vesicle release.[44,45] Taken together with the fact that vesicles which remain in the mossy fiber terminal are near active synaptic zones and that there are more mitochondria in the terminals of stressed rats, this suggests that CRS increases the release of glutamate.[44]

Extension of Stress Effects to Amygdala and Prefrontal Cortex

Besides the hippocampus, the amygdala and prefrontal cortex are targets of stress and display structural plasticity after both acute and chronic stress (Figure 4). Neurons in the basolateral amygdala (BlA) expand dendrites after chronic immobilization stress and increase spine density,[46] whereas neurons in medial amygdala show reduced spine density after chronic stress.[47] The latter changes are dependent on tissue plasminogen activator released by CRF,[48] based on a tPA-ko mouse, whereas stress effects in BlA are not so dependent.[47] These stress-induced changes are accompanied by increases in anxiety-like behavior[46,49] and suggest that stress causes a reorganization and dysfunction of circuits within the amygdala.

Glucocorticoids and excitatory amino acids are involved in the mechanism for dendritic expansion in the BlA with chronic stress, along with brain derived neurotrophic factor (BDNF)[50,51] and, indeed, a single bolus of corticosterone (CORT) mimics the effects of 10 days of chronic immobilization to cause BlA dendrite expansion.[52] Overexpression of BDNF in mice increases dendritic length in both CA3 and BlA and occludes the effects of chronic stress to decrease dendritic branching in CA3 and increase it in BlA.[53] Without such overexpression, chronic stress causes a down-regulation of BDNF in CA3 hippocampus and an up-regulation of BDNF in the BlA, and the effect in BlA persists after 21-day post-stress while that in CA3 has normalized[50]; moreover, acute stress with a 10-day delay, which causes BlA to develop increased anxiety and increased density of spines in BlA neurons,[54] caused BDNF expression to rise and stay elevated for 10 days while that in CA3 fell after acute stress, but did so only transiently.[50] CORT levels increased after both acute and chronic stress and remained elevated after chronic, but not after acute stress.

Although some of the immediate consequences of stress—elevated glucocorticoids and glutamate—are similar in amygdala and hippocampus, they lead to contrasting patterns of BDNF expression and structural plasticity. This implies that signaling mechanisms more downstream of the initial changes in glucocorticoids and glutamate, but upstream of BDNF, may hold the key to the differential impact of stress in these brain areas. Importantly, BDNF infusion into the hippocampus of stressed rodents helped to protect against the deleterious effects of stress despite high levels of circulating CORT. This suggests that BDNF could be a final point of convergence for the stress induced effects in the hippocampus. BDNF-mediated signaling is involved in stress response but the direction and nature of signaling is region specific, stress specific, and is influenced by epigenetic modifications along with post-translational modifications.[50,55]

Concurrently, with changes in the amygdala, neurons in the medial prefrontal cortex show reversible dendritic shrinkage after chronic stress,[56,57] with spine loss,[58] that can be inhibited by blocking N-methy-D-aspartate (NMDA) receptors,[59] similar to stress-induced atrophy of neurons in the CA3 hippocampus (see above). This chronic stress-induced atrophy is associated with deficits in executive function and cognitive flexibility,[60,61] and the stressors that cause this to happen also include circadian disruption.[62] While medial prefrontal cortical neurons atrophy with chronic stress, neurons in the orbitofrontal cortex show hypertrophy[60] similar to what happens in the BlA.[46]

TRANSLATION TO THE HUMAN BRAIN

Studies of the human hippocampus have demonstrated in many, but not all studies,[63] shrinkage of the hippocampus not only in mild cognitive impairment and Alzheimer's,[64] but also in Type 2 diabetes,[65] prolonged major depression,[66] Cushing's disease,[67] and PTSD.[68] Moreover, in nondisease conditions, such as chronic stress,[69] chronic inflammation,[70] lack of physical activity,[71] and jet lag,[72] smaller hippocampal or temporal lobe volumes have been reported. In the human brain, MRI has shown amygdala enlargement and overactivity, as well as hippocampal and prefrontal cortical shrinkage in a number of mood disorders.[66,73] Moreover, amygdala reactivity to sad and angry faces is enhanced by sleep deprivation and living in an urban environment, and excessive amygdala reactivity is associated with early signs of cardiovascular disease.[26]

These changes may not be due to neuron loss but rather to volume reduction in dentate gyrus due to inhibited neuronal replacement, as well as dendritic shrinkage and glial cell loss. Autopsy studies on depression-suicide have indicated loss of glial cells and smaller neuron soma size,[74] which is indicative of a smaller dendritic tree. With regard to Type 2 diabetes, it should be emphasized that the hippocampus has receptors for, and the ability to take up and respond to, insulin, ghrelin, insulin-like growth factor-1 (IGF1), and leptin; importantly, IGF-1 mediates exercise-induced neurogenesis.[14] Thus, besides its response to glucocorticoids, the hippocampus is an important target of metabolic hormones that have a variety of adaptive actions in the healthy brain which is perturbed in metabolic disorders, such as diabetes.[14] It should also be noted that amygdala volume is reported to decrease in anxiety-disorder subjects in relation to successful anxiety reduction using cognitive-behavioral therapy (CBT).[75] Conversely, longitudinal increases in medial prefrontal cortex volume have been reported after CBT in chronic fatigue patients.[76]

One surprising finding from animal models and human studies concerning PTSD is that a timed elevation of glucocorticoids can prevent PTSD symptoms. This comes from epidemiologic studies showing that low cortisol is a risk factor for PTSD[77] and is also supported by interventions for possible PTSD with cortisol supplementation for low normal cortisol levels during surgery.[78] Another finding is that glucocorticoid administration within an hour or so after a traffic accident reduces incidence of PTSD symptoms.[79] Two animal models support this,[79,80] the latter of which showed that elevation of CORT at the time of trauma prevented the increased anxiety and spine density of BlA neurons in a rat model 10 days later.[80]

SEX DIFFERENCES IN STRESS RESPONSIVENESS AND WHAT THIS MEANS FOR THE REST OF THE BRAIN

All of the animal model studies of stress effects summarized above and below were carried out on male rodents. Thus, it is very important to note before proceeding further by discussing sex differences in how the brain responds to stressors. Indeed, female rodents do not show the same pattern of neural remodeling after chronic stress as do males. The first realization of this was for the hippocampus, in which the remodeling of CA3 dendrites did not occur in females after CRS, even though all the measures of stress hormones indicated that the females were experiencing the stress as much as males.[81] Females and males also differ in the cognitive consequences of repeated stress, with males showing impairment of hippocampal dependent memory, whereas females do not.[82–84]

In contrast, acute tail shock stress during classical eyeblink conditioning improves performance in males, but suppresses it in females[85] by mechanisms influenced by gonadal hormones in development and in adult life.[86,87] However, giving male and female rats control over the shock abolishes both the stress effects and the sex differences.[88] These findings suggest that sex differences involve brain systems that mediate how males and females interpret stressful stimuli and that a sense of control is paramount to coping with those stimuli.

Female rats fail to show the mPFC dendritic remodeling seen in males after CRS in those neurons that do not project to amygdala. Instead, they show an expansion of the dendritic tree in the subset of neurons that project to the basolateral amygala.[89] Moreover, ovariectomy prevented these CRS effects on dendritic length and branching. Furthermore, estradiol treatment of ovariectomized (OVX) females increased spine density in mPFC neurons, irrespective of where they were projecting.[89]

Taken together with the fact that estrogen, as well as androgen, effects are widespread in the central nervous system, these findings indicate that there are likely to be many more examples of sex × stress interactions related to many brain regions and multiple functions, as well as developmentally programmed sex differences that affect how the brain responds to stress (e.g., in the locus ceruleus).[90,91] Clearly, the impact of sex and sex differences has undergone a revolution and much more is to come,[92–96] including insights into X and Y chromosome contributions to brain sex differences.[97] In men and women, neural activation patterns to the same tasks are quite different between the sexes even when performance is similar.[98] This leads to the concept that men and women often use different strategies to approach and deal with issues in their daily lives, in part because of the subtle differences in brain architecture. Nevertheless, from the standpoint of gene expression and epigenetic effects, the principles of what we have learned in animal models regarding plasticity, damage, and resilience, are likely to apply to both males and females.

LESSONS FROM GENE EXPRESSION

Even when the healthy brain and associated behavior appears to have recovered from a stressful challenge, studies of gene expression have revealed that the brain is not the same, just as the morphology after recovery appears to be somewhat different from what it was before stress.[99] Transcriptional profiling of the mouse hippocampus has revealed that after a recovery period from chronic stress, which is equivalent to the duration of the stressor (21 days) and is sufficient to restore anxiety-like behaviors to pre-stress baselines, the expression levels of numerous genes remained distinct from the stress naïve controls.[100] Further, exposure to a novel swim stress 24 h after chronic stress or after a 21-day recovery period from the chronic stress, produced distinct gene expression profiles from mice that experienced a swim stress but had no history of chronic stress. Together, these findings suggest that gene expression patterns after recovery from stress do not reflect a return to the stress naïve baseline (even when the behaviors have recovered) and chronic stress alters reactivity to future stressors. Studies examining longer recovery periods, as well as how intermittent stress during recovery might alter gene expression will be necessary to answer whether these seemingly lasting changes might eventually reverse, or if an additional stressor can compound certain changes. These changes in transcriptome reactivity represent one molecular signature for resilience that are themselves likely to be driven by epigenetic changes discussed in the next section.

Importantly, recent evidence has suggested that the in vivo transcriptional changes in response to stress represent a synthesis of multiple cellular pathways, not simply CORT activation of GR-dependent transcription. Chronic stress increases inflammatory tone and this release of cytokines can activate other signaling pathways, such as NF-κB-dependent transcription.[100] Microarray studies have found that glucocorticoid injections produce distinct gene expression profiles from naïve acute stress and that the gene expression response to a glucocorticoid injection changes after exposure to chronic stress.[100,101] In support of these findings, in vitro studies have demonstrated that simultaneous activation of GR and NF-κB-dependent transcription results in a unique pattern of gene expression that is distinct from the predicted sum of either pathway activated alone.[102] These findings illustrate that gene expression changes in response to stress are not solely the product of glucocorticoid activity. Increasingly, research into stress resilience is looking beyond GR-dependent transcription in order to capture the complexity of the cellular response to stress.

LESSONS FROM EPIGENETIC INFLUENCES

Functional insights into the ever-changing brain come from studies of epigenetic regulation. The term "epigenetics" now extends beyond its original definition[103] to include the continuous, seamless interaction between genes and the factors which regulate gene expression over the life course. The core of the genomic response to those environmental factors such as hormones, cytokines, and chemokines and other neuromodulators involves modification of histones,[104] methylation of cytosine residues on DNA, noncoding RNAs that modify expression of mRNA molecules, and retrotransposon DNA elements[10] (Figure 5).

In our studies of stress neurobiology, acute restraint stress was shown to increase expression in the dentate gyrus of a repressive histone mark, H3K9me3, and this was accompanied by the repression of certain retrotransposon elements of as yet unidentified function; this repressive response habituates with repeated stress raising the possibility of increased genomic instability. Such instability may manifest itself in terms of genomic activity that is no longer responsive to environmental influences or lead to genomic activity that is increased as a result of chronic stress, as in accelerated aging.[106,107]

"Epigenetics"

Emergence of individual/species characteristics during development (Waddington 1942) .

Now means " above the genome"—not changing DNA sequence

Refers to the gene-environment interactions that bring about the phenotype of an individual.

- Modifications of histones—unfolding/folding of chromatin to expose or hide genes

- Binding of transcription regulators to DNA response elements on genes

- Methylation of cytosine bases in DNA without changing genetic code

- MicroRNA's—regulate mRNA survival and translation to protein

----

--Transposons and retrotransposons—DNA rearrangements and insertions

FIGURE 5 "Epigenetics" in its present definition refers to mechanisms for regulation of gene expression that do not change the genetic code.[10] Transposons and retrotransposons are somewhat of an exception in that pieces of DNA are removed and inserted into other locations "jumping genes."[105]

Loss of reversal of stress induced structural plasticity, as seen in aging rats[108] is one example, and increased expression of inflammatory mediators together with loss of cholinergic and dopaminergic function[109] is another.

In contrast, there are examples of epigenetic activation of neural activity. Indeed, acute swim stress as well as novelty exposure induce an activational histone mark in dentate gyrus, namely, acetylation of lysine residue 14 and phosphorylation of the serine residue on histone H3, which is dependent on both GR and NMDA activation and is associated with c-fos induction among other genes.[110] Acetylation of another lysine residue, K27 on histone H3, is associated with increased expression of metabotropic glutamate receptor, mGlu2, in hippocampus of Flinders Sensitive Line (FSL) rats as shown by chromatin immunoprecipitation.[111] mGlu2 is known to exert an inhibitory tone on glutamate release from synapses. The acetylating agent L-acetylcarnitine (LAC), a naturally occurring substance, behaves as an antidepressant, at least in part by the epigenetic up-regulation of mGlu2 receptors via this epigenetic mechanism. LAC caused a rapid and long-lasting antidepressant effect in both FSL rats and in mice exposed to chronic unpredictable stress, which, respectively, model genetic and environmentally induced depression. Beyond the epigenetic action on the acetylated H3K27 bound to the Grm2 promoter, LAC also increased acetylation of NF-κB-p65 subunit, thereby enhancing the transcription of Grm2 gene encoding for the mGlu2 receptor in hippocampus and prefrontal cortex. The involvement of NF-κB in LAC antidepressant-like effects supports a growing literature that shows depression may be associated with a chronic inflammatory response.[112] Importantly, LAC reduced the immobility time in the forced swim test and increased sucrose preference as early as 3 days of treatment, whereas 14 days of treatment were needed for the antidepressant effect of chlorimipramine.[111] This suggests LAC is important for stress resilience.

A recent study from our laboratory has shown that hippocampal expression of mGlu2, is also a marker of individual susceptibility to mood disorders. Interestingly, mGlu2 is the same receptor regulating inhibitory glutamate tone that has been shown to be elevated by treatment with LAC in FSL rats to reverse depressive-like behavior.[111] Using a novel and acute approach for rapidly screening an inbred population of laboratory animals, it has been shown that both chronic unpredictable stress and acute restraint stress results in individual behavioral and molecular differences in wild-type mice that are more (high sensitivity, HS) or less (low sensitivity, LS) susceptible to stress-induced mood abnormalities. At the molecular level, HS and LS mice differ in the ability of stress to induce a decrease of mGlu2 receptor expression in hippocampus. Mapping the steps of this intricate dance that allow some individuals to face adverse life experience, the HS subset of mice was associated with higher baseline levels of MR genes than the LS subset, showing an MR-dependent down-regulation of mGlu2 receptors in hippocampus. These findings led to the introduction of the "epigenetic allostasis model," which incorporates an epigenetic core into the allostasis-allostatic load model of stress and adaptation to emphasize the gene-environment interactions. In particular, the epigenetic allostasis model helps to elucidate how nonshared experience early in life can epigenetically set each individual, via MR genes or other genes, to a somewhat different trajectory of development as far as responses to subsequent stressful life experiences.[113] In agreement, juvenile stress was associated with increased hippocampal MR mRNA levels and anxiety-like behavior in adulthood.[114]

THE LIFE COURSE AND THE EPIGENETICS OF INDIVIDUAL DIFFERENCES

The individual traits that allow these adaptive or maladaptive outcomes depend upon the unique neurological capacity of each individual, which is built upon experiences in the life course, particularly those early in life.[115] These influences can result in healthy or unhealthy brain architecture and in epigenetic regulation that either promotes or fails to promote gene expression responses to new challenges. Genetically similar or identical individuals differ in many ways ranging from length of dendrites in the prefrontal cortex[116] to differences in MR levels in hippocampus,[113] locomotor activity, and neurogenesis rates,[117] and the influences that lead to those differences begin early in life. For example, identical twins diverge over the life course in patterns of CpG methylation of their DNA reflecting the influence of "nonshared" experiences.[118]

Early life events related to maternal care in animals, as well as parental care in humans, play a powerful role in later mental and physical health, as demonstrated by the adverse childhood experiences studies,[119] and recent work that will be noted below. Animal models have contributed enormously to our understanding of how the brain and body are affected, starting with the "neonatal handling" studies of Levine and Denenberg[120] and the more recent, elegant work of Meaney and Syzf[121] involving methylation of CpG residues in DNA. Such epigenetic, transgenerational effects transmitted by maternal care are central to these findings. Besides the amount of maternal care, the consistency over time of that care and the exposure to novelty are also very important not only in rodents,[122,123] but also in monkey models.[124] Prenatal stress impairs hippocampal development in rats, as does stress in adolescence.[125] Insufficient maternal care in rodents[126] and the surprising attachment shown by infant rats to their less-attentive mothers appears to involve an immature amygdala,[127] activation of which by glucocorticoids causes an aversive conditioning response to emerge. Maternal anxiety in the variable foraging demand model in rhesus monkeys leads to chronic anxiety in the offspring, as well as signs of metabolic syndrome.[128,129]

INTERVENTIONS

What can be done to remediate the effects of chronic stress over the life course at both individual and societal levels? For the individual, the complexity of interacting, nonlinear and biphasic actions of the mediators of stress and adaptation, as described above, emphasizes behavioral, or "top-down," interventions (i.e., interventions that involve integrated CNS activity) that include CBT, mindfulness-based stress reduction, including meditation, physical activity, and programs such as the Experience Corps that promote social support and integration and meaning and purpose in life.[14,130,131] In contrast, pharmacological agents, which are useful in many circumstances to redress chemical and molecular imbalances, nevertheless run the risk of dysregulating other adaptive pathways (i.e., no pharmaceutical is without side effects). It should also be noted that many interventions that are intended to promote plasticity and slow decline with age, such as physical activity and positive social interactions that give meaning and purpose, are also useful for promoting "positive health" and "eudaimonia"[132–134] independently of any notable disorder and within the range of normal behavior and physiology.

A powerful "top down" therapy (i.e., an activity, usually voluntary, involving activation of integrated nervous system activity, as opposed to pharmacological therapy which has a more limited target) is regular physical activity, which has actions that improve prefrontal and parietal cortex blood flow and enhance executive function.[135] Moreover, regular physical activity, consisting of walking an hour a day, 5 out of 7 days a week, increases hippocampal volume in previously sedentary adults.[136] This finding complements work showing that fit individuals have larger hippocampal volumes than sedentary adults of the same age-range.[71] It is also well known that regular physical activity is an effective antidepressant and protects against cardiovascular disease, diabetes, and dementia.[137–141] Moreover, intensive learning has also been shown to increase volume of the human hippocampus.[142]

Social integration and support and finding meaning and purpose in life are known to be protective against allostatic load and overload[143] and dementia,[144] and programs such as the Experience Corps that promote these along with increased physical activity, have been shown to slow the decline of physical and mental health and to improve prefrontal cortical blood flow in a similar manner to regular physical activity.[145,146]

Depression and anxiety disorders are examples of a loss of resilience, in the sense that changes in brain circuitry and function, caused by the stressors that precipitate the disorder, become "locked" in a particular state and thus need external intervention. Indeed, prolonged depression is associated with shrinkage of the hippocampus[66,147] and prefrontal cortex.[148] While there appears to be no neuronal loss, there is evidence for glial cell loss and smaller neuronal cell nuclei,[74,149] which is consistent with a shrinking of the dendritic tree described above after chronic stress. Indeed, a few studies indicate that pharmacological treatment may reverse the decreased hippocampal volume in unipolar[150] and bipolar[151] depression, but the possible influence of concurrent CBT in these studies is unclear. One useful model is the reversal of amblyopia (lazy eye; monocular visual field created by closing one eye early in life) via plasticity-inducing interventions such as fluoxetine, caloric restrictions, and cortisol combined with binocular visual stimulation in amblyopic adult animals.[152,153] These studies support the concept that a combination of a pharmaceutical intervention, like fluoxetine, or physical activity, that opens up a "window of plasticity" might improve the efficacy of targeted behavioral therapies.

In this connection it is important to reiterate that successful behavioral therapy, which is tailored to individual needs, can produce volumetric changes in both prefrontal cortex in the case of chronic fatigue,[76] and in amygdala, in the case of chronic anxiety,[75] as measured in same subjects longitudinally. This reinforces the notion that plasticity-facilitating treatments should be given within the framework of a positive behavioral or physical therapy intervention. On the other hand, negative experiences during the window of enhanced plasticity may have negative consequences, such as a person going back into a bad family environment that may have precipitated anxiety or depression in the first place.[154] In that connection, it should be noted that BDNF, a plasticity-enhancing class of molecules, also has the ability to promote pathophysiology, as in seizures.[155–157]

At the societal level, the most important top-down interventions are the policies of government and the private sector that not only improve education but also allow people to make choices that improve their chances for a healthy life.[130] This point was made by the Acheson report of the British Government in 1998[158] which recognized that no public policy of virtually any kind should be enacted without considering the implications for health of all citizens. Thus basic education, housing, taxation, setting of a minimum wage, and addressing occupational health and safety and environmental pollution regulations are all likely to affect health via a myriad of mechanisms. At the same time, providing higher quality food and making it affordable and accessible in poor, as well as affluent neighborhoods, is necessary for people to eat better, providing they also learn what types of food to eat.[159] Likewise, making neighborhoods safer and more congenial and supportive[160,161] can improve opportunities for positive social interactions and increased recreational physical activity. However, governmental policies are not the only way to reduce allostatic load. For example, businesses that encourage healthy lifestyle practices among their employees are likely to gain reduced health insurance costs and possibly a more loyal workforce.[162–164]

Acknowledgments

This review is dedicated to current and former postdoctoral fellows and students, as well as collaborators, who, in many cases, catalyzed the research discussed in this review, many of whose names are noted in the text and references. I am very proud of their accomplishments!

References

1. McEwen BS. Protective and damaging effects of stress mediators. *N Engl J Med*. 1998;338:171–179.
2. McEwen BS, Wingfield JC. The concept of allostasis in biology and biomedicine. *Horm Behav*. 2003;43:2–15.
3. Dhabhar F, McEwen B. Enhancing versus suppressive effects of stress hormones on skin immune function. *Proc Natl Acad Sci U S A*. 1999;96:1059–1064.
4. Yudt MR, Cidlowski JA. The glucocorticoid receptor: coding a diversity of proteins and responses through a single gene. *Mol Endocrinol*. 2002;16:1719–1726.
5. Mrosovsky N. *Rheostasis: The Physiology of Change*. New York, NY: Oxford University Press; 1990.
6. Selye H. A syndrome produced by diverse nocuous agents. *Nature*. 1936;138:32.
7. Selye H, ed. *The Stress of Life*. New York, NY: McGraw Hill; 1956.
8. Mason J. Psychological influences on the pituitary-adrenal cortical system. In: Pincus G, ed. *Recent Progress in Hormone Research*. New York, NY: Academic Press; 1959:345–389.
9. Koolhaas JM, Bartolomucci A, Buwalda B, et al. Stress revisited: a critical evaluation of the stress concept. *Neurosci Biobehav Rev*. 2011;35:1291–1301.
10. Mehler MF. Epigenetic principles and mechanisms underlying nervous system functions in health and disease. *Prog Neurobiol*. 2008;86:305–341.
11. Sterling P, Eyer J. Allostasis: a new paradigm to explain arousal pathology. In: Fisher S, Reason J, eds. *Handbook of Life Stress, Cognition and Health*. New York, NY: John Wiley & Sons; 1988:629–649.
12. McEwen BS. Protective and damaging effects of stress mediators: central role of the brain. *Dialogues Clin Neurosci*. 2006;8:367–381.
13. Koob GF, Le Moal M. Drug abuse: hedonic homeostatic dysregulation. *Science*. 1997;278:52–58.
14. McEwen BS. Physiology and neurobiology of stress and adaptation: central role of the brain. *Physiol Rev*. 2007;87:873–904.
15. Schulkin J, McEwen BS, Gold PW. Allostasis, amygdala, and anticipatory angst. *Neurosci Biobehav Rev*. 1994;18:385–396.
16. McEwen BS, Weiss J, Schwartz L. Selective retention of corticosterone by limbic structures in rat brain. *Nature*. 1968;220:911–912.
17. Reul JM, DeKloet ER. Two receptor systems for corticosterone in rat brain: microdistribution and differential occupation. *Endocrinology*. 1985;117:2505–2511.
18. Joels M. Corticosteroid effects in the brain: U-shape it. *Trends Pharmacol Sci*. 2006;27:244–250.
19. Margineanu D-G, Gower AJ, Gobert J, Wulfert E. Long-term adrenalectomy reduces hippocampal granule cell excitability in vivo. *Brain Res Bull*. 1994;33:93–98.
20. Diamond DM, Bennett MC, Fleshner M, Rose GM. Inverted-U relationship between the level of peripheral corticosterone and the magnitude of hippocampal primed burst potentiation. *Hippocampus*. 1992;2:421–430.
21. Pavlides C, Kimura A, Magarinos AM, McEwen BS. Type I adrenal steroid receptors prolong hippocampal long-term potentiation. *NeuroReport*. 1994;5:2673–2677.
22. Pavlides C, Kimura A, Magarinos AM, McEwen BS. Hippocampal homosynaptic long-term depression/depotentiation induced by adrenal steroids. *Neuroscience*. 1995;68:379–385.
23. Pavlides C, Watanabe Y, Magarinos AM, McEwen BS. Opposing role of adrenal steroid Type I and Type II receptors in hippocampal long-term potentiation. *Neuroscience*. 1995;68:387–394.
24. Pugh CR, Tremblay D, Fleshner M, Rudy JW. A selective role for corticosterone in contextual-fear conditioning. *Behav Neurosci*. 1997;111:503–511.
25. Okuda S, Roozendaal B, McGaugh JL. Glucocorticoid effects on object recognition memory require training-associated emotional arousal. *Proc Natl Acad Sci U S A*. 2004;101:853–858.
26. McEwen BS, Gianaros PJ. Stress- and allostasis-induced brain plasticity. *Annu Rev Med*. 2011;62:431–445.
27. McEwen BS. Stress and hippocampal plasticity. *Annu Rev Neurosci*. 1999;22:105–122.
28. Sousa N, Lukoyanov NV, Madeira MD, Almeida OFX, Paula-Barbosa MM. Reorganization of the morphology of hippocampal neurites and synapses after stress-induced damage correlates with behavioral improvement. *Neuroscience*. 2000;97:253–266.
29. McKittrick CR, Magarinos AM, Blanchard DC, Blanchard RJ, McEwen BS, Sakai RR. Chronic social stress reduces dendritic arbors in CA3 of hippocampus and decreases binding to serotonin transporter sites. *Synapse*. 2000;36:85–94.
30. Magarinos AM, McEwen BS, Flugge G, Fuchs E. Chronic psychosocial stress causes apical dendritic atrophy of hippocampal CA3 pyramidal neurons in subordinate tree shrews. *J Neurosci*. 1996;16:3534–3540.
31. Vollmann-Honsdorf GK, Flugge G, Fuchs E. Chronic psychosocial stress does not affect the number of pyramidal neurons in tree shrew hippocampus. *Neurosci Lett*. 1997;233:121–124.
32. Stewart MG, Davies HA, Sandi C, et al. Stress suppresses and learning induces plasticity in CA3 of rat hippocampus: a three-dimensional ultrastructural study of thorny excrescences and their postsynaptic densities. *Neuroscience*. 2005;131:43–54.
33. Magarinos AM, McEwen BS, Saboureau M, Pevet P. Rapid and reversible changes in intrahippocampal connectivity during the course of hibernation in European hamsters. *Proc Natl Acad Sci U S A*. 2006;103:18775–18780.
34. Sunanda Rao MS, Raju TR. Effect of chronic restraint stress on dendritic spines and excrescences of hippocampal CA3 pyramidal neurons—a quantitative study. *Brain Res*. 1995;694:312–317.
35. Magarinos AM, Li CJ, Gal Toth J, et al. Effect of brain-derived neurotrophic factor haploinsufficiency on stress-induced remodeling of hippocampal neurons. *Hippocampus*. 2011;21:253–264.
36. Conrad CD, Magarinos AM, LeDoux JE, McEwen BS. Repeated restraint stress facilitates fear conditioning independently of causing hippocampal CA3 dendritic atrophy. *Behav Neurosci*. 1999;113:902–913.
37. Popov VI, Bocharova LS. Hibernation-induced structural changes in synaptic contacts between mossy fibres and hippocampal pyramidal neurons. *Neuroscience*. 1992;48:53–62.

38. Popov VI, Bocharova LS, Bragin AG. Repeated changes of dendritic morphology in the hippocampus of ground squirrels in the course of hibernation. *Neuroscience*. 1992;48:45–51.
39. Arendt T, Stieler J, Strijkstra AM, et al. Reversible paired helical filament-like phosphorylation of tau is an adaptive process associated with neuronal plasticity in hibernating animals. *J Neurosci*. 2003;23:6972–6981.
40. Magarinos AM, Deslandes A, McEwen BS. Effects of antidepressants and benzodiazepine treatments on the dendritic structure of CA3 pyramidal neurons after chronic stress. *Eur J Pharm*. 1999;371:113–122.
41. Sousa AR, Lane SJ, Cidlowski JA, Staynov DZ, Lee TH. Glucocorticoid resistance in asthma is associated with elevated in vivo expression of the glucocorticoid receptor b-isoform. *J Allergy Clin Immunol*. 2000;105:943–950.
42. McEwen BS. Stress, sex, and neural adaptation to a changing environment: mechanisms of neuronal remodeling. *Ann N Y Acad Sci*. 2010;1204(suppl):E38–E59.
43. Wood GE, Young LT, Reagan LP, Chen B, McEwen BS. Stress-induced structural remodeling in hippocampus: prevention by lithium treatment. *Proc Natl Acad Sci U S A*. 2004;101:3973–3978.
44. Magarinos AM, Verdugo Garcia JM, McEwen BS. Chronic restraint stress alters synaptic terminal structure in hippocampus. *Proc Natl Acad Sci U S A*. 1997;94:14002–14008.
45. Grillo CA, Piroli GG, Wood GE, Reznikov LR, McEwen BS, Reagan LP. Immunocytochemical analysis of synaptic proteins provides new insights into diabetes-mediated plasticity in the rat hippocampus. *Neuroscience*. 2005;136:477–486.
46. Vyas A, Mitra R, Rao BSS, Chattarji S. Chronic stress induces contrasting patterns of dendritic remodeling in hippocampal and amygdaloid neurons. *J Neurosci*. 2002;22:6810–6818.
47. Bennur S, Shankaranarayana Rao BS, Pawlak R, Strickland S, McEwen BS, Chattarji S. Stress-induced spine loss in the medial amygdala is mediated by tissue-plasminogen activator. *Neuroscience*. 2007;144:8–16.
48. Matys T, Pawlak R, Matys E, Pavlides C, McEwen BS, Strickland S. Tissue plasminogen activator promotes the effects of corticotropin releasing factor on the amygdala and anxiety-like behavior. *Proc Natl Acad Sci U S A*. 2004;101:16345–16350.
49. Pawlak R, Magarinos AM, Melchor J, McEwen B, Strickland S. Tissue plasminogen activator in the amygdala is critical for stress-induced anxiety-like behavior. *Nat Neurosci*. 2003;6:168–174.
50. Lakshminarasimhan H, Chattarji S. Stress leads to contrasting effects on the levels of brain derived neurotrophic factor in the hippocampus and amygdala. *PLoS ONE*. 2012;7:e30481.
51. McEwen BS, Chattarji S. *Neuroendocrinology of stress. Handbook of Neurochemistry and Molecular Neurobiology*. 3rd ed. New York, NY: Springer-Verlag; 2007 pp. 572–593.
52. Mitra R, Sapolsky RM. Acute corticosterone treatment is sufficient to induce anxiety and amygdaloid dendritic hypertrophy. *Proc Natl Acad Sci U S A*. 2008;105:5573–5578.
53. Govindarajan A, Rao BSS, Nair D, et al. Transgenic brain-derived neurotrophic factor expression causes both anxiogenic and antidepressant effects. *Proc Natl Acad Sci U S A*. 2006;103: 13208–13213.
54. Mitra R, Jadhav S, McEwen BS, Vyas A, Chattarji S. Stress duration modulates the spatiotemporal patterns of spine formation in the basolateral amygdala. *Proc Natl Acad Sci U S A*. 2005;102: 9371–9376.
55. Gray JD, Milner TA, McEwen BS. Dynamic plasticity: the role of glucocorticoids, brain-derived neurotrophic factor and other trophic factors. *Neuroscience*. 2013;239:214–227.
56. Radley JJ, Sisti HM, Hao J, et al. Chronic behavioral stress induces apical dendritic reorganization in pyramidal neurons of the medial prefrontal cortex. *Neuroscience*. 2004;125:1–6.
57. Radley JJ, Rocher AB, Janssen WGM, Hof PR, McEwen BS, Morrison JH. Reversibility of apical dendritic retraction in the rat medial prefrontal cortex following repeated stress. *Exp Neurol*. 2005;196:199–203.
58. Radley JJ, Rocher AB, Rodriguez A, et al. Repeated stress alters dendritic spine morphology in the rat medial prefrontal cortex. *J Comp Neurol*. 2008;507:1141–1150.
59. Martin KP, Wellman CL. NMDA receptor blockade alters stress-induced dendritic remodeling in medial prefrontal cortex. *Cereb Cortex*. 2011;21:2366–2373.
60. Liston C, Miller MM, Goldwater DS, et al. Stress-induced alterations in prefrontal cortical dendritic morphology predict selective impairments in perceptual attentional set-shifting. *J Neurosci*. 2006;26:7870–7874.
61. Dias-Ferreira E, Sousa JC, Melo I, et al. Chronic stress causes frontostriatal reorganization and affects decision-making. *Science*. 2009;325:621–625.
62. Karatsoreos IN, Bhagat S, Bloss EB, Morrison JH, McEwen BS. Disruption of circadian clocks has ramifications for metabolism, brain, and behavior. *Proc Natl Acad Sci U S A*. 2011;108:1657–1662.
63. Fink G. Stress controversies: post-traumatic stress disorder, hippocampal volume, gastroduodenal ulceration. *J Neuroendocrinol*. 2011;23:107–117.
64. de Leon MJ, George AE, Golomb J, et al. Frequency of hippocampus atrophy in normal elderly and Alzheimer's disease patients. *Neurobiol Aging*. 1997;18:1–11.
65. Gold SM, Dziobek I, Sweat V, et al. Hippocampal damage and memory impairments as possible early brain complications of type 2 diabetes. *Diabetologia*. 2007;50:711–719.
66. Sheline YI. Neuroimaging studies of mood disorder effects on the brain. *Biol Psychiatry*. 2003;54:338–352.
67. Starkman MN, Giordani B, Gebrski SS, Berent S, Schork MA, Schteingart DE. Decrease in cortisol reverses human hippocampal atrophy following treatment of Cushing's disease. *Biol Psychiatry*. 1999;46:1595–1602.
68. Gurvits TV, Shenton ME, Hokama H, et al. Magnetic resonance imaging study of hippocampal volume in chronic, combat-related posttraumatic stress disorder. *Biol Psychiatry*. 1996;40:1091–1099.
69. Gianaros PJ, Jennings JR, Sheu LK, Greer PJ, Kuller LH, Matthews KA. Prospective reports of chronic life stress predict decreased grey matter volume in the hippocampus. *NeuroImage*. 2007;35:795–803.
70. Marsland AL, Gianaros PJ, Abramowitch SM, Manuck SB, Hariri AR. Interleukin-6 covaries inversely with hippocampal grey matter volume in middle-aged adults. *Biol Psychiatry*. 2008;64:484–490.
71. Erickson KI, Prakash RS, Voss MW, et al. Aerobic fitness is associated with hippocampal volume in elderly humans. *Hippocampus*. 2009;19:1030–1039.
72. Cho K. Chronic 'jet lag' produces temporal lobe atrophy and spatial cognitive deficits. *Nat Neurosci*. 2001;4:567–568.
73. Drevets WC, Raichle ME. Neuroanatomical circuits in depression: implications for treatment mechanisms. *Psychopharmacol Bull*. 1992;28:261–274.
74. Stockmeier CA, Mahajan GJ, Konick LC, et al. Cellular changes in the postmortem hippocampus in major depression. *Biol Psychiatry*. 2004;56:640–650.
75. Holzel BK, Carmody J, Evans KC, et al. Stress reduction correlates with structural changes in the amygdala. *Soc Cogn Affect Neurosci*. 2010;5:11–17.
76. de Lange FP, Koers A, Kalkman JS, et al. Increase in prefrontal cortical volume following cognitive behavioural therapy in patients with chronic fatigue syndrome. *Brain*. 2008;131:2172–2180.
77. Yehuda R, McFarlane AC, Shalev AY. Predicting the development of posttraumatic stress disorder from the acute response to a traumatic event. *Biol Psychiatry*. 1998;44:1305–1313.
78. Schelling G, Roozendaal B, De Quervain DJ-F. Can posttraumatic stress disorder be prevented with glucocorticoids? *Ann N Y Acad Sci*. 2004;1032:158–166.
79. Zohar J, Yahalom H, Kozlovsky N, et al. High dose hydrocortisone immediately after trauma may alter the trajectory of PTSD:

interplay between clinical and animal studies. *Eur Neuropsychopharmacol*. 2011;21:796–809.
80. Rao RP, Anilkumar S, McEwen BS, Chattarji S. Glucocorticoids protect against the delayed behavioral and cellular effects of acute stress on the amygdala. *Biol Psychiatry*. 2012;72:466–475.
81. Galea LAM, McEwen BS, Tanapat P, Deak T, Spencer RL, Dhabhar FS. Sex differences in dendritic atrophy of CA3 pyramidal neurons in response to chronic restraint stress. *Neuroscience*. 1997;81:689–697.
82. Luine V, Villegas M, Martinez C, McEwen BS. Repeated stress causes reversible impairments of spatial memory performance. *Brain Res*. 1994;639:167–170.
83. Luine VN, Beck KD, Bowman RE, Frankfurt M, MacLusky NJ. Chronic stress and neural function: accounting for sex and age. *J Neuroendocrinol*. 2007;19:743–751.
84. Bowman RE, Zrull MC, Luine VN. Chronic restraint stress enhances radial arm maze performance in female rats. *Brain Res*. 2001;904:279–289.
85. Wood GE, Shors TJ. Stress facilitates classical conditioning in males, but impairs classical conditioning in females through activational effects of ovarian hormones. *Proc Natl Acad Sci U S A*. 1998;95:4066–4071.
86. Wood GE, Shors TJ, Beylin AV. The contribution of adrenal and reproductive hormones to the opposing effects of stress on trace conditioning in males versus females. *Behav Neurosci*. 2001;115:175–187.
87. Shors TJ, Miesegaes G. Testosterone in utero and at birth dictates how stressful experience will affect learning in adulthood. *Proc Natl Acad Sci U S A*. 2002;99:13955–13960.
88. Leuner B, Mendolia-loffredo S, Shors TJ. Males and females respond differently to controllability and antidepressant treatment. *Biol Psychiatry*. 2004;56:964–970.
89. Shansky RM, Hamo C, Hof PR, Lou W, McEwen BS, Morrison JH. Estrogen promotes stress sensitivity in a prefrontal cortex-amygdala pathway. *Cereb Cortex*. 2010;20:2560–2567.
90. Bangasser DA, Curtis A, Reyes BA, et al. Sex differences in corticotropin-releasing factor receptor signaling and trafficking: potential role in female vulnerability to stress-related psychopathology. *Mol Psychiatry*. 2010;15:877–904.
91. Bangasser DA, Zhang X, Garachh V, Hanhauser E, Valentino RJ. Sexual dimorphism in locus coeruleus dendritic morphology: a structural basis for sex differences in emotional arousal. *Physiol Behav*. 2011;103:342–351.
92. Cahill L. Why sex matters for neuroscience. *Nat Rev Neurosci*. 2006;7:477–484.
93. McEwen BS, Lasley EN. *The end of sex as we know it. Cerebrum: The Dana Forum on Brain Science*; vol. 7. Washington, DC: Dana Press; 2005.
94. McEwen BS. Introduction: the end of sex as we once knew it. *Physiol Behav*. 2009;97:143–145.
95. Laje G, Paddock S, Manji H, et al. Genetic markers of suicidal ideation emerging during citalopram treatment of major depression. *Am J Psychiatry*. 2007;164:1530–1538.
96. Meites J. Short history of neuroendocrinology and the International Society of Neuroendocrinology. *Neuroendocrinology*. 1992;56:1–10.
97. Carruth LL, Reisert I, Arnold AP. Sex chromosome genes directly affect brain sexual differentiation. *Nat Neurosci*. 2002;5:933–934.
98. Derntl B, Finkelmeyer A, Eickhoff S, et al. Multidimensional assessment of empathic abilities: neural correlates and gender differences. *Psychoneuroendocrinology*. 2010;35:67–82.
99. Goldwater DS, Pavlides C, Hunter RG, et al. Structural and functional alterations to rat medial prefrontal cortex following chronic restraint stress and recovery. *Neuroscience*. 2009;164:798–808.
100. Gray JD, Rubin TG, Hunter RG, McEwen BS. Hippocampal gene expression changes underlying stress sensitization and recovery. *Mol Psychiatry*. 2014;19:1171–1178.
101. Datson NA, van den Oever JM, Korobko OB, Magarinos AM, de Kloet ER, McEwen BS. Previous history of chronic stress changes the transcriptional response to glucocorticoid challenge in the dentate gyrus region of the male rat hippocampus. *Endocrinology*. 2013;154:3261–3272.
102. Rao NA, McCalman MT, Moulos P, et al. Coactivation of GR and NFKB alters the repertoire of their binding sites and target genes. *Genome Res*. 2011;21:1404–1416.
103. Waddington CH. The epigenotype. *Endeavour*. 1942;1:18–20.
104. Maze I, Noh KM, Allis CD. Histone regulation in the CNS: basic principles of epigenetic plasticity. *Neuropsychopharmacology*. 2013;38:3–22 PubMed PMID: 22828751.
105. Baillie JK, Barnett MW, Upton KR, et al. Somatic retrotransposition alters the genetic landscape of the human brain. *Nature*. 2011;479:534–537.
106. Hunter RG, Murakami G, Dewell S, et al. Acute stress and hippocampal histone H3 lysine 9 trimethylation, a retrotransposon silencing response. *Proc Natl Acad Sci U S A*. 2012;109:17657–17662.
107. Hunter RG, McEwen BS, Pfaff DW. Environmental stress and transposon transcription in the mammalian brain. *Mob Genet Elements*. 2013;3:e24555.
108. Bloss EB, Janssen WG, McEwen BS, Morrison JH. Interactive effects of stress and aging on structural plasticity in the prefrontal cortex. *J Neurosci*. 2010;30:6726–6731.
109. Bloss EB, Hunter RG, Waters EM, Munoz C, Bernard K, McEwen BS. Behavioral and biological effects of chronic S18986, a positive AMPA receptor modulator, during aging. *Exp Neurol*. 2008;210:109–117.
110. Reul JMHM, Chandramohan Y. Epigenetic mechanisms in stress-related memory formation. *Psychoneuroendocrinology*. 2007;32: S21–S25.
111. Nasca C, Xenos D, Barone Y, et al. L-Acetylcarnitine causes rapid antidepressant effects through the epigenetic induction of mGlu2 receptors. *Proc Natl Acad Sci U S A*. 2013;110:4804–4809.
112. Dantzer R, O'Connor JC, Freund GG, Johnson RW, Kelley KW. From inflammation to sickness and depression: when the immune system subjugates the brain. *Nat Rev Neurosci*. 2008;9:46–56.
113. Nasca C, Bigio B, Zelli D, Nicoletti F, McEwen BS. Mind the gap: glucocorticoids modulate hippocampal glutamate tone underlying individual differences in stress susceptibility. *Mol Psychiatry*. 2015;20:755–763.
114. Brydges NM, Jin R, Seckl J, Holmes MC, Drake AJ, Hall J. Juvenile stress enhances anxiety and alters corticosteroid receptor expression in adulthood. *Brain Behav*. 2014;4:4–13.
115. Halfon N, Larson K, Lu M, Tullis E, Russ S. Lifecourse health-development: past, present and future. *Matern Child Health J*. 2014;18:344–365.
116. Miller MM, Morrison JH, McEwen BS. Basal anxiety-like behavior predicts differences in dendritic morphology in the medial prefrontal cortex in two strains of rats. *Behav Brain Res*. 2012; 229:280–288.
117. Freund J, Brandmaier AM, Lewejohann L, et al. Emergence of individuality in genetically identical mice. *Science*. 2013;340:756–759.
118. Fraga MF, Ballestar E, Paz MF, et al. Epigenetic differences arise during the lifetime of monozygotic twins. *Proc Natl Acad Sci U S A*. 2005;102:10604–10609.
119. Felitti VJ, Anda RF, Nordenberg D, et al. Relationship of childhood abuse and household dysfunction to many of the leading causes of death in adults. The adverse childhood experiences (ACE) study. *Am J Prev Med*. 1998;14:245–258.
120. Levine S, Haltmeyer G, Kara G, Denenberg V. Physiological and behavioral effects of infantile stimulation. *Physiol Behav*. 1967;2:55–59.
121. Meaney MJ, Szyf M. Environmental programming of stress responses through DNA methylation: life at the interface between a dynamic environment and a fixed genome. *Dialogues Clin Neurosci*. 2005;7:103–123.
122. Akers KG, Yang Z, DelVecchio DP, et al. Social competitiveness and plasticity of neuroendocrine function in old age: influence of

neonatal novelty exposure and maternal care reliability. *PLoS ONE*. 2008;3:e2840.
123. Tang AC, Akers KG, Reeb BC, Romeo RD, McEwen BS. Programming social, cognitive, and neuroendocrine development by early exposure to novelty. *Proc Natl Acad Sci U S A*. 2006;103:15716–15721.
124. Parker KJ, Buckmaster CL, Sundlass K, Schatzberg AF, Lyons DM. Maternal mediation, stress inoculation, and the development of neuroendocrine stress resistance in primates. *Proc Natl Acad Sci U S A*. 2006;103:3000–3005.
125. Isgor C, Kabbaj M, Akil H, Watson SJ. Delayed effects of chronic variable stress during peripubertal-juvenile period on hippocampal morphology and on cognitive and stress axis functions in rats. *Hippocampus*. 2004;14:636–648.
126. Rice CJ, Sandman CA, Lenjavi MR, Baram TZ. A novel mouse model for acute and long-lasting consequences of early life stress. *Endocrinology*. 2008;149:4892–4900.
127. Moriceau S, Sullivan R. Maternal presence serves as a switch between learning fear and attraction in infancy. *Nat Neurosci*. 2006;8:1004–1006.
128. Kaufman D, Smith ELP, Gohil BC, et al. Early appearance of the metabolic syndrome in socially reared bonnet macaques. *J Clin Endocrinol Metab*. 2005;90:404–408.
129. Coplan JD, Smith ELP, Altemus M, et al. Variable foraging demand rearing: sustained elevations in cisternal cerebrospinal fluid corticotropin-releasing factor concentrations in adult primates. *Biol Psychiatry*. 2001;50:200–204.
130. Juster RP, McEwen BS, Lupien SJ. Allostatic load biomarkers of chronic stress and impact on health and cognition. *Neurosci Biobehav Rev*. 2010;35:2–16.
131. Gard T, Taquet M, Dixit R, et al. Fluid intelligence and brain functional organization in aging yoga and meditation practitioners. *Front Aging Neurosci*. 2014;6:76.
132. Ryff CD, Singer B. The contours of positive human health. *Psychol Inq*. 1998;9:1–28.
133. Singer B, Friedman E, Seeman T, Fava GA, Ryff CD. Protective environments and health status: cross-talk between human and animal studies. *Neurobiol Aging*. 2005;26S:S113–S118.
134. Fredrickson BL, Grewen KM, Coffey KA, et al. A functional genomic perspective on human well-being. *Proc Natl Acad Sci U S A*. 2013;110:13684–13689.
135. Colcombe SJ, Kramer AF, Erickson KI, et al. Cardiovascular fitness, cortical plasticity, and aging. *Proc Natl Acad Sci U S A*. 2004;101:3316–3321.
136. Erickson KI, Voss MW, Prakash RS, et al. Exercise training increases size of hippocampus and improves memory. *Proc Natl Acad Sci U S A*. 2011;108:3017–3022 PubMed PMID: 21282661, Pubmed Central PMCID: 3041121.
137. Babyak M, Blumenthal JA, Herman S, et al. Exercise treatment for major depression: maintenance of therapeutic benefit at 10 months. *Psychosom Med*. 2000;62:633–638.
138. Kahle EB, Zipf WB, Lamb DR, Horswill CA, Ward KM. Association between mild, routine exercise and improved insulin dynamics and glucose control in obese adolescents. *Int J Sports Med*. 1996;17:1–6.
139. Bonen A. Benefits of exercise for Type II diabetics: convergence of epidemiologic, physiologic, and molecular evidence. *Can J Appl Physiol*. 1997;20(3):261–279.
140. Rovio S, Kareholt I, Helkala EL, et al. Leisure-time physical activity at midlife and the risk of dementia and Alzheimer's disease. *Lancet Neurol*. 2005;4:705–711.
141. Larson EB, Wang L, Bowen JD, et al. Exercise is associated with reduced risk for incident dementia among persons 65 years of age or older. *Ann Intern Med*. 2006;144:73–81.
142. Draganski B, Gaser C, Kempermann G, et al. Temporal and spatial dynamics of brain structure changes during extensive learning. *J Neurosci*. 2006;26:6314–6317.
143. Seeman TE, Singer BH, Ryff CD, Dienberg G, Levy-Storms L. Social relationships, gender, and allostatic load across two age cohorts. *Psychosom Med*. 2002;64:395–406.
144. Boyle PA, Buchman AS, Barnes LL, Bennett DA. Effect of a purpose in life on risk of incident Alzheimer disease and mild cognitive impairment in community-dwelling older persons. *Arch Gen Psychiatry*. 2010;67:304–310.
145. Fried LP, Carlson MC, Freedman M, et al. A social model for health promotion for an aging population: Initial evidence on the experience corps model. *J Urban Health Bull NY Acad Med*. 2004;81:64–78.
146. Carlson MC, Erickson KI, Kramer AF, et al. Evidence for neurocognitive plasticity in at-risk older adults: the experience corps program. *J Gerontol A: Biol Med Sci*. 2009;64:1275–1282.
147. Sheline YI. Hippocampal atrophy in major depression: a result of depression-induced neurotoxicity? *Mol Psychiatry*. 1996;1:298–299.
148. Drevets WC, Price JL, Simpson Jr. JR, et al. Subgenual prefrontal cortex abnormalities in mood disorders. *Nature*. 1997;386:824–827.
149. Rajkowska G. Postmortem studies in mood disorders indicate altered numbers of neurons and glial cells. *Biol Psychiatry*. 2000;48:766–777.
150. Vythilingam M, Vermetten E, Anderson GM, et al. Hippocampal volume, memory, and cortisol status in major depressive disorder: effects of treatment. *Biol Psychiatry*. 2004;56:101–112.
151. Moore GJ, Bebehuk JM, Wilds IB, Chen G, Manji HK. Lithium-induced increase in human brain grey matter. *Lancet*. 2000;356:1241–1242.
152. Vetencourt JFM, Sale A, Viegi A, et al. The antidepressant fluoxetine restores plasticity in the adult visual cortex. *Science*. 2008;320:385–388.
153. Spolidoro M, Baroncelli L, Putignano E, Maya-Vetencourt JF, Viegi A, Maffei L. Food restriction enhances visual cortex plasticity in adulthood. *Nat Commun*. 2011;2:320.
154. Castren E, Rantamaki T. The role of BDNF and its receptors in depression and antidepressant drug action: reactivation of developmental plasticity. *Dev Neurobiol*. 2010;70:289–297.
155. Heinrich C, Lahteinen S, Suzuki F, et al. Increase in BDNF-mediated TrkB signaling promotes epileptogenesis in a mouse model of mesial temporal lobe epilepsy. *Neurobiol Dis*. 2011;42:35–47.
156. Kokaia M, Ernfors P, Kokaia Z, Elmer E, Jaenisch R, Lindvall O. Suppressed epileptogenesis in BDNF mutant mice. *Exp Neurol*. 1995;133:215–224.
157. Scharfman HE. Hyperexcitability in combined entorhinal/hippocampal slices of adult rat after exposure to brain-derived neurotrophic factor. *J Neurophysiol*. 1997;78:1082–1095.
158. Acheson SD. *Independent Inquiry into Inequalities in Health Report*. London: The Stationary Office; 1998.
159. Drewnowski A, Specter SE. Poverty and obesity: the role of energy density and energy costs. *Am J Clin Nutr*. 2004;79:6–16.
160. Kawachi I, Kennedy BP, Lochner K, Prothrow-Stith D. Social capital, income inequality, and mortality. *Am J Public Health*. 1997;87:1491–1498.
161. Sampson RJ, Raudenbush SW, Earls F. Neighborhoods and violent crime: a multilevel study of collective effects. *Science*. 1997;277:918–924.
162. Whitmer RW, Pelletier KR, Anderson DR, Baase CM, Frost GJ. A wake-up call for corporate America. *JOEM*. 2003;45:916–925.
163. Pelletier KR. A review and analysis of the clinical- and cost-effectiveness studies of comprehensive health promotion and disease management programs at the worksite: 1998-2000 update. *Am J Health Promot*. 2001;16:107–115.
164. Aldana SG. Financial impact of health promotion programs: a comprehensive review of the literature. *Am J Health Promot*. 2001;15:296–320.

CHAPTER

6

Behavior: Overview

R. Dantzer

MD Anderson Cancer Center, Houston, TX, USA

OUTLINE

Abstract

When confronted with stressors, individuals adjust their behavior and physiology to cope with the situation. Behavioral responses to stress are not reflex-like. Their nature depends on the appraisal of the eliciting situation along the dimensions of novelty, uncertainty, and controllability. Coping strategies are associated with different forms of behavioral responses depending in particular on the ability or inability to control the situation. Social settings are often seen as a source of stress leading to agonistic behavior. However, stress can also be alleviated by the different forms of social support that are available in the group. Successful coping strategies decrease physiological arousal, measured by cortisol and catecholamines. Stress does not necessarily originate only from the outside. It can also originate from the inside, as a result of the demands to which individuals expose themselves, depending on their personality.

BEHAVIORAL RESPONSES TO STRESS

The stress response was initially described by Hans Selye in the context of catastrophic events such as hemorrhage, surgical injury, burns, septic shock, and poisoning. These events were claimed to cause a nonspecific reflex-like response aiming at restoring homeostasis and mediated by activation of the hypothalamic-pituitary-adrenal axis.[1] However, more common stressors, such as life events, do not act this way. The stress response to life events depends on the evaluation by the subject of the potential threat these events represent and the way he or she can cope with them based on personal resources and previous experience.[2–4] The transactional model of stress that emphasizes cognitive and emotional components of the stress response is therefore more appropriate than the stimulus-response model of stress to describe what is occurring in individuals exposed to external stressors[5] (Figure 1). Novelty, uncertainty, and the absence or loss of control, are important dimensions in the evaluation process. An emotional response occurs if the event is recognized as discrepant, and this emotional response involves both physiological and physiological adjustments.[6]

Novelty occurs when the situation contains no or very few familiar elements. Subjects exposed to novelty engage in exploring the situation in order to identify potential sources of danger that need to be avoided and potentially interesting objects that can be safely approached. When novelty is too intense, fear predominates over exploration, and escape attempts replace investigatory behavior (Figure 2).

Uncertainty refers to the inability of predicting what is going to happen when the subject is exposed to a potentially dangerous or rewarding situation.[10] In many cases, harmful or rewarding events do not happen suddenly. They are preceded by warning signals that can be used to predict the occurrence of what will happen, via a process of classical or Pavlovian conditioning. A tone that occurs before a painful electric shock becomes a conditioned fear signal. Inversely, a tone that rings only when there is no electric shock becomes a conditioned safety signal. Distinct signals are not always available and the subject may have to rely on contextual cues to predict what will happen. Response to contextual cues involves the hippocampus while response to conditioned fear signals involve the amygdala.[11] In an uncertain situation,

Stress: Concepts, Cognition, Emotion, and Behavior
http://dx.doi.org/10.1016/B978-0-12-800951-2.00006-6

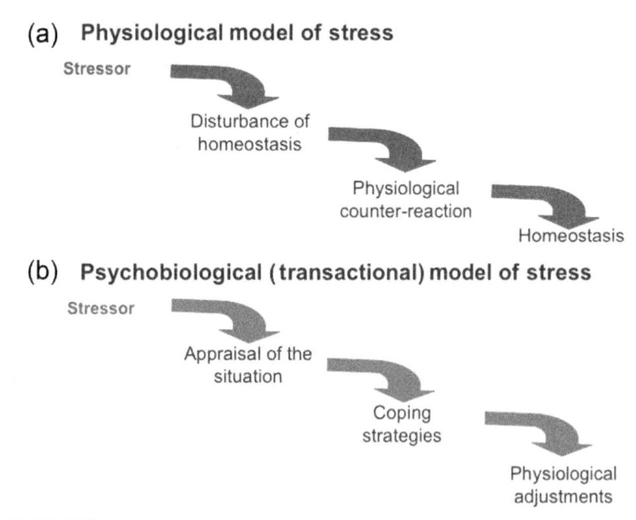

FIGURE 1 Differences between the homeostatic and transactional models of stress. (a) In the homeostatic model of stress, the behavioral response to stress is reflex-like and is a necessary conduit for reestablishment of homeostasis. (b) In the transactional model of stress, behavioral adjustments take place in response to cognitive and emotional appraisal of the situation. They allow the subjects to cope with the stress situation, which ultimately results in decreases in physiological arousal.

the subject tries to find regularities in the succession of events to which he or she is exposed, so as to make the situation more predictable. The behavioral response to a fear signal is an anticipatory response that helps the individual prepare for coping with the danger both physiologically (the alarm response) and behaviorally (the fight/flight response). The response to a safety signal is a decrease in the tension or anxiety induced by the possible occurrence of the harmful situation. Behavioral responses to a fear signal depend on the subject's proximity to the danger. When the subject is directly in contact with an uncontrollable threat, he or she tries to move away from the danger (active avoidance or escape). When the subject is at a close distance from the danger and cannot move away, he or she usually refrains from getting in contact with it (passive avoidance). This response can be associated with active attempts to hide oneself, or immobility and freezing when hiding is not possible.

When the situation is rewarding rather than aversive, events that regularly precede the occurrence of the reward become positively conditioned stimuli or secondary reinforcers. They induce the development of conditioned behavioral responses that involve approach and other appetitive or instrumental components of the normal behavioral response to the reward. If the secondary reinforcer is no longer followed by the occurrence of the expected reward, it induces a state of frustration that is characterized by agitation, aggression, and other displacement activities (see below).[12]

When individuals are confronted with a potentially harmful situation, they usually try to do something in order to avoid the danger. Coping refers to the efforts undertaken to resolve the situation.[13] Active and passive avoidance responses are examples of coping responses. Behavioral manifestations of coping vary according to the situation and the way the threat can be controlled. In laboratory animals, for instance, coping is typically studied by enabling the subjects to escape or avoid painful electric shocks when they emit a specific instrumental response, such as a lever press, a wheel turn, or a move from one compartment to the other in a two-compartment cage of which the floor is electrified. While animals in the escape/avoidance group are permitted to terminate electric shocks or delay their occurrence, animals in another group can serve as yoked controls to assess the consequences of being exposed to painful electric shocks without having the possibility to control them. Yoked controls receive the same electric shocks as the

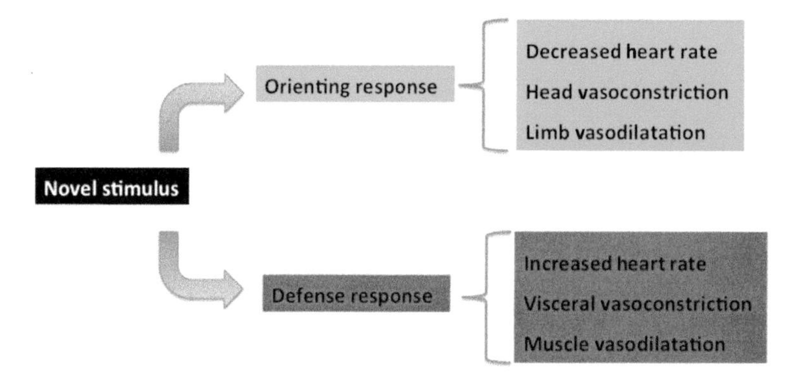

FIGURE 2 Coupling of behavioral responses to novelty with physiology. Exposure to novelty can lead to orienting or defense responses, with different physiological correlates. An orienting response usually develops in response to a mildly novel stimulus that occurs in familiar surroundings. A defense response typically develops in response to a sudden novel stimulus especially when it occurs in unfamiliar surroundings. The exact nature of the relationship between the physiological signature of the behavioral response to novelty that was initially described by Sokolov[7] and the ability to process environmental stimuli was the object of heated disputes between Lacey and Obrist. Lacey made the physiological response a necessary condition for the ability of the cortex to process novel information[8] whereas Obrist saw it as a simple correlate of the somatic component of the behavioral response to novelty.[9]

animals that are actively escaping or avoiding electric shocks, but their behavioral response has no effect on the occurrence and duration of electric shocks. As a further control, other animals are placed in the same apparatus as the animals in the other two groups, but they are never exposed to electric shocks. This triadic design has been widely used to study the behavioral and physiological consequences of controllability versus uncontrollability.[14] Controllability can also be studied in human subjects by asking them to press a button to terminate an intense noise. As in the animal experiment, the yoked group has no control over the noise.

Helplessness refers to an individual's inability to control the situation when he or she has the opportunity to do so. When this happens after a previous experience of uncontrollable stress, it is referred to as learned helplessness.[15] Learned helplessness is often used as a model of depression and it has cognitive, motivational, and emotional components.

Coping attempts can be classified according to the strategy that is employed to confront the situation. Active coping refers to all ways of dealing behaviorally directly with the situation (e.g., avoidance or attack). In human subjects, active coping can be cognitive (logical analysis and positive reappraisal) or behavioral. The latter case includes seeking guidance and support and taking concrete actions to deal directly with the situation or its aftermath. In an aversive context, cognitive coping comprises mental processes aimed at denying or minimizing the seriousness of the situation or its consequences, as well as accepting the situation as it is. Behavioral coping includes seeking alternative strategies to directly confront the problem.

In general, active coping enables the subject to deal directly with the problem and is therefore characteristic of what is referred to as problem-oriented coping. In cases in which the problem cannot be directly dealt with, it might be preferable to engage in passive coping that consists mainly of managing the emotions induced by the situation, a coping strategy that is referred to as emotion-oriented coping. In the case of passive coping, the behavioral activities that are normally elicited by the stimuli present in the situation can be redirected toward more easily accessible targets. A typical example is the redirected aggression that develops toward a subordinate when an animal in a social group is attacked by the dominant member of the group. Whereas redirected activities belong to the same behavioral repertoire as the thwarted behavioral activity, displacement activities belong to a different behavioral repertoire.[16] This is the case for aggressive behavior that is targeted toward peers or other objects in the environment when a subject does not get the reward it was expecting. A well-studied displacement behavior is schedule-induced behavior.[17,18] Schedule-induced behaviors, also known as adjunctive activities, typically occur when hungry animals are exposed to intermittent food rewards, delivered in very small amounts, independently of the animals' behavior. Immediately after animals have eaten their tiny food ration, they vigorously engage in other activities depending on what is available in the situation. They drink exaggerated amounts of liquid if a drinking tube is available (Figure 3), attack a congener when present, or gnaw a piece of wood if available. To qualify as an adjunctive behavior, the behavioral activity must occur as an adjunct to the reinforcement

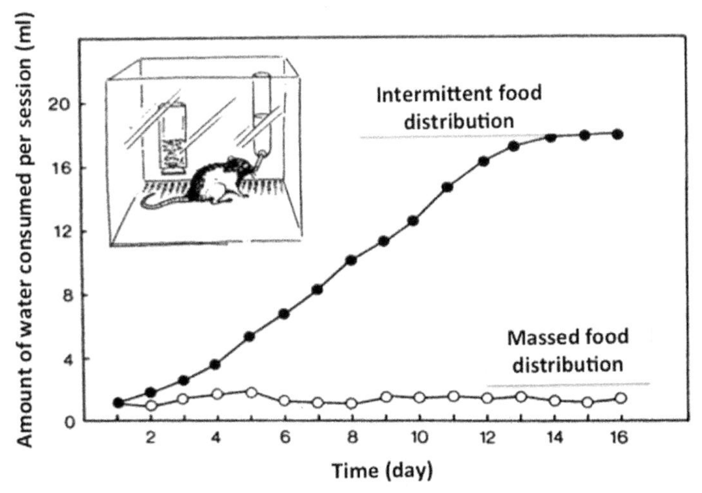

FIGURE 3 Development of schedule-induced polydipsia in rats submitted to an intermittent schedule of food reinforcement. Rats maintained at approximately 80% of their free feeding body weight were placed daily in a Skinner box in which they received a food pellet every minute during a 30 min session. They had free access to a tube connected to a water bottle. Control rats received the same amount of food but in one distribution at the beginning of the session. Note that excessive drinking (polydipsia) developed only in rats submitted to the intermittent delivery of food. Modified from Ref. 19.

schedule, not be directly involved in, or maintained by, the reinforcement contingency, and be persistent and excessive. Independently of the object toward which they are directed, adjunctive behaviors occur at a maximum rate immediately after the reinforcement, and decrease in probability when food deprivation is alleviated.

Most animals, including humans, are social and their response to stressful events depends on the social context. Congeners can be perceived as dangers or as protective. Threatening social stimuli induce agonistic behavior. Agonistic behavior refers to the complex of aggression, threat, appeasement, and avoidance behaviors that occur during encounters between members of the same species. Attack is a highly ritualized behavior when it occurs in property-protective fights. The offender attacks its opponent at very specific parts of the body, such as the neck in rats, whereas the defender tries to protect these targets from bites. When defeated, the defender adopts a submissive posture, exposing vulnerable body sites to the attacker. Submission usually puts an end to the fight.[20]

When stress events occur in a social situation, the outcome depends on the degree of familiarity between members of the group, as individual subjects are very sensitive to behavioral signals from congeners. Social support refers to the various types of support that individuals receive from others. It can be emotional, instrumental, or informational. Social modulation of stress responses has mainly been studied in the context of mother-infant bonding, but also applies to adult social bonds.[21] Social relationships and social support improve stress responses both directly and as stress buffers. Mechanisms are multiple and involve social comparison, social control, role-based purpose and meaning, self-esteem, sense of control, belonging and companionship, and perceived support ability.[22]

There is a vast continuum of individual differences in the way subjects react to stress. Individual differences are present even in genetically homogenous populations living in the same environment. For instance, exposure of rats of the same genetic strain to a novel environment allows distinguishing high responders from low responders. Both groups show a habituation response to novelty over time but the high responder rats are twice more active during the test compared to the low responders.[23] This difference in response to novelty is associated with differences in self-administration of addictive drugs such as amphetamines. Genetic differences in behavioral responses to stress have also been described, and selection experiments have even been carried out in order to select different lines of rats and test further their coping abilities.[24] Genetic manipulations of specific target genes in mouse models have now replaced genetic selection experiments although the genetic background of the strain in which genes are overexpressed or knocked out plays an important role in the resulting behavioral phenotype.

All individual differences in response to stress are not genetically determined. Some of these differences find their origin in previous life experiences. In particular, living in an enriched environment allowing repeated exposure to novelty, renewed social contacts, and opportunities for physical exercise enhances sensory, cognitive, and motor experience. Compared to isolated rats, rats with an enriched environment experience better adapt to stressful events and show increased resilience possibly by stress inoculation.[25] The effects of experience are more marked during infancy and in adolescence. Inversely, aging is associated with decreased resilience to stress.[26] Early experiences of stress can also have deleterious effects at adulthood depending on the nature of the stressors and their severity. The three-hit concept of vulnerability and resilience postulates that genetic predispositions (hit 1) interact with early life environment (hit 2) and with later life environment (hit 3) in a cumulative manner. Mild early life adversity prepares for the future and promotes resilience to similar challenges later in life, whereas mismatch between early life and later life experience compromises coping and enhances vulnerability.[27]

KEY POINTS

- Behavioral responses to stress are not reflexes but depend on the appraisal of the stressful situation and in particular, the ability to control it.
- Lack of control is associated with the development of displacement activities that serve to dissipate physiological arousal induced by the stressful situation.
- For animals living in social groups, congeners are not only a possible cause of stress leading to agonistic behavior, but also a resource for social support.
- There are important individual differences in the way individuals adapt behaviorally to a stressful situation. These differences are based on genetic features and previous experience.
- Behavioral adjustments shape the physiological component of the stress response. In addition, baseline activity of physiological stress response systems modulates the type of coping strategy that is deployed in the face of stressors.
- Stress does not necessarily occur independently of the subject's behavior. In many cases, stress is a direct consequence of the way individuals behave in their environment.

INTERACTIONS BETWEEN BEHAVIORAL AND PHYSIOLOGICAL COMPONENTS OF THE STRESS RESPONSE

The physiological response to stress is mediated by activation of the hypothalamic-pituitary-adrenal axis, resulting in higher level of glucocorticoids, and activation of the sympathetic nervous system and the adrenal medulla, resulting in higher levels of catecholamines, adrenaline, and noradrenaline. Circulating levels of glucocorticoids (cortisol in most species, corticosterone in rodents) reflect the level of arousal experienced by subjects when exposed to stress. Compared to uncertainty and lack of control, predictability and controllability have deactivating effects on the pituitary-adrenal response evoked by the stress situation (Figure 4). In social conflict situations, activation of stress hormones is in general more pronounced in subordinate animals that behave defensively or passively than in offensive animals that remain active and become dominant. However, the social role is more important than the social status since dominance is associated with a reduced arousal response only when dominant animals are recognized as such by their peers and do not need to reaffirm their status each time they are challenged by subdominant animals.

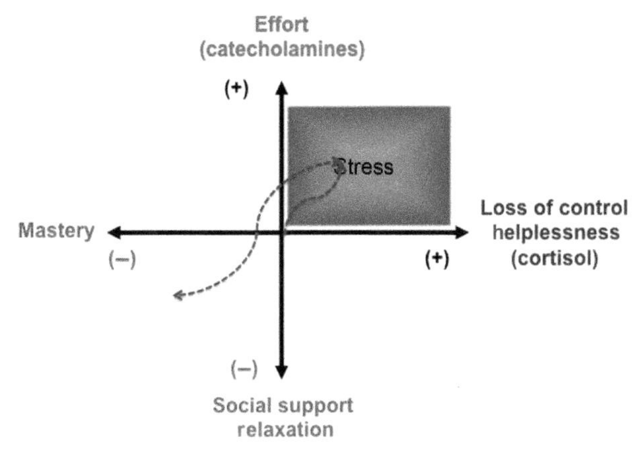

FIGURE 4 Bidirectional relationships between coping strategies and stress hormonal responses. Each axis represents a behavioral trajectory together with its neuroendocrine correlates. The diagram contrasts trajectories on the effort/relaxation axis associated with activation of the sympathetic nervous system and the adrenal medulla to trajectories on the loss of control/mastery axis associated with activation of the hypothalamic-pituitary-adrenal axis. The intersection of the two axes corresponds to the lack of challenge. In response to a stressful challenge, individuals will move toward the upper right of the diagram and lean toward the horizontal or the vertical axis depending on their fear of losing control and the amount of effort they need to deploy to control the situation. Mastering the situation will bring them to the left side of the diagram and even further down vertically if they obtain positive feedback on the success of their strategy and congratulations from their peers. The blue dotted line represents such a hypothetical trajectory. Note that the upper right quadrant corresponds to stress in its usual acceptation and that individuals with a higher than normal baseline neuroendocrine activity will not start their trajectory from the origin but from some position on the vertical or horizontal axis that corresponds to their hormonal status. Adapted from Ref. 28.

Schedule-induced polydipsia is associated with lowered plasma levels of glucocorticoids. This feature is shared by other motor activities that involve sensory overloading from the oral cavity (e.g., chewing or licking) or the limbs (e.g., pacing). The dearousal properties of coping behavior are not always that clear. Coping attempts are associated with increased activation of the sympathetic nervous system, and they result in physiological deactivation only when they have been successful and the situation is fully mastered.

It is tempting to see the response of the autonomic nervous system to stress in a dichotomic manner. Stress activates the sympathetic branch, which results in increased metabolic activity and redirection of blood from the splanchnic cavity to the muscles and brain. At the same time, the parasympathetic branch of the autonomic nervous system that promotes conservation of metabolic activity is inhibited. However, this simplified view does not account for the existence of a myelinated branch of the vagus nerves. Activation and deactivation of the myelinated vagus is a powerful way of rapidly regulating cardiac output to foster engagement and disengagement with the environment.[29] More specifically, releasing the myelinated vagal brake on the heart allows to respond in a well-controlled manner to a challenging situation without activating the energetically demanding sympathetic nervous system. Inversely, re-engagement of the myelinated branch of the vagus nerve by positive social sensory stimulation (e.g., touch) helps to buffer the effect of stress.

The relationship between behavioral and physiological components of the stress response is not unidirectional, with behavior influencing physiological responses, but truly bidirectional. The likelihood of a given behavioral response is modulated by the subject's initial physiology. Individuals with high activity of the sympathetic-adrenal medullary axis are more likely to try to cope in an active way with a threat than individuals with a low activity of the sympathetic-adrenal medullary axis. Inversely, individuals with a relatively high activity of the hypothalamic-pituitary-adrenal axis are more likely to resign and display helplessness than individuals with a normal or lower activity of this axis.

BEHAVIORAL SOURCES OF STRESS

Stress does not always reside outside of the subject. In many occasions, stress stems from inside depending on the way individuals perceive and respond to their environment. The type A behavior pattern and the personality trait of sensation seeking are good examples of this.

Type A behavior is defined as an action-emotion complex that individuals use to confront challenges.[30] This pattern of behavior involves behavioral dispositions such as aggressiveness, competitiveness, and impatience; specific behaviors such as muscle tenseness, alertness, rapid and emphatic vocal stylistics, and accelerated pace of activities; and emotional responses, such as irritation, covert hostility, and above-average potential for anger. Type B individuals are characterized by the relative absence of type A behaviors and confront every challenge with placid nonchalance. The distinction between type A and type B behavior has been proposed by cardiologists to account for individual differences in risk of coronary heart disease. Type A individuals have a higher risk of developing coronary heart disease than type B individuals. The deleterious effects of type A behavior on cardiovascular morbidity and mortality are mediated by the higher physiological reactivity to external demands, especially when there are elements of competition. These findings corroborate the previously mentioned influences of behavioral factors on the physiological response to stress. In accordance with the bidirectional nature of hormone-behavior relationships, type A behavior does not occur at will. It is determined at least in part by biological factors and autonomic hyperactivity. In addition, type A individuals actively contribute to their fate by having a higher tendency to put themselves in situations of competition and take all possible steps to transform the events they are confronted with in challenges that need to be met in a more hostile way than their peers would be inclined to do. Although the type A behavior pattern appears to be the quintessential psychobiological construct in stress theory, unfortunately it has a number of drawbacks. The psychological scales that have been developed to assess the type A behavior pattern are fraught with serious psychometric problems.[31] In addition, the importance of type A behavior in epidemiological studies on the relationship between psychosocial factors and health has seen a progressive decline in favor over time. It is now apparent that funding from the tobacco industry aiming at countering concerns regarding smoking and health helped to inflate the only few positive results that were reported during the golden age of research on type A behavior pattern.[32]

Sensation seeking is part of the extraversion domain of personality and involves the basic need to seek stimulation and thrill independently of the associated risk.[33] It contributes to impulsive behavior and has been mainly studied in the context of addiction. There is evidence that sensation seekers display enhanced responses to novelty and danger. The probability for a given individual to be a sensation seeker is determined by innate and experiential factors. However, what is characteristic of sensation seekers is their strong preference for sensation-inducing situations rather than their increased response to such situations.

Independently of their construct validity, the examples of type A behavior pattern and sensation seeking illustrate very well the notion that we are not passively exposed to stress; we actually create our own stress by the way we perceive the environment and ultimately act upon it. This can be reverberating as the experience of stress impacts on decision-making processes by tending to increase risk taking behavior, although the exact outcome depends on whether the risk leads to gains, losses, or mixed options.[34,35] Part of these reverberating effects could be due to previous experience as stress biases decision-making toward habitual behavior that is driven by past experience and relatively insensitive to current stimulus-response contingencies.[36]

References

1. Selye H. The general adaptation syndrome and the diseases of adaptation. *Am J Med*. 1951;10(5):549–555.
2. Lazarus RS. From psychological stress to the emotions: a history of changing outlooks. *Annu Rev Psychol*. 1993;44:1–21.
3. Mason JW. A historical view of the stress field. *J Hum Stress*. 1975;1 (1):6–12. Contd.
4. Mason JW. A historical view of the stress field. *J Hum Stress*. 1975;1 (2):22–36.
5. Folkman S, Lazarus RS. The relationship between coping and emotion: implications for theory and research. *Soc Sci Med*. 1988;26 (3):309–317.
6. Grandjean D, Scherer KR. Unpacking the cognitive architecture of emotion processes. *Emotion*. 2008;8(3):341–351.
7. Sokolov EN. Higher nervous functions; the orienting reflex. *Annu Rev Physiol*. 1963;25:545–580.
8. Lacey BC, Lacey JI. Two-way communication between the heart and the brain. Significance of time within the cardiac cycle. *Am Psychol*. 1978;33(2):99–113.
9. Obrist P. The cardiovascular-behavioral interaction: as it appears today. *Psychophysiology*. 1976;13:95–107.
10. Hennessy JW, King MG, McClure TA, Levine S. Uncertainty, as defined by the contingency between environmental events, and the adrenocortical response of the rat to electric shock. *J Comp Physiol Psychol*. 1977;91(6):1447–1460.
11. LeDoux JE. Emotion circuits in the brain. *Annu Rev Neurosci*. 2000;23:155–184.
12. Mineka S, Hendersen RW. Controllability and predictability in acquired motivation. *Annu Rev Psychol*. 1985;36:495–529.
13. Lazarus R, Folkman S. *Stress, Appraisal, and Coping*. New York: Springer; 1984.
14. Drugan RC, Basile AS, Ha JH, Healy D, Ferland RJ. Analysis of the importance of controllable versus uncontrollable stress on subsequent behavioral and physiological functioning. *Brain Res Protoc*. 1997;2(1):69–74.
15. Seligman ME. Learned helplessness. *Annu Rev Med*. 1972;23: 407–412.
16. McFarland DJ. On the causal and functional significance of displacement activities. *Z Tierpsychol*. 1966;23(2):217–235.
17. Wallace M, Singer G. Schedule induced behavior: a review of its generality, determinants and pharmacological data. *Pharmacol Biochem Behav*. 1976;5(4):483–490.
18. Wetherington CL. Is adjunctive behavior a third class of behavior? *Neurosci Biobehav Rev*. 1982;6(3):329–350.

19. Brett LP, Levine S. Schedule-induced polydipsia suppresses pituitary-adrenal activity in rats. *J Comp Physiol Psychol.* 1979;93(5):946–956.
20. Blanchard RJ, Blanchard DC. Bringing natural behaviors into the laboratory: a tribute to Paul MacLean. *Physiol Behav.* 2003;79(3):515–524.
21. DeVries AC, Glasper ER, Detillion CE. Social modulation of stress responses. *Physiol Behav.* 2003;79(3):399–407.
22. Thoits PA. Mechanisms linking social ties and support to physical and mental health. *J Health Soc Behav.* 2011;52(2):145–161.
23. Piazza PV, Deminiere JM, Le Moal M, Simon H. Factors that predict individual vulnerability to amphetamine self-administration. *Science.* 1989;245(4925):1511–1513.
24. Armario A, Nadal R. Individual differences and the characterization of animal models of psychopathology: a strong challenge and a good opportunity. *Front Pharmacol.* 2013;4:137.
25. Crofton EJ, Zhang Y, Green TA. Inoculation stress hypothesis of environmental enrichment. *Neurosci Biobehav Rev.* 2015;49C:19–31.
26. McEwen BS, Morrison JH. The brain on stress: vulnerability and plasticity of the prefrontal cortex over the life course. *Neuron.* 2013;79(1):16–29.
27. Daskalakis NP, Bagot RC, Parker KJ, Vinkers CH, de Kloet ER. The three-hit concept of vulnerability and resilience: toward understanding adaptation to early-life adversity outcome. *Psychoneuroendocrinology.* 2013;38(9):1858–1873.
28. Henry JP. Biological basis of the stress response. *Integr Physiol Behav Sci.* 1992;27(1):66–83.
29. Porges SW. The polyvagal theory: phylogenetic substrates of a social nervous system. *Int J Psychophysiol.* 2001;42(2):123–146.
30. Friedman M, Rosenman RH. Overt behavior pattern in coronary disease. Detection of overt behavior pattern A in patients with coronary disease by a new psychophysiological procedure. *JAMA.* 1960;173:1320–1325.
31. Edwards JR, Baglioni Jr AJ, Cooper CL. Examining the relationships among self-report measures of the Type A behavior pattern: the effects of dimensionality, measurement error, and differences in underlying constructs. *J Appl Psychol.* 1990;75(4):440–454.
32. Petticrew MP, Lee K, McKee M. Type A behavior pattern and coronary heart disease: Philip Morris's "crown jewel". *Am J Public Health.* 2012;102(11):2018–2025.
33. Zuckerman M, Kuhlman DM. Personality and risk-taking: common biosocial factors. *J Pers.* 2000;68(6):999–1029.
34. Pabst S, Schoofs D, Pawlikowski M, Brand M, Wolf OT. Paradoxical effects of stress and an executive task on decisions under risk. *Behav Neurosci.* 2013;127(3):369–379.
35. Buckert M, Schwieren C, Kudielka BM, Fiebach CJ. Acute stress affects risk taking but not ambiguity aversion. *Front Neurosci.* 2014;8:82.
36. Radenbach C, Reiter AM, Engert V, et al. The interaction of acute and chronic stress impairs model-based behavioral control. *Psychoneuroendocrinology.* 2015;53:268–280.

CHAPTER

7

Conservation of Resources Theory Applied to Major Stress

S.E. Hobfoll, V. Tirone, L. Holmgreen, J. Gerhart

Rush University Medical Center, Chicago, IL, USA

OUTLINE

Abstract

Conservation of resources (COR) theory informs our understanding of how individuals cope with major stress and trauma. COR theory asserts that traumatic stress occurs when events threaten and erode the basic resources human beings need for survival or self-integrity. This process occurs within an ecological framework, meaning that patterns of risk and resilience in the face of resource loss are intimately tied to an individual's family, community, and culture. The basic principles and corollaries of COR theory are reviewed to illustrate patterns of post-trauma adjustment over time. Finally, the use of COR theory in guiding individual and collective post-trauma interventions is explored.

Traumatic events such as war, terrorism, rape, natural disasters, and life-threatening illness tax people's coping resources and can result in severe psychological distress. Trauma is ubiquitous, and nearly all humans are eventually exposed to events that lead to significant loss. Approximately 22% of Americans are exposed to natural disasters at some point during their lives.[1] Over one-third of American women are raped, stalked, or physically assaulted by an intimate partner at some point in their lives.[2] As many as 20% of service members returning from Iraq and Afghanistan suffer from post-traumatic stress disorder (PTSD), leaving veterans' families to shoulder the burden of war on a daily basis.[3] Because these events are widespread, it is important to examine why some people can emerge from trauma unscathed, whereas others are almost completely debilitated. This chapter describes conservation of resources (COR) theory, an ecological approach to understanding individual differences in response to stress and trauma.

COR theory begins with the basic premise that human beings are motivated to acquire and preserve resources needed for survival. Through personal experience, people come to recognize what they need in order to acquire and retain what is important directly, indirectly, and symbolically for success within their culture and for sheer survival. These resources commonly include external resources like food, shelter, social relationships, and attachments, along with internal resources such as optimism, self-esteem, energy, and hope. Given the necessity of these resources for survival, cultures have evolved such that the major activities of human life (e.g., work, family, and leisure activities) are organized around their development and preservation. COR is broad in scope and is applicable to the study of occupational burnout

Stress: Concepts, Cognition, Emotion, and Behavior
http://dx.doi.org/10.1016/B978-0-12-800951-2.00007-8

(see Halbesleben et al.[4] for review), chronic stress, and psychopathology. This chapter focuses specifically on COR as a model for responses to traumatic stress.

KEY POINTS

- *Conservation of resources (COR) theory* states that stress has central environmental, social, and cultural bases in terms of the demands on people to acquire and protect the circumstances that insure their well-being and distance themselves from threat to well-being.
- *Resources* are objects, personal characteristics, conditions, or energies that are valued by the individual or that serve as the means for attainment of other objects, personal characteristics, conditions, or energies.
- *Caravans* are groups of resources or risk factors that tend to co-occur.
- *Caravan passageways* are environmental conditions that support or undermine the aggregation of resources.
- *Stress* occurs in circumstances that represent a threat of loss or actual loss of the resources required to sustain the individual-nested-in-family-nested-in-social-organization. Because people will invest what they value to gain further, stress also occurs when individuals do not receive reasonable gain following resource investment.
- *Traumatic stressors* are severe, typically infrequent, and unexpected events that usually include serious threat to life and well-being.

THE INTERDEPENDENCE OF RESOURCES

Resources can be organized hierarchically from the most basic or primary to the most abstract or tertiary. Primary resources such as ample food, shelter, safety, and clothing are those that directly promote health and survival. Secondary resources such as social support, marriage, and optimism are those that contribute indirectly to survival by providing access to primary resources. Tertiary resources are socially or culturally constructed and include social status, money, and credit that are symbolically related to primary and secondary resources. As resources become more distal from basic survival their relative importance may vary cross-culturally.

Resources may be acquired piecemeal, but there is a high level of interdependence and statistical correlation among resources, referred to as *resource caravans*. The presence of resource caravans is explained in part by the sociocultural structuring of resources such that access to one resource begets access to others. The environmental conditions that render particular resource caravans possible are known in COR theory as *caravan passageways*.[5] These passageways represent the factors which support or undermine the aggregation of resources. Often, passageways occur as a matter of circumstance as opposed to personal choice or will. For example, a key caravan passageway in the resource caravans outlined above involves the US mechanism of public education funding, whereby schools are financed largely by local property taxes. This environmental condition has the effect of concentrating resources in certain groups and restricting access to them in others.

Many caravan passageways involve inheritance processes, of which there are three broad types.[5] One type of inheritance passageway consists of *cultural capital*, by which particular habits and proclivities associated with status are passed along.[6] For example, the child of a preacher in an impoverished rural community may be well equipped to become a preacher as an adult because he is revered for being from a religious family and has learned important skills from spending time in the church (e.g., conflict resolution, public speaking, and community organizing). As a result, the child may have preferential access to one of few means of income and status in the community. Alternatively, an individual with a stereotypically "White name" who grows up attending preparatory schools and country clubs will benefit from caravan passageways affording her greater preparation for college admissions processes, job interviews, and business networking.

Another passageway consists of *inter vivos* transfers,[5] or benefits exchanged between living individuals. These benefits may take the form of childrearing bestowals (such as parents paying for college or a car), gifts to other adults (such as giving a sibling or friend a loan), one family sponsoring the immigration of another family to a new country, or less tangible exchanges such as facilitating important educational, social, or business connections. Finally, passageways may consist of simple testamentary inheritance, whereby one's wealth is transferred to others upon death. According to some estimates, up to 80% of US wealth changes hands through such passageways.[7] The converse of such inheritance passageways, of course, represents a lack of resources and access to resource-rich environments, which further impedes the acquisition and preservation of resources.

The fact that any given resource is embedded within a caravan is critical to understanding how that resource predicts psychological distress or promotes resilience in the face of trauma. This premise suggests that it will be more fruitful to examine the sociocultural processes that explain the existence of related groups of resources than to examine any single resource in isolation. Further, consistent with COR theory's emphasis on the role of individuals-nested-in-families, nested-in-communities, resource caravans frequently represent assets that people share with each other.

THE ORIGINS OF TRAUMATIC STRESS

Within the COR framework, stress and trauma both emerge when resources are lost or threatened. While low-intensity stressors produce effects like burnout or negative affect, trauma can lead to profound distress because loss is rapid, life-threatening, and life-altering. War, mass disasters, and sexual assault are vastly different events, but all gravely threaten primary resources of safety, secondary resources of hope, energy, and social support, and tertiary resources related to socioeconomic status (SES). The core symptoms of PTSD (i.e., intrusion, negative cognition and affect, avoidance, and hypervigilance) function to increase awareness of ongoing threat and prevent further harm and loss. However, these symptoms frequently become exaggerated and overgeneralized and persist into settings that are objectively safe. As such, even when the objective threat to primary resources is low, symptoms will erode secondary and tertiary resources related to energy, hope, and social support.

Just as resources tend to nest within resources caravans, *risk factor caravans* exist in the co-occurrence of conditions such as unsafe neighborhoods, minority status, low SES, low-quality education, childhood trauma, and health risk. For example, in the context of childhood trauma, the physical or sexual abuse that directly threatens safety is also accompanied by psychological maltreatment that contributes to cycles of self-blame and the erosion of self-efficacy and hope. Moreover, due to intergenerational violence, poor policing, and other sociopolitical factors, abuse and interpersonal violence are highly nested within neighborhood communities.[8,9] As such, the pool of well-adjusted, healthy, and engaged others to rely on for social support may be limited, and community health agencies may be overwhelmed with caseloads of the severely traumatized. Given the principles of COR theory detailed below, the aggregation of loss in risk factor caravans is often too socially engrained and overwhelming for one individual or family to escape.

PRINCIPLES OF COR THEORY

Principle 1

There are two major principles that follow from the basic proposition of COR theory. The first principle is that resource loss is disproportionately more salient than resource gain. According to COR theory, it is the loss and the threat of loss of resources that principally define stress. This principle has also been developed in work by Kahneman and Tversky in their prospect theory, which states that the gradient for loss is steeper than the gradient for gain, which results in a salience bias in favor of loss. In experiments they have shown that when problems are framed or worded in terms of loss, greater risk will be taken to try to preserve resources than if the same situation is framed in terms of potential gain.[10]

If loss is more salient than gain, resource loss should have a greater impact on psychological distress than gain. This was examined empirically with students and community members who indicated whether they had lost or gained each of 74 resources recently as well as during the past year.[11] They repeated this process twice, separated by 3 weeks. They also completed well-known anxiety and depression scales. Results indicated that neither recent resource gain nor resource gain during the past year had any direct impact on psychological distress for either students or community members. Recent resource loss and resource loss during the past year, in contrast, had major negative effects on psychological distress. People were deeply negatively affected when they lost resources, but were hardly impacted when they experienced resource gain.

Researchers have applied COR theory to study the impact of both Hurricane Hugo, which affected South Carolina in 1990,[12] and the Sierra Madre earthquake that hit Los Angeles County in 1991.[13] They examined how resource loss influenced the mobilization of resources and also how loss affected mental health. It was found that the greater the resource loss, the more coping individuals engaged in and the more psychologically distressed they became. They also found that the influence of resource loss was both of greater magnitude and independent of the influence of positive coping responses. Supporting COR theory, resource loss was more important in predicting psychological distress than were personal characteristics and coping behavior.

The impact of resource loss on coping, physical health, and psychological well-being was examined further after Hurricane Andrew struck South Florida in 1992.[14] Comparing the impact of resource loss to other factors, it was found that resource loss had the single most profound influence. The greater the resources people lost, the greater was their psychological distress and the worse their immune resistance. Resource loss spirals also predicted post-traumatic distress among Virginia Tech students following the mass shooting on the university's campus in 2007.[15]

Principle 2

The second principle that follows from COR theory is that people must invest resources in order to protect against resource loss, recover from loss, and gain resources. Stress occurs when resources are lost or threatened, and people use resources to prevent or offset loss and to make other resource gains. This investment of resources occurs by several mechanisms. The first mechanism of resource investment deals with the total investment of a resource. The second kind of resource investment involves risking the resource without total investment. Resource investment may also occur directly or through substitution. Resource investment may counterbalance loss, protect against threat of loss, or contribute toward resource gain.

When self-esteem, social support, or finances are put into place to offset a major stressor, the importance of principle 2 is underscored. First, the resource must be valuable enough to be used effectively. Low self-esteem, an unreliable support system, and a nearly depleted savings account will do little to offset losses. The second principle of COR theory also suggests a more subtle point. Specifically, because people typically have little experience with major stressors, they may not know how to employ their resources in these special circumstances. Just as combat troops must learn to apply their skills under increasing pressure, people will learn and adjust to even catastrophic circumstances. However, the cost of resource investment will accelerate as people will at first misuse resources, hold back on resource use to preserve resource integrity, and will need to allow for a quick depletion of resources in some instances.

COROLLARIES OF COR THEORY

Corollary 1

The first corollary of COR theory, along with the facts of resource caravans and passageways, has crucial implications for chronic and traumatic stress, as well as resiliency. In the case of stress, COR theory predicts that those people with fewer resources will be more deeply impacted by a major crisis (e.g., exposure to a traumatic event) or by chronic demands (e.g., as poverty, discrimination, and work stress) given their smaller initial resource reserves, lesser access to shared resources, and inability to invest for future gain. Therefore, initial setbacks of stress can be devastating and result in immediate and rapid loss spirals. These individuals are at greater risk of developing psychological distress, including symptoms consistent with PTSD following exposure to traumatic events.[16–18]

With respect to resiliency to trauma and stress, COR theory posits that greater initial resources and more-intact caravan passageways (that is, passageways less impacted by a trauma) will result in greater resiliency to traumatic or chronic stress. For example, a family which experiences a traumatic hurricane will be expected to be more resilient if they had more money to begin with (greater initial resources) and if their house was not destroyed in the hurricane (intact caravan passageway in the form of continuing shelter). Such a family can be generally predicted to recover from an acute traumatic response more rapidly than a family who was poorer to begin with and who lost their home in the storm.

Corollary 2

The second corollary of COR theory states that initial loss begets future loss. People rely on resources to apply to losses, and therefore with each loss there are fewer resources that can be utilized or invested in gains that might influence the occurrence of stress. With a depleted resource pool, future challenges are decreasingly likely to be met, and a downward spiral increases in momentum. This results in depleted resource reserves that are less capable of being mobilized to defend against future challenges. This further suggests that loss cycles will have initially higher velocity for resource-poor individuals or groups, as they are from the outset in a resource-challenged state characterized by their resources being already stretched in protection of the self, family, or social system. Therefore, this corollary predicts that loss cycles will have progressing momentum and strength. An illustrated example of resource loss and gain spirals following the rape of a college student appears in Figure 1. As can be seen, the trauma of rape activates both an individual's normative stress response and stored resources. The severity of the trauma, intensity of the traumatic response, and availability of existent resources predict distress and impairment in the immediate aftermath. Moving forward, the cumulative impact of resource loss leads to ongoing distress, an effect that can be partially mitigated by resource gain. Because these dynamic processes are imbedded in layers of social context they are impacted by factors such as a family's sexual values and institutional responses to sexual victimization.

When individuals are resource-poor, especially if they are in a loss spiral, they are likely to be particularly vulnerable. This is illustrated in the ongoing hardships of Somali Refugees fleeing to the United States.[19] Although these Somalians escaped the trauma and loss of war, many continued to experience resource loss in America including loss of support from extended family, failure of American employers to recognize credentials and education, and loss of mastery over the dominant language. These resource loss spirals cause Somali-Americans continued distress despite their removal from a war-torn environment. Again highlighting the importance of culture and community, established Somali-Americans play an important role in recent refugees' postmigration

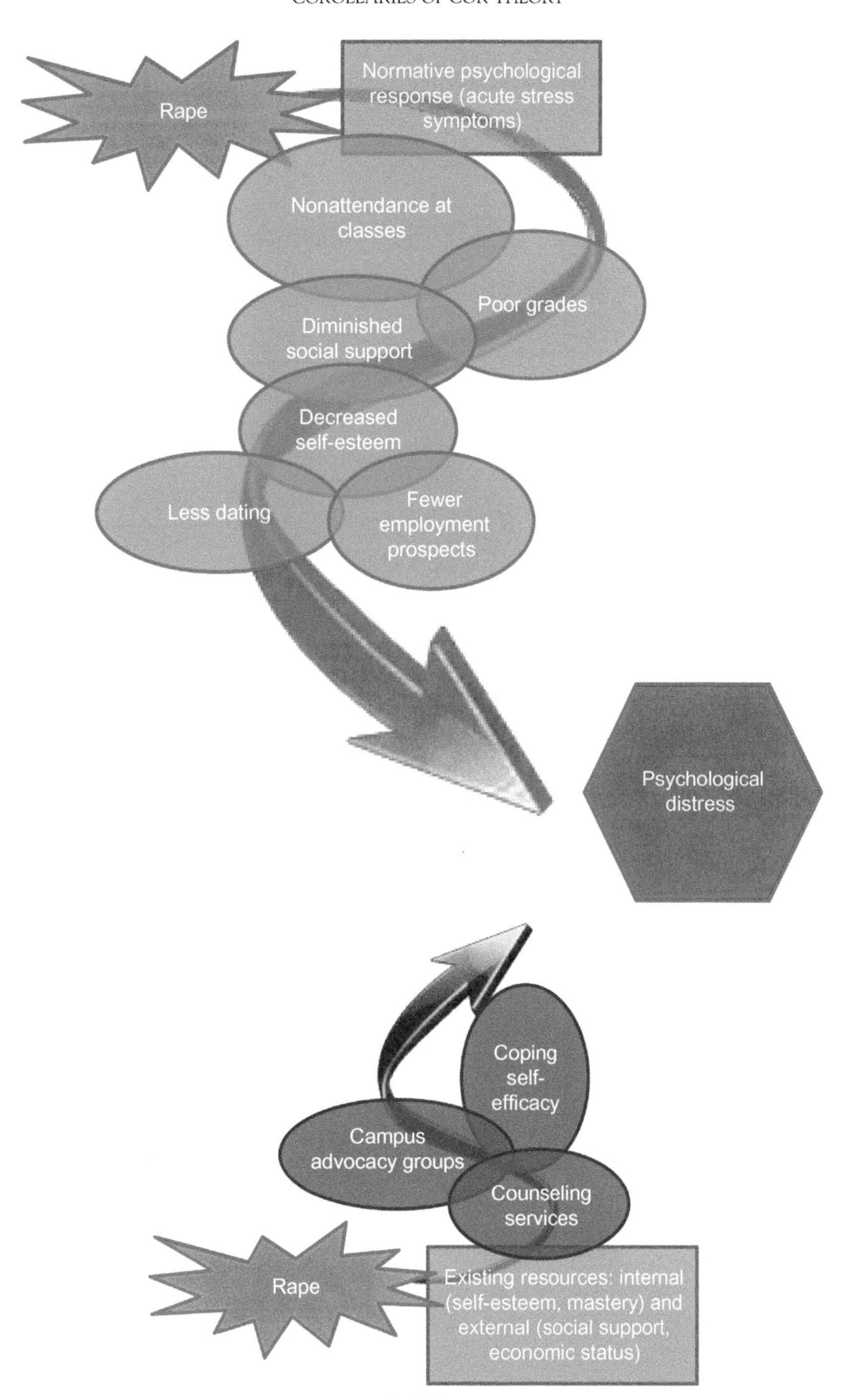

FIGURE 1 Conceptual illustration of resource loss and gain spirals following a rape.

adjustment by acting as conduits to community resources, thereby helping to initiate gain spirals or at least mitigate loss spirals.

Corollary 3

The third corollary of COR theory states both that those who possess resources are more capable of gain and that initial resource gain begets further gain. When initial gains are achieved, additional resources become available for investment, rendering individuals and social systems less vulnerable to loss and loss spirals. In addition, individuals with greater resources do not necessarily need to rely on these resource surpluses for daily survival, so they can benefit further by investing them toward future resource gains. However, because resource loss is more potent than resource gain, gain cycles will have less momentum or speed and less impact.

In the case of major and traumatic stress, gain cycles will often only occur well after the event's initial impact.

Corollary 4

The final corollary affirms that those who lack resources adopt a defensive posture to guard their resources. For those with few assets, the demands placed on resources to preserve basic survival in the face of trauma exceed existent resources, making the individual or organization vulnerable. Investment of resources in order to obtain additional resources is highly unlikely because the system is overtaxed. For example, a migrant farm worker might benefit from the interest earned from a savings account to offset the fact that their income is unstable; however, their initial pay is so low that they cannot afford to invest. A defensive posture keeps a maximum of resources readily available in case the person needs to offset a future loss.

Those with weak resources have been found to use their resources best when challenged by everyday stressors, but to have the least success under high-stress conditions. Among terrorism victims, PTSD and denial coping are positively associated with outgroup bias.[20] Thus, people are more conscious of threat from others after trauma has depleted their psychological resources. In studying victims of Hurricane Andrew, researchers found that resource loss resulted in a marked increase in use of denial coping. This, in turn, placed victims at increased risk for PTSD. To expend resources from a depleted system may be too risky a venture. Instead, a shutdown response may best preserve limited resource pools until the storm blows over.

PRACTICAL IMPLICATIONS

As an ecological and multilevel theory of stress and trauma, COR theory has many practical implications for the prevention and treatment of traumatic stress. COR theory informs interventions that equip people to weather trauma (i.e., primary prevention), reduce vulnerability to long-term distress among trauma exposed individuals (i.e., secondary prevention) and to relieve distress among those who have developed post-traumatic psychopathology (i.e., tertiary prevention). Empirical evidence and consensus of trauma experts suggest that interventions across at all levels focus on helping people and communities build and retain the resources of sense of safety, calming, sense of self and community efficacy, and hope.[21] At the individual level of intervention, COR suggests that clinicians help individuals adopt greater flexibility in their pursuit of these resources and strategically use resources to stop cycles of loss in times of acute trauma. In a complementary fashion, organizations that provide traditional psychological services to trauma survivors could provide social interventions. According to COR, individuals are more likely to use resources like psychological coping skills effectively when their basic survival needs are met. COR also has practical applications for intervening on sociocultural processes more broadly. Public policy interventions that include social programming, mental health education, establishment of resource reserves and violence reduction are highly beneficial because they create resource caravan passageways which can make communities and their members more resilient before and after trauma occurs.

CONCLUSION

This chapter reviews COR theory, an ecological and multilevel theory of stress and trauma. Individuals are highly motivated and play an active role in the acquisition and preservation of the resources needed for health, survival, and well-being. However, sociocultural forces have a significant impact on the degree of flexibility one has to pursue and preserve their resources. COR theory defines psychological stress as a reaction to the environment in which there is the threat of a net loss of resources, an actual net loss of resources, or a lack of resource gain following investment. In high-risk situations such as war, natural disasters, or sudden illness, many if not all of these conditions can occur. Therefore, COR theory explains distress following a high-risk situation not as an individualized response, but as one that occurs because of the threats to resources as shared within a community.

Although loss is more salient than gain, this is not to imply that resource gain is not important. Resource gain is important because it can prevent, counterbalance, or forestall loss. However, in a high-risk situation, many factors may prevent resource gain and therefore limit the role of gain in forestalling loss cycles. Because of the long tail of resource loss that follows major stressful circumstances, follow-up to protect the individual and social system are needed on a long-term basis.

References

1. Briere J, Elliott D. Prevalence, characteristics, and long-term sequelae of natural disaster exposure in the general population. *J Trauma Stress*. 2000;13(4):661–679.
2. Black MC, Basile KC, Breiding MJ, et al. *The National Intimate Partner and Sexual Violence Survey (NISVS): 2010 Summary Report*. Atlanta, GA: National Center for Injury Prevention and Control, Centers for Disease Control and Prevention; 2011.
3. Ramchand R, Schell TL, Karney BR, Osilla KC, Burns BM, Caldarone LB. Disparate prevalence estimates of PTSD among

service members who served in Iraq and Afghanistan: possible explanations. *J Trauma Stress*. 2010;23(1):59–68.
4. Halbesleben JRB, Neveu JP, Paustian-Underdahl SC, Westman M. Getting to the "COR": understanding the role of resources in conservation of resources theory. *J Manage*. 2014;40(5):1334–1364.
5. Hobfoll SE. Conservation of resource caravans and engaged settings. *J Occup Organ Psychol*. 2011;84(1):116–122.
6. Miller RK, McNamee SJ. *The Inheritance of Wealth in America*. New York, NY: Springer; 1998.
7. Kalmijn M, Kraaykamp G. Race, cultural capital, and schooling: an analysis of trends in the United States. *Sociol Educ*. 1996;69(1):22–34.
8. Stevens NR, Gerhart J, Goldsmith RE, Heath NM, Chesney SA, Hobfoll SE. Emotion regulation difficulties, low social support, and interpersonal violence mediate the link between childhood abuse and posttraumatic stress symptoms. *Behav Ther*. 2013;44(1):152–161.
9. Espino SR, Fletcher J, Gonzalez M, Precht A, Xavier J, Matoff-Stepp S. Violence screening and viral load suppression among HIV-positive women of color. *AIDS Patient Care STDS*. 2015;29(suppl 1):S36–S41.
10. Kahneman D, Tversky A. Prospect theory: an analysis of decision under risk. *Econometrica*. 1979;47(2):263–291.
11. Hobfoll SE, Lilly RS. Resource conservation as a strategy for community psychology. *J Commun Psychol*. 1993;21(2):128–148.
12. Freedy JR, Shaw DL, Jarrell MP, Masters CR. Towards an understanding of the psychological impact of natural disasters: an application of the conservation of resources stress model. *J Trauma Stress*. 1992;5(3):441–454.
13. Freedy JR, Saladin ME, Kilpatrick DG, Resnick HS, Saunders BE. Understanding acute psychological distress following natural disaster. *J Trauma Stress*. 1994;7(2):257–273.
14. Ironson G, Wynings C, Schneiderman N, et al. Posttraumatic stress symptoms, intrusive thoughts, loss, and immune function after Hurricane Andrew. *Psychosom Med*. 1997;59(2):128–141.
15. Littleton H, Grills-Taquechel A, Axsom D. Resource loss as a predictor of posttrauma symptoms among college women following the mass shooting at Virginia Tech. *Violence Vict*. 2009;24(5):669–686.
16. DeSalvo KB, Hyre AD, Ompad DC, Menke A, Tynes LL, Muntner P. Symptoms of posttraumatic stress disorder in a New Orleans workforce following Hurricane Katrina. *J Urban Health*. 2007;84(2):142–152.
17. Adeola FO. Mental health and psychosocial distress sequelae of Katrina: an empirical study of survivors. *Hum Ecol Rev*. 2009;16(2):195–210.
18. Lee EO, Shen C, Tran TV. Coping with Hurricane Katrina: psychological distress and resilience among African American evacuees. *J Black Psychol*. 2009;35(1):5–23.
19. Betancourt TS, Abdi S, Ito BS, Lilienthal GM, Agalab N, Ellis H. We left one war and came to another: resource loss, acculturative stress, and caregiver-child relationships in Somali refugee families. *Cultur Divers Ethnic Minor Psychol*. 2015;21(1):114–125.
20. Hall BJ, Hobfoll SE, Canetti D, Johnson RJ, Galea S. The defensive nature of benefit finding during ongoing terrorism: an examination of a national sample of Israeli Jews. *J Soc Clin Psychol*. 2009;28(8):993–1021.
21. Hobfoll SE, Watson P, Bell CC, et al. Five essential elements of immediate and mid-term mass trauma intervention: empirical evidence. *Psychiatry*. 2007;70(4):283–315. discussion 316–369.

CHAPTER

8

Control and Stress

A. Steptoe, L. Poole
University College London, London, UK

OUTLINE

Abstract

Control over aversive experiences is a central construct in stress research, and can come in many guises, from behavioral control over the source of stimulation, through perceptions of control, to cognitive control as a form of coping response. Control modulates the neurobiological and health consequences of stress exposure, and can also be harnessed in clinical and other situations to ameliorate stress responses.

INTRODUCTION

Stressors come in many guises, from confrontation with a predator through death of a close relative to conflict at work or living in crowded, noisy conditions. When we try to understand what aspects of situations make them more or less threatening, the concept of control emerges as a crucial feature. Stressors differ in the extent to which they can be controlled, and behavioral control is typically associated with a diminution of stress responses. Sometimes the perception of control is sufficient to reduce stress reactions, even though control itself is never actually exercised. For example, studies of people who are anxious about dentistry have shown that telling people they can give a signal that will stop the procedure reduces pain and anxiety, even though they never use the signal. Humans and other animals also seek out information about impeding threatening events and about ways of coping with the situation. This phenomenon is known as cognitive control and is an important weapon in the armory of psychological coping. Finally, there are control beliefs or the sense of control that people have in specific areas or over their lives, and destiny more generally. A sense of control is generally adaptive for health, but the need for control may under some circumstances be maladaptive and lead to an inflexibility in coping with difficult problems. Since control has so many facets, there is a danger that the construct is misused and disguises careless reasoning.[1] In this chapter, we describe the different aspects of control and stress, along with evidence related to physiological responses, well-being, and health. We also discuss the ways in which control can be enhanced to improve quality of life, ameliorate pain, and promote rehabilitation.

Stress: Concepts, Cognition, Emotion, and Behavior
http://dx.doi.org/10.1016/B978-0-12-800951-2.00008-X

KEY POINTS

- Control is a key concept in stress research, with greater control typically resulting in reduced stress responses.
- Control takes many forms, from direct behavioral control (the availability of direct actions that can modify stress stimulation), through perceived control (the sense that one has control over events), to cognitive control (the capacity to control one's reactions to events), and self-control (ability to regulate emotions and delay gratification).
- Control is relevant in human development, the experience of aging, in potentially stressful situations such as work, in health settings, and in the broader societal context in relation to autonomy and self-determination.

BEHAVIORAL CONTROL AND PHYSIOLOGICAL STRESS RESPONSES

Behavioral control can be defined as having at one's disposal a behavioral response that can prevent, reduce, or terminate stressful stimulation. Natural stressors vary across the entire spectrum of controllability. Some events, such as the unexpected death of a family member, are outside personal control. But in many situations there is an element of behavioral control. People who regularly find themselves stuck in traffic jams on their morning journey to work may feel helpless, but actually have some control; they might choose to use some other form of transport; to start their journey at a different time; to select a different route; stop, and take a break until the traffic clears; and so on.

Study of the impact of behavioral control over stress is complicated by the fact that many of the most aversive life events are uncontrollable. Comparisons between controllable and uncontrollable events may therefore be confounded with the aversiveness of the experience. Experimental studies provide the best opportunity to examine the effects of behavioral control, since aversiveness can be held constant. The yoked design championed by Jay Weiss[2] and later by Martin Seligman[3] and others has served to illustrate the effects of behavioral control most clearly. In this paradigm, a pair of rodents is exposed to identical intermittent aversive stimuli (electric shocks or loud noise). One animal is able to make a behavioral response which either terminates the stimuli or delays their onset, while the second animal has no response available. The amount of aversive stimulation experienced by both depends on the effectiveness of the performance of the one in the behavioral control condition, and any differences in physiological responses or pathology will result from the availability of behavioral control.

TABLE 1 Adverse Effects of Uncontrollable Stress in Comparison with Exposure to Matched Controllable Aversive Stimulation

Decreased food/water consumption
Greater weight loss
Higher plasma corticosterone
Increased gastric lesions
Reduced production of specific antibodies
Reduced lymphocyte reactivity
Decreased cytotoxic activity of natural killer cells
Decreased tumor rejection
Increased susceptibility to malignancy

Table 1 summarizes some of the many physiological and pathological advantages conferred by control in this design.[4] It can be seen that behavioral control is associated with amelioration of physiological stress responses and pathological outcomes. These range from reduced levels of neuroendocrine response to slower proliferation of experimentally implanted malignancies. The regions of the brain responsible for these effects have been identified. Studies have highlighted the role of infralimbic and prelimbic regions of the ventral medial prefrontal cortex. Controllable stressors lead to the inhibition of stress-induced activation of the dorsal raphe nucleus by the ventral medial prefrontal cortex, blocking the behavioral sequelae of uncontrollable stress. Stress-induced neural activity in brainstem nuclei is inhibited by control, which is not consistent with previous research which has shown that such activity is induced by a lack of control.[5] But not all the biological responses to uncontrollable stress are detrimental. For example, uncontrollable conditions tend to lead to greater stress-induced analgesia than matched controllable conditions,[6] and this mechanism may permit organisms to endure stressful stimulation at a reduced level of physical discomfort.

Research in humans has gone some way to duplicating these effects, with differences in the magnitude of acute cardiovascular and endocrine responses. However, an important caveat concerns the effort or response cost associated with maintaining behavioral control.

Control and Effort

Effects of the type shown in Table 1 have generally emerged in studies in which behavioral control is easy to exert. For example, electric shock may be avoided by a lever press on a simple schedule of reinforcement. However, control may require great effort to maintain and can be associated with a degree of uncertainty as to whether it has been successful. Under these circumstances, physiological activation may be greater than that elicited in uncontrollable situations. This was illustrated in the classic human experimental studies carried by Paul Obrist and colleagues. They randomized human volunteers to three conditions in a shock avoidance reaction time study.[7] All participants had to respond rapidly to aversive stimuli and were informed that successful performance would lead to the avoidance of electric shock. Three criteria were employed: an "easy" condition in which even slow reaction times would be successful, a "hard" condition in which participants had to maintain or improve performance over time, and an "impossible" condition in which the reaction time criterion was too fast for most people. The impossible condition is comparable to uncontrollable stress, while the other two conditions represent easy and effortful behavioral control, respectively. The number of shocks administered was held constant across conditions. They found that the greatest physiological activation was present not in the impossible or uncontrollable condition, but in the hard condition. This physiological reaction pattern was sustained by selective activation of β-adrenergically mediated sympathetic nervous system responses. Later studies have established that the enhanced physiological reactions associated with control under these circumstances are related to the effort expended, with greater effort being correlated with heightened norepinephrine responses.

Control and Fear

Lack of control over aversive stimuli potentiates fear responses. Using the yoked control design, it has been found that rats subsequently placed in a different environment and given two brief shocks so as to generate conditioned fear, showed greater freezing responses if they had previously had no control over shock compared with those that had had control. These effects are long-lasting, but can be blocked by benzodiazepines.[8] Fear responses associated with uncontrollable stress are again related to release of serotonin in the dorsal raphe nucleus and amygdala. Experiments using in vivo microdialysis have shown that uncontrollable stress leads to increased levels of extracellular serotonin in both these sites, while animals exposed to controllable stress show no such elevation.[9]

Research on humans also suggests that control is relevant to fear responses. Fear of pain during medical and dental procedures is strongly associated with perceptions of uncontrollability. Fear of social threats (for example, when the individual is threatened with physical or sexual assault) is greater when people believe the situation is uncontrollable.

Control and Work

Control is important to understanding the physiological and health consequences of stressful events in human life. One area in which control has emerged as a key variable is in the study of work stress. Jobs vary on many dimensions of control. Some work is self-paced, while other tasks are externally paced by machines or other people. In some jobs, workers have choice about their posture, when they can take rest breaks, and how the work should be done. Flexibility in work hours, participation in decision-making, and autonomy at work are all factors that vary across jobs. In psychophysiological studies, it has been found that cardiovascular and neuroendocrine activation is greater during externally paced than self-paced work, even when the actual pace of work is held constant.[10] The same patterns are present in the real work environment. Figure 1 illustrates the pattern of ambulatory blood pressure measured every 20 min over the working day in middle-aged men and women divided into low and high job control on a standard questionnaire measure. Systolic and diastolic ambulatory blood pressure was significantly greater throughout the day in people working in low-control than in high-control jobs (means 125.7/81.5 vs. 122.4/78.6 mmHg, $P < 0.05$), independently of age, gender, employment grade, body mass index, smoking, and physical activity.[11]

A model of work stress has been developed in which high demands coupled with low control is regarded as particularly toxic (the demand/control or job strain model). Several meta-analyses and large-scale prospective epidemiological observational studies have shown that low control at work, in combination with high demands, predicts future coronary heart disease in previously healthy adults[12,13]; other prospective studies indicate that job strain is a risk factor for stroke and diabetes, but not for cancer.[14,15] Control over work pace

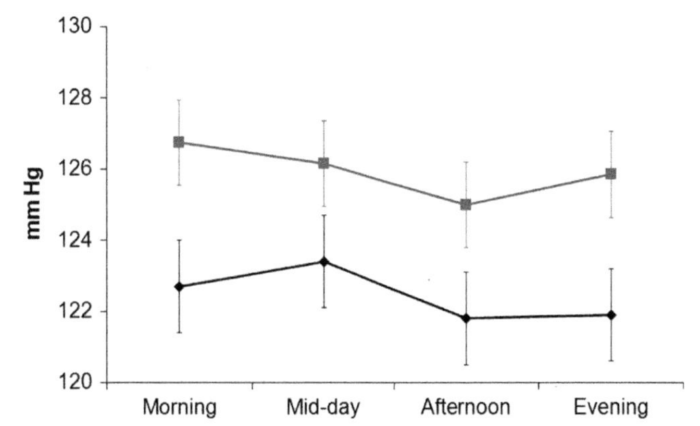

FIGURE 1 Mean systolic blood pressure in the morning, mid-day, afternoon, and evening periods in high (black line) and low (blue line) job control groups, adjusted for gender, employment grade, age, body mass index, smoking status, and physical activity. Error bars are standard errors of the mean (s.e.m.). From Steptoe and Willemsen.[11]

influences health during pregnancy, and women with lower control jobs may be at greater risk for preeclampsia and low back pain, after statistical adjustment for age, parity, education, smoking, and type of work.[16]

Conversely, greater autonomy and participation at work promote higher levels of self-reported job satisfaction, commitment to work, better performance, and reduced levels of emotional distress, staff turnover, and absenteeism. However, these studies of work illustrate a complication in human research in control, since it is clear that control cannot be seen solely as an objective characteristic of the environment, but depends on subjective perceptions as well.

PERCEIVED CONTROL

Perceived control is the sense that one has control over aversive events. Objective differences in behavioral control are presumably mediated by perceived control. Thus the benefits of having control over a particular aspect of work (such as being able to make a personal telephone call if necessary) will only be apparent if the worker realizes that this option is available. Perceived control is the result of a subjective judgment, based on the individual's appraisal of the situation. It has two elements. The first is a perception that the situation is potentially controllable, and the second is that the individual can take the correct action. There may be circumstances in which events are perceived as controllable, but no action is taken since people lack (or believe they lack) appropriate behavioral competence. Psychologists such as Albert Bandura[17] have postulated a distinction between outcome expectations and self-efficacy expectations to characterize these two aspects.

It is generally found that distress is associated with a perception that events are uncontrollable, and this phenomenon may operate independently of objective behavioral control. Thus two individuals may experience the same event, such as the breakdown of an important personal relationship. One may perceive that the event was outside their control, due to the callousness of their partner, the circumstances in which they lived, or the predatory behavior of a third party. Another person may perceive themselves to have been responsible; if they had not been preoccupied with their work or had responded better to the needs of their partner, then the relationship might not have floundered. One student who has failed an academic examination may attribute this to their own lack of preparation and effective revision, while another may blame arbitrary examiners or bad luck in the selection of questions.

Human beings seek to understand their experiences in life by developing explanations for aversive events and often construe events retrospectively as having been controllable (for example, "if I had not stayed later than I planned at that party, and had not decided to walk down that street, I would not have been assaulted"). Such cognitive attributions may be maladaptive in leading to self-blame for occurrences over which one has little control, but also may help to impose sense and order on a world which may otherwise appear cruel and arbitrary.

Sense of Control

Perception of control is not restricted to individual aversive experiences, but also relates to domains of life or even to life in general. The term "locus of control" is used to describe these generalized control beliefs. This construct was first introduced by Rotter,[18] who argued that people vary in the extent to which they believe that important outcomes are determined by their own internal abilities and activities or by external factors. Locus of control has since been elaborated in two important ways. First, locus of control may be domain specific, with different levels of belief in personal control over different aspects of life (health, finances, work, personal relationships, etc.). Thus an individual may feel a strong sense of control over their work, but believe that their living environment (traffic, difficult neighbors, etc.) is out of their own control. Second, it may be useful to understand locus of control as consisting of three dimensions: belief in internal control, chance, and powerful others. For example in the domain of health, people can be categorized according to the extent to which they believe that their health is determined by their own actions (internal control), that whether they remain healthy is a matter of fate or luck (belief in chance), and that their health is influenced by the competence of doctors and other health professionals (belief in powerful others). A dimensional perspective implies that people may simultaneously hold strong beliefs about more than one set of influences. They may believe, for instance, that they have quite a lot of control over their health, while at the same time believing that doctors are also very influential.

Sense of control is an individual belief pattern that also has developmental and social determinants. Animal studies indicate that monkeys provided with experiences of controllable (contingent) events early in life are subsequently less reactive to stressful events in adulthood.[19] The mechanism for this effect is thought to involve changes to stress sensitive brain regions, namely the vermis, dorsomedial prefrontal cortex, and dorsal anterior cingulate cortex.[20] The experience of a major uncontrollable event in childhood, such as parental separation, is associated with increased likelihood of major depression, generalized anxiety disorder, and drug and alcohol dependence in adult life.[21] Differences in sense of control may also relate to socioeconomic position, with economically more deprived people facing greater external obstacles and fewer opportunities to influence events than do the better off. One study assessed sense of mastery (e.g., agreement with the statement "whether I am able to get what I want is in my own hands") and perceptions of constraints (e.g., "there is little I can do to change many of the important things in my life") in three large population surveys.[22] It was found that sense of mastery was greater in higher than lower income sectors of the sample, while perceived constraints were less. The financially better off also reported greater life satisfaction and less depression and rated their health as better. However, not all the lower income respondents had low mastery and high constraint ratings. Some individuals in the low-income sectors had high mastery ratings, and they reported greater life satisfaction, less depression, and health levels comparable with those of the high-income group. Moreover, sense of control is also thought to buffer against the negative effects of social adversity. For example, a 14-year prospective analysis of the Midlife in the United States (MIDUS) study showed that baseline sense of control attenuated the negative impact of low education on mortality risk.[23]

The association between sense of control and coronary heart disease was established in the INTERHEART STUDY, a study involving 11,119 patients who had suffered an acute myocardial infarction from 52 countries and 13,648 controls.[24] Ratings of sense of control on six items were averaged and divided into quartiles. Figure 2 shows that in both men and women, greater sense of control was associated with reduced odds of being a cardiac case.

Sense of control has cultural and political determinants as well.[25] Surveys in Eastern Europe during the first decade after the collapse of communism identified a profound sense of lack of control over life, as people witnessed the destruction of familiar institutions and ways of life, and experienced greater difficulty in fulfilling daily needs. These years have been characterized by a rapid deterioration in health and well-being in many sectors of the population of Russia and several Central-Eastern European countries.[26]

COGNITIVE CONTROL AND COPING

Another closely related phenomenon is cognitive control, which can be defined as the belief that one can control one's reactions to events. It is a common characteristic to seek out information about events even when they cannot be controlled behaviorally. The person who has an abnormal result on a medical test and is scheduled for further examination may try to find out about the test and its consequences from acquaintances, the Internet, or other sources of information. The purpose of this information-seeking is not to control the outcome, but to feel "in control" by reducing uncertainty and increasing the predictability of the situation and to manage reactions to the result by anticipatory coping. This type of control is closely allied with the repertoire of other coping responses that people mobilize in an effort to regulate their responses to aversive experiences. It has important uses in medical settings, where the procedures for psychological preparation for surgery and stressful medical procedures have the effect of increasing cognitive control by providing information and helping patients engage in a range of coping responses. In the self-regulation model of health developed by Leventhal,[27] control is a key dimension in the cognitive representation of health threats, and variations in controllability influence effective adaptation.

People vary in the extent to which they exert effort to establish control over the situations with which they are confronted, and some individuals appear to have a high need for control. This may be maladaptive in some circumstances. People with a high need of control may find it particularly difficult to come to terms with experiences

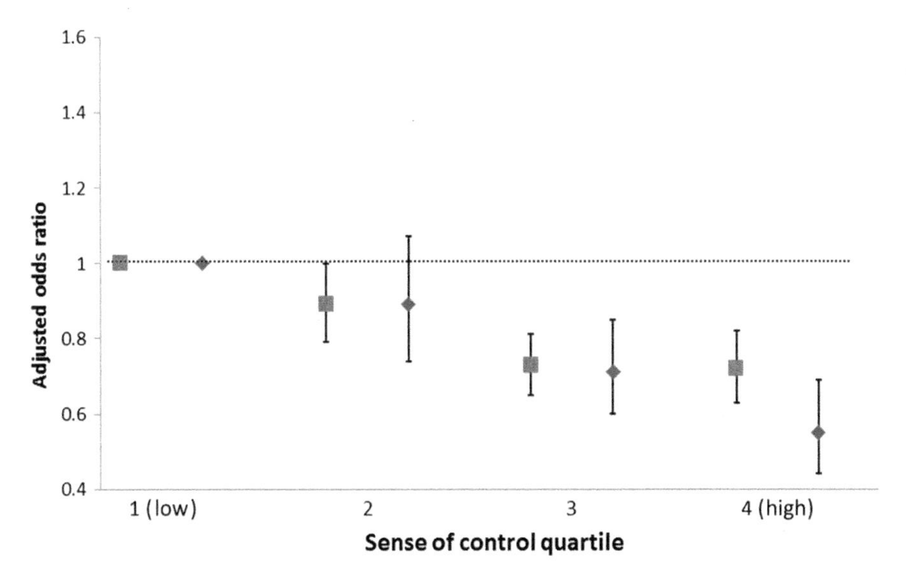

FIGURE 2 Adjusted odds of being an acute myocardial infarction case rather than control in four quartiles of sense of control in men (red) and women (blue). Data are adjusted for age, geographic region, and smoking. The lowest sense of control quartile is the reference group, and values below 1 indicate a protective effect. Error bars are 95% confidence intervals. From the INTERHEART study. From Rosengren et al.[24]

over which they have no control, such as the injury of a family member in a traffic accident.

SELF-CONTROL

Self-control research is an emerging field that focuses on the person's ability to suppress impulses (these could be thoughts, emotions, or behaviors) and delay gratification. It is linked to other concepts such as conscientiousness, self-regulation, and willpower. Several birth cohort studies have begun to examine the emergence of self-control during childhood and its influence on later health and well-being. Moffitt and colleagues[28] studied 1000 children from birth and showed that childhood self-control predicts physical health, substance dependence, income, and criminal offenses at 32-years of age. Children low in self-control made mistakes in adolescence that set them on a path of disadvantage and underachievement. For example, they were more likely than those with high self-control to smoke, leave school without any qualifications, and have unplanned pregnancies. Further work using data from the UK have added to this literature showing that low self-control in childhood predicts unemployment throughout adulthood.[29] Moreover, a bi-directional association exists, whereby low socioeconomic position in childhood predicts low self-control later in life.[30]

Given that self-control is linked to adversity, it is not surprising that self-control is also relevant to stress. Self-control is thought to be linked to stress in three main ways.[31] First, the direct hypothesis proposes that self-control directly predicts exposure to fewer stressors, perhaps due to superior anticipatory coping or avoidance skills. For example, someone with high self-control would be less likely to face unemployment because they remain vigilant to work opportunities. Second, self-control might prevent impulses from turning into a stressor, because of an ability to pay heed to warning signs. For instance, a person high in self-control would anticipate that staying for an extra drink after work would lead to an argument with their spouse causing extra stress at home, so they refuse. Third, self-control can act as a buffer against stressors by altering stress appraisals. In this scenario, those high in self-control feel less stressed as they have the resources to combat challenging situations. For example, an individual with high self-control will go into an examination well prepared, and so feel less stressed about the risk of failure.

CONTROL AND WELL-BEING

Control and Aging

Growing old is associated with a diminution of control. When people retire from work, their economic power is reduced, they have less authority than before, and their social networks shrink. Impaired physical health may increase the person's sense of vulnerability, while they are exposed more often to uncontrollable events such as the death of contemporaries. A number of studies have documented the links between aging and the loss of control, though there are exceptions. "Successful aging," or the maintenance of well-being and effective functioning despite chronological progression, is coupled with maintenance of beliefs in personal control. People may adapt by lowering their valuation of the importance of areas of life over which they lose control and by investing in social relationships more selectively than before.[32,33] Cognitive control becomes more important at older ages when primary behavioral control is attenuated. The impact of enhancing sense of control and autonomy on the well-being of elderly population has also been investigated. Randomized trials of increased involvement in decision-making in institutionalized elderly groups have demonstrated improvements in mood and activity levels associated with enhanced control.[34–36]

Control and Pain

The effects on pain of having control over the source of painful stimulation have been studied extensively. These experiments suggest that having behavioral control leads to a reduction in the subjective and physiological impact of the painful stimulus and may also increase pain tolerance.[37] However, the benefits of control depend on the confidence that people have in their response efficacy, so perceptions are again important. It is interesting that the favorable effects of control over acute pain are observed despite the fact that stress-induced analgesia is elicited when there is a lack of control. It is likely that control allows behavioral and cognitive coping methods to be deployed that in turn reduce distress. Brain imaging studies have shown that when pain is perceived to be controllable, there is an attenuation of activation in the anterior cingulate, insular, and secondary somatosensory cortex, areas of the brain that are associated with pain processing. However, there is a large degree of interindividual variability. Those individuals who showed greater activation in the pregenual anterior cingulate cortex, periaqueductal gray, and posterior insula/SII in response to uncontrollable pain also reported higher levels of pain during the uncontrollable versus controllable conditions. On the other hand, greater activation in the ventral lateral prefrontal cortex in anticipation of uncontrollable pain is associated with a lower pain response to uncontrollable pain compared with controllable pain.[38] Interestingly, placebo analgesia results in alterations in brain activity in pain-sensitive brain regions, indicating that the control associated with beliefs leads to changes in the experience of pain.[39]

Chronic clinical pain is often accompanied by feelings of helplessness, lack of control, and depression. Some authorities have argued that there may be a progression from feelings of personal helplessness in the early phases of a chronic pain problem to general helplessness (the belief that no one is able to control the pain) in later stages. The impact of chronic pain on well-being is that of an uncontrollable stressor, since by definition a pain that becomes chronic has not been alleviated and so has not been controlled. The generalization of distress about pain to disability and depression mirrors the phenomenon of learned helplessness.

The links between control and pain can be harnessed for patient care. As noted earlier, providing patients with control over sources of potential pain (for example, during dental examinations) can increase tolerance. Patient-controlled analgesia is used in a variety of medical settings with both adults and children, and has benefits both in terms of emotional well-being and analgesic use. Several studies have found that when patients can control the level of analgesia, they experience lower pain intensity than if it is provided by the physician.[40] Rehabilitation programs for patients with chronic pain have a strong element of increasing perceived control by providing patients with greater and more effective coping options.

Control and Illness

Most serious illnesses bring with them strong feelings of loss of control. When people have heart attacks or are diagnosed with cancer, they feel that they are no longer in control of their bodies, and their future plans are put in jeopardy so that they lose a sense of control over their destiny. Loss of control is often accompanied by other facets of helplessness, including hopelessness, distress, and passivity. A large-scale study of older people found perceived control was associated with several precursors to ill-health including greater grip strength (an indicator of physical function) and with lower cardio-metabolic risk.[41] Similar patterns occur in chronic health problems. For example, high levels of perceived control over asthma are associated with better self-management skills, reduced risk of hospitalization, and less restricted activity, independent of the severity of asthma itself.[42] Some studies have found that recovery from serious illness such as stroke and major fractures is positively correlated with high internal control beliefs.[43]

Many procedures for improving the emotional state of patients have the effect of enhancing perceived control. For example, the methods used to prepare women for childbirth increase perceived control by providing a repertoire of coping responses. Stress management procedures during cardiac rehabilitation enhanced perceived control and patient self-efficacy by increasing confidence in the resumption of normal activities. A particular area in which the issue of control has been contentious surrounds patient choice. To what extent should patients be involved in medical decision-making? The literature indicates that although patients desire information about issues surrounding medical decisions, they sometimes prefer to delegate responsibility for the decision itself to the doctor.[44] This should not be seen as too surprising since it is the typical pattern found in any society with specialist roles. While we might want to know all the facts, we would expect the car mechanic to decide how to fix an engine or the chef to use judgement and skill to prepare the complicated dish we have ordered. Patients often wish to be involved in decision-making, particularly when there is a genuine choice of more than one effective alternative, but seldom wish to take authority completely away from the physician.

CONCLUSIONS

Control is a powerful explanatory variable in stress research. It is relevant to a wide variety of settings, from basic experimental research on neuroendocrinology and neurochemistry to the experience of people in different socioeconomic sectors of society. There is a danger such an attractive integrating notion will be misused and employed as a global explanation for completely unrelated phenomena. However, if care is taken not to confuse the different meanings of control, the construct has considerable value in understanding stress.

References

1. Skinner EA. A guide to constructs of control. *J Pers Soc Psychol*. 1996;71(3):549–570.
2. Weiss JM. Effects of coping behavior in different warning signal conditions on stress pathology in rats. *J Comp Physiol Psychol*. 1971;77(1):1–13.
3. Seligman ME, Beagley G. Learned helplessness in the rat. *J Comp Physiol Psychol*. 1975;88(2):534–541.
4. Steptoe A. The significance of personal control in health and disease. In: Steptoe A, Appels A, eds. *Stress, Personal Control and Health*. Chichester: John Wiley; 1989:309–318.
5. Amat J, Baratta MV, Paul E, Bland ST, Watkins LR, Maier SF. Medial prefrontal cortex determines how stressor controllability affects behavior and dorsal raphe nucleus. *Nat Neurosci*. 2005;8(3):365–371.
6. Sutton LC, Lea SE, Will MJ, et al. Inescapable shock-induced potentiation of morphine analgesia. *Behav Neurosci*. 1997; 111(5):1105–1113.
7. Light KC, Obrist PA. Cardiovascular response to stress: effects of opportunity to avoid, shock experience, and performance feedback. *Psychophysiology*. 1980;17(3):243–252.

8. Bouton ME, Kenney FA, Rosengard C. State-dependent fear extinction with two benzodiazepine tranquilizers. *Behav Neurosci.* 1990;104(1):44–55.
9. Forster GL, Feng N, Watt MJ, et al. Corticotropin-releasing factor in the dorsal raphe elicits temporally distinct serotonergic responses in the limbic system in relation to fear behavior. *Neuroscience.* 2006; 141(2):1047–1055.
10. Steptoe A, Fieldman G, Evans O, Perry L. Control over work pace, job strain and cardiovascular responses in middle-aged men. *J Hypertens.* 1993;11(7):751–759.
11. Steptoe A, Willemsen G. The influence of low job control on ambulatory blood pressure and perceived stress over the working day in men and women from the Whitehall II cohort. *J Hypertens.* 2004; 22(5):915–920.
12. Kivimäki M, Nyberg ST, Batty GD, et al. Job strain as a risk factor for coronary heart disease: a collaborative meta-analysis of individual participant data. *Lancet.* 2012;380(9852):1491–1497.
13. Steptoe A, Kivimäki M. Stress and cardiovascular disease: an update on current knowledge. *Annu Rev Public Health.* 2013;34:337–354.
14. Nyberg ST, Fransson EI, Heikkilä K, et al. Job strain as a risk factor for type 2 diabetes: a pooled analysis of 124,808 men and women. *Diabetes Care.* 2014;37(8):2268–2275.
15. Heikkilä K, Nyberg ST, Theorell T, et al. Work stress and risk of cancer: meta-analysis of 5700 incident cancer events in 116,000 European men and women. *BMJ.* 2013;346:f165.
16. Wergeland E, Strand K. Work pace control and pregnancy health in a population-based sample of employed women in Norway. *Scand J Work Environ Health.* 1998;24(3):206–212.
17. Bandura A. *Self-Efficacy: The Exercise of Control.* New York, NY: Freeman; 1997.
18. Rotter JB. Generalized expectancies for internal versus external control of reinforcement. *Psychol Monogr Gen Appl.* 1966;80(1):1–28.
19. Suomi SJ. Early stress and adult emotional reactivity in rhesus monkeys. *Ciba Found Symp.* 1991;156:171–183.
20. Spinelli S, Chefer S, Suomi SJ, Higley JD, Barr CS, Stein E. Early-life stress induces long-term morphologic changes in primate brain. *Arch Gen Psychiatry.* 2009;66(6):658–665.
21. Otowa T, York TP, Gardner CO, Kendler KS, Hettema JM. The impact of childhood parental loss on risk for mood, anxiety and substance use disorders in a population-based sample of male twins. *Psychiatry Res.* 2014;220(1–2):404–409.
22. Lachman ME, Weaver SL. The sense of control as a moderator of social class differences in health and well-being. *J Pers Soc Psychol.* 1998;74(3):763–773.
23. Turiano NA, Chapman BP, Agrigoroaei S, Infurna FJ, Lachman M. Perceived control reduces mortality risk at low, not high, education levels. *Health Psychol.* 2014;33(8):883–890.
24. Rosengren A, Hawken S, Ounpuu S, et al. Association of psychosocial risk factors with risk of acute myocardial infarction in 11119 cases and 13648 controls from 52 countries (the INTERHEART study): case–control study. *Lancet.* 2004;364(9438):953–962.
25. Bobak M, Pikhart H, Hertzman C, Rose R, Marmot M. Socioeconomic factors, perceived control and self-reported health in Russia. A cross-sectional survey. *Soc Sci Med.* 1998;47(2):269–279.
26. Cockerham WC. *Health and Social Change in Russia and Eastern Europe.* London: Routledge; 1999.
27. Cameron LD, Leventhal H. *The Self-Regulation of Health and Illness Behaviour.* London: Routledge; 2003.
28. Moffitt TE, Arseneault L, Belsky D, et al. A gradient of childhood self-control predicts health, wealth, and public safety. *Proc Natl Acad Sci U S A.* 2011;108(7):2693–2698.
29. Daly M, Delaney L, Egan M, Baumeister RF. Childhood self-control and unemployment throughout the life span: evidence from two British cohort studies. *Psychol Sci.* 2015;26(6):709–723.
30. Hostinar CE, Ross KM, Chen E, Miller GE. Modeling the association between lifecourse socioeconomic disadvantage and systemic inflammation in healthy adults: the role of self-control. *Health Psychol.* 2015;34(6):580–590.
31. Galla BM, Wood JJ. Trait self-control predicts adolescents' exposure and reactivity to daily stressful events. *J Pers.* 2015;83(1):69–83.
32. Scheibe S, Carstensen LL. Emotional aging: recent findings and future trends. *J Gerontol B Psychol Sci Soc Sci.* 2010; 65B(2):135–144.
33. Brandtstädter J, Rothermund K. Self-percepts of control in middle and later adulthood: buffering losses by rescaling goals. *Psychol Aging.* 1994;9(2):265–273.
34. Rodin J, Langer EJ. Long-term effects of a control-relevant intervention with the institutionalized aged. *J Pers Soc Psychol.* 1977; 35(12):897–902.
35. Rodin J. Aging and health: effects of the sense of control. *Science.* 1986;233(4770):1271–1276.
36. Grönstedt H, Frändin K, Bergland A, et al. Effects of individually tailored physical and daily activities in nursing home residents on activities of daily living, physical performance and physical activity level: a randomized controlled trial. *Gerontology.* 2013;59 (3):220–229.
37. Rosenbaum M. Individual differences in self-control behaviors and tolerance of painful stimulation. *J Abnorm Psychol.* 1980;89 (4):581–590.
38. Salomons TV, Johnstone T, Backonja M-M, Shackman AJ, Davidson RJ. Individual differences in the effects of perceived controllability on pain perception: critical role of the prefrontal cortex. *J Cogn Neurosci.* 2007;19(6):993–1003.
39. Wager TD, Rilling JK, Smith EE, et al. Placebo-induced changes in FMRI in the anticipation and experience of pain. *Science.* 2004; 303(5661):1162–1167.
40. McNicol ED, Ferguson MC, Hudcova J. Patient controlled opioid analgesia versus non-patient controlled opioid analgesia for postoperative pain. *Cochrane Database Syst Rev.* 2015;6 CD003348.
41. Infurna FJ, Gerstorf D. Perceived control relates to better functional health and lower cardio-metabolic risk: the mediating role of physical activity. *Health Psychol.* 2014;33(1):85–94.
42. Calfee CS, Katz PP, Yelin EH, Iribarren C, Eisner MD. The influence of perceived control of asthma on health outcomes. *Chest.* 2006; 130(5):1312–1318.
43. Partridge C, Johnston M. Perceived control of recovery from physical disability: measurement and prediction. *Br J Clin Psychol.* 1989;28(1):53–59.
44. Levinson W, Kao A, Kuby A, Thisted RA. Not all patients want to participate in decision making. A national study of public preferences. *J Gen Intern Med.* 2005;20(6):531–535.

CHAPTER

9

Effort-Reward Imbalance Model

J. Siegrist

University of Duesseldorf, Duesseldorf, Germany

OUTLINE

Abstract

The effort-reward imbalance model was developed to identify health-adverse effects of stressful psychosocial work and employment conditions in developed and rapidly developing countries. It posits that exposure to the recurrent experience of failed reciprocity at work "high cost/low gain" increases the risk of incident stress-related disorders, such as depression or coronary heart disease. Evidence from prospective epidemiological investigations, as well as from experimental and naturalistic studies focusing on psychobiological mechanisms, supports this notion. Distinct measures of prevention can be derived from available scientific knowledge, addressing the organizational level as well as the level of national labor and social policies. Recent extensions of the model beyond the context of paid work highlight the far-reaching importance of fair procedures of social exchange for human health and well-being.

BACKGROUND

Stress concepts that focus on the social environment and its effects on exposed people often address work and employment. This is due to the significance of work in adult life in modern societies. Having a job is a prerequisite for a continuous income and, more so than any other social circumstance, employment characteristics determine adult socioeconomic status. Beyond economic livelihood, a person's occupation is an important aim of long-term socialization, thus providing opportunities for personal growth and development. While a good quality of work including stable employment may contribute to employees' health and well-being, poor jobs and precarious work adversely affect their health. Traditionally, research on associations of work with health was the task of occupational medicine. Most often, noxious effects of physical, chemical, and biological hazards were analyzed. However, with the advent of advanced technologies including automation, the growth of the service sector of employment and the expansion of computer-based, information processing jobs in postindustrial modern societies, the spectrum of occupational exposures with potential impact on health has changed rather dramatically. While traditional hazards still prevail in certain sectors of the labor market, the majority of employed people are now confronted with a variety of mental and emotional demands, threats, or conflicts, rather than with toxic substances and environments. This is exactly the place where stress concepts developed by social and behavioral sciences are needed to cope with this challenge. Effort-reward imbalance is one such concept identifying a stressful psychosocial work environment. This chapter describes its theoretical basis and its measurement, and it summarizes a substantial amount of epidemiological and experimental research documenting elevated risks of stress-related mental and physical diseases as well as psychobiological pathways underlying these associations. In the final part, implications for prevention are discussed, and future developments of this line of research are outlined.

Stress: Concepts, Cognition, Emotion, and Behavior
http://dx.doi.org/10.1016/B978-0-12-800951-2.00009-1

HOW DOES STRESSFUL WORK GET UNDER THE SKIN?

Work and employment in modern societies underwent significant changes in the recent past. Large parts of the workforce are no longer confronted with physical demands and traditional occupational hazards, but are exposed to mental and emotional demands, high work pressure, and job insecurity. If experienced over a longer period of time, these psychosocial conditions can adversely affect the working people's health. To identify respective "toxic" features within diverse and variable work environments, a theoretical model is needed that offers a stress-theoretical explanation of the links between adverse work and reduced health.

"Effort-reward imbalance" is one such model.[1] It asserts that the recurrent experience of failed reciprocity between high cost spent at work and low gain received in turn, activates sustained negative emotions of reward frustration and associated circuits of the brain reward system, including nucleus accumbens, anterior cingulated cortex, and insula. This sustained activation is due to the violation of a basic principle of social exchange, the equivalence of "give" and "take" in costly transactions. As the brain's reward circuitry seems to be sensitive to the experience of inequality in social exchange[2] it may recurrently arouse the stress axes, in particular the hypothalamic-pituitary-adrenocortical axis, and thus trigger states of allostatic load within several regulating systems of the body. In consequence, the risk of developing a stress-related disorder, such as coronary heart disease or depression, is increased in working people exposed to effort-reward imbalance at work.

This hypothesis was tested in a number of epidemiological investigations and in experimental and naturalistic studies. A solid body of evidence supports the notion that failed reciprocity in costly transactions at work is associated with elevated relative risks of depression and coronary heart disease, after adjusting for relevant confounders. In addition, potential psychobiological pathways were identified. Importantly, rewards at work include salary and wage, job promotion, and job security, as well as recognition and esteem of well-performed work. Thus, the model does not only cover problematic aspects of modern work and employment in a globalized economy, but it also offers entry points for investments in healthy work and for the prevention of stress-related disorders.

THEORY AND MEASUREMENT

The Sociological Basis

Given the complexities and varieties of modern work and employment conditions, it is not easy to identify common features that account for elevated health risks among people exposed to these conditions. Yet, identifying such common features and demonstrating their contribution toward explaining health and well-being is the task of a stress-theoretical model. The notion of theory implies that stressful features of work are delineated at a level of generalization that allows for their identification in a wide range of different occupations and contexts. To this end, a theoretical model selectively reduces the complexities of the real world by introducing some core analytical notion.

The "effort-reward imbalance" model is concerned with stressful features of the work contract, with a selective focus on the analytical notion of social reciprocity in costly transactions.[1] Social reciprocity has been identified as a fundamental, evolutionary stable principle of collaborative human exchange.[3] According to this principle, any costly transaction provided by person A to person B that has some utility to B is expected to be returned by person B to A. Exchange expectancy does not implicate full identity of the service in return, but it is essential that this activity meets some agreed-upon standard of equivalence. Failed reciprocity results from situations where service in return is either denied or does not meet the agreed-upon level of equivalence. To secure equivalence of return in crucial types of costly transactions, social contracts have been established as a universal societal institution. The work contract (or contract of employment) is one such type where efforts are expected to be delivered by employees in exchange for rewards provided by the employer. Three basic types of rewards are transmitted in this case: salary or wage (financial reward), career promotion or job security (status-related reward), and esteem or recognition (socioemotional reward). Importantly, contracts of employment do not specify efforts and rewards in all details, but provide some room of flexibility and adaptation.

The model of effort-reward imbalance at work asserts that lack of reciprocity in terms of high cost spent and low gain received in turn occurs frequently under specific conditions. "Dependency" is one such condition, defined by situations where workers have no alternative choice in the labor market. For instance, unskilled or semiskilled workers, elderly employees, or those with restricted mobility or reduced work ability may be susceptible to unfair contractual transaction. "Dependency" is relatively frequent in modern economies with a globalized labor market where parts of the workforce are exposed to job instability or job loss due to mergers, organizational downsizing, rapid technological change, and growing economic competition. This latter observation points to a second condition of failed reciprocity at work, "strategic choice." Here, people accept the experience of "high cost/low gain" in their employment for a certain time, often without being forced to do so, because they tend to improve their chances of career promotion in a highly competitive job market.

The notion of effort at work implies both an extrinsic demand to which the working person responds as well as a subjective motivation to match the demand. In most instances, matching the demands is part of the control structures established in organizations, thus leaving little room for variations of subjective motivation. Yet, demands are likely to be exceeded in situations of strong informal pressure exerted by a competing work team (e.g., group piece work). Similarly, demands are likely to be exceeded if people are characterized by a motivational pattern of excessive work-related "overcommitment." Consciously or unconsciously, they may strive toward continuously high achievement because of their underlying need for approval and esteem at work. This motivation contributes to "high cost/low gain" experience at work even in the absence of extrinsic pressure. To summarize, the model of effort-reward imbalance at work maintains that failed contractual reciprocity in terms of high cost and low gain is often experienced by people who have no alternative choice in the labor market, by those exposed to heavy job competition, and by those who are overcommitted to their work (see Figure 1).

The Stress-Theoretical Basis

As the conditions identified by this model are expected to occur across different sectors of employment, in a variety of jobs and in different socioeconomic and sociocultural contexts the model's claim may be relevant for working populations in several parts of the world. Experiencing effort-reward imbalance at work is particularly distressing as it frustrates basic expectations of equivalence of return in costly transactions. Most importantly, the recurrent experience of failed reciprocity is expected to afflict the health and well-being of working people by compromising their self-esteem and by eliciting negative emotions with special propensity to elicit sustained autonomic and neuroendocrine activation of the organism.

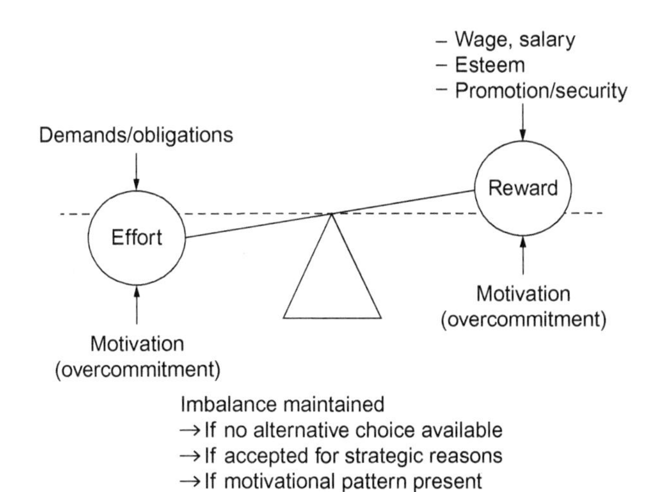

FIGURE 1 The model of effort-reward imbalance at work. Modified from Ref. 1.

As a long-term consequence, recurrent activation of the main stress axes within the organism may precipitate the development of manifest stress-related mental and physical disorders.[4,5]

The experience of effort-reward imbalance at work due to unfair exchange, trust violation, or broken promise is assumed to activate distinct areas in the brain reward circuits, including nucleus accumbens, anterior cingulated cortex, and insula.[6] This activation suppresses the production of dopamine and oxytocin (i.e., neurotransmitters associated with pleasurable emotions and stress-buffering properties). Moreover, activation of the insula is associated with the experience of physical and emotional pain, and with strong visceral and somatic sensations.[7,8] If combined with the occurrence of threats to a person's self or social status, these processes of sustained activation may trigger states of allostatic load within several regulating systems of the body, driven by an extensively aroused hypothalamic-pituitary-adrenocortical stress axis.[4] Neuroscience research has only recently demonstrated that insular activation is modulated by the magnitude of loss following effort, and that the intensity of positive stimulation of the brain reward circuits depends on the amount of effort previously expended.[9] It seems that the brain's reward circuitry is sensitive to the experience of disadvantageous inequality in social exchange.[2] While this recent evidence from neuroscience research is in accordance with basic assumptions of the effort-reward imbalance model, further studies are needed to unravel the links between sustained experience of reward deficiency at work and the development of stress-related disorders, triggered by the described psychobiological processes (see below).

"Threat or loss of reward" is not the only analytical notion within a stress-theoretical framework that has been applied to the study of work and health. "Threat or loss of control" is a complementary powerful notion underlying the well-established job demand-control model.[10] This model claims that stressful experience is elicited by jobs which fail to offer control and decision latitude to working people, especially so under conditions of high demands and work pressure. Lack of control threatens a person's sense of mastery and autonomy, thus evoking negative emotions of anger or anxiety and related psychobiological responses.[11] In conclusion, the two models of demand-control and effort-reward imbalance complement each other, and their combined effects may aggravate the working people's health.[12] It should also be noted that there is some similarity between the latter model and the notion of distributive injustice inherent in social-psychological inequity theory.[13,14] However, effort-reward imbalance addresses inequity of exchange within a core contractual relationship embedded in the labor market, whereas the notion of distributive injustice applies to a group of people who assess the fairness of their individual share by comparing it with the share obtained by significant others.

Measurement

As the dimensions depicted in Figure 1 reflect, the working people's lived experiences they can only partially be assessed by external observation. Rather, self-reported information based on psychometrically validated scales has evolved as the leading methodological approach in this field of inquiry. To be scientifically valid, respective questionnaires or interviews must meet established quality criteria, such as internal consistency of scales, sensitivity to change over time, and content or criterion validity [i.e., the ability to predict or explain an expected outcome (see below)]. Moreover, there should be a satisfactory fit between the structure of the scales reflecting the theoretical model and the observed empirical data. Several additional procedures are available to further improve the validity of subjective data. They include the control of reporting bias by adjusting for distinct personality characteristics, the combination of subjective data with externally assessed information, the construction of aggregated measures (e.g., at the level of work units), or the application of ecological momentary assessment where data is recorded in real time within critical environments. Extensive information on the scales measuring effort-reward imbalance at work is beyond the scope of this chapter and can be found elsewhere[12] (see www.uniklinik-duesseldorf.de/medsoziologie). As the questionnaire is available in a number of languages cross-country comparisons can be conducted on the basis of respective research.

SELECTED EMPIRICAL EVIDENCE

There are three study designs that allow a test of causal associations of models of a stressful psychosocial work environment with health. The first one is the prospective cohort study, considered a gold standard in this area of research. In a cohort study, work stress is assessed as exposure at study entry among an otherwise healthy working population. This cohort is followed over time, where the occurrence of a defined health outcome is monitored and analyzed in relation to the exposure. The respective hypothesis states that exposed people suffer from a significantly higher relative risk of disease onset as compared to nonexposed people, and that this holds true after adjustment for relevant confounders. Moreover, a dose-response relation should be observed where higher levels of exposure result in higher relative risks of disease. The second design concerns the assessment of psychobiological or behavioral pathways that are assumed to link the exposure with the disease outcome, either by classical stress experiments or by "naturalistic" studies of working people's physiological activity in everyday life (e.g., ambulatory blood pressure assessment). As a third approach, intervention studies are used to demonstrate that a reduction of exposure is followed by a subsequent reduction of disease risk. For instance, a large company may manage to reduce overall levels of work stress among employees by implementing health-conducive organizational changes, thus diminishing demands and increasing options of reward and control at work. According to the hypothesis, subsequent health levels of employees should be significantly better compared to those of employees in a similar company where work stress levels remained unchanged. Obviously, a strict observation of all criteria of scientific accuracy may be difficult in the context of such complex interventions.

All three designs were applied with regard to the model of effort-reward imbalance at work. Several prospective epidemiological investigations documented elevated relative risks of depression[15,16] and of coronary heart disease[11,17] among employees exposed to high effort/low reward conditions at work. In addition, elevated relative risks were found for hypertension,[18] disability pension due to depression,[19] type 2 diabetes,[20] and alcohol dependence.[21] In most studies, the range of increased relative risks or odds ratios varied from 50% to 100%. Some findings were restricted to men, and although the results hold true for a wide variety of occupational and professional groups, workers in low socioeconomic positions are generally more vulnerable to these health-adverse effects of stressful work.[15,22] Importantly, the findings are not restricted to modern western societies, but hold equally true for rapidly developing countries (e.g., China[23]).

Experimental and naturalistic investigations were conducted in order to explore pathways linking effort-reward imbalance at work with stress-related disorders. These studies used cardiovascular reactivity,[24,25] hormonal release,[26] or markers of the immune system[27] as psychobiological indicators. In general, study participants suffering from work-related reward frustration demonstrated marked deviations in their psychobiological responses from those of participants who were free from this type of stress. Fewer reports based on the third study design, intervention studies, are available so far, but their findings point to favorable outcomes among employees of work settings undergoing health-promoting reorganization.[28,29]

In summary, despite some inconsistencies and methodological challenges, a robust body of scientific evidence is now available demonstrating associations of an adverse psychosocial work environment, as assessed by the effort-reward imbalance model, with a range of stress-related disorders.

IMPLICATIONS FOR PREVENTION AND FUTURE RESEARCH DIRECTIONS

Scientific advances, as the ones reported here, may strengthen efforts to improve health-conducive working conditions. Yet, the gap between available scientific knowledge and the practical use of this information for investments into health-promoting working conditions remains a serious concern, specifically in view of the negative sides of economic globalization and neoliberal policies.[30,31] At two levels, practical implications of available scientific evidence for primary and secondary prevention are particularly convincing.

The first level concerns the prevention or reduction of stress in companies and organizations. In addition to efforts toward increasing health-promoting behavior, interventions are required that tackle the organizational and personnel development. Measures include leadership training, strengthening social capital, redistribution of workload and work time control, and balancing efforts at work with equitable material and nonmaterial rewards. Special emphasis should be put on secondary prevention, resulting in high return-to-work rates among people with disability, ill health, or unemployment. The second level relates to the development of national social and labor policies as well as distinct international regulations. One such policy concerns the implementation of regulations dealing with monitoring and surveillance of work-related stress. In addition, investments into a well-trained workforce of occupational safety and health professionals and efficient information infrastructure are needed. A further policy is directed toward providing appropriate measures of social protection and active labor market programs. Countries with a strong development of integrative employment policies, including continued education, benefit from significantly improved quality of work which in turn reduces the burden of work-related disease.[32] To conclude, models of good practice and evidence of favorable cost-benefit relations are available from several pioneering countries and organizations.[33]

Whereas evidence derived from research on effort-reward imbalance at work can be useful for preventive purposes, its further scientific development defines an additional challenge. Broadening the scope beyond organizational-level aspects by including macrosocial indicators and embedding the model in a life course perspective are two such developments.[32] In theoretical terms, an extension of its basic notion of failed reciprocity in costly social transactions beyond paid work is of particular interest. For instance, recent research applied the model to house and family work[34] or schoolwork,[35] testifying the far-reaching importance of fair procedures of social exchange for human health and well-being. In conclusion, recent research based on the model of effort-reward imbalance has provided innovative knowledge that can enrich both science and policy.

References

1. Siegrist J. Adverse health effects of high effort—low reward conditions at work. *J Occup Health Psychol.* 1996;1:27–43.
2. Tricomi E, Rangel A, Camerer CF, O'Doherty JP. Neural evidence for inequality-averse social preferences. *Nature.* 2010;463 (7284):1089–1091. http://dx.doi.org/10.1038/nature08785.
3. Gouldner AW. The norm of reciprocity: a preliminary statement. *Am Sociol Rev.* 1960;25(2):161–178. http://dx.doi.org/10.2307/2092623.
4. Chrousos GP. Stress and disorders of the stress system. *Nat Rev Endocrinol.* 2009;5(7):374–381. http://dx.doi.org/10.1038/nrendo.2009.106.
5. McEwen B. Protective and damaging effects of stress mediators. *N Engl J Med.* 1998;338:171–179.
6. Schultz W. Behavioral theories and the neurophysiology of reward. *Annu Rev Psychol.* 2006;57:87–115.
7. Singer T, Seymour B, O'Doherty J, Kaube H, Dolan R, Frith CD. Empathy for pain involves the affective but not sensory components of pain. *Science.* 2004;303:1157–1162.
8. Baumgartner T, Fischbacher U, Feierabend A, Lutz K, Fehr E. The neural circuitry of a broken promise. *Neuron.* 2009;64:756–770.
9. Hernandez Lallement J, Kuss K, Trautner P, Weber B, Falk A, Fliessbach K. Effort increases sensitivity to reward and loss magnitude in the human brain. *Soc Cogn Affect Neurosci.* 2014;9(3):342–349. http://dx.doi.org/10.1093/scan/nss147.
10. Karasek R, Theorell T. *Healthy Work: Stress, Productivity, and the Reconstruction of Working Life.* New York, NY: Basic Books; 1990.
11. Steptoe A, Kivimaki M. Stress and cardiovascular disease. *Nat Rev Cardiol.* 2012;9(6):360–370. http://dx.doi.org/10.1038/nrcardio.2012.45.
12. Siegrist J. Job control and reward: effects on well-being. In: Cartwright S, Cooper CL, eds. *The Oxford Handbook of Organizational Well-Being.* Oxford u.a: Oxford University Press; 2009:109–132. http://www.loc.gov/catdir/enhancements/fy0907/2008024470-b.html.
13. Greenberg J. Organizational injustice as an occupational health risk. *Acad Manag Ann.* 2010;4:205–243.
14. Adams JS. Inequity in social exchange. In: Berkowitz L, ed. *Advances in Experimental Social Psychology;* vol. 2. New York: Academic Press; 1965:267–299.
15. Rugulies R, Aust B, Madsen IEH, Burr H, Siegrist J, Bultmann U. Adverse psychosocial working conditions and risk of severe depressive symptoms. Do effects differ by occupational grade? *Eur J Public Health.* 2013;23:415–420. http://dx.doi.org/10.1093/eurpub/cks071.
16. Siegrist J, Lunau T, Wahrendorf M, Dragano N. Depressive symptoms and psychosocial stress at work among older employees in three continents. *Glob Health.* 2012;8(1):27. http://dx.doi.org/10.1186/1744-8603-8-27.
17. Backé E, Seidler A, Latza U, Rossnagel K, Schumann B. The role of psychosocial stress at work for the development of cardiovascular diseases: a systematic review. *Int Arch Occup Environ Health.* 2012;85(1):67–79. http://dx.doi.org/10.1007/s00420-011-0643-6.

18. Gilbert-Ouimet M, Trudel X, Brisson C, Milot A, Vézina M. Adverse effects of psychosocial work factors on blood pressure: systematic review of studies on demand-control-support and effort-reward imbalance models. *Scand J Work Environ Health*. 2014;40(2):109–132. http://dx.doi.org/10.5271/sjweh.3390.
19. Juvani A, Oksanen T, Salo P, et al. Effort-reward imbalance as a risk factor for disability pension: the Finnish Public Sector Study. *Scand J Work Environ Health*. 2014;40(3):266–277. http://dx.doi.org/10.5271/sjweh.3402.
20. Kumari MHJ, Marmot M. Prospective study of social and other risk factors for incidence of type II diabetes in Whitehall 2 study. *Ann Intern Med*. 2004;164:1873–1880.
21. Head J, Stansfeld S, Siegrist J. Psychosocial work environment and alcohol dependence. *Occup Environ Med*. 2004;61:219–224.
22. Hoven H, Siegrist J. Work characteristics, socioeconomic position and health: a systematic review of mediation and moderation effects in prospective studies. *Occup Environ Med*. 2013;70(9):663–669. http://dx.doi.org/10.1136/oemed-2012-101331.
23. Xu W, Hang J, Cao T, et al. Job stress and carotid intima-media thickness in Chinese workers. *J Occup Health*. 2010;52(5):257–262.
24. Falk A, Menrath I, Verde P, Siegrist J. *Cardiovascular consequences of unfair pay*. IZA discussion paper no. 5720, 2011.
25. Jarczok MN, Jarczok M, Mauss D, et al. Autonomic nervous system activity and workplace stressors—a systematic review. *Neurosci Biobehav Rev*. 2013;37:1810–1823.
26. Steptoe A, Siegrist J, Kirschbaum C, Marmot M. Effort-reward imbalance, overcommitment, and measures of cortisol and blood pressure over the working day. *Psychosom Med*. 2004;66:1899–1903.
27. Nakata A, Takahashi M, Irie M. Effort-reward imbalance, overcommitment, and cellular immune measures among white-collar employees. *Biol Psychol*. 2011;88(2-3):270–279. http://dx.doi.org/10.1016/j.biopsycho.2011.08.012.
28. Bourbonnais R, Brisson C, Vézina M. Long-term effects of an intervention on psychosocial work factors among healthcare professionals in a hospital setting. *Occup Environ Med*. 2011;68(7):479–486. http://www.scopus.com/inward/record.url?eid=2-s2.0-79958273161&partnerID=40&md5=e325639261a47f41819fd8dd06a752e9.
29. Limm H, Gündel H, Heinmüller M, et al. Stress management interventions in the workplace improve stress reactivity: a randomised controlled trial. *Occup Environ Med*. 2011;68(2):126–133. http://dx.doi.org/10.1136/oem.2009.054148.
30. Schnall PL, Dobson M, Rosskam E. *Unhealthy Work: Causes, Consequences, Cures*. Amityville, NY: Baywood Publishing; 2009.
31. Benach J, Muntaner C, Santana V. *Employment conditions and health inequalities*. Final report to the WHO Commission on Social Determinants of Health (CSDH), Geneva, 2007. www.who.int. Accessed 04.03.14.
32. Wahrendorf M, Siegrist J. Proximal and distal determinants of stressful work: framework and analysis of retrospective European data. *BMC Public Health*. 2014;14:849.
33. WHO. *Review of social determinants and the health divide in the WHO European Region*. Final report, Copenhagen: WHO; 2014. *www.euro.who.int*.
34. Sperlich S, Peter R, Geyer S. Applying the effort-reward imbalance model to household and family work: a population-based study of German mothers. *BMC Public Health*. 2012;12:12. http://dx.doi.org/10.1186/1471-2458-12-12.
35. Li J, Shang L, Wang T, Siegrist J. Measuring effort-reward imbalance in school settings: a novel approach and its association with self-rated health. *J Epidemiol*. 2010;20:111–118.

CHAPTER

10

Environmental Factors

W.R. Avison[1,2]

[1]The University of Western Ontario, London, ON, Canada

[2]Children's Health Research Institute, Lawson Health Research Institute, London, ON, Canada

OUTLINE

Abstract

Environmental factors are social or economic conditions that affect exposure to stressors. The stress process model has been used widely as a conceptual framework for understanding how environmental factors influence exposure to a wide array of stressors. Three major environmental determinants of stress are social status, social roles, and the ambient environment. Research on the stress process has extensively studied the relative importance of differential exposure and differential vulnerability to stress as explanations of differences in mental health outcomes by various statuses and roles.

HISTORICAL BACKGROUND

There is a vast body of research literature on the impact of social and economic factors on physical and mental health problems. In general, these studies have demonstrated that individuals' social locations in society, in terms of their social statuses and social roles, are important determinants of both the kinds of stressors to which they are exposed and the levels of exposure that they experience. Indeed, one of the most important contributions of the sociology of mental health has been to demonstrate conclusively that stressors are not experienced randomly by individuals but, rather, that there is a social distribution of stressors. This perspective has emerged out of a number of theoretical and empirical developments.

Much of the work on the social distribution of stressors has its intellectual origins in the work of Bruce and Barbara Dohrenwend[1–3] and of Leonard Pearlin[4–6] in the latter part of the twentieth century. They observed that exposure to stressful life events was positively correlated with symptoms of mental illness. They argued that the higher rates of psychological impairment found among members of socially disadvantaged groups might be explained by their greater exposure to life events. The underlying assumption of this explanatory model was that social causation processes were operating: differences in social statuses and social roles produce variations in the experience of life events that result in different rates of mental health problems among these social groups.

Stress: Concepts, Cognition, Emotion, and Behavior
http://dx.doi.org/10.1016/B978-0-12-800951-2.00010-8

KEY POINTS

- *Environmental factors* are social or economic circumstances that affect exposure to stressors. They can be classified as social statuses, social roles, or ambient social conditions.
- *The stress process paradigm* is a model that describes how environmental factors expose individuals to stressors that manifest themselves in distress or disorder. It postulates the existence of three factors that either mediate or moderate the effects of stressors on distress or disorder: psychosocial resources, coping resources and strategies, and social supports. The model also specifies how environmental factors influence all other variables in the paradigm.
- *The differential exposure hypothesis* contends that specific statuses or roles are associated with greater or lesser levels of stressors that arise out of the conditions of life and which, in turn, affect levels of psychological distress.
- By contrast, the *differential vulnerability hypothesis* argues that elevated levels of distress among individuals in disadvantaged statuses reflect their greater responsiveness to stressors. That is, among two social groups exposed to the same levels of stressors, distress will be higher among those people in the disadvantaged statuses.

In subsequent decades, social scientists exhaustively studied the relationships among environmental factors, stressful life events, and psychological distress and disorder. At least three major themes emerged from this work. First, a conceptual paradigm, the stress process, was developed to examine the interplay among environmental factors, stressors, psychosocial mediators and moderators, and mental health outcomes. Second, it became clear that stressful life events constituted only one kind of stressor and that research needed to identify and study other types. Third, there was general agreement with the notion that the distribution of stressors is socially patterned; that is, it is importantly influenced by environmental factors such as social statuses and social roles.

The Stress Process

Researchers working independently in the 1980s[4,5,7] began to develop models that enable us to better understand how socially induced stressors manifest themselves in psychological distress, symptoms of psychiatric disorders, or in other dysfunctions or health problems. Perhaps the most fully articulated model is *the stress process*, developed over three decades ago by Leonard Pearlin and his colleagues[4] and elaborated by Pearlin[8–10] in a series of conceptual updates and elaborations.

According to Pearlin, the foundation of the stress process rests on three key assumptions. First, the process is a dense causal web that involves dynamic interconnections among the components of the model. Changes in one set of factors produce changes in others. Second, consistent with Selye's[11] classic statement about stress as a "normal" experience, social stress is a typical experience of ordinary life; it is not unusual or abnormal. Third, the origins of stress are essentially social in nature: stressors emerge out of individuals' experiences in the social environment. This directs sociological stress researchers toward more proximal than distal environmental sources of social stress and to a greater emphasis on social context than on history or biology.

Pearlin's model of the stress process contains four major components. *Sources of stress* include stressful life events as well as other dimensions such as role-related strains, daily hassles, and life traumas. These sources of stress are potentially interactive in their effects on health outcomes; that is, life events can intensify or exacerbate existing strains and vice versa. In addition, stressful life experiences can create new role strains and, conversely, role strains can generate new stressful life events.

Manifestations of stress include an array of possible health outcomes. Sociologists have tended to focus on symptoms of mental illness or measures of psychological distress. These measures typically include symptoms of depressive illness; however, some sociologists have examined the effects of stressors on diagnosable disorders[12–14] and others have extended the study of stress to include alcohol consumption and drug dependence,[15,16] and physical illness.[17]

The third component of the stress process model, *mediators of stress*, refers to a broad range of factors including social support, psychosocial resources such as mastery, self-esteem, mattering, interpersonal dependency, and coping strategies. These mediators are hypothesized to function as pathways that connect exposure to stress to its manifestations. So, for example, individuals exposed to stress experience the erosion of their sense of control over their lives. In turn, this decline in mastery manifests itself in symptoms of distress or depression.

These mediating factors may also operate as *moderators* of the stress-distress relationship. The classic examples are those in which the impact of stress on mental health outcomes is reduced in the presence of higher levels of psychosocial resources, social support, or coping resources. These moderating effects, often referred to as stress-buffering effects, typically reflect processes in which the consequences of stress are mitigated by individuals' abilities to cope or otherwise deal with stress.[18–21]

The interplay among sources of stress, mediators and moderators of stress, and outcomes all occurs in a social context that is defined by the social and economic statuses and roles that individuals occupy.

The Stress Universe

Throughout the 1970s and early 1980s, most of the research on environmental sources of stressors focused on life changing events. The most widely used measures of life stress were events checklists that usually contained from 30 to 100 events that represented potential changes in peoples' lives. These included events such as the death of a loved one, marital separation or divorce, job loss, and geographical moves. Over time, however, two developments substantially altered the study of environmental stressors. First, it became clear to researchers studying the stress process that the associations between eventful stressors and distress were modest at best. For many researchers, this suggested that stressors were not being measured comprehensively. Accordingly, they expanded life events inventories to include events that were not life changes, but which included other difficulties such as family conflicts, work conflicts, and financial difficulties. In part, the inclusion of these "events" reflected the realization among many researchers that stress often arises in the context of individuals' work and family roles. Second, investigations of intraevent variability in stressful life events revealed that life events inventories included ongoing or chronic stressors (such as family conflict, marital difficulties, and financial problems) as well as eventful stressors.

These considerations led researchers to develop more extensive strategies to measure a broader array of life stressors than had previously been the case. Perhaps the most comprehensive consideration of the many dimensions of life stress has been presented by Wheaton.[22] He distinguishes among various dimensions of stressors, including chronic stress, daily hassles, macro events, and traumatic events. In doing so, he presents a scheme for arraying these different stressors on a continuum ranging from sudden traumatic experiences and life change events at the discrete end to chronic stressors that are more continuous in nature. Wheaton demonstrates two important properties of these various dimensions of the stress universe. First, they are relatively independent of each other and therefore are unlikely to be empirically confounded with one another. Second, they each have significant, independent effects on various measures of physical and mental health outcomes. These findings suggest that any comprehensive attempt to estimate the impact of life stressors on health outcomes requires that one measure an array of stressors that includes both discrete life experiences and more chronic or ongoing stressors.

The Social Distribution of Stressors

Early research on life events clearly established that socially induced stressors exerted important influences on various measures of mental health outcomes. What seemed particularly interesting was that those socially disadvantaged groups with the highest levels of distress and disorder also appeared to be exposed to the most stressful life events. This led to the hypothesis that social differences in distress and disorder might be accounted for by variations in either exposure or vulnerability to social stressors.

Subsequently, in a seminal article on the sociological study of stress, Pearlin[6] argued for the need to systematically investigate the ways in which social structure affects individuals' exposure to stressors. Essentially, Pearlin asserted that the roles and statuses that people occupy in their everyday lives have important consequences for the kinds of stressors to which they are exposed and the frequency with which this occurs. In short, he argued that accounting for the impact of environmental factors is crucial for understanding the ways in which individuals experience social stressors.

This point has been made even more salient by Turner et al.[14] in their paper on the epidemiology of social stress. Their findings from a large community study reveal that younger people are more exposed to a variety of stressors than are older respondents. They also find this pattern among women compared to men, unmarried compared to married people, and individuals in lower compared to higher socioeconomic status (SES) positions in society. In their view, this clearly indicates that stressors are not experienced randomly in the population. Quite the contrary, they assert that their results reveal that there is a social distribution of stressors that is characterized by greater exposure among members of disadvantaged social groups.

THE IMPACT OF ENVIRONMENTAL FACTORS ON STRESS

In thinking about the social distribution of stressors, social scientists have focused primarily on three major environmental determinants: social statuses, social roles, and the ambient environment. Most formulations of the stress process model take the view that individuals' locations in the structure of society place them at greater or lesser risk of encountering stressors. These locations are defined by the various statuses that individuals hold and by the various social roles they occupy. Additionally, there are more ambient characteristics of the social environment that are not specific to statuses or roles, but which may generate stressful experiences for individuals.

Social Statuses

In the literature, individuals' positions in the structure of society are often defined by six major status characteristics: age, gender, marital status, race/ethnicity, employment status, and SES. Each of these environmental factors is associated with differential exposure to stressful experiences.

Age

There is widespread agreement among stress researchers that exposure to stressful experiences declines with age. Schieman et al.[23] have conducted a systematic assessment of this issue. Whether the stressors in question are life events or chronic role strains, younger people report significantly more stress than do the elderly. Consistent with these patterns, Mirowsky and Ross[24] note that economic hardship tends to decline with age as does marital conflict among marriages that stay intact. Nevertheless, it remains inconclusive whether the impact of age on exposure to stressors is a function of maturation processes, birth cohort or generational effects, or life-cycle processes.

Gender

Differences between men and women in exposure to stressors have generated inconsistent findings. Some research concludes that women experience more stressful life events than men, while other research finds no differences. Turner and Avison[25] find that women report more stressful life events than men, but they experience fewer lifetime traumatic events and less discrimination stress. Women and men do not differ in their exposure to chronic stressors. When these various dimensions of stress are considered cumulatively, they find that men experience greater exposure compared to women, but this difference is relatively small. Turner and Avison conclude that these small gender variations in exposure to stressors are unlikely to account for gender differences in psychological distress. A similar conclusion has been proposed by Meyer et al.[26]

It seems clear that many of these observed gender differences in exposure to stressors are a function of the gendered roles that women and men play in the workplace and in the household division of labor. Indeed, one of the lessons of stress research in the sociology of mental health is that gender conditions the effects of a variety of roles and statuses on the exposure to stress.

Marital Status

One of the more robust findings in the study of social stress is the observation that married individuals experience considerably fewer stressors than either never married or formerly married individuals.[27] This pattern can be observed for chronic strains as well as for stressful life events. Unlike age and gender, however, the causal direction of the relationship between marital status and stressful experience is open to competing interpretations. A social selection interpretation suggests that people with high levels of stress in their lives are less likely to marry or, if they do so, they are more likely to separate and divorce. A social causation interpretation suggests that persons who have never married and those who have separated or divorced are more likely to experience an array of life events and chronic strains than are the married.

Some longitudinal research indicates that divorce leads to elevated levels of depression and that this change is accounted for by a decline in living standard, economic difficulties, and a reduction in social support.[28] Even if the end of a marriage provides some escape from a stressful situation, divorce is accompanied by life stress that has depressive consequences. Other studies also report significant increases in psychological distress among the maritally disrupted over and above their predivorce levels, but find virtually no evidence that changes in financial stressors, changes in role demands, or changes in geographic location mediated the relationship between marital disruption and distress. These are reflective of longitudinal studies on marital disruption and remarriage insofar as they generate few consistent findings other than the observations that marital disruption is associated with elevated psychological distress and that remarriage results in only a partial reduction in this elevation.

Race and Ethnicity

Studies of racial and ethnic variations exposure to stressors are relatively recent in the literature. There is general agreement that experiences of discrimination constitute an important set of stressors for African Americans, Hispanic Americans, and American Indians.[29] In addition, stressors arising out of the social disadvantages of these groups have also been observed. Among Asian Americans, the experience of discrimination and the stress of migration have been noted as important threats to their mental health.

Turner and Avison[25] have systematically examined differences between African Americans and whites in the United States in exposure to stressors. Across five different dimensions (recent life events, chronic stressors, total lifetime major events, lifetime major discrimination, and daily discrimination), African Americans experience significantly more stress than do whites. Indeed, the cumulative difference between African Americans and whites in exposure to stress is more than one-half standard deviation.

It is interesting to note, however, that these elevated levels of stressors among racial and ethnic minority members do not necessarily translate into higher rates of distress or disorder for all groups. In their review of this issue, Williams and Harris-Reid[30] conclude that African Americans have lower prevalence rates of psychiatric disorders than do whites. The data are inconsistent when comparing whites with Hispanic Americans or Asian Americans. What little data exists about the epidemiology of mental health among American Indians suggests that their rates of disorder, especially depression and alcohol abuse, are elevated.

It is difficult to interpret the meaning of these findings. Lower rates of disorder in the face of elevated levels of distress suggests that some racial and ethnic groups may be less vulnerable than others to these stressors or that countervailing effects of psychosocial resources may reduce the impact of stressors on their mental health. Turner and Avison[25] have suggested that African Americans may exhibit a response tendency in which they underreport infrequent or mild experiences of distress, thus leading to an underestimation of their levels of distress. They demonstrate that once this tendency is taken into account, there is clear evidence that African Americans' elevated exposure to stress manifests itself in significantly higher distress.

Employment Status

Research leaves little doubt that the unemployed experience more negative mental health outcomes than do the employed. The evidence of this correlation is most clear for outcomes such as symptoms of depression and anxiety and measures of psychological distress. Longitudinal studies support the conclusion that job loss results in higher levels of mental health symptoms.

Studies of the factors that intervene between individuals' job losses and their health problems have identified at least two major sources of environmental stressors that mediate this relationship. Some researchers have shown that job loss and unemployment create financial strains that lead to mental health problems.[31] Others have examined the mediating role of marital and family conflict.[12] These studies report that unemployment leads to increasing conflicts between the unemployed worker and other family members. Some researchers have suggested that the elevated levels of distress observed among women whose husbands are experiencing job-related stress may be consistent with the "costs of caring" hypothesis described earlier.

Socioeconomic Status

Although there is general agreement that SES and mental illness are inversely correlated and that exposure to stressors are a major determinant of mental health problems, there has been surprisingly little consensus among researchers about the SES-stress relationship. Whether SES is measured by some combination of education and income or by occupational prestige among those with jobs, contradictory results emerge. These inconsistent findings have led some researchers to argue that it is not stress exposure that produces higher rates of mental illness among individuals from lower SES circumstances. Rather, they suggest that the impact of stressors on mental illness is more substantial among low SES than high SES individuals. In other words, they argue that lower class individuals are differentially vulnerable to stressors. This issue of differences in exposure and vulnerability is addressed later; nevertheless, it seems clear that this debate with respect to the influence of SES has not been resolved. Nevertheless, more recent studies suggest that when stressors are comprehensively measured, there is a significant social class gradient in exposure to stress.[32]

Some of the most compelling evidence of the effects of economic hardship on marital relationships has been presented in studies of families during the Great Depression of the 1930s.[33] This work clearly indicates that economic difficulties increase marital tensions in most families and especially in those that were most vulnerable prior to the economic hardship. Similar findings have been reported in studies of families whose lives have been affected by the farm crises of the 1980s.

Social Roles

A central focus of much research on the stress process has been on the ways in which the social roles that individuals occupy expose them to stressors. Pearlin[34] has described these role strains in rich detail. For Pearlin, several types of stress may arise from role occupancy: (a) excessive demands of certain roles; (b) inequities in rewards; (c) the failure of reciprocity in roles; (d) role conflict; (e) role captivity; and (f) role restructuring. These various types of stress are important sources of stress that may manifest themselves in symptoms of distress or disorder. These kinds of environmental stressors have been studied most intensively in investigations of family roles and work roles.

Family Roles

Recent research on the family and mental health has focused on family structure in terms of the intersection of marital status and parenthood.[28] In this context, there has been intense interest in the impact of single parenthood on symptoms of distress. Results consistently show higher levels of distress among single compared to married mothers. These studies of single parenthood clearly reveal that family structure and the roles imbedded in that structure are important determinants of women's mental health.

Research indicates that separation and divorce trigger chronic stressors such as income reduction and housing relocation. In addition, the divorced experienced more life events than the married, particularly negative events involving loss. When children are involved, there may be additional strains associated with separation or divorce. The custodial parent, usually the mother, assumes many household, financial, and emotional responsibilities previously shared by two parents.

The Work Role

In addition to examining differences in employment or work status influence exposure to stressors and subsequent mental health outcomes, social scientists have become aware of the importance of understanding how experiences in the work role are related to stress and health. Despite the observation that being employed generally has positive psychosocial consequences for individuals, not all employment circumstances are the same. Indeed, there are important variations in the stressors associated with any particular work situation. Work exposes individuals to various kinds of stressful experiences and provides different kinds of rewards for different people—financial rewards, self-esteem, a sense of control over one's life, etc. It seems, then, that the net effect of paid employment for any individual will depend on the balance of these costs and benefits.

A number of recent contributions to the study of work and mental health have provided some useful models for this kind of research. In her comprehensive review of the literature on the interplay between work and family, Menaghan[35] has demonstrated clearly that work-related stressors are importantly associated with the mental health of family members. Other researchers have convincingly documented how role overload and the sense of personal power have important implications for the effects of women's employment on their mental health. Some investigations have shown how job characteristics such as full-time versus part-time work, and substantive complexity have significant effects on mental health.

The Intersection of Work and Family Roles

Research on work and family stress among women suggests a number of ways in which work and family roles interact in their effects upon psychological distress and depression. These studies highlight the importance of considering both family stressors to which women are exposed and work-related stress.

Relatively few studies have examined how single parenthood and paid employment interact in their impact on mental health problems. Those studies that have investigated this issue conclude that differences between single-parent families and two-parent families in role obligations and opportunities may have significant effects on the balance between work and family responsibilities.[36] For married mothers, paid employment may be more easily integrated into daily family life. The presence of a spouse provides the opportunity to share some of the child care responsibilities. Alternatively, dual income families have more funds for outside child care or paid assistance in the home.

Employment for single-parent mothers may represent more of a pressing responsibility than an opportunity for achievement and development. While income from employment may address a number of financial burdens of single-parent mothers, it nevertheless is the case that most single-parent families live in poverty or near poverty. In such circumstances, paid employment may alleviate some of the most pressing financial strains but other strains persist. Moreover, when single-parent mothers obtain employment, it is common for them to also bear continued sole responsibility for the care and nurturance of their children. Such dual demands may generate a cost to employment—role overload.

Under these circumstances, it seems probable that fewer psychosocial rewards associated with employment (greater self-esteem, self-efficacy, or social support) will accrue to single-parent mothers. This is all the more likely to be the case because single parents may be constrained to select jobs that are not their first choice but which are near their homes, have hours of work that fit with their children's schedules or with the availability of child care.

Ambient Strains

Not all stressors are associated with statuses and roles. Ambient strains are not attributable to a specific role but, rather, are diffuse in nature and have a variety of sources. These include experiences such as living through an economic recession, living in unsafe housing, or living in a dangerous neighborhood.

Some studies have demonstrated how ambient strains associated with the neighborhoods in which adolescents live magnify the impact of other stressors on their mental health.[37] Others make the same point with respect to the ambient effect of economic environments.[32]

DIFFERENTIAL EXPOSURE AND VULNERABILITY TO ENVIRONMENTAL STRESSORS

An important issue has been to test the relative importance of differential exposure and differential vulnerability to stressors as explanations of differences in psychological distress by social status or social role.[38,39] The differential exposure hypothesis contends that specific statuses or roles are associated with greater or lesser levels of stressors that arise out of the conditions of life and which, in turn, affect levels of psychological distress. For example, applying this argument to marital status, the hypothesis is that the transition from marriage to separation or divorce brings with it significant increases in exposure to financial strains, role overload, and other types of stressors. This increase in the burden of stress experienced by the separated or divorced translates into elevated levels of mental health problems. Conversely, individuals who remain married experience far fewer stressful circumstances and, accordingly, have lower levels of psychological distress.

The competing explanation, the differential vulnerability hypothesis, argues that elevated levels of distress among individuals in certain statuses reflect their greater responsiveness to stressors. Such increased responsiveness or vulnerability to stressors has been attributed to a number of different sources. These include dimensions of personal and social competence such as self-efficacy that contribute to individuals' greater or lesser resilience in the face of stress. They also involve access to coping resources that moderate stressful experience.

The most recent work on this debate has been conducted by Turner and colleagues.[14,25] Their investigation of age, gender, marital status, and SES reveals little evidence that any of the differences by these social statuses in distress or disorder can be attributed to differential vulnerability. Instead, they find that observed variations in mental health outcomes are due largely to different levels of exposure to stressors. They and others argue that the use of a more comprehensive measure of stressors that includes live events and chronic strains is likely to better estimate actual exposure and to avoid the error of attributing unmeasured differential exposure to differential vulnerability.

References

1. Dohrenwend BS. Social status and stressful life events. *J Pers Soc Psychol*. 1973;28:225–235.
2. Dohrenwend BP, Dohrenwend BS. Social and cultural influences on psychopathology. *Annu Rev Psychol*. 1974;25:417–452.
3. Dohrenwend BP, Dohrenwend BS. Socioenvironmental factors, stress, and psychopathology. Part I: quasi-experimental evidence on the social causation-social selection issue posed by class differences. *Am J Community Psychol*. 1981;9:129–146.
4. Pearlin LI, Lieberman MA, Menaghan EG, Mullan JT. The stress process. *J Health Soc Behav*. 1981;22:337–356.
5. Pearlin LI, Schooler C. The structure of coping. *J Health Soc Behav*. 1978;19:2–21.
6. Pearlin LI. The sociological study of stress. *J Health Soc Behav*. 1989;30:241–256.
7. Lazarus RS. *Psychological Stress and the Coping Process*. New York, NY: McGraw-Hill; 1966.
8. Pearlin LI, Skaff MM. Stress and the life course: a paradigmatic alliance. *Gerontologist*. 1996;36:239–247.
9. Pearlin LI. The stress concept revisited: reflections on concepts and their interrelationships. In: Aneshensel CS, Phelan JC, eds. *Handbook of the Sociology of Mental Health*. New York, NY: Plenum; 1999:395–415.
10. Pearlin LI, Bierman A. Current issues and future directions in research into the stress process. In: Aneshensel CS, Phelan JC, Bierman A, eds. *Handbook of the Sociology of Mental Health*. 2nd ed. New York, NY: Springer; 2013:325–340.
11. Selye H. *The Stress of Life*. New York, NY: McGraw-Hill; 1956.
12. Avison WR. Unemployment and its consequences for mental health. In: Marshall VW, Heinz W, Krueger H, Verma A, eds. *Restructuring Work and the Life Course*. Toronto: University of Toronto Press; 2001:177–200.
13. Brown GW, Harris T. *Social Origins of Depression*. New York, NY: Free Press; 1978.
14. Turner RJ, Wheaton B, Lloyd DA. The epidemiology of social stress. *Am Sociol Rev*. 1995;60:104–125.
15. Aneshensel CS, Rutter CM, Lachenbruch PA. Competing conceptual and analytic models: social structure, stress, and mental health. *Am Sociol Rev*. 1991;56:166–178.
16. Turner RJ, Lloyd DA, Taylor J. Physical disability and mental health: an epidemiology of psychiatric and substance disorders. *J Rehabil Psychol*. 2006;51:214–223.
17. Brown GW, Harris T. *Life Events and Illness*. New York, NY: Guilford Press; 1989.
18. Avison WR, Cairney J. Social structure, stress, and personal control. In: Zarit SH, Pearlin LI, Schaie KW, eds. *Personal Control in Social and Life Contexts*. New York, NY: Springer; 2003:127–164.
19. Thoits PA. Stress and health: major findings and policy implications. *J Health Soc Behav*. 2010;51:S41–S53.
20. Turner JB, Turner RJ. Social relations, social integration, and social support. In: Aneshensel CS, Phelan JC, Bierman A, eds. *Handbook of the Sociology of Mental Health*. 2nd ed. New York, NY: Springer; 2013:341–356.
21. Wheaton B. Models for the stress-buffering functions of coping resources. *J Health Soc Behav*. 1985;26:352–365.
22. Wheaton B. Sampling the stress universe. In: Avison WR, Gotlib IH, eds. *Stress and Mental Health: Contemporary Issues and Prospects for the Future*. New York, NY: Plenum; 1994:77–114.
23. Schieman S, Van Gundy K, Taylor J. Status, role, and resource explanations for age patterns in psychological distress. *J Health Soc Behav*. 2001;42:80–96.
24. Mirowsky J, Ross CE. Age and depression. *J Health Soc Behav*. 1992;33:187–205.
25. Turner RJ, Avison WR. Status variations in stress exposure: implications for the interpretation of research on race, socioeconomic status, and gender. *J Health Soc Behav*. 2003;44:488–505.
26. Meyer IH, Schwartz S, Frost DM. Social patterning of stress and coping: does disadvantaged social status confer more stress and fewer coping resources? *Soc Sci Med*. 2008;67:368–379.
27. Amato PA. The consequences of divorce for adults and children. *J Marriage Fam*. 2000;62:1269–1287.
28. Umberson D, Thomeer MB, Williams K. Family status and mental health: recent advances and future directions. In: Aneshensel CS, Phelan JC, Bierman A, eds. *Handbook of the Sociology of Mental Health*. 2nd ed. New York, NY: Springer; 2013:405–432.
29. Brown TN, Donato KM, Laske MT, Duncan EM. Race, nativity, ethnicity, and cultural influences in the sociology of mental health. In: Aneshensel CS, Phelan JC, Bierman A, eds. *Handbook of the Sociology of Mental Health*. 2nd ed. New York, NY: Springer; 2013:255–276.
30. Williams DR, Harris-Reid M. Race and mental health: emerging patterns and promising approaches. In: Horwitz AV, Scheid TL, eds. *A Handbook for the Study of Mental Health: Social Contexts, Theories, and Systems*. New York, NY: Cambridge University Press; 1999:295–314.
31. Kessler RC, Turner JB, House JS. Unemployment, reemployment, and emotional functioning in a community sample. *Am Sociol Rev*. 1989;54:648–657.
32. McLeod JR. Social stratification and inequality. In: Aneshensel CS, Phelan JC, Bierman A, eds. *Handbook of the Sociology of Mental Health*. 2nd ed. New York, NY: Springer; 2013:229–254.
33. Elder GH, Liker JK. Hard times in women's lives: historical influences across forty years. *Am J Sociol*. 1982;88:241–269.

34. Pearlin LI. Role strains and personal stress. In: Kaplan HB, ed. *Psychosocial Stress: Trends in Theory and Research*. New York, NY: Academic Press; 1983:3–32.
35. Menaghan EG. The daily grind: works stressors, family patterns, and intergenerational outcomes. In: Avison WR, Gotlib IH, eds. *Stress and Mental Health: Contemporary Issues and Prospects for the Future*. New York, NY: Plenum Press; 1994:115–147.
36. Ali J, Avison WR. Employment transitions and psychological distress: the contrasting experiences of single and married mothers. *J Health Soc Behav*. 1997;38:345–362.
37. Hill TD, Maimon D. Neighborhood context and mental health. In: Aneshensel CS, Phelan JC, Bierman A, eds. *Handbook of the Sociology of Mental Health*. 2nd ed. New York, NY: Springer; 2013:479–501.
38. Kessler RC. A strategy for studying differential vulnerability to the psychological consequences of stress. *J Health Soc Behav*. 1979; 20:100–108.
39. Kessler RC, McLeod JD. Sex differences in vulnerability to undesirable life events. *Am Sociol Rev*. 1984;49:620–631.

CHAPTER

11

Evolutionary Origins and Functions of the Stress Response System

R.M. Nesse[1], *S. Bhatnagar*[2], *B. Ellis*[3]

[1]Arizona State University School of Life Sciences, Tempe, AZ, USA
[2]Children's Hospital of Philadelphia and the University of Pennsylvania School of Medicine, Philadelphia, PA, USA
[3]University of Arizona Norton School of Family and Consumer Sciences, Tucson, AZ, USA

OUTLINE

Abstract

Evolution is the process in which traits such as physiological stress response systems (SRSs) are shaped by natural selection. A full understanding of any trait requires knowing its evolutionary history, how it has given a selective advantage, and the trade-offs and costs involved. Stress-related mechanisms emerged early in the history of life. Like all traits, they have costs as well as benefits. Because the stress response is so often associated with negative events, its utility has often been neglected. This chapter reviews the phylogeny and functional significance of the SRS, with a special focus on how selection has shaped the mechanisms that process environmental information to regulate the stress response, and how the stress response influences other traits such as risk-taking and sexual behavior.

UTILITY OF THE STRESS RESPONSE SYSTEM

The vast bulk of research on stress has investigated its causes, mechanisms, and effects. An evolutionary approach instead addresses two very different and relatively neglected questions: (1) How does the stress response system (SRS) give a selective advantage? and (2) What is the evolutionary history of the SRS? The answers to these questions provide a foundation in Darwinian medicine[1] for understanding why the stress response is the way it is and why it causes so much suffering and disease. The first and most important contribution of an evolutionary perspective on stress is a clear focus on its utility. The SRS is a complex, sophisticated, and carefully regulated adaptation that has been shaped by natural selection because its advantages that must be substantial in order to outweigh its huge costs.[2] The idea that stress is useful is by no means new. In fact, the very phrase Hans Selye chose to describe it, "The General Adaptation Syndrome," emphasizes its utility.[3] Despite this early emphasis on its benefits, as the idea of "stress" has entered the popular imagination, there has been a tendency to emphasize its dangers so that the fundamental fact of the utility of the stress response is often neglected.

Stress: Concepts, Cognition, Emotion, and Behavior
http://dx.doi.org/10.1016/B978-0-12-800951-2.00011-X

KEY POINTS

- The stress response system was shaped by natural selection to adjust physiology and behavior to changing circumstances, especially regarding energy usage and environmental threats and opportunities.
- Selection shaped the stress regulation system to express the stress response whenever the benefits are greater than the costs.
- In many situations, the costs of expressing a stress response are low compared to the costs of not expressing the response if a major threat is present, so false alarms are expected in the normal system (the "smoke detector principle").
- Selection has shaped mechanisms that adjust the threshold and magnitude of stress responses as a function of prior experience.

Stress and Other Defenses

Other defenses are also often confused with the problems they protect against. The capacities for pain, fever, vomiting, cough, and inflammation are often thought of as medical problems, although a moment's thought reveals that they are useful protective reactions. The ubiquity of the illusion that defenses are abnormalities arises from several sources. First, defenses are often associated with some kind of suffering and therefore seem maladaptive. Unfortunately, however, discomfort is itself probably one aspect of a mechanism that makes it useful. Second, defenses are reliably associated with disadvantageous situations, so the association bias makes it seem as if they are the problem. Finally, it is often possible to use drugs to block the expression of many defenses with very little harm, fostering the illusion that defenses are useless. In fact, blocking a defense can be harmful. For instance, suppressing cough for a patient with pneumonia makes it harder to clear the infection and may lead to death. Stopping the diarrhea of a person with a serious intestinal infection may lead to complications. Blocking fever, however, usually has little effect on the speed of recovery from a cold. When blocking a defense is not dangerous, this is because the body has back-up protective mechanisms and because the regulation mechanism seems to be set, for reasons we revisit, to a hair-trigger that expresses the defense at the slightest hint of threat.[4]

Situations in which Stress is Useful

Stress responses, like fever and pain, are useful only in certain situations. These responses have low basal activation levels until aroused by the particular circumstances in which they are useful. This means that the evolutionary explanation for such traits cannot be summarized in a single function. Instead, inducible defenses give advantages by changing multiple aspects of the body that increase its ability to cope effectively with the adaptive challenges that arise in a particular kind of situation. One defense may have many aspects that serve many functions, so, the first step in understanding the adaptive value of stress is not to specify its function, but to understand the situations in which the stress response is useful.

To do that, we need to go back to the origins of complex life forms 600 million years ago. If a very primitive organism had only two states, what would they be? The answer is quite straightforward: activity and rest. This is a fundamental divide, one that is maintained even in our biochemical and nervous systems. Biochemical pathways are divided into the catabolic, in which energy is used, and the anabolic, in which energy is stored and tissues are repaired. Parallel to this division are the two arms of the autonomic nervous system. The sympathetic system, which is activated as part of the stress response, increases arousal, blood pressure, heart rate, respiratory rate, and physical activity and institutes other endocrine and physiological changes necessary for action. The other half of the autonomic nervous system, the parasympathetic, inhibits muscular activity, stores energy, and shunts blood to digestion and bodily repair. Is stress, then, the same as arousal for action? Not exactly. As soon as a generic state of arousal was established, natural selection likely began to differentiate it into subtypes to better meet different kinds of challenges. Here again, the main bifurcation is clear. Arousal is useful in two different situations: threats and opportunities. This division is also represented in our nervous systems. As Gray and others have pointed out, the brain seems to have moderately distinct systems for behavioral inhibition and for reward-seeking.[5] The corresponding behaviors are said to be defensive or appetitive and are associated with feelings of fear/pain or pleasure. In psychology, the same division is recognized in the distinct cognitive states described by "promotion" as compared to "prevention."

PHYLOGENY OF THE STRESS RESPONSE

Cross-Species Comparisons

Comparisons among different species can help to reconstruct the phylogeny of the stress response. All vertebrates have the proopiomelanocortin molecule that gives rise not only to adrenocorticotropic hormone (ACTH), but also to opiate-like peptides. It is intriguing to note that these molecules, with their related functions, are derived from the same parent molecule. All vertebrates also make corticosteroids. Peptide sequences very similar to those of human ACTH are found not only in mammals, but also in amphibians, reptiles, and even in insects, mollusks, and marine worms. Interestingly, these peptides are usually associated with immune cells, equivalent to macrophages, where they set defensive processes in motion. ACTH has long been closely associated with other signaling molecules such as CRH (corticotrophin releasing hormone), biogenic amines such as epinephrine and norepinephrine, steroids such as cortisol, cytokines such as interleukin-1, and nitric oxide. All are crucial to defensive systems. The remarkable thing is that genetic sequences for these molecules have not only been conserved over hundreds of millions of years, but they continue to serve closely related defensive functions. Why have they changed so little? If a single molecule has several essential functions, this will create a strong selective force against mutations that change the sequence. By contrast, mutations that result in differentiation of different classes of receptors in target tissues can slowly specialize the responses of that tissue to the signal molecule. And they have, judging from the proliferating classes and subclasses of receptors that are now being discovered.

Cost-Benefit Trade-offs

Why isn't the SRS better? It could provide more effective protection against danger, but only at a still greater cost. Soldiers undergoing high-intensity military survival training show increases in the sympathetic neural transmitter neuropeptide Y (NPY) following interrogations.[6] This increase plays a functional role in adjusting to high-stress conditions: soldiers who experienced greater increases in NPY remained more interactive with their environment and were rated as exhibiting greater mental alertness during the interrogations.[6] The trade-off is that up-regulation of the NPY system mediates stress-induced obesity and metabolic syndrome.[7] Like everything else in the body, stress responses are shaped by trade-offs, sometimes with benefits and costs occurring in different parts of the life cycle.

The mechanisms that regulate the responsiveness of the SRS are shaped by the trade-off between the long-term costs versus the immediate benefits. Individuals who have smaller stress responses may be less vulnerable to stress-related diseases,[8] but they may be less able to cope with some stressors. The SRS responds not only to threats and challenges, but also to novel stimuli and positive social opportunities (e.g., unexpected or exciting rewards, opportunities for status enhancement, potential sexual partners). For example, in a naturalistic study on a Caribbean island, significantly elevated cortisol levels among children were documented during the 2 days prior to Christmas, compared with a control period, but only among children who had high expectations for presents or other exciting activities.[9] More generally, the SRS appears to mediate the effects of environmental influences, operating as an amplifier (when highly responsive) or filter (when unresponsive).[10] This dual function of the SRS is captured by the concept of *biological sensitivity to context*.[11] Lack of response to adversity and stress-related disorders may be associated with inability to take advantage of opportunities.

Resilience is a dynamic concept[12,13] that to some extent must be environment and stressor specific. An individual resilient to one type of stressor may be vulnerable to another. More interestingly, a response profile that was resilient in one environmental context may now produce vulnerability because the response profile is optimal for another environment.[14] Thus, systems that adjust the responsiveness of the SRS give a selective advantage.

Other trade-offs reflects the benefits and costs of habituation (down-regulation) and sensitization/facilitation (up-regulation) of SRS parameters.[15] For instance, rats habituate to mild cognitive stressors.[16] Habituation conserves energy when the stressor is known and can be easily coped with. However, habituation to unpredictable dangerous situations can be maladaptive, so sensitization may be useful, despite the risks of positive feedback dysregulation. These trade-offs have shaped the brain mechanisms that regulate the SRS.

In the longer term, activation of autonomic, neuroendocrine, metabolic, and immune system responses during ontogeny provides information about threats and opportunities likely to be encountered throughout life. Over time, this information becomes embedded in set points and reactivity patterns of an individual's SRS.[2] Thus, understanding the long-term impact of stress during development requires attention to the changes in an individual's environment from adolescence to adulthood. The impact of a specific kind of stress on SRS regulation mechanisms during development may be advantageous when the environment stays the same, but disadvantageous if the environment shifts.[14,17] Mismatch between early life experience and the adult environment may cause excessive or deficient responses to stressors experienced later in life.

Habituation and sensitization of the SRS is different for the sexes in rats, with males showing greater tendencies to habituation.[18] Selection forces acting on males and females may have differed enough to shape different patterns of habituation, a topic of current research. Humans show consistent sex differences in the type of events that elicit a SRS response and in the physiological and behavioral correlates of the response. Men tend to show more hypothalamic-pituitary-adrenal (HPA) activation than women in achievement-related tasks (which may elicit status-related motives), whereas women show larger HPA activation in situations involving social rejection.[19]

Stress responses in adult animals are profoundly affected by prenatal stress and variations in maternal care. The effects of variations in maternal care are transmitted across generations with offspring who experience high maternal care exhibiting lower stress responses and providing high maternal care themselves.[20] Such effects would be adaptive when offspring experience an environment similar to their parents. Mothers providing low maternal care tend to have high-stress responsiveness, as do their offspring when they become adults. However, offspring cross-fostered to other mothers show patterns of stress responsivity more similar to that of their foster mothers. Such results suggest that stress responsivity and maternal care are influenced by early experiences as well as genetic factors. Such regulation is seen in other mammals and even plants.[20] Some transmission across generations may be mediated by facultative epigenetic mechanisms that evolved to adjust the system based on early life experiences, and some may arise from more general learning mechanisms.

Difficulties in Defining Stress

The human mind seems wired to try to make neat categories with sharp boundaries, perhaps because we communicate with words and this requires dividing the world up into categories even when that is unnatural. This leads to a tendency to try to make sharp distinctions between different states that may, in fact, overlap considerably. States of defensive arousal, for instance, are different from states of arousal for seeking food, but this differentiation is not complete. For instance, cortisol secretion is aroused by opportunities as well as threats. In fact, cortisol is even involved in reward mechanisms. Thus, any attempt to define the SRS in terms of cortisol arousal is doomed. For that matter, any attempt to define stress or the SRS is liable to be an exercise in frustration, for the evolutionary reason that the system does not have sharp boundaries or a single function. The closest we can come to a defining characteristic is the kinds of situations in which stress responses have given a selective advantage, and those situations are not sharply defined. The SRS was, after all, not designed by an engineer, but shaped by a process of tiny tinkering changes. The long unsatisfying history of attempts to define stress, and the wish expressed by many researchers that the term would go away, arise from this difficulty. Even after defensive arousal was differentiated considerably from appetitive arousal, there were undoubtedly advantages from further differentiating subtypes of stress responses to match specific challenges. Thus, different situations—a predator, a high place, injury, infection, starvation, loss of a status battle, and speaking in public—all seem to have shaped somewhat different defensive responses.[21] These responses are only partially differentiated from a more generic response so they have overlapping characteristics with functions in common. For the same reasons, attempts to sharply distinguish different kinds of anxiety disorders are as frustrating as attempts to define stress itself. New attempts to study anxiety disorders in the context of normal anxiety will be helpful.

HOW DOES THE SRS HELP?

Immediate Response

A stress response is a coordinated pattern of changes that is useful in situations in which the organism is faced with possible damage or a loss/gain of resources. The next question is, "How is it useful?" Even before Selye, Walter Cannon provided some answers. In situations that might require "fight or flight," he observed the utility of increased heart rate and contractility to speed circulation, increased rate and depth of breathing to speed gas exchange, sweating to cool the body and make it slippery, increased glucose synthesis to provide energy, shunting of blood from gut and skin to muscles, increased muscle tension to increase strength and endurance, and increased blood clotting in preparation for possible tissue damage.[22] More recently, others have demonstrated faster reaction times and cognitive benefits as a result of sympathetic arousal. These immediate responses are mostly mediated by the sympathetic nervous system and the associated release of epinephrine from the adrenal medulla.

Adrenal Cortical Response

The SRS also includes release of cortisol from the adrenal cortex, a more delayed response, although one with some rapid effects such as fast negative feedback that are likely mediated by putative membrane steroid receptors.[23] Cortisol release is initiated by neural signals to the hypothalamus that releases CRH, which in turn results in secretion of ACTH from the anterior pituitary gland. ACTH induces cortisol synthesis and release from the adrenal gland. The whole system is called the HPA system because the signal acts via the hypothalamus, the pituitary, and the adrenal glands. Many actions of the HPA system seem, like those of the sympathetic system, well designed for acute action. It changes physiology so the liver breaks glycogen down into glucose, and it increases the entry of glucose into cells.[24] CRH not only releases ACTH, it also directly increases anxiety and arousal and activates cells in the locus coeruleus, the brain center where the cell bodies for most noradrenergic neurons are located. All in all, the system seems admirably designed to get the organism ready for action. Indeed, both branches of the system are readily aroused by exercise, and trained athletes, far from having low levels of cortisol, have chronic high levels, as is appropriate to cope with high levels of exertion.

Association with Negative Events

So, why should the SRS as a whole be associated so closely with bad events instead of positive ones? To answer that question, we need to understand why the components of the SRS are carefully packaged. If stress responses are so useful, why aren't stress responses expressed all the time? There are at least three good reasons. First, it is calorically expensive. No organism can afford to waste energy. Second, it interferes with other adaptive behavior. A vigilant organism has less time for finding food and eating, to say nothing of mating. Finally, some changes that give an advantage in the face of threats also cause tissue damage. For this reason, they need to be carefully sequestered, except in circumstances in which the costs are outweighed by the benefits. This helps to explain why some aspects of the SRS are associated more with negative than positive arousal. The benefits of a stress response that increases the likelihood of catching prey may sometimes be worth the costs, but a response that prevents being caught, as prey will almost always be worth almost any costs. An optimal regulatory mechanism will express a stress response whenever, on average, the benefits are greater than the costs. This "smoke detector principle," based on a signal detection analysis of how selection shapes mechanisms that regulate defense responses explains why so many instances of stress response seem excessive or unnecessary.[4] The global conclusion is that the damage caused by stress responses is not necessarily from "abnormal" stress. Some components of the SRS may be a part of the response specifically because they are too damaging to be expressed except when they protect against great danger. Normal stress, like every other bodily trait, has costs as well as benefits. This idea is expressed in the concepts of allostasis and allostatic load, as proposed by McEwen,[25] which emphasizes the short-term benefits and the long-term costs.

The idea that the normal stress response is crucial for optimal functioning has implications for proposed pharmacological therapies for reducing stress responses. For example, a CRH inhibitor that blocks stress responses can be expected to disrupt the body's adaption to situations that call for increased energy use. Disrupting the normal operation of the HPA system is also likely to interfere with counter-regulatory systems, such as regulation of insulin release. Furthermore, human and animal studies provide many examples of mismatches between perceived stress and behavioral indices of stress and HPA activation. The extreme of an absent stress response is, of course, Addison's disease. Thus, although the cognitive nature of many current human stressors results in costs disproportionate to actual threats, from an evolutionary point of view, general inhibition of stress responses is by no means optimal. It would be an irony as well as a tragedy if the history of excessive use of cortisone were to repeat itself with a new generation of drugs that block the SRS.

Cortisol as Protection Against Other Aspects of Stress

If some aspects of the stress response cause harm, has selection shaped systems to protect against this damage? In 1984, Munck and colleagues reviewed the actions of cortisol and said, "We propose that stress-induced increases in glucocorticoids levels protect, not against the source of stress itself, but rather against the body's normal reactions to stress, preventing those reactions from overshooting and themselves threatening homeostasis."[26] They noted that many inflammatory diseases had been attributed to overproduction of cortisol until 1941, when adrenal steroids were shown to decrease inflammation. Subsequent demonstrations showed that steroids inhibit production of cytokines, prostaglandins, and other mediators of the immune response, thus decreasing immune function. This is just the opposite of what would make sense as protection from danger, but is entirely consistent with a role in protecting against damage from immune system activation induced by other SRS induced changes. The effects of glucocorticoids on immune function are much more complicated than originally thought. For instance, recent findings regarding the effects of glucocorticoids on the brain versus the body on regulation of energy balance and fat deposition[27] suggest continuing challenges to our assumptions about the physiological and neural functions of glucocorticoids.

ADAPTIVE REGULATION OF STRESS RESPONSIVENESS

The bimodal pattern of stress response in some organisms, with some individuals responding far more quickly and strongly than others, may result from frequency dependent selection for fast response "hawk" patterns that are optimal in crowded situations, while a "dove" pattern is otherwise better.[28] Maternal effects on an offspring's stress responsiveness may also be adaptive. Mothers exposed to stressful environments give birth to offspring with especially responsive stress systems that may give an advantage in harsh environments,[20] a finding that may help explain the connection between early abuse and increased stress vulnerability.

Several other lines of thinking also address the adaptive significance of individual differences in SRS functioning.[29] Integrating and extending these past theories, the Adaptive Calibration Model is an evolutionary theory of developmental programing focusing on calibration of SRSs and associated life history strategies to local environmental conditions. It attempts to move beyond the primarily descriptive science that now dominates SRS studies to more explanatory models that seek to account for individual differences and their adaptive significance. For example, exposures to danger, unpredictable or uncontrollable contexts, and social evaluation generate sustained activation of the HPA axis.[30] Because HPA responses track the key environmental variables, they can feed into mechanisms that adjust life history strategies via changes in defensive behaviors, competitive risk-taking, learning, attachment, affiliation, and reproduction.[31]

MISMATCH BETWEEN ANCESTRAL AND MODERN ENVIRONMENTS

Much has been made of the differences between our environment and that of our ancestors.[32] In the case of stress, this argument comes in several flavors. Some suggest that life is more stressful now than it was for our predecessors. Special aspects of our environment do cause new kinds of stress. Working in a bureaucracy is tedious and political at best. Driving to work, living in a ghetto, running a corporation, working in a factory—these all arouse the SRS. Despite the amount of stress we experience, however, our ancestors almost certainly experienced more. With no police, no food reserves, no medicine, no laws, rampant infections, and prevalent predators, danger could come at any time. True, social groups were closer, kin networks were stronger, and people spent all their time with each other, none of it alone reading books. Still, life was hard. Perhaps in that environment, where stressors were more often physical, the SRS was more useful than it is now. Today, we mainly face social and mental threats, so the actions of the HPA system may more often yield net costs. This plausible hypothesis supports efforts to reduce stress and to find drugs that block the SRS.

This brings us back to the general concept of stress as aroused when demands are greater than an individual's ability to meet them. We think of these demands as coming from the outside, and sometimes they do, as when we are attacked by the proverbial tiger. But most stresses in modern life arise not from physical dangers or deficiencies, but from our tendency to commit ourselves to personal goals that are too many and too high, and to ruminate about why we cannot achieve them all. When our efforts to accomplish these goals are thwarted, or when we cannot pursue all the goals at once and must give something up, the SRS is activated. In short, much stress arises, ultimately, not from a mismatch between our abilities and the environment's demands, but from a mismatch between what we desire and what we can have.

Glossary

Defense A trait that is latent until aroused by threatening situations in which it is useful.

Evolutionary medicine The application of evolutionary biology to address problems in medicine and public health.

Natural selection The process by which genes that give a fitness advantage become more common from generation to generation and those that decrease fitness become less common, thus shaping adaptive traits, including defenses.

Phylogeny The evolutionary history of a trait or a species.

Trade-offs The fitness costs and benefits of a trait whose net effects yield a selective advantage.

References

1. Nesse RM, Williams GC. *Why We Get Sick – The New Science of Darwinian Medicine.* New York, NY: Times Books; 1994.
2. Del Giudice M, Ellis BJ, Shirtcliff EA. The adaptive calibration model of stress responsivity. *Neurosci Biobehav Rev.* 2011;35(7): 1562–1592.
3. Selye H. *The Stress of Life.* rev. ed. New York: McGraw-Hill; 1978.
4. Nesse RM. Natural selection and the regulation of defenses: a signal detection analysis of the smoke detector principle. *Evol Hum Behav.* 2005;26:88–105.
5. Gray JA. *Fear and Stress.* 2nd ed. Cambridge: Cambridge University Press; 1987.
6. Morgan CA, Wang S, Southwick SM, et al. Plasma neuropeptide-Y concentrations in humans exposed to military survival training. *Biol Psychiatry.* 2000;47(10):902–909.
7. Kuo LE, Kitlinska JB, Tilan JU, et al. Neuropeptide Y acts directly in the periphery on fat tissue and mediates stress-induced obesity and metabolic syndrome. *Nat Med.* 2007;13(7):803–811.
8. Charney DS. Psychobiological mechanisms of resilience and vulnerability. *FOCUS.* 2004;2(3):368–391.
9. Flinn MV, Nepomnaschy PA, Muehlenbein MP, Ponzi D. Evolutionary functions of early social modulation of hypothalamic-pituitary-adrenal axis development in humans. *Neurosci Biobehav Rev.* 2011;35 (7):1611–1629.
10. Ellis BJ, Del Giudice M, Shirtcliff EA. Beyond allostatic load: the stress response system as a mechanism of conditional adaptation. In: Beauchaine TP, Hinshaw SP, eds. 2nd ed. Child and Adolescent Psychopathology; vol. 2:Hoboken, NJ: Wiley and Sons; 2013:251–284.
11. Boyce WT, Ellis BJ. Biological sensitivity to context: I. An evolutionary-developmental theory of the origins and functions of stress reactivity. *Dev Psychopathol.* 2005;17(02):271–301.

12. Bracha HS, Ralston TC, Matsukawa JM, Williams AE, Bracha AS. Does "fight or flight" need updating? *Psychosomatics*. 2004;45(5):448–449.
13. Rutter M. Implications of resilience concepts for scientific understanding. *Ann N Y Acad Sci*. 2006;1094(1):1–12.
14. Wood SK, Bhatnagar S. Resilience to the effects of social stress: evidence from clinical and preclinical studies on the role of coping strategies. *Neurobiol Stress*. 2015;1:164–173.
15. Frankenhaeuser M. *The role of peripheral catecholamines in adaptation to understimulation and overstimulation. Psychopathology of Human Adaptation*. New York: Springer; 1976. pp. 173-191.
16. Grissom N, Bhatnagar S. Habituation to repeated stress: get used to it. *Neurobiol Learn Mem*. 2009;92(2):215–224.
17. Daskalakis NP, Diamantopoulou A, Claessens SE, et al. Early experience of a novel-environment in isolation primes a fearful phenotype characterized by persistent amygdala activation. *Psychoneuroendocrinology*. 2014;39:39–57.
18. Bhatnagar S, Lee TM, Vining C. Prenatal stress differentially affects habituation of corticosterone responses to repeated stress in adult male and female rats. *Horm Behav*. 2005;47(4):430–438.
19. Stroud LR, Salovey P, Epel ES. Sex differences in stress responses: social rejection versus achievement stress. *Biol Psychiatry*. 2002;52(4):318–327.
20. Zhang TY, Parent C, Weaver I, Meaney MJ. Maternal programming of individual differences in defensive responses in the rat. *Ann N Y Acad Sci*. 2004;1032:85–103.
21. Marks IM, Nesse RM. Fear and fitness: an evolutionary analysis of anxiety disorders. *Ethol Sociobiol*. 1994;15(5-6):247–261.
22. Cannon WB. *Bodily Changes in Pain, Hunger, Fear, and Rage. Researches into the Function of Emotional Excitement*. New York: Harper and Row; 1929.
23. Dallman MF. Fast glucocorticoid actions on brain: back to the future. *Front Neuroendocrinol*. 2005;26(3):103–108.
24. Warne JP, Akana SF, Ginsberg AB, Horneman HF, Pecoraro NC, Dallman MF. Disengaging insulin from corticosterone: roles of each on energy intake and disposition. *Am J Physiol Regul Integr Comp Physiol*. 2009;296(5):R1366–R1375.
25. McEwen BS. Interacting mediators of allostasis and allostatic load: towards an understanding of resilience in aging. *Metabolism*. 2003;52:10–16.
26. Munck A, Guyre PM, Holbrook NJ. Physiological functions of glucocorticoids in stress and their relation to pharmacological actions. *Endocr Rev*. 1984;5(1):25–44.
27. Dallman MF, Pecoraro N, Akana SF, et al. Chronic stress and obesity: a new view of "comfort food". *Proc Natl Acad Sci U S A*. 2003;100(20):11696–11701.
28. Korte SM, Koolhaas JM, Wingfield JC, McEwen BS. The Darwinian concept of stress: benefits of allostasis and costs of allostatic load and the trade-offs in health and disease. *Neurosci Biobehav Rev*. 2005;29(1):3–38.
29. Ellis BJ, Jackson JJ, Boyce WT. The stress response systems: universality and adaptive individual differences. *Dev Rev*. 2006;26(2):175–212.
30. Dickerson SS, Kemeny ME. Acute stressors and cortisol responses: a theoretical integration and synthesis of laboratory research. *Psychol Bull*. 2004;130(3):355.
31. Ellis BJ, Del Giudice M. Beyond allostatic load: rethinking the role of stress in regulating human development. *Dev Psychopathol*. 2014;26(1):1–20.
32. Gluckman PD, Hanson M. *Mismatch: Why Our World No Longer Fits Our Bodies*. New York: Oxford University Press; 2006.

CHAPTER

12

Life Events Scale

E. Wethington

Cornell University, Ithaca, NY, USA

OUTLINE

Abstract

There is convincing evidence that exposure to stress in daily life is related to physical and mental health. A life events scale is a comprehensive list of external events and situations (stressors) that are hypothesized to place demands that tend to exceed the capacity of the average person to adapt (Cohen et al.[4]). Difficulty in adaptation leads to physical and psychological changes or dysfunction, creating risk for disorders and disease. The naturalistic measurement of life events has a long history in health research, primarily in observational, nonexperimental, population-based descriptive studies. Life events scales have proliferated, applying different theories of why stress exposure affects health, and addressing different levels of analysis, from momentary exposure to stress during the day, to the accumulation of stressors and their impact across many years of life. Four types of stress measurement are described: major life events, chronic stressors, stress appraisal, and daily hassles, or events.

THE IMPORTANCE OF ASSESSING LIFE EVENTS AND STRESS EXPOSURE

It is now generally recognized that exposure to stressors in daily life, or over the course of life, may be one of the most critical components of health and well-being. Research related to stress exposure, health behavior, and physical and mental health outcomes has increasingly been applied to understanding serious social and public health problems, such as the existence and persistence of health disparities.[1,2] A new synthesis of theory and research from psychoimmunology, neuroscience, psychology, sociology, and public health has emerged which focuses on how the accumulation, or "piling up" of stressful experiences across the course of life may lead to physiological damage, and in burdened populations, contribute to health disparities between social groups.[1,3]

However, there remains ambiguity about the dimensions of stress that are involved in this process, specifically the types of stressors that have the more deleterious effects on health, both immediately and in the long term, and which types of stressors are the most potent risk factors that predispose an individual to develop physical or psychological illness.[4,5] Variations in stressor assessment reflect different theories of how, when, and why stressor exposure poses a risk to health.

Life event scales remain one of the most commonly utilized means of naturalistic assessment of exposure to stressors, in distinction to the manipulation of stressor exposure in experimental research. Assessment of stressor exposure has been dependent on two well-established research approaches in stress assessment. The first approach is the environmental perspective, which focuses on characteristics of events, such as severity, timing in the life course, and chronicity as the predisposing risks for developing disease. The second approach is the psychological perspective, which focuses on individual perceptions and appraisals of the threats posed by events and the capacity to manage or cope with them.[4] Both perspectives have virtues and each is represented in the different

Stress: Concepts, Cognition, Emotion, and Behavior
http://dx.doi.org/10.1016/B978-0-12-800951-2.00012-1

KEY POINTS

- It is now generally recognized that exposure to stressors in daily life, or over the course of life, is one of the most critical components of health and well-being.
- The life events scale, first developed in the 1960s, remains among the most frequently used measures of stressor exposure, despite questions about the method's overall validity and reliability and its reliance on retrospective recall of stressful experiences.
- Life event scales are inexpensive to use and are invaluable for exploratory studies in new areas where the probable influence of stress as a risk factor has not been highly developed.
- This chapter reviews alternative ways in which measures of life events and exposure to stressors have been conceived and developed, across many different types of studies, and prospects for their continued use in studies that combine them with physiological measures of stress.

types of life events scales that have been developed for particular types of research questions and designs. In order to assess the physical impact of stress exposure on the body over longer periods of time, measures developed from both of these perspectives are now being combined in studies with physiological measures of stress arousal and biomarkers indicating accumulated damage to physical systems.

Four types of naturalistic stressor assessment have predominated in the research literature: *Life events*, or exposure to out-of-the-ordinary, demanding events such as divorce or sexual abuse during childhood, that have the capacity to change the patterns of life or arouse very unpleasant feelings; *Stress appraisals*, which are self-reports of perceived degree of stressfulness of the demands and threat posed by events; *Chronic stressors*, also known as long-term difficulties, which are enduring or recurrent difficulties and strains in an area of life, or affecting many areas of life; and *Hassles*, also known as daily events, which are smaller, relatively minor, and normally less emotionally arousing events whose effects fade in a day or two. Although conceived primarily as comprehensive lists of exposure to discrete events outside the organism, naturalistic life events assessments often combine two or more of these four approaches, often including appraisals of stressors, measuring dimensions such as severity, frequency, controllability, and expectedness of events. Some researchers have advocated for the measurement of multiple types of stressor exposure in order to assess exposure to stress more comprehensively in studies of health disparities.[6]

Frequently, the goal of research on stress exposure is to establish the relationship between exposure to stressors and specific health outcomes or changes in health behavior; these studies evaluate stress as a possible risk factor for disease. For example, if the research question is to estimate the size of the association between stressor exposure and distressed mood, then the usual procedure is to use a retrospective count of life events occurring in the months prior to the measurement of mood. However, if a researcher is interested primarily in the relationship between stress and the onset or timing of a major clinical disorder, then the stressor exposure measure shifts to a determination of the type, severity, and timing of events and preexisting chronic stressors that preceded the disorder, sometimes using a prospective or longitudinal design of data collection. Studies of daily variation in mood, or the onset of more transient physical disorders, such as colds or body aches, use measures of short-term exposure to relatively minor stressors, daily events, or hassles, but also chronic stressors over a period of time that are hypothesized to make the organism more vulnerable to infection.[7] Studies focusing on diseases that develop over long periods of time, such as heart disease, tend to use theoretically-based measures of exposure to long-term, recurrent, and persistent chronic stressors, such as the famous longitudinal Whitehall Studies of civil servants in England.[8] In practice, researchers often combine many methods.

MAJOR LIFE EVENTS

Two contrasting methods of life events measurement have developed: checklist measures and personal interview measures.[9] The latter uses qualitative probes in order to specify more precisely the characteristics of life events believed to produce illness as well as the timing of events in relationship to the health outcome. Because of their ease of use and the amount of exploratory health research conducted, checklist measures predominate in studies of stressor exposure. This method has been used in thousands of studies. Moreover, researchers still find the life event checklist to be a valuable research tool[10] especially for descriptive preliminary studies where a stress-health relationship has not yet been fully established; or in studies where researchers must assess multiple risk factors, such as stress, social support, coping, and personality factors and must use relatively brief, easy to administer measures.

Checklist measures were derived from an environmental perspective on stress proposing that the basis of experienced stress is an event that brings about a need for social, physical, or psychological readjustment. The earliest of these perspectives was the life change-readjustment paradigm developed by Holmes and

Rahe[11] which was further refined into successive versions of the Social Readjustment Rating Scale.[10,12] Other theoretical perspectives on stress, such as those developed by Lazarus and Folkman,[13] Dohrenwend,[14,15] and Brown and colleagues,[16,17] have elaborated on Holmes and Rahe's approach, and also have introduced other theoretical perspectives on the characteristics of stress that affect health.

A typical checklist measure consists of a series of yes/no questions, asking participants to report if any event like the one described has occurred over a past period of time. The time of recall can range from a month to a year or more. Some checklist measures have included more detailed information about timing and severity. For example, it is relatively easy to ask respondents to report the date something occurred, although it is often hard for them to remember those dates precisely.[18] Other methods based on the checklist have asked for a brief written description of "other events that may have occurred."[14] Still others include follow-up questions asking respondents to rate the relative stressfulness of the event (from "not at all" to "very") or even its frequency.

Self-report descriptive questions estimate the severity of the event and its likely relationship to an outcome. However, the typical checklist measure does not rely on descriptive information. A typical measure sums the absolute number of events or sums investigator-assigned severity ratings for each type of event report. These severity ratings, termed "normative," are based on responses in larger surveys where participants were asked to rate the relative severity of stressors such as deaths, job losses, marital problems, and residential moves.[12,14] The summary score, however derived, is used as an estimate of the stressfulness of changes and events experienced over a period of time. Related to the latter point, checklist measures tend to be used in studies that utilize continuous symptom or "mood" outcome measures, although they are also used in exploratory studies of the onset of illness.

The ancestor of most current measures is the Holmes and Rahe Social Readjustment Rating Scale.[11,12] This measure included both positive and negative events because Holmes and Rahe theorized that change per se was associated with changes in health status. Over time, some checklist life events scales have moved away from the change-readjustment paradigm to an emphasis on negative and undesirable events. There are much-replicated findings that undesirable events are more predictive of health problems than the average positive event.[19] Life event scales intended for use in the general population have become more comprehensive and inclusive of events occurring to women, minorities, and other populations. Researchers have also developed comprehensive event lists for age groups and special populations, including children, adolescents, and older people.[19]

Despite their popularity, checklist measures are frequently criticized. Most of this criticism questions their reliability and validity as measures of stressor exposure; in a recent example, Ayalon questioned whether older adult respondents can report accurately about events that happened to them during childhood and adolescence,[20] a strategy that was implemented in the U.S. Health and Retirement Survey. This criticism is not new. Dohrenwend[15] recently analyzed the many methodological issues that threaten the reliability and validity of checklist measures. His criticisms also include the vagueness and generality of life events questions, which may lead respondents to misreport or overreport the occurrence of events or to report minor events as "major." Life events scales are typically administered in retrospective studies, where recall can be faulty even over relatively short periods of time, let alone over a lifetime. Life events scales need to be comprehensive and to include a wide variety of possible stressors to represent the experiences of the entire population; but often shorter measures are preferred and in practice, life events scales are cut when used in many studies. Life events scales also do not distinguish acute time-limited events, such as a divorce decree, from longer-term difficulties that came before it, such as increasingly severe marital arguments and the period of separation. Dohrenwend also noted that some life events scales include events that are confounded with the feelings and emotional state of psychiatric illness. In addition to these threats to reliability and validity, there is also considerable evidence that life events scales suffer from recency bias: respondents are more likely to recall events that occur in the last few months than those that occurred a years ago.[21]

The early development of *personal interview methods* for assessing life events utilized a theoretical perspective distinct from the change-readjustment paradigm, which informed life events checklists. The major developers of the interview methods, George W. Brown, Tirrill Harris, and other colleagues, proposed that social and environmental changes that threaten the most strongly held goals, plans, commitments, and social roles are the basis for experienced severe stress.[16] Brown's perspective also holds that severe stressors, rather than minor, threaten health, distinguishing it from both change-readjustment and hassles paradigms (see Daily Hassles).

Interview measures are not used commonly, primarily because of their greater cost and complexity.[22] Investigators tend to use interviews under three circumstances. First, some investigators prefer to use interview measures when more precise severity ratings are required.[18] Second, other researchers advocate for the use of interviews (or similar more complicated and expensive measurement procedures) when the relative timing of stressor exposure and disease onset is critical to a study,[23–25] and third, when the occurrence of an event, or series of events, may be related to respondent illness or behavior.[23,26] Promoters of interview methods also claim that they are more comprehensive, reliable, and valid than checklist measures, although there is considerable debate on this point.[22]

However, the complexity of interview methods discourages their use in lower budget or more exploratory studies. Interview methods are very expensive, requiring extensive interviewer training periods, long interviews which need to be recorded and transcribed for analysis, and the coding and rating of stressful situations by trained raters. In contrast, most uses of life event checklists require simply summing up a series of yes/no questions or life event weights.

The life events and difficulties schedules (LEDS) developed by Brown and Harris,[16] is the most widely used personal interview methods; and several other researchers in the United States and elsewhere have developed interviews based on the LEDS for other populations such as HIV patients, teens, and low income mothers. These instruments include the UCLA Life Stress Interview[23] and the Stressful Life Events and Difficulties Interview.[26] The LEDS and these newer American instruments are semi-structured interview survey instruments that assess a wide variety of stressors in specific contexts.

Life event interviews consist of a series of questions asking whether certain types of events and chronic stressors have occurred over a period of time and a set of guidelines for probing about objective features of the events. The probes are used by investigators to rate the long-term contextual threat of events; contextual threat is an estimate of the severity of a stressor based on the objective demands posed for a particular individual who undergoes a more or less demanding stressor of this type. The contextual threat ratings for the LEDS are documented in dictionaries that have been developed over many studies over time and in different populations. Similarly, other investigators have also developed dictionaries documenting contextual threat ratings for special populations in the United States.[23,26] Contextual rating is the key component of these methods, because the occurrence of a severely threatening event is believed to be a major risk factor for the onset of illness.

There are two important categories of life events scales that researchers believe are particularly important for predicting health across the life course, as well as the causes of group health disparities in the population. The first category focuses on stressful experiences during childhood. The types of events consist of persistent poverty, separation from parents because of death or divorce, serious physical illness, mental disorders, and substance abuse among members of the childhood family, and physical and sexual abuse. These scales are now included in many surveillance surveys[27] and longitudinal studies.[20,28] Interview measures also include an assessment of childhood living conditions and stressors.[16] The second category is measurement of instances of racial or other types of discrimination, lifetime and recent occurrences.[29] Discrimination measures are important for studies that examine the relationship of lifetime and recent stressor exposure to group health disparities. Stressful events in childhood and discrimination measures are designed as self-report measures, similar to standard life events checklist scales.

STRESS APPRAISAL

Appraisal plays an important part in life events scales, as well as theories of the stress process, most notably the transactional model of stress of Lazarus and Folkman.[13] Appraisal of stressors is believed to underlie the emotional experience of stress. Measures of appraisal focus on the degree to which an event threatens well-being (primary appraisal, or assessment of threat) or overwhelms resources to cope (secondary appraisal).[13]

Life events scales differ in whether they explicitly include appraisal as a component or explicitly exclude appraisal. Several measures of stressor exposure include assessments of how well one is coping with stressors or perceptions of the level of demand in different social roles.[6] Checklist and interview measures of life events, for the most part, aim to exclude appraisal from stressor exposure assessments, although some explicitly do not: judging an occurrence as a discriminatory event involves appraisal of threat.[29] Those that exclude appraisal do so because of concerns that stress appraisal is confounded with the health and psychological outcomes stressor exposure is hypothesized to predict.[14] Indeed, researchers have speculated that some stress appraisals are "caused" by underlying persistent mood disturbance or ill health, rather than vice versa.[14] The transactional model of stress posits that appraisals can be measured separately and objectively.[13]

CHRONIC STRESSORS

More recent research on stress and illness has turned toward emphasizing the role of persistent, continuous, or regular exposure to stressors as important risk factors for the development of disease. There are three methods by which researchers measure chronic stress: (a) multi-item self-report interviews and questionnaires covering ranges of situations believed to produce chronic stress exposure in a given setting, such as work role demands, conflicts, and lack of control; (b) personal interview (contextual) measurement in life event interview instruments such as the LEDS; and (c) third-party observation of those believed to be at risk for chronic stress exposure. The most widely used method is the self-report interview or questionnaire, administered in person or (increasingly) online. Multi-item measures of chronic stressors are an important class of life events scales.[30]

A typical measure of chronic stressors is a set of questions designed to assess either the frequency or severity (or both) of the more frequently occurring stressors in important life roles, such as work, marriage, and parenting. The strength of this approach is its grounding in theory, which has produced detailed, multidimensional assessment of environmental factors known to produce chronic stress, such as work or family stressors.[8] A weakness of the approach for some investigators is that the questions may be confounded with stressor appraisal, particularly those scales assessing stressor severity.

DAILY HASSLES

The research perspective in which measures of hassles were developed, the transactional model of stress,[13] differs significantly from the research perspective in which major life event scales were developed. The hassles paradigm focuses attention on the potentially deleterious ways in which minor stressors, even those whose effects are relatively fleeting, can have long-term negative impacts on health. Measures of hassles are life events scales, but assess exposure to minor hassles such as traffic tie-ups, missed appointments, lost pets, and quarrels with small children. Although such events might in fact be major events for some people under some circumstances, they are made more or less severe by the ways in which the individual appraises and copes with the event.[13]

Methods assessing daily events or hassles were developed using diary methods of data collection.[31] Diary methods require participants to keep records of small events occurring over a given period of time, usually 1 day or a 24-h period, although weekly diaries are not unknown. However, it is now has become more common for researchers to use a technique called ecological momentary experience sampling, where participants are contacted regularly or at random on the phone to respond to questions about their experiences at that time or the very recent past.[32]

Styles of measurement vary by research question and data collection method: open-ended approaches, which ask respondents to describe aspects of bothersome events of the day (not unlike personal interview measures)[33,34]; and structured question approaches, simple yes or no response questions modeled on life events checklists.[35] Many hassles and daily events questions are administered in tandem with measures of appraisal. Daily event or hassles measurement has also been combined with measures of chronic stress,[36] to assess whether minor events that arise from more major chronic stressors have greater impact on daily well-being and physical symptoms than isolated minor stressors.

Current hassle scales share the strengths and weaknesses of related approaches to the measurement of more major life events. Because diary methods of data collection rely on written and self-report, there is concern that they may confound objective event occurrence with unmeasured appraisal processes. Instantaneous collection of experiences on phones, however, aims to reduce bias in event measurement that may be brought about by the demands of retrospective recall.[32]

FUTURE DIRECTIONS IN RESEARCH INVOLVING LIFE EVENTS SCALES

A number of major longitudinal cohort studies in the United States and other countries, including the U.S. Health and Retirement Surveys, English Longitudinal Survey of Aging, the Midlife in the United States Studies, and others have incorporated physiological measures of stressor exposure, including biomarkers indicating the accumulation of significant damage to biological systems.[18,37] The purpose of including these biomarkers and indicators is to combine them with the more naturalistic measures of stress exposure also included in those studies. But these are studies of aging, focusing on middle-aged and older adults. The newly emerging interdisciplinary biopsychosocial life course perspective on health suggests that longitudinal surveys through early childhood, adolescence, and young adulthood should also incorporate age-appropriate biomarkers.[2] Combining physiological stress research with naturalistic assessment will advance the field of stress measurement considerably, while also providing new clues (or confirmation) about which types of stressors or stressor characteristics are associated with physical and mental health across the life course.

References

1. Wethington E, Glanz K, Schwartz MD. Stress, coping and health behavior. In: Glanz K, Rimer BK, Viswanath K, eds. *Health Behavior: Theory, Research, and Practice*. 5th ed. San Francisco, CA: John Wiley & Sons; 2015:223–242.
2. Chen E, Miller GE. Socioeconomic status and health: mediating and moderating factors. *Annu Rev Clin Psychol*. 2013;9:723–749.
3. Ganzel B, Morris P, Wethington E. Allostasis and the human brain: integrating models of stress from the social and life sciences. *Psychol Rev*. 2010;117(1):134–174.
4. Cohen S, Kessler RC, Gordon LU. Strategies for measuring stress in studies of psychiatric and physical disorders. In: Cohen S, Kessler RC, Gordon LU, eds. *Measuring Stress: A Guide for Health and Social Scientists*. New York: Oxford University Press; 1995:3–26.
5. Contrada RJ. Stress, adaptation, and health. In: Contrada RJ, Baum A, eds. *Handbook of Stress Science*. New York: Springer; 2010:3–9.

6. Turner RJ, Wheaton B, Lloyd DA. The epidemiology of social stress. *Am Sociol Rev*. 1995;60:104–125.
7. Cohen S, Tyrell DA, Smith AP. Psychological stress and susceptibility to the common cold. *N Eng J Med*. 1991;325(9):606–612.
8. Aboa-Éboulé C, Brisson C, Maunsell E, et al. Effort-reward imbalance at work and recurrent coronary heart disease events: a 4-year prospective study for post-myocardial infarction patients. *Psychosom Med*. 2011;73:436–447.
9. Wethington E, Brown GW, Kessler RC. Interview measurement of stressful life events. In: Cohen S, Kessler RC, Gordon LU, eds. *Measuring Stress: A Guide for Health and Social Scientists*. New York: Oxford University Press; 1995:59–79.
10. Scully JA, Tosi H, Banning K. Life event checklists: revisiting the social readjustment rating scale after 30 years. *Educ Psychol Meas*. 2000;60(6):864–876.
11. Holmes TH, Rahe RH. The social readjustment rating scale. *J Psychosom Res*. 1967;4:1–39.
12. Masuda M, Holmes TH. Magnitude estimates of social readjustments. *J Psychosom Res*. 1967;11:219–225.
13. Lazarus RS, Folkman S. *Stress, Appraisal and Coping*. New York: Springer; 1984.
14. Dohrenwend BS, Krashnoff L, Ashkenasy AR, Dohrenwend BP. Exemplification of a method for scaling life events: the PERI life events scale. *J Health Soc Behav*. 1978;19:205–229.
15. Dohrenwend BP. Inventorying stressful life events as risk factors for psychopathology: toward resolution of the problem of intracategory variability. *Psychol Bull*. 2006;132:477–495.
16. Brown GW, Harris TO. *Social Origins of Depression: A Study of Depressive Disorder in Women*. New York: The Free Press; 1978.
17. Brown GW, Harris TO, eds. *Life Events and Illness*. New York: Guilford; 1989.
18. Kessler RC, Wethington E. The reliability of life event reports in a community survey. *Psychol Med*. 1991;21:723–738.
19. Turner RJ, Wheaton B. Checklist measurement of stressful life events. In: Cohen S, Kessler RC, Gordon LU, eds. *Measuring Stress: A Guide for Health and Social Scientists*. New York: Oxford University Press; 1995:29–58.
20. Ayalon L. Retrospective reports of negative early life events over a 4-year period: a test of measurement invariance and response consistency. *J Gerontol B Psychol Sci Soc Sci*. http://dx.doi.org/10.1093/geronb/gbv087 [published online September 24, 2015].
21. Herbert TB, Cohen S. Measurement issues in research on psychosocial stress. In: Kaplan HB, ed. *Psychosocial Stress: Perspectives on Structure, Theory, Life-Course, and Methods*. New York: Academic; 1996:295–332.
22. Anderson B, Wethington E, Kamarck TW. Interview assessment of stressor exposure. In: Contrada R, Baum A, eds. *Handbook of Stress Science: Biology, Psychology, and Health*. New York: Springer; 2010:565–582.
23. Hammen C, Mayo A, Demy R, Marks T. Initial symptoms levels and the life event-depression relationship. *J Abnorm Psychol*. 1986;95:112–114.
24. Luhmann M, Orth U, Specht J, Kandler C, Lucas RE. Studying changes in life circumstances and personality: it's about time. *Eur J Personal*. 2015;28:256–266.
25. Wethington E. The relationship of work turning points to perceptions of psychological growth and change. *Adv Life Course Res*. 2002;7:111–131.
26. Lesserman J. HIV disease progression: depression, stress and probable mechanisms. *Biol Psychiatry*. 2003;54:295–306.
27. Anda RF, Butchart A, Felitti J, Brown DW. Building a framework for global surveillance of the public health implications of adverse childhood experiences. *Am J Prev Med*. 2010;39(1):93–98.
28. Umberson D, Williams K, Thomas PA, Liu H, Thomeer MB. Race, gender, and chains of disadvantage: childhood adversity, social relationships, and health. *J Health Soc Behav*. 2014;55(1):20–38.
29. Williams DR. Race, socioeconomic status, and health: the added effects of racism and discrimination. *Ann N Y Acad Sci*. 1999;896:173–188.
30. Lepore SJ. Measurement of chronic stressors. In: Cohen S, Kessler RC, Gordon LU, eds. *Measuring Stress: A Guide for Health and Social Scientists*. New York: Oxford University Press; 1995:102–120.
31. Stone AA, Neale JM. Development of a methodology for assessing daily experiences. In: Baum A, Singer JE, eds. *Advances in Environmental Psychology: Environment and Health*; vol. 4, Hillsdale, NJ: Erlbaum; 1982:49–83.
32. Kamarck TW, Shiffman S, Wethington E. Measuring psychosocial stress using ecological momentary assessment. In: Contrada R, Baum A, eds. *Handbook of Stress Science: Biology, Psychology, and Health*. New York: Springer; 2010:597–618.
33. Almeida DM, Wethington E, Kessler RC. The daily inventory of stressful events (DISE): an investigator-based approach for measuring daily stressors. *Assessment*. 2002;9:41–55.
34. Almeida DM, Stawski RS, Cichy KE. Combining checklist and interview approaches for assessing daily stressors: the daily inventory of stressful events. In: Contrada RJ, Baum A, eds. *Handbook of Stress Science*. New York: Springer; 2010:583–595.
35. Kanner AD, Coyne JC, Schaefer C, Lazarus RS. Comparison of two modes of stress measurement: daily hassles and uplifts versus major life events. *J Behav Med*. 1981;4:1–39.
36. Serido J, Almeida DM, Wethington E. Chronic stressors and daily hassles: unique and interactive relationships with psychological distress. *J Health Soc Behav*. 2004;45:17–33.
37. Jodczyk S, Fergusson DM, Horwood J, Pearson JF, Kennedy MA. No association between mean telomere length and life stress observed in a 30 year birth cohort. *PLoS One*. 2014;9(5):e97102.

CHAPTER

13

Psychological Stressors: Overview

S.M. Monroe[1], G.M. Slavich[2]

[1]University of Notre Dame, Notre Dame, IN, USA [2]University of California, Los Angeles, CA, USA

OUTLINE

Abstract

Psychological stressors are social and physical environmental circumstances that challenge the adaptive capabilities and resources of an organism. These circumstances represent an extremely wide and varied array of different situations that possess both common and specific psychological and physical attributes. The challenge for theory, research, and practice is to abstract and understand the specific qualities and characteristics of environmental exposures that most strongly elicit noxious psychological and biological responses, which in turn can lead to serious mental and physical health problems over the life course. In the present article, historical perspectives and conceptual considerations are addressed first, which provides the context for the subsequent discussion of key issues for defining and assessing psychological stressors. Susceptibility to psychological stressors is subject to individual differences, which can alter the impact and adverse consequences of such environmental exposures, necessitating a discussion of these moderating influences as well.

HISTORICAL AND GENERAL CONSIDERATIONS

Historical Matters

The term "stress" has become very popular in contemporary society, and is commonly invoked to explain a wide variety of psychological and medical problems. Along with this fashionable trend, it is commonly assumed that psychological stressors represent a concern of particularly present-day origins, or at least that they have become much more prominent and pervasive with advances in modern technologies and the apparent quickening pace of life. As a consequence of these perceived pressures, it is also commonly believed that with this accelerating progress of civilization, more people are succumbing to mental and physical disorders than ever before.

Historical accounts, however, caution against such limited perspectives and suggest that similar ideas about stressors, civilization, and disease have been common for quite some time. Sir Clifford Allbutt[1] expressed such sentiments quite clearly well over 100 years ago, writing:

> To turn now...to nervous disability, to hysteria...to the frightfulness, the melancholy, the unrest due to living at a high pressure, the world of the railway, the pelting of telegrams, the strife of business...surely, at any rate, these maladies or the causes of these maladies are more rife than they were in the days of our fathers?

Stress: Concepts, Cognition, Emotion, and Behavior
http://dx.doi.org/10.1016/B978-0-12-800951-2.00013-3

KEY POINTS

- The terms "stress" and "psychological stressors" have a long and varied semantic history.
- Contemporary usage of these terms often reflects these varied, and often, vague meanings.
- More precise definitions and scientifically useful measures of psychological stressors have come from animal laboratory research as well as human experimental and field studies.
- Important themes for better understanding psychological stressors include the specificity of effects associated with particular types of psychological stressors and individual differences in response to psychological stressors.

The tendency to view life as being challenging or stressful may be even more basic to human cognition than is readily apparent. The Greek myth of Sisyphus is enlightening in this regard. The perpetual work of pushing a boulder up a mountain—only to have gravity bring it back down after each and every effort—captures some of the qualities and characteristics linked to modern views of psychological stressors and the challenges of everyday life. Perhaps there is something fundamental about the human condition and psyche that fosters a perception of the world as a place rife with unrelenting demands that can never be fully met, resulting in common subjective states of fatigue and distress that can lead to ill health. Each era may bring its unique colorations to such perceptions, and its own attributions regarding their origins.

It is against this perhaps universal psychological backdrop of belief and bias in thinking that modern work on psychological stressors must be critically examined. Psychological stressors and related concepts have been popular explanatory devices throughout recent, and not-so-recent, history. As a result of their phenomenological allure and tempting explanatory power, these ideas have often been loosely formulated and accepted at "face value." Owing to conceptual fuzziness and ambiguity, not only has progress in science been slowed, but nonscientific issues and ideas are permitted to masquerade as scientific truths.

The concept of psychological stressors is rich with possibilities for shedding light on important matters in adaptation, dysfunction, and disease. The concept is paralleled, though, by the potential pitfalls that may accompany its intuitive, yet potentially misleading, appeal. The challenge is to translate the fertile ideas about psychological stressors into more precise concepts, definitions, and operational procedures. With more sound definitional and methodological procedures in place, the utility of stress concepts for understanding adaptation and maladaptation in relation to mental and physical disorders will be better understood.

Early Ideas and Research

A broad template for understanding the organism's reactions to challenging environmental circumstances was laid down by Claude Bernard and Charles Darwin during the nineteenth century. Each of these influential individuals in his own way touched on the tension resulting from the ongoing adaptation of the organism to changing and challenging environmental circumstances.[2] Yet it was not until the early-to-mid twentieth century that such generality and complexity was translated into more specific terminology and technologies. These efforts can be traced to at least three different lines of thought and research.

The early work of Walter Cannon dealt with ideas about common emotions and their physiological consequences, particularly with respect to the body's maintenance of homeostasis.[2] This line of study was complemented shortly thereafter by the animal laboratory studies of Hans Selye, wherein acute and severe stressors were systematically investigated. It was in Selye's work that the concept of stress most forcefully emerged. Stress was defined in terms of "the nonspecific response of the body to any demand" (Ref. 3, p. 74). Stressors, in turn, were defined as "that which produces stress" (Ref. 3, p. 78). Yet from another vantage point, Adolph Meyer popularized the "life chart" methodology. This approach emphasized the importance of the dynamic interplay between biological, psychological, and social factors, such that important life events became focal points for studying health and disease. Collectively, these activities, and the multiple lines of research they generated, served to initiate specific awareness of, and interest in, psychological stressors.[4]

Other developments arising outside of science also contributed to the emerging idea that psychological stressors could lead to both mental and physical disorders. Prior to World War II, psychopathology was predominantly attributed to genetic factors or to acquired biological propensities; so-called "normal" people without such taints were thought to be largely invulnerable to serious mental illness. However, World War II dramatically altered thinking in medical and psychiatric circles to incorporate the idea that severe stress could precipitate breakdowns in previously healthy individuals.[2,4] Once this conceptual shift began, it underscored the multiplicity of health consequences associated with severe stressors. It also opened the door for enlarging conceptual perspectives on psychological stressors for considering how less severe, yet still aversive, aspects of the social and physical environment might also promote pathology.

CONCEPTUAL DEVELOPMENTS

Upon the foundations of stress research and theory laid down by Selye, Cannon, and Meyer, along with the influences of experiences of World War II, contemporary inquiry into the effects of psychological stressors became a topic of increasing interest and, eventually, of extensive empirical inquiry. Two general themes can be discerned that underpinned advances in theory: first, characteristics of psychological stressors, and second, individual differences in response to psychological stressors.

Stressor Characteristics

Despite general agreement about the importance of psychological stressors for health and well-being, determining exactly what it "is" about stressful circumstances that is deleterious has proven challenging. An initial question of considerable theoretical importance involved the basic nature of psychological stressors: Are they better viewed in a unitary manner as "nonspecific demands" on the organism (as Selye postulated) or as a class of conditions that harbor specific bodily demands? Investigators from two traditions—animal and human research—addressed this issue, with parallel and sometimes intersecting developments. Although considerable progress was made, stressor characteristics remain one of central topics of importance in current thinking on psychological stressors.[4–6]

Animal Laboratory Research

A great deal of work in the 1960s and 1970s addressed whether specific psychological characteristics of stressors possess qualitatively distinct implications for the organism. Initially this work revealed how particular features associated with environmental stressors might be important for adverse outcomes (as opposed to the more psychologically neutral general, or nonspecific, adaptive demands). Such research went on to probe different types of psychological stressors and their effects. It became of central interest to understand in a more differentiated way the effects of diverse psychological stressors.

Animal laboratory studies adopted ingenious ways to differentiate psychological components associated with environmental stressors. The findings from these studies demonstrated that distinctive psychological characteristics were responsible for many immediate behavioral or physiological responses. For example, specific psychological characteristics of stressors, such as undesirability or controllability, were important for the development of various disorders.[7] It became clear, too, that other characteristics of stressors were pertinent. For example, different parameters of shock administration (acute, intermittent, or chronic) produced distinctive physiological effects in animals. Further, such differences could increase, decrease, or not influence the development of particular diseases.[2] Lastly, psychological stressors could not only influence immediate psychobiologic functioning, but also have long-term effects by permanently altering the psychobiological characteristics of the organism.[2]

As the importance of specificity of stressor or "stimulus" characteristics became apparent, questions about the specificity of stress *responses* also arose. What were the implications of specific stressor characteristics for different facets of psychological and physiological functioning? Such theoretical developments greatly extended the framework for inquiry, requiring attention to multiple characteristics of stressors in relation to multiple psychological and biological processes and outcomes. Relatively simple, singular response indices (e.g., corticosteroids, catecholamines) were replaced by more complex patterns of behavioral and biological effects, or profiles of neuroendocrine responses. Other intriguing levels of conceptualization have been proposed. For example, psychological stressors may promote fundamental disruptions in oscillatory regulation of basic biological functions, or reversions to earlier modes of functioning.[2]

Overall, research on psychological stressors from animal research has moved beyond unidimensional and linear concepts of stressors and their effects. More recent thinking has adopted a larger framework for understanding the diverse characteristics of stressors that influence particular response systems of the organism. The systems of interest have expanded from single systems to patterns or profiles of response across multiple indices.

Human Experimental and Field Studies

Investigators of psychological stressors in humans also conducted innovative laboratory and field studies.[8] The early work focused on the aversive subjective attributes, particularly perception or appraisal, of psychological stressors as evaluated in an experimental setting.[8] Research on stressful life events also began around this time, and it is in this area that research on psychological stressors perhaps reached its pinnacle in terms of both productivity and popular interest.

Extrapolating from animal laboratory studies on the one hand, and integrating with Meyer's life chart procedures on the other, Thomas Holmes and Richard Rahe first formulated the idea that distinctive changes in one's life circumstances—specific and documentable life events—could be defined and assessed in an objective manner. The work was initially based on case histories of some 5000 tuberculosis patients, from which they derived a list of 43 life events "empirically observed to occur just prior to the time of onset of disease, including, for example, marriage, trouble with the boss, jail term, death of spouse, change in sleeping habits, retirement, death in the family, and vacation" (Ref. 9, p. 46). The Schedule of Recent Experiences (SRE) was developed and published,[10] and by 1978 alone, over 1000 publications had utilized this convenient method for probing questions pertaining to stress and illness.[9]

The common feature associated with these disparate life changes—the stressor characteristic of primary concern—was thought to be the degree of social readjustment caused by the event: "The relative importance of each item is determined not by the item's desirability, by the emotions associated with the item, nor by the meaning of the item for the individual; it is the amount of change that we are studying and the relationship of the amount of change to the onset of illness" (Ref. 9, p. 47). This viewpoint is consistent with Selye's ideas about stressors and stress (i.e., stress as the nonspecific response of the body to any demand). Hence, the psychologically neutral notion of the "readjustment" required of life changes was conceptualized as the characteristic responsible for vulnerability to a wide variety of psychological and physical maladies.

Much as the emphasis in animal laboratory studies shifted from psychological neutral concepts of "any demand," the emphasis in the stressful life events literature shifted from the neutral concept of "readjustment" to concepts involving the undesirable social-psychological characteristics of events. Human studies of life events consequently began to focus on the particular characteristics of psychological stressors and their potentially unique effects. The principle of specificity also was extended from the characteristics of stressors to the specific consequences of such experiences, elaborating theory about the importance of specific psychological stressors for specific responses and eventually for specific types of disorder or disease.[5] A vast literature on this topic exists, with diverse conceptualizations of psychological stressors and myriad methods to measure them.[6,11]

Most recently, researchers have focused on interpersonal loss and social rejection as psychological characteristics that may make some experiences particularly deleterious for health and well-being.[12] Life events with these qualities, called "targeted rejection" events, are among the strongest precipitants of depression.[13] Additionally, there is some evidence that targeted rejection events uniquely trigger biological responses that promote disease.

Research on the desirability of events, along with the more general issue involving stressor characteristics, brought into focus another important topic in the study of psychological stressors and their impact on health and well-being: individual differences. What one person might experience as being undesirable, another person could experience as being desirable. As discussed next, a variety of considerations come into play for explaining variability in the effects of psychological stressors on health.

Individual Differences

There is considerable variability in response to psychological stressors across individuals. Even under extremely stressful conditions, not all animals or individuals breakdown. Additional factors are useful to effectively model the variability in effects attributable to psychological stressors. Progress in understanding this matter has again come from both the basic laboratory and human studies of psychological stressors.

Animal Laboratory Research

Although there were characteristic features of physiologic responses to the stressors employed in the early paradigm adopted by Selye, not all animals responded to stress in an identical manner. Further, individual differences in response were even more pronounced when the less severe types of stressors were used.

Factors such as prior experience, availability of "coping" responses, and attributes of the social and experimental context (e.g., social ties) were found to moderate the influence of psychological stressors. For example, when rats are exposed to electric shock, animals that cannot predict shock occurrence (via warning tones) develop a sixfold increase in gastric ulceration compared to their yoked counterparts (who receive the warning tones).[2,7] Additional research demonstrated the delicate and often subtle interplay between stressor, social context, and resources available to the organism in moderating response outcomes. These lines of study, too, suggested that individual differences in susceptibility also could be viewed within a dynamic and developmental framework over time. For instance, laboratory animals repeatedly exposed to severe psychological stressors can become neurobiologically sensitized to the stressors, such that progressively less severe degrees of stress acquire the capability of triggering the pathogenic responses.[14,15] Moreover, considerable animal research has now demonstrated that exposure to stressors early in life can have long-lasting effects on stress reactivity over the life course.[16]

Human Life Stress Research

The importance of individual differences is perhaps most apparent in studies of human life stress and its consequences. A consistent criticism of life events research was the relatively weak association between psychological stressors and disorder. It was assumed that many considerations moderated stress effects, and the elucidation of such factors would increase the predictive strength of the association between psychological stressors and disorder. Again, a number of factors were believed to influence the impact of psychological stressors, ranging from environmental factors, such as the availability of social support, to more individual factors, such as prior experience and coping abilities. Developmental considerations also are important in recent theorizing about individual differences in reaction to psychological stressors, with the idea that prior exposure to severe psychological stressors renders individuals more sensitive and thereby susceptible to increasingly lower levels of psychological stress.[14]

A major arena for understanding individual differences in response to psychological stressors has been the topic of perception. The early and elegant laboratory studies of human stress indicated the importance of such individual differences in perception, or appraisal, of stressors, and such thinking was readily incorporated into theory and method.[8] Studies of life events, for example, would use subjective weights of events experienced by the study participants. Once this avenue of inquiry was opened, it also brought to the forefront a variety of influences on perception, along with other factors that might influence stress responsivity. Thus, research not only began to focus on appraisal of stressors, but also on coping, social support, personality, and other considerations that in theory could moderate the effects of psychological stressors.

Most recently, research has examined genetic factors that may shape health risk following psychological stressors. Some candidate polymorphisms have been identified, but empirical support for these factors has been mixed, likely due in part to poor stress measurement and the fact that polymorphisms exert effects only when genes are "turned on" by certain environmental influences.[17] As a result, a new field of research on "human social genomics" has emerged that examines how different psychological stressors activate genes that are relevant for health.[18]

As research progressed, it became clear that making some distinctions was easier in theory than in practice. Although it made good sense to consider an individual's subjective perception of psychological stressors, for example, employing such information in scientifically sound manner was difficult. When it came to measurement, serious problems became apparent. For example, owing to depression-based perceptual biases, a depressed person might have a skewed perception of events and rate them as particularly negative (irrespective of the objectively stressful qualities per se). Such concerns raised a paradox for investigations. Namely, while a large part of what one wants to know about pertains to the individual's personal appraisal of psychological stressors, methodological concerns caution against direct use of such information. Consequently, alternative approaches were developed to avoid the pitfalls of using subjective reports and associated problems with these methods.

METHODOLOGICAL CONSIDERATIONS AND RECENT DEVELOPMENTS

While concepts and methods intertwine and, united, nurture progress, at times one or the other component may unduly influence development (for good or for bad). This comment is applicable to research on psychological stressors, where the methods adopted in animal laboratory research have constrained theory, and where methods adopted in human life stress research have misled theory on psychological stressors.

Animal Laboratory Research

The original work of Selye typically employed situations that were overpowering or unavoidable for animals. Such conditions did not permit an evaluation of behavioral responses or of other moderating influences that could influence an animal's adaptation to stressors. Further, it was realized that this paradigm did not provide information about responses to stressors of high ecological and evolutionary relevance, such as those found in the animal's natural environment and evolutionary history. Thus, such an approach masked the implications of less severe but more normative psychological stressors on physiology and behavior, which in turn might represent a more fertile area of inquiry into stressor effects.[2] Finally, the nature of the stressor employed in the early animal laboratory studies, too, contributed to the aforementioned difficulty in differentiating physical from psychological effects, which inhibited progress in the arena of conceptual development.

Overall, the range of psychological stressors was constrained by the methods adopted. Theory, in turn, was constrained to account for the consequences of stressors under such restricted and relatively unnatural environmental conditions. More recent research has benefited from methods involving the assessment of a more diverse array of psychological stressors that incorporate the assessment of a wider variety of behavioral and biological response possibilities. Current perspectives based on these broader methodological approaches suggest that the organism's responses are often "exquisitely specific" nuances of stressors encountered.[2]

Human Studies of Stressful Life Events

The bulk of empirical work on human life stress has been based on self-report checklist methods. The prototype of this approach is the SRE, the instrument that catalyzed research on the topic. The popularity of the SRE was due to the combination of the intuitive appeal of the stress concept, the ease and apparent objectivity of the method, and the overall impression of scientific legitimacy.

The methodological paradigm launched by the SRE, however, embodied several problems. It became clear that subjects did not report life events in a reliable manner over time, and that investigators did not adequately control for the directionality of effects in research designs (e.g., being depressed could bring about life events such as "trouble at work," "difficulties with spouse," and so on). Indeed, many of the initial items on the SRE were direct indicators of disorder or illness. For example, some of the key criteria for defining clinical depression were represented in the original SRE ("Major change in eating habits," "Major change in sleeping habits"). If measures of life events were directly confounded with the presence of disorder, or contaminated by the effects of pre- or coexisting disorder, then clearly general theory about psychological stressors, as well as theory about the characteristics of psychological stressors, rested on flawed information.[5]

In response to these methodological concerns, investigators designed semistructured interview protocols and developed explicit guidelines, decision rules, and operational criteria for defining and rating life events.[4,5] These developments further highlighted serious problems with self-report checklist methods. For example, there is too much subjective leeway permitted in defining what constitutes an "event" with self-report procedures, resulting in unacceptable variability of content within ostensibly uniform categories of events.[4] To have a more firm methodological foundation, more elaborate and extensive interview and rater-based procedures were employed, helping to standardize measurement across individuals.

In general, interview and rater-based approaches enhance the reliability of life event assessments and provide stronger predictions of particular kinds of disorders following the occurrence of psychological stressors.[5,6,19,20] Procedures such as these, too, provide a solid foundation upon which to build in terms of developing taxonomies of psychological stressors and their effects.[4] Although such approaches are more time-intensive and costly, they represent the current-day gold standard for assessing psychological stressors.

Human Studies Employing Other Measures of Psychological Stressors

Other methods have been developed for assessing psychological stressors. None of these approaches has received the degree of attention devoted to the work on life events, yet each may have useful properties for the study of psychological stressors. Two lines of investigation are noteworthy.

First, many investigations have targeted people who experience a specific life event and compared them to controls who have not experienced the event. For example, individuals who become unemployed are compared to individuals who do not experience this event in relation to a variety of psychological and physical processes and outcomes. Such work is useful for examining a potentially more homogenous process with more readily identifiable outcomes. On the other hand, these studies may oversimplify the psychological stressors associated with an event, and not specifically articulate the different components within the general event that are most pernicious for health. For example, the effects can be partitioned into a variety of stressful themes that, although often intercorrelated, may not have uniform effects. Thus, although people who become unemployed in general may experience as loss of self-esteem, loss of income, loss of daily schedule, and so on, each particular situation may pull more or less for heightened responses along these different dimensions. Research sensitive to variability in the component characteristics will be most useful for research on psychological stressors.

Finally, there also have been efforts to measure psychological stressors through questionnaire or diary methods, inquiring about less major but common daily experiences, chronic conditions, appraisal processes, and other indicators or correlates of psychological stressors.[11] A promising recent avenue of research involves ecological momentary assessment, where subjects can be prompted throughout the day to respond to queries about their circumstances and psychological states. Such procedures help minimize problems with standard retrospective methods, although may still pose challenges for reliably assessing major types of life events.[21]

In closing, it is appropriate to return to the concerns and caveat with which the discussion began. The specter of biases in the measurement of psychological stressors consistently must be borne in mind, and methods employed must be rigorously attentive to such concerns, to provide a solid empirical foundation upon which theory and research can build for this important area of investigation.

References

1. Allbutt C. Nervous diseases and modern life. *Contemp Rev*. 1895;67:217.
2. Weiner H. *Perturbing the Organism: The Biology of Stressful Experience*. Chicago, IL: The University of Chicago Press; 1992.
3. Selye H. *The Stress of Life*. 2nd ed. New York, NY: McGraw-Hill; 1976.
4. Dohrenwend BP, ed. *Adversity, Stress, and Psychopathology*. New York, NY: Oxford University Press; 1998.

5. Brown GW, Harris TO, eds. *Life Events and Illness*. London: Guilford Press; 1989.
6. Monroe SM. Modern approaches to conceptualizing and measuring life stress. *Annu Rev Clin Psychol*. 2008;4:33–52.
7. Weiss JM. Psychological factors in stress and disease. *Sci Am*. 1972;226:104–113.
8. Lazarus RS. *Psychological Stress and the Coping Process*. New York, NY: McGraw-Hill; 1966.
9. Holmes TH. Development and application of a quantitative measure of life change magnitude. In: Barrett JE, ed. *Stress and Mental Disorder*. New York, NY: Raven Press; 1979:37–53.
10. Holmes TH, Rahe RH. The social readjustment rating scale. *J Psychosom Res*. 1967;11:213–218.
11. Cohen S, Kessler RC, Gordon LU, eds. *Measuring Stress: A Guide for Health and Social Scientists*. New York, NY: Oxford University Press; 1995.
12. Slavich GM, Irwin MR. From stress to inflammation and major depressive disorder: a social signal transduction theory of depression. *Psychol Bull*. 2014;140:774–815.
13. Monroe SM, Slavich GM, Georgiades K. The social environment and depression: the importance of life stress. In: Gotlib IH, Hammen CL, eds. *Handbook of Depression*. 3rd ed. New York, NY: Guilford Press; 2014:296–314.
14. Monroe SM, Harkness KL. Life stress, the 'kindling' hypothesis, and the recurrence of depression: considerations from a life stress perspective. *Psychol Rev*. 2005;112:417–445.
15. Post RM. Transduction of psychosocial stress into the neurobiology of recurrent affective disorder. *Am J Psychiatr*. 1992;149:999–1010.
16. Zhang T-Y, Meaney MJ. Epigenetics and the environmental regulation of the genome and its function. *Annu Rev Psychol*. 2010;61:439–466.
17. Monroe SM, Reid MW. Gene-environment interactions in depression: genetic polymorphisms and life stress polyprocedures. *Psychol Sci*. 2008;19:947–956.
18. Slavich GM, Cole SW. The emerging field of human social genomics. *Clin Psychol Sci*. 2013;1:331–348.
19. Dohrenwend BP. Inventorying stressful life events as risk factors for psychopathology: toward resolution of the problem of intracategory variability. *Psychol Bull*. 2006;132:477–495.
20. Hammen C. Stress and depression. *Annu Rev Clin Psychol*. 2005;1:293–319.
21. Shiffman S, Stone AA, Hufford MR. Ecological momentary assessment. *Annu Rev Clin Psychol*. 2008;2005(4):1–32.

CHAPTER

14

Remodeling of Neural Networks by Stress

B. Czéh[1], E. Fuchs[2]

[1]MTA—PTE, Neurobiology of Stress Research Group, Szentágothai Research Center, University of Pécs, Pécs, Hungary
[2]German Primate Center, Göttingen, Germany

OUTLINE

Abstract

Challenging events in an individual's environment initiate a cascade of adaptive physiological and psychological processes that constitute the stress response. Reactions to stress are mediated through a concerted activity of many brain areas and peripheral endocrine organs. There is experimental evidence that various forms of stress, especially when severe or repeated, can induce diverse structural and neurochemical/neurophysiological changes in neuronal and glial networks. These findings demonstrate that even the adult and differentiated brain is a plastic organ. Many of these stress-induced structural and functional changes in the brain are thought to contribute to the development of various types of neuropsychiatric disorders, but animal experiments provide evidence that in fact many of these alterations are reversible given sufficient recovery periods.

EARLY IDEAS ABOUT THE STRESS RESPONSE

When Selye formulated his stress theory nearly 80 years ago, stress was thought to be merely a endocrine character, and noxious stimuli of physical or chemical nature were the primary stressors discussed at that time.[1] However, subsequent research demonstrated that stress-induced activation of the endocrine system leads to increased release of glucocorticoids (cortisol in primates including man, corticosterone in rodents) from the adrenal cortex and that these peripheral "stress hormones" profoundly influence brain function.[2] Within the last decades, great progress has been made in understanding the biology of stress reactions and the neurobiological alterations that occur as a consequence of exposure to challenging events in the environment.

ACTIVATION OF BRAIN SYSTEMS

Stress does not have a "global" effect on all brain areas. Rather, depending on the type of stressor, different neuronal circuits are activated. Limbic brain regions comprising, for example, the hippocampus and amygdala, as well as the prefrontal cortex, are sensitive to stressors such as restraint or anxiogenic external stimuli, for example, a novel environment.[3] Common to these stressors is that they require, prior to initiation or inhibition of the stress response, a central nervous processing of information coming from different sensory organs. This information processing involves limbic and cortical regions, and its outcome depends on previous experience. By contrast, physiological threats, such as exposure to ether, result in a direct activation of efferent visceral pathways, in part mediated by the paraventricular nucleus of the hypothalamus.[4] In this case, the rapid activation of the brain stem

Stress: Concepts, Cognition, Emotion, and Behavior
http://dx.doi.org/10.1016/B978-0-12-800951-2.00014-5

and hypothalamus circumvents cognitive-emotional processing via the "higher" brain regions.

KEY POINTS

- Within the last decades, our view of the mature mammalian brain has changed. It is far from being fixed and immutable, because a number of factors such as environmental stimulation, stress, adrenocortical and gonadal hormones, aging, and learning can rearrange neuronal networks including the production of new neurons.
- This capacity has forced us to look anew at plasticity of the brain not only on the neuronal but also on the glial level.

The neuronal crosstalk that finally leads to a well-organized stress reaction is brought about by neurotransmitter systems.[5] Rapid neuronal responses to stress are elicited by, for example, glutamate (the primary excitatory neurotransmitter in the brain) and gamma-amino butyric acid (GABA, the primary inhibitory neurotransmitter in the brain).[5] These neurotransmitters act on specific receptors and ion channels in neuronal membranes to directly control neuronal firing. This fast neurotransmission is modulated by slow-acting neurotransmitters such as the monoamines (e.g., the catecholamines dopamine, noradrenaline, adrenaline, and the indolamine serotonin) and various neuropeptides.[5] Activation of brain monoamine systems is a major component of the stress response, and turnover of monoamines is increased during stressful experiences. Stress-induced alterations in the monoamine system are regarded as the basis for stress-related behavior and are of special interest because the hyperactivity of catecholaminergic and serotonergic neurons that occurs during stress may induce an imbalance between the different neurotransmitter systems.[6,7] These imbalances are thought to contribute to stress-related psychopathologies, such as major depressive disorder.[6,8]

Another example of substances that modulate neuronal activity and are involved in the mediation of stress responses are neuropeptides, for example, corticotropin-releasing factor, which is thought to play a major role in the stress response.[9] Neuropeptides and monoamines exert their actions via specific receptors located in the membranes of neurons. Some of these receptors are also found in glia cells.[10] Many of these receptors belong to the family of G-protein-coupled receptors (GPCRs) that are linked to intracellular second messenger systems via guanosine 5′-triphosphate-binding proteins. The intracellular signal transduction pathways that are regulated by GPCRs bring about many of the short- and long-term effects of stress on neuronal activity and regulate most aspects of neuronal functioning, including metabolism and gene expression.[11,12] Thus, intracellular signal transduction systems play a key role with respect to stress-mediated regulation of neuronal functioning and structure and, on the cellular level, they represent the first step in a cascade of events in the brain that leads to changes in neuronal cells including alterations in gene transcription and morphology.[8] Adverse effects of stress are thought to result partly from (i) sustained activation or inhibition of certain neurotransmitter systems and intracellular pathways and/or from (ii) chronic exposure to neuroactive substances, such as the glucocorticoids.[2,13–15] Glucocorticoids are steroid hormones that interact with intracellular receptors forming complexes that bind to specific sites on DNA to regulate transcription.[2,13]

STRUCTURAL PLASTICITY OF THE MATURE BRAIN

The classic view on the neuroanatomy of the adult mature brain was that it has a static structure and electrical and chemical information is processed by the fixed circuits of neurons. Ramón y Cajal stated that "In adult centers the nerve paths are something fixed, ended, immutable. Everything may die, nothing may be regenerated."[16] However, in recent years, this view has been gradually revised on the basis of studies demonstrating that neuronal circuits and connections between cells in the brain are subject to lifelong modification and reorganization. An impressive example of the consequences of alterations in sensory input and changes in circulating hormones comes from research in songbirds. In the early 1980s, a substantial neurogenesis was demonstrated in a vocal control nucleus of the adult canary brain, and a functional link between behavior, song learning, and the production of new neurons was established.[17] The finding that in songbirds (canaries, zebra finches) males have larger song control nuclei in their brains compared with females indicated that the number of neurons in those adult birds may change with the season.[18] Indeed, the neuron number in song control nuclei increases in spring when male zebra finches begin to sing, and new-born neurons are found in the hyperstriatum ventrale pars caudalis (HVC) of adult canaries.[19] Studies on the HVC in birds show that steroid hormones play important roles in these processes of neuroplasticity, in particular the gonadal hormone testosterone. In ground squirrels, a hibernating mammalian species, it was shown that synaptic contacts of mossy fibers with CA3 hippocampal pyramidal neurons are altered in many aspects of their structure during different stages of the torpor-activity cycle.[20] These findings clearly demonstrate the capacity of the adult brain to adjust its neuronal circuits to altered hormonal and/or sensory input. Based on such findings, it was hypothesized that similar structural modifications may also occur during stressful life events when hormonal levels are altered (e.g., elevated glucocorticoids, reduced gonadal steroids) and when sensory input is changed. It has transpired that this view is accurate.

This ability of the brain to perform functionally relevant adaptations following various challenges is called plasticity. *Neuronal plasticity* is mandatory for adequate functioning of an individual in a continuously changing environment.[21] The dynamic processes of neuroplasticity are based on the capability of neurons to adapt to alterations in the internal and/or external environment by modifying specific structures and function, and they do this primarily by changing their synaptic contacts with other neurons eventually changing their axon projections and dendritic architecture.[21]

Adaptive or experience-dependent plasticity reveals itself in many forms ranging from changes in gene expression to changes in neurotransmitter release, and in behavior. Even under normal or undisturbed conditions, contacts between neurons are continuously replaced and renewed in the adult and differentiated brain.[21] Enhanced axonal outgrowth and collateral sprouting on the presynaptic site may lead to the formation of new synapses, and existing contacts between neurons may be eliminated by terminal retrograde degeneration. The number of postsynaptic sites on a neuron can be increased or decreased by alterations in the size of its dendritic tree or the spine density on the dendrites.[22]

REMODELING OF BRAIN CELLS BY STRESS

Probably the most thoroughly investigated stress-induced change in neuronal morphology is the regression of apical dendrites of pyramidal neurons, which was first demonstrated in the hippocampus (Figure 1).[23] Dendritic remodeling of CA3 pyramidal neurons has been repeatedly documented after various chronic stress protocols, and after corticosterone administration.[24] It is assumed that as a result of the reduced surface area of the neurons, synaptic contacts are also diminished.[25] Indeed, a significant loss of synapses on CA3 pyramidal cells and profound changes in the morphology of the afferent mossy fibers that terminate on these neurons (rearrangements of synaptic mitochondria and vesicles at the presynaptic terminals) were detected in chronically stressed or corticosterone-treated animals.[24,25] However, if the stress exposure is terminated, then the complexity and length of the dendritic arbor recover.[26] Interestingly, even a brief social defeat stress with a long time delay thereafter can significantly *reduce* the length of apical dendrites, but *increase* the length and complexity of the primary dendrites at the basis of the pyramidal neurons.[27] These data suggest that a brief social conflict is sufficient to drive a dynamic reorganization of neuronal networks with site-selective elimination and *de novo* growth of dendritic branches, changes that may persist for several weeks after the acute stress exposure.

Further studies demonstrated that pyramidal cells of the medial prefrontal cortex also respond to chronic stress with reduced apical dendritic complexity and spine loss (Figure 1).[28–30] Dendritic spines are of particular relevance because they are critically involved in the storage of information.[31] In addition to changes in spine density, chronic unpredictable stress reduces the density of synapses in the rat prefrontal cortex.[29,30,32]

Consistent with circuit-specific effects of stress, increases were also found in the dendritic arborization, for example, in the orbital frontal cortex,[33] in the basolateral amygdala and nucleus accumbens, where stress generally results in hypertrophy of dendritic arborization and increases in spine density (Figure 1).[34] Notably, it appears that the amygdala displays differential structural and functional responses to stress that are dependent on a variety of factors, including the type of stressor and the duration of the stress paradigm.[35]

Contrary to conclusions from initial studies proposing that increased glucocorticoid levels resulting from severe or chronic stress can induce apoptotic cell death of pyramidal neurons in the hippocampus, recent stereological investigations failed to detect a statistically significant loss of these neurons following prolonged hypercortisolism resulting from stress exposure, corticosteroid administration, or aging.[36] However, it should be noted that the number of hippocampal interneurons or of neurons in the hilus was not specifically examined in these studies. Recently, evidence was provided that long-term stress may reduce the number of specific types of GABAergic interneurons especially the parvalbumin-immunoreactive GABAergic basket cells.[37] Thus, the possibility that long-term stress may induce loss of interneurons without affecting the number of principal cells in the hippocampus cannot be excluded.

STRESS SUPPRESSES NEUROGENESIS IN THE ADULT DENTATE GYRUS

It has long been a central hypothesis in neuroscience that, in the mammalian brain, the production of neurons occurs only during development and stops before puberty, and that new neurons cannot be formed in the adult brain. This widely held belief has been overturned in recent years by the extensive evidence from many mammalian species, including nonhuman primates and humans, showing that even in the adult brain, certain areas retain the capacity to generate new neurons.[38] Neurogenesis in the adult hippocampus was proposed to contribute to various forms of cognitive functions and disturbances in this form of cellular plasticity have been

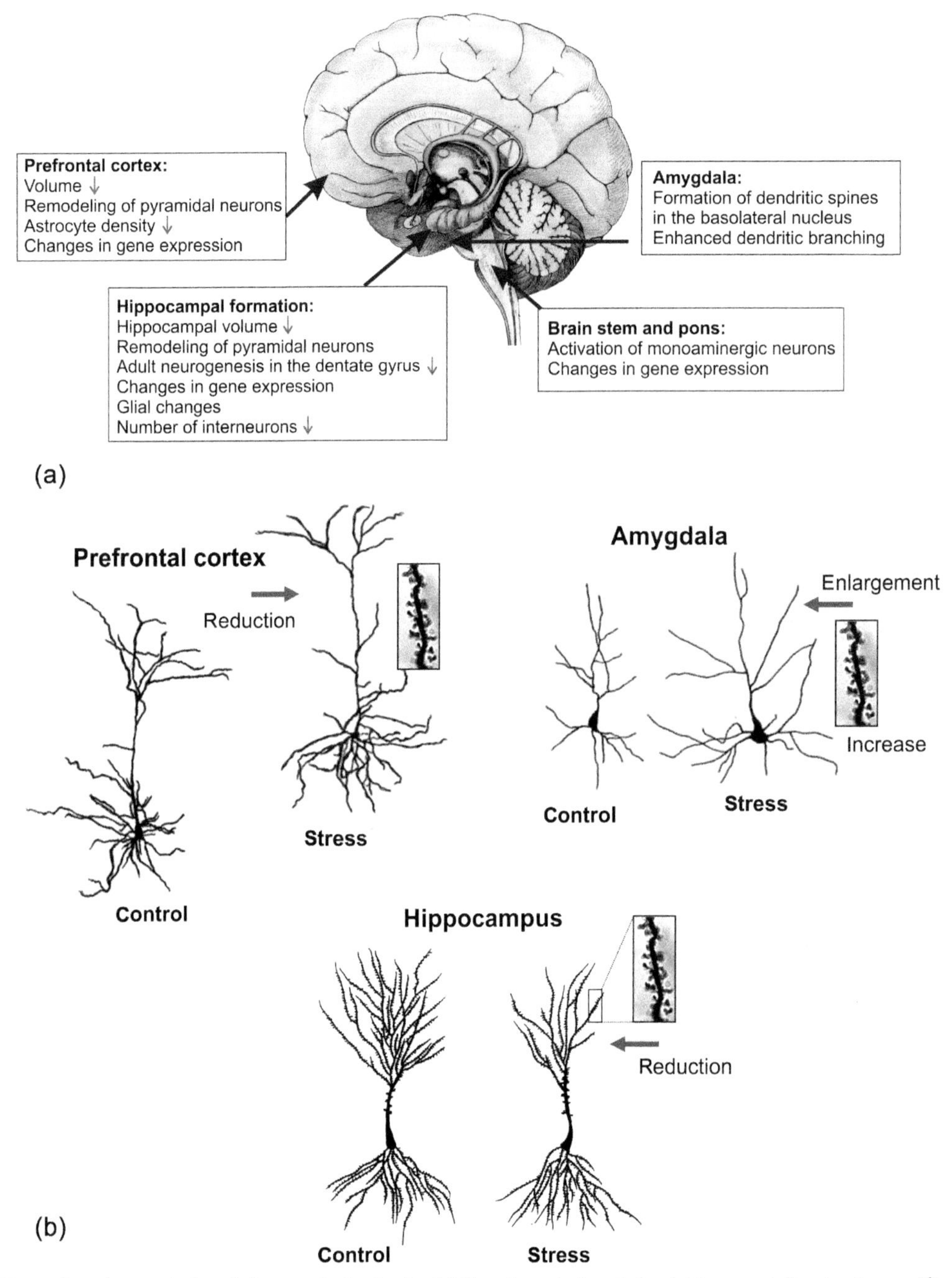

FIGURE 1 (a) Examples of stress-induced changes in the brain. (b) The stress-induced dendritic remodeling is site specific. Hippocampal formation: remodeling of pyramidal neurons characterized by shrinkage of apical dendrites and loss of spines of pyramidal neurons in region CA3. Medial prefrontal cortex: remodeling of the apical dendritic tree of pyramidal neurons together with the loss of dendritic spines. Amygdala: in the basolateral nucleus, formation of new spines on the dendrites of spiny neurons and increased complexity of the dendritic arbor.

implicated in various forms of neuropsychiatric disorders including depression, schizophrenia, epilepsy, and Alzheimer's disease.[39]

Spontaneous neurogenesis in adults takes place only in certain regions, such as the subgranular zone of the dentate gyrus in the hippocampal formation and the subventricular zone at the lateral ventricle. In the dentate gyrus, newly generated cells become incorporated into the granule cell layer and attain the morphological and biochemical characteristics of mature granule neurons (Figure 2a).[40] The neuronal nature of these cells is demonstrated by the formation of synapses on the cell bodies and dendrites, the extension of axons into the CA3 region, and the generation of action potentials.[40] Immature neurons exhibit specific electrophysiological features that make them more capable of acquiring a cellular form of learning, so-called long-term potentiation.[41] The demonstration of adult neurogenesis was facilitated by the

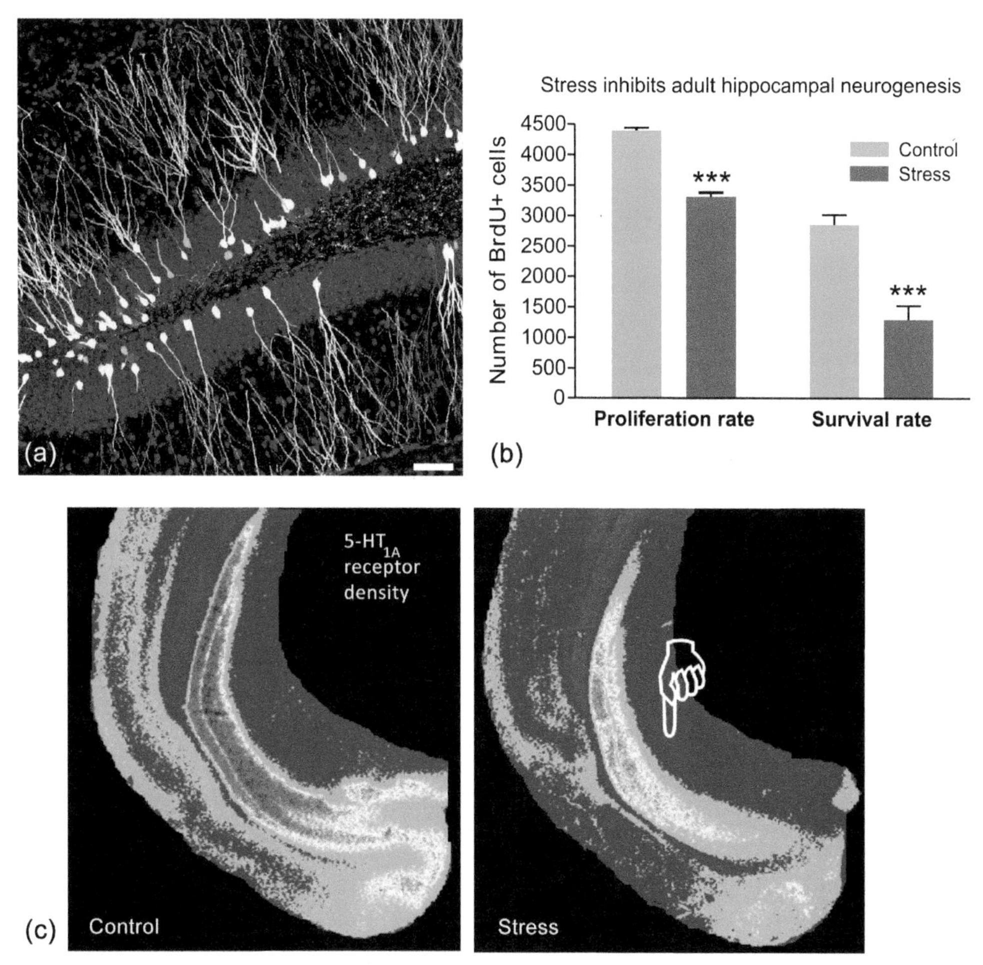

FIGURE 2 Chronic stress inhibits neurogenesis in the adult hippocampal dentate gyrus and alters serotonergic transmission in the hippocampus. (a) Representative confocal images of newborn neurons in the hippocampus (retroviral labeling). Scale bar 50 μm. (b) Chronic stress inhibits both the proliferation rate and the survival rate of newly generated cells in the hippocampal dentate gyrus. (c) Chronic stress down regulates the density of 5-HT_{1A} receptors in the hippocampus. Male tree shrews were submitted to subordination stress for 28 days, and brain 5-HT_{1A} receptors were quantified by *in vitro* receptor autoradiography. For details see Ref. 77.

advent of a novel immunohistochemistry method, BrdU labeling, and currently intensive research is being undertaken to develop sensitive and, preferably noninvasive methods, to detect the formation of new neurons not only in the postmortem brain, but also in the living organism.[42] Such methods range from targeted delivery of reporter genes by the use of viral vectors to less invasive methods such as NMR imaging, NMR spectroscopy, or measuring the concentration of nuclear bomb test-derived ^{14}C in genomic DNA.[42–44] The exact amount of newborn neurons in the adult brain is a crucial point–and an issue that is the subject of intense debate–mainly because different methods detect significantly different amounts. We still do not know whether the current estimates of the amount of newborn neurons in the human brain significantly underestimate (or perhaps overestimate) the true situation. Currently, it seems that the new neurons comprise a relatively small proportion of the total neuronal population; however, their continuous addition over an entire lifespan implies considerable structural changes within the neural network.[40,45]

Stress is a very potent suppressive factor for adult neurogenesis (Figure 2b), although a large number of environmental and endogenous parameters also have an impact on the formation of new neurons in the adult brain.[46] The suppression of cell turnover by stress predicts that the age of the cell population, the connectivity of the neurons, and the resulting properties of neuronal circuits in a stressed individual might be substantially different from control situations, with potentially important functional consequences. Several different types of stressful experiences inhibit neurogenesis in the dentate gyrus in several species, including nonhuman primates.[46] Furthermore, it appears that the susceptibility of neurogenesis to stress is age-dependent in that the effects of stress on cell proliferation in the dentate gyrus are more pronounced in old animals than in young ones.[47]

The exact underlying cellular mechanisms mediating the inhibitory effect of stress on neurogenesis are largely unknown. Adrenal glucocorticoids have been suggested as key players and mineralocorticoid, glucocorticoid, and

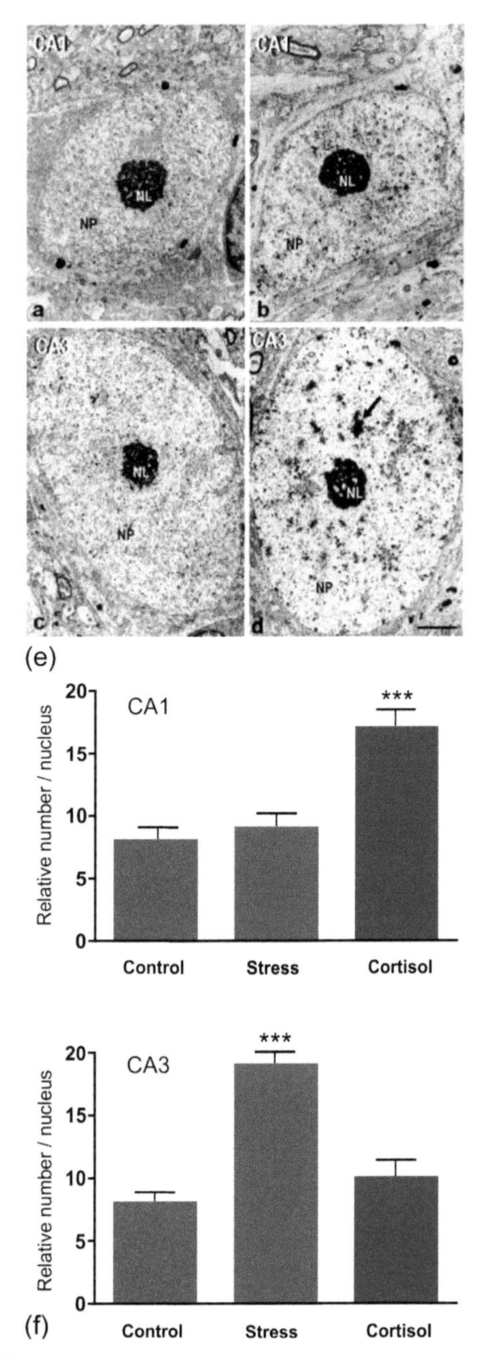

FIGURE 3 Stress and glucocorticoids affect the formation of heterochromatin in the hippocampus in a site and treatment specific way. Electron micrographs of pyramidal neuron nuclei in the hippocampus of control and stressed male tree shrews: (a) control CA1; (b) stress CA1; (c) control CA3; (d) stress CA3. Note the homogeneous nucleoplasma (NP) in the controls and the large number of heterochromatin clusters (arrow) in nuclei of CA3 pyramidal neurons in stressed animals. NL, nucleolus. Calibration bar: 2 μm. (e–f) Relative number of heterochromatin clusters in electron micrographs from nuclei of pyramidal cells in CA1 and CA3. With modifications from Ref. 48.

NMDA receptors have been identified on progenitor cells.[49] At the same time, there are several examples of the long-lasting inhibition of neurogenesis after an initial stressor, despite later normalized glucocorticoid concentrations.[50] Notably, even prenatal stress has been shown to reduce the number of newborn neurons in the hippocampus of adult offspring.[51] This suggests that while glucocorticoids may be involved in the initial suppression of cell proliferation, particularly in early life, when neurogenesis is abundant, they are not always necessary for the maintenance of this effect.

Stress also affects levels of various neurotransmitters implicated in the regulation of neurogenesis, including GABA, serotonin (Figure 2c), noradrenalin, dopamine, cannabinoids, opioids, and nitric oxide.[49] Stress further reduces the expression of several growth factors, such as brain-derived neurotrophic factor, insulin-like growth factor-1, nerve growth factor, epidermal growth factor, and vascular endothelial growth factor, which can all influence neurogenesis, while gonadal steroids should not be neglected.[49] The proximity of the precursors to blood vessels further suggests a strong interaction with the vasculature and it is this population that is particularly sensitive to stress.[52]

PLASTICITY OF GLIAL CELLS

Glia are the most abundant type of cell within the central nervous system.[53] In the adult mammalian brain, there are 10-50 times more glial cells than neurons.[53] Astrocytes are the most prominent type of glial cells, accounting for about one-third of brain mass.[54] Recent studies have revealed that, in addition to their housekeeping functions, astrocytes are dynamic regulators of synaptogenesis and synaptic strength.[55,56] Thus, abnormalities in glial functioning are likely to impair the structural plasticity of the brain. Notably, a considerable proportion of astrocytes express glucocorticoid receptors in the rodent and human hippocampus, and similarly to neurons, these cells are stress responsive.[54] In line with these ideas, it was demonstrated that long-term stress significantly decreased both the number and somatic volumes of astrocytes in the hippocampus and in several other brain areas.[57] Although these data suggest that astrocytes may die because of stress, a recent study suggests the opposite. By comparing the results of different astrocyte labeling methods, it was found that chronic stress was associated with a decrease in GFAP-positive cell numbers; however, there was no indication for astrocytic cell loss based on Nissl staining or S100β immunoreactivity.[58] This later study also showed that astrocytes respond to chronic stress by reorganizing their cellular morphology and reducing the length, complexity, and volume of their processes.

Chronic stress can also reduce the proliferation rate of glial cells. This was shown in the medial prefrontal cortex of rats subjected to 5 weeks of social defeat,[59] or to chronic unpredictable stress[60] or after chronic corticosterone administration.[61] Chronic stress also promotes significant structural remodeling of microglia, and can enhance the release of proinflammatory cytokines from microglia.[62]

Morphological changes in astrocytes most probably have functional significance with respect to neuron-glia interaction, and because this interaction supports neuronal functioning, neuronal communication is also affected by the stress-induced changes. The reduced number or weakened activity of astrocytes could change levels of extracellular glutamate, thus leading to high concentrations of this excitatory neurotransmitter that may have excitotoxic effects. An upregulation of the glial glutamate transporter (GLT-1) in the hippocampus has been reported after chronic stress, and it has been suggested that this might be a compensatory mechanism to control for the increased extracellular glutamate concentrations caused by stress.[63] Therefore, the modulation of glial GLT-1 expression in hippocampal region CA3 can be considered a regional neurochemical correlate of dendritic remodeling.

Finally, glial cells, especially astrocytes, are key components of the "neurogenic niche" that provides the necessary local microenvironment for generation of neurons in specific brain areas.[64] They support maturation and integration of newborn neurons, both physically and by releasing a cocktail of growth factors and cytokines.[64] Together, this implies that GCs not only are influenced by stress, but also stimulate interactions between astrocytes and neural progenitors.

PLASTIC CHANGES OF BRAIN CELLS ARE REFLECTED BY ALTERATIONS IN GENE EXPRESSION

The stress-induced changes in neuronal structure imply that brain cells adjust their biosynthetic pathways to the new requirements (Figure 3). For example, concomitant with retraction of dendrites, there is probably a reduced need for membrane proteins. Related adaptational processes take place, partly at the level of gene transcription. Various methods, namely cDNA microarrays, serial analysis of gene expression, and subtractive hybridization, are currently used to identify genes that are differentially regulated by stress, and such genes appear to be rather heterogeneous.[65] However, it is already clear that some observations from gene transcription studies are consistent with findings using other methods.

In the hippocampal formation, the expression of several genes known to be involved in neuronal differentiation was found to be downregulated by chronic social stress.[66] Such genes included those encoding the membrane glycoprotein M6a, the CDC-like kinase 1, and a gene encoding a distinct subunit of certain G-proteins, GNAQ. All these genes are known to be involved in neurite outgrowth and neuronal differentiation, supporting the view that alterations in neuronal morphology and/or formation of neurons are primary effects of stress, at least in the hippocampal formation. Furthermore, the expression of the neural cell adhesion molecule (NCAM) was found to be downregulated in the hippocampus after chronic restraint stress.[67] NCAM is known to regulate neurite outgrowth and target recognition in the developing nervous system by mediating cell adhesion and regulating signal transduction.

Neuronal plasticity is accompanied by dynamic changes in elements of the cytoskeleton. Alpha-tubulin, the major component of microtubules, can be post-translationally modified; tyrosinated as well as acetylated alpha-tubulin are considered markers of dynamically changing or stable forms of microtubules, respectively. Restraint stress for 4 days decreased the expression of tyrosinated alpha-tubulin and increased expression of the acetylated form in the hippocampus.[68] This is consistent with the view that stress alters the morphology of pyramidal neurons.

Plastic changes are not restricted to forebrain areas. In a brain stem region, the rat dorsal raphe nucleus, which contains a large number of serotonergic neurons innervating the forebrain, long-term social stress increased the expression of genes involved in regulation of neurotransmitter release (synaptosomal-associated protein 25 and synaptic vesicle glycoprotein 2b).[69] These data support the view that stress increases the activity of neurons in the dorsal raphe nucleus.

In the current state of the art, data from mRNA or protein expression studies sometimes appear contradictory. However, when interpreting such data, it should be considered in which brain region, on which cellular level (neurons vs. glia), and in which cellular compartment (neuronal cell body versus area of the dendrites versus axon terminal areas) the stress-induced changes were detected. Besides the sex and age of the investigated species, both the type and the duration of the stressor play a crucial role with respect to effects of stress on gene expression.

Finally, it should be noted that the more recent studies investigating the effect of stress on gene expressional changes have increasingly focused on epigenetic mechanisms[70] and on the regulatory role of microRNAs.[71]

STRUCTURAL CHANGES ARE REVERSIBLE

As shown in animal studies, apical dendrites of hippocampal pyramidal neurons that were shortened by restraint stress or glucocorticoid exposure regained their normal length within 3 weeks after the treatment.[72,73] Several *in vivo* MRI studies in humans suggest a correlation between hippocampal volume reduction and cumulative glucocorticoid exposure, although exceptions have also been reported. In Cushing's patients who suffer from hypercortisolism, smaller hippocampal volumes returned to normal after successful surgery that led to normal glucocorticoid levels.[74] Thus, up to a certain point, structural changes may be at least partially reversible and may permit restoration of normal functioning. However, it appears that appropriate intervention should take place within a certain time before the changes become irreversible.

The mechanisms that are responsible for hippocampal volume loss have not yet been identified.[75] Massive neuronal loss following exposure to repeated episodes of hypercortisolemia can be excluded, because no major cell loss was apparent in experimental animals or in human postmortem brain tissue.[36,76] Neurogenesis has been regarded as another mechanism that might affect the volume of the hippocampus. However, hippocampal neurogenesis adds relatively few neurons per day and of those, only to the granule cell layer of the dentate gyrus where an equivalent number of neurons also die.[75] An alternative explanation is that hippocampal volume loss is the result of alterations in amounts of the dendritic, axonal, and synaptic components or to changes in levels of glia.[75] Furthermore, one may also speculate that stress may induce a shift in fluid balance between the ventricles and brain tissue. This assumption is supported by numerous clinical studies reporting lower volumes of different brain structures in conjunction with enlarged ventricles in patients with affective disorders.[75] However, it remains to be determined whether changes in fluid content cause the neuroplastic changes described above or just accompany them.

CONCLUSIONS

The diverse forms of stress-induced changes in neural cells are of particular interest. They show that even the adult and differentiated brain is a plastic organ. This lifelong plasticity is a prerequisite for allowing the brain to adapt to environmental changes. Plasticity of the neuronal networks is the basis for learning and memory. Extensive neuroplastic changes induced by stress may lead to imbalances in neural networks of the central nervous system. A promising aspect with regard to therapy is that many of these processes appear to be reversible. Reversibility of structural and functional plasticity has already been demonstrated, for example, in response to pharmacological treatments. Thus, progress in our understanding of neural plasticity has profound implications for the treatment of a number of neurodegenerative and psychiatric disorders.

References

1. Selye H. A syndrome produced by diverse noxious agents. *Nature (London)*. 1936;138:32.
2. Reul JM, Sutanto W, van Eekelen JA, Rothuizen J, de Kloet ER. Central action of adrenal steroids during stress and adaptation. *Adv Exp Med Biol*. 1990;274:243–256.
3. McEwen BS. Brain on stress: how the social environment gets under the skin. *Proc Natl Acad Sci U S A*. 2012;109(suppl 2): 17180–17185.
4. Herman JP, Cullinan WE. Neurocircuitry of stress: central control of the hypothalamo-pituitary-adrenocortical axis. *Trends Neurosci*. 1997;20(2):78–84.
5. Joels M, Baram TZ. The neuro-symphony of stress. *Nat Rev Neurosci*. 2009;10(6):459–466.
6. Stanford S. Monoamines in response and adaptation to stress. In: Stanford S, Salmon P, eds. *Stress: From Synapse to Syndrome*. London: Academic Press; 1993:281–331.
7. Flugge G, Van Kampen M, Mijnster MJ. Perturbations in brain monoamine systems during stress. *Cell Tissue Res*. 2004;315(1):1–14.
8. Krishnan V, Nestler EJ. Linking molecules to mood: new insight into the biology of depression. *Am J Psychiatry*. 2010;167(11):1305–1320.
9. Binder EB, Nemeroff CB. The CRF system, stress, depression and anxiety-insights from human genetic studies. *Mol Psychiatry*. 2010;15(6):574–588.
10. Volterra A, Meldolesi J. Astrocytes, from brain glue to communication elements: the revolution continues. *Nat Rev Neurosci*. 2005;6(8): 626–640.
11. Lopez de Maturana R, Sanchez-Pernaute R. Regulation of corticostriatal synaptic plasticity by G protein-coupled receptors. *CNS Neurol Disord Drug Targets*. 2010;9(5):601–615.
12. Hazell GG, Hindmarch CC, Pope GR, et al. G protein-coupled receptors in the hypothalamic paraventricular and supraoptic nuclei—serpentine gateways to neuroendocrine homeostasis. *Front Neuroendocrinol*. 2012;33(1):45–66.
13. De Kloet ER, Vreugdenhil E, Oitzl MS, Joels M. Brain corticosteroid receptor balance in health and disease. *Endocr Rev*. 1998;19(3): 269–301.
14. Sapolsky RM. Glucocorticoids, stress, and their adverse neurological effects: relevance to aging. *Exp Gerontol*. 1999;34(6):721–732.
15. Joels M, Sarabdjitsingh RA, Karst H. Unraveling the time domains of corticosteroid hormone influences on brain activity: rapid, slow, and chronic modes. *Pharmacol Rev*. 2012;64(4):901–938.
16. Cajal S. *Degeneration and Regeneration of the Nervous System*. London: Oxford University Press; 1928.
17. Goldman SA, Nottebohm F. Neuronal production, migration, and differentiation in a vocal control nucleus of the adult female canary brain. *Proc Natl Acad Sci U S A*. 1983;80(8):2390–2394.
18. Nottebohm F, Arnold AP. Sexual dimorphism in vocal control areas of the songbird brain. *Science*. 1976;194(4261):211–213.
19. Nottebohm F. From bird song to neurogenesis. *Sci Am*. 1989;260 (2):74–79.
20. Popov VI, Bocharova LS. Hibernation-induced structural changes in synaptic contacts between mossy fibres and hippocampal pyramidal neurons. *Neuroscience*. 1992;48(1):53–62.
21. Sale A, Berardi N, Maffei L. Environment and brain plasticity: towards an endogenous pharmacotherapy. *Physiol Rev*. 2014; 94(1):189–234.
22. Sjostrom PJ, Rancz EA, Roth A, Hausser M. Dendritic excitability and synaptic plasticity. *Physiol Rev*. 2008;88(2):769–840.
23. Watanabe Y, Gould E, McEwen BS. Stress induces atrophy of apical dendrites of hippocampal CA3 pyramidal neurons. *Brain Res*. 1992;588(2):341–345.
24. McEwen BS. Stress and hippocampal plasticity. *Annu Rev Neurosci*. 1999;22:105–122.
25. Conrad CD. What is the functional significance of chronic stress-induced CA3 dendritic retraction within the hippocampus? *Behav Cogn Neurosci Rev*. 2006;5(1):41–60.
26. Hoffman AN, Krigbaum A, Ortiz JB, et al. Recovery after chronic stress within spatial reference and working memory domains: correspondence with hippocampal morphology. *Eur J Neurosci*. 2011;34 (6):1023–1030.
27. Kole MH, Costoli T, Koolhaas JM, Fuchs E. Bidirectional shift in the cornu ammonis 3 pyramidal dendritic organization following brief stress. *Neuroscience*. 2004;125(2):337–347.

28. Wellman CL. Dendritic reorganization in pyramidal neurons in medial prefrontal cortex after chronic corticosterone administration. *J Neurobiol*. 2001;49(3):245–253.
29. Holmes A, Wellman CL. Stress-induced prefrontal reorganization and executive dysfunction in rodents. *Neurosci Biobehav Rev*. 2009;33(6):773–783.
30. Shansky RM, Morrison JH. Stress-induced dendritic remodeling in the medial prefrontal cortex: effects of circuit, hormones and rest. *Brain Res*. 2009;1293:108–113.
31. Kasai H, Fukuda M, Watanabe S, Hayashi-Takagi A, Noguchi J. Structural dynamics of dendritic spines in memory and cognition. *Trends Neurosci*. 2010;33(3):121–129.
32. Kang HJ, Voleti B, Hajszan T, et al. Decreased expression of synapse-related genes and loss of synapses in major depressive disorder. *Nat Med*. 2012;18(9):1413–1417.
33. Drevets WC. Functional anatomical abnormalities in limbic and prefrontal cortical structures in major depression. *Prog Brain Res*. 2000;126:413–431.
34. Mitra R, Jadhav S, McEwen BS, Vyas A, Chattarji S. Stress duration modulates the spatiotemporal patterns of spine formation in the basolateral amygdala. *Proc Natl Acad Sci U S A*. 2005;102(26):9371–9376.
35. Wilson M, Grillo C, Fadel J, Reagan L. Stress as a one-armed bandit: differential effects of stress paradigms on the morphology, neurochemistry and behavior in the rodent amygdala. *Neurobiol Stress*. 2015;1:195–208.
36. Lucassen PJ, Heine VM, Muller MB, et al. Stress, depression and hippocampal apoptosis. *CNS Neurol Disord Drug Targets*. 2006;5(5):531–546.
37. Czeh B, Varga ZK, Henningsen K, Kovacs GL, Miseta A, Wiborg O. Chronic stress reduces the number of GABAergic interneurons in the adult rat hippocampus, dorsal-ventral and region-specific differences. *Hippocampus*. 2015;25(3):393–405.
38. Gross CG. Neurogenesis in the adult brain: death of a dogma. *Nat Rev Neurosci*. 2000;1(1):67–73.
39. Eisch AJ, Cameron HA, Encinas JM, Meltzer LA, Ming GL, Overstreet-Wadiche LS. Adult neurogenesis, mental health, and mental illness: hope or hype? *J Neurosci*. 2008;28(46):11785–11791.
40. Aimone JB, Li Y, Lee SW, Clemenson GD, Deng W, Gage FH. Regulation and function of adult neurogenesis: from genes to cognition. *Physiol Rev*. 2014;94(4):991–1026.
41. Schmidt-Hieber C, Jonas P, Bischofberger J. Enhanced synaptic plasticity in newly generated granule cells of the adult hippocampus. *Nature*. 2004;429(6988):184–187.
42. Ho NF, Hooker JM, Sahay A, Holt DJ, Roffman JL. In vivo imaging of adult human hippocampal neurogenesis: progress, pitfalls and promise. *Mol Psychiatry*. 2013;18(4):404–416.
43. van Praag H, Schinder AF, Christie BR, Toni N, Palmer TD, Gage FH. Functional neurogenesis in the adult hippocampus. *Nature*. 2002;415(6875):1030–1034.
44. Spalding KL, Bergmann O, Alkass K, et al. Dynamics of hippocampal neurogenesis in adult humans. *Cell*. 2013;153(6):1219–1227.
45. Kempermann G. New neurons for 'survival of the fittest'. *Nat Rev Neurosci*. 2012;13(10):727–736.
46. Lucassen PJ, Meerlo P, Naylor AS, et al. Regulation of adult neurogenesis by stress, sleep disruption, exercise and inflammation: implications for depression and antidepressant action. *Eur Neuropsychopharmacol*. 2010;20(1):1–17.
47. Simon M, Czeh B, Fuchs E. Age-dependent susceptibility of adult hippocampal cell proliferation to chronic psychosocial stress. *Brain Res*. 2005;1049(2):244–248.
48. Fuchs E, Flugge G. Adult neuroplasticity: more than 40 years of research. *Neural Plast*. 2014;2014:541870.
49. Balu DT, Lucki I. Adult hippocampal neurogenesis: regulation, functional implications, and contribution to disease pathology. *Neurosci Biobehav Rev*. 2009;33(3):232–252.
50. Czeh B, Welt T, Fischer AK, et al. Chronic psychosocial stress and concomitant repetitive transcranial magnetic stimulation: effects on stress hormone levels and adult hippocampal neurogenesis. *Biol Psychiatry*. 2002;52(11):1057–1065.
51. Coe CL, Kramer M, Czeh B, et al. Prenatal stress diminishes neurogenesis in the dentate gyrus of juvenile rhesus monkeys. *Biol Psychiatry*. 2003;54(10):1025–1034.
52. Heine VM, Zareno J, Maslam S, Joels M, Lucassen PJ. Chronic stress in the adult dentate gyrus reduces cell proliferation near the vasculature and VEGF and Flk-1 protein expression. *Eur J Neurosci*. 2005;21(5):1304–1314.
53. Kandel ER. Glial cells are support cells. In: Kandel ERSJ, Jessell T, eds. *Principles of Neural Science*. 5th ed. New York, NY: McGraw-Hill; 2012:20–21.
54. Sofroniew MV, Vinters HV. Astrocytes: biology and pathology. *Acta Neuropathol*. 2010;119(1):7–35.
55. Kimelberg HK. Functions of mature mammalian astrocytes: a current view. *Neuroscientist*. 2010;16(1):79–106.
56. Kimelberg HK, Nedergaard M. Functions of astrocytes and their potential as therapeutic targets. *Neurotherapeutics*. 2010;7(4):338–353.
57. Czeh B, Simon M, Schmelting B, Hiemke C, Fuchs E. Astroglial plasticity in the hippocampus is affected by chronic psychosocial stress and concomitant fluoxetine treatment. *Neuropsychopharmacology*. 2006;31(8):1616–1626.
58. Tynan RJ, Beynon SB, Hinwood M, et al. Chronic stress-induced disruption of the astrocyte network is driven by structural atrophy and not loss of astrocytes. *Acta Neuropathol*. 2013;126(1):75–91.
59. Czeh B, Muller-Keuker JI, Rygula R, et al. Chronic social stress inhibits cell proliferation in the adult medial prefrontal cortex: hemispheric asymmetry and reversal by fluoxetine treatment. *Neuropsychopharmacology*. 2007;32(7):1490–1503.
60. Banasr M, Valentine GW, Li XY, Gourley SL, Taylor JR, Duman RS. Chronic unpredictable stress decreases cell proliferation in the cerebral cortex of the adult rat. *Biol Psychiatry*. 2007;62(5):496–504.
61. Wennstrom M, Hellsten J, Ekstrand J, Lindgren H, Tingstrom A. Corticosterone-induced inhibition of gliogenesis in rat hippocampus is counteracted by electroconvulsive seizures. *Biol Psychiatry*. 2006;59(2):178–186.
62. Walker FR, Beynon SB, Jones KA, et al. Dynamic structural remodelling of microglia in health and disease: a review of the models, the signals and the mechanisms. *Brain Behav Immun*. 2014;37:1–14.
63. Reagan LP, Rosell DR, Wood GE, et al. Chronic restraint stress upregulates GLT-1 mRNA and protein expression in the rat hippocampus: reversal by tianeptine. *Proc Natl Acad Sci U S A*. 2004;101(7):2179–2184.
64. Horner PJ, Palmer TD. New roles for astrocytes: the nightlife of an 'astrocyte'. La vida loca! *Trends Neurosci*. 2003;26(11):597–603.
65. Karssen AM, Her S, Li JZ, et al. Stress-induced changes in primate prefrontal profiles of gene expression. *Mol Psychiatry*. 2007;12(12):1089–1102.
66. Alfonso J, Frasch AC, Flugge G. Chronic stress, depression and antidepressants: effects on gene transcription in the hippocampus. *Rev Neurosci*. 2005;16(1):43–56.
67. Bisaz R, Conboy L, Sandi C. Learning under stress: a role for the neural cell adhesion molecule NCAM. *Neurobiol Learn Mem*. 2009;91(4):333–342.
68. Bianchi M, Heidbreder C, Crespi F. Cytoskeletal changes in the hippocampus following restraint stress: role of serotonin and microtubules. *Synapse*. 2003;49(3):188–194.
69. Abumaria N, Rygula R, Havemann-Reinecke U, et al. Identification of genes regulated by chronic social stress in the rat dorsal raphe nucleus. *Cell Mol Neurobiol*. 2006;26(2):145–162.
70. Mifsud KR, Gutierrez-Mecinas M, Trollope AF, Collins A, Saunderson EA, Reul JM. Epigenetic mechanisms in stress and adaptation. *Brain Behav Immun*. 2011;25(7):1305–1315.

71. Schouten M, Aschrafi A, Bielefeld P, Doxakis E, Fitzsimons CP. microRNAs and the regulation of neuronal plasticity under stress conditions. *Neuroscience*. 2013;241:188–205.
72. Conrad CD, LeDoux JE, Magarinos AM, McEwen BS. Repeated restraint stress facilitates fear conditioning independently of causing hippocampal CA3 dendritic atrophy. *Behav Neurosci*. 1999;113(5): 902–913.
73. Alfarez DN, De Simoni A, Velzing EH, et al. Corticosterone reduces dendritic complexity in developing hippocampal CA1 neurons. *Hippocampus*. 2009;19(9):828–836.
74. Bourdeau I, Bard C, Noel B, et al. Loss of brain volume in endogenous Cushing's syndrome and its reversibility after correction of hypercortisolism. *J Clin Endocrinol Metab*. 2002;87 (5):1949–1954.
75. Czeh B, Lucassen PJ. What causes the hippocampal volume decrease in depression? Are neurogenesis, glial changes and apoptosis implicated? *Eur Arch Psychiatry Clin Neurosci*. 2007;257(5):250–260.
76. Lucassen PJ, Pruessner J, Sousa N, et al. Neuropathology of stress. *Acta Neuropathol*. 2014;127(1):109–135.
77. Flügge G, Kramer M, Rensing S, Fuchs E. $5HT_{1A}$-receptors and behaviour under chronic stress: selective counteraction by testosterone. *Eur J Neurosci*. 1998;10(8):2685–2693.

CHAPTER

15

Epigenetics, Stress, and Their Potential Impact on Brain Network Function

V.A. Diwadkar

Wayne State University School of Medicine, Detroit, MI, USA

OUTLINE

Abstract

What "causal" relationships might link epigenetics, stress, and human brain function? Understanding these interrelationships can significantly enhance how we view the role of life events in shaping how the brain "works." Behavior reflects the brain in "action," and a mechanistic understanding of these relationships may also illuminate why so many neuropsychiatric illnesses are characterized by disordered stress-responses and behavior. This chapter explores these questions. We briefly review the origins of the epigenetics revolution in biology. We then discuss findings linking hypothalamic-pituitary-adrenal axis function and cortical metrics drawn from neuroimaging. We then briefly summarize known epigenetic mediation of molecular brain function at the level of the synapse. Finally we provide an overview of current and forward-looking approaches toward understanding macroscopic brain *network* function. This preliminary synthesis is intended to motivate integrative future approaches grounded in previous experimental and theoretical work in neurobiology, psychosocial studies, and neuroimaging.

INTRODUCTION

Biological stress typically reflects the system's response to an external challenge or "stressor."[1] The stress response is a ubiquitous biological defense mechanism, evident at all of life's time points, and with relevance to both the central and peripheral nervous systems.[2] Though inherently normative, disordered stress-responses are associated with an array of psychopathology in humans, suggesting a delicate (and perhaps finely-tuned) balance between stress and behavioral responses within the "typical" range. Stress also has clear physiological signatures: by inducing the release of corticotrophin releasing factor, it ultimately activates the hypothalamic-pituitary-adrenal (HPA) axis to produce cortisol, and the sympathetic nervous system to produce norepinephrine and epinephrine. Stress also has clear autonomic psychological signatures. For example, the "fight-or-flight" response to stress causes the organism to oscillate between competing psychological states before returning to homeostasis.[3] Chronic stress also has pathological signatures even in groups not characterized by overt psychopathology: for example, chronic stress (indexed by higher levels of blood cortisol) occurs frequently in individuals with intense professional lifestyles, results in hyporeactivity of the autonomic nervous system following acute mental stressors, and impairs function in key neurocognitive domains such as cognitive control.[4] Stress also plays a facilitatory role in the behavioral economy. In addition to mediating adaptive flight responses, experimentally induced social stress has been associated with higher levels of pro-social behavior.[5]

Stress: Concepts, Cognition, Emotion, and Behavior
http://dx.doi.org/10.1016/B978-0-12-800951-2.00015-7

KEY POINTS

- There is an increasing focus on how the stress response is mediated by epigenetic effects. The stress response is an experienced and/or manifested behavioral phenotype distributed in the general population, and in some cases (particularly neuropsychiatric illness) pathological.
- Fundamentally, behavior emerges from causal interactions between the brain's array of corticocortical and corticosubcortical networks, an idea that is increasingly the focus of analytic approaches toward in vivo neuroimaging signals. Therefore, a highly plausible "causal" explanatory arc may link epigenetics with brain network function (and dysfunction).
- Here we present logical extensions to this model incorporating ideas drawn from epigenetic mediation of psychiatric disease, and the effects of epigenetics on the emergence of brain network function and dysfunction in adolescence.
- We discuss how exploring relationships between epigenetics and brain network function can enhance understanding of the relationship between stress, epigenetics and functional neurobiology.
- We extend this exploration to the schizophrenia diathesis. This extension grounds our framework on highly debilitating neuropsychiatric illnesses that have been called an "epigenetic puzzle," and that has been robustly associated with both pathological stress-responses, and impairments in brain network function.
- Stress is a biological process that has mechanistic relationships with, and effects on, other processes such as gene expression and brain structure and function.
- Future work is needed to (a) create a theoretical framework that motivates the search for these relationships and (b) help devise and operationalize a research agenda for the future.

Stress and the stress response clearly shape overt behavior, but overt behavior emerges from a complex pattern of functional brain interactions.[6] Logically, stress must act to shape the emergence of brain networks, interactions within which are increasingly seen as the macroscopic level at which behavior might be explained. Two questions of central interest have been relatively understudied in the literature: (1) what is the role of genetics, but more specifically, epigenetics in shaping responses to stress, and (2) consequently, how might these factors combine to mediate the emergence of the brain's functioning networks?

The aim of this chapter is to introduce a framework for thinking about the relationship between stress and brain network function, and more fundamentally about the role of how epigenetic mechanisms might mediate this "causal" relationship. We will also allude to potential relationships in the context of neuropsychiatric illnesses. Neuropsychiatric illnesses are the most behaviorally devastating of medical phenotypes and the most poorly understood from the perspectives of mechanisms. Therefore such an extension helps inform an explanatory model linking epigenetics and brain network function through the lens of a highly disruptive neuropsychiatric illness.

THE EPIGENETICS REVOLUTION IN BIOLOGY

Conrad Waddington introduced the term "epigenetics" in the 1940s.[7] It is notable that his introduction of the term predates the genetic revolution in biology (which did not arrive until the 1950s) and before much was known about mechanisms of genetic transmission and heredity. However, the semantics of Waddington's usage closely foreshadows the thrust of this chapter. He was particularly interested in conceptualizing how genetic interactions with the environment produce phenotypes, particularly in morphology. Thus per Waddington,

> The ...important part of the task (i.e., understanding the relationship between genes and phenotypes) is to discover the *causal mechanisms* (emphasis added) at work, and to relate them as far as possible to what experimental embryology has already revealed of the mechanics of development. We might use the name 'epigenetics' for such studies, thus emphasizing their relation to the concepts, so strongly favorable to the classical theory of epigenesis, which have been reached by the experimental embryologists.

Waddington's initial focus was in providing a framework for the diversity in biological development, and ultimately resulted in his influential "canalization" model of development, a model that over time has been formally realized in dynamical systems-based approaches toward development.[8]

Born from the terms "genetics" and "epigenesis," the term epigenetics while historically used in the context of studying causal relationships between genes and their phenotypic effects,[9] now has a more focused meaning. It is increasingly associated with the study of *changes in gene activity* independent of the DNA sequence, that may or may not be heritable, and that may also be modified through the life span, thus reflecting Waddington's idea of an epigenetic landscape in development. Epigenetic mechanisms include (a) DNA methylation which in vertebrates typically involves the addition of a methyl group to cytosine where cytosine and guanine occur on the same DNA strand, (b) histone modifications, involving the addition (or removal) of chemical groups to the core proteins around which DNA is wound, and (c) noncoding RNAs such as microRNAs (miRNAs), which bind to mRNAs to suppress gene expression post-transcriptionally.[10] For example, glucocorticoids (GCs) such as cortisol induce epigenetic, DNA methylation changes in HPA axis genes (e.g., FK506 binding protein 5, *FKBP5*), in the brain,[11,12] in the periphery[11,13,14], and in pituitary cells.[12] Furthermore, GC-induced methylation changes persist long after cessation of exposure,[11,13,14] suggesting that stress-induced GC cascades have *long lasting consequences for HPA axis function*.[11,15]

THE HPA AXIS AND MEASURES OF THE CENTRAL NERVOUS SYSTEM

The relationship with HPA axis functionality provides a point of contact with measures of the central nervous system in the brain. For example, structural neuroimaging studies have focused on the size of the pituitary gland, particularly in disorders associated with impaired stress-responses. The pituitary gland is located at the base of the brain and is closely associated (along with structures such as the hypothalamus) with autonomous and endocrine function. The release of GCs from the adrenal gland results in rapid ultradian oscillations of hormone levels both in the blood and in the brain.[16] Notably, the timing of the stressor relative to the phase of the underlying ultradian rhythm has been shown as critical in mediating the magnitude of the corticosterone defensive response.[17,18]

GCs themselves affect fundamental changes in the structure and function of multiple brain regions, particularly those in the medial temporal lobe, such as the amygdala and the hippocampus, most closely associated with fight-or-flight responses.[19,20] There is ample evidence of stress-related effects on pituitary and medial temporal lobe morphology, particularly in neuropsychiatric illnesses including schizophrenia and post-traumatic stress disorder (PTSD).[21] Hypertrophic pituitary volumes are characteristic of the schizophrenia diathesis, observed in chronic and patients at genetic risk for the illness as well, which is highly suggestive of a heightened activation of the hormonal stress response and to an increase in the size and number of corticotroph cells producing the adrenocorticotropic hormone.[22–25]

Disordered stress reactivity may play a key role in amplifying disposition for psychosis in the risk-state[26] and provides a direct link between stress and measures of central nervous system structures. Patients who are at clinical risk for schizophrenia (termed "prodromal") show heightened sensitivity to interpersonal interaction, a measure that indirectly assesses heightened stress responsivity.[27] Moreover, a significant percent of individuals who have experienced trauma in their lives convert to psychosis[28] and evince the previously noted increases in volume of the pituitary. These HPA axis-related findings reverberate beyond the endocrine system, and are (at least) correlated with cortical impairments. For instance, dopamine synthesis is increased in prodromal subjects, and the degree of synthesis is positively associated with the severity of sub-threshold clinical symptoms.[29] Heightened dopamine sensitivity, itself a leading intermediate phenotype of the "sociodevelopmental cognitive" model of schizophrenia,[30,31] has substantial implications for cortical network function. Dopamine, particularly in the prefrontal cortex, has been strongly associated with modulation of long-term potentiation of pyramidal neurons,[32,33] themselves essential to working memory.[34] Elevated dopamine release during acute stress[35] adversely affects prefrontal pyramidal cells leading to a series of degenerative molecular events. The resultant dendritic spine loss in the infra-granular prefrontal cortex results in reductions in prefrontal-based network connectivity, particularly on prefrontal efferent pathways.[36] In part, the strength of synapses (and hence connectivity at the microscopic scale) is modulated by molecular signaling within slender spines, resulting in a pattern that has been termed "dynamic network connectivity."[37,38] Excessive dopamine stimulation of D1 receptors impairs prefrontal function via cyclic adenosine monophosphate (cAMP) intracellular signaling, leading to disconnection of prefrontal networks[39]. Moreover, chronic stress exacerbates dopamine release.[40] This formal framework, linking stress, molecular mechanisms (particularly dopamine signaling), and ultimately central nervous system function has been suggested to be a fundamental causal pathway underlying schizophrenia and risk for schizophrenia.[41] The idea of stress reactivity potentially impacting brain network function is a particular extension of the seminal concept of "allostatic load"[1,42] (i.e., morphologic degeneration) as a response to repeated responses to stress.

EPIGENETIC MEDIATION OF STRESS-RELATED PROCESSES

Independent evidence is beginning to suggest how these CNS (Central Nervous System) processes might be under epigenetic mediation. Peripheral DNA methylation measurements appear to predict brain endophenotypes. For example, leukocyte DNA methylation in the serotonin transporter locus (*SLC6A4*) is higher among adult males who experienced high childhood-limited physical aggression, and *SLC6A4* DNA methylation was negatively correlated with serotonin synthesis in the orbitofrontal cortex.[43] Similarly, leukocyte DNA methylation in the promoter region of the *MAOA* gene—whose product metabolizes monoamines such as serotonin and dopamine—is negatively associated with brain MAOA levels as measured by PET in healthy male adults.[44] Structural imaging data analyses in relation to the *FKBP5* locus discussed above have identified a negative association between DNA methylation in peripheral blood and volume of the right (but not left) hippocampal head.[11] This suggests that lower *FKBP5* DNA methylation in peripheral blood is associated not only with altered stress sensitivity (as indexed by a GC receptor sensitivity assay within the same study), but also with structural brain differences in the medial temporal lobe, a brain region particularly sensitive to stress.[11,45] Other genetic loci have also been investigated. Investigations of the Catechol-O-methyltransferase, a gene encoding enzyme critical for degradation of dopamine and other catecholamines, has shown that the Val/Val genotype with higher stress scores have reduced DNA methylation at a CpG site located in the promoter region of the gene.[46] Moreover, DNA methylation at this site was positively correlated with working memory accuracy (see previous discussion relating to the emergence of prefrontal-centric brain networks): greater methylation predicting a greater percentage of correct responses (with results again limited to analysis of the Val/Val subjects). Finally, though limited in number, functional magnetic resonance imaging (fMRI) studies suggest an interaction between methylation and stress scores on bilateral prefrontal activity during working memory: greater stress, when combined with lower methylation, is associated with greater activity,[46] a pattern suggestive of greater "resource" consumption.[47–50]

PATHWAYS I: STRESS, EPIGENETICS AND SYNAPTIC PLASTICITY

Long lasting impacts have been particularly salient in the case of prenatal stress. Epigenetics and stress in the prenatal neurodevelopment stage have a profound impact on synaptic plasticity,[51,52] impairing long-term potentiation of cells to afferent stimulation.[53] Synaptic plasticity is a process through which synapses undergo activity-driven changes in their strength. It is an essential characteristic underpinning complex cognitive and memorial processes, particularly prefrontal and hippocampal-based associative learning and memory, as is evident from both animal and human studies.[54,55] Moreover, synaptic facilitation mechanisms that drive memory formation in simple systems may also be similarly expressed in more complex organisms.[56] Thus, mechanisms of memory are a fertile domain in which to explore epigenetic and stress-related effects on memory formation and brain networks, particularly as neuroimaging studies have provided a relatively robust understanding of the macroscopic network architecture of hippocampal-based memory formation.[57–59]

Several studies have investigated whether *epigenetic marking* of the genome has effects on the induction of synaptic plasticity. During cell differentiation, a process critical to pre- and post-natal neurodevelopment, epigenetic tags associated with genomal marking are preserved, rendering epigenetic cellular memory unchanged. Thus, epigenetic marking of the genome can be thought of as a persistent form of cellular memory, where differentiated cells remember their phenotype.[60] This is an important mechanism that might protect against memory extinction in the face of constant neuronal turnover through adulthood (but may also confer vulnerability if stress impacts memory formation at critical neurodevelopmental periods). Animal models have documented distinctive behavioral deficits; mice exposed to stress show increased levels of anxiety and impairment in spatial learning tasks[61] and in long-term memory.[62] These effects are unsurprising: synaptic plasticity and long-term potentiation are essential elements of long-term memory formation.

Because synaptic plasticity occurs during memory acquisition,[63] it has long been suspected that synaptic plasticity is itself under epigenetic influence and evidence suggests that the regulation of the epigenome occurs during induction of synaptic plasticity in the mammalian brain. For example, the induction of LTP (Long-Term Potentiation) requires the activation of *N*-methyl-D-aspartate receptors (NMDAR).[64] The direct activation of NMDAR's in the hippocampus results in an extracellular signal kinase-dependent increase in acetylation of histone H3,[65] implicated in long-term memory formation. These and other studies have established that changes in the epigenome affect the induction of synaptic plasticity in the mammalian and invertebrate brain[66,67] ultimately affecting memory formation.[68]

PATHWAYS II: TOWARD BRAIN NETWORK FUNCTION

The influence of stress on DNA methylation on HPA axis genes in blood is well established.[11,13,14] As GC hormones produced by the HPA axis are dispersed throughout the body through blood, it results in the regulation of gene expression in virtually all cell types.[2] Thus, the broad reach of HPA axis activity, together with evidence that blood-derived DNA methylation in HPA axis genes is altered through stress,[11,69] indicates that stress sensitivity, measured in the periphery, affects CNS mechanisms.[70] As previously noted, the concept of allostatic load[71] has been used primarily to refer to plasticity of brain structures such as the hippocampus. However, increasingly the search for the "neural" correlates of behavior is turning toward the basis of these not only in individual brain structures but also in the dynamics of macroscopic brain networks.[6]

As noted, the epigenome mediates both the stress response, and long-term memory relevant mechanisms of synaptic plasticity. An open question is how this mediation might exert effects at the level of *macroscopic brain network function*. To explore this, we must first define what we specifically mean by macroscopic brain networks, distinguishing this notion from the concept of brain networks at the molecular scale.[36] Secondly, we must consider how extant approaches toward in vivo neuroimaging data[72] permit the process of "network discovery,"[73] or the estimation of hidden "neural states" that give rise to observable neuroimaging time-series data.

A fundamental challenge for brain research is to unravel the precise relationships that relate emergent brain function to its underlying structure. As noted,[6,74] this is a nontrivial problem for several reasons. Firstly, it has become relatively easy to collect spatiotemporal signals from outside and inside the brain, yet it is substantially more challenging to discover the hidden "mechanistic" brain interactions that give rise to these signals.[75] This problem is further compounded by the fact that within the brain, neurophysiological and hemodynamic signals have highly complex and uncertain interrelationships.[76] A significant additional challenge is that the relationship between behavior and the underlying structures that subserve that behavior is "regressive": multiple complex behaviors can emerge dynamically from the same structural architecture.[77] Thus, whereas a particular cognitive function might load on, and deterministically activate a brain region, it is impossible to use observed activation in a brain region to infer the *specific* task that generated it. This fundamental indeterminacy reflects the limitations of relying on models of regional brain function to infer the bases of complex behavior. Rather, in humans, complex overt behavior (the kind that has been typically measured to assess dysfunctional stress-responses) is more likely to be associated with macroscopic brain networks at the level of brain regions. These networks have been estimated using a multiplicity of techniques.

Electroencephalography is time-series data at very high temporal resolutions acquired at multiple surface (and in surgical cases, cortical) leads, that following frequency decomposition, provide frequency components with distinct functional properties.[78,79] These electrophysiological signals result in large-scale oscillations that have successfully been analyzed to uncover brain networks at a relatively "high" spatial scale,[80] (i.e., lower resolution). The relationship between these neurophysiological measures and molecular/synaptic assessments of brain function is still obscure. Evidence suggests that phase-locked firing to slow gamma oscillations (typically 20-80 Hz) is mediated by synaptic plasticity dependent on the AMPA (α-amino-3-hydroxy-5-methyl-4-isoxazole propionate) receptors in the hippocampus,[81] suggesting that molecular mechanisms such as synaptic plasticity (considered in the previous section) have correlates with signals collected at a broader spatial scale.[82] These higher order spatial scales, provide information on agglomerated neural signals, and are deemed "macroscopic." As noted, they may provide a more proximate and/or tractable correlate of complex behaviors.[6] This is particularly so, as recovering brain network function at higher spatial resolutions is considered intractable.[83] While scalp electrophysiology provides striking insights on patterns of macroscopic coherence and the relationship of this coherence to behavior, a typical challenge with electrophysiological recordings is one of spatial resolution. Spatial resolution is relatively indeterminate, given that in most human studies, signals must be recorded from the scalp and cannot easily be resolved with respect to distant cortical loci. In that sense, electrophysiology is more confined to fine-grained analyses of the brain signals in the frequency domain. In vivo techniques such as fMRI are becoming more widely used in the estimate of macroscopic brain network function.

fMRI is a relatively young discipline,[84] yet it has become the most dominant methodological technique used in in vivo assessments of brain function. Multiple reasons for its status include the non-invasiveness of the method, but also the relatively favorable balance of spatial and temporal resolution.[85] This balance is important because it allows assessment of both agglomerative "neural" signals in circumscribed areas of tissue, and temporal information on signal fluctuations. These signal fluctuations principally arise from dynamics in the brain's network states induced by perturbations to the system (e.g., cognitive processing).[85] In this sense, cognitive tasks can be seen as "stressors" to the intrinsic connectivity of the brain's macro-networks; their role is to induce dynamics that can be recovered and modeled using causal generative models of data.[86] The induction of dynamics, and the recovery and modeling of signals is particularly important for probing the operating range of brain networks. In other words, cognitive stressors exert systemic loads that in the case of health are appropriately supported by brain network function, but in the case of disease, reveal impaired brain network interactions assessed by techniques of effective connectivity,[87–90] where effective connectivity is broadly defined as causal effects between brain regions.[75] In this way, cognitive tasks play the same functional role as exercise in "stress" test assessments of cardiovascular health and risk.[91]

More fundamentally, analyses of fMRI signals are moving forward from assessments of regional specialization of function (what is a specific brain region relatively specialized for?) to assessments of functional integration of information across brain regions (i.e., networks).[72] This change in focus is more than academic: it reflects advances in the understanding of modeling spatiotemporal fMRI signals in an effort to recover causal dynamics between brain regions,[86,92,93] providing a more realistic model of how the brain "works." It is this level of network interaction that is most likely to be sensitive to perturbations from the epigenome and stress.[41]

The field is slowly orienting towards this methodological framework, though little attention has been paid to assessing macroscopic brain dynamics as outlined above. Studies have begun to address epigenetic mediation of the intrinsic connectivity of brain networks estimated from low frequency fluctuations in the fMRI signal. This work in the area of resting state functional connectivity (rsFC)[94] examines statistical correlations in the time-series drawn from multiple brain regions in an effort to quantify synchronization of macroscopic brain networks in the basal state. A recent study examined rsFC related to SLC6A4 promoter methylation. SLC6A4 is a protein-coding gene that has been associated with multiple disorders, including obsessive-compulsive disorder and depression. A positive correlation between SLC6A4 methylation and rsFC between the amygdala and its targets, including the insula and the dorsal anterior cingulate cortex, provided evidence of epigenetic changes effect macroscopic brain coherence.[95] The effects of stress have been more widely addressed in the literature.[96] PTSD which provides a powerful clinical model of dysfunctional stress-responses is characterized by decreased rsFC in default-mode and salience brain networks[97,98] that may result from GC alterations of brain structures that mediate connectivity.[99] However, these system-level changes induced by stress exposure in human clinical models, have uncertain correlates with the animal literature. In rodents, exposure to chronic stress leads to increased correlation in resting state fMRI signals in default-mode networks.[100]

In experimentally induced contexts, exposure to traumatic memories results in increased functional connectivity between the amygdala and its targets, the hippocampus, prefrontal cortex, and the insula, evidence that stress reactivity may alter dynamics in functioning brain networks.[101] There is also some preliminary evidence of relationships between macroscopic brain network connectivity and measures of HPA axis function. For instance, connectivity within hippocampal-based networks appears predictive of stimulated cortisol concentration in healthy subjects, suggesting that estimates of macroscopic brain network integrity may be systematically related to measures of pituitary function.[102]

There has been virtually no assessment of the role of psychosocial stress on dynamic changes in network connectivity using previously discussed techniques of effective connectivity, a glaring lacuna in the field. Using heroin as a pharmacologic stressor, a recent study showed compelling reductions in fear-induced effective connectivity between the amygdala and the orbitofrontal cortex,[103] though the focus of this work was on the "normalization" of effective connectivity patterns following placebo.

Assessment of epigenomic mediation of stress and its effects on brain network function is unquestionably in its infancy. However, this is a research vista that is overdue for focused investigation, particularly as multiple research tools in multiple domains begin to mature.[104] Population studies that employ a motivated focus on specific genes, coupled with sophisticated fMRI task design and analyses will be the standard for future studies. Understanding the relationship between epigenetics, stress, and brain network function will provide needed articulation of plausible mechanisms underlying multiple neuropsychiatric diseases, and the "causal" pathways that plausibly link the molecular and the macroscopic scales of the brain.

Acknowledgments

This work was supported by the National Institute of Mental Health (MH 059299), the National Association for Research on Schizophrenia and Depression (NARSAD, now Brain Behavior Research Fund), the Prechter World Bipolar Foundation, the Children's Research Center of Michigan, the Cohen Neuroscience Endowment, the Children's Hospital of Michigan Foundation, and the Lyckaki Young Fund from the State of Michigan. The agencies played no role in the shaping of the ideas presented herein. The author thanks Monica Uddin, Paolo Brambilla, and Steven Bressler for helpful discussions concerning the ideas presented herein.

References

1. McEwen BS. Effects of adverse experiences for brain structure and function. *Biol Psychiatry*. 2000;48(8):721–731.
2. Irwin MR, Cole SW. Reciprocal regulation of the neural and innate immune systems. *Nat Rev Immunol*. 2011;11(9):625–632.
3. Goldstein DS. Adrenal responses to stress. *Cell Mol Neurobiol*. 2010;30(8):1433–1440.
4. Teixeira RR, Diaz MM, Santos TV, et al. Chronic stress induces a hyporeactivity of the autonomic nervous system in response to acute mental stressor and impairs cognitive performance in business executives. *PLoS One*. 2015;10(3):e0119025.
5. von Dawans B, Fischbacher U, Kirschbaum C, Fehr E, Heinrichs M. The social dimension of stress reactivity: acute stress increases prosocial behavior in humans. *Psychol Sci*. 2012;23(6):651–660.

6. Park HJ, Friston K. Structural and functional brain networks: from connections to cognition. *Science (New York, NY)*. 2013;342(6158). http://dx.doi.org/10.1126/science.1238411.
7. Waddington CH. The epigenotype. 1942. *Int J Epidemiol*. 2012;41 (1):10–13.
8. Davila-Velderrain J, Martinez-Garcia JC, Alvarez-Buylla ER. Modeling the epigenetic attractors landscape: toward a post-genomic mechanistic understanding of development. *Front Genet*. 2015;6:160.
9. Van Speybroeck L. From epigenesis to epigenetics: the case of C. H. Waddington. *Ann N Y Acad Sci*. 2002;981:61–81.
10. Weaver IC. Integrating early life experience, gene expression, brain development, and emergent phenotypes: unraveling the thread of nature via nurture. *Adv Genet*. 2014;86:277–307.
11. Klengel T, Mehta D, Anacker C, et al. Allele-specific FKBP5 DNA demethylation mediates gene-childhood trauma interactions. *Nat Neurosci*. 2013;16(1):33–41.
12. Yang X, Ewald ER, Huo Y, et al. Glucocorticoid-induced loss of DNA methylation in non-neuronal cells and potential involvement of DNMT1 in epigenetic regulation of Fkbp5. *Biochem Biophys Res Commun*. 2012;420(3):570–575.
13. Lee RS, Tamashiro KL, Yang X, et al. Chronic corticosterone exposure increases expression and decreases deoxyribonucleic acid methylation of Fkbp5 in mice. *Endocrinology*. 2010;151(9):4332–4343.
14. Lee RS, Tamashiro KL, Yang X, et al. A measure of glucocorticoid load provided by DNA methylation of Fkbp5 in mice. *Psychopharmacology*. 2011;218(1):303–312.
15. Mehta D, Binder EB. Gene x environment vulnerability factors for PTSD: the HPA-axis. *Neuropharmacology*. 2012;62(2):654–662.
16. Spiga F, Walker JJ, Terry JR, Lightman SL. HPA axis-rhythms. *Compr Physiol*. 2014;4(3):1273–1298.
17. Windle RJ, Wood SA, Lightman SL, Ingram CD. The pulsatile characteristics of hypothalamo-pituitary-adrenal activity in female Lewis and Fischer 344 rats and its relationship to differential stress responses. *Endocrinology*. 1998;139(10):4044–4052.
18. Windle RJ, Wood SA, Shanks N, Lightman SL, Ingram CD. Ultradian rhythm of basal corticosterone release in the female rat: dynamic interaction with the response to acute stress. *Endocrinology*. 1998;139(2):443–450.
19. Shirazi SN, Friedman AR, Kaufer D, Sakhai SA. Glucocorticoids and the brain: neural mechanisms regulating the stress response. *Adv Exp Med Biol*. 2015;872:235–252.
20. LeDoux J. The emotional brain, fear, and the amygdala. *Cell Mol Neurobiol*. 2003;23(4–5):727–738.
21. Bremner JD. Long-term effects of childhood abuse on brain and neurobiology. *Child Adolesc Psychiatr Clin N Am*. 2003;12(2): 271–292.
22. Pariante CM. Pituitary volume in psychosis: the first review of the evidence. *J Psychopharmacol (Oxford, England)*. 2008;22(suppl 2): 76–81.
23. Mondelli V, Dazzan P, Gabilondo A, et al. Pituitary volume in unaffected relatives of patients with schizophrenia and bipolar disorder. *Psychoneuroendocrinology*. 2008;33(7):1004–1012.
24. Nordholm D, Krogh J, Mondelli V, Dazzan P, Pariante C, Nordentoft M. Pituitary gland volume in patients with schizophrenia, subjects at ultra high-risk of developing psychosis and healthy controls: a systematic review and meta-analysis. *Psychoneuroendocrinology*. 2013;38(11):2394–2404.
25. Pariante CM, Vassilopoulou K, Velakoulis D, et al. Pituitary volume in psychosis. *Br J Psychiatry*. 2004;185:5–10.
26. Walker EF, Diforio D, Baum K. Developmental neuropathology and the precursors of schizophrenia. *Acta Psychiatr Scand Suppl*. 1999;395:12–19.
27. Masillo A, Day F, Laing J, et al. Interpersonal sensitivity in the at-risk mental state for psychosis. *Psychol Med*. 2012;42(9):1835–1845.
28. Bechdolf A, Thompson A, Nelson B, et al. Experience of trauma and conversion to psychosis in an ultra-high-risk (prodromal) group. *Acta Psychiatr Scand*. 2010;121(5):377–384.
29. Howes OD, Bose SK, Turkheimer F, et al. Dopamine synthesis capacity before onset of psychosis: a prospective [18F]-DOPA PET imaging study. *Am J Psychiatry*. 2011;168(12):1311–1317.
30. Howes OD, Murray RM. Schizophrenia: an integrated sociodevelopmental-cognitive model. *Lancet*. 2014;383(9929): 1677–1687.
31. Howes OD, Kapur S. The dopamine hypothesis of schizophrenia: version III—the final common pathway. *Schizophr Bull*. 2009;35(3): 549–562.
32. Sheynikhovich D, Otani S, Arleo A. Dopaminergic control of long-term depression/long-term potentiation threshold in prefrontal cortex. *J Neurosci*. 2013;33(34):13914–13926.
33. Otani S, Daniel H, Roisin MP, Crepel F. Dopaminergic modulation of long-term synaptic plasticity in rat prefrontal neurons. *Cereb Cortex*. 2003;13(11):1251–1256.
34. Lewis DA, Gonzalez-Burgos G. Neuroplasticity of neocortical circuits in schizophrenia. *Neuropsychopharmacology*. 2008;33(1):141–165.
35. Pruessner JC, Champagne F, Meaney MJ, Dagher A. Dopamine release in response to a psychological stress in humans and its relationship to early life maternal care: a positron emission tomography study using [11C]raclopride. *J Neurosci*. 2004;24 (11):2825–2831.
36. Arnsten AF. Prefrontal cortical network connections: key site of vulnerability in stress and schizophrenia. *Int J Dev Neurosci*. 2011;29(3):215–223.
37. Arnsten AF, Paspalas CD, Gamo NJ, Yang Y, Wang M. Dynamic network connectivity: a new form of neuroplasticity. *Trends Cogn Sci*. 2010;14(8):365–375.
38. Goto Y, Yang CR, Otani S. Functional and dysfunctional synaptic plasticity in prefrontal cortex: roles in psychiatric disorders. *Biol Psychiatry*. 2010;67(3):199–207.
39. Hains AB, Arnsten AF. Molecular mechanisms of stress-induced prefrontal cortical impairment: implications for mental illness. *Learn Mem*. 2008;15(8):551–564.
40. Mizrahi R, Addington J, Rusjan PM, et al. Increased stress-induced dopamine release in psychosis. *Biol Psychiatry*. 2012;71(6): 561–567.
41. Diwadkar VA, Bustamante A, Rai H, Uddin M. Epigenetics, stress, and their potential impact on brain network function: a focus on the schizophrenia diatheses. *Front Psychiatry*. 2014;5:71.
42. McEwen BS. Stress, adaptation, and disease. Allostasis and allostatic load. *Ann N Y Acad Sci*. 1998;840:33–44.
43. Wang D, Szyf M, Benkelfat C, et al. Peripheral SLC6A4 DNA methylation is associated with in vivo measures of human brain serotonin synthesis and childhood physical aggression. *PLoS One*. 2012; 7(6):e39501.
44. Shumay E, Logan J, Volkow ND, Fowler JS. Evidence that the methylation state of the monoamine oxidase A (MAOA) gene predicts brain activity of MAO A enzyme in healthy men. *Epigenetics*. 2012;7(10):1151–1160.
45. Brown ES, Jeon-Slaughter H, Lu H, et al. Hippocampal volume in healthy controls given 3-day stress doses of hydrocortisone. *Neuropsychopharmacology*. 2015;40(5):1216–1221.
46. Ursini G, Bollati V, Fazio L, et al. Stress-related methylation of the catechol-*O*-methyltransferase Val 158 allele predicts human prefrontal cognition and activity. *J Neurosci*. 2011; 31(18):6692–6698.
47. Carpenter PA, Just MA. Modeling the mind: very-high-field functional magnetic resonance imaging activation during cognition. *Top Magn Reson Imaging*. 1999;10(1):16–36.
48. Toepper M, Gebhardt H, Bauer E, et al. The impact of age on load-related dorsolateral prefrontal cortex activation. *Front Aging Neurosci*. 2014;6:9.
49. Engstrom M, Landtblom AM, Karlsson T. Brain and effort: brain activation and effort-related working memory in healthy participants and patients with working memory deficits. *Front Hum Neurosci*. 2013;7:140.

50. Bakshi N, Pruitt P, Radwan J, et al. Inefficiently increased anterior cingulate modulation of cortical systems during working memory in young offspring of schizophrenia patients. *J Psychiatr Res.* 2011;45(8):1067–1076.
51. Bock J, Wainstock T, Braun K, Segal M. Stress in utero: prenatal programming of brain plasticity and cognition. *Biol Psychiatry.* 2015;78(5):315–326.
52. Provencal N, Binder EB. The effects of early life stress on the epigenome: from the womb to adulthood and even before. *Exp Neurol.* 2015;268:10–20.
53. Avital A, Segal M, Richter-Levin G. Contrasting roles of corticosteroid receptors in hippocampal plasticity. *J Neurosci.* 2006;26(36):9130–9134.
54. Stephan KE, Friston KJ, Frith CD. Dysconnection in schizophrenia: from abnormal synaptic plasticity to failures of self-monitoring. *Schizophr Bull.* 2009;35(3):509–527.
55. Silva AJ. Molecular and cellular cognitive studies of the role of synaptic plasticity in memory. *J Neurobiol.* 2003;54(1):224–237.
56. Kandel ER, Tauc L. Mechanism of prolonged heterosynaptic facilitation. *Nature.* 1964;202:145–147.
57. Buchel C, Coull JT, Friston KJ. The predictive value of changes in effective connectivity for human learning. *Science (New York, NY).* 1999;283:1538–1541.
58. Banyai M, Diwadkar VA, Erdi P. Model-based dynamical analysis of functional disconnection in schizophrenia. *Neuroimage.* 2011;58(3):870–877.
59. Diwadkar VA, Flaugher B, Jones T, et al. Impaired associative learning in schizophrenia: behavioral and computational studies. *Cogn Neurodyn.* 2008;2(3):207–219.
60. Levenson JM, Sweatt JD. Epigenetic mechanisms: a common theme in vertebrate and invertebrate memory formation. *Cell Mol Life Sci.* 2006;63(9):1009–1016.
61. Avital A, Ram E, Maayan R, Weizman A, Richter-Levin G. Effects of early-life stress on behavior and neurosteroid levels in the rat hypothalamus and entorhinal cortex. *Brain Res Bull.* 2006;68(6):419–424.
62. Bohacek J, Farinelli M, Mirante O, et al. Pathological brain plasticity and cognition in the offspring of males subjected to postnatal traumatic stress. *Mol Psychiatry.* 2015;20(5):621–631.
63. Ekstrom AD, Meltzer J, McNaughton BL, Barnes CA. NMDA receptor antagonism blocks experience-dependent expansion of hippocampal "place fields" *Neuron.* 2001;31(4):631–638.
64. Bliss TV, Collingridge GL. A synaptic model of memory: long-term potentiation in the hippocampus. *Nature.* 1993;361(6407):31–39.
65. Levenson JM, O'Riordan KJ, Brown KD, Trinh MA, Molfese DL, Sweatt JD. Regulation of histone acetylation during memory formation in the hippocampus. *J Biol Chem.* 2004;279(39):40545–40559.
66. Guzman-Karlsson MC, Meadows JP, Gavin CF, Hablitz JJ, Sweatt JD. Transcriptional and epigenetic regulation of Hebbian and non-Hebbian plasticity. *Neuropharmacology.* 2014;80:3–17.
67. Rahn EJ, Guzman-Karlsson MC, Sweatt David J. Cellular, molecular, and epigenetic mechanisms in non-associative conditioning: implications for pain and memory. *Neurobiol Learn Mem.* 2013;105:133–150.
68. Puckett RE, Lubin FD. Epigenetic mechanisms in experience-driven memory formation and behavior. *Epigenomics.* 2011;3(5):649–664.
69. Oberlander TF, Weinberg J, Papsdorf M, Grunau R, Misri S, Devlin AM. Prenatal exposure to maternal depression, neonatal methylation of human glucocorticoid receptor gene (NR3C1) and infant cortisol stress responses. *Epigenetics.* 2008;3(2):97–106.
70. Millan MJ. An epigenetic framework for neurodevelopmental disorders: from pathogenesis to potential therapy. *Neuropharmacology.* 2013;68:2–82.
71. McEwen BS. Plasticity of the hippocampus: adaptation to chronic stress and allostatic load. *Ann N Y Acad Sci.* 2001;933:265–277.
72. Friston KJ. Models of brain function in neuroimaging. *Annu Rev Psychol.* 2005;56:57–87.
73. Friston KJ, Li B, Daunizeau J, Stephan KE. Network discovery with DCM. *Neuroimage.* 2012;56(3):1202–1221.
74. Noppeney U, Friston KJ, Price CJ. Degenerate neuronal systems sustaining cognitive functions. *J Anat.* 2004;205(6):433–442.
75. Friston KJ. Functional and effective connectivity: a review. *Brain Connect.* 2011;1(1):13–36.
76. Singh KD. Which "neural activity" do you mean? fMRI, MEG, oscillations and neurotransmitters. *Neuroimage.* 2012;62(2):1121–1130.
77. Price CJ, Friston KJ. Functional ontologies for cognition: the systematic definition of structure and function. *Cogn Neuropsychol.* 2005;22(3):262–275.
78. Singer W. Coherence as an organizing principle of cortical functions. *Int Rev Neurobiol.* 1994;37:153–183. discussion 203–207.
79. Hipp JF, Engel AK, Siegel M. Oscillatory synchronization in large-scale cortical networks predicts perception. *Neuron.* 2011;69(2):387–396.
80. Uhlhaas PJ, Roux F, Rodriguez E, Rotarska-Jagiela A, Singer W. Neural synchrony and the development of cortical networks. *Trends Cogn Sci.* 2010;14(2):72–80.
81. Kitanishi T, Ujita S, Fallahnezhad M, Kitanishi N, Ikegaya Y, Tashiro A. Novelty-induced phase-locked firing to slow gamma oscillations in the hippocampus: requirement of synaptic plasticity. *Neuron.* 2015;86(5):1265–1276.
82. Stepp N, Plenz D, Srinivasa N. Synaptic plasticity enables adaptive self-tuning critical networks. *PLoS Comput Biol.* 2015;11(1):e1004043.
83. Papo D, Buldu JM, Boccaletti S, Bullmore ET. Complex network theory and the brain. *Philos Trans R Soc Lond.* 2014;369(1653).
84. Ogawa S, Lee TM, Kay AR, Tank DW. Brain magnetic resonance imaging with contrast dependent on blood oxygenation. *Proc Natl Acad Sci U S A.* 1990;87:9868–9872.
85. Logothetis NK. What we can do and what we cannot do with fMRI. *Nature.* 2008;453(7197):869–878.
86. Stephan KE, Roebroeck A. A short history of causal modeling of fMRI data. *Neuroimage.* 2012;62(2):856–863.
87. Diwadkar VA. Adolescent risk pathways toward schizophrenia: sustained attention and the brain. *Curr Top Med Chem.* 2012;12:2339–2347.
88. Diwadkar VA, Bakshi N, Gupta G, Pruitt P, White R, Eickhoff SB. Dysfunction and dysconnection in cortical-striatal networks during sustained attention: genetic risk for schizophrenia or bipolar disorder and its impact on brain network function. *Front Psychiatry.* 2014;5:50.
89. Diwadkar VA, Burgess A, Hong E, et al. Dysfunctional activation and brain network profiles in youth with obsessive-compulsive disorder: a focus on the dorsal anterior cingulate during working memory. *Front Hum Neurosci.* 2015;9:149.
90. Diwadkar VA, Ofen N. Disordered brain network function in adolescence: impact on thought, language and vulnerability for schizophrenia. In: Brambilla P, Marini A, eds. *Brain Evolution, Language and Psychopathology in Schizophrenia.* Oxford: Taylor and Francis Group; 2014:73–95.
91. Helbing WA, Luijnenburg SE, Moelker A, Robbers-Visser D. Cardiac stress testing after surgery for congenital heart disease. *Curr Opin Pediatr.* 2010;22(5):579–586.
92. Bressler SL, Seth AK. Wiener-Granger causality: a well established methodology. *Neuroimage.* 2011;58(2):323–329.
93. Friston K, Moran R, Seth AK. Analysing connectivity with Granger causality and dynamic causal modelling. *Curr Opin Neurobiol.* 2013;23(2):172–178.
94. Biswal BB, Mennes M, Zuo XN, et al. Toward discovery science of human brain function. *Proc Natl Acad Sci U S A.* 2010;107(10):4734–4739.
95. Muehlhan M, Kirschbaum C, Wittchen HU, Alexander N. Epigenetic variation in the serotonin transporter gene predicts resting state functional connectivity strength within the salience-network. *Hum Brain Mapp.* 2015;36(11):4361–4371.

96. Soares JM, Sampaio A, Ferreira LM, et al. Stress impact on resting state brain networks. *PLoS One*. 2013;8(6):e66500.
97. Zhang Y, Liu F, Chen H, et al. Intranetwork and internetwork functional connectivity alterations in post-traumatic stress disorder. *J Affect Disord*. 2015;187:114–121.
98. Brown VM, LaBar KS, Haswell CC, et al. Altered resting-state functional connectivity of basolateral and centromedial amygdala complexes in posttraumatic stress disorder. *Neuropsychopharmacology*. 2014;39(2):351–359.
99. Hall BS, Moda RN, Liston C. Glucocorticoid mechanisms of functional connectivity changes in stress-related neuropsychiatric disorders. *Neurobiol Stress*. 2015;1:174–183.
100. Henckens MJ, van der Marel K, van der Toorn A, et al. Stress-induced alterations in large-scale functional networks of the rodent brain. *Neuroimage*. 2015;105:312–322.
101. Cisler JM, Steele JS, Lenow JK, et al. Functional reorganization of neural networks during repeated exposure to the traumatic memory in posttraumatic stress disorder: an exploratory fMRI study. *J Psychiatr Res*. 2014;48(1):47–55.
102. Kiem SA, Andrade KC, Spoormaker VI, Holsboer F, Czisch M, Samann PG. Resting state functional MRI connectivity predicts hypothalamus-pituitary-axis status in healthy males. *Psychoneuroendocrinology*. 2013;38(8):1338–1348.
103. Schmidt A, Walter M, Gerber H, et al. Normalizing effect of heroin maintenance treatment on stress-induced brain connectivity. *Brain*. 2015;138(Pt 1):217–228.
104. Sweatt JD. The emerging field of neuroepigenetics. *Neuron*. 2013; 80(3):624–632.

PART 2

COGNITION, EMOTION, AND BEHAVIOR

CHAPTER

16

Cognition and Stress

M.G. Calvo, A. Gutiérrez-García
University of La Laguna, Tenerife, Spain

OUTLINE

Abstract

For a cognitive conceptualization, stress arises when environmental demands are perceived as taxing or potentially exceeding one's own capacity or resources to manage them, and there is threat to well-being if coping responses do not satisfy such demands. A cognitive vulnerability factor enhancing stress is trait anxiety, through hypervigilant processing styles involving selective orienting to threat cues, biased negative interpretation of ambiguous stimuli, and focusing attention on unfavorable thoughts. Short-term episodes of mild-intensity stress can facilitate cognitive functions, mainly encoding and memory consolidation of task-relevant stimuli, and in implicit memory or simple declarative tasks. However, exposure to high-intensity stress impairs the formation and retrieval of explicit memories and cognitive processes requiring complex or flexible reasoning. Long-term stress, particularly during childhood and adolescence, consistently undermines cognitive mechanisms. This can be due to chronic elevations of glucocorticoids inhibiting neurogenesis, which damages important functions in the hippocampus and, possibly, the prefrontal cortex.

ENVIRONMENTAL, BIOLOGICAL, AND COGNITIVE COMPONENTS OF STRESS

The major adaptive goal of evolved organisms is to survive and also to thrive. Such adaptive functions, nevertheless, have to be achieved in a complex and hostile environment. Complex, because there are multiple stimuli to attend to and multiple demands that must be met. Hostile, because many of those stimuli and demands involve physical dangers (e.g., harms to health or well-being) or psychological threats (e.g., social rejection or loss), if responses are not satisfactory. Individuals thus need to be able, first, to detect demands and also to prioritize them depending on their relative adaptive importance; and, second, coping resources must be used to respond to the demands in a way that harm is minimized and benefit is maximized. In such conditions, stress develops with a general adaptive purpose, by potentiating an alerting mechanism in cognitive and neural systems, and by recruiting and mobilizing resources in behavioral and physiological systems.

But, what is stress, and how can it be at the core of crucial adaptive functions for organisms? There are three major components in stress. First, the environmental, objective stimulus conditions that constitute the demands on the individual, which are called stressors. By restricting our consideration of them to human environments, multiple classifications can be made such as physical (e.g., accidents, natural disasters, noise, extreme temperatures, crowding, etc.), biogenic (e.g., illness, injury, pain, hunger, etc.) stressors, and psychosocial stressors (e.g., divorce, aggression, loss of job, exams, failures, etc.). A common characteristic of all stressors is that they alter a current biological or psychological homeostatic state or a desired goal, and thus threaten well-being.

Second, stress is characterized by a biological response to the environmental demands. Essentially, the brain initiates a course of action that releases neurotransmitters, peptides, and hormones throughout the

Stress: Concepts, Cognition, Emotion, and Behavior
http://dx.doi.org/10.1016/B978-0-12-800951-2.00016-9

body. Particularly, two systems are mobilized: the fast acting sympathetic nervous system (SNS) and the slower hypothalamus-pituitary-adrenal (HPA) axis. SNS responses include the release of the catecholamines adrenaline and noradrenaline from the adrenal medulla, which cause, for example, increases in heart rate and enhanced blood flow to skeletal muscles, and thus prepare the organism for a "fight-or-flight" behavioral response. Activation of the HPA-axis leads to the release of glucocorticoids (cortisol, in humans) from the adrenal cortex, causing, for example, an increase in the availability of energy substrates in different parts of the body, to support response resistance to environmental demands.

A third major component is the interpretation assigned by the individual to the objective stressor. The same objective stimulus, situation, or event may be more or less stressful, or not stressful, for different individuals, and even for the same individual in different contexts. In addition, the same physiological "stress response" can also be produced by nonstressful events. For example, components of the neuroendocrine response to appetitive, rewarding stimuli (e.g., sexual), events (e.g., social success), or physical activity (e.g., exercise) can be as large as when reacting to (aversive) stressful stimuli. This implies that stress cannot be defined only in terms of objective stimulus and response properties. Rather, it is commonly accepted that stress occurs when environmental demands tax or exceed the natural regulatory capacity of an organism, particularly when there is perceived uncertainty about the occurrence and the personal control of potentially threatening outcomes.

This introduction serves as a context for two cognitive approaches to stress: the cognitive origins and the cognitive consequences of the stress process. First, although some events (e.g., major life events or disasters) are objectively stressful for most individuals, the way they are (subjectively) perceived makes them more or less stressful. Also, for most daily hassles of mild intensity, the way they are perceived may determine their real impact, as to become stressful or not. The cognitive mechanisms—both the immediate triggers and the predisposing vulnerability factors—leading to perceive an event as stressful will be considered in section "Cognition and Stress: Cognitive Origins of Stress." Second, the catecholamines and glucocorticoids released during the biological stress response can afterwards enter the brain. This occurs particularly for the amygdala, the hippocampus, and the prefrontal cortex (PFC). Through feedback mechanisms, stress can thus exert important effects on the cognitive processes performed by such neural structures, involving interpretation, learning, memory, and decision. Such effects and mechanisms will be considered in section "Stress and Cognition: Cognitive Consequences of Stress."

KEY POINTS

- Stress is an adaptive biological mechanism that enhances alertness to environmental demands and mobilizes coping resources.
- The stress mechanism is activated by the perception of potential threats to well-being because of excessive demands.
- Individual differences in cognitive styles (mainly, attentional and interpretive biases) account for differences in vulnerability to stress.
- Short-term stress can facilitate cognitive encoding and learning of simple tasks, but impairs memory retrieval and complex cognitive processes.
- Long-term stress worsens a wide range of cognitive functions, possibly due to chronic elevations of glucocorticoids damaging brain structures.

COGNITION AND STRESS: COGNITIVE ORIGINS OF STRESS

Cognitive Appraisal

From a psychological perspective,[1,2] stress was defined by Lazarus and Folkman[3] as a kind of relationship between the person and the environment, in which the environmental demands are appraised as taxing the personal capacities and endangering well-being. Stress would occur when people confront circumstances that are perceived to exceed one's own ability to manage them. In this conceptualization, a central construct is appraisal, as a cognitive process that evaluates an event in terms of its significance for the person's well-being. Thus, the origin of stress would be cognitive, in that an environmental demand becomes stressful and activates the biological stress response to the extent that it is perceived and interpreted as threatening. This explains why the same objective situation or event (e.g., talking in front of a group of people) can be very stressful for one person or it may be less stressful, neutral, or even appetitive for others who could see it as an opportunity for benefits rather than potential harm. Nevertheless, while this view can generally be applied to most psychological stressors, the contribution of objective properties is probably greater for physical and biogenic stressors (e.g., noise, illness, etc.).

There are two classes of appraisal: primary and secondary.[1,2] *Primary appraisal* is an evaluation of what is at stake in the encounter, that is, what the environmental demands are and their implications. Three major outcomes of primary appraisal have been described. First, circumstances can be appraised as *irrelevant* to the personal well-being, if the situation does not concern the person's needs or goals, in which case no stress develops. Second, the situation can be appraised as positive or *benign*, if it is appraised as preserving or enhancing the personal well-being, with no stressful effects, either. Third, the circumstances can be appraised as *stressful* if the personal needs or goals are implicated in the situation. It is in this third case that *secondary appraisal* intervenes, as an evaluation of the personal resources or capacity and the available options to deal with the demands. Depending on the outcome of secondary appraisal, four major types of stress can develop: threat, harm, loss, and challenge. *Threat* appraisal means that the person perceives an impending event that may have negative consequences. *Harm* appraisal is the perception that something bad has already happened. *Loss* appraisal constitutes a specific kind of harm appraisal in which something that is positively valued becomes inaccessible. All these three kinds of appraisal lead to *distress*, which is the typical form of stress. In contrast, *challenge* appraisal leads to a different form of stress that has been termed as *eustress*. Challenge appraisal still involves seeing the situation as demanding (and taxing and potentially exceeding capacity, and therefore stressful), but also as an opportunity for obtaining benefits, which positively motivates effort and approach behavior.

The previous conceptualization of appraisal is formulated in a way that seems to entail strategic cognitive activity, as conscious, thoughtful, and voluntary analysis of environmental demands and personal resources. However, Lazarus[4] himself—as the proponent of the appraisal conceptualization—admits that appraisal can also be performed through automatic cognitive processes. In any case, within either view, perception of *unpredictability* and *uncontrollability* are critical ingredients of appraisal in current cognitive views of stress.[5] Unpredictability varies as a function of the degree of uncertainty about whether, when, where, how, or with what intensity a potential, threat, harm, or loss can *occur*. Uncontrollability varies as a function of the degree of uncertainty about whether and to what extent a person feels able to *cope* with the demands and minimize or avoid threat, harm, or loss. Unpredictability and uncontrollability will thus make environmental demands stressful. Furthermore, while uncertainty of control leads to an appraisal of threat, and perception of uncontrollability leads to appraisals of harm or loss, predictability of control leads to an appraisal of challenge.

Cognitive Vulnerability

As indicated above, the probability and intensity of the stress response significantly varies for different individuals, even for the same objective stressor. In addition, for a cognitive conceptualization of the stress process, the response is initiated and maintained depending greatly on the person's appraisal of the stressor. This implies that there must be a vulnerability factor involving relatively stable personality characteristics, and that such factor must be related to cognitive styles in the processing of environmental demands. Three such cognitive vulnerability mechanisms may enhance perception of threatening demands. First, a hypervigilant style that leads observers to selectively attend to threat-related cues in the environment, among other multiple stimuli, with low-thresholds for detection of threat cues of minimal intensity. Second, a tendency to preferentially interpret in a threatening way ambiguous stimuli that can, otherwise, also have nonthreat meanings (e.g., a neutral facial expression, or even a smile, which can be interpreted as anger or even as contempt). And, third, once a situation has been appraised as stressful, the proneness to focus on worrisome thoughts, with anticipation of dangers and negative evaluations of one's own coping capacity to face the demands. Understandably, selective attentional orienting toward threat cues, as well as biased negative interpretations of ambiguous stimuli, will enhance the probability of perceiving stressful demands. Relatedly, maintained rumination of internal representations of threat, harm, or loss, will increase the intensity and duration of the stress response.

Is there any type of personality characterized by such cognitive vulnerability processing styles? Research on trait anxiety and anxiety disorders (generalized anxiety disorder, obsessive-compulsive disorder, panic and post-traumatic stress disorder, social phobia, and simple phobia) has provided abundant evidence that people high in anxiety typically use all three types of processing mechanisms (for integrative reviews and meta-analytic studies, see Refs. 6–11). Hypervigilance and selective attention to threat cues in trait anxious people and anxious patients have been demonstrated with various tasks and measures, including gaze direction, attentional capture of covert attention, and lowered perceptual thresholds for threat stimuli. Similarly, interpretive bias or the preferential negative interpretation of ambiguous stimuli (words, sentences, and facial expressions) has proved to characterize trait anxious individuals and anxious patients, who also exhibit an enhanced risk estimation and predictive inferences of aversive consequences of ambiguous events. Relatedly, trait anxious individuals and anxious patients typically show a bias in the allocation of processing resources to threat-related internal representations, as evidenced by difficulties in inhibiting attention to such thoughts once threat has been detected. Given that a major biological function of anxiety is to facilitate the anticipation of threat detection, it is understandable that highly anxious people are especially prone to perceive stressful demands and to react with stress accordingly. What makes anxious people more able to detect and prioritize demands that may affect well-being, on one hand, also makes them more vulnerable to stress, on the other.

There is a neural substrate underlying the cognitive vulnerability styles of anxious individuals.[9,12–14] A critical brain mechanism that has been proposed to account for all three cognitive biases in anxiety entails an amygdala hyperresponsivity to threat signals along with PFC hyporesponsivity. The amygdala, a subcortical neural structure, is typically engaged with emotional processing and is particularly sensitive to threat. The prefrontal cortices, particularly the lateral PFC, are associated with conscious emotion regulation and executive processes. Neuroimaging studies have shown enhanced amygdala activity and reduced PFC activity in high-anxiety individuals. Enhanced amygdala activation would lead to hypervigilance for threat and a bias toward negative interpretations of stimuli. Reduced recruitment of PFC mechanisms would impoverish control of negative thoughts and anticipation of aversive consequences, thus delaying disengagement from worrisome internal representations. In the absence of such inhibitory mechanisms, which would allow for the persistence of the biased attention and interpretation, stress perception and reactivity would increase.

STRESS AND COGNITION: COGNITIVE CONSEQUENCES OF STRESS

Cognitive Effects of Acute Stress

Acute stress refers to episodes of limited duration (generally occurring within minutes or hours; e.g., a speech in public or an exam), as compared to long-term stress (lasting for weeks or months; e.g., looking after an ill person or undergoing academic promotion). The duration of stress is a relevant factor affecting cognitive mechanisms and activities. In general, while acute stress can have both a beneficial and a detrimental cognitive influence, prolonged stress generally undermines cognition. Acute stress impinges upon cognition in many ways, with either facilitating or impairing effects, depending on a number of factors. Among cognitive factors, the specific cognitive operation (e.g., implicit vs. explicit memory, long-term vs. working memory, and goal-directed vs. habit learning) and information processing phase (e.g., encoding, consolidation, and retrieval) are particularly important (see reviews in Refs. 15–18).

More specifically, in humans, psychosocial stress occurring just *before* learning may have no effects on *encoding*, or it can either impair or improve learning.[16,17] Memory for emotional information is generally enhanced by prior stress exposure, whereas memory for neutral information is less likely to be affected or it is even reduced by prior stress.[19] The facilitating effect of stress on memory for emotional words has been observed particularly for negative rather than positive words.[20] In addition, administration of cortisol before learning facilitates memory for emotionally arousing pictures.[21] Stress or glucocorticoid administration just *after* learning generally has shown beneficial effects on memory *consolidation* processes.[22] These effects appear to be particularly strong for emotionally arousing stimuli. In contrast, the exposure to stress or the administration of glucocorticoids shortly *before* memory testing of previously acquired information reduces *retrieval* performance.[23] Again, these effects are most pronounced for emotionally arousing material.[24]

Models have been proposed to explain the time-dependent effects of stress on memory.[17] As indicated above, there are two main stress hormones, namely, glucocorticoids (cortisol) and catecholamines (epinephrine and norepinephrine). The glucocorticoids directly access the brain where they bind to receptors in three brain areas involved in learning and memory, such as the hippocampus, amygdala, and frontal lobes. Also, epinephrine can act on the brain via vagal afferents, noradrenergic cells in the nucleus of the solitary tract and the locus ceruleus, which in turn, stimulate the basolateral amygdala. This way, if an individual is stressed shortly before, during, or shortly after learning, rapidly acting catecholamine and glucocorticoid effects facilitate attentional and other encoding processes. In addition, delayed glucocorticoid actions suppress competing information processing immediately after learning and hence promote memory consolidation of the recently encoded information. If, however, an individual is exposed to stress a considerable time before learning, and glucocorticoid action is already active during learning, stress can impede new learning and memory processes. Such a stress-induced retrieval impairment would be due to the stressful episode competing with, or directly suppressing, the concurrent cognitive activities required for the retrieval of previously learned information.

In addition to the influence of acute stress on the learning and memory functions associated with the amygdala and hippocampus, higher-order cognitive functions supported by the PFC are affected. In general, under high stress, the more flexible higher-order "cognitive" functions tend to be replaced by more rigid "habit" memory functions in the control of learning and response.[16,25,26] Nevertheless, the effects of stress are varied, with both impairment and improvement of working memory function,[27] cognitive flexibility,[25] and decision-making.[28] Differences across studies in stress intensity, cognitive load, and timing can probably account for the discrepant findings. In support of this view, first, mild increases in monoamines, such as dopamines associated to moderate stress exposure, enhance functional connectivity within PFC networks.[29] Second, in contrast, under highly stressful situations, an excessive release of neurotransmitters, such as norepinephrine, undermines PFC function and associated behaviors.[29] Third, a moderate increase in catecholamines boosts decision-making (leads to less risky decisions) while elevated cortisol impairs it (more risky decisions).[28] Fourth, delayed—but not immediate—effects of glucocorticoid administration in humans improve working memory performance and increase neuronal activity during performance in the dorsolateral PFC depending on task load.[30] This suggests that glucocorticoid receptors (GRs) in the PFC regulate stress-evoked dopamine efflux and the associated working memory impairment.[31]

Cognitive Effects of Prolonged Stress

Unlike acute stress, which, at least when it is of mild intensity, can yield some favorable effects on various cognitive processes (mainly encoding and memory consolidation), repeated or prolonged stress typically causes functional cognitive deterioration, in addition to structural damage of some brain structures, such as the hippocampus. Available evidence for the impact of chronic stress on memory in healthy humans is, nevertheless, scarce, mainly due to the ethical constraints inherent to intentionally exposing humans to repeated stress.

Indirect evidence of the effects of stress, however, has been obtained from various sources. Considerable available data come from studies focusing on stress-related neuropsychiatric disorders (see reviews in Refs. 16, 18, 32–34). This approach has provided information about a link between accumulated exposure to stress and impaired hippocampal memory function in humans. Mental disturbances mimicking mild dementia, such as performance decrements in simple and complex attentional tasks, verbal and visual memory, encoding, storage, and retrieval, have been described in depressed patients with hypercortisolism, and also in steroid psychosis following glucocorticoid treatment. There is also evidence that chronic exposure to stress and/or glucocorticoids worsens cognition and neuropathology in humans with Alzheimer's disease. Cognitive deficits are also reported in patients suffering from Cushing's disease, a medical condition in which endogenous levels of glucocorticoids are chronically elevated. Relatedly, during human aging, a significant proportion of elderly individuals present an endogenous increase of glucocorticoid levels, and this increase has been related to decreased memory performance.

Importantly, such cognitive deficits have been attributed to reduced hippocampal volume due to chronic elevations of glucocorticoids. Presumably, long-lasting elevated levels of glucocorticoids inhibit neurogenesis (i.e., the formation of new neurons or synapses), which damages important memory functions in the hippocampus. Studies have revealed the presence of smaller hippocampal volumes in various psychiatric disorders such as depression, post-traumatic stress disorder, and schizophrenia, all of which involve glucocorticoid level dysregulations.[33] For example, in major depressive disorder, a clinical syndrome that is highly sensitive to stress, the greatest degree of cognitive impairment occurs for memory dependent on hippocampal function, and it is the case that major depressive disorder patients have reduced hippocampal volume.[35] Hippocampal atrophy associated with chronic exposure to high levels of glucocorticoids is also reported in Cushing's patients and in elderly individuals. Hippocampal volume, however, increases and restores in part in such patients after treatment that diminishes cortisol levels to normal concentrations. After cortisol levels decline, structural volumetric increase in hippocampal volume is also accompanied by functional improvement in memory performance.[18] This is a significant finding since it implies the possibility of functional reorganization of the hippocampus once the chronic stress has been taken away.

A special issue regarding the cognitive effects of long-term stress is concerned with early-life stress (ELS; for a review, see Ref. 33). ELS refers to prolonged stressful experiences during infancy (e.g., abuse or maltreatment, early institutionalization, neglect, etc.). Children undergoing such stressful experiences exhibit global cognitive deficits, including decreased intellectual performance, less academic success, as well as poorer language abilities, and deterioration of various aspects of executive functioning (e.g., attention, inhibitory control, planning), and such children require greater individualized education programs.[36,37] These ELS cognitive deficits are also associated with smaller intracranial volume, reduced hemispheric integration and a smaller corpus callosum, and reduced hippocampus size.[38,39] In addition, the PFC has a high density of GRs and dopaminergic projections that are stress-sensitive. This makes the PFC especially susceptible of being damaged by ELS, given that PFC networks undergo critical development during childhood and adolescence. Importantly, the fact that the PFC is critically involved in high-level executive cognitive processes implies that relevant cognitive functions are negatively affected. Altogether, these findings suggest that the cognitive-deficit correlates of ELS are due to the ELS detrimental influence on critical brain structures that are necessary for cognitive development. As a consequence, the early stress experiences, if prolonged, can be especially harmful on cognition.

References

1. Carver CS. Coping. In: Contrada RJ, Baum A, eds. *The Handbook of Stress Science: Biology, Psychology, and Health.* New York, NY: Springer; 2011:195–208.

2. Smith CA, Kirby LD. The role of appraisal and emotion in coping and adaptation. In: Contrada RJ, Baum A, eds. *The Handbook of Stress Science: Biology, Psychology, and Health*. New York, NY: Springer; 2011:221–229.
3. Lazarus RS, Folkman S. *Stress, Appraisal, and Coping*. New York, NY: Springer; 1984.
4. Lazarus RS. Vexing research problems inherent in cognitive-mediational theories of emotion and some solutions. *Psychol Inq*. 1995;3:183–196.
5. Koolhaas JM, Bartolomucci A, Buwalda B, et al. Stress revisited: a critical evaluation of the stress concept. *Neurosci Biobehav Rev*. 2011;35:1291–1301.
6. Armstrong T, Olatunji BO. Eye tracking of attention in the affective disorders: a meta-analytic review and synthesis. *Clin Psychol Rev*. 2012;32:704–723.
7. Bar-Haim Y, Lamy D, Pergamin L, Bakermans-Kranenburg MJ, van IJzendoorn MH. Threat-related attentional bias in anxious and non-anxious individuals: a meta-analytic study. *Psychol Bull*. 2007;133:1–24.
8. Blanchette I, Richards A. The influence of affect on higher-level cognition: a review of research on interpretation, judgment, decision making, and reasoning. *Cogn Emot*. 2010;24:561–595.
9. Cisler JM, Koster EHW. Mechanisms of attentional biases towards threat in anxiety disorders: an integrative review. *Clin Psychol Rev*. 2010;30:203–216.
10. Eysenck MW, Derakshan N, Santos R, Calvo MG. Anxiety and cognitive performance: attentional control theory. *Emotion*. 2007;7:336–353.
11. Ouimet AJ, Gawronsky B, Dozois DJA. Cognitive vulnerability to anxiety: a review and integrative model. *Clin Psychol Rev*. 2009;29:459–470.
12. Bishop SJ. Neurocognitive mechanisms of anxiety: an integrative account. *Trends Cogn Sci*. 2007;11:307–316.
13. Bishop SJ. Trait anxiety and impoverished prefrontal control of attention. *Nat Neurosci*. 2009;12:92–98.
14. Hofmann SG, Ellard KK, Siegle GJ. Neurobiological correlates of cognitions in fear and anxiety: a cognitive-neurobiological information-processing model. *Cogn Emot*. 2012;26:282–299.
15. Marin MF, Lupien SJ. Stress and glucocorticoid effects on learning and memory: human studies. In: Conrad CD, ed. *Handbook of Stress: Neuropsychological Effects on the Brain*. Chichester: Wiley-Blackwell; 2011:248–265.
16. Sandi C. Stress and cognition. *Wiley Interdiscip Rev Cogn Sci*. 2013;4:245–261.
17. Schwabe L, Joëls M, Roozendaal B, Wolf OT, Oitzl MS. Stress effects on memory: an update and integration. *Neurosci Biobehav Rev*. 2012;36:1740–1749.
18. Van Stegeren AH. Imaging stress effects on memory: a review of neuroimaging studies. *La Rev Can Psychiatr*. 2009;64:16–27.
19. Payne JD, Jackson ED, Ryan L, Hoscheidt S, Jacobs JW, Nadel L. The impact of stress on neutral and emotional aspects of episodic memory. *Memory*. 2006;14:1–16.
20. Schwabe L, Bohringer A, Chatterjee M, Schachinger H. Effects of pre-learning stress on memory for neutral, positive and negative words: different roles of cortisol and autonomic arousal. *Neurobiol Learn Mem*. 2008;90:44–53.
21. Buchanan TW, Lovallo WR. Enhanced memory for emotional material following stress-level cortisol treatment in humans. *Psychoneuroendocrinology*. 2001;26:307–317.
22. Smeets T, Otgaar H, Candel I, Wolf OT. True or false? Memory is differentially affected by stress-induced cortisol elevations and sympathetic activity at consolidation and retrieval. *Psychoneuroendocrinology*. 2008;33:1378–1386.
23. Kuhlmann S, Piel M, Wolf OT. Impaired memory retrieval after psychosocial stress in healthy young men. *J Neurosci*. 2005;25:2977–2982.
24. Smeets T, Wolf OT, Giesbrecht T, Sijstermans K, Telgen S, Joëls M. Stress selectively and lastingly promotes learning of context-related high arousing information. *Psychoneuroendocrinology*. 2009;34:1152–1161.
25. Schwabe L, Wolf OT. Stress and multiple memory systems: from 'thinking' to 'doing'. *Trends Cogn Sci*. 2013;17:60–68.
26. Schwabe L, Wolf OT, Oitzl MS. Memory formation under stress: quantity and quality. *Neurosci Biobehav Rev*. 2010;3:514–591.
27. Beste C, Yildiz A, Meissnera TW, Wolf OT. Stress improves task processing efficiency in dual-tasks. *Behav Brain Res*. 2013;252: 260–265.
28. Pabst S, Brand M, Wolf OT. Stress and decision making: a few minutes make all the difference. *Behav Brain Res*. 2013;250:39–45.
29. Arnsten AF. Stress signalling pathways that impair prefrontal cortex structure and function. *Nat Rev Neurosci*. 2009;10:410–422.
30. Henckens MJ, van Wingen GA, Jöels M, Fernández G. Time-dependent corticosteroid modulation of prefrontal working memory processing. *Proc Natl Acad Sci U S A*. 2011;108:5801–5806.
31. Butts KA, Weinberg J, Young AH, Phillips AG. Glucocorticoid receptors in the prefrontal cortex regulate stress-evoked dopamine efflux and aspects of executive function. *Proc Natl Acad Sci U S A*. 2011;108:18459–18464.
32. Marin MF, Lord C, Andrews J, et al. Chronic stress, cognitive functioning and mental health. *Neurobiol Learn Mem*. 2011;96: 583–595.
33. Pechtel P, Pizzagalli DA. Effects of early life stress on cognitive and affective function: an integrated review of human literature. *Psychopharmacology (Berl)*. 2011;214:55–70.
34. Wolf OT, Buss C. Effect of chronic stress on cognitive function through life. In: Cooper CL, Field J, Goswami U, Jenkins R, Sahakian BJ, eds. *Mental Capital and Wellbeing*. Chichester: Wiley-Blackwell; 2010:233–241.
35. MacQueen G, Frodl T. The hippocampus in major depression: evidence for the convergence of the bench and bedside in psychiatric research? *Mol Psychiatry*. 2011;16:252–264.
36. De Bellis MD, Hooper SR, Spratt EG, Woolley DP. Neuropsychological findings in childhood neglect and their relationships to pediatric PTSD. *J Int Neuropsychol Soc*. 2009;15:868–878.
37. Loman MM, Wiik KL, Frenn KA, Pollak SD, Gunnar MR. Postinstitutionalized children's development: growth, cognitive, and language outcomes. *J Dev Behav Pediatr*. 2009;30:426–434.
38. Noble KG, Tottenham N, Casey BJ. Neuroscience perspectives on disparities in school readiness and cognitive achievement. *Future Child*. 2005;15:71–89.
39. Teicher M, Dumont N, Ito Y, Vaituzis C, Giedd J, Andersen S. Childhood neglect is associated with reduced corpus callosum area. *Biol Psychiatry*. 2004;56:80–85.

CHAPTER

17

Stress, Memory, and Memory Impairment

S. Musić[1], S.L. Rossell[1,2,3]

[1]Swinburne University of Technology, Melbourne, VIC, Australia

[2]The Alfred Hospital and Monash University Central Clinical School, Melbourne, VIC, Australia

[3]St. Vincent's Mental Health, Melbourne, VIC, Australia

OUTLINE

Abstract

Stress can have impairing, enhancing, or no influence on memory systems and processes. The present chapter reviews, albeit selectively, the impact of stress on human memory with a particular focus on clinical populations. The consequence of stress on memory systems will be reviewed focusing on working memory, implicit and explicit memory, episodic, and autobiographical memory. This is followed by the review of the impact of stress on memory phases including encoding, consolidation, retrieval, and reconsolidation. Throughout the review, the effects of stress on memory are linked to psychiatric disorders, and clinical implications are considered.

INTRODUCTION

When faced with imminent stress, our body initiates a number of physiological reactions, including the activation of the fast-acting autonomic nervous system and the slower hypothalamic-pituitary-adrenal axis (HPA, as measured by an increase in glucocorticoid (GC)). This stress system activation is mostly adaptive and beneficial, aiding survival. However, recurrent stress exposure, particularly if experienced early in life, can induce chronic stimulation of the HPA axis, thus leading to modifications in the consequent reactivity to acute stressors. There is accumulating evidence that such abnormalities in the stress response can influence the quality and quantity of memory processes,[1] and have deleterious implications for several psychiatric disorders such as post-traumatic stress disorder (PTSD[2]), depression,[3] and generalized anxiety disorder (GAD[4]). The impact and direction of these maladaptive patterns is modulated by a number of factors including: type, timing, intensity and frequency of the stressor, type of paradigm (valence and arousal dependent), as well as demographic variables such as sex and age of the person at the time of the experienced stress.[5]

Stress does not impair all memory systems equally, nor does it affect all areas of the brain thought to be associated with memory functions.[6,7] We review the present advances on the impact of stress on human memory with a particular focus on clinical populations. We will discuss the most promising findings, and their implication for the pathogenesis and treatment of several psychiatric conditions.

Stress: Concepts, Cognition, Emotion, and Behavior
http://dx.doi.org/10.1016/B978-0-12-800951-2.00017-0

KEY POINTS

- Chronic stimulation of the HPA axis can affect the quality and quantity of memory.
- The effects of stress on memory systems and processes are dependent on a number of factors including: type, timing, intensity, and frequency of the stressor, type of paradigm (valence and arousal dependent), as well demographic variables such as sex and age of the person at the time of the experienced stress.
- The impact of stress on memory performance is studied in experimental settings through psychopharmacological administrations of stress hormones or exposure to psychosocial stressors.
- Memory performance under stress appears to be most sensitive to emotional stimuli, with this information most often remembered.
- Working memory difficulties have been linked to the hallmark features of depression.
- Information-processing bias may be a vulnerability and maintenance factor in the symptomatology of psychiatric conditions such as PTSD and depression.
- Stress-related memory impairments have been observed in a number of psychiatric conditions including OCD, GAD, PTSD, social anxiety, substance disorders, spider phobia, and depression.
- Implicit memory bias in PTSD is thought to predict PTSD symptomatology.
- Information-processing modification treatment trials have commenced with some showing promising results.
- Overgeneralized autobiographical memory has been reported in a number of clinical populations, including PTSD.
- Depending on the time interval interposed between learning and memory testing, the effects of stress on encoding may confound the process of memory retrieval or consolidation. Thus it is difficult to study the complex interaction.
- Overconsolidation of trauma memories fails effective autobiographical memory integration and thus promotes the development of intrusive memories in PTSD.
- Implicitly and explicitly biased memory retrieval has been observed in a number of psychiatric conditions including social anxiety disorder and GAD.
- Interference with the reconsolidation process of different memory systems using various pharmacological (e.g., beta blockers) and behavioral manipulations (e.g., stress exposure) has been attempted with promising results.
- Postretrieval plasticity of reconsolidation carries significant treatment implications for many psychiatric conditions characterized by pathological memories, including psychosis and substance abuse disorders.
- β-Adrenergic blocking agent has been used in reconsolidation disruption of traumatic memories, and has demonstrated attenuated fear and startle responses in PTSD patients.

MEMORY

Memory is thought to consist of multiple anatomically and functionally distinct, yet interacting, entities that influence behavioral performance, with some studies suggesting a compensatory and others a competitive interaction between the memory systems.[8] The switching between these systems has been attributed to corticosteroid stress hormones.[9] Given the number of memory systems and processes to be discussed in the proceeding review, working definitions are presented in Table 1.

For the purpose of this chapter, memory systems will be divided hierarchically as presented in Figure 1. The impact of stress on the memory subsystem will be discussed, with the understanding that some subsystems share neural structures (Figure 2). For example, memory-related limbic structures such as the amygdala and hippocampus have been implicated in explicit and implicit emotional memory[10] and are involved in the emotional modulation of multiple memory systems.[6]

Working Memory and Stress

While the effects of stress on memory have been the focus of extensive research (for review, Ref. 8), only recent studies have started to consider the consequences on working memory (WM[11]). Findings have been somewhat inconsistent with some research showing enhanced,[12] while other studies showing detrimental effects of stress on WM.[43] Duncko and colleagues,[12] for example, postulated that WM function is stress-dose dependent with acute stress exposure improving WM performance. Heterogeneous neuroimaging findings have also been reported; with stress-induced increase and decrease of activity in the lateral and dorsolateral prefrontal cortex (PFC) and intraparietal sulcus.[7] Sex differences have been

TABLE 1 Working Definitions of Memory Systems and Processes

Autobiographical memory	Autobiographical memories are semantic memories of biographical facts that store some of the episodic memory content with its spatiotemporal context, and allow for extensions, modifications, integrations, and de-emotionalizations of knowledge through the semantic memory system.
Consolidation	Memory consolidation is the slow protein synthesis dependant stabilization process of novel events that occurs post initial encoding, and leads to the formation of a memory trace.
Encoding	Encoding refers to the first critical step in creating a new memory. Perceived items of interest are converted into a construct that it is stored in the brain. If adequately "encoded" it can be recalled at a later date.
Episodic memory	Episodic memory refers to memory of an event or episode. The episodic memory is a storage system that receives highly detailed multimodal information about past personal experiences including their spatiotemporal context, internal state of the person's emotions, perceptions and thoughts, and the ability to engage in episodic future thinking.
Explicit memory	Explicit memory (sometimes referred to as declarative) consists of systems that involve conscious recollections of previous experiences.[8] Declarative memory is divided into episodic and semantic memory.
Implicit memory	Implicit memory (sometimes termed nondeclarative) is a nonintentional form of memory that impacts on the person without their conscious awareness. Priming tasks are most often employed to investigate implicit memory.[8]
Procedural memory	Procedural memory is a memory storage system of skills usually acquired through repetition and practice.
Reconsolidation	Recalled or reactivated consolidated memories can, under specific conditions, reenter a labile state requiring protein synthetic restabilization. This process is termed reconsolidation.
Semantic memory	Semantic memory refers to our general knowledge and facts, and is shared with others.
Sensory memory	Sensory memory allows the retention of incoming sensory information for a short period of time after incoming stimuli have ceased.
Working memory	Working memory refers to a temporary storage (several seconds) system that maintains and manipulates information.[3]

noted in a recent study undertaken by Schoofs et al.,[13] where stress improved WM performance in males, and impaired it in females.

There are several lines of evidence that support the notion that stress can impair WM performance[11,43]; particularly when WM is engaged in demanding executive processing, and/or emotionally charged tasks (e.g., Refs. 14 but see 13). For example, to engender greater chance of survival, emotionally charged information recruits attentional resources that enable us to rapidly detect and respond to potentially threatening changes in the environment. However, these emotional stressors are distracting and impair WM performance for other task-relevant information.[14] In accordance with these observations,[43] investigated the immediate impact of emotion on WM using human faces coupled with emotional and neutral pictorial distracters. Their functional magnetic resonance imaging findings unveiled an increased amygdala and hippocampal, and reduced dorsolateral PFC activity in response to emotional distracters. Although such interference is a normal response in healthy individuals, this information-processing bias may induce and maintain symptoms such as hypervigilance in PTSD, or rumination in depression,[3] thus leading to WM impairments. With this in mind, some authors have speculated that many hallmark features of depression are linked to WM processes and the mood-congruent information bias that interferes with this process.[3]

Implicit Memory and Stress

Stress appears to mostly facilitate, rather than disrupt, aspects of the learned, nonintentional, and automatic implicit memory system. Although this relationship is understudied, cognitive theories of emotional disorders posit that biased processing of information are vulnerability and maintenance factors for these conditions.[15] For example, depressed individuals show increased hippocampus, amygdala, and anterior inferotemporal cortex activity to implicitly presented sad faces,[10] and people with GAD have shown an implicit memory bias (IMB) for threat words specific to their domains of worry.[4]

IMB for traumatic cues has been observed in several groups with PTSD (e.g., car accident, interpersonal violence survivors) using supra and subliminal word completing paradigms, free recall and word-stem completion tasks.[2] Moreover, Amir and colleagues[2] showed that individuals with PTSD exhibited an IMB for negative and trauma-relevant pictorial stimuli. Trauma exposed persons (without PTSD) demonstrated a greater IMB for negative images relative to trauma and neutral pictorial stimuli suggesting a differentiated memory mechanism for PTSD compared to that of general distress and trauma history. The authors proposed that it is this faulty IMB that predicts PTSD symptomatology. IMB is thought to facilitate the cue-driven involuntary retrieval of traumatic memories which play a crucial role in the onset and maintenance of the PTSD symptoms.[16] Therefore,

FIGURE 1 Hierarchical memory system subdivision.

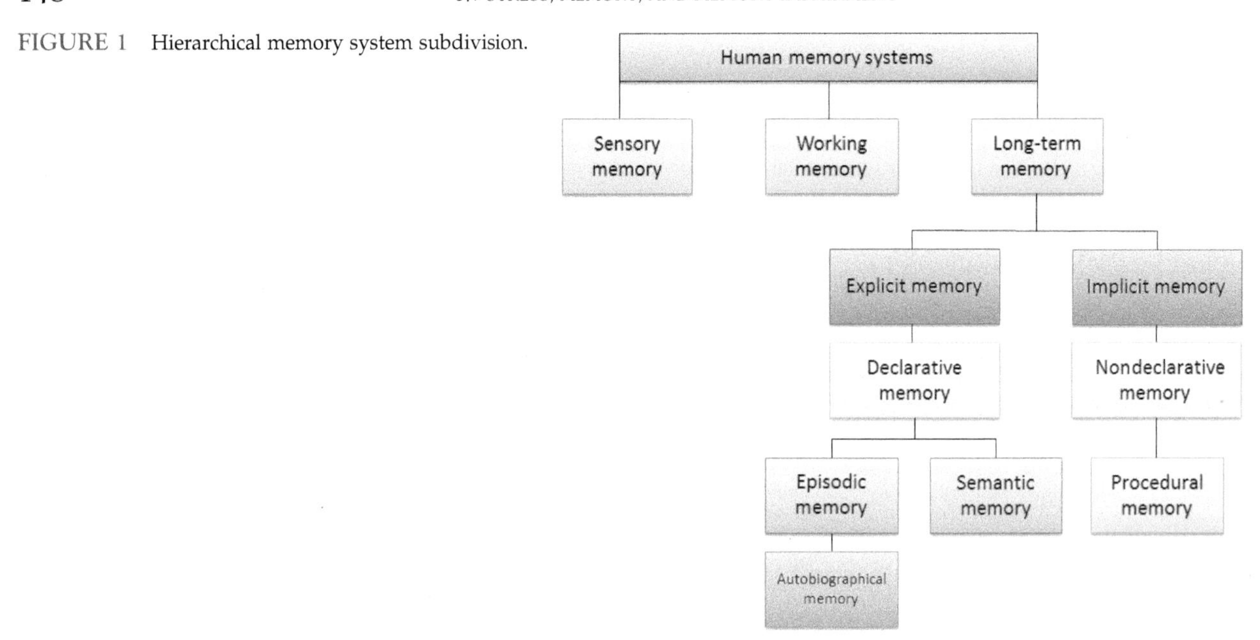

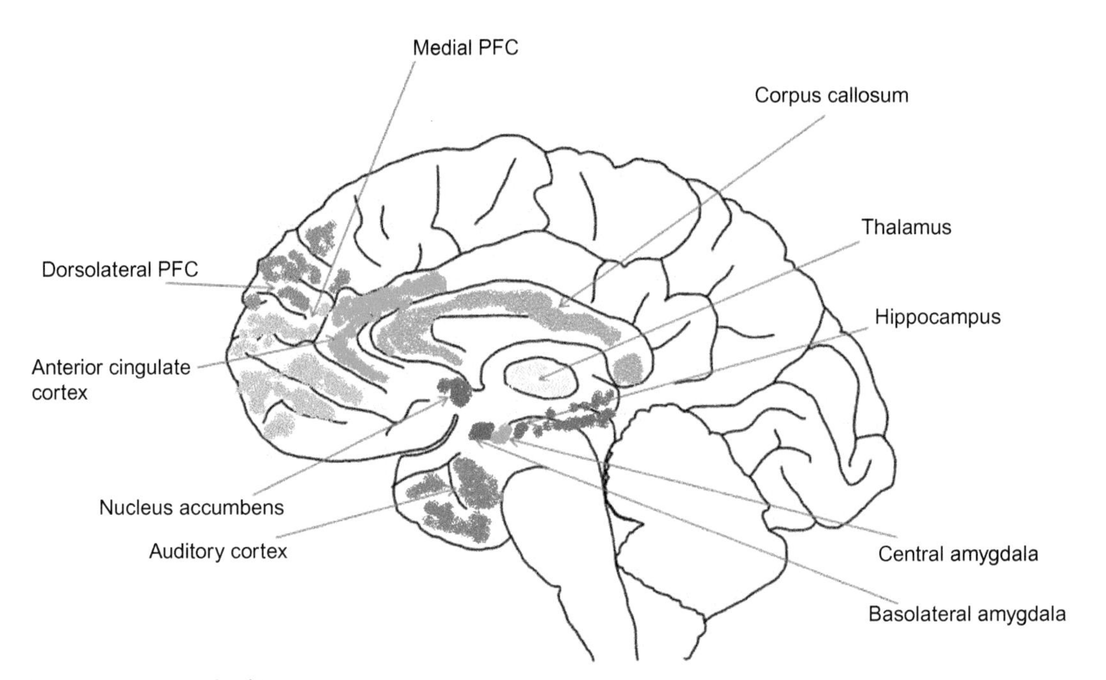

FIGURE 2 Brain structures involved in memory.

understanding the automatic biases in emotional disorders, as discussed in the aforementioned studies, has implications for information-processing modification treatments of these debilitating psychiatric conditions.[4,15]

Explicit Memory and Stress

The effect of stress and GC administration has been thoroughly investigated in terms of explicit memory (EM).[2,17] The direction of this observed effect is dependent on several factors, including: the memory phase (discussed below) and emotionality of the presented material[17,18] among many others that are beyond the scope of our review. EM impairments have been reported in healthy adults when treated with cortisone or exposed to psychosocial stress,[19] and in emotional conditions including PTSD, depression, and obsessive compulsive disorder (OCD[18,20]: for meta-analytic review). For example, cognitive models of OCD postulate that sufferers show deficits in cued-recall and recognition memory with dysfunctions particularly evident for threat-related and neutral material.[20] Recent studies indicate that beta blockers[21] and nonarousing testing, and congruent learning and testing environments[22,23] eliminate the stress-induced retrieval of EM deficits.

Emotional disorders have also shown an EM enhancement in experimental settings. A meta-analytic study of anxiety disorders, for instance, has demonstrated superior retrieval abilities of threatening material. Superior free and cued recall findings have been observed in PTSD, GAD, OCD, depression, and panic disorder with agoraphobia.[4] For instance, using various word paradigms, individuals with panic disorder exhibited enhanced retrieval related to threat, physical-threat, panic with and without the imagining of personal scenes in comparison to healthy controls.[24] This EM bias is thought to maintain the vicious cycle of panic attacks.[26] Enhanced recall has also been reported in OCD for contaminated objects and previously experienced threat-related actions. Similarly, EM enhancements have also been observed in depression with individuals reporting more overgeneralized and negatively valanced memories in comparison to nondepressed controls.[25] It has been argued that EM bias in individuals with emotional disorders is due to increased attention for disorder-congruent information, thus the evaluation and subsequent interpretation of the situation is guided by (at least partially) the retrieval of memories of disorder-congruent stressful and threatening experience.[24]

It is also important to note that numerous studies have found no enhanced memory retrieval in emotional disorders, such as for threat-relevant information in social anxiety, or disorder-congruent information in spider phobia (Ref. 26 for review). The discrepancy in findings may be due to methodological differences between the studies, but also because of the sample characteristics, such as comorbidity with other conditions. For example, PTSD is often comorbid with other emotional disorders, thus the participant samples are never pure. Some authors have also argued that the memory impairments and enhancements in emotional disorders are not global affecting disorder-congruent and unrelated material differently (Ref. 15 for review).

Episodic Memory and Stress

The hippocampus is crucial for the encoding, consolidation, and retrieval of episodic memories,[5] and the dysfunctional retrieval of these memories has been linked to the etiology of anxiety disorders.[27] In their population-based study of episodic memory deficits in anxiety disorders, Airaksinen and colleagues[28] found that dysfunctional patterns for emotionally irrelevant stimuli are similar for most anxiety disorders, with most significant deficits evident in social phobia and panic disorder. In line with this finding, Zlomuzica et al.[27] proposed that emotional valence of presented information affects the storage and retrieval of episodic memory and this has been supported by increased amygdala and hippocampus activity while viewing emotionally provoking information.

In contrast, excessive spontaneous episodic memory retrieval of stressful experiences increases re-experiencing symptomatology in PTSD and phobia patients, thus continuously increasing the reconsolidation of the episodic trauma memory (Ref. 27 for review). Overconsolidation of trauma memories is linked to increased arousal, intense intrusive memories, and psychological and physiological reactivity to trauma reminders.[44] This hyperconsolidation may be linked to the interaction and simultaneous elevation of cortisol and noradrenalin which consequently leads to increased hippocampal and amygdala activity.

Studies have suggested that memory function could be employed as an indicator of treatment prognosis and guide treatment planning, as deficits in episodic memory have been associated with reduced daily functioning. In a recent review, Zlomuzica et al.[27] argued that these dysfunctions might contribute to the etiology of anxiety disorders, proposing the active use of training programs to support therapeutic intervention of memory dysfunction in anxiety disorders.

Autobiographical Memories and Stress

Enhanced recall for emotional and threatening autobiographical memories (AM) has been shown in numerous psychiatric conditions, including in panic disorder and social anxiety. In contrast, Young et al.[29] demonstrated AM retrieval deficits in unmedicated individuals with depression, and found reduced activity in the right hippocampus and parahippocampus in response to emotionally valanced cue words. In addition, AM deficits have been linked to poor problem solving capacity and self-efficacy, and reduced psychological treatment outcomes.

A recent study showed that stress can have detrimental effects on the reconsolidation of AMs in healthy controls.[30] In PTSD, deficits in contextualization of the traumatic memory into the autobiographical system due to inefficient encoding may contribute to the distortion in trauma memory and development of PTSD symptomatology.[16] This may also intensify the retrieval deficits of personal memory in PTSD.[31]

Overgeneralized autobiographical memory (OGM) refers to difficulties retrieving the specific spatiotemporal autobiographical events,[31] with a recent study showing that clinical populations exhibit memory specificity dysfunctions in the generation of future events too. The OGM bias has been observed in individuals with depression[29] and PTSD.[31] For example, neuroimaging findings indicate spatial and temporal differences in the neural networks in PTSD compared to healthy controls.[31] It has been argued that OGM is a protective mechanism linked to the avoidance of conscious, specific cognitive processing of aversive autobiographical experiences as memory specificity of that event may lead to increased re-experiencing, physiological reactivity, and negative affect.[25] Although, OGM biases may temporarily dampen distress, they continue to preserve weakened emotional regulation and lead to enhanced vulnerability to future emotional deficits.[45]

This section has reviewed, albeit selectively, the most important findings of the impact of stress on memory systems. Memory can also be phase dependant, and theoretically further divided into encoding, consolidation, and retrieval.[1] In the section that follows the effect of stress on memory processes will be discussed.

STRESS AND MEMORY ENCODING

It is difficult to isolate the effects of stress on the encoding process as stress should be administered prior to learning.[1] Depending on the time interval interposed between learning and memory testing, the effects of stress on encoding may confound the process of memory retrieval or consolidation.[32] Thus, unsurprisingly the results of encoding studies are heterogeneous with some reporting enhancing,[33] others impairing,[34] or no effect[35] of stress. For example, Weymar et al.[36] found that for a subset of participants, emotional words encoded in the context of anticipatory anxiety prompted better memory performance, and increased event-related potential activity over the centroparietal region in the 500-700 ms window. Future research is required to provide insight into the understanding of the effects of stressful experiences (e.g., witnessing a car accident, rape) on the initial acquisition and encoding of trauma memories.[33] This may lead to a better understanding of the complex interaction and its implications in the pathogenesis and treatment of psychiatric conditions such as PTSD.

STRESS AND MEMORY CONSOLIDATION

A considerable body of literature has demonstrated that components of the stress reaction and stress hormones, such as GC and catecholamines, can affect memory consolidation.[1,5] Neural circuits of memory consolidation in animal and human studies indicate the involvement of the lateral and central amygdala, hippocampus, nucleus accumbens, auditory cortex, and medial PFC.[37] Accumulating evidence postulates that the direction of the effect is dependent on the timing of the induced stress (pre, during, or post) particularly for hippocampal and striatum dependent learning and memory processes.[32] For instance, if the stress is induced prior to learning, some studies report memory enhancing effects, while stress prior to retention testing appears to reduce memory retrieval.[19] These effects were particularly evident for emotionally charged content or pharmacologically elevated GC concentrations.[37]

The overconsolidation hypothesis of PTSD states that extreme arousal caused by traumatic events creates overconsolidated trauma memories that fail effective AM integration.[16,44] The intense arousal triggers excessive adrenergic activity resulting in enhanced encoding of environmental stimuli, thus promoting the development of intrusive and debilitating trauma memories that are more easily triggered.[44] These overconsolidated memories facilitate the development of PTSD (e.g., intrusive memories, reactivity to trauma reminders.[44] Recent pharmacological studies have unveiled the possibility of trauma memory consolidation disruption shortly after the trauma event by preventing the increase of emotional arousal required for consolidation to occur, or inhibiting the retrieval of the trauma memory and enabling integration into autobiographic memory systems.[44]

STRESS AND MEMORY RETRIEVAL

Stress or GCs administration, can have deleterious effects on memory retrieval particularly if stressed prior to retention testing or emotionally arousing material is presented.[38] A recent study tested memory retrieval of previously learnt material during the experience of stress and found no performance effect.[38] Similarly, some studies reported that nonarousing testing environments do not have an effect on EM retrieval.[22] However, declarative memory retrieval deficits have been reported in a number of studies,[35] and retrieval impairments have been observed in episodic memory for emotionally irrelevant stimuli in persons with clinical anxiety.[28] In line with this view, PTSD patients exhibited reduced hippocampal functional responses during retrieval of negative word paradigms.

Similar to the reconsolidation studies, β-adrenergic blocking agents appear to abolish the effects of stress on memory retrieval.[21] It is important to note that stress can also enhance memory retrieval performance.[21] This is particularly significant if psychiatric conditions are to be treated with GCs, where stress-related deficits in memory retrieval are expected.

RECONSOLIDATION

Reconsolidation is not a simple repetition of consolidation (for review, Ref. 39), as it offers a unique opportunity to enhance, impair, or update the memory to incorporate new information under specific conditions. To date, researchers have attempted to interfere with the reconsolidation process of different memory systems using various pharmacological (e.g., beta blockers) and behavioral manipulations (e.g., stress exposure).[40] For instance, in a study of abstinent heroin addicts, Zhao et al.[41] demonstrated that social stress appeared to disrupt the reconsolidation of memories for addiction-congruent, but not neutral words. Similar findings of the role of behavioral stress and stress hormones have been observed using a variety of paradigms in different memory systems.[30] Furthermore, neuroimaging studies have implied that the disruption of the reconsolidation process is amygdala, hippocampus, insular cortex, and medial PFC dependant.[32]

The postretrieval plasticity of reconsolidation carries significant treatment implications for many psychiatric conditions characterized by pathological memories, including substance abuse disorders,[41] phobias,[42] and PTSD[40] as it may lead to "erasure," or at least reduced emotionality of the memory traces. Although still in its infancy, a number of clinical trials using β-adrenergic blocking agent in reconsolidation disruption of traumatic memories, have demonstrated attenuated fear and startle responses in PTSD patients.[40] The generalization of fear and the fear response were affected by the administration of β-adrenergic blocker propranolol in healthy controls, but not the cognitive components of that response.[32] As has been noted throughout this review, the literature surrounding the impact of stress on memory is heterogeneous, with studies reporting impairing, enhancing and no effect in healthy controls and clinical populations. This is of clinical importance as blunting the emotionality of traumatic memories raises significant ethical concerns.

CONCLUSION

The interactions between stress and memory are complex, and one must understand the limitations of the studies presented in this chapter. The effect of stress on memory appears to depend on a number of factors including the paradigm; the timing, dose, and type of stressor; and the participant sample. Notwithstanding these limitations, the effect of stress on memory systems and processes has significant implications for the understanding and treatment of many debilitating psychiatric conditions. The recent advances in the psychopharmacological approach to the treatment of intrusive memories, for example, may assist many individuals with PTSD who might not be able to, or willing to engage in psychological therapy. Understanding the implications of the administration of GCs in clinical populations and the effects of it is a challenge for future research studies.

References

1. Schwabe L, Wolf OT, Oitzl MS. Memory formation under stress: quantity and quality. *Neurosci Biobehav Rev*. 2010;34:584–591. http://dx.doi.org/10.1016/j.neubiorev.2009.11.015.
2. Amir N, Leiner AS, Bomyea J. Implicit memory and posttraumatic stress symptoms. *Cogn Ther Res*. 2010;34:46–58. http://dx.doi.org/10.1007/s10608-008-9211-0.
3. Baddeley A. Working memory and emotion: ruminations on a theory of depression. *Rev Gen Psychol*. 2013;17(1):20–27. http://dx.doi.org/10.1037/a0030029.
4. Coles ME, Turk CL, Heimberg RG. Memory bias for threat in generalized anxiety disorder: the potential importance of stimulus relevance. *Cogn Behav Ther*. 2007;36(2):65–73. http://dx.doi.org/10.1080/16506070601070459.
5. Wolf OT. Stress and memory in humans: twelve years of progress? *Brain Res*. 2009;1293:142–154. http://dx.doi.org/10.1016/j.brainres.2009.04.013.
6. McGaugh JL. Memory—a century of consolidation. *Science*. 2000;287:248. http://dx.doi.org/10.1126/science.287.5451.248.
7. van Ast VA, Spicer J, Smith EE, et al. Brain mechanisms of social threat effects on working memory. *Cereb Cortex*. 2014; http://dx.doi.org/10.1093/cercor/bhu206.
8. Squire LR, Dede AJO. Conscious and unconscious memory systems. *Cold Spring Harb Perspect Biol*. 2015;7:a021667. http://dx.doi.org/10.1101/cshperspect.a021667.
9. Schwabe L, Oitzl MS, Richter S, Schachinger M. Modulation of spatial and stimulus-response learning strategies by exogenous cortisol in healthy young women. *Psychoneuroendocrinology*. 2009;34:358–366. http://dx.doi.org/10.1016/j.psyneuen.2008.09.018.
10. Victor TA, Furey ML, Fromm SJ, Bellgowan PSF, Ohman A, Drevets WC. The extended functional neuroanatomy of emotional processing biases for masked faces in major depressive disorder. *PLoS ONE*. 2012;7:10. http://dx.doi.org/10.1371/journal.pone.0046439.
11. Luethi M, Meier B, Sandi C. Stress effects on working memory, explicit memory, and implicit memory for neutral and emotional stimuli in healthy men. *Front Behav Neurosci*. 2009;2:5. http://dx.doi.org/10.3389/neuro.08.005.2008.
12. Duncko R, Johnson L, Merikangas K, Grillon C. Working memory performance after acute exposure to the cold pressor stress in healthy volunteers. *Neurobiol Learn Mem*. 2009;91:377–381. http://dx.doi.org/10.1016/j.nlm.2009.01.006.
13. Schoofs D, Pabst S, Brand M, Wolf OT. Working memory is differentially affected by stress in men and women. *Behav Brain Res*. 2013;241:144–153. http://dx.doi.org/10.1093/cercor/bhu206.
14. García-Paciosa J, Del Ríob D, Maestú F. State anxiety in healthy people can increase their vulnerability to neutral but not to unpleasant distraction in working memory. *Clín Salud*. 2014;25:181–185. http://dx.doi.org/10.1016/j.clysa.2014.10.002.
15. Mathews A, MacLeod C. Cognitive vulnerability to emotional disorder. *Annu Rev Clin Psychol*. 2005;1:167–195.
16. Ehlers A, Clark DM. A cognitive model of posttraumatic stress disorder. *Behav Res Ther*. 2000;38:319–345. http://dx.doi.org/10.1016/S0005-7967(99)00123-0.
17. Buchanan TW, Tranel D. Stress and emotional memory retrieval: effects of sex and cortisol response. *Neurobiol Learn Mem*. 2008;89:134–141. http://dx.doi.org/10.1016/j.nlm.2007.07.003.

18. Brewin CR, Kleiner JS, Vasterling JJ, Field AP. Memory for emotionally neutral information in posttraumatic stress disorder: a meta-analytic investigation. *J Abnorm Psychol.* 2007;116(3):448–463. http://dx.doi.org/10.1037/0021-843X.116.3.448.
19. Kuhlmann S, Piel M, Wolf OT. Impaired memory retrieval after psychosocial stress in healthy young men. *J Neurosci.* 2005;25:2977–2982. http://dx.doi.org/10.1523/JNEUROSCI.5139-04.2005.
20. Tuna S, Tekcan AI, Topcuoglu V. Memory and metamemory in obsessive-compulsive disorder. *Behav Res Ther.* 2005;43:15–27. http://dx.doi.org/10.1016/j.brat.2003.11.001.
21. Schwabe L, Romer S, Richter S, Dockendorf S, Bilak B, Schachinger H. Stress effects on declarative memory retrieval are blocked by ab-adrenoceptor antagonist in humans. *Psychoneuroendocrinology.* 2009;34:446–454. http://dx.doi.org/10.1016/j.psyneuen.2008.10.009.
22. Kuhlmann S, Wolf OT. A non-arousing test situation abolishes the impairing effects of cortisol on delayed memory retrieval in healthy women. *Neurosci Lett.* 2006;399:268–272. http://dx.doi.org/10.1016/j.neulet.2006.02.007.
23. Schwabe L, Wolf OT. The context counts: congruent learning and testing environments prevent memory retrieval impairment following stress. *Cogn Affect Behav Neurosci.* 2009;9(3):229–236. http://dx.doi.org/10.3758/CABN.9.3.229.
24. Cloitre M, Shear MK, Cancienne J, Zeitlin SB. Implicit and explicit memory for catastrophic associations to bodily sensation words in panic disorder. *Cogn Ther Res.* 1994;19(3):225–240.
25. Williams JMG, Barnhofer T, Crane C, et al. Autobiographical memory specificity and emotional disorder. *Psychol Bull.* 2007;133(1):122–148. http://dx.doi.org/10.1037/0033-2909.133.1.122.
26. Coles ME, Heimberg RG. Memory biases in the anxiety disorders: current status. *Clin Psychol Rev.* 2002;22:587–627.
27. Zlomuzica A, Dere D, Machulska A, Adolph D, Dere E, Margraf J. Episodic memories in anxiety disorders: clinical implications. *Front Behav Neurosci.* 2014;8:131. http://dx.doi.org/10.3389/fnbeh.2014.00131.
28. Airaksinen E, Larsson M, Forsell Y. Neuropsychological functions in anxiety disorders in population-based samples: evidence of episodic memory dysfunction. *J Psychiatr Res.* 2005;39:207–214. http://dx.doi.org/10.1016/j.jpsychires.2004.06.001.
29. Young KD, Bellgowan PSF, Bodruka J, Drevets WC. Behavioral and neurophysiological correlates of autobiographical memory deficits in patients with depression and individuals at high risk for depression. *JAMA Psychiatry.* 2013;70(7):698–708. http://dx.doi.org/10.1001/jamapsychiatry.2013.1189.
30. Schwabe L, Wolf OT. Stress impairs the reconsolidation of autobiographical memories. *Neurobiol Learn Mem.* 2010;94:153–157. http://dx.doi.org/10.1016/j.nlm.2010.05.001.
31. St. Jacques PL, Kragel PA, Rubin DC. Neural networks supporting autobio-graphical memory retrieval in posttraumatic stress disorder. *Cogn Affect Behav Neurosci.* 2013;13:554–566. http://dx.doi.org/10.3758/s13415-013-0157-7.
32. Schwabe L, Joels M, Roozendaal B, Wold OY, Oitzl MS. Stress effects on memory: an update and integration. *Neurosci Biobehav Rev.* 2012;36:1740–1749. http://dx.doi.org/10.1016/j.neubiorev.2011.07.002.
33. Payne JD, Jackson ED, Hoscheidt S, Ryan L, Jacobs WJ. Stress administered prior to encoding impairs neutral but enhances emotional long-term episodic memories. *Learn Mem.* 2007;14:861–868. http://dx.doi.org/10.1101/lm.743507.
34. Dickie WW, Brunet A, Akerib V, Armony JL. An fMRI investigation of memory encoding in PTSD: influence of symptom severity. *Neuropsychologia.* 2008;46:1522–1531. http://dx.doi.org/10.1016/j.neuropsychologia.2008.01.007.
35. Smeets T, Otgaar H, Candel I, Wolf OT. True or false? Memory is differentially affected by stress-induced cortisol elevations and sympathetic activity at consolidation and retrieval. *Psychoneuroendocrinology.* 2008;33:1378–1386. http://dx.doi.org/10.1016/j.psyneuen.2008.07.009.
36. Weymar M, Bradley MM, Hamm AO, Lamg PJ. When fear forms memories: threat of shock and brain potentials during encoding and recognition. *Cortex.* 2013;49:819–826. http://dx.doi.org/10.1016/j.cortex.2012.02.012.
37. Roozendaal B, McEwen BS, Chattarji S. Stress, memory and the-amygdala. *Nat Rev Neurosci.* 2009;10:423–433. http://dx.doi.org/10.1038/nrn2651.
38. Schonfeld P, Ackerman K, Schwabe L. Remembering under stress: different roles of autonomic arousal and glucocorticoids in memory retrieval. *Psychoneuroendocrinology.* 2014;39:249–256. http://dx.doi.org/10.1016/j.psyneuen.2013.09.020.
39. Alberini CM. Mechanisms of memory stabilization: are consolidation and reconsolidation similar or distinct processes? *Trends Neurosci.* 2005;28(1):51–56. http://dx.doi.org/10.1016/j.tins.2004.11.001.
40. Brunet A, Orr SP, Tremblay J, Robertson K, Nader K, Pitman RK. Effect of post-retrieval propranolol on psychophysiologic responding during subsequent script-driven traumatic imagery in post-traumatic stress disorder. *J Psychiatr Res.* 2008;42:503–506. http://dx.doi.org/10.1016/j.jpsychires.2007.05.006.
41. Zhao L-Y, Zhang X-L, Shi J, Epstein D, Lin L. Psychosocial stress after reactivation of drug-related memory impairs later recall in abstinent heroin addict. *Psychopharmacology (Berl).* 2009;203(3):599–608. http://dx.doi.org/10.1007/s00213-008-1406-2.
42. Sevenster D, Beckers T, Kindt M. Retrieval per se is not sufficient to trigger reconsolidation of human fear memory. *Neurobiol Learn Mem.* 2012;97:338–345. http://dx.doi.org/10.1016/j.nlm.2012.01.009.
43. Dolcos F, Iordan AD, Kragel J, et al. Neural correlates of opposing effects of emotional distraction on working memory and episodic memory: an event-related fMRI investigation. *Front Psychol.* 2013;4. http://dx.doi.org/10.3389/fpsyg.2013.00293.
44. Pitman RK, Delahanty DL. Conceptually driven pharmacologic approaches to acute trauma. *CNS Spectr.* 2005;10:99–106.
45. Watkins ER. Constructive and unconstructive repetitive thoughts. *Psychol Bull.* 2008;134:163–206. http://dx.doi.org/10.1037/0033-2909.134.2.16.

CHAPTER

18

Effects of Stress on Learning and Memory

M. Lindau[1], O. Almkvist[1,2], A.H. Mohammed[2,3]

[1]Stockholm University, Stockholm, Sweden
[2]Karolinska Institutet, Stockholm, Sweden
[3]Linnaeus University, Växjö, Sweden

OUTLINE

Abstract

Stress activates the hypothalamus-pituitary-adrenal axis, which causes the release of glucocorticoids, a class of adrenal steroid hormones. Stress also activates the sympathetic nervous system and thereby, the release of the transmitters adrenaline and noradrenaline. Stress has a memory-modulatory effect in humans as well as in animals.

In humans, the hippocampus, prefrontal cortex, and amygdala are rich in cortisol receptors. Acute and tolerable stress may increase memory performance, while excessive levels and chronic stress may have negative effects, thereby mimicking the pattern in animals. Stress in humans seems to have different effects on the various stages of memory (the memory process: encoding, consolidation, and retrieval) and can be enhanced by emotional arousal.

Animals learn to associate events in their environment. Studies of the effects of manipulation of corticosterone levels in animals have helped to disentangle the influences of stress on memory and learning, and indicated that low levels enhance spatial learning, whereas higher levels impair performance.

EFFECTS OF STRESS ON LEARNING AND MEMORY IN HUMANS

Introduction

Etymologically, the term stress originates in the Latin word *strictus*, signifying something that narrows or a tightening. It has the meaning of a "force, pressure (or) strain"[1] exerting psychological or physiological pressure upon the organism. In current scientific terminology, the word *stressor* is used to designate the stimuli that puts strain on the organism. Stressors may be either exteroceptive, that is represent external conditions that potentially evoke aversive reactions in the organism, such as extreme noise, timekeeping, or traumatic situations, or interoceptive, that is represent real or imaginary processes going on inside the body. A *stressed state* is experienced when

Stress: Concepts, Cognition, Emotion, and Behavior
http://dx.doi.org/10.1016/B978-0-12-800951-2.00018-2

the organism perceives that the demands of a situation exceed its resources to cope with it.[2] Whether a situation is perceived as stressful or not depends on how the individual subjectively interprets the potentially stressful stimuli. To evaluate the situation and take a stand on the ability to cope with it, the individual mostly intuitively makes a cognitive appraisal. "A cognitive appraisal is an evaluation of the potential significance of a situation, along with one's ability to control it."[3] Psychologically, stress is associated with perceptions of *"novelty"*, *"unpredictability"*, *"lack of control"*, and *"social-evaluative judgment"*,[4] as, for example, during public speeches, which are highly demanding situations. All these dimensions of perceived stress have been closely associated with activation of the hypothalamus-pituitary-adrenal (HPA) axis,[4] further described below. If the situation is perceived as positive, the stimuli may be conceived as a challenge, stimulate efforts, and possibly ameliorate performance. On the contrary, if stimuli are deemed as negative, the situation is understood as potentially harmful, efforts will be disturbed or decrease and, as a consequence, performance may be lowered. The same pattern applies for fight-or-flight reactions, also named acute stress responses, which involve the sympathetic limb of the autonomous nervous system. Flight or fight is determined by central brain structures starting with the sensory cortex, thalamus, hypothalamus, amygdala, locus coeruleus and other brain stem nuclei, the limbic, and neocortex. The sympathetic-adrenal-medullary (SAM) and HPA systems are activated by higher brain centers in response to stress. To be properly interpreted, these types of reactions must be understood as the result of an instinctive experience of a situation as being harmful, but all the same positive, in the sense that the individual finds herself or himself being able to cope with it, either by a fight or a flight. A negative evaluation risks crippling the efforts. Neuroendocrinologically, the HPA and SAM are the main brain outflow systems that in response to stress act synergistically to release glucose from the liver, and in the case of SAM, increase cardiac output and shunt blood from skin and gut to skeletal muscle, all of which facilitates fight or flight. The glucorticoids released by HPA action and the catecholamines released by the SAM do affect the brain, but the brain response to stress precedes the HPA and SAM responses. In addition, research has demonstrated that the immune and inflammatory defense systems are involved in the response to stress.

It is commonly recognized that stress may have both positive as well as detrimental effects on memory. It is also widely known that acute, as well as chronic stress, affect some aspects of memory, as demonstrated in psychologically traumatic events (e.g., being an eyewitness to an accident) or prolonged mental overload (e.g., time pressure and high task demands). In these situations, the recall is often inaccurate, particularly in details.

KEY POINTS

- Stress is associated with novelty, unpredictability, lack of control, and social-evaluative judgment (for review, see Ref. 1).
- Two brain regions important in regulating learning and memory are the hippocampus and amygdala, which contain receptors for stress hormones.
- Stress activates the hypothalamic-pituitary-adrenal axis (HPA axis) which causes a release of hormones called glucocorticoids: cortisol in humans, corticosterone in rodents and birds. This output of hormones is found to have memory-modulatory effects through its influence on the hippocampus and amygdala.
- Memory may be analyzed as a process, in encoding, consolidation and retrieval, and as systems, in episodic, semantic, working, procedural, and perceptual memory.
- Stress may be advantageous as well as detrimental for memory, and seems to have different effects on different processes and systems, depending on the stress load and the time when it occurs. Moderate levels of stress may enhance encoding and consolidation, but impair retrieval from long-term memory. High stress load usually has negative effects for memory.
- Stress has an impact on hippocampal function as it reduces levels of the neurotrophin Brain-Derived Neurotrophic Factor (BDNF), suppresses neurogenesis in the dentate gyrus of the hippocampus, and impairs cognitive function.
- The amygdala exerts neuromodulatory influences in memory processing across diverse brain regions, including the hippocampus, entorhinal cortex, striatum, medial frontal cortex, and anterior cingulate cortex.

The Conceptual and Anatomical Basis in Humans of Learning and Memory

The contemporary conceptual frameworks of memory and learning emanate from experimental cognitive psychology and research in cognitive neuroscience.[5,6] According to this model, human memory is not a unitary system, but composed of a series of interdependent systems characterized by different behavioral and brain features. The behavioral features of these systems vary in terms of time for registration of new information (fractions of a second to seconds and longer), type of information (sensory features vs. symbolic information), availability of consciousness (yes vs. no), activity at encoding (elaborating on target information or not), amount of information stored (limited or large), delay of retrieval (yes vs. no) and modes of retrieval (recognition vs. recall). The brain characteristics of these systems vary in terms of anatomical structures/networks involved.[7] The processes, mapping, encoding, storage, and retrieval are related and undergo temporary physiological changes and long-term molecular structural changes.

It is commonly considered that there are two important aspects of memory: systems and processes, the latter sometimes also called stages. At least five systems are recognized: episodic, semantic, working, procedural, and perceptual. Three processes are recognized: encoding, consolidation, and retrieval. First stimuli are encoded into internal representations, which may be followed by consolidation of the memory trace and finally, there is the possibility to retrieve the information in memory. Below, the brain networks and mechanisms subserving the memory systems and processes are briefly presented. Recent research has shown that both processes and systems are affected by stress.

One major distinction of memory is that between declarative and nondeclarative memory. Declarative memory (also denoted as explicit) includes episodic (events related to time, location, and person), semantic memory (general facts), and short-term/working memory (limited information on-line). Nondeclarative memory (often denoted as implicit knowledge) includes procedural memory (motor and cognitive skills), perceptual representation memory (sensory features), conditioning, and habituation.

The episodic memory is primarily related to medial temporal lobe structures such as the hippocampus and adjacent regions, in which the stimuli are thought to be linked together as a memory trace. This trace may be consolidated as mediated by long-term potentiation (LTP), which refers to the reinforcement of synaptic connections in the nervous system during the memory process. The molecular changes that generate LTP are thought to play a key role in the encoding and consolidation of memories, but not in retrieval according to recent research. LTP helps understand how synaptic strength can be enhanced through repeated activation of signal transmission.[8] The process of encoding and storage is influenced by conscious thought and effortful elaboration of the stimuli in focus. This conscious thinking is considered to be related primarily to dorsolateral prefrontal cortex of the brain. The medial temporal lobe and frontal areas are also involved in retrieval of the information.

The semantic memory system seems to rely on temporal lobe and/or other neocortical structures, but the extent to which semantic memory traces are dependent on hippocampal learning is still under consideration as exemplified by the complementary learning systems theory. There may be alternative ways of learning as suggested by individuals who have demonstrated semantic memory capacity, in the presence of hippocampal lesions. Studies are lacking regarding the effect of stress on this type of memory.

Short-term/working memory refers to information that is attended to, and possibly responded to, in real time in the conscious individual. The memory is continuously updated across time and the amount of temporarily stored information is highly limited (a matter of seconds). Without efforts to keep the information in mind, items will be lost. Certain prefrontal brain regions in collaboration with the parietal regions seem to be responsible for the temporary storage and processing of memory.[7]

Procedural memory is primarily related to basal ganglia structures (striatum) in addition to prefrontal regions of the brain and the cerebellum (i.e., regions involved in motor performance). The acquisition of new skills (encoding) rely on the appropriate sequence of actions which involve repeated trials until the action has become satisfactory. Both acquisition and retrieval appear to operate in an unconscious manner. This feature stands in contrast to what is typical for declarative memory. So far, studies on human striatal-dependent learning and memory are sparse. However studies indicate that the hippocampal- and striatal-dependent learning and memory systems are integrated.[9]

The perceptual memory system is another example of implicit memory and this system is primarily related to cortical regions in the brain subserving perceptual processes (i.e., primary projections areas in the brain).

Conditioning may be viewed as learning a connection between, for example, threatening stimuli and the emotion of fear and behavioral reactions, which seem to involve parts of amygdala and other parts of the brain.[10] This system may be seen as encoding and consolidation of recent experience and it is involved in many kinds of human psychopathology, for instance posttraumatic stress disorder (PTSD), anxiety disorder, and phobias.[11,12]

Mechanisms Behind the Relationship Between Stress, Learning, and Memory

Most of the literature in the field concerns the effects of stress on memory, which from a definitional point of view is something different from learning. Memory is a question of remembering previous information, whereas learning has to do with the elaboration, generalization, and application of the acquired information to pertinent new situations. Thus, the following discussion will focus on the relationship between stress and memory.

During the last decade, knowledge about the mechanisms behind stress and memory has been considerably enhanced. The relationships between these parameters are complex, and depend on the influence of the hormones released during stress, the amount and type of stress (acute, chronic, mild, moderate, or severe), the timing of stress during the memory process, as well as the type of learning stimuli (neutral or emotionally arousing).

Stress in humans influences memory formation (i.e., the process from encoding to storage) through the activation of the HPA axis, whose principal task is to mobilize energy.[4] Activation of the HPA axis causes the release of stress hormones and reaches its peak in the secretion of glucocorticoid (steroid) hormones (cortisol in humans, and corticosterone in rodents and birds) from the adrenal cortex. The existence of the HPA axis may be said to have evolutionary and adaptive functions since it enables the organism to mobilize glucose and thereby, energy. Recollection is better for emotionally charged experiences than for neutral. From an evolutionary point of view agreeable (e.g., sexual) as well as disagreeable (e.g., aversive) stimuli are said to be adaptive in that they are more important for reproduction than neutral stimuli.[13]

The hippocampus is a key structure for many memory systems. Together with other structures crucial for memory, (such as prefrontal cortex and amygdala), the hippocampus is known to be rich in receptors for glucocorticoids. The hippocampus also contains receptors for noradrenaline, a key stress hormone released by the sympathetic nervous system in the periphery and by the locus coeruleus and other brain stem nuclei.[14] During stress, the levels of glucocorticoids, as well as noradrenaline, are increased. Together with the paraventricular nucleus, basolateral amygdala, and locus coeruleus, glucocorticoids and noradrenaline act synergistically to facilitate hippocampal receptivity for memory and learning modulation.[15,16]

The hippocampus interacts with the amygdala, bilateral structures which are located in the anterior temporal lobe in close topographical relationship with the hippocampus. The amygdala plays an important role in memories linked to emotional arousal. It has foremost been associated with negative feelings, such as fear, and negative memories, but involvement in memories from positive emotional reinforcement is also evident.

Memory is most delicate and responsive to modulation immediately after encoding. During this phase it may be reinforced or weakened.[17] Stress may be advantageous as well as detrimental for memory. Usually, amelioration of memory performance under acute stress follows the pattern of an inverted-U dose-response curve, so that the amount of stress corresponding to the memory performance at the top of the inverted U-shape may reflect the optimal stress level.[15] Too low or too high levels of glucocorticoids may impair declarative memory. The phase under which memories are stabilized is the synaptic or cellular consolidation phase,[17,18] during which memories become more and more independent of hippocampus. During consolidation, the information will instead gradually be transformed to long lasting representations through the involvement of other brain structures committed to memory. With time, memory usually turns pale and loses its sharpness for contextual details. However, stress intensity during learning may modulate the quality of memories. Exposure to moderate stress during the encoding phase predicts a more detailed memory, rather than too high levels of stress which risk just shaping schematic or diaphanous representations. According to the Multiple Trace Transformation Model, memories are rich in contextual details during the hippocampus-dependent phase, but become less sharp when they enter the hippocampus-independent state. Thus, exposure to moderate stress directly after acquisition will prolong the hippocampus-dependent phase and with this, the acuity of mental representations.[17]

Vulnerability of Human Memory

The current concept of human memory posits that the order of development during ontogenesis is reversed when the brain is exposed to various adverse influences such as aging, somatic (e.g., metabolic syndrome) and psychiatric (e.g., depression) disease, toxins, inflammatory agents, and stress (e.g., acute stress, PTSD). The implication of this hypothesis is that more sophisticated memory systems, such as declarative memory, are hit more severely and earlier than other memory systems, when an individual is exposed to chronic stress and the brain is affected. In line with this hypothesis, the majority of studies on stress and memory have dealt with episodic memory and few studies have investigated the effect of stress on the perceptual memory and conditioning. However, there are some recent studies on procedural memory.

Encoding, Consolidation, and Retrieval in Stressful, and Emotionally Loaded Situations

The research about the effects of stress on memory is dominated by two types of experiments, one that studies the effects of cortisol, and another that focuses on the consequences for memory of the interaction between glucocorticoids and adrenergic activation, mostly under the influence of acute stress. Chronic stress is foremost associated with a PTSD.[19]

Acute stress affects memory formation in another way than chronic stress. Most indications are that the effects of acute stress are stage-dependent (acquisition, consolidation, and retrieval). Each part of the memory complex seems to have its specific neurobiological profile, meaning that stress, brought about by the release of cortisol or cortisone, may have different effects on the various phases of the memory process, and consequently, with these phases, interconnected memory systems. Chronic stress may have more widespread implications for cognition, by causing a cumulative and enduring overload of hippocampus, with implications for its functioning and morphology. When the stress is chronic, the effects for encoding and retrieval are reported to be negative.[20–23]

During the memory process, there appears to be some kind of reallocation of energy resources: Stress-induced cortisol secretion may ameliorate encoding, but hinder memory recall from long-term memory. An explanation for this is that cortisol, in order to make the brain ready for the encoding of new material, impedes the brain from the recall of old material.[24,25] This raises the intriguing question whether it is possible to utilize several memory systems at the same time, or if the brain must focus on one memory stage/system at the time. Cortisol has also been reported to enhance memory consolidation.[12,26] One hook in cortisol studies is that it may be difficult to isolate the consequences of cortisol for distinct phases/systems since cortisol is not stabilized until 90 min after inducement.[14]

High levels of stress may lower hippocampus-dependent memory performance, whereas emotionally arousing tasks may increase the remembrance for positively or negatively emotionally salient stimuli. Findings indicate that bilateral amygdala activation during memory acquisition is connected with an amelioration of episodic recognition memory for emotionally arousing visual stimuli, agreeable, as well as disagreeable, as compared to neutral material. According to the memory modulating hypothesis, memories of aversive episodes are better remembered because of the interaction between noradrenaline and cortisol in the basolateral nucleus of the amygdala during memory acquisition. Emotional arousal during encoding has also been found to reflect a reinforcement of the interconnection from the hippocampus to the amygdala, which supports the memory-modulation hypothesis emanating from animal studies, that amygdala controls the mediation of information in the hippocampus.[13,27] An efficient encoding is an absolute prerequisite for remembrance whatsoever, and a deficient encoding will prevent the transfer of the information to the short-term memory and other memory systems in the memory sequence.

Since different types of stress—with or without emotional arousal—are reported in the literature, the results are not always comparable, although it seems as if the results merge in a stage dependency for stress on memory, with or without emotional load.

Summary

Memory is created during a process of encoding, consolidation, and retrieval of information, and is stored in explicit or implicit memory systems. Stress influences memory through the activation of the HPA axis, which release cortisol. The effects of this hormone seem to be stage- (encoding, consolidation, and retrieval) and system-dependent, and it has been found that acute and tolerable stress may enhance encoding and consolidation, but impair retrieval from long-term memory. One mechanism behind this irregularity is that while cortisol operates in preparation for the encoding, it shuts down the recall. Emotionally arousing memories under stress are more readily remembered than neutral, through the interaction between noradrenaline and cortisol in the basolateral nucleus of the amygdala during memory acquisition. Chronic stress may impair memory as well as cognition.

EFFECTS OF STRESS ON LEARNING AND MEMORY IN ANIMALS

Introduction

Animals learn about the relationship or association between two events. In classical conditioning the animal learn to associate a stimulus such as a tone (conditioned stimulus) with food (unconditioned stimulus). In operant or instrumental conditioning, the animal learns to associate a response. For instance, pressing a lever (instrumental response) is associated with food reward (reinforcement). It has been suggested by some researchers[28] that associative learning mechanisms have been shaped by evolution to enable animals to detect and store information about real causal relationships in their environment.

Stress Hormones and Learning

The impact of stress hormones in modulating associative learning and cognitive function in animals has been a subject of continued investigations. The studies that have documented the relationship of levels of stress hormones with cognitive function, such as attention, perception, and memory have been done predominantly in rodents where investigators have examined the influence of stress hormones on acquisition, consolidation and retrieval of information. Glucocorticoids are the stress hormones secreted by the adrenal glands in animals. In primates and dogs, the naturally occurring glucocorticoid is cortisol, whereas in rodents and birds, it is corticosterone.

In response to emotional arousal, the stress hormones adrenalin and corticosterone are secreted in the adrenal glands with the concomitant release of several neurotransmitters and neuropeptides in the brain. Fluctuations in corticosterone levels can be said to reflect emotional states related to stress. In experimental animals, changes in hormonal levels can be achieved by pharmacological means or environmental manipulations and their effects assessed. Several studies show that stress and glucocorticoids influence cognitive function. The administration of low levels of corticosterone improves performance in learning tasks in animals.[29] However, whereas short-term exposure to low levels of corticosterone can enhance cognitive function, it is known that short-term exposure to higher levels, as well as long-term effects of high levels of corticosterone (e.g., through sustained exposure to restraint stress), has deleterious effects on learning, memory, and cognitive function.[30]

The effects of corticosterone on cognitive function are mediated through binding of the stress hormones to specific receptors in the brain. These receptors, known as glucocorticoid receptors (GRs), have been found to be abundant in the hippocampus, a brain region that is critically involved in modulating learning and memory. Another region in the brain that participates in cognitive function is the amygdala, which also has moderate amounts of receptors for corticosterone. The actions of corticosterone in the hippocampus and amygdala can be induced by the administration of selective drugs that interact with the GRs. Drugs include those that enhance the effects of corticosterone (glucocorticoid agonists) or those that block or attenuate the effect of corticosterone (glucocorticoid antagonists). These drugs have provided powerful tools in dissecting the role of stress hormones in the hippocampus and the amygdala in modulating cognitive function in rodents. Pharmacological or environmental manipulation of the GRs influences cognitive function in rats. For example, administration of the GR antagonist RU38486 impairs fear memory reconsolidation in rats[31]; and neonatal stimulation (handling) of rat pups causes increased density of GRs in the hippocampus during aging and enhances spatial learning.[30]

Hippocampus and Learning in Animals

Glucocorticoids exert numerous effects on the central nervous system that regulate the stress response, mood, learning and memory, and various neuroendocrine functions. The actions of corticosterone in the brain are mediated via two receptor systems: the mineralocorticoid receptor (MR) and the GR, both of which are highly localized in the hippocampus. The hippocampus is characterized by, among other things, the presence of GR and MR receptors, neurotrophins such as Brain-Derived Neurotrophic Factor (BDNF), and neurogenesis in the dentate gyrus region, all of which have been found to be affected by stress. For example, exposure to stress reduces levels of BDNF in the hippocampus (e.g., Ref. 32) and suppresses neurogenesis in the dentate gyrus.[33] Corticosterone effects on GRs and MRs, hippocampal BDNF and neurogenesis have an impact on learning, including LTP, which is thought to be or mimic the neurophysiological process associated with learning and memory. Thus, exposure to stress has been shown to impair LTP in the CA1 region of the hippocampus.

Considerable evidence shows that the hippocampus mediates spatial learning in rodents. Spatial learning in rodents can be examined by the use of a water maze known as the Morris maze. The procedure involves placing an experimental animal in a large pool of water of about 1.5 m in diameter. The animal has to use spatial cues in the experimental room to locate an invisible escape platform that is hidden 2 cm below the surface of the opaque water. After a few days of testing, the animals learn to locate the platform within seconds. It has been demonstrated that rats perform poorly in this test if hippocampal function is impaired. Thus, aged rats with pathological changes in the hippocampus show impaired spatial learning in the water maze compared to aged rats that do not show pathological hippocampal changes. Aged rats that have been exposed to repeated restrained stress have a reduction of GRs in the hippocampus. This results in, among other things, an impaired spatial learning ability. In contrast, animals that have higher levels of GRs in the hippocampus show an increased ability for spatial learning. Administration of a GR antagonist in the brain appears to impair information processing during spatial learning.

To sum up, spatial memory in animals is dependent on hippocampal function. Through their action in the hippocampus, stress hormones facilitate spatial information processing so that low levels of corticosterone improve learning, whereas higher levels impair performance.

Amygdala and Learning in Animals

The amygdala is a region of the brain that plays a crucial role in acquisition and expression of fear responses. Binding of GRs in the basolateral nucleus is known to affect memory storage. It was found that administration of glucocorticoid agonist in this area immediately after training enhanced retention of a passive avoidance response. Many studies have examined the effects glucocorticoids on the acquisition of new information. In a test of memory retrieval of long-term spatial learning, it was found that stress caused impaired performance, and this impairment could be related to circulating levels of corticosterone. These findings point to a critical involvement of the amygdala in regulating stress hormone effects on learning and memory.

The amygdala is thought to influence memory storage processes in other brain regions, such as the hippocampus and cortex. Accumulating evidence from the studies by McGaugh and Roozendaal[16] has established the amygdala to be critically involved in mediating stress-related modulation of hippocampal function. These investigators have provided evidence for the crucial involvement of amygdala—via its interaction with other brain regions—in regulating stress hormone effects on such memory processing phenomena as memory retrieval, memory extinction, and working memory. An interaction of stress hormones and amygdala nuclei appears to be important in modulating memory consolidation for emotional events. Roozendaal and McGaugh[16] have presented convincing evidence that activation of noradrenaline in the amygdala is crucial in enhancing memory consolidation. Emotional arousal experience induces release of noradrenaline into the basolateral amygdala. And this noradrenergic activity in the amygdala influences other hormones and neurotransmitters in the hippocampus on memory consolidation.

For example, pharmacological blockade of noradrenaline receptors in the amygdala can impair LTP (the neurophysiological substrate of memory) in the dentate gyrus of the hippocampus.

What has now emerged from intensive research is that the amygdala exerts neuromodulatory influences in memory processing across diverse brain regions, including the hippocampus, entorhinal cortex, striatum, medial frontal cortex, and anterior cingulate cortex. To conclude, the amygdala is a region of the brain known to be importantly involved in emotions. Stress hormones can affect the memory storage of emotional events by their action on the amygdala, which can interact with other brain regions to promote memory processes.

Summary

Two brain regions important in regulating learning and memory are the hippocampus and amygdala, which contain receptors for stress hormones. The effects of stress hormones on learning are mediated by these receptors in the hippocampus and amygdala. However, research in rodents points to the involvement of other brain regions in stress-related learning. Emotional arousal activates the amygdala which exerts neuromodulatory influences on memory processing across diverse brain regions, including the hippocampus, entorhinal cortex, striatum, and medial frontal cortex.

Glossary

Amygdala A group of nuclei involved in the medial anterior part of the temporal lobe concerned with fear and the regulation of emotion and certain types of learning.

Corticosterone A steroid hormone in many species, including rodents, produced by the adrenal cortex and involved in response to stress and immune reaction.

Cortisol A steroid hormone in humans produced by the adrenal cortex, involved in response to stress and that has direct influence on the nervous system.

Emotion Mental state of arousal.

Hippocampus A region of the cerebral cortex in the basal medial part of the temporal lobe thought to be important for learning and memory.

Learning A relatively permanent change in cognition as a consequence of the acquisition of new information and experience, directly influencing behavior.

Memory Encoding of information, storage, and recall.

Stress hormones Hormones of the adrenal glands that serve to maintain bodily homeostasis in response to challenging external and emotional environment. These hormones are the glucocorticoids and adrenaline.

Stressor Stimulus that puts an extra demand on the organism.

References

1. Goodnite PM. Stress: a concept analysis. *Nurs Forum*. 2014;49:71–74.
2. Sandi C. Memory impairments associated with stress and aging. In: Bermúdez-Rattoni F, ed. *Neural Plasticity and Memory: From Genes to Brain Imaging*. Boca Raton (FL): CRC Press; 2007. Frontiers in Neuroscience [chapter 12].
3. White JB. Fail or flourish? Cognitive appraisal moderates the effect of solo status on performance. *Pers Soc Psychol Bull*. 2008;34:1171–1184.
4. Wirth MM. Hormones, stress, and cognition: the effects of glucocorticoids and oxytocin on memory. *Adapt Human Behav Physiol*. 2015;1:177–201.
5. Giovanello KS, Verfaellie M. Memory systems of the brain: a cognitive neuropsychological analysis. *Semin Speech Lang*. 2001;22: 107–116.
6. Nadel L, Hardt O. Update on memory systems and processes. *Neuropsychopharmacology*. 2011;36:251–273.
7. Cabeza R, Nyberg L. Imaging cognition II: an empirical review of 275 PET and fMRI studies. *J Cogn Neurosci*. 2000;12:1–47.
8. Kumar A. Long-term potentiation at CA3-CA1 hippocampal synapses with special emphasis on aging, disease, and stress. *Front Aging Neurosci*. 2011;3:7.
9. Schwabe L. Stress and the engagement of multiple memory systems: integration of animal and human studies. *Hippocampus*. 2013;23:1035–1043.
10. Van Stegeren AH. Imaging stress effects on memory: a review of neuroimaging studies. *Can J Psychiatry*. 2009;54:16–27.
11. Finsterwald C, Alberini CM. Stress and glucocorticoid recepetor-dependent mechanisms in long-term memory: from adaptive responses to psychopathologies. *Neurobiol Learn Mem*. 2014;112:17–29.
12. Wingenfeld K, Wolf OT. Stress, memory, and the hippocampus. *Front Neurol Neurosci*. 2014;34:109–120.
13. Hamann SB, Ely TD, Grafton ST, Kilts CD. Amygdala activity related to enhanced memory for pleasant and aversive stimuli. *Nat Neurosci*. 1999;2:289–293.
14. Payne JD, Jackson ED, Hoscheidt S, Ryan L, Jacobs WJ, Nadel L. Stress administered prior to encoding impairs neutral but enhances emotional long-term episodic memories. *Learn Mem*. 2007;14: 861–868.
15. Osborne DM, Pearson-Leary J, McNay EC. The neuroenergetics of stress hormones in the hippocampus and implications for memory. *Front Neurosci*. 2015;9:1–16.
16. Roozendaal B, McGaugh JL. Memory modulation. *Behav Neurosci*. 2011;125:797–824.
17. Pedraza LK, Sierra RO, Boos FZ, Haubrich J, Quillfeldt JA, de Oliveira Alvares L. The dynamic nature of systems consolidation: stress during learning as a switch guiding the rate of the hippocampal dependency and memory quality. *Hippocampus*. 2015; http://dx.doi.org/10.1002/hipo.22527 [Epub ahead of print].
18. McGaugh JL. Memory—a century of consolidation. *Science*. 2000;287:248–251.

19. Ahmadian A, Mirzaee J, Omidbeygi M, Holsboer-Trachsler E, Brand S. Differences in maladaptive schemas between patients suffering from chronic and acute posttraumatic stress disorder and healthy controls. *Neuropsychiatr Dis Treat*. 2015;11: 1677–1684.
20. de Quervain DJ, Roozendaal B, Nitsch RM, McGaugh JL, Hock C. Acute cortisone administration impairs retrieval of long-term declarative memory in humans. *Nat Neurosci*. 2000;3:313–314.
21. de Quervain DJ, Henke K, Aerni A, et al. Glucocorticoid-induced impairment of declarative memory retrieval is associated with reduced blood flow in the medial temporal lobe. *Eur J Neurosci*. 2003;17:1296–1302.
22. de Quervain DJ, Aerni A, Schelling G, Roozendaal B. Glucocorticoids and the regulation of memory in health and disease. *Front Neuroendocrinol*. 2009;30:358–370.
23. de Quervain DJ, McGaugh JL. Stress and the regulation of memory: from basic mechanisms to clinical implications. Neurobiology of learning and memory special issue. *Neurobiol Learn Mem*. 2014;112:1.
24. Stauble MR, Thompson LA, Morgan G. Increases in cortisol are positively associated with gains in encoding and maintenance working memory performance in young men. *Stress*. 2013;16:402–410.
25. Roozendaal B. Stress and memory: opposing effects of glucocorticoids on memory consolidation and memory retrieval. *Neurobiol Learn Mem*. 2002;78:578–595.
26. Tops M, van der Pompe G, Baas D, et al. Acute cortisol effects on immediate free recall and recognition of nouns depend on stimulus valence. *Psychophysiology*. 2003;40:167–173.
27. Fastenrath M, Coynel D, Spalek K, et al. Dynamic modulation of amygdala-hippocampal connectivity by emotional arousal. *J Neurosci*. 2014;34:13935–13947.
28. Dickinson A. *Contemporary Animal Learning Theory*. Cambridge: Cambridge University Press; 1980.
29. Sandi C, Rose SP. Corticosterone enhances long-term retention in one-day-old chicks trained in a weak passive avoidance learning paradigm. *Brain Res*. 1994;647:106–112.
30. Meaney MJ, Aitken DH, van Berkel C, Bhatnagar S, Sapolsky RM. Effect of neonatal handling on age-related impairments associated with the hippocampus. *Science*. 1988;239:766–768.
31. Nikzad S, Vafaei AA, Rashidy-Pour A, Haghighi S. Systemic and intrahippocampal administrations of the glucocorticoid receptor antagonist RU38486 impairs fear memory reconsolidation in rats. *Stress*. 2011;14:459–464.
32. Lakshminarasimhan H, Chattarji S. Stress leads to contrasting effects on the levels of brain derived neurotrophic factor in the hippocampus and amygdala. *PLoS One*. 2012;7:30481.
33. Schoenfeld TJ, Gould E. Differential effects of stress and glucocorticoids on adult neurogenesis. *Curr Top Behav Neurosci*. 2013;15:139–164.

CHAPTER

19

Trauma and Memory

B. Iffland, F. Neuner

Bielefeld University, Bielefeld, Germany

OUTLINE

Abstract

During a traumatic event, mainly emotional, sensory, and perceptual information is stored in an interconnected neural network. This network can be activated by environmental stimuli and internal cues later at any given time. An activation of the entire network is thought to be a flashback, which is one of the cardinal symptoms of post-traumatic stress disorder (PTSD). Additionally, the PTSD symptom of avoidance can be seen as a consequence of the activation of the trauma network. In contrast to the extensive memory of sensory-perceptual information, memories of traumatic events are characterized by a disconnectedness from temporal and spatial information about the general event and cannot be clearly positioned in a lifetime period. Furthermore, traumatic stress and the distribution of stress hormones during a traumatic event cause strong structural and functional alterations of brain structures involved in memory processing, like the hippocampus and the amygdala.

CHARACTERISTICS OF TRAUMATIC MEMORY

Personally important and emotional life events leave strong and durable memory representations that allow a person to remember these events in detail even years later. Memories of traumatic events, however, are qualitatively different from memories of every day events. The specific characteristics of traumatic memories contribute to the onset and maintenance of trauma-related psychopathology, most of all post-traumatic stress disorder (PTSD). The most distinct symptom of PTSD is the re-experiencing of traumatic events in the form of intrusive recollections such as flashbacks and nightmares. These involuntary intrusions can be triggered by cues that remind a person of the traumatic situation and have been described as being brief and perceptually detailed,[1,2] with visual details predominating.[1,3] During a flashback, victims feel that the traumatic event is happening again which implies that they are not fully aware that their experiences are memories from the past, but think they are back in the situation as memories of traumatic events do not seem to be fixed in the context of time and space in which they actually occurred.[4] It has been shown that this sense of nowness is more strongly present in trauma memories of individuals with PTSD than in memories that are not related to traumatic events.[5]

Foa and Rothbaum[6] proposed that the emotional intensity of a traumatic event interferes with encoding processes of attention and memory, leading to a disjointed and fragmented narration that is relatively brief, simplistic, and poorly articulated. Similarly, Ehlers and Clark[4] suggested that difficulties in conceptually processing the events lead to poorly elaborated and inadequate integration of traumatic events into autobiographical knowledge. Accordingly, greater disorganization or fragmentation in trauma memories is associated with the onset and maintenance of symptoms of acute stress disorder and PTSD.[7–9]

Thus, memories of a traumatic event are characterized by very vivid recollections of the event including many sensory details and, at the same time, difficulties in facing the memories and in learning to put the details into

http://dx.doi.org/10.1016/B978-0-12-800951-2.00019-4

coherent speech and chronological order. Because traumatic events are stored differently than memories of everyday events, this pathological representation of traumatic memories is suggested to be responsible for the core symptoms of PTSD.[4,10,11]

In order to explain this dichotomy of traumatic memories, it is necessary to refer to knowledge about the principles of memories of life events. According to Tulving,[12] there is a specific store of memories about past events called "episodic memory." Episodic memory involves events in particular places at particular times and covers context information about "what," "where," and "when." A unique feature that distinguishes episodic memory from other memory systems is the possibility of consciously re-experiencing previous events in the form of recollective experiences. Furthermore, most theories of memory separate at least two different bases of episodic knowledge: nondeclarative memory and declarative memory. Nondeclarative memory covers skills, habits, emotional associations, and conditioned responses. In contrast, declarative memory consists of memory of personal events, as well as memory of facts and knowledge of the world.[13] These memory systems differ in the form of retrieval of information: declarative (explicit) memory can be retrieved deliberately and accessed verbally, whereas nondeclarative (implicit) memory is activated automatically by environmental or internal cues and affects a person's behavior and experience, similar to the memories of trauma survivors.[13] Therefore one must distinguish between a nondeclarative episodic memory system ("sensory-perceptual representation") and a declarative episodic memory system ("autobiographical representation"). A detailed theoretical account of different memory systems involved in the retrieval of information is provided by the dual representation theory of PTSD.[10,14]

SENSORY-PERCEPTUAL REPRESENTATION

Thinking about past life events, a person may not only retrieve abstract knowledge about what has happened, but also sometimes can directly access visual and other sensory information about the past event. The retrieval of this sensory information is fundamentally different from the retrieval of autobiographic information. Whereas the contextual facts stored in autobiographic memory are retrieved as verbally accessible knowledge, the retrieval of sensory information is perceived as an experience of the information itself. For example, a person who has a vivid memory of his first day at a beach actually might see the water in front of his eyes and re-experience the excitement of that moment when thinking back on the event. This type of information is provided by a sensory-perceptual representation of the event (other authors have used such terms as "hot memory,"[11] "situationally accessible memory," (SAM[10]), or "event-specific knowledge."[15] According to a later version of the dual representation theory of PTSD,[14] these perceptual or image-based memories consist of sensation-near representations (S-reps) that are a product of processing in the dorsal visual stream, insula, and amygdala and that are specialized for action on the environment.

The sensory-perceptual representation itself does not contain any context information about single events, but a close tie to the corresponding autobiographical representations offers a spatial and temporal context for this memory. As a consequence, the activation of the sensory-perceptual details of an event usually is accompanied by the activation of autobiographical knowledge about the sequence of the event and the location of the event in lifetime periods. However, a vivid and detailed recollection is not possible for all events. It was suggested that the storage depends on the significance of an event to a person.[16] An enduring storage of sensory-perceptual representations only happens for events stored in a highly emotional state, as this means that they are significant for the achievement or failure of individual goals.[15]

The interconnections and the interaction between sensory-perceptual representations, their embedding emotions, and external or internal stimuli triggering memories of traumatic events are described in Lang's bioinformational theory of emotion.[17] Here, emotions are considered "response propositions," (i.e., modes in neural networks—neuronal connections within the brain) that initiate distinct sets of actions. These representations consist of sensory-perceptual information about the stimuli present in the past situation in different modalities (visual, auditory, olfactory, etc.), as well as cognitive or emotional information in response to the stimuli. Additionally, the body is also responding with a set of corresponding physiological responses. Furthermore, Lang[17] suggested that motor and physiological responses to the stimulus or memory represent an additional set of "response propositions." It is assumed that all elements of this network of brain activity are connected such that the activation of a single item (such as a sensory experience) leads to an activation of other elements (see Figure 1). These sensory-perceptual representations of traumatic events have also been called *fear structures* or *fearnetworks.*[17] This network model predicts that recollections of an event would occur simultaneously with physiological and emotional responses similar to those that occurred when the event happened. For instance, if someone experienced great levels of fear and terror accompanied by heart palpitations while watching a horrible event, this fear, terror, and the heart palpitations would be relived upon remembering the event. The impact of

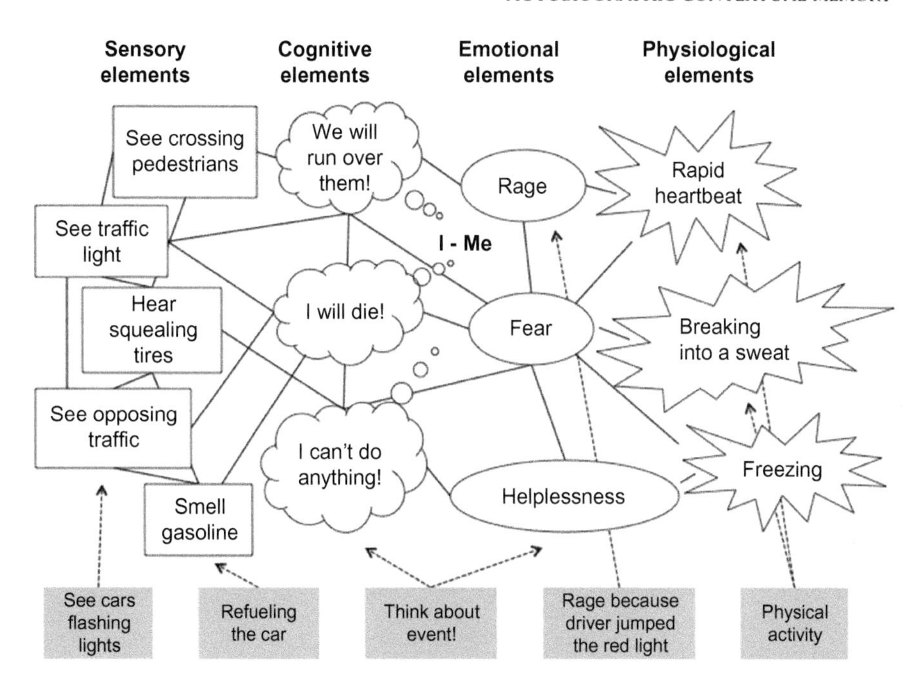

FIGURE 1 Schematic presentation of a hypothetical sensory-perceptual representation—a so-called fear network, including sensory, cognitive, emotional, and physiological elements. The network represents a casualty's memory of a car accident. The boxes indicate environmental stimuli with the potential to activate the representation. Modified from Ref. 18.

emotional structures on behavior, however, is not limited to the deliberate retrieval of memories of past events, but elements of the structure can also be automatically retrieved by a cue, and lead to emotional behavior.

Emphasizing the differences between representations of normal and traumatic events, Foa and Kozak[19] have related the network theory to PTSD. They suggested that fear structures encoded during a traumatic event are unusually large and cover a wide variety of single elements being coded, such as sight, sound, emotions, physical sensations. This implies that nodes in the trauma structures can be easily activated. Hence, many stimuli in the environment can act as cues as they may have similarities to one or the other element of the structure. Furthermore, Foa and Kozak[19] argued that interconnections between single elements of the network are unusually powerful. As a consequence, the activation of only a single element may be sufficient to activate the whole trauma structure, causing all of the related memories to return. Accordingly, an environmental stimulus or internal cue, such as thinking about the event or an intense heartbeat, can cause the full firing of the sensory information. Moreover, the associated emotional, physiological, and motor responses stored in the trauma structure will also be triggered. With respect to this theory, a flashback is the activation of the entire trauma network.

AUTOBIOGRAPHIC CONTEXTUAL MEMORY

The declarative part of the episodic memory has been called "autobiographical memory,"[15] other authors have used the terms "verbally accessible memory (VAM),"[10] or "cold memory."[11] Alternatively, the updated dual representation theory of PTSD proposes parallel contextualized representations (C-reps) that are a product of processing in the ventral visual stream and medial temporal lobe.[14]

In contrast to the sensory-perceptual representations, autobiographical memory is a highly developed and structured memory system that allows filing the abundant knowledge about past events in a highly efficient way. It is the principal resource for the retrieval of information about one's life and is the main base for the narration of events and life periods. To allow a rapid and organized access of information, this autobiographic memory is structured in a hierarchical way. At the top of the hierarchy, memories of *lifetime periods* represent general knowledge of persons, places, actions, plans, or goals that characterize a special period covering a distinct time-period with an identifiable beginning and ending. These memories are typically organized along major themes of persons, including relationships, occupations, and places one lived.

One step beyond lifetime periods is the memory of *general events*. General events can contain both, repeated events like "Having lunch at the cafeteria" and single events like "My first day at school." These knowledge bases organize the chronologic sequence of single events. However, not all events a person experiences are represented with the same accuracy. In particular, events that demarcate beginnings and ends of lifetime periods are suggested to be represented in more detail and accessed more easily.[20] On the contrary, memories of activities that cover a series of events can be connected to form so-called mini-histories.[21]

The lowest category of the hierarchy of autobiographic memory has been referred to as *event-specific knowledge*[15] and corresponds to the sensory-perceptual representations described above. Within the hierarchy of memory, the different categories have strong interconnections and are able to activate each other. For instance, the activation of the sensory-perceptual details of an event is usually accompanied by the activation of knowledge about the sequence of the event and the location of the event in lifetime periods.

On the basis of this conceptualization of memory, it may be assumed that patients who suffer from PTSD have a significant distortion of their autobiographic memory. In particular, there is evidence to suggest that in persons with PTSD, the traumatic event is not clearly represented as a specific event. Furthermore, the traumatic event does not seem to be clearly positioned in a lifetime period. Even though one's memories and the sensory-perceptual representations of traumatic events are very strong and long-lasting, the corresponding autobiographical structure is unreliable and distorted.

According to the dual representation theory of PTSD,[14,22] the encoding of S-reps (perceptual memories) is strengthened during a traumatic event, while the encoding of C-reps (contextualized episodic memories), and the connections between S-reps and C-reps are weakened. As a consequence, traumatized subjects are able to voluntarily retrieve C-reps of the event and communicate them, although they are fragmented and disorganized.[22] Additionally, reminders of the event initiate the automatic retrieval of S-reps with vivid, decontextualized images. In line with this, it was shown that individuals with weaker contextual memory abilities were more vulnerable to developing involuntary memories[23] and a lack of temporal context was associated with naturally occurring intrusive memories.[24] In terms of PTSD, flashbacks are more likely when encoding of traumatic events into the SAM system is unimpeded or enhanced and when encoding or re-encoding into the VAM system is degraded or reduced.[10] Accordingly, it has been suggested that memories of traumatic events are less well integrated with autobiographical material in PTSD patients than memories of distressing but nontraumatic events.[25]

SUMMARY

During a traumatic event, subjects are in a highly emotional state which has strong implications for the storage of this event in memory. Mainly emotional, sensory, and perceptual information about the event is stored in an interconnected neural network from which a trauma network is established. The trauma network includes sensory, cognitive, physiological, and emotional experiences including the action disposition (i.e., fighting or running away) related to the traumatic experience (=sensory-perceptual representations). Environmental stimuli (e.g., a smell or noise) and internal cues (e.g., a thought) can activate this trauma structure later at any given time. At this juncture, the activation of only a few elements in the network is sufficient to activate the whole structure. An activation of the entire network is thought to be a flashback, which is one of the cardinal symptoms of PTSD. Additionally, the PTSD symptom of avoidance of reminders of the traumatic event can be seen as a consequence of the activation of the trauma network because it implies the frightening and painful recollection and reliving of the event. In contrast to the extensive memory of sensory-perceptual information, storage of traumatic events in the autobiographic memory is dramatically impeded in patients who suffer from PTSD. Memories of traumatic events are characterized by a disconnectedness of sensory-perceptual representations from temporal and spatial information about the general event and cannot be clearly positioned in a lifetime period.

NEUROBIOLOGICAL BASIS OF MEMORY AND TRAUMA

Neuroimaging studies have demonstrated significant neurobiological changes in subjects who experienced traumatic events. Several areas of the brain in particular are different in patients with PTSD compared with those in control subjects: the hippocampus, the amygdala, and the medial prefrontal cortex, which includes the anterior cingulate cortex.[26] The medial temporal lobe and the connected hippocampus are the key brain structures in the consolidation of memories that contain autobiographical information, including the temporal and spatial context of an event. The hippocampus is associated with the construction of a meaningful episodic representation of specific events in a spatio-temporal context,[27] also during fear conditioning. Furthermore, the hippocampus is involved in the coding of information that contradicts previously learned knowledge.[28] In terms of trauma, this is especially important as much of what individuals experience during times of trauma contradicts basic assumptions that have been learned in life about security, trust, and human nature.

In the storage of sensory-perceptual representations, a complex involvement of areas that are responsible for visuospatial processing and emotion (the limbic structures, especially the posterior cingulated cortex, including the surrounding cortex and the occipital and parietal cortex) and for preparation for action (motor cortex) has been suggested.[29] In particular, the amygdala and cingulate cortex are relevant for these memories. The amygdala is involved in the processing of highly arousing events,[30,31] the assessment of threat-related stimuli and the acquisition and expression of conditioned fear.[32,33] In receiving information from all sensory modalities and projecting to subcortical structures that are involved in mediating specific symptoms of fear and anxiety (e.g., facial expression of fear, stress hormone release, galvanic skin response, blood pressure elevation, hypoalgesia, and freezing), the amygdala is the cardinal brain structure in mediating stress-related effects on behavior and modulating hippocampal function. The medial prefrontal cortex inhibits activation of the amygdala and is involved in the extinction of conditioned fear. It includes the anterior cingulate cortex, which is implicated in evaluating the emotional significance of stimuli and in attentional function.[34]

In the context of traumatic stress, the amygdala appears to be hyperreactive to trauma-related stimuli. This increased functioning has been suggested to represent the formation of overly strong implicit memories that are related to autonomic conditioning and fear, including priming.[35,36] Similarly, amygdala activation is associated with subjective experience of a vivid memory and recognition of more event details, but not recognition of all event details.[37,38] Furthermore, the amygdala is a key structure for the recall of flashbulb memories.[39] In contrast with the amygdala, higher brain regions, such as the hippocampus and the medial frontal cortex, are affected as they are unable to handle the overload of stimuli that results from symptoms such as an exaggerated startle response and flashbacks. In these regions, exposure to *glucocorticoids* (stress hormones that are released during traumatic events) impairs memory for facts and events in the way that the encoding of events into memory is facilitated but the retrieval is impaired.[40,41] Based on research with rodents (for a review, see Ref. 42), there is evidence that stress hormones affect both the function and the structure of the hippocampus. Normally, exposure to glucocorticoids increases the activity of the hippocampus. Under very high levels of stress, however, the functioning of the hippocampus becomes severely impaired, and the hippocampus, along with the medial frontal cortex, is unable to mediate the exaggerated symptoms of arousal and distress that occur in the amygdala in response to reminders of the traumatic event.[43] In animals, chronic stress induces dendritic atrophy (decrease in size) in the hippocampus whereas the dendritic arborization of neurons in the amygdala was enhanced.[44] In humans, several studies examined structural abnormalities of the hippocampus and other brain regions in subjects with various trauma. Several studies using volumetric magnet resonance imaging reported that the hippocampus is smaller in individuals with PTSD than in comparison subjects.[45] The effect of traumatic events on the hippocampal volume could be confirmed in meta-analyses.[45–48] However, these analyses emphasized that volumetric reduction of the hippocampus was not specific for individuals suffering from PTSD, but was also present in trauma survivors without PTSD. Therefore, it was speculated whether volumetric reduction is caused by a genetic determination which may serve as pre-existing risk factor for the development of PTSD, whether it is due to experiencing a traumatic event, or whether it is a marker of PTSD. However, studies that compared Vietnam veterans with PTSD to their healthy twin brothers showed that reduction of size in the hippocampus, and as well the anterior cingulate cortex, was particularly present in individuals with PTSD but not their healthy twins.[49] Accordingly, it was found that smaller hippocampal volume is not a pre-existing risk factor for the development of PTSD, but that it later occurs in individuals with chronic or complicated PTSD.[50] Recent studies reported bilateral reduction in hippocampal volume in traumatized individuals with and without PTSD. However, these groups showed differences in the size of the right hippocampus.[51] Additionally, male and female subjects differed in reduction of hippocampal volume.[52] Furthermore, it became increasingly evident that volumetric hippocampal reduction depends on the developmental period when the traumatic stress was first experienced. It was speculated that when studies were controlled for drug intake, traumatic stress in adults would not result in a reduction of hippocampal volume.[53] With respect to functional changes, findings are mixed, ranging from no or lower activation of the hippocampus during cognitive tasks, to increased hippocampal activation when at rest or across tasks.[54]

Compared to findings about the hippocampus, data reporting volumetric changes in the amygdala are twofold. Karl et al.[46] reported that PTSD diagnosis influences the size of the left amygdala. In contrast, another meta-analysis found no differences in the volume of the amygdala in individuals with PTSD and traumatized and nontraumatized controls.[55] However, a recent meta-analysis showed size reduction in bilateral amygdala volume when PTSD subjects were compared to healthy controls.[48] No significant differences, though, were found in amygdala volume between PTSD subjects and trauma-exposed controls. Both, in PTSD subjects and traumatized controls alterations were accompanied by symptoms of hypervigilance and increased propensity in acquisition of conditioned fear memories.

In addition, structural alterations of the cortex in subjects with PTSD (and to a lesser extent, trauma-exposed controls) have been identified.[53] Reduced gray matter volumes were found in the right inferior parietal cortex, the left rostral middle frontal cortex, the bilateral lateral orbitofrontal cortex, and the bilateral isthmus of the cingulate. As these regions are particularly involved in episodic memory, emotional processing, and executive control, these changes might be part of the physiological substrate of PTSD symptoms.

Additional to these structural changes, functional alterations in individuals who experienced traumatic events were observed in left temporal brain regions, with peak activities in the region of the insula, as well as in right frontal areas.[56] The insula, as a site of multimodal convergence, was suggested to account for a reduced ability to identify, express, and regulate emotional responses to reminders of traumatic events. Furthermore, differences in activity in right frontal areas may indicate a dysfunctional prefrontal cortex, which may lead to diminished extinction of conditioned fear and reduced inhibition of the amygdala.

SUMMARY

Traumatic stress and the distribution of stress hormones cause strong structural and functional alterations of brain structures important for memory coding, like the hippocampus and the amygdala. When exceeding a certain level, traumatic stress impairs the function of the hippocampus and related neural networks. Hence, during times of severe stress, the key brain structure in the consolidation of the autobiographical memory, including integration of the temporal and spatial context of an event, is affected. On the contrary, activation of the amygdala, which is the structure that prepares the body for danger, and the interconnected frontal lobe regions is increased when exposed to traumatic stress or reminders of the traumatic event. This may explain the exaggerated emotional responses to fear situations in PTSD patients. In addition to functional alterations, it has been reported that experiences of traumatic stress induce alterations of gray matter volumes in brain structures involved in memory processing.

References

1. Ehlers A, Steil R. Maintenance of intrusive memories in posttraumatic stress disorder: a cognitive approach. *Behav Cogn Psychother*. 1995;23(3):217–249.
2. van der Kolk BA, Fisler R. Dissociation and the fragmentary nature of traumatic memories: overview and exploratory study. *J Trauma Stress*. 1995;8:505–525.
3. Hackmann A, Ehlers A, Speckens A, Clark DM. Characteristics and content of intrusive memories in PTSD and their changes with treatment. *J Trauma Stress*. 2004;17(3):231–240.
4. Ehlers A, Clark DM. A cognitive model of posttraumatic stress disorder. *Behav Res Ther*. 2000;38:319–345.
5. Rubin DC, Feldman ME, Beckham JC. Reliving, emotions, and fragmentation in the autobiographical memories of veterans diagnosed with PTSD. *Appl Cogn Psychol*. 2004;18(1):17–35.
6. Foa EB, Rothbaum BO. *Treating the Trauma of Rape: Cognitive-Behavioral Therapy for PTSD*. New York, NY: Guilford Press; 1998.
7. Halligan SL, Michael T, Clark DM, Ehlers A. Posttraumatic stress disorder following assault: the role of cognitive processing, trauma memory, and appraisals. *J Consult Clin Psychol*. 2003;71(3):419–431.
8. Harvey AG, Bryant RA. A qualitative investigation of the organization of traumatic memories. *Br J Clin Psychol*. 1999;38:401–405.
9. Salmond CH, Meiser-Stedman R, Glucksman E, Thompson P, Dalgleish T, Smith P. The nature of trauma memories in acute stress disorder in children and adolescents. *J Child Psychol Psychiatry*. 2011;52(5):560–570.
10. Brewin CR, Dalgleish T, Joseph S. A dual representation theory of posttraumatic stress disorder. *Psychol Rev*. 1996;103(4):670–686.
11. Metcalfe J, Jacobs WJ. A hot-system/cool-system view of memory under stress. *PTSD Res Q*. 1996;7:1–3.
12. Tulving E. Episodic memory and common sense: how far apart? *Philos Trans R Soc Lond B Biol Sci*. 2001;356:1505–1515.
13. Squire LR. Declarative and nondeclarative memory: multiple brain systems supporting learning and memory. In: Schacter DL, Tulving E, eds. *Memory Systems 1994*. Cambridge, MA: MIT Press; 1994:207–228.
14. Brewin CR, Gregory JD, Lipton M, Burgess N. Intrusive images in psychological disorders: characteristics, neural mechanisms, and treatment implications. *Psychol Rev*. 2010;117(1):210–232.
15. Conway MA, Pleydell-Pearce CW. The construction of autobiographical memories in the self-memory system. *Psychol Rev*. 2000;107(2):261–288.
16. Conway MA. Sensory-perceptual episodic memory and its context: autobiographical memory. *Philos Trans R Soc Lond B Biol Sci*. 2001;356:1375–1384.
17. Lang PJ. A bio-informational theory of emotional imagery. *Psychophysiology*. 1979;16(6):495–512.
18. Schauer M, Neuner F, Elbert T. *Narrative Exposure Therapy: A Short-Term Treatment for Traumatic Stress Disorders*. 2nd ed. Cambridge, MA: Hogrefe Publishing; 2011.
19. Foa EB, Kozak MJ. Emotional processing of fear: exposure to corrective information. *Psychol Bull*. 1986;99(1):20–35.
20. Shum MS. The role of temporal landmarks in autobiographical memory processes. *Psychol Bull*. 1998;124(3):423–442.
21. Robinson JA. First experience memories: contexts and function in personal histories. In: Conway MA, Rubin DC, Spinnler H, Wagenaar WA, eds. *Theoretical Perspectives on Autobiographical Memories*. Dodrecht: Kluwer Academics; 1992:223–239.
22. Brewin CR. Episodic memory, perceptual memory, and their interaction: foundations for a theory of posttraumatic stress disorder. *Psychol Bull*. 2014;140(1):69–97.
23. Bisby JA, King JA, Brewin CR, Burgess N, Curran HV. Acute effects of alcohol on intrusive memory development and viewpoint dependence in spatial memory support a dual representation model. *Biol Psychiatry*. 2010;68(3):280–286.
24. Glazer D, Mason O, King J, Brewin C. Contextual memory, psychosis-proneness, and the experience of intrusive imagery. *Cogn Emot*. 2013;27:150–157.
25. Kleim B, Wallott F, Ehlers A. Are trauma memories disjointed from other autobiographical memories in posttraumatic stress disorder? An experimental investigation. *Behav Cogn Psychother*. 2008;36(2):221–234.
26. Kolassa IT, Elbert T. Structural and functional neuroplasticity in relation to traumatic stress. *Curr Dir Psychol Sci*. 2007;16(6):321–325.
27. Shastri L. Episodic memory and cortico-hippocampal interactions. *Trends Cogn Sci*. 2002;6:162–168.
28. McClelland JL, McNaughton BL, McNaughton BL, O'Reilly RC. Why there are complementary learning systems in the hippocampus and neocortex: insights from the successes and failures of connectionist models of learning and memory. *Psychol Rev*. 1995;102(3):419–457.
29. Bremner JD. *Does Stress Damage the Brain? Understanding Trauma Related Disorders from a Mind-Body Perspective*. New York, NY: W.W. Norton; 2002.
30. Canli T, Zhao Z, Brewer J, Gabrieli JD, Cahill L. Event-related activation in the human amygdala associates with later memory for individual emotional experience. *J Neurosci*. 2000;20:RC99.
31. Kensinger EA. Remembering emotional experiences: the contribution of valence and arousal. *Rev Neurosci*. 2004;15(4):241–251.

32. LeDoux JE. Emotion: clues from the brain. *Annu Rev Psychol.* 1995;46:209–235.
33. LeDoux JE. Emotion circuits in the brain. *Annu Rev Neurosci.* 2000;23:155–184.
34. Cardinal RN, Parkinson JA, Hall J, Everitt BJ. Emotion and motivation: the role of the amygdala, ventral striatum, and prefrontal cortex. *Neurosci Biobehav Rev.* 2002;26:321–352.
35. Metcalfe J, Jacobs WJ. Emotional memory the effects of stress on "Cool" and "Hot" memory systems. *Psychol Learn Motiv.* 1998;38:187–222.
36. Pitman RK, Shalev AY, Orr SP. Posttraumatic stress disorder: emotion, conditioning and memory. In: Cambridge, MA: MIT Press; 2000:1133–1147. Gazzaniga MS, ed. The New Cognitive Neurosciences; vol. 2.
37. Holland AC, Kensinger EA. Emotion and autobiographical memory. *Phys Life Rev.* 2010;7:88–131.
38. Kensinger EA, Garoff-Eaton RJ, Schacter DL. Memory for specific visual details can be enhanced by negative arousing content. *J Mem Lang.* 2006;54(1):99–112.
39. Sharot T, Martorella EA, Delgado MR, Phelps EA. How personal experience modulates the neural circuitry of memories of September 11. *Proc Natl Acad Sci U S A.* 2007;104(1):389–394.
40. de Quervain DJF, Aerni A, Schelling G, Roozendaal B. Glucocorticoids and the regulation of memory in health and disease. *Front Neuroendocrinol.* 2009;30:358–370.
41. McEwen BS, Sapolsky RM. Stress and cognitive function. *Curr Opin Neurobiol.* 1995;5:205–216.
42. McEwen BS. *The End of Stress as We Know It.* Washington, DC: Joseph Henry Press/Dana Press; 2002.
43. Nutt DJ, Malizia AL. Structural and functional brain changes in posttraumatic stress disorder. *J Clin Psychiatry.* 2004;65(suppl 1):11–17.
44. Vyas A, Mitra R, Shankaranarayana Rao BS, Chattarji S. Chronic stress induces contrasting patterns of dendritic remodeling in hippocampal and amygdaloid neurons. *J Neurosci.* 2002;22(15):6810–6818.
45. Smith ME. Bilateral hippocampal volume reduction in adults with post-traumatic stress disorder: a meta-analysis of structural MRI studies. *Hippocampus.* 2005;15(6):798–807.
46. Karl A, Schaefer M, Malta LS, Dörfel D, Rohleder N, Werner A. A meta-analysis of structural brain abnormalities in PTSD. *Neurosci Biobehav Rev.* 2006;30:1004–1031.
47. Kitayama N, Vaccarino V, Kutner M, Weiss P, Bremner JD. Magnetic resonance imaging (MRI) measurement of hippocampal volume in posttraumatic stress disorder: a meta-analysis. *J Affect Disord.* 2005;88(1):79–86.
48. O'Doherty DCM, Chitty KM, Saddiqui S, Bennett MR, Lagopoulos J. A systematic review and meta-analysis of magnetic resonance imaging measurement of structural volumes in posttraumatic stress disorder. *Psychiatry Res Neuroimaging.* 2015;232(1):1–33.
49. Kasai K, Yamasue H, Gilbertson MW, Shenton ME, Rauch SL, Pitman RK. Evidence for acquired pregenual anterior cingulate gray matter loss from a twin study of combat-related posttraumatic stress disorder. *Biol Psychiatry.* 2008;63(6):550–556.
50. Bonne O, Brandes D, Gilboa A, et al. Longitudinal MRI study of hippocampal volume in trauma survivors with PTSD. *Am J Psychiatry.* 2001;158(8):1248–1251.
51. Woon FL, Sood S, Hedges DW. Hippocampal volume deficits associated with exposure to psychological trauma and posttraumatic stress disorder in adults: a meta-analysis. *Prog Neuropsychopharmacol Biol Psychiatry.* 2010;34(7):1181–1188.
52. Woon F, Hedges DW. Gender does not moderate hippocampal volume deficits in adults with posttraumatic stress disorder: a meta-analysis. *Hippocampus.* 2011;21(3):243–252.
53. Eckart C, Stoppel C, Kaufmann J, et al. Structural alterations in lateral prefrontal, parietal and posterior midline regions of men with chronic posttraumatic stress disorder. *J Psychiatry Neurosci.* 2011;36(3):176–186.
54. Shin LM, Rauch SL, Pitman RK. Amygdala, medial prefrontal cortex, and hippocampal function in PTSD. *Ann N Y Acad Sci.* 2006;1071:67–79.
55. Woon FL, Hedges DW. Amygdala volume in adults with posttraumatic stress disorder: a meta-analysis. *J Neuropsychiatr Clin Neurosci.* 2009;21(1):5–12.
56. Kolassa I-T, Wienbruch C, Neuner F, et al. Altered oscillatory brain dynamics after repeated traumatic stress. *BMC Psychiatry.* 2007;7:56.

CHAPTER

20

Stress, Trauma, and Memory in PTSD

J. Nursey, A.J. Phelps

Phoenix Australia, Centre for Posttraumatic Mental Health, Carlton, VIC, Australia

OUTLINE

Abstract

PTSD is a debilitating disorder that is precipitated by exposure to a traumatic stressor. A hallmark feature of PTSD that sets it apart from other mental health disorders is the presence of recurring intrusive memories of the traumatic event. However, PTSD also impacts memory systems in other ways. Episodes of dissociation during and after the event can lead to memory gaps. Fear extinction, learning, and recall are impaired and episodic memory or recall of daily events can be compromised. This chapter explores the neurobiological, psychological, and cognitive underpinnings of these memory disturbances in PTSD and the implications for treatment.

INTRODUCTION

PTSD is a psychological disorder that can develop following exposure to a traumatic stressor. The traumatic stressor can be a single event, such as rape or a serious motor vehicle accident, or repeated events, such as war trauma, childhood sexual abuse, or domestic violence. The symptoms of PTSD are grouped into four clusters: reexperiencing symptoms, avoidance symptoms, negative alterations in cognition and mood, and increased arousal. The diagnosis requires that the symptoms have been present for at least 1 month and cause clinically significant distress or impairment in social, occupational, or other important areas of functioning. For the first time in DSM-5,[7] PTSD has been reclassified from an anxiety disorder to a new category of trauma and stressor related disorders. This highlights the central importance of the precipitating stressor to the diagnosis and locates PTSD on a continuum of stress reactions from normal distress to short-term and diffuse stress reactions, to more severe and chronic psychopathology.

With extreme stress leading to the development of PTSD, an understanding of the neurobiology of extreme stress is helpful to our understanding of PTSD. The impact of extreme stress on neurocircuitry implicated in memory is particularly relevant because various aspects of memory disturbance are core to PTSD. These memory disturbances have been the subject of both neuropsychological investigations and psychological theories of PTSD.

This chapter begins with an overview of the neurobiology of stress, and the impacts of extreme stress on brain circuitry including the implications for various types of memory. It then presents what is known about the neuropsychological features of PTSD as they pertain to

Stress: Concepts, Cognition, Emotion, and Behavior
http://dx.doi.org/10.1016/B978-0-12-800951-2.00020-0

memory. Finally, the chapter presents the dominant psychological theories of PTSD that highlight the centrality of memory for the traumatic event in the etiology, maintenance, and treatment of the disorder.

KEY POINTS

- The hallmark feature of PTSD is the recurrence of intrusive memories of the traumatic event.
- Chronic and/or extreme stress leads to dysregulation of the hypothalamic-pituitary-adrenal-axis stress system resulting in changes to the production, and uptake of neurotransmitters and hormones such as cortisol, noradrenaline, dopamine, and serotonin.
- Long-term changes in these neuroendocrine and neurohormonal systems impact the structural and functional integrity of brain structures and neurocircuits involved in regulation of the fear response and through which symptoms of PTSD manifest.
- Neuroimaging studies demonstrate increased dendritic connectivity and hyperfunctionality of the amygdala along with reduced volume and hypofunctionality in the hippocampus and parts of the medial prefrontal cortex. The net effect is that the "braking system" on the amygdala no longer works.
- The dual action of neurochemical imbalances and concomitant changes to connectivity in the limbic-prefrontal circuits of the brain are implicated in the expression and maintenance of PTSD symptoms including intrusive memories, hypervigilance and hyperarousal, dissociation, reduced fear extinction, and emotional dysregulation.
- These same structural and functional changes in these critical brain circuits also account for the cognitive deficits associated with PTSD, including impairments in working memory, attention, speed of processing, encoding, and retrieval of episodic memories and executive functions such as inhibition and attentional control.
- Four different psychological theories of PTSD seek to explain the primary nature of intrusive memories in PTSD.
- Metcalfe and Jacobs[1] suggest that in PTSD our "hot memory system" (emotional-fear system driven by the amygdala) takes precedence over our "cool memory system" (cognitive system driven by the hippocampus).
- Foa and Rothbaum[2,3] describe a "trauma memory network" system, where traumatic memories are stored in an unprocessed and fragmented form but can be triggered and brought into awareness by unconscious recall of any aspect of the trauma memory, resulting in a reexperiencing of the traumatic event.
- Brewin et al.[4] propose a "dual representation theory" where trauma is stored in two separate memory systems—verbally accessible memories (VAMs) that contain conscious autobiographical memories, and situationally accessible memories (SAMs) that contain involuntary intrusive memories.
- Finally, Ehlers and Clark[5] submit that a person's appraisal of the trauma and poorly integrated trauma memories that are easily triggered results in the person feeling under constant threat.
- The cornerstone of PTSD treatment involves confronting the traumatic memory to allow for effective emotional processing of the trauma. On the strength of high quality treatment trials across a range of traumatized populations, Australian and international treatment guidelines[6] recommend trauma focused cognitive behavioral therapy (TF-CBT) for the treatment of PTSD.

OVERVIEW OF THE NEUROBIOLOGY OF STRESS

As with most species in the animal kingdom, the human brain is highly attuned to threat detection and preparing our bodies for the "flight-or-fight" response. Operating through a relay loop called the hypothalamic-pituitary-adrenal-axis (HPA axis), the system is designed to be instantly activated on threat detection and quickly closed down once the threat has passed. Incoming sensory information is filtered through the thalamic and limbic centers of our brain, with the amygdala and the hippocampus, in particular, primed to detect threatening stimuli. Once a threat is detected, the hypothalamus is activated to release corticotropin-releasing hormone (CRH), which in turn signals the pituitary gland to produce adrenocorticotropic hormone (ACTH). ACTH is carried through the bloodstream to the adrenal gland that sits above the kidney and signals it to release catecholamines including the neurotransmitters adrenaline and noradrenaline (also known as epinephrine and norepinephrine) and stress hormones, including glucocorticoid and cortisol. Adrenaline and noradrenaline act in the short term to prepare the body for responding to the stressor by activating the sympathetic nervous system. This induces a range of physiological changes including, for example, increased heart rate, opening up of the airways, diluting pupils, slowing down digestion, converting glucose to energy, and sending energy and oxygen to the brain and muscles. Adrenaline also stimulates the ongoing production of CRH and ACTH to ensure the HPA axis and sympathetic nervous system remain activated throughout the threat. Cortisol is released more slowly and helps to ensure a steady availability of glucose to facilitate a sustained response to the threat. It also assists in tissue repair and suppresses the immune system. Another important function of cortisol is to provide a negative feedback to the hypothalamus to stop production of CRH and ACTH. This facilitates the activation of the parasympathetic nervous system and helps the body to return to its normal resting state once the threat has passed.[8]

IMPACTS OF EXTREME STRESS ON BRAIN CIRCUITRY

Chronic or extreme stress has been shown to change the sensitivity of the HPA axis stress response system.[8–11] The negative feedback loop of cortisol on the hypothalamus can become desensitized, resulting in overproduction of CRH and ACTH and concomitant imbalances in the other neurotransmitters and hormones that they help to regulate, including cortisol, noradrenaline, dopamine, and serotonin. These chemical imbalances can result not only from changes at a neuronal level in terms of cell receptivity or uptake of the neurotransmitters, but also from changes in their production and availability. If these neurotransmitter and hormonal changes are prolonged they can lead to both structural and functional changes in critical brain structures. These structural changes in turn impact the neurocircuits involved in a range of cognitive functions implicated in PTSD, including fear acquisition and fear extinction, attention, working memory, memory encoding and retrieval, planning and emotional, and behavioral regulation.[9,10,12–14] Both structural and functional imaging studies consistently demonstrate changes in the limbic-prefrontal circuits in people suffering from PTSD.[15] In particular they show increased dendritic connectivity and hyperfunctionality of the amygdala and reduced volume and hypofunctionality in the hippocampus and parts of the medial prefrontal cortex.[16] Disruptions to these circuits account for many of the symptoms associated with PTSD, including hyperarousal, intrusive memories, avoidance, and poor emotional regulation.[17,18]

IMPLICATIONS FOR MEMORY IN PTSD

The different types of memory and processes of memory formation referred to in this section are described in Boxes 1 and 2, respectively.

The brain structures that are highly sensitive to the negative impacts of excess cortisol and catecholamines arising from extreme stress, including amongst others, the medial prefrontal cortex, the anterior cingulate, the dorsolateral prefrontal cortex, the hippocampus and medial temporal gyrus, the insula and the amygdala all play a part, either directly, or through network connections, in the encoding and retrieval of episodic memories and fear extinction learning as well as in aspects of working memory and executive functions. In addition, the altered functionality of the frontal-limbic network in PTSD contributes to information processing biases evident in PTSD with sufferers orienting towards and failing to disengage from stimuli perceived as threatening to the detriment of attending to less threatening information in their environment. It is not surprising then that there are various features of memory disturbance apparent in PTSD.

ENHANCED RECALL OF TRAUMA-RELATED INFORMATION

First and foremost, PTSD is associated with enhanced recall of trauma-related information. Reexperiencing symptoms is a hallmark of the PTSD diagnosis, and takes many forms including: intrusive (unwanted and uninvited) thoughts, images, or sensations related to the trauma; distress triggered by trauma-related sights, sounds, smells and sensations; nightmares; and flashbacks. Flashbacks are distinct in that the person is awake but feels as though they are back in the traumatic situation and may even behave as though they are.

DISSOCIATION

Another distinctive element of PTSD with particular relevance to memory is dissociation. Dissociation is characterized by a disruption of and/or discontinuity in the normal integration of consciousness, memory, identity, emotion, perception, motor control, and behavior.[7] It can include symptoms such as depersonalization, derealization, and dissociative amnesia, where the person is unable to recall autobiographical details either related to specific events or more generalized personal information. Lanius et al.[19] note that while not everyone with a diagnosis of PTSD will have dissociative symptoms, most people who have high levels of dissociative symptoms meet criteria for PTSD.[19] Dissociative symptoms are likely to be most prominent in the first few days following trauma exposure and then will slowly dissipate over the coming weeks. However, a minority of people will continue to demonstrate high levels of dissociation in an ongoing way and often demonstrate poor response to

BOX 1

DIFFERENT TYPES OF MEMORY

Sensory memory is the perception of sight, hearing, smell, taste, and touch information entering through the sensory cortices of the brain and relaying through the thalamus. It lasts only milliseconds and is mostly outside conscious awareness.[27]

Immediate or short-term memory is our capacity to hold a very limited amount of information in a temporary buffer for a short period of time (from a few seconds to a couple of minutes). The information enters through auditory or visual channels and captures our attention. The average adult can hold between seven and nine pieces of information in short-term memory before it becomes overloaded.

Working memory refers to our ability to manipulate and make use of the information held in short-term memory, for example calculating the change you should get back from a shopkeeper or reading and recalling a phone number while you dial. Information only stays in our working memory buffer for a couple of minutes. Working memory is comprised of two parts, the phonological loop that processes verbal information and the visuospatial sketch pad that handles visual information, coordinated by a central executive system.[34] Working memory relies on a network of interconnected brain regions, however the dorsolateral prefrontal cortex plays a fundamental role in working memory.

Long-term memory refers to information that is stored over a period of days, months, or years. There are two streams of long-term memory—declarative and nondeclarative.

Declarative or explicit memory stores facts and events that are available to our conscious recall. It relies on the medial temporal lobe including the hippocampus and its connections to adjacent structures such as the parahippocampal gyrus, the entorhinal cortex, and the limbic system to facilitate the encoding and consolidation of new information. Declarative memory can be further broken down into two subsystems known as episodic and semantic memory.

Episodic memory contains memories of events and facts in our daily life, for example what we had for dinner last night. Episodic memory stores the autobiographical details of our life and is always self-referential. *Semantic memory* stores knowledge that we acquire about the world, including things such as word meanings, general knowledge, rules, concepts, and customs. Semantic memory involves structures such as the cerebellum, basal ganglia, amygdala, and the neocortex.[35]

Nondeclarative or implicit memory includes behaviors, skills, and knowledge learned through unconscious processes such as conditioning, habituation, priming, and procedural learning. It is not available for conscious recall. Examples of implicit memory include fear conditioning and procedural memory.

Emotional memory: Emotional experiences enhance both the encoding of new memories as well as their ability to be retrieved. Memories associated with strong emotions are known as emotional memory. Emotional memories can be both implicit (unconsciously encoded and retrieved) or explicit (consciously mediated). The amygdala is central to both the encoding and retrieval of emotional memories, however both the hippocampus and prefrontal cortex also play important roles.

BOX 2

PHASES OF MEMORY PROCESSING INVOLVED IN NEW LEARNING

Information that is stored in long-term memory has to be learned in the first instance. There are a number of phases involved in the learning of new information.

Encoding involves dual processes of acquisition and consolidation. In order to acquire new information we have to pay attention to it, assign it meaning, link it to other related information and assign it a context. Consolidation involves the embedding of this information into long-term storage. Repeated retrieval of newly acquired information helps to consolidate it over time.

Storage refers to the permanent holding of information that has been well encoded and consolidated.

Retrieval is the recollection of memories held in long-term storage. Retrieval can be facilitated either through an active recall of the information where we voluntarily bring it into our conscious awareness or via a recognition process where when information is presented to us we make a decision as to whether or not we have seen or heard the information previously.

traditional evidence-based treatments for PTSD. For example, high levels of dissociation are thought to interfere with the emotional processing of and habituation to trauma memories during exposure therapy.[20] While symptoms of hyperarousal and reexperiencing in PTSD are commonly seen as being related to the hypoactivation of the medial prefrontal cortex and anterior cingulate and concomitant hyperactivation of the amygdala, dissociative symptoms in contrast are thought to result from a hyperactivation of the anterior cingulate and medial prefrontal cortex resulting in an overmodulation or inhibition of the amygdala and hippocampal reactivity.[17,19] This cortico-limbic inhibition model of emotional regulation posits that reexperiencing and hyperarousal symptoms and numbing and dissociation represent different sides of the same coin and reflect different subtypes of PTSD symptoms. Dissociative states prevent effective encoding and consolidation of episodic memories, but have also been shown to interfere with nondeclarative memory functions including conditioning and habituation.[21,22]

IMPAIRED NEW LEARNING

Aside from intrusive memories and dissociation, PTSD can also impact other memory and new learning processes. Fear extinction learning refers to the gradual reduction of a conditioned fear response by repeatedly exposing the person to the conditioned stimulus without the associated unconditioned stimulus. In PTSD this would involve exposing people to a reminder of a traumatic event without the associated aversive consequences. However, people with PTSD demonstrate a persistent exaggerated fear response that is difficult to extinguish. It is hypothesized that this reflects a failure to consolidate extinction learning and therefore an inability to recall the memory of the extinction learning over time (extinction recall), rather than a failure in extinction learning per se.[10] Milad et al.[36] provide support for this hypothesis and demonstrate that the deficit in extinction recall is directly associated with the reduced activation of the ventromedial prefrontal cortex (vmPFC) and the hyperactivation of the amygdala during extinction learning as well as the hypoactivation of the vmPFC and the hippocampus during recall.

Another way that PTSD impacts memory processing is to reduce the sufferer's ability to learn and recall verbally mediated episodic memories. To lay down new memories we need to be able to attend to incoming information, ignore information that is not relevant, interpret the incoming information and give it meaning and context, associate it with other information that we have stored away, and be motivated to rehearse it or repeatedly retrieve it in order to effect consolidation. Thus our ability to encode, consolidate, and retrieve episodic and semantic memories is highly dependent on the robustness of a range of other cognitive processes including attention, working memory, and executive functions such as inhibition and attentional switching, as well as emotional regulation and language and visual processing. The neurobiological processes that underlie the symptoms of PTSD also lead to disruption in many of these neuropsychological processes with the result that the sufferer's ability to learn and retain new nontrauma-related information is impaired.

NEUROPSYCHOLOGICAL CORRELATES OF PTSD

Much work has been done over the past 10 years to elucidate the exact nature and extent of the neuropsychological impairments associated with PTSD.[23–25] A recent metaanalysis of over 60 studies examining the cognitive deficits associated with PTSD found consistent evidence for moderate impairments in verbal learning and verbal recall, speed of information processing, attention, working memory, and executive function alongside weaker, but still evident, impairments in language, visuospatial functions, and visual learning and recall.[26] This profile of deficits differs from that seen in people diagnosed with depression and other anxiety disorders suggesting that the neural mechanisms underlying them are unique to PTSD. There is ongoing debate about whether some of these deficits might be preexisting and therefore represent a vulnerability or risk factor for PTSD rather than a consequence of it. Scott and colleagues[26] argue that while some of the documented cognitive deficits (for example, reduced verbal recall) can be identified as risk factors for developing PTSD, they have also been shown to worsen after developing the disorder. Regardless, the profile of cognitive deficits associated with PTSD has significant implications not only for the sufferer's capacity to maintain occupational, social, and daily life functions, but also for their capacity to engage in and benefit from evidence-based psychological treatments for the disorder.

PSYCHOLOGICAL THEORIES OF PTSD

Diagnostic Criteria for PTSD

As noted above, PTSD comprises four groups of symptoms. The reexperiencing symptoms have already been described. The avoidance symptoms of PTSD can be understood as ways in which the individual, consciously and unconsciously, attempts to keep the reexperiencing symptoms at bay. They may avoid any reminders of the trauma including people, places, events, news, and TV programs that remind them of the event. They may withdraw from social activities and avoid any situation that leads to distress or feelings of anxiety. The negative alterations in cognitions and mood cluster includes the range of negative mood and cognition states associated with PTSD. Individuals may experience persistent and distorted blame of self or others, persistent negative emotional states, and estrangement from others. They may also find it difficult to remember key aspects of the traumatic event. Symptoms related to alterations in arousal and reactivity comprise the final cluster. When a person has been exposed to a highly threatening or traumatic situation, they can be left with an ongoing increased level of arousal. This may include feeling constantly irritable or angry, engaging in reckless or destructive behavior, feeling constantly "on guard" and on the lookout for signs of danger, startle reactions to sudden noises or other unexpected stimuli during the day, problems with concentration and memory, and sleep disturbance. The arousal symptoms of PTSD can be seen to serve the purpose of keeping the person alert to any external or internal threats.

The clinical syndrome of PTSD reflects the interplay between these symptom groups. Intrusive recollections that arise from the memory of trauma are associated with high distress, an exacerbation of the already elevated arousal. The natural response is to reduce this distress through active avoidance of the memory or any potential reminders. The result of avoidance, however, is that the traumatic memory remains unprocessed and so continues to give rise to intrusive recollections and negative cognitions and mood, thus perpetuating the cycle.

CENTRALITY OF MEMORY TO PSYCHOLOGICAL THEORIES OF PTSD

Thus, involuntary and intrusive memories of the traumatic event, in the form of flashbacks, nightmares, and intrusive recollections, are at the core of PTSD. In an apparent paradox, people with PTSD also report gaps in their memory for important parts of the traumatic event, described by Brewin[27] as an inability to retrieve ordinary episodic memories of the trauma.

From the earliest theories of PTSD, memory for the traumatic event has been central to our understanding of the disorder. Metcalfe and Jacobs[1] seek to explain the impact of stress on memory with reference to two subsystems of the brain; the "cool" cognitive hippocampal memory system that records wellelaborated autobiographical events in their proper spatio-temporal context; and the "hot" emotional-fear amygdala system that records unintegrated fragmentary and fear-provoking features of events. Under stressful conditions, as arousal increases, the hippocampus becomes less functional while, conversely, the amygdala becomes progressively more responsive. Thus, at traumatic levels of stress, the individual will focus exclusively on the fear-evoking features of the situation and not attend to the broader spatiotemporal context, creating memories that are fragmentary and fear-evoking. Subsequent researchers[28] have highlighted the coexistence of seemingly inconsistent memories of the traumatic event (i.e., vivid and detailed sensory memories alongside vague and incomplete autobiographical memories) as specific to PTSD.

Foa and Rothbaum[2,3] proposed that high levels of physiological arousal at the time of the trauma contributes to memory for the event being stored in a different form to memory for everyday events because it interferes with the individual's capacity to emotionally process or come to terms with the experience. Nontraumatic experiences are processed and integrated into episodic memory where they are available for deliberate recall. In contrast, memory for traumatic experience in individuals who develop PTSD, is stored in an unprocessed way—an encapsulation of stimuli, response, and meanings experienced at the time of the trauma. Foa and Rothbaum call the resulting memory structure, the *trauma memory network*. The trauma memory network comprises stimulus information (sensory details of the traumatic event), response information (cognitive, emotional, behavioral, physiological responses at the time of the trauma), and meaning information (assumptions about the self and others arising from the traumatic experience). According to the model, the three elements of the network are linked through classical conditioning but, importantly, those links are not necessarily available to the person's conscious mind. Rather the elements are fragmented in the individual's awareness and in deliberate recall of the experience. However, the entire trauma memory network can be activated by any one of the elements—trauma stimuli, responses, or meaning. When activation occurs, the trauma is reexperienced in the form of daytime intrusions, psychological and physiological reactivity to reminders, flashbacks, and trauma-related nightmares during sleep.

Brewin et al.[4] propose a "dual representation" theory of PTSD to better account for the apparently disparate representations of memory for trauma observed in PTSD. Dual representation theory argues that, rather than being stored in a single trauma memory network, memory for trauma is stored in two memory systems, distinguished according to whether the memory is accessible verbally or situationally. Verbally accessible memories are described as ordinary autobiographical memories that can be retrieved at will, reflect conscious appraisal of the trauma and its consequences, and interact with other information in autobiographical memory. Memories stored thus can be verbalized and do not evoke strong emotional responses. Situationally accessible memories, on the other hand, are described as involuntary intrusive memories that were stored rapidly with minimum conscious attention, and when triggered by internal or external cues, involve intense reliving of the trauma.

The central tenant of Ehlers and Clark's[5] cognitive model of PTSD is that PTSD persists when the traumatic experience and its sequelae are processed in a way that leads to a sense of current threat. They propose that two key processes can lead to the sense of current threat. The first of these is the individual's appraisal of the trauma and its consequences, for example, "nowhere is safe" and "I'll never get over this," respectively. The second process proposed by Ehlers and Clark relates to the memory for the traumatic event. They describe traumatic memory as poorly elaborated and poorly integrated into normal autobiographical memory with contextual information such as time and place. The content of trauma memory contains strong associations between trauma-related stimuli and between stimuli and responses and there is strong perceptual priming such that there is a low threshold for the memory to be triggered—vaguely similar cues are all that is needed. Thus, memory for trauma is triggered involuntarily, experienced as happening again "now," and involves reexperiencing the original senses and emotions felt during the trauma.[29–31]

PSYCHOLOGICAL TREATMENT OF PTSD

Within each of these theoretical models, the nature of traumatic memory, containing unprocessed sensory memory and contributing to a sense of ongoing threat, is conceptualized as the core of the disorder. It is not surprising then, that the cornerstone of PTSD treatment involves confronting the traumatic memory in order to help the individual to emotionally process the memory and in so doing, come to terms with their traumatic experience.

On the strength of high quality treatment trials across a range of traumatized populations, Australian and international treatment guidelines[6] recommend trauma focused cognitive behavioral therapy (TF-CBT) for the treatment of PTSD. TF-CBT encompasses a range of interventions that assist the individual to confront their memory of the traumatic event, including the meaning of the event and subsequent interpretations which impact their beliefs about themselves, other people, and the world. One of the key TF-CBT interventions is prolonged exposure.

Exposure therapy has long been established as an effective treatment for a range of anxiety disorders, with the key objective of confronting, rather than avoiding the object of fears.[32] In PTSD, this can include in vivo exposure to external stimuli (e.g., people, places, conversations that serve as trauma reminders), as well as imaginal exposure to internal stimuli—the traumatic memory itself.

Two mechanisms underpin exposure therapy: habituation and emotional processing. Habituation is based on the principle that if a person stays in contact with the anxiety-provoking stimulus for long enough, their anxiety will inevitably subside. In this way, the conditioned association between stimulus (the feared object) and response (anxiety) is broken or unlearned. Habituation may occur within a session or between sessions, with between-session habituation particularly important in imaginal exposure.[20]

However, conditioning alone cannot explain the acquisition and maintenance of all fears; vicarious exposure and information transmission are recognized as alternative pathways.[33] Thus, in the treatment of PTSD, the meaning of the stimulus and response elements also needs to be addressed. According to Foa and Kozak,[37] for fear reduction to occur, two conditions are required. The fear memory needs to be activated and cognitive and affective information that is incompatible with what exists in the fear memory must be made available and integrated into the information structure so that a new memory can be formed. This change in the fear memory structure is what is termed emotional processing.

SUMMARY

PTSD is a stress-related disorder, which features disturbances in a range of normal memory processes. Our understanding of the disorder is informed by the neurobiology of stress and the impact of extreme stress on brain circuitry that is implicated in memory. Neuropsychological investigations have identified a number of specific cognitive deficits in PTSD related to attention and memory. Psychological theories of the disorder also emphasize the importance of memory of the traumatic event in the etiology, maintenance, and treatment of the disorder. There is agreement across each of these fields of study that memory is central to our understanding of PTSD. However, further work is needed to integrate the findings from each field to create a comprehensive model of the disorder.

References

1. Metcalf J, Jacobs WJ. An interactive hot system/cool system view of memory under stress. *PTSD Res Q*. 1996;7:1–6.
2. Foa EB, Rothbaum BO. Behavioural psychotherapy for post-traumatic stress disorder. *Int Rev Psychiatry*. 1989;1(3):219–226.
3. Foa EB, Rothbaum BO. *Treating the Trauma of Rape: Cognitive-Behavioral Therapy for PTSD*. New York: Guilford Press; 1998.

4. Brewin CR, Dalgleish T, Joseph S. A dual representation theory of posttraumatic stress disorder. *Psychol Rev*. 1996;103(4):670–686.
5. Ehlers A, Clark DM. A cognitive model of posttraumatic stress disorder. *Behav Res Ther*. 2000;38(4):319–345.
6. Forbes D, Creamer M, Bisson JI, et al. A guide to guidelines for the treatment of PTSD and related conditions. *J Trauma Stress*. 2010;23 (5):537–552. http://dx.doi.org/10.1002/jts.20565.
7. American Psychiatric Association. *Diagnostic and Statistical Manual of Mental Disorders*. 5th ed. Arlington, VA: American Psychiatric Association; 2013.
8. Smith SM, Vale WW. The role of the hypothalamic-pituitary-adrenal axis in neuroendocrine responses to stress. *Dialogues Clin Neurosci*. 2006;8(4):383–395.
9. Arnsten AF. Stress signalling pathways that impair prefrontal cortex structure and function. *Nat Rev Neurosci*. 2009;10(6):410–422. http://dx.doi.org/10.1038/nrn2648.
10. Bremner JD. Traumatic stress: effects on the brain. *Dialogues Clin Neurosci*. 2006;8(4):445–461.
11. Shin LM, Liberzon I. The neurocircuitry of fear, stress, and anxiety disorders. *Neuropsychopharmacology*. 2010;35(1):169–191.
12. Bremner JD, Elzinga B, Schmahl C, Vermetten E. Structural and functional plasticity of the human brain in posttraumatic stress disorder. *Prog Brain Res*. 2008;167:171–186.
13. Samuelson KW. Post-traumatic stress disorder and declarative memory functioning: a review. *Dialogues Clin Neurosci*. 2011;13 (3):346–351.
14. Sylvester CM, Corbetta M, Raichle ME, et al. Functional network dysfunction in anxiety and anxiety disorders. *Trends Neurosci*. 2012;35(9):527–535. http://dx.doi.org/10.1016/j.tins.2012.04.012.
15. Rauch SL, Shin LM, Segal E, et al. Selectively reduced regional cortical volumes in post-traumatic stress disorder. *Neuroreport*. 2003;14(7):913–916. http://dx.doi.org/10.1097/01.wnr.0000071767.24455.10.
16. Sherin JE, Nemeroff CB. Posttrumatic stress disorder: the neurobiological impact of psychological trauma. *Dialogues Clin Neurosci*. 2011;13(3):263–277.
17. Hopper JW, Frewen PA, van der Kolk BA, Lanius RA. Neural correlates of reexperiencing, avoidance, and dissociation in PTSD: symptom dimensions and emotion dysregulation in responses to script-driven trauma imagery. *J Trauma Stress*. 2007;20(5):713–725. http://dx.doi.org/10.1002/jts.20284.
18. Pitman RK, Rasmusson AM, Koenen KC, et al. Biological studies of post-traumatic stress disorder. *Nat Rev Neurosci*. 2012;13 (11):769–787. http://dx.doi.org/10.1038/nrn3339.
19. Lanius RA, Brand B, Vermetten E, Frewen PA, Spiegel D. The dissociative subtype of posttraumatic stress disorder: rationale, clinical and neurobiological evidence, and implications. *Depress Anxiety*. 2012;29(8):701–708. http://dx.doi.org/10.1002/da.21889.
20. Jaycox LH, Foa EB, Morral AR. Influence of emotional engagement and habituation on exposure therapy for PTSD. *J Consult Clin Psychol*. 1998;1:185.
21. Schmahl C, Lanius R, Pain C, Vermetten E. Biological framework for traumatic dissociation related to early life trauma. In: Lanius RA, Vermetten E, Pain C, eds. *The Impact of Early Life Trauma on Health and Disease: The Hidden Epidemic*. Cambridge: Cambridge University Press; 2010:178–188.
22. Staniloiu A, Markowitsch HJ. Review: dissociative amnesia. [Review article]. *Lancet Psychiatry*. 2014;1:226–241. http://dx.doi.org/10.1016/s2215-0366(14)70279-2.
23. Cohen BE, Neylan TC, Yaffe K, Samuelson KW, Li Y, Barnes DE. Posttraumatic stress disorder and cognitive function: findings from the mind your heart study. *J Clin Psychiatry*. 2013;74(11):1063–1070. http://dx.doi.org/10.4088/JCP.12m08291.
24. Vasterling J, Brewin CR. *Neuropsychology of PTSD: Biological, Cognitive, and Clinical Perspectives*. New York, NY: Guilford Press; 2005.
25. Vasterling J, Verfaellie M. Posttraumatic stress disorder: a neurocognitive perspective. *J Int Neuropsychol Soc*. 2009;15(826–829). http://dx.doi.org/10.1017/S1355617709990683.
26. Scott JC, Wrocklage KM, Cmich C, et al. A quantitative meta-analysis of neurocognitive functioning in posttraumatic stress disorder. *Psychol Bull*. 2015;141(1):105–140. http://dx.doi.org/10.1037/a0038039.
27. Brewin CR. Episodic memory, perceptual memory, and their interaction: foundations for a theory of posttraumatic stress disorder. *Psychol Bull*. 2014;140(1):69–97. http://dx.doi.org/10.1037/a0033722.
28. Brewin CR, Holmes EA. Psychological theories of posttraumatic stress disorder. *Clin Psychol Rev*. 2003;23:339–376.
29. Ehlers A, Hackmannb A, Steilb R, Clohessyb S, Wenningerb K, Winterb H. The nature of intrusive memories after trauma: the warning signal hypothesis. *Behav Res Ther*. 2002;40(9): 995–1002.
30. Michael T, Ehlers A, Halligan SL, Clark DM. Unwanted memories of assault: What intrusion characteristics are associated with PTSD? *Behav Res Ther*. 2005;43(5):613–628.
31. Steil R, Ehlers A. Dysfunctional meaning of posttraumatic intrusions in chronic PTSD. *Behav Res Ther*. 2000;38(6):537–558.
32. Mowrer OH. *Learning Theory and Behaviour*. New York, NY: Wiley; 1960.
33. Rachman S. The conditioning theory of fear acquisition: a critical examination. *Behav Res Ther*. 1997;15(5):375–387.
34. Baddeley AD. *Working Memory*. Oxford/New York: Clarendon Press/Oxford University Press; 1986.
35. Gazzaniga MS, Ivry RB, Mangun GR. *Cognitive Neuroscience: The Biology of the Mind*. 4th ed. New York, NY: WW Norton and Company; 2014.
36. Milad MR, Pitman RK, Ellis CB, et al. Neurobiological basis of failure to recall extinction memory in posttraumatic stress disorder. *Biol Psychiatry*. 2009;66:1075–1082.
37. Foa EB, Kozak MJ. Emotional processing of fear:exposure to corrective information. *Psychol Bull*. 1986;99(1):20–35.

CHAPTER

21

Adolescent Cognitive Control: Brain Network Dynamics

D.B. Dwyer[1], *B.J. Harrison*[1], *M. Yücel*[1,2], *S. Whittle*[1], *A. Zalesky*[1,3], *C. Pantelis*[1], *N.B. Allen*[3], *A. Fornito*[2]

[1]The University of Melbourne and Melbourne Health, Melbourne, VIC, Australia
[2]Monash University, Clayton, VIC, Australia
[3]The University of Melbourne, Melbourne, VIC, Australia

OUTLINE

Abstract

Adolescence is a time when the ability to consciously control thoughts, actions, and emotions is linked to successful outcomes later in life. Despite a historical focus on the prefrontal cortex, modern neuroimaging research suggests that these cognitive control abilities are mediated by the activity and interactions of brain regions that collectively form distributed networks. This chapter outlines research and methods that have highlighted the importance of brain network dynamics (i.e., the spatiotemporal interactions between brain regions) during cognitive control in adolescence. On the basis of current research, we suggest that transient reconfigurations of brain networks that occur in specific contexts are critical to cognitive control in adolescence. In particular, we suggest that the spatiotemporal dynamics of brain networks play a crucial role in governing adolescent cognitive control across different contexts (i.e., they are generalizable). We suggest that further research of generalizable principles will help adolescents with low cognitive control.

INTRODUCTION

During adolescence, the ability to overcome impulsive tendencies, contextually inappropriate behavior, and general distraction is an important component of adaptive outcomes in later life. For example, low levels of self-control ability have been linked to risk-taking behavior, substance abuse, and mental disorders,[1] while high levels have been associated with academic and vocational success.[2] Understanding the neural basis of adolescent self-control is thus an important goal of developmental neuroscience.

In this chapter, we consider the neural basis of an aspect of self-control known as cognitive control. We review how neurobiological models of cognitive control have shifted from an early emphasis on functional

http://dx.doi.org/10.1016/B978-0-12-800951-2.00021-2

localization, principally within the prefrontal cortex (PFC), to more recent theories that consider control as arising from complex interactions between elements of distributed neural systems. In particular, we will focus on the contribution that network science is making to promote these theoretical developments.

KEY POINTS

- Cognitive control is the ability to consciously overcome habitual thoughts, actions, and emotions to achieve goals.
- Adolescent cognitive control has historically been linked to the structure and functioning of the prefrontal cortex.
- More recent functional magnetic resonance imaging (fMRI) research shows that cognitive control involves antagonistic interactions between two large-scale neural systems: the cognitive control network (CCN) and the default-mode network (DMN).
- Studies of context-dependent functional interactions between these systems also show that transient reconfigurations of connections between regions across the brain are associated with performance on cognitive control tasks.
- These findings suggest that the maturation of cognitive control relies on the integration of distributed neural systems, rather than the development of any single brain region.

EARLY RESEARCH IMPLICATING THE FRONTAL LOBES

Research investigating the neural basis of cognitive control began with seminal investigations of brain lesion patients. One famous case was Phineas Gage,[3] who suffered a large, focal lesion to his left PFC. The result was a marked change in personality characterized by increased impulsivity without regard for future consequences. Similar behaviors, including maladaptive control over impulses and habitual reactions, were observed when corresponding areas of the frontal lobes were experimentally lesioned in animals.[4] Additionally, seminal neuropsychological investigations further demonstrated the link between prefrontal regions and the regulation of human behavior.[5,6] These early studies suggested a causal link between the ability to exert control over behavior to fulfill long-term goals and the structure and functioning of the PFC.

The role of more specific regions of the PFC in cognitive control later came from neuroimaging work, in which increasingly specific psychological probes of different cognitive processes were used to map prefrontal specialization on increasingly finer spatial scales. A well-known task in this area is the Stroop interference task,[7] which measures the speed and accuracy in which an individual can overcome a habitual reaction to read a word in order to produce an alternate response (naming the color that the word is printed in). There are now a multitude of cognitive control tasks that involve different aspects of the ability to overcome habitual or automatic responses, such as switching between tasks or stopping behavior when it is inappropriate.[8] Over time, the concept of cognitive control grew to include more diverse cognitive processes that are also thought to be important, such as planning, organization, and the ability to hold and manipulate information in-mind (i.e., working memory[9]). In this chapter, we focus principally on results from tasks assessing more basic, or fundamental, control processes assessed in interference paradigms, such as the Stroop task.

Early neuroimaging studies using positron emission tomography (PET) indicated that Stroop-like interference specifically activated a region of the PFC called the anterior cingulate cortex (ACC).[10] This research was supported by later functional magnetic resonance imaging (fMRI) studies showing activation of the dorsal ACC (dACC),[11] in addition to the dorsolateral prefrontal cortex (DLPFC).[12] On the basis of these studies, it was theorized that the dACC monitors and detects conflict between competing responses (e.g., a habitual response and a goal-oriented response) before sending this signal to the DLPFC. This signal indicates a need for controlled processing, which the DLPFC implements by amplifying or suppressing neural pathways to perform the task appropriately.[13] Thus, these studies were the first to propose that interactions within a small PFC network were important to cognitive control.

The PFC network hypothesis was also useful in explaining reports of a reduced capacity for cognitive control in adolescents compared to adults.[14] It was discovered that the PFC undergoes pronounced maturation during adolescence, associated with enhanced functional differentiation, synaptogenesis, myelination, and neural pruning.[15] Structural neuroimaging studies confirmed these reports[16] and longitudinal functional neuroimaging research showed changes in PFC activation during cognitive control tasks during the transition from adolescence to adulthood.[17] These studies implied that the maladaptive control over impulses and habitual reactions observed in adolescence was due to the immaturity of a prefrontal network of brain regions.

BRAIN NETWORKS OF COGNITIVE CONTROL

Recent research has extended the PFC network model by highlighting the importance of activation of other regions distributed across the brain. Simultaneously, there is a growing recognition of the importance of reductions in neural activity from a baseline state (i.e., deactivations) to the engagement of cognitive control within a spatially disparate network of brain regions. Quantitative assessment of regional interactions in different experimental contexts has further extended brain activation studies by suggesting that cognitive control is supported by complex interactions within and between large-scale networks.

Network Activations and Deactivations

The PFC network model was first extended by research demonstrating the importance of other regions within the frontal cortex. For example, Aron and colleagues[8] demonstrated a critical role of the inferior frontal junction during stop-signal tasks, which assess the ability to overcome habitual responses. Other studies also highlighted the importance of activation in additional brain regions, such as the anterior insula cortex, premotor cortex, posterior parietal lobe, and subcortical regions.[18–22] Combined, these regions have been proposed to fulfill separate functions during task performance; for example, vigilance is supported by the anterior insula cortex and spatial attention mediated by the posterior parietal lobes.[19] This research led to the hypothesis of a cognitive control network (CCN; Figure 1), characterized by

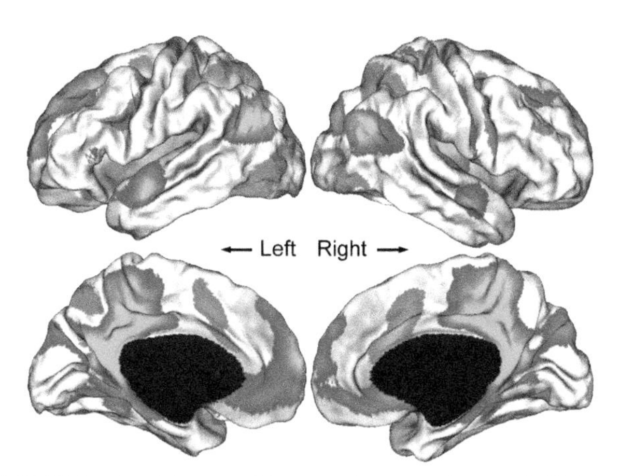

FIGURE 1 Spatial maps of the cognitive control network (CCN; red) and the default-mode network (DMN; blue) in adolescence. The maps were derived from a sample of 16-year-olds who performed the Multi-Source Interference Task.[23] The CCN was derived from the comparison of a cognitive control condition with a control condition; it is putatively involved in attention to external stimuli and includes frontal, parietal, cingulate, and subcortical regions (not shown here) that often activate across a range of cognitive control tasks.[20] The DMN was derived from the comparison of a cognitive control condition with a resting condition; it has been linked to self-related mental processes and includes the posterior cingulate, medial frontal, and lateral parietal regions that widely deactivate during cognitive control tasks.[24] Figure adapted and reprinted with permission from the Society for Neuroscience.[23]

cooperation between anatomically distributed regions to exert conscious control over actions.[18]

In addition to the CCN, the default-mode network (DMN; Figure 1) plays a complementary role in supporting cognitive control. This network contains regions that exhibit coordinated decreases in brain activity (or "deactivations") as measured by PET or fMRI when a task condition involving attention to external environmental stimuli is contrasted with a task-free, so-called resting-state, where mental activity is hypothesized to be largely self-referential (e.g., thinking about oneself in the past or future, or making moral judgements).[25] The relevance of DMN deactivation to cognitive control has been demonstrated by studies showing that failures of deactivation are associated with poorer cognitive control performance,[26] and that deactivation scales with task difficulty.[27]

The regions that comprise the DMN include the medial PFC, medial parietal/posterior cingulate cortex, inferior parietal, and middle temporal.[28] Similar to prefrontal regions of the CCN, areas such as the posterior cingulate cortex are defined by their high degree of connectivity with regions distributed across the brain.[29] However, instead of exhibiting associations with psychological functions that mainly involve focused attention to external events (e.g., during a Stroop task), the psychological functions of DMN have been linked to self-referential processing, such as thinking about the past, planning for the future, assimilating information, and semantic processing.[30,31] Combined, these processes are commonly reported while subjects are quietly resting while their mind wanders,[32] and as such, it has been proposed that deactivations of the DMN during task performance reflect a need to suppress self-referential thoughts in order to fulfill task goals—otherwise known as "losing one's self in one's work."[33]

These results suggest that optimal cognitive control performance is associated with activation of the CCN, which engages cognitive processes needed to perform the task, and deactivation of the DMN, which disengages self-referential processing. These observations led to the hypothesis that effective cognitive control during the engagement of cognitive control involves dynamic changes of the CCN and DMN, which are possibly mediated by communication both within and between the network.[34]

Interactions Within and Between Networks

The research outlined thus far has predominantly inferred network connectivity based on the simultaneous activation or deactivation of localized regions in response to task demands. However, coactivation (or codeactivation) between regions may occur in the absence of direct communication between them. To properly characterize these regions as forming part of a distributed network, we must quantify their functional interactions. To this end, we can examine statistical dependencies in their activity time courses—so-called functional connectivity.[35]

Studies of functional connectivity during task performance have suggested that cognitive control performance is associated with enhanced integration between regions that comprise the CCN.[18,36–38] However, it remains unclear whether the network can be divided into functionally segregated subnetworks that putatively fulfill different psychological cognitive functions. This heterogeneity may be influenced by the context of the task under investigation (e.g., whether it employs visual or auditory stimuli).

Substantially more research has been directed toward investigating functional connectivity of CCNs during task-free, resting-state conditions. In these conditions, functional connectivity is assessed using the time-series of low-frequency (e.g., <0.08 Hz) spontaneous fluctuations of the fMRI blood oxygenation level dependent (BOLD) signal between brain regions. Biswal et al.[39] first demonstrated the validity of this technique by showing that the low-frequency fMRI BOLD signal from the left sensorimotor cortex at rest was highly correlated with the right sensorimotor cortex and other regions involved in the motor control system.[39] Later research showed that resting-state networks recapitulate well-known activation patterns found during the performance of other tasks, including those related to speech, hearing, emotion, and cognitive control.[40] Importantly, while some degree of resting-state connectivity may be associated with conscious, self-referential, mental processes discussed above (e.g., autobiographical memory), similar patterns are found under anesthesia.[41] These observations imply that transient changes in functional connectivity arising from self-referential thoughts occur against a background of continuous "intrinsic" connectivity that is shaped by brain anatomy, maturation, and learning.[42,43]

Resting-state fMRI has been very useful in specifically assessing the functional connectivity of the CCN and DMN. For example, by extracting the signal from key regions of each of the networks and correlating it with the remainder of the brain, research has found a high degree of connectivity within the CCN and within the DMN.[44] Functional segregation of the networks, indicative of a separation between psychological functions, has also been shown in research demonstrating that the network signals are highly negatively correlated.[45] Moreover, the degree of functional segregation between the networks (i.e., the degree of negative correlation) is associated with a simple measure of cognitive control (performance stability).[46] This research indicates that the strength of connections within each network, and between the networks, may be important for cognitive control in adolescence.

LARGE-SCALE NETWORK DYNAMICS

The research described thus far indicates that cognitive control is supported by interactions within and between large-scale networks. However, the contribution of individual regions within the networks has not been addressed, questions remain regarding the presence of subnetworks, and most research to date has been conducted on adults rather than adolescents where developmental changes in network organization may play a role. These factors are beginning to be addressed with graph theoretical approaches that facilitate the quantitative assessment of complex relationships between multiple brain regions.

Using Graph Theory to Investigate Brain Networks

Graph theory is the mathematical study of systems of interacting elements. The elements are modeled as nodes in a graph, and their connections are represented as edges. These edges could represent physical (e.g., an axon between neurons) or statistical (e.g., a correlation between time-series) relationship.[47] By representing brain regions in graph form as nodes connected by edges, the connections of each node can be tested for relationships with behavior, subnetworks can be identified, and a range of other metrics can be used to investigate properties of the graph. For example, modularity analyses can be used to explore the large-scale organization of nodes and their connections into subsets of regions (i.e., "modules") that are more highly connected with each other than they are with regions from other modules (Figure 2).

When applied to task and resting-state neuroimaging data, graph theoretic approaches have revealed a more complex relationship between CCN/DMN nodes than previously found and have challenged the hypothesis that the segregation between the networks results in better task performance. For example,[66] suggest that the CCN can be further subdivided into a cingulo-opercular subnetwork that interacts with the DMN to enable efficient performance. It has also been demonstrated that the dorsal posterior cingulate cortex acts as an integrative hub between the CCN and DMN during memory retrieval,[49] enabling efficient integration between these two networks when required. Moreover, this research

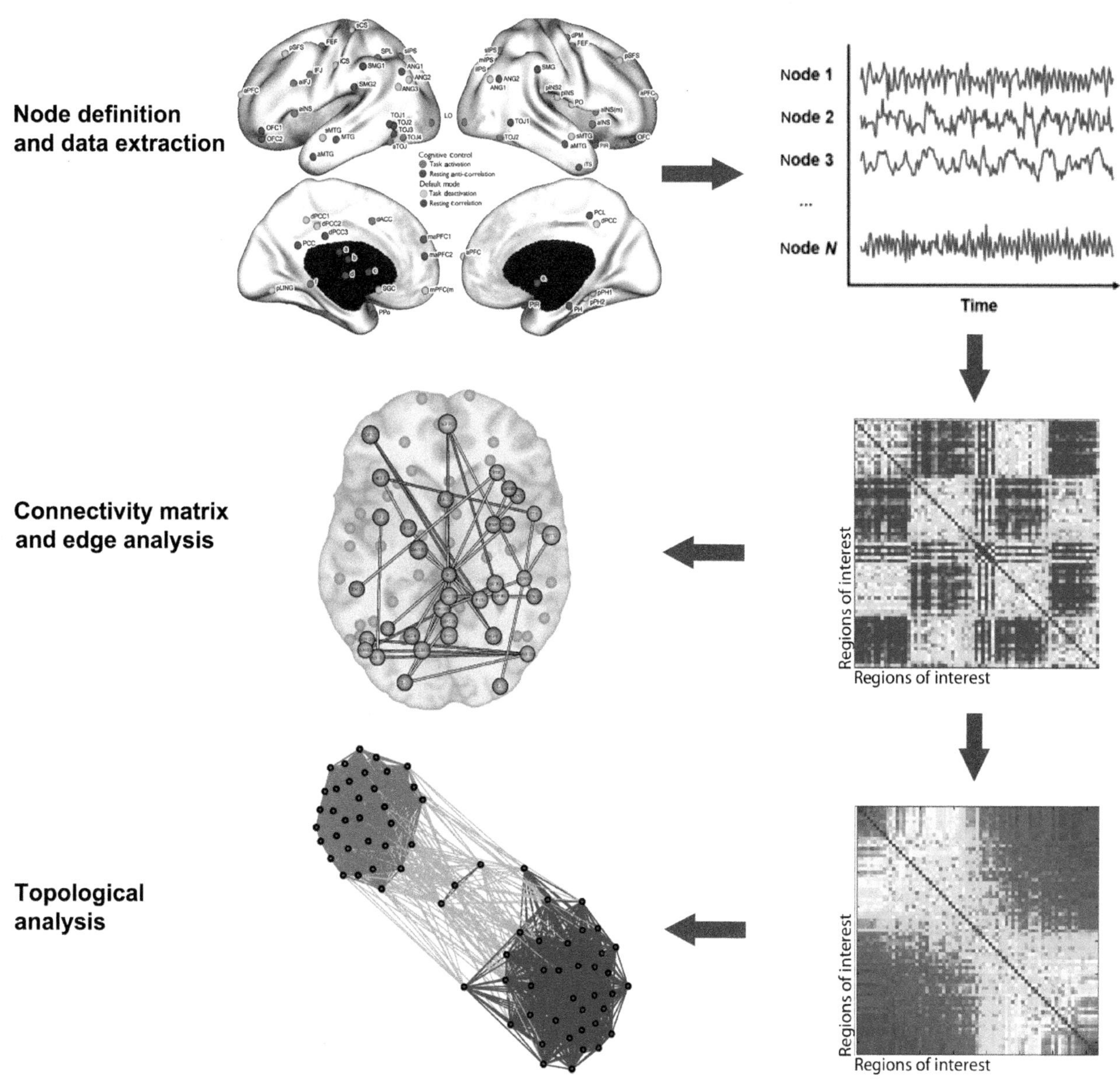

FIGURE 2 The main steps involved in graph analysis of human functional neuroimaging data. Top row: network nodes are first defined by parcellating the brain or defining regions-of-interest (left). Time-series during rest or task paradigms for each region are then extracted (right). Second row: functional connectivity is measured between every pair of time-series to generate a connectivity matrix (right; hot colors represent strong connections), and statistically significant edges are defined by applying a statistical threshold accounting for multiple comparisons or through the use of statistics that identify significant network components (left).[48] Third row: connectivity matrices can be analyzed with a variety of topological metrics, such as modularity, where the connectivity matrix is decomposed into subsets of regions that are more highly connected to one another than they are to other subsets of regions (right). Results from modularity analyses can be graphically represented using topological techniques reflecting the degree of integration or segregation of the regions (left; integrated regions are highly clustered). Figure adapted and reprinted with permission from the Society for Neuroscience.[23]

has been supported by other work showing that the degree of integration between the dorsal posterior cingulate cortex and other brain regions is influenced by the task difficulty.[29] These studies suggest that integration between specific subnetworks or regions is important in certain task contexts (i.e., they are context-dependent).

Other research has also highlighted the transience of context-dependent relationships between regions from the CCN and DMN.[49,50] For example, the anterior insula from the CCN has been proposed to act as a switch that toggles activity in the CCN and DMN when employing cognitive control to orient to salient auditory events.[50] Combined, observations such as these have informed theoretical models positing that transitory, context-dependent reconfigurations of individual connections between specific regions and subnetworks are critical for efficient cognitive control.[51] Within such models, reorganization of connections distributed across the brain is flexible and putatively occurs in response to changes in sensory input, task context, task difficulty, or learning.[52]

Developmental Network Changes

Research to-date has been largely conducted in adults, where it is assumed that functional brain networks have reached full maturity. However, in adolescence, this is not the case as the task-related activity and connectivity of brain networks continue to mature until the late twenties.[53–55] For example, studies have found that maturation is associated with progressively increased CCN activation in core brain areas of the network during cognitive control tasks,[53,55] while other research has found progressively decreased DMN activity possibly indicating a better ability to suppress distracting self-related thoughts.[56]

The organization of large-scale functional and structural brain networks also develops during adolescence. Beginning in infancy,[57] functional networks mature in a pattern that has been characterized as a progression from short-range connectivity based on anatomical proximity (e.g., between motor cortices) to the formation of long-range connections associated with specific psychological functions (e.g., language networks comprised by connections between parietal and frontal cortices).[58] The changes are hypothesized to be due to the maturation and strengthening of white matter connections in the brain through learning and environmental interactions that mold networks, such as the CCN and DMN.[53–55] In comparison with the task-related dynamic changes that occur over the course of milliseconds to seconds, dynamic developmental changes evolve over the course of months to years.

Brain Network Dynamics During Adolescence

Together with the maturation of brain activation, developmental studies indicate that the transient, context-dependent, interactions of brain regions reported in adults may be different in adolescence. However, there has been very little research to date investigating transient network dynamics during the engagement of cognitive control during this specific developmental stage. Although studies have specifically investigated PFC connections,[59] work investigating broader network changes across the CCN and DMN is still very limited.[23]

In one study of large-scale network dynamics of cognitive control in adolescence, 16-year-olds were assessed in two contexts. They first underwent a resting-state protocol to assess intrinsic network connectivity, and then performed a cognitive control task assessing the ability to overcome habitual responses and distractions (the Multi-Source Interference Task[60]). Task-related activations, task-related connectivity, and resting-state connectivity were then assessed in 73 different regions representing core areas of the CCN and DMN (identified through task and resting-state mapping). In all analyses, the behavioral significance of brain measures was directly quantified by correlating the strength of activation, deactivation, or functional connectivity with task performance (reaction-time).

The results of the study demonstrated that cognitive control behavior in adolescence was associated with brain activation during task performance predominantly in CCN regions, as well as with to task-dependent network connectivity in a distributed system that included both CCN and DMN regions (Figure 3). Better cognitive control ability was associated with greater connectivity between CCN regions, reduced connectivity between DMN regions, and greater connectivity between CCN-DMN regions. These results agreed with previous adult research by suggesting that cognitive control is facilitated by the engagement of interdependent processes related to external attention mediated by the CCN (e.g., vigilance, monitoring, spatial attention) and the disengagement of processes normally related to self-related thoughts as mediated by the DMN (e.g., autobiographical memory). However, they also agreed with other research in suggesting that strong connections between the networks were also important, perhaps to regulate the switch between the functions of either network.[61]

Importantly, correlations between task performance and functional connectivity were only found during the task and not for resting-state connectivity measures. This result suggests that performance was driven by transient, context-dependent network recruitment. To further investigate dynamic connectivity between task and rest states, modularity analyses was conducted within subsamples of participants with high and low cognitive control abilities (Figure 3). The results of this analysis showed that high performers exhibited two primary modules corresponding to the CCN and DMN across both conditions, whereas the low performing group demonstrated the presence of a third intermediary module consisting of regions that are not considered to be core components of either network[20]). In the high performing group, the regions comprising this intermediary module were switched between modules from the rest to the task states. These results demonstrate the behavioral importance of large-scale, dynamic network reorganization to cognitive control during adolescence.

Collectively, these results suggest that adolescent cognitive control is not merely a property of PFC function, or of the CCN as a whole. Instead, they support previous

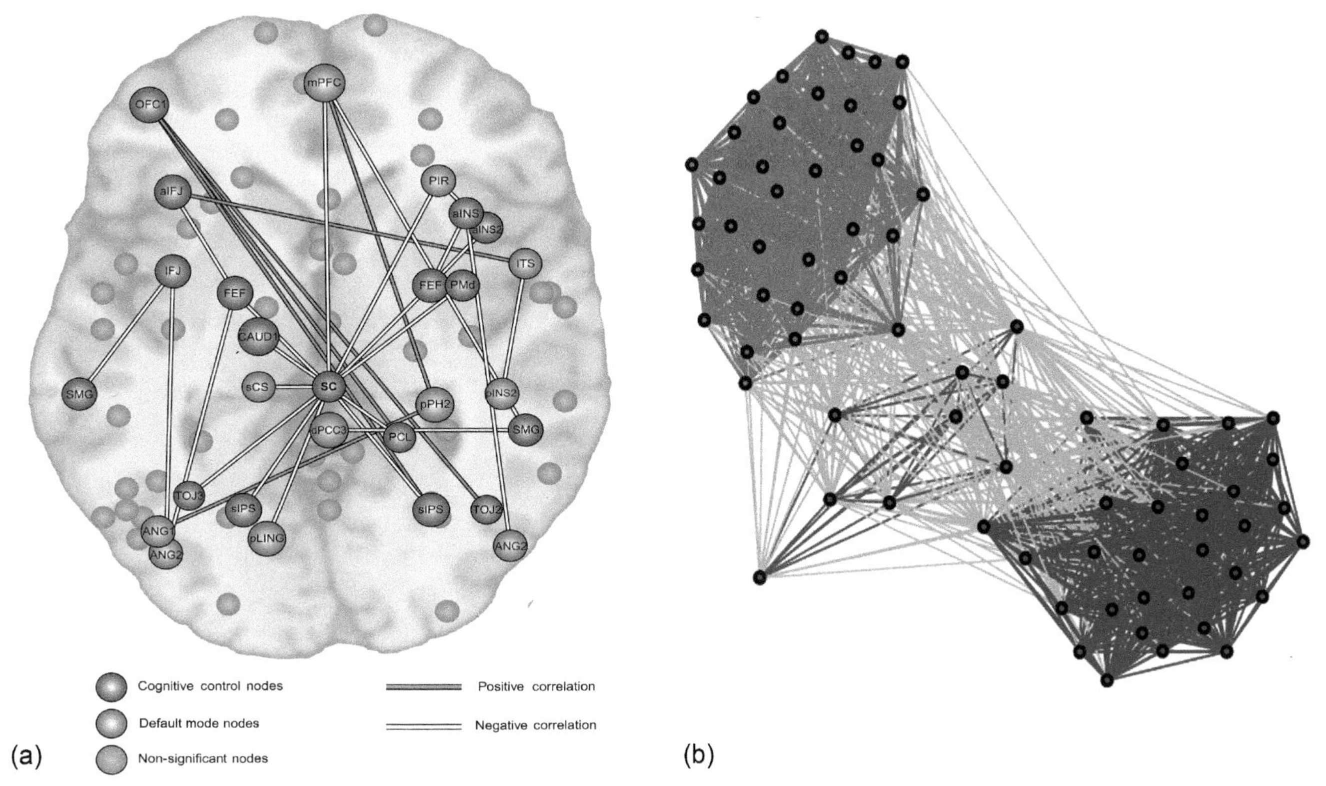

FIGURE 3 Adolescent brain network engaged during the performance of a cognitive control task. (a) Increased connectivity within the cognitive control network (CCN) and decreased connectivity within the default-mode network (DMN) was associated with better cognitive control—putatively reflecting focussed engagement to external tasks and suppression of self-referential processes. Further, increased connectivity between specific nodes of the CCN and DMN was also associated with better performance, possibly reflecting internetwork coordination or modulation. (b) Modularity analysis demonstrated the presence of two main modules reflecting the CCN (red) and DMN (blue) divisions. The efficient reallocation of regions between modules from rest to task states was found in participants who exhibited higher cognitive control, whereas the presence of a third intermediary module (purple) was associated with low performing individuals. Figure adapted and reproduced with permission from the Society for Neuroscience.[23]

findings in adults by showing that adolescent cognitive control involves a transient, context-dependent balance between functional segregation and integration of specific CCN and DMN elements.[49,51] Overall, this study points to the complex neural basis of cognitive control abilities in adolescents and the large-scale reconfigurations of network architecture required to support optimal task performance. In agreement with previous work,[54] these novel findings suggest that cognitive control deficits in adolescence are unlikely to be simply the result of immature PFC structural development. Rather, they are more likely to be associated with the differences in whole-brain network dynamics.

FUTURE DIRECTIONS AND CONCLUDING REMARKS

This chapter has discussed recent neuroimaging studies that suggest transient, context-dependent network changes of brain activation and connectivity underlies adolescent cognitive control. However, much of this research has been linked to comparisons of brain activity or connectivity across different task conditions (e.g., rest and task states). Although instructive, these comparisons represent a small fraction of the neural dynamics that are unfolding during the course of a task across different temporal scales (e.g., changes over milliseconds, seconds, or minutes) and do not address the dynamics that occur over the course of development (i.e., months and years). Thus, further research may be directed toward investigating large-scale network connectivity during cognitive control tasks across multiple temporal scales with the aim of better understanding the neural basis of cognitive control during adolescence.

Another limitation of the research discussed in this chapter is that most of the network dynamics have been studied using laboratory tasks that might not reflect real-world circumstances where cognitive control is influenced by both individual differences associated with state and trait factors.[62] For example, a state related factor is the reduction in cognitive control found with acute stress,[63] while personality traits such as impulsivity are also associated with differences in large-scale brain network organization.[64] Thus, further research may be directed toward uncovering generalizable principles of brain network dynamics associated with both state and trait variability in cognitive control and self-control more broadly.

By understanding the temporal (i.e., seconds, minutes, months, and years), as well as psychological state and trait principles of adolescent cognitive control, it is expected that further insight will be gained into psychiatric disorders that emerge during this time. For example, substance disorders that arise during adolescence are clearly linked to reduced ability to control behavior, while other conditions such as schizophrenia and depression are often associated with cognitive control deficits as well.[1] Within this context, further investigations of disruptions to network maturation may be particularly important to identify adolescents who may be at risk of psychiatric disorders with the aim to rehabilitate or enhance network connectivity in order to prevent illness onset.[65] Additionally, given the association of self-control with good functional outcomes in life more generally (e.g., academic and vocational success),[2] understanding how to enhance the functioning of CCNs would also be beneficial to all adolescents and not just those who may be at risk of mental illness.

In conclusion, adolescent cognitive control has been proposed to be mediated by interactions encompassing multiple brain regions, including the PFC. In addition to providing the opportunity to discover fundamental principles of brain and behavior, further advances will also help to identify, understand, and assist adolescents who are at risk of physical or psychological harm associated with low levels of self-control.

References

1. Paus T, Keshavan M, Giedd JN. Why do many psychiatric disorders emerge during adolescence? *Nat Rev Neurosci*. 2008;9(12):947–957.
2. Moffitt TE, Arseneault L, Belsky D, et al. A gradient of childhood self-control predicts health, wealth, and public safety. *Proc Natl Acad Sci U S A*. 2011;108(7):2693–2698.
3. Harlow JM. Recovery from the passage of an iron bar through the head. *Publ Mass Med Soc*. 1868;2:327–347.
4. Benton AL. The prefrontal region: its early history. In: *Frontal Lobe Function and Dysfunction*. USA: Oxford University Press; 1991: 3–32.
5. Luria AR. *Higher Cortical Functions in Man*. 1st ed. Oxford: Oxford University Press; 1966.
6. Stuss DT, Benson DF. *The Frontal Lobes*. 1st ed. MI: Raven Press; 1986.
7. Stroop JR. Studies of interference in verbal reactions. *J Exp Psychol*. 1935;18(6):3–662.
8. Aron AR, Robbi TW. Inhibition and the right inferior frontal cortex. *Trends Cogn Sci (Regul Ed)*. 2004;8(4):170–177.
9. Baddeley A. Working memory: theories, models, and controversies. *Annu Rev Psychol*. 2012;63(5044):1–29.
10. Pardo JV, Pardo PJ, Janer KW, Raichle ME. The anterior cingulate cortex mediates processing selection in the Stroop attentional conflict paradigm. *Proc Natl Acad Sci U S A*. 1990;87(1):256.
11. Botvinick M, Braver T, Barch D, Carter C, Cohen J. Conflict monitoring and cognitive control. *Psychol Rev*. 2001;108(3):624.
12. Carter CS, Braver TS, Barch DM, et al. Anterior cingulate cortex, error detection, and the online monitoring of performance. *Science*. 1998;280(5364):747–749.
13. Miller E, Cohen J. An integrative theory of prefrontal cortex function. *Neuroscience*. 2001;24(1):167.
14. Anderson VA, Anderson P, Northam E, et al. Development of executive functions through late childhood and adolescence in an Australian sample. *Dev Neuropsychol*. 2001;20(1):385–406.
15. Rice D, Barone S. Critical periods of vulnerability for the developing nervous system: evidence from humans and animal models. *Environ Health Perspect*. 2000;108(suppl 3):511–533.
16. Giedd J, Blumenthal J, Jeffries N, et al. Brain development during childhood and adolescence: a longitudinal MRI study. *Nat Neurosci*. 1999;2:861–862.
17. Casey B, Trainor RJ, Orendi JL, et al. A developmental functional MRI study of prefrontal activation during performance of a go-no-go task. *J Cogn Neurosci*. 1997;9(6):835–847.
18. Cole M, Schneider W. The cognitive control network: integrated cortical regions with dissociable functions. *NeuroImage*. 2007;37(1):343–360.
19. Dosenbach NU, Visscher KM, Palmer ED, et al. A core system for the implementation of task sets. *Neuron*. 2006;50(5):799–812.
20. Houde O, Sandrine R, Lubin A, Joliot M. Mapping numerical processing, reading, and executive functions in the developing brain: an fMRI meta-analysis of 52 studies including 842 children. *Dev Sci*. 2010;13(6):876–885.
21. Nee D, Wagner TD, Jonides J. Interference resolution: insights from a meta-analysis of neuroimaging tasks. *Cogn Affect Behav Neurosci*. 2007;7(1):1–17.
22. Roberts KL, Hall DA. Examining a supramodal network for conflict processing: a systematic review and novel functional magnetic resonance imaging data for related visual and auditory stroop tasks. *J Cogn Neurosci*. 2008;20(6):1063–1078.
23. Dwyer DB, Harrison BJ, Yucel M, et al. Large-scale brain network dynamics supporting adolescent cognitive control. *J Neurosci*. 2014;34(42):14096–14107.
24. Shulman G, Fiez J, Buckner R, Raichle E. Common blood flow changes across visual tasks: II. Decreases in cerebral cortex. *J Cogn Neurosci*. 1997;95:648–663.
25. Harrison B, Pujol J, López-Solà M, et al. Consistency and functional specialization in the default mode brain network. *Proc Natl Acad Sci U S A*. 2008;105(28):9781–9786.
26. Weissman DH, Roberts KC, Visscher KM, Woldorff MG. The neural bases of momentary lapses in attention. *Nat Neurosci*. 2006;9(7):971–978.
27. Persson J, Lustig C, Nelson JK, Reuter-Lorenz PA. Age differences in deactivation: a link to cognitive control? *J Cogn Neurosci*. 2007;19(6):1021–1032.
28. Raichle M, MacLeod A, Snyder A, et al. A default mode of brain function. *Proc Natl Acad Sci U S A*. 2001;98(2):676–682.
29. Leech R, Sharp DJ. The role of the posterior cingulate cortex in cognition and disease. *Brain*. 2014;137(Pt 1):12–32.
30. Binder J, Desai R, Graves W, Conant L. Where is the semantic system? A critical review and meta-analysis of 120 functional neuroimaging studies. *Cereb Cortex*. 2009;19(12):2767–2796.
31. Gusnard DA, Akbudak E, Shulman GL, Raichle ME. Medial prefrontal cortex and self-referential mental activity: relation to a default mode of brain function. *Proc Natl Acad Sci*. 2001;98(7):4259.
32. Binder J, Frost J, Hammeke T, et al. Conceptual processing during the conscious resting state: A functional MRI study. *J Cogn Neurosci*. 1999;11(1):80–93.
33. Raichle ME. The brain's default mode network. *Annu Rev Neurosci*. 2015;38:433–447.
34. Sonuga-Barke E, Castellanos F. Spontaneous attentional fluctuations in impaired states and pathological conditions: a neurobiological hypothesis. *Neurosci Biobehav Rev*. 2007;31(7):977–986.
35. Friston KJ. Functional and effective connectivity in neuroimaging: a synthesis. *Hum Brain Mapp*. 1994;2(1–2):56–78.
36. Egner T, Hirsch J. The neural correlates and functional integration of cognitive control in a Stroop task. *NeuroImage*. 2005;24(2):539–547.

37. Harrison B, Shaw M, Yücel M, et al. Functional connectivity during Stroop task performance. *NeuroImage*. 2005;24(1):181–191.
38. Peterson BS, Skudlarski P, Gatenby JC, et al. An fMRI study of Stroop word-color interference: evidence for cingulate subregions subserving multiple distributed attentional systems. *Biol Psychiatry*. 1999;45(10):1237–1258.
39. Biswal B, Zerrin Yetkin F, Haughton VM, Hyde JS. Functional connectivity in the motor cortex of resting human brain using echo-planar MRI. *Magn Reson Med*. 1995;34(4):537–541.
40. Smith SM, Fox PT, Miller KL, et al. Correspondence of the brain's functional architecture during activation and rest. *Proc Natl Acad Sci U S A*. 2009;106(31):13040–13045.
41. Vincent JL, Patel GH, Fox MD, et al. Intrinsic functional architecture in the anaesthetized monkey brain. *Nature*. 2007;447(7140):83–86.
42. Deco G, Jirsa V, Mcintosh A. Emerging concepts for the dynamical organization of resting-state activity in the brain. *Nat Rev Neurosci*. 2011;12(1):43–56.
43. Honey C, Kötter R, Breakspear M, Sporns O. Network structure of cerebral cortex shapes functional connectivity on multiple time scales. *Proc Natl Acad Sci*. 2007;104(24):10240.
44. Fox M, Snyder A, Vincent J, et al. The human brain is intrinsically organized into dynamic, anticorrelated functional networks. *Proc Natl Acad Sci U S A*. 2005;102(27):9673.
45. Greicius M, Krasnow B, Reiss A, Menon V. Functional connectivity in the resting brain: a network analysis of the default mode hypothesis. *Proc Natl Acad Sci U S A*. 2003;100(1):253.
46. Kelly AMC, Uddin LQ, Biswal BB, Xavier Castellanos F, Milham MP. Competition between functional brain networks mediates behavioral variability. *NeuroImage*. 2008;39(1):527–537.
47. Bullmore E, Sporns O. Complex brain networks: graph theoretical analysis of structural and functional systems. *Nat Rev Neurosci*. 2009;10(3):186–198.
48. Zalesky A, Cocchi L, Fornito A, Murray M, Bullmore E. Connectivity differences in brain networks. *NeuroImage*. 2012;60(2):1055–1062.
49. Fornito A, Harrison BJ, Zalesky A, Simons JS. Competitive and cooperative dynamics of large-scale brain functional networks supporting recollection. *Proc Natl Acad Sci U S A*. 2012;109(31):12788–12793.
50. Sridharan D, Levitin DJ, Menon V. A critical role for the right fronto-insular cortex in switching between central-executive and default-mode networks. *Proc Natl Acad Sci*. 2008;105(34):12569–12574.
51. Cocchi L, Zalesky A, Fornito A, Mattingley JB. Dynamic cooperation and competition between brain systems during cognitive control. *Trends Cogn Sci (Regul Ed)*. 2013;17(10):493–501.
52. Cole MW, Reynolds JR, Power JD, et al. Multi-task connectivity reveals flexible hubs for adaptive task control. *Nat Neurosci*. 2013;16(9):1348–1355.
53. Casey B, Getz S, Galvan A. The adolescent brain. *Dev Rev*. 2008;28:62–77.
54. Fair DA, Dosenbach NUF, Church JA, et al. Development of distinct control networks through segregation and integration. *Proc Natl Acad Sci U S A*. 2007;104(33):13507.
55. Luna B, Padmanabhan A, O'Hearn K. What has fMRI told us about the development of cognitive control through adolescence? *Brain Cogn*. 2010;72(1):101–113.
56. Velanova K, Wheeler M, Luna B. Maturational changes in anterior cingulate and frontoparietal recruitment support the development of error processing and inhibitory control. *Cereb Cortex*. 2008;18(11):2505–2522.
57. Kiviniemi V, Jauhiainen J, Tervonen O, et al. Slow vasomotor fluctuation in fMRI of anesthetized child brain. *Magn Reson Med*. 2000;44(3):373–378.
58. Fair DA, Cohen AL, Power JD, et al. Functional brain networks develop from a "local to distributed" organization. *PLoS Comput Biol*. 2009;5(5):e1000381.
59. Hwang K, Velanova K, Luna B. Strengthening of top-down frontal cognitive control networks underlying the development of inhibitory control: a functional magnetic resonance imaging effective connectivity study. *J Neurosci*. 2010;30(46):15535–15545.
60. Bush G, Shin L, Holmes J, Rosen B, Vogt B. The multi-source interference task: validation study with fMRI in individual subjects. *Mol Psychiatry*. 2003;8(1):60–70.
61. Seghier ML, Friston KJ. Network discovery with large DCMs. *NeuroImage*. 2013;68:181–191.
62. Yücel M, Harrison B, Wood S, et al. State, trait and biochemical influences on human anterior cingulate function. *NeuroImage*. 2007;34(4):1766–1773.
63. Mueller SC, Maheu FS, Dozier M, et al. Early-life stress is associated with impairment in cognitive control in adolescence: An fMRI study. *Neuropsychologia*. 2010;48(10):3037–3044.
64. Davis F, Knodt A, Sporns O, et al. Impulsivity and the modular organization of resting-state neural networks. *Cereb Cortex*. 2012;23(6):1444–1452.
65. Gogtay N, Vyas NS, Testa R, Wood SJ, Pantelis C. Age of onset of schizophrenia: perspectives from structural neuroimaging studies. *Schizophr Bull*. 2011;37(3):504–513.
66. Bressler S, Menon V. Large-scale brain networks in cognition: emerging methods and principles. *Trends Cogn Sci*. 2010;14(6):277–290. http://dx.doi.org/10.1016/j.tics.2010.04.004.

CHAPTER

22

The Behavioral, Cognitive, and Neural Correlates of Deficient Biological Reactions to Acute Psychological Stress

D. Carroll[1], *A.T. Ginty*[2], *A.C. Phillips*[1]

[1]University of Birmingham, Birmingham, UK

[2]University of Pittsburgh, Pittsburgh, PA, USA

OUTLINE

Abstract

Blunted or deficient cardiovascular and cortisol reactions to acute psychological stress are associated with a range of adverse behavioral and health outcomes: depression, obesity, bulimia, substance, and nonsubstance addictions. What links these diverse outcomes is that they all reflect suboptimal functioning in the face of challenge to the fronto-limbic systems in the brain that regulate motivated behavior. Available brain imaging data support this. Deficient stress reactivity is also associated with other manifestations of impaired motivation, including lower cognitive ability and poorer performance on motivation-dependent tests of lung function. In addition, deficient stress responding is typical of stable behavioral characteristics, such as neuroticism, impulsivity, and lack of perseverance, and is common in various behavioral disorders. We amend the reactivity hypothesis to include deficient stress reactivity, as well as sketching out research priorities for the future.

INTRODUCTION

Most stressful encounters are short-lived, lasting seconds, minutes, hours at most, and what is now abundantly clear is that individuals differ markedly in their cardiovascular and cortisol reactions to such acute stress exposures. This biological variability has proved fertile territory for researchers over the years. For the most part, research has been guided by the reactivity hypothesis[1] which argues compellingly that those who consistently show exaggerated cardiovascular reaction to acute stress will be at increased risk of subsequent cardiovascular disease, particularly hypertension. There is now substantial

Stress: Concepts, Cognition, Emotion, and Behavior
http://dx.doi.org/10.1016/B978-0-12-800951-2.00022-4

evidence in favor, with population studies attesting to a link between heightened cardiovascular reactions to laboratory stress exposures and hypertension,[2,3] atherosclerosis,[4] increased left ventricular mass,[5] and increased cardiovascular disease mortality.[6] Meta-analyses and more qualitative reviews confirm the contention that exaggerated cardiovascular reactivity presages poorer cardiovascular health.[7]

KEY POINTS

- Deficient cardiovascular and cortisol reactions to acute stress are associated with a range of adverse behavioral and health outcomes, such as depression, obesity, and addiction.
- Deficient stress reactivity would appear to reflect dysregulation, during stress exposure, of brain areas involved in both and autonomic regulation motivation.
- Deficient stress reactivity is also associated with other manifestations of impaired motivation, including lower cognitive ability and poorer performance on motivation-dependent tests of lung function.
- Deficient stress responding is also typical of stable behavioral characteristics, such as neuroticism, impulsivity, and lack of perseverance, and is common in various behavioral disorders.
- The corollaries of deficient stress reactivity are strikingly similar to those associated with deficient biological responses to reward.
- The original reactivity hypothesis linking exaggerated cardiovascular reactivity with cardiovascular pathology requires radical revision to incorporate deficient stress responding and its consequences.

DEFICIENT STRESS REACTIVITY

By implication, low or blunted cardiovascular reactivity might be presumed to be benign or even protective. However, recent evidence strongly indicates that this is far from the case. What now can rightly be regarded as deficient cardiovascular, as well as cortisol, stress reactivity is implicated in a range of adverse behavioral and health outcomes. The detailed evidence has been marshaled elsewhere[8,9] and so we can be necessarily brief. Smokers and those with alcohol and other substance dependencies have been consistently found to exhibit deficient cardiovascular and cortisol reactivity. Pathological gamblers and those showing symptoms of exercise dependence are also characterized by blunted cardiovascular and cortisol stress reactions. This suggests that the link between addiction and deficient stress reactivity does not primarily reflect the chronic or acute effects of ingested toxins on autonomic function. Along with the finding that blunted stress reactivity is also typical of the adolescent offspring of alcoholics, who are not themselves alcohol users and/or abusers, the data imply that deficient stress reactivity may predate dependencies and, indeed, may be a marker of susceptibility. The observation that those with blunted stress reactions are more likely to relapse from a smoking cessation program also suggests that deficient reactivity may have considerable prognostic value. Both cross-sectional and prospective analyses affirm a link between deficient cardiovascular and/or cortisol stress reactivity and both depression and obesity. Finally, in this context, blunted cardiovascular and cortisol reactivity is also characteristic of women with symptoms of bulimia.

MOTIVATIONAL DYSFUNCTION

The association between exaggerated stress reactivity and adverse cardiovascular health outcomes is conceptually plausible and, in retrospect, even self-evident, given mechanisms such as metabolic uncoupling, auto-regulation, and shear stress. Much less intuitively obvious is the link between deficient biological stress reactivity and the diverse behavioral and health outcomes summarized above. In an attempt at reconciliation, what we and others have proposed is that these various outcomes in their different ways reflect motivational dysregulation, by which we mean impaired functioning of areas of the brain essential for motivation and behavioral regulation.[8,10] From this perspective, deficient stress reactions are markers of suboptimal functioning in key fronto-limbic brain systems in the face of stressful and challenging encounters. Tellingly, those fronto-limbic brain areas that are concerned with motivated behavior are also implicated in autonomic regulation. However, before proceeding, it is worth considering a fairly parsimonious explanation of the link between the outcomes listed above and low stress reactivity; that it reflects attenuated conscious psychological engagement/effort with the stress task by the depressed, obese, or otherwise dysfunctional participants, which in turn results in reduced biological responding. Evidence in favor comes from studies examining reactivity and stress task difficulty.[11] Essentially, within-subject cardiovascular reactivity rises with increasing task difficulty but eventually decreases when the task becomes so difficult that success is impossible; this is attributed to conscious disengagement. However, in our studies and those of others, we have adjusted for subjective stress, self-reported stress task engagement and/or objective task performance or have found no differences between high and low reactors in these variables. Indeed, in one study, we found a complete dissociation between ratings of stressfulness and cardiovascular and cortisol stress reactivity; participants high in neuroticism and low on openness showed blunted cardiovascular and cortisol reactions to a battery of stress tasks but rated themselves as significantly more stressed by the experience.[12] It would appear, then, that blunted stress reactivity largely reflects something more nuanced: an unconscious biological disengagement in the face of stressful psychological challenges.

COGNITIVE ABILITY, TESTS OF RESPIRATORY FUNCTION, AND DEFICIENT STRESS REACTIVITY

If this conceptual framework is anywhere close to correct, we would expect to observe associations between tasks that perceptibly rely on intact motivation for optimal performance and blunted stress reactions. Two quite different sorts of tasks suggested themselves: cognitive ability tasks and tests of respiratory function. In two large scale population studies we have shown that deficient cardiovascular and cortisol reactivity is related to poorer general cognitive ability performance scores as well as more rapid cognitive decline over time in older participants.[13,14] Given the imperative of predicting susceptibility to age-related cognitive decline, the size of the latter association clearly recommends further scrutiny of deficient biological stress reactivity in this context. The most commonly applied test of lung function is forced expiratory volume in one second (FEV_1). Aside from lung function per se, a major contributor to FEV_1 is intrinsic motivation.[15] In the same two studies, we found positive associations between FEV_1 and cardiovascular and cortisol reactions in middle-aged men and women; those with smaller stress reactions achieved significantly lower FEV_1 scores, even after adjusting for smoking.[16,17] It is also worth pointing out that in addition to smoking, we adjusted statistically for a range of demographic and health behavior variables including those already known to be associated with deficient biological stress reactivity. The same was true for our analyses of stress reactivity and cognitive ability. Overall, the associations between blunted stress reactivity and outcomes such as depression, obesity, addiction, and bulimia would all appear to be largely independent of one another.

DEFICIENT STRESS REACTIONS AND NEURAL HYPOACTIVATION

A number of functional magnetic resonance imaging (fMRI) studies afford evidence that many of the outcomes associated with deficient stress reactivity are associated with relatively reduced activation in the frontal and subcortical limbic regions.[18] Only a few studies have examined the neural correlates of diminished peripheral reactions to acute stress exposure. Smaller cardiovascular reactions to stress have been associated with reduced neural activity in the midanterior and posterior cingulate cortex, insula, and amygdala.[19,20] However, for the most part, these studies have been concerned with characterizing the neural reactions to stress of exaggerated cardiovascular reactors. In a recent study, we shifted the focus to unambiguously blunted cardiovascular reactors and an explicit comparison of their neural reactions to a standard stress exposure with those who were unambiguously exaggerated cardiovascular reactors.[21] Blunted and exaggerated cardiac stress reactors were selected from previous studies, on the basis of being at least two standard deviations below and above the mean heart rate reactivity for their respective cohorts. Neural activity, using fMRI, and heart rate were recorded at baseline rest and during exposure to a standard laboratory stress task, as well as to a nonstressful control task. For blunted reactors heart rate change from baseline did not differ between the stress and control tasks, whereas for exaggerated reactors, heart rate change was substantially and significantly greater during stress than control. Blunted reactors also exhibited reduced activation in the anterior midcingulate cortex and insula, as well as a greater deactivation in the amygdala and posterior cingulate. These outcomes are not substantially dissimilar from the results of earlier less focused studies. Not only are the amygdala, insula, and cingulate cortex involved in autonomic regulation, but they comprise a network vital for motivated behavioral responses and adaptation to threat and challenges.[22] Accordingly, the results of this study very much underpin our contention that deficient stress reactivity is a marker of the under-recruitment of pertinent brain systems during behavioral challenges that require motivated action.

DEFICIENT STRESS REACTIVITY AND OTHER BEHAVIORAL MANIFESTATIONS OF POOR MOTIVATIONAL REGULATION

We have already eluded to the link between deficient stress reactivity and personality characteristics, such as high levels of neuroticism and low levels of openness. However, there is emerging evidence that blunted cardiovascular reactions to stress may be typical of other stable behavioral characteristics. In a study of preadolescent children, impulsivity, assessed by two standard behavioral inhibition tasks (one where participants had to draw a circle at their natural pace and then as slowly as possible, and a Go-NoGo task where they had to inhibit prepotent responses) and maternal self-report was associated with blunted heart rate reactions to a mental arithmetic stress.[23] It is also worth pointing out that the self-perceived stressfulness of the stress task was not related to any of our measures of impulsivity. We recently replicated this result in a study of undergraduates where deficient and exaggerated heart rate reactors were selected from stress-testing a large screening sample, again using a standard mental arithmetic challenge.[24] Selected participants were subsequently exposed to inhibitory control (stop signal) and motor impulsivity (circle drawing) tasks, analogous to those used in our earlier study of school children. Deficient stress reactors showed greater impulsivity on both behavioral assessments (see Figure 1). Again, self-reported stressfulness, self-reported stress task engagement, and mental arithmetic performance did not differ between the stress reactor groups.

Perseverance is another characteristic that could conceivably be associated with the deficient stress reactivity phenotype. As indicated above, smokers who display the most blunted stress reactions are most likely to relapse in smoking cessation programs. We examined this further in a study in of nearly 176 high school students who were stress-tested, using a challenging mental arithmetic task, and cardiac reactivity recorded; a year later, when they were at university, all participants were contacted to complete a simple and undemanding online assessment.[25] Twenty-five percent of participants failed to complete the online assessment despite repeated prompts from the same researcher who had conducted the original stress testing session. Crucially, the noncompleters were characterized by blunted heart rate and cardiac output reactions to the earlier stress task exposure (see Figure 2, which illustrates the findings for heart rate reactivity). Once more, the association was independent of stress task performance. Our results add to the idea that individual differences in stress reactivity may have implications for likely completion of multisession intervention programs, as well as for selection bias in longitudinal

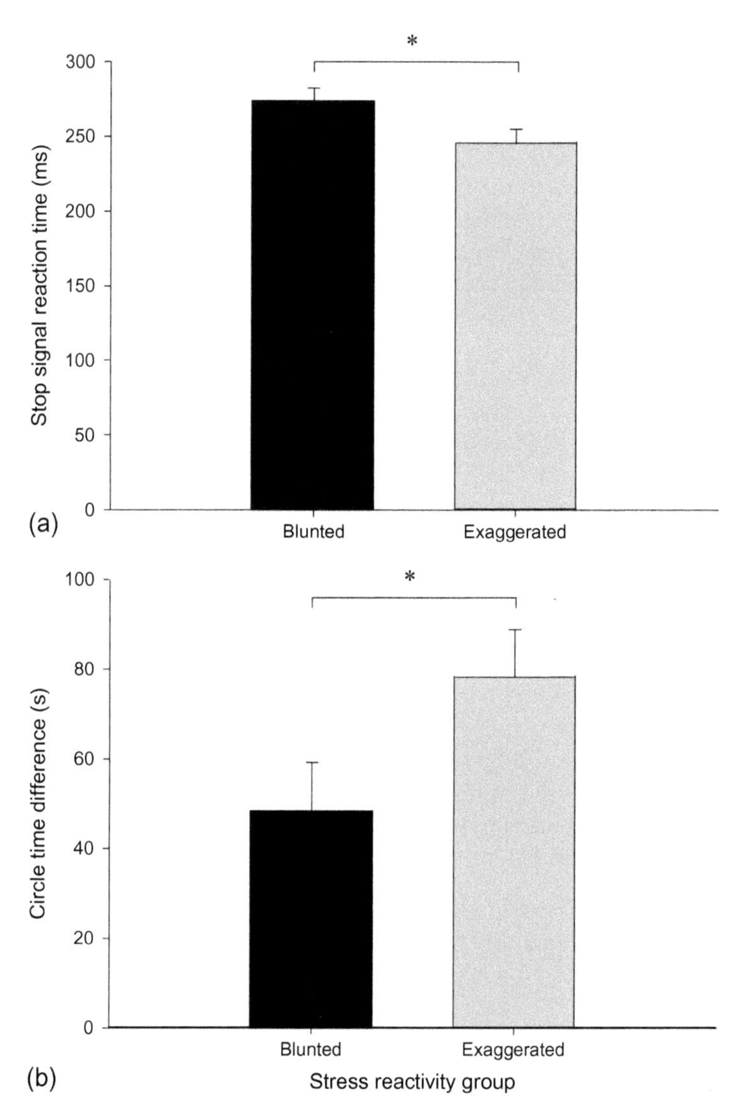

FIGURE 1 Mean (SE) (a) stop signal reaction time, and (b) circle time difference for the deficient and exaggerated stress reactivity groups, $*p \leq 0.05$. Deficient stress reactors show greater impulsivity.

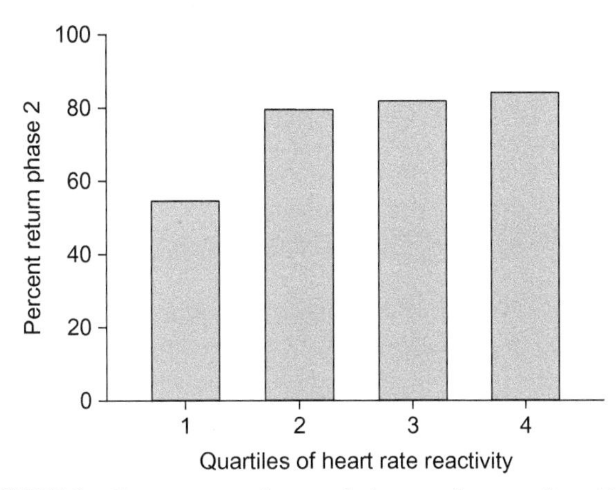

FIGURE 2 Percentage study completion rate by quartiles of heart rate reactivity. The quartile with the smallest heart rate reactions were significantly less likely to complete the study.

stress reactivity studies, as loss to follow-up would not appear to be arbitrary, but rather skewed toward those with deficient stress responses, with all the attendant adverse health and behavioral corollaries. Overall, the results of these studies of impulsivity and caprice fit neatly with our and others' contention[26] that deficient stress reactivity is a marker of motivational dysfunction and consequently poor behavioral regulation. That model also proposes an association stress reactivity and behavioral disorders and there is now accumulating evidence that blunted stress reactions typify individuals who consistently display antisocial behavior,[27] have a conduct disorder,[28] or have been diagnosed with attention deficit hyperactive disorder.[29]

THE ORIGINS OF DEFICIENT STRESS REACTIVITY

Why some individuals show deficient biological reactions to acute stress exposures remains to be determined. Nevertheless, on the basis of preliminary evidence we can usefully speculate on two broad classes of cause worthy of further scrutiny: genotype and early childhood adversity. A metaanalysis of twin studies of cardiovascular stress reactivity[30] revealed pooled heritability estimates ranging from 0.26 to 0.43, indicating a far from negligible contribution from our genes. However, as the authors plaintively observed, "... in spite of significant heritability emerging from twin studies, independently and consistently replicated gene variants that explain this heritability are still at large" (p. 66). The most likely polymorphisms are those involved in the regulation of neurotransmitters, such as noradrenalin, serotonin, monoamine oxidase, and dopamine. We have briefly reviewed their candidature elsewhere,[8] and, although promising, substantial further concerted research is required before definitive pronouncements can be made.

Although counter evidence exists, recent analyses suggest a possible link between early childhood adversity and deficient stress reactivity. Perhaps the most persuasive evidence to date comes from the Oklahoma Family Health Patterns Project.[26] The project's participants comprised a cohort of 426 healthy young adults with or without a family history of alcoholism. Regardless of family history, participants who had experienced high levels of psychological adversity before the age of 16 showed blunted heart rate and cortisol reactions to mental arithmetic and public speaking stress. They also showed poorer cognitive ability, higher behavioral impulsivity, and greater antisocial tendencies. A model is proposed whereby early adversity affects fronto-limbic function, which in turn is manifest in reduced stress reactivity, lower cognitive ability, and unstable affect regulation; this then increases the likelihood of impulsivity and risk taking, which in turn contributes to unhealthy behavior and addiction risk. While promising, the model is yet to encompass all the corollaries of deficiencies, such as depression, which again has been strongly linked to childhood adversity.

THE REACTIVITY HYPOTHESIS REVISITED: INVERTED-U OR ORTHOGONAL PROCESSES

The reactivity hypothesis has shown remarkable resilience and durability. As indicated, its primary postulate is that exaggerated cardiovascular reactions to acute psychological stress will contribute to the subsequent development of hypertension and, by implication, other manifestations of end-point cardiovascular disease. Few fairly circumspect hypotheses in the biobehavioral sciences can have generated such a volume of empirical and conceptual interest over the last 40 years. However, it is now abundantly clear that the reactivity hypothesis requires radical revision. Although, it is now beyond contention that exaggerated cardiovascular reactions to stress predict future cardiovascular pathology, it is now equally apparent that blunted or deficient cardiovascular, and cortisol, reactions to stress are association with a range of adverse health and behavioral outcomes, from depression and obesity to addiction and behavioral disorders to poorer cognitive ability. How might we reconceptualize the reactivity hypothesis? The data conform to an inverted-U model where high and low reactivity can be considered maladaptive depending on the outcome in question. The inverted-U has a substantial pedigree in psychology. Originally characterized as the Yerkes-Dodson law,[31] the inverted-U proposed to describe the relationship between motivation and performance on one hand and physiological arousal on the other, such that there is an optimal midpoint in the arousal continuum where performance is best served. Accordingly, by conceiving of stress reactivity in terms a unitary system that for some operates in a biased state, either over-responding or under-responding, we may simply be putting a few new clothes on a much loved but rather old doll. However, a model which conceives of continuous positive associations between stress reactivity and some outcomes and continuous negative associations between reactivity and other outcomes can also fit the results. Indeed, increasingly this model has much to commend it. For the most part, the negative associations between stress reactivity and adverse health and behavioral outcomes involve deficient cardiac and cortisol reactions. There are exceptions, but very few consistent negative associations emerge for blood pressure reactivity, particularly for diastolic blood pressure reactivity. In contrast, the positive relationships between cardiovascular stress reactivity and cardiovascular pathology are largely confined to associations involving exaggerated blood pressure responding. We might speculate that the former involve impaired parasympathetic withdrawal and/or insufficient β-adrenergic drive,[32] whereas the latter are more vascular in origin, relying on enhanced α-adrenergic activation. We depict these two models in Figure 3; time will tell which best serves revision of the reactivity hypothesis.

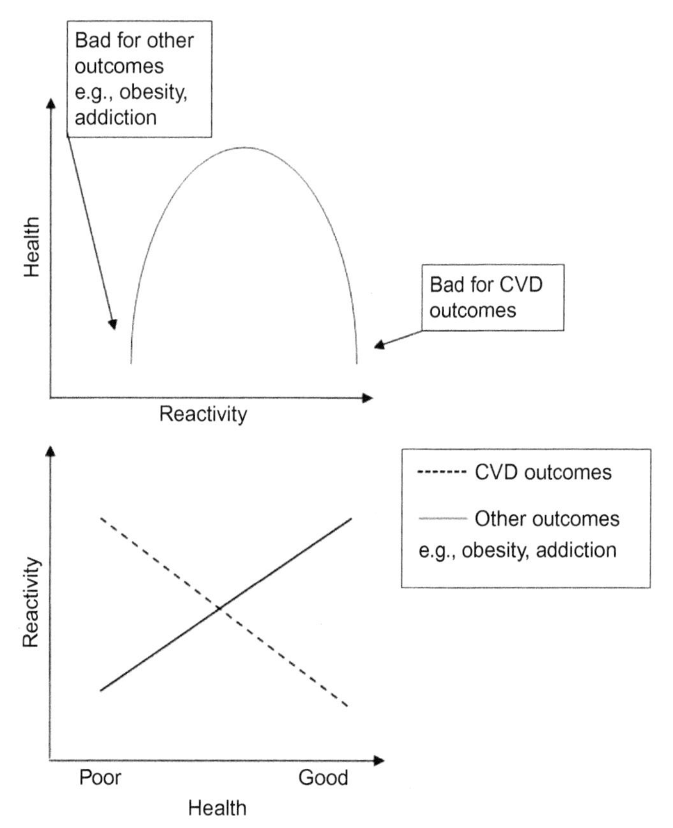

FIGURE 3 Alternative models, inverted-U and distinct mechanisms, linking stress reactivity to health and behavioral outcomes.

BIOLOGICAL DISENGAGEMENT: A WIDER PHENOMENON

Research on the antecedents and consequences of deficient biological reactions to acute psychological stress is still in its infancy, but is gathering momentum. On the other hand, research into deficient biological responses to reward has a much longer pedigree. What is initially striking is that individuals differ markedly in their peripheral biological and central neural responses to reward, with some showing blunted responding. Further, what really catches the eye is that the behavioral and health correlates of deficient reward responding are remarkably similar to those associated with deficient stress responding: depression, obesity, bulimia, addiction, including gambling addiction, attention deficit hyperactive disorder, and antisocial behavior. Further, the neural mechanisms proposed to underlie deficient stress reactivity and what has come to be called reward deficiency syndrome (RDS)[33] are noticeably similar; as with our notion of central motivational dysregulation, RDS would seem to be a manifestation of hypoactivation to reward stimuli in the same fronto-limbic network. In addition, RDS has been attributed to a dopamine deficiency, specifically at the D2 receptor. The dopamine system has also been attributed a key role in regulating autonomic activity, as well as in motivated behavior and stimulus processing.[10] A much fuller account of the parallels and their implications can be found elsewhere[34]; the present account is necessarily brief. However, the major implication is that deficient stress reactivity and RDS may be two sides of the same coin and both may reflect a more general failure in biological responding to the active challenges in life, reflecting suboptimal functioning in motivational and emotional processing systems.

CONCLUDING REMARKS AND FUTURE DIRECTIONS

It is broadly accepted that exaggerated cardiovascular reactions to acute psychological stress are implicated in cardiovascular pathology. What has proved more contentious until late is the notion that low or deficient cardiovascular and cortisol stress responding might also be associated with adverse behavioral and health outcomes. However, the strength and consistency of the empirical findings is now compelling, and deficient biological stress reactivity has been linked to depression, obesity, substance and nonsubstance addiction, bulimia, impulsivity and behavioral disorders, and poorer cognitive ability. Our thesis is that what links these diverse outcomes is that they all reflect, either directly or indirectly, suboptimal functioning in those fronto-limbic systems in the brain that are normally engaged during behavioral challenges, such as acute stress, that required motivation and adaptation. Those same systems are also involved in autonomic regulation, suggesting that deficient peripheral biological stress reactivity is a readily accessible marker of such central neural dysfunction. The available fMRI research supports this contention, but more research that simultaneously examines peripheral autonomic and central neural reactions to acute stress is clearly warranted. What we also require is a model fitting together the various pieces of what is becoming quite a complex jigsaw. The antecedents of central motivational dysregulation are likely to be found in genes and in childhood exposures. Although twin studies strongly implicate heredity, the precise polymorphisms involved have yet to be determined. We would argue that genes involved in neurotransmission, particularly dopamine transmission, would be a good starting point. Although preliminary evidence is emerging, there needs to be much more focus on the association between deficient stress responding and early life adversity, particularly emotional neglect and abuse. Downstream from central neural dysfunction are stable behavioral and emotional characteristics such as neuroticism, impulsivity, lack of perseverance, depression, low cognitive ability. These in turn increase the likelihood of addiction and other unhealthy outcomes such as obesity and bulimia. This is the barest outline of a model based on the current data and is illustrated in Figure 4; however, much more has to be done by way of confirmation and refinement. The parallels between deficient stress reactivity and the RDS also need further investigation; it is an essential next step to establish whether who underreact to stress also underreact to reward both peripherally and centrally. Given the prognostic importance of cognitive decline for states of dementia, further attention needs to be paid to the association between blunted stress reactivity and age-related deterioration in cognitive ability. Finally, almost all of the corollaries of deficient stress responding are targets for intervention; we would argue that stress testing and the determination of deficient reactors could be informative in identifying participants likely to require more aggressive treatment.

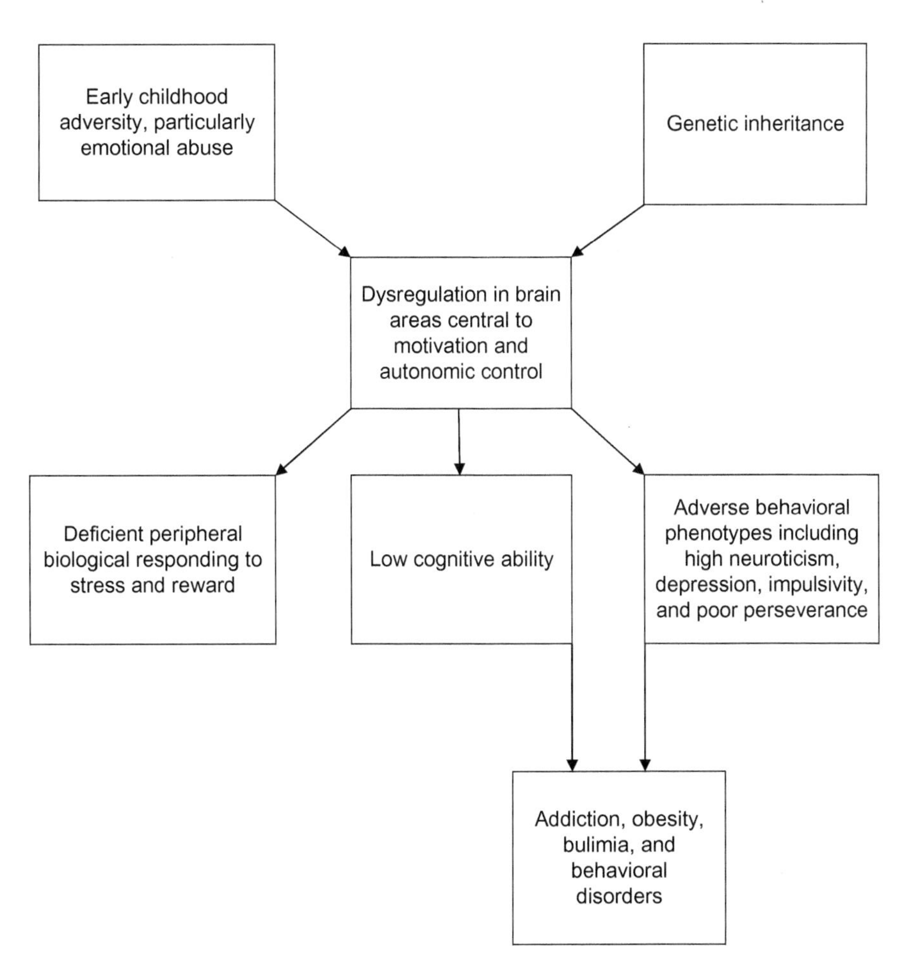

FIGURE 4 Provisional model of deficient biological stress reactivity; from this perspective, deficient reactivity is regarded as a marker rather than a cause of endpoint behavioral and health outcomes.

References

1. Obrist PA. The cardiovascular-behavioral interaction—as it appears today. *Psychophysiology*. 1976;13:95–107.
2. Carroll D, Phillips AC, Der G, Hunt K, Benzeval M. Blood pressure reactions to acute mental stress and future blood pressure status: data from the 12-year follow-up of the West of Scotland Study. *Psychosom Med*. 2011;73:737–742.
3. Carroll D, Ring C, Hunt K, Ford G, Macintyre S. Blood pressure reactions to stress and the prediction of future blood pressure: effects of sex, age, and socioeconomic position. *Psychosom Med*. 2003;65:1058–1064.
4. Barnett PA, Spence JD, Manuck SB, Jennings JR. Psychological stress and the progression of carotid artery disease. *J Hypertens*. 1997;15:49–55.
5. Kapuku GK, Treiber FA, Davis HC, Harshfield GA, Cook BB, Mensah GA. Hemodynamic function at rest, during acute stress, and in the field: predictors of cardiac structure and function 2 years later in youth. *Hypertension*. 1999;34:1026–1031.
6. Carroll D, Ginty AT, Der G, Hunt K, Benzeval M, Phillips AC. Increased blood pressure reactions to acute mental stress are associated with 16-year cardiovascular disease mortality. *Psychophysiology*. 2012;49:1444–1448.
7. Chida Y, Steptoe A. Greater cardiovascular responses to laboratory mental stress are associated with poor subsequent cardiovascular risk status: a meta-analysis of prospective evidence. *Hypertension*. 2010;55:1026–1032.
8. Carroll D, Phillips AC, Lovallo WR. Are large physiological reactions to acute psychological stress always bad for health? *Soc Personal Psychol Compass*. 2009;3:725–743.
9. Phillips AC, Ginty AT, Hughes BM. The other side of the coin: blunted cardiovascular and cortisol reactivity are associated with negative health outcomes. *Int J Psychophysiol*. 2013;90:1–7.
10. Lovallo WR. Do low levels of stress reactivity signal poor states of health? *Biol Psychol*. 2011;86:121–128.
11. Richter M, Friedrich A, Gendolla GH. Task difficulty effects on cardiac activity. *Psychophysiology*. 2008;45:869–875.
12. Bibbey A, Carroll D, Roseboom TJ, Phillips AC, de Rooij SR. Personality and physiological reactions to acute psychological stress. *Int J Psychophysiol*. 2013;90:28–36.
13. Ginty AT, Phillips AC, Der G, Deary IJ, Carroll D. Cognitive ability and simple reaction time predict cardiac reactivity in the West of Scotland Twenty-07 study. *Psychophysiology*. 2011;48:1022–1027.
14. Ginty AT, Phillips AC, Roseboom TJ, Carroll D, de Rooij SR. Cardiovascular and cortisol reactions to acute psychological stress and cognitive ability in the Dutch Famine Birth Cohort study. *Psychophysiology*. 2012;49:391–400.
15. Crim C, Celli B, Edwards LD, et al. Respiratory system impedance with impulse oscillometry in healthy and COPD subjects: ECLIPSE baseline results. *Respir Med*. 2011;105:1069–1078.
16. Carroll D, Bibbey A, Roseboom TJ, Phillips AC, Ginty AT, de Rooij SR. Forced expiratory volume is associated with cardiovascular and cortisol reactions to acute psychological stress. *Psychophysiology*. 2012;49:866–872.
17. Carroll D, Phillips AC, Der G, et al. Low forced expiratory volume is associated with blunted cardiac reactions to acute psychological stress in a community sample of middle-aged men and women. *Int J Psychophysiol*. 2013;90:17–20.
18. Stice E, Spoor S, Bohon C, Veldhuizen MG, Small DM. Relation of reward from food intake and anticipated food intake to obesity: a functional magnetic resonance imaging study. *J Abnorm Psychol*. 2008;117:924–935.
19. Gianaros PJ, May JC, Siegle GJ, Jennings JR. Is there a functional neural correlate of individual differences in cardiovascular reactivity? *Psychosom Med*. 2005;67:31–39.
20. Gianaros PJ, Sheu LK, Matthews KA, Jennings JR, Manuck SB, Hariri AR. Individual differences in stressor-evoked blood pressure reactivity vary with activation, volume, and functional connectivity of the amygdala. *J Neurosci*. 2008;28:990–999.
21. Ginty AT, Gianaros PJ, Derbyshire SWG, Phillips AC, Carroll D. Blunted cardiac stress reactivity relates to neural hypoactivation. *Psychophysiology*. 2013;50:219–229.
22. Gianaros PJ, Onyewuenyi IC, Sheu LK, Christie IC, Critchley HD. Brain systems for baroreflex suppression during stress in humans. *Hum Brain Mapp*. 2012;33:1700–1716.
23. Bennett C, Blissett J, Carroll D, Ginty AT. Rated and measured impulsivity in children is associated with diminished cardiac reactions to acute psychological stress. *Biol Psychol*. 2014;102:68–72.
24. Bibbey A, Ginty A, Brindle RC, Phillips AC, Carroll D. Blunted cardiac stress reactors exhibit relatively high levels of behavioral impulsivity. *Physiol Behav*. Under review.
25. Ginty A, Brindle RC, Carroll D. Cardiac stress reactions and perseverance: diminished reactivity is associated with study non-completion. *Biol Psychol*. 2015;109:200–205.
26. Lovallo WR. Early life adversity reduces stress reactivity and enhances impulsive behavior: implications for health behaviors. *Int J Psychophysiol*. 2013;90:8–16.
27. De Vries-Bouw M, Popma A, Vermeiren R, Doreleijers TA, Van De Ven PM, Jansen LM. The predictive value of low heart rate and heart rate variability during stress for reoffending in delinquent male adolescents. *Psychophysiology*. 2011;48:1597–1604.
28. Fairchild G, van Goozen SH, Stollery SJ, et al. Cortisol diurnal rhythm and stress reactivity in male adolescents with early-onset or adolescence-onset conduct disorder. *Biol Psychiatry*. 2008;64:599–606.
29. Pesonen AK, Kajantie E, Jones A, et al. Symptoms of attention deficit hyperactivity disorder in children are associated with cortisol responses to psychosocial stress but not with daily cortisol levels. *J Psychiatr Res*. 2011;45:1471–1476.
30. Wu T, Snieder H, de Geus E. Genetic influences on cardiovascular stress reactivity. *Neurosci Biobehav Rev*. 2010;35:58–68.
31. Yerkes RM, Dodson JD. The relation of strength of stimulus to rapidity of habit-formation. *J Comp Neurol Psychol*. 1908;18:459–482.
32. Brindle RC, Ginty AT, Phillips AC, Carroll D. A tale of two mechanisms: a meta-analytic approach toward understanding the autonomic basis of cardiovascular reactivity to acute psychological stress. *Psychophysiology*. 2014;51:964–976.
33. Blum K, Braverman ER, Holder JM, et al. Reward deficiency syndrome: a biogenetic model for the diagnosis and treatment of impulsive, addictive, and compulsive behaviors. *J Psychoactive Drugs*. 2000;32(suppl i–iv):1–112.
34. Ginty AT. Blunted responses to stress and reward: reflections on biological disengagement? *Int J Psychophysiol*. 2013;90:90–94.

CHAPTER

23

When the Work Is Not Enough: The Sinister Stress of Boredom

A. Weinberg
University of Salford, Salford, Lancashire, UK

Abstract

Boredom as a state and understimulation as a stressor are examined in the context of working, with consideration also given to the potential role played by these factors in evolutionary terms. It is suggested that the waste of mental capital engendered by understimulation inside and outside of work can have wider negative consequences—a high proportion of employees already report experiences of boredom. Links are drawn between work design and optimal levels of variety and arousal, both of which have implications for psychological functioning and mental health. These relations are located within discussion of the influential roles of attention, personality characteristics—including sensation-seeking—and brain-related factors, highlighting the interplay between internal and external variables which lead to the experience of boredom. Reference is made to a range of individual coping strategies—some deliberate and some instinctive—as well as much-needed social approaches to these widespread phenomena.

"Are we there yet?" is a familiar prompt spoken in hope by a traveller trying to be patient. In all likelihood he or she is trying to stave off feelings of boredom invoked by a seeming eternity sitting in the car while time passes. The journey may not be any easier for the child either! That period of waiting for something and being aware of this is characteristic of feeling bored. From this boredom is defined as "the aversive experience of wanting, but being unable to engage in satisfying activity."[1] Enforced unemployment is therefore a daunting prospect, carrying with it an uncertain endpoint and in the meantime has the potential to sap positive energy from the would-be worker. This is compounded as a protracted span of wasted opportunity is often shaped by accompanying financial constraints. Indeed, working itself does not guarantee a qualitatively stimulating experience, as many jobs involve monotonous tasks through repetition or underuse of skills which can occupy significant proportions of our waking time. Experiences of ill health and recovery feature a waiting game, shaped by a lack of physical or psychological resources to relieve periods of inactivity. Depending on personal resources, choosing to retire or remain at work is influenced by a desire to do something stimulating with our time, but how easy is it to maintain? For that matter, how desirable is it that we avoid a state of boredom (Figure 1)?

Underpinning these examples is an assumption that life should hold something more interesting. However this may be motivated by a more than an existentialist imperative. Our survival as a species depends on the ability to perceive what stands out in our vicinity, so our visual processes accordingly analyze salient aspects of the physical world.[2] Considered from this perspective, the experience of boredom or understimulation means that our abilities and opportunities to identify and therefore classify new stimuli as a threat or asset to survival are enhanced. In evolutionary terms, it can be argued that boredom is good for us! So have we become complacent as perceiving potential threats remains a vital survival mechanism? Maslow's hierarchy of needs highlights how our expectations of existence are raised in conjunction with the fulfillment of more basic functioning. Nevertheless, consideration of boredom as undesirable may be a luxury afforded by having more time to focus on self-actualizing activities. How much of this is "spare" capacity which can be redirected to survival mode as necessary varies with circumstances. Our ability to build lives dependent on this "spare" resource is particularly evident in the world of work. This chapter explores how boredom is promoted by external factors such as the absence of meaningful activity and also by internal states linked to the focus of our attention. Each reflects a relationship with psychological states, including stress.

Fundamentally, work serves motivations which are variously owned by the job-holder, their employer, and wider society. The notion that we add value to society

Stress: Concepts, Cognition, Emotion, and Behavior
http://dx.doi.org/10.1016/B978-0-12-800951-2.00023-6

FIGURE 1 Are we there yet?

if we are contributing via the use of labor and skills relates clearly to survival of communities and ultimately the species. This in turn can feed back into feelings of self-worth: hence the significance of the consideration that we are doing something useful with our time. However, 43% of employees in the European Union rate their work as monotonous[3] which suggests that there is a lot of time which could be better spent maximizing the use of mental capital. Up to 87% of employees report boredom at work at least sometimes, which in turn promotes a host of negative consequences for the individual and the organization, including low job satisfaction and poorer performance.[4] Furthermore there are clear dangers in work environments characterized by high levels of monotony, as evidenced by error-proneness in the nuclear military personnel[5] and airline pilots.[6] The understimulation inherent in monotonous work means many spend time doing something which may not enhance a sense of value—either perceived or real. Taken to a logical conclusion, this means the chances of creating or discovering something inherently useful are clearly reduced. In other words, poorly designed work takes away from, rather than adds to, our survival potential. In addition to monotonous work there are job situations which reduce opportunities to utilize or develop individual skills. These can be manual, clerical, or professional, as the lack of variety can impact negatively on our inbuilt need to find something new or stimulating on which to focus. One of the publicized benefits of the development of technology in the workplace has been its potential to relieve humans of boring tasks. A less than flattering appraisal of this aim suggests it has been less than forthcoming and created more monotony as we spend more time monitoring technology instead, and some innovations have resulted in the removal of livelihoods. Both of these scenarios are linked with increased prevalence of poorer psychological health.

KEY POINTS

- Boredom is an undesirable state determined by freedom, motivation, and capacity to engage in a rewarding task.
- We are designed to detect new stimuli in our environment, but there is likely to be an optimum level which is naturally influenced by individual differences.
- High levels of boredom at work are linked with negative outcomes for employees and organizations.
- Wasted human capital threatens to impair health as well as progress.
- There are internal biological and psychological motivations which appear to engage spare attention and psychological energy, for better or worse.
- Gaining insight into our mental functioning—including how well we attend and our emotional state—can help individuals gain better control of the psychological resources available to us.
- Individual coping strategies, combined with workplace and job design that are informed by psychological considerations, can help to avert negative boredom states.

Some may take a hard-nosed species level evolutionary perspective and suggest the need to find new opportunities—in existing or new jobs—has actually led to the kinds of invention and creativity essential to our continued survival. However such advances are far from the norm in the world of work as the figures show. Occupational psychology theories pinpoint understimulation as only one stressor for the individual[38] and a century of research has demonstrated the contributory roles of factors such as boredom in the work environment,[7] the individual[8] and the interaction between them.[9] However if we remind ourselves that earning a living is based on the premise that we are safe in our environments and that our capacity for paid work is largely additional to the attention we therefore need to devote to perceiving threats from our surroundings, it is possible that the prevalence of stress-related disorders across workplaces is doubly damaging to our long-term survival prospects. Firstly, our mental resources to cope with threats to well-being are likely to be finite and secondly, if we expend these in work environments designed without great consideration to psychosocial factors such as variety and growth, we are potentially trapped into acting counter to our survival needs. In other words, boredom at work disables effective functioning and taken to an evolutionary extreme threatens successful adaptation.

Naturally the societal need for financial security suggests that paid work, including that which is less than stimulating, is important for survival and we would rather be employed than not. However history has shown that workplaces are not fundamentally infused with a desire to maximize the psychosocial benefits of job design (i.e., how work impacts on psychological health has not been a focus driven by financial considerations). That is perhaps, until now.

It is entirely possible that we have reached a "tipping point." The concept of stress sits squarely at the fulcrum of the debate. In a posteconomic collapse scenario, the financial imperative for organizations to consider costs is even more salient. Not surprisingly these include sickness absence as a warning sign about organisational performance and the leading reasons for it are stress-related and musculoskeletal.[10] Coincident with these, national policy directives, such as those in the United Kingdom, have begun to emphasize not simply the importance of work,[11] but also of well-designed work which can actively promote positive psychological health through flexible working and more considered management styles[12,13]; in particular this means maximizing the benefits of work by recognizing the mental capital of the workforce.[14] Boredom at work clearly runs counter to this favored approach. The idea that human resources are the key to the survival of organizations is nothing new, however the manner in which people at work are treated and the work environments to which they are exposed are designed, has assumed a new significance. This highlights the need to avoid wasting human potential in the workplace and to identify a way forward based on supracapitalist rather than subcapitalist considerations. Indeed the notion of humans as "resources" hints at where things may have gone badly wrong, as arguably it implies a relegation of considering employees' psychological needs.[15] This challenged evolutionary intuition, so how could things be better (Figure 2)?

FIGURE 2 Is this a stimulating work environment? It worked for Rembrandt.

Work which is designed to engage is recognized far more in higher performing organizations[16] and better employee health outcomes seem to exist in workplaces designed to enhance variety, improve control over work tasks, and promote positive relationships.[17] It is important to acknowledge the link between these factors, for it is unlikely that adjusting one psychosocial variable at work can or will happen in isolation from its impact on others. A range of models has been designed in the field of occupational psychology to capture the essence of such variables and to describe their impact on workers. Warr's Vitamin Model[18] is one such approach which posits a curvilinear relationship between aspects of the work environment—including variety (absence of boredom)—and psychological health. There has been mixed empirical support for this intuitively attractive relationship, which otherwise appears to meet our expectations of the need for an optimum level of certain job characteristics. Equally, one might anticipate that a lack of variety—which we would recognize as boredom—would have a similar relationship with well-being, but perhaps the absence of consistent data in support of Warr's broad model to date is that individuals experience their optimal exposures to aspects of the environment at differing levels (even if some features appear to others as positives or negatives). In order to understand boredom, it is therefore important to consider internal as well as external processes (i.e., the "why" and the "how").

So how do we experience boredom? It is logical to assume that exposure to stimuli in the external environment, contributes proportionally to a sense of feeling understimulated. From a biological perspective, identifying novelty is part of a successful survival strategy, however the lengths to which this capacity is predetermined has both physiological and psychological roots. For example, the ultimate state of boredom is represented in stimulus deprivation experiments.[19] These erect an effective barrier between the environment and the individual by removing the usual sensory cues which provide information about the world around us. For example, blindfolding and being placed in a darkened room effectively takes away sight, while cushioning the body in padding, including particular sense-gathering areas such as fingers and toes, prevents meaningful touch as experienced in everyday life. What has been recorded in such experimental conditions illustrates the potential for the brain to relieve this unnatural state by creating its own stimuli. Individuals participating in such studies have reported hallucinations which indicate our ability to generate stimuli internally. This type of compensatory experience enables the individual's physiological system to remain alert for when it is next required and provides a focus for our attention as the individual is clearly awake—both suggest we need a certain level of sensory arousal for optimal functioning. Although we might consider that our perception of boredom is a social construction which enables us to label that emotional state in which we are understimulated, it also highlights insufficient sensory activity to meet our internal needs. All of which gives rise to the question, how much is the right level of stimulation for each of us? A common feature of pertinent answers is an optimum determined by individual considerations, suggesting that at certain frequencies of incoming stimuli we may perceive the demands on our resources as either too limited or indeed excessive, which invokes a stress response. For example, it is claimed that the Internet has resulted in the huge increase of information to which we are exposed, estimated daily to be equivalent to reading dozens of newspapers back to back.[20] This suggests we are more likely to be over-than understimulated in a world infused with new technology. Yet the quality of this information and its capacity to distract our mental focus at a given time can have negative implications for how stimulated or bored we feel, as extraneous information such as that signaled by bleeping communication devices, can degrade our ability to remain absorbed in an otherwise rewarding task.[1] Again this appears to underline the importance of taking control over the stimuli to which we are exposed.

Theoretical explanations of the use of attention have provided the basis for recent considerations of boredom. "Menton" theory suggests the existence of so-called units of mental energy equivalent to a resource which permits the execution of everyday activities.[21] According to this approach, available resources are unevenly spread across brain and when boredom occurs, it is due to a surplus of mentons and the "feeling of boredom is your mind pushing you to find new challenges" (Ref. 21, p. 134). In this way daydreaming and doodling,[22] or even chewing gum,[23] are activities using excess mental energy and thereby topping up demands on available capacity which in turn facilitate maintenance of optimal arousal levels. Similarly the presence of factors which are not tied to the task at hand, for example while eating, might include the hubbub of conversation or images on a computer screen together provide "noise" which more fully occupies the mind. It is possible to extend the principles of Menton Theory further in a way which links with the kinds of problems previously outlined in this chapter. The concept of an optimal level of "mind-working" is logical given our physiological mechanisms governed by homeostatic regulation. Whether the experience of understimulation aids the processing of other information which is relevant to our survival, or provides spare capacity which feeds creative processes such as new ideas, it is also the case that an experience of boredom warns our resources are not being optimally employed. This means there is clear scope for negative outcomes too which feed on preexisting internal states.

Internal psychological factors which require and consume attentional resources therefore need to be considered. For example the presence of a chronic anxiety disorder occupies a proportion of mental energy and one method of dealing with this is to provide a different and hopefully, adaptive focus in order for the individual to cope. However during the enacting of the task which is occupying the individual's attention, anxiety is likely to remain in the background and if attention is not fully engaged by the activity—thus providing surplus mental energy—there is a risk that the negative feelings associated with the anxiety state will proliferate given the strength of the associated emotions and interfere with the primary focus of attention. Indeed such a mechanism can be considered relevant to a range of psychological states. Consistent with this, Davies and Fortney[21] suggest attention deficit hyperactivity disorder may be indicative of an abundance of attentional capacity or mentons and this may explain why certain types of background noise may actually prove beneficial for those who experience it.[24] Similarly if one considers the human population on a continuum of capacity for attention maintenance—whether motivated by physiology or psychology or both—it is not surprising that chronic experiences and perceptions of understimulation are linked to a range of psychological health problems, including depression and anxiety.[25] It is possible that these examples tend to support the possibility that negative psychological experiences can be directly associated with spare mental energy which we are unable to direct.

So how does the relationship between boredom and negative psychological states function? Eastwood et al.[1] highlight the relevance of attention and our capacity to direct, control, or sustain it. In this way, they identify a range of contributory cognitive and behavioral factors for boredom, including distraction of attention that "disrupts adequate engagement with information pertaining to the current activity" (p. 484), problems with maintaining a focus once it has been established, an awareness that attention is wavering, reduced arousal linked to insufficient external stimulation and an awareness that time is passing slowly.[1]

However, their review also highlights factors that hark back to the start of this chapter, namely the situational experience of being prevented from doing something meaningful and the dispositional tendency to feel the external world determines our own experiences. These are familiar themes within the literature on psychological health in which control—whether actual or perceived—is linked to better outcomes, and an absence of it leads to the potential for learned helplessness which predicts increased likelihood of distress and disorder.[26] Eastwood et al. suggest that the use of attention and direction of energy related to it underpin the mechanism which describes boredom: "the ability to successfully exert control to utilize attention provides the foundation for our elaborated sense of agency and, conversely, the inability to engage attention results in a self that is blocked or inarticulate"[1] (p. 489). In other words, the battle for control of our attention is likely to be a battle for control of our well-being, too (Figure 3).

How much control do we really have over what we do with these attentional processes? Research into personality yields part of the answer to the first of these questions. According to the five-factor model, trait theory has us all placed on a continuum between introversion and extraversion which describes the level of our desire to seek stimulation from interactions with others. Within this domain, the concept of sensation-seeking highlights further subcomponent traits including "boredom susceptibility," which refers to an aversion to routine, repetition, and predictability.[27] An impressive body of findings confirms that those scoring as high on the Sensation Seeking Scale[28] (SSS) are more likely to take risks which are reflected in daily behaviors such as breaking driving laws, gambling, drug taking, or indeed, in choices of careers characterized by higher levels of risk.[29] A sobering study of mortality statistics has shown that those reporting increased boredom levels tend to die younger.[30] This suggests a link between more extreme outcomes and psychological restlessness among those who describe themselves as bored, not to mention the psychological strain placed on others close to them. However this does not mean we are without conscious control over our attention and choices in behavior. The remaining part of the answer to the question which opens this paragraph suggests there are chemical and structural mechanisms linked to our arousal systems which shape our behaviors too (Figure 4).

Those who score high on the SSS tend to respond with greater levels of electrical brain activity than low-scorers to new stimuli.[27] Additionally, the cortical activity of those who register higher scores on the disinhibition subscale of the SSS tends to increase in the presence of novel stimuli.[31] Both findings indicate that there are motivations to seek out more daring (e.g., bungee-jumping) and high arousal experiences (e.g., dancing) of which the individual may not be aware, which clearly has implications for the threshold at which we might take notice of a state of boredom. Additionally, it has been suggested that the neurotransmitters dopamine and monoamine oxidase, which influence bodily movements and readiness to act, have been identified as promoting certain behaviors including sensation-seeking, to compensate for or reflect lower levels of these chemicals in the bodily system. Similarly reinforcement sensitivity theory[32] links the function of the locus coeruleus, located in the lower part of the brain, to sensation-seeking or avoidant behavior. It regulates sensitivity to new stimuli, as well as those we link with punishment, in such a way that the degree to

FIGURE 3 Time stands still in the street where Einstein lived—is the speed at which time passes relative to our emotions?

FIGURE 4 Would you find it hard to watch a cricket match for 5 days?

which the locus coeruleus responds influences decisions about whether to approach or avoid particular stimuli. Ongoing developments in neuroimaging promise further insights, such as the role of fluctuations in brain activity during the kinds of shift in attention associated with mind-wandering and experiences of boredom[33] and also the suggested existence of a "default network" active when attention is devoted to internal thought processes which provides a more rewarding experience than when focusing on externally driven factors.[34] It would appear that such mechanisms have clear implications for behaviors linked to boredom. Consistent with this, theorists have recognized the behavioral activation system—broadly leading more extraverted and less anxious individuals more readily toward positive emotional states—and the behavioral inhibition system which quickly alerts more introverted and anxious individuals to the presence of potentially threatening stimuli. While it would be unwise to overgeneralize the outcomes of such systems to every instance of an individual's behavior, or to overstate consideration of either physiological or psychological motivations, it is clear that there is the potential for internal states to play an influential role in our experiences of boredom and understimulation. In terms of stress, it is not hard to see that where the chemical or structural systems of an individual promote extremes in response, or patterns of behavior become dysfunctional within a given environmental context, that the likelihood of challenging psychological and behavioral outcomes is raised.

This begs the question of how do we cope with boredom? From the issues considered thus far, it would appear that while in a state of understimulation, the brain has the capacity for other activities to become the focus for our attention, even if these are spontaneous or more conscious examples of exercising the mind, such as doodling. Where boredom is a stressor characterized by a lack of control over our environment, it is increasingly recognized that employees will craft their job so that it provides them with opportunities to take back some level of autonomy[4] (i.e., ensuring a better fit between personal and job requirements). Additionally, taking control over one's internal drivers such as attention or negative psychological state presents challenges, but gaining awareness of these is a key step toward availing ourselves of cognitive strategies which make positive outcomes more likely. Meditation is viewed as a successful method for avoiding negative sources of mental distraction although given that scheduling can present logistical and practical challenges, it is perhaps unsurprising why mindfulness has gained in popularity. By emphasizing an ability to bring "attention to the experiences occurring in the present moment, in a nonjudgemental or accepting way" (Ref. 35, p. 27), it is possible in everyday nonclinical contexts to exert greater control over the psychological processes linked with experiences of boredom by encouraging a more meaningful experience. By focusing attention in such a deliberate manner there is scope for absorbing excess or negative psychological energy that in turn facilitates improved appraisal of stressful stimuli and job satisfaction.[36] However where options to become engaged in "satisfying activity" appear constrained, for example by enforced circumstances such as unemployment, then taking control over unpaid aspects of work holds promise not simply for improving how well we cope, but also for protecting health. For example, volunteering reduces the incidence of hypertension for both employed and unemployed individuals.[37] However, creative coping strategies are not always so positive in intention, as counterproductive work behaviors are linked to experiences of monotony[4]; accordingly one may argue that some criminal behavior exhibits creativity motivated by boredom. Whatever the outcomes, an alternative focus for one's attention while working tends to be viewed with disparagement, but without this capacity, Einstein, who worked in a patent office, Darwin who was keener on collecting plants than medicine and Gainsborough who preferred to paint rather than study, would not be familiar names. It has been argued here that inside and out of many modern workplaces, there is a considerable waste of mental capital due to poor job design and a general lack of understanding of our cognitive capabilities—stressors which hold back individuals as well as organizations. If boredom as an outcome is linked with higher risk of accidents and sabotage, as well as poorer psychological health, then understimulation threatens progress at a sinister level, befitting the notion of the "daily grind."

References

1. Eastwood JD, Frischen A, Fenske MJ, Smilek D. The unengaged mind: defining boredom in terms of attention. *Perspect Psychol Sci.* 2012;7(5):482–495.
2. Hubel DH, Wiesel TN. *Brain and Visual Perception: The Story of a 25-Year Collaboration.* Oxford: Oxford University Press; 2005.
3. Eurofound. *European working conditions survey results.* Available at: http://www.eurofound.europa.eu/surveys/smt/ewcs/ewcs2010_07_06.htm; 2010.
4. Van Hooff MLM, van Hooft EAJ. Boredom at work: proximal and distal consequences of affective work-related boredom. *J Occup Health Psychol.* 2014;19(3):348–359.
5. Dumas L. Why mistakes happen even when the stakes are high: the many dimensions of human fallibility. *Med Glob Surviv.* 2001;7:12–19.
6. Bhana H. Correlating boredom proneness and automation complacency in modern airline pilots. *Coll Aviat Rev.* 2010;28:9–24.
7. Münsterberg H. *Psychology and Industrial Efficiency.* Boston, MA: Houghton, Mifflin and Company; 1913.
8. Caplan R, Cobb S, French J, Harrison R, Pinneau S. *Job Demands and Worker Health.* Washington, DC: HEW Publication, NIOSH; 1975.
9. French JRP, Caplan RD, Harrison RV. *Mechanisms of Job Strain and Stress.* New York, NY: John Wiley; 1982.
10. Cooper CL, Dewe P. Well-being—absenteeism, presenteeism, costs and challenges. *Occup Med.* 2008;58(8):522–524.

11. *Working for a healthier tomorrow.* Black report, London: TSO; 2008. Accessed at: www.workingforhealth.gov.uk/documents/working-for-a-healthier-tomorrow-tagged.pdf.
12. NICE (2013) Workplace policy and management practices to improve the health of Employees. Accessed at: https://www.nice.org.uk/guidance/ng13, 2015.
13. NICE. *Promoting Mental Wellbeing Through Productive and Healthy Working Conditions: Guidance for Employers.* London: NICE; 2009.
14. Foresight Project. *Mental capital and wellbeing.* Accessed at: www.foresight.gov.uk/Ourwork/ActiveProjects/Mental%20Capital/Welcome.asp; 2008.
15. Torrington D, Hall L, Taylor S. *Human Resource Management.* 5th ed. Harlow: Pearson Education Limited; 2002.
16. McLeod D, Clarke N. *Engaging for Success: Enhancing Performance Through Employee Engagement.* London: Department for Business, Innovation and Skills; 2009. Accessed at: www.engageforsuccess.org/wp-content/uploads/2012/09/file52215.pdf.
17. Warr PB. Jobs and job-holders: two sources of happiness and unhappiness. In: David SA, Boniwell I, Ayers AC, eds. *The Oxford Handbook of Happiness.* Oxford: Oxford University Press; 2013:733–750.
18. Warr PB. *Work, Unemployment and Mental Health.* Oxford: Clarendon Press; 1987.
19. Heron W. The pathology of boredom. *Sci Am.* 1957;196:52–69.
20. Levitin DJ. *The Organized Mind.* Hialeah, FL: Dutton; 2014.
21. Davies J, Fortney M. The Menton theory of engagement and boredom. In: *First Annual Conference on Advances in Cognitive Systems. Cognitive Systems Foundation*; 2012.
22. Andrade J. What does doodling do? *Appl Cogn Psychol.* 2010;24:100–106.
23. Scholey A, Haskell C, Robertson B, Kennedy D, Milne A, Wetherell M. Chewing gum alleviates negative mood and reduces cortisol during acute laboratory psychological stress. *Physiol Behav.* 2009;97:301–312.
24. Söderlund G, Sikström S, Smart A. Listen to the noise: noise is beneficial for cognitive performance in ADHD. *J Child Psychol Psychiatry.* 2007;48:840–847.
25. Goldberg YK, Eastwood JD, LaGuardia J, Danckert J. Boredom: an emotional experience distinct from apathy, anhedonia, or depression. *J Soc Clin Psychol.* 2011;30:647–666.
26. Weinberg A, Cooper CL. *Stress in Turbulent Times.* London: Palgrave-Macmillan; 2012.
27. Zuckerman M. *Psychobiology of Personality.* 2nd ed. New York: Cambridge University Press; 2005.
28. Zuckerman M. Dimensions of sensation seeking. *J Consult Clin Psychol.* 1971;36:45–52.
29. Zuckerman M. *Sensation Seeking: Beyond the Optimal Level of Arousal.* Hillsdale, NJ: Erlbaum; 1979.
30. Britton A, Shipley MJ. Bored to death? *Int J Epidemiol.* 2010;39:370–371.
31. Brocke B, Beauducel A, John R, Debener S, Heilemann H. Sensation seeking and affective disorders: characteristics in the intensity dependence of acoustic evoked potentials. *Neuropsychobiology.* 2000;41:24–30.
32. Gray JA, McNaughton N. *The Neuropsychology of Anxiety: An Enquiry into the Functions of the Septohippocampal System.* 2nd ed. Oxford: Oxford University Press; 2000.
33. Leech R, Kamourieh S, Beckmann CF, Sharp DJ. Fractionating the default mode network: distinct contributions of the ventral and dorsal posterior cingulated cortex to cognitive control. *J Neurosci.* 2011;31(9):3217–3224.
34. Spreng RN, Mar R, Kim ASN. The common neural basis of autobiographical memory, prospection, navigation, theory of mind, and the default mode: a quantitative meta-analysis. *J Cogn Neurosci.* 2009;21:489–510.
35. Baer RA, Smith GT, Hopkins J, Krietemeyer J, Toney L. Using self-report assessment methods to explore facets of mindfulness. *Assessment.* 2006;13:27–45.
36. Hülsheger UR, Alberts HJ, Feinholdt A, Lang JW. Benefits of mindfulness at work: the role of mindfulness in emotion regulation, emotional exhaustion and job satisfaction. *J Appl Psychol.* 2013;98(2):310–325.
37. Griep Y, Hyde M, Vantilburgh T, Bidee J, de Witte H. Voluntary work and the relationship with unemployment, health and well-being: a two-year follow-up study contrasting a materialistic and psychosocial pathway perspective. *J Occup Health Psychol.* 2014;20(2):190–204. http://dx.doi.org/10.1037/a0038342.
38. Warr PB. *Work, Happiness and Well-Being.* New York: Lawrence Erlbaum; 2007.

CHAPTER

24

Anxiety Disorders

V. Starcevic[1], D.J. Castle[2,3]

[1]University of Sydney, Sydney, NSW, Australia

[2]University of Melbourne, Melbourne, VIC, Australia

[3]St Vincent's Mental Health, Melbourne, VIC, Australia

OUTLINE

Abstract

Anxiety disorders are the most common group of psychiatric disorders in the general population. They are also important because of their association with significant impairment in functioning and with high direct and indirect costs. Anxiety disorders are often associated with depressive and substance use disorders and may have other complications. These conditions are often unrecognized, misdiagnosed, or trivialized, which is unfortunate because their timely recognition and treatment are beneficial to the sufferers, their families, and society. This chapter provides a brief review of each of the following anxiety disorders in adults: panic disorder, agoraphobia, social anxiety disorder, specific phobia, generalized anxiety disorder, and separation anxiety disorder. Clinical features, diagnostic issues, epidemiological data, etiological factors, and treatments of these disorders are summarized, with the similarities and differences between individual disorders being highlighted.

INTRODUCTION

Anxiety disorders are very common, with the lifetime prevalence in various countries around the world ranging from 9.2% to 28.7%. The best-estimate lifetime prevalence rate for all anxiety disorders was calculated at 16.6%.[1] These conditions are primarily characterized by pathological anxiety, although other emotions, such as disgust, irritability, and shame may also be prominent. Compared to normal anxiety, pathological anxiety is usually experienced as more intense, persistent, and overwhelming, with the person having little or no control over it; also, pathological anxiety often arises in the absence of any real danger or such danger, as perceived by the individual, is grossly exaggerated. Pathological anxiety is associated with significant distress or impairment. Although the diagnostic and classification systems such as the *Diagnostic and Statistical Manual of Mental Disorders, Fifth Edition* (DSM-5) recognize the diagnostic categories of substance/medication-induced anxiety disorder and anxiety disorder due to another medical condition, these are usually not subsumed within the construct of anxiety disorders per se. Likewise, pathological anxiety in the context of schizophrenia, another psychotic disorder or bipolar affective disorder, which is not a part of a specific anxiety disorder, is not considered to denote a distinct anxiety disorder.

According to the DSM-5, anxiety disorders comprise the following conditions: panic disorder, agoraphobia, social anxiety disorder (social phobia), specific phobia, generalized anxiety disorder (GAD), separation anxiety

Stress: Concepts, Cognition, Emotion, and Behavior
http://dx.doi.org/10.1016/B978-0-12-800951-2.00024-8

disorder, and selective mutism. Given that this chapter focuses on the anxiety disorders in adults, selective mutism will not be reviewed and separation anxiety disorder will be considered insofar as it pertains to adults. Obsessive-compulsive disorder and post-traumatic stress disorder have close relationships with many anxiety disorders, but in the DSM-5 they are classified elsewhere.

KEY POINTS

- Anxiety disorders are the most common group of psychiatric disorders in the general population and they are associated with significant functional impairment and high direct and indirect costs.
- Anxiety disorders are often unrecognized or misdiagnosed at the primary point of contact with health care providers and are often complicated by depressive or substance use disorders.
- Panic disorder is common in primary care, specialized medical settings, and hospital emergency departments and requires a thorough diagnostic workup to check for the presence of underlying medical conditions.
- Agoraphobia is conceptually independent from panic disorder; when its sufferers become homebound, agoraphobia is incapacitating.
- Social anxiety disorder is arguably the most disabling anxiety disorder, with its frequent complications further contributing to treatment difficulties.
- Specific phobia is a heterogeneous diagnosis; avoidance is the main coping mechanism that often prevents the sufferers from seeking professional help.
- Generalized anxiety disorder has a close relationship with depression and should be better conceptualized so that it could be readily recognized and treated.
- Adult separation anxiety disorder needs to be more clearly delineated from other psychopathology.
- Pharmacological and psychological treatments, especially cognitive-behavioral therapy, have improved the outcomes for anxiety disorders, but there is a need to further refine treatments.

PANIC DISORDER

The concept of panic disorder includes recurrent panic attacks and the consequent anticipatory anxiety or panic-related behavioral changes. Panic attacks are sudden surges of an intense fear of impending catastrophe, such as dying, losing consciousness and passing out, losing control, or "going crazy." These episodes are accompanied by bodily symptoms, such as palpitations, chest discomfort or pain, shortness of breath or other respiratory disturbance, dizziness, sweating, trembling or shaking, numbness or tingling sensations, hot or cold flushes, and/or nausea or abdominal distress. Many of these symptoms reflect hyperarousal of the autonomic nervous system. The peak of a panic attack is reached very quickly, within minutes. While the first panic attack may be experienced as occurring without any reason ("unexpected" or "spontaneous" panic attack), the subsequent panic attacks are usually "situational," that is, more likely to occur in situations that the person has associated with panic attacks. Panic attacks do not have to lead to panic disorder and may appear in the context of various other mental disorders; in the latter case, they usually denote a more severe form of the disorder and a less favorable course.

Anticipatory anxiety refers to a persistent and often disabling fear of another panic attack. More specifically, the anticipated, though unlikely somatic (heart attack), psychological (loss of control, madness), and/or social (embarrassment, humiliation) consequences of panic attacks are the focus of this fear. Panic-related behavioral changes include avoidance of activities or situations that may precipitate a panic attack (e.g., exercising or crowded places). If this avoidance pertains to agoraphobic situations and becomes prominent and disabling, a diagnosis of agoraphobia may also be considered.

Because of the nature of their symptoms, individuals with panic attacks often present to emergency departments and request urgent help. Bodily symptoms that are a part of the clinical presentation during panic attacks often require a thorough diagnostic workup to check for the presence of an underlying heart disease (e.g., cardiac arrhythmias), neurological condition (e.g., epilepsy), endocrinological disease (e.g., thyroid disease), or other (e.g., pulmonary, vestibular, gastrointestinal) illness.

Table 1 shows epidemiological findings for panic disorder. A striking feature of panic disorder is that it is commonly encountered in primary care, specialized medical settings and hospital emergency departments. Occurrence of the first panic attack after the age of 50 is rare and may suggest "organic" etiology. The course of panic disorder is variable, with at least one third of sufferers reporting a complete recovery and about one half having a chronic, fluctuating course. Common associated conditions are substance misuse and depression. The links between panic disorder and a higher risk of suicide attempts and suicide[2] and cardiovascular disease[3] are not well understood.

A genetic predisposition to developing panic attacks has been proposed, but the nature and specificity of

TABLE 1 Epidemiological Data for Anxiety Disorders

	Panic disorder	Agoraphobia	Social anxiety disorder	Specific phobia	Generalized anxiety disorder	Adult separation anxiety disorder
Lifetime prevalence (community)	• 0.4-3.8% • Best-estimate: 1.2% • Panic attacks: 7.3-28.3%	• 0.73-10.8% • Best-estimate: 3.1%	• 0.5-16% • Best-estimate: 3.6%	• 0.6-12.5% • Best-estimate: 5.3%	• 1.9-31.1% • Best-estimate: 6.2%	6.6%
Gender ratio	F:M = 2-3:1	F:M = 2.5-4:1	F:M = 1.5:1 (community) F:M = 1:1 (clinical)	F:M = 2-2.5:1 F:M = 1:1 for blood-injection-injury phobia	F:M = 2:1	F:M = 1:1 (?) F > M for childhood separation anxiety disorder
Age of onset (in years)	• Typical: 20s • Mean: 25	• Typical: mid-teens to early 20s • Mean: 17	• Typical: early teens to early 20s • Mean: 15-16	• Typical: childhood • Mean: 10 • Animal phobia: early childhood • Natural environment phobia: childhood • Blood-injection-injury phobia: late childhood • Situational phobia: mid-teens to early 20s	• Typical: late teens to late 20s • Can begin at any age	• Typical: early childhood • Can begin in adulthood
Demographic correlates	Strong association with • Being separated, divorced, widowed • Lower education • Urban residence	Strong association with • Work disability	Strong association with • Being single • Lower education • Unemployment • Belonging to lower socio-economic group		Strong association with • Being separated, divorced, widowed • Unemployment • Lower income • Being a homemaker • Urban residence	Strong association with • Being unmarried, divorced (?) • Lower education (?) • Unemployment (?) • Work disability (?)

such predisposition remain unclear. An abnormally sensitive, anxiety-regulating mechanism that originates in the amygdala may be involved in panic disorder and other anxiety disorders. The more specific mechanisms implicated in the pathogenesis of panic attacks include hyperventilation, hypersensitivity to carbon dioxide of the brain-stem chemoreceptors,[4] lower threshold for activating the suffocation alarm mechanism,[5] and failure of the gamma-aminobutyric acid (GABA) system to inhibit the locus coeruleus. With regards to the psychological mechanisms, an exaggerated perception of threat has been proposed to characterize all anxiety disorders.[6] In panic disorder, this threat is perceived to originate within one's body; other factors that are relatively specific for panic disorder include a heightened anxiety sensitivity (fear of anxiety and its bodily symptoms based on the beliefs about their dangerousness)[7] and misinterpretation of bodily sensations as a sign of an impending catastrophe.[8]

AGORAPHOBIA

Agoraphobia shares several features with other phobic disorders: social anxiety disorder and specific phobia (Table 2). It refers to a fear of multiple, interrelated situations that can be grouped into three "clusters." Situations in the first cluster are those of which the person is afraid because it might be difficult or impossible to immediately escape; they include being in crowded or enclosed places (e.g., shopping centers, cinemas, tunnels) and using public transportation (e.g., buses, trains, planes). The second cluster is represented by situations (e.g., traveling far away from home, staying at home alone) in which the person is alone or outside of their safety zone, so immediate medical or other help might not be available. Situations in the third cluster are those from which it might be awkward or embarrassing to escape immediately (e.g., standing in line, sitting in the middle of a row in a theater).

TABLE 2 General Characteristics of the Phobic Disorders (Agoraphobia, Social Anxiety Disorder, Specific Phobia)

- The fear pertains to the known objects, situations, activities, or phenomena ("phobic stimuli")
- The fear is out of proportion to the actual threat posed by the phobic stimuli and to the sociocultural context
- Insight that the fear is irrational or excessive is usually preserved, but may be absent in children
- Exposure to phobic stimuli elicits an immediate fearful response, sometimes in the form of a panic attack
- Phobic stimuli are avoided or endured with great distress and/or fear if avoidance is not possible
- The fear is persistent and lasts for months and years (minimum 6 months according to the DSM-5)
- The fear or fear-related avoidance causes significant distress or impairment in functioning

These situations are often feared because of the anticipation of panic attacks or symptoms, and the purpose of the subsequent avoidance is to prevent them. This is the reason for the frequent co-occurrence of agoraphobia and panic disorder and for the view that agoraphobia is actually a part of panic disorder. However, in many cases of agoraphobia, the fear is unrelated to panic[9,10] and sufferers believe that their avoidance is due to fears of falling, being incontinent, getting lost, having an accident, or being mugged. Unlike other phobic disorders, agoraphobia is characterized by reliance on a "phobic companion" (i.e., a person who accompanies the affected individual to a variety of agoraphobic situations so that full-scale avoidance is averted). In the absence of a phobic companion, avoidance is usually extensive and often leads to the person becoming homebound and incapacitated.

Many of the epidemiological findings on agoraphobia (Table 1) are derived from the data on panic disorder with agoraphobia and some may relate more to panic disorder than to agoraphobia. Higher prevalence rates of agoraphobia than those of panic disorder and earlier mean age of onset of agoraphobia than that of panic disorder also suggest the conceptual independence of agoraphobia from panic disorder. It has been consistently reported that agoraphobia is much more likely to occur in women, which gave rise to the hypotheses about gender-related psychological and social factors in the development of agoraphobia. However, the dominant explanatory model is that agoraphobic fear is a consequence of learning (i.e., of associating unpleasant anxiety symptoms or traumatic experiences with certain situations); avoidance, in turn, maintains this fear.

SOCIAL ANXIETY DISORDER (SOCIAL PHOBIA)

The key characteristic of social anxiety disorder is excessive and persistent fear of situations in which the person might be exposed to the scrutiny by others. There are two types of these situations. The first involves performance, typically giving a talk to a number of people or doing something (e.g., eating, writing, working) in front of others. The second type pertains to a broad variety of interpersonal interactions, from simple conversations with unfamiliar people and asking for directions to assertiveness-testing communication with authority figures or expressing disagreement. These situations are usually avoided; if it is not possible to resort to avoidance, the person endures feared situations with substantial distress (Table 2). While some people fear and avoid only performance-type situations, others are troubled by almost any social interaction.

The key underlying cognition in social anxiety disorder is the fear of negative evaluation and humiliation, which is often manifested through concerns about coming across as "stupid" or incompetent, making a mistake or performing poorly, and subsequent rejection. In addition, individuals with this disorder may have a prominent fear of having visible bodily symptoms (such as blushing, sweating, or trembling) that would reveal their anxiety. There is a close relationship between shyness and social anxiety disorder, but the latter denotes a greater severity of social anxiety and decreased ability to habituate in social situations and to respond to positive social cues.

Table 1 shows epidemiological data on social anxiety disorder. Many individuals with the condition do not seek professional help because of shame and embarrassment. Those who do seek help often do so after many years and in the context of a complication (e.g., depression) or important life change (e.g., going to a different school or starting a new job). Professional help is usually sought from school counselors, psychologists, or psychotherapists. Social anxiety disorder typically has a chronic course, with significant impairment and disability in many areas of functioning. Depression and substance use disorders, especially alcohol abuse, are frequent complications. The relationship between social anxiety disorder and avoidant personality disorder is important, but remains poorly understood.[11,12]

There may be a genetic predisposition to social anxiety disorder, but it is likely to be shared with other anxiety and depressive disorders. Behavioral inhibition to the unfamiliar[13] represents a predisposition to social anxiety disorder, but also to other anxiety disorders; it refers to the manifestations of inborn temperament observable during early life, which include difficulty sleeping in unfamiliar surroundings, irritability in novel situations and avoidance of contacts with unfamiliar people, places, and objects. According to the cognitive models,[14,15] social anxiety disorder is a consequence of the negative assumptions and beliefs about oneself, others and social situations. These assumptions and beliefs result in appraisals of social situations as threatening and/or perception of the social environment as hostile. Assumptions and beliefs specific for social anxiety disorder and expectations to be evaluated negatively in social situations are maintained through avoidance of social situations, use of safety behaviors and biases in information processing.

SPECIFIC PHOBIA

Specific phobia is a heterogeneous group of conditions that includes phobia of animals, blood-injection-injury phobia (i.e., phobia of the sight of blood, injured tissues, mutilation of the body, or needle penetrating the skin), situational phobia (i.e., phobia of driving or flying, claustrophobia), natural environment phobia (i.e., phobia of water, heights, storms) and other "unclassified" phobias such as dental phobia and phobia of choking or vomiting. The affected individuals are excessively afraid of particular objects, situations, activities, or phenomena (i.e., "phobic stimuli") because of the perceived threat posed by them. The threat is based on the specific dangers associated with phobic stimuli (e.g., a danger of suffocation while the person is in a small, enclosed place) or disgust. The feeling of disgust is experienced vis-à-vis certain animals, especially insects, and stimuli that serve as reminders of the animal origin and mortality of the humans (i.e., the sight of blood, wounds, or needle penetrating the skin). The blood-injection-injury phobia is a unique type of phobia as it is characterized by a vasovagal reaction, with bradycardia, hypotension, and fainting; unlike all other phobias, it is as common in males as it is in females (Table 1). In all specific phobias, phobic stimuli are avoided as much as possible; if avoidance is not possible, phobic stimuli are endured with much fear or distress (Table 2).

Although specific phobia is common in the community (Table 1), its frequency is lower in the treatment-seeking populations, suggesting that specific phobia may cause less impairment than most other anxiety disorders and/or that individuals with specific phobia are less likely to seek professional help. Indeed, only about 8% of individuals with specific phobia may seek treatment,[16] and most use avoidance as their coping strategy. When they seek professional help, that is usually because changes in their life circumstances prevent them from continuing to avoid their phobic stimuli (e.g., commencing a job that involves frequent travel by plane makes avoidance of flying impossible).

Similar to other anxiety disorders, genetic predisposition to specific phobia is not specific. Many phobias develop as a result of learning. This can occur through a traumatic conditioning (direct aversive experience with the phobic stimulus), vicarious learning (observation of the fear in others) or transmission of the information on the dangerousness of certain objects or situations. Some types of specific phobia (e.g., phobia of heights or water, spider phobia) have been posited to represent "innate" fears, which are not a product of learning; these phobias may have a survival or evolutionary value.[17]

GENERALIZED ANXIETY DISORDER

The main features of GAD are pathological worry and symptoms of tension. Autonomic arousal symptoms may also be present, but they are usually less prominent in GAD than in panic disorder and other anxiety disorders.

Pathological worry relates to several topics or issues, including relationships, family, health, work, or finances, but its cardinal feature is an anxiety-amplifying, uncontrollable cascade of the "what if" pattern of thinking.[18] Pathological worry is largely driven by a difficulty in coming to terms with uncertainty, where worrying could perhaps cease only with an unlikely attainment of complete certainty. The associated doubt prevents a closure, which results in constant anticipation of further problems and no solution in sight. Thus, pathological worrying denotes overthinking that is almost incessant and rather fruitless; as such, it interferes with problem-solving and decision-making.

Tension is usually experienced or expressed as nervousness, feeling "keyed up" or "on edge" or irritability, often with an exaggerated startle response, hypervigilance, restlessness, and inability to relax. Many individuals with GAD complain of muscle tension, which is manifested as tension headache or as tightness, stiffness, or pain in the neck, shoulder, or back. Muscle spasms, tic-like movements, jerks, fine tremor, and difficulty swallowing may also be encountered. Largely as a consequence of pathological worry and tension, symptoms of GAD may also include problems with concentration, sleep disturbance, fatigue, and/or exhaustion.

GAD is a chronic disorder closely related to depression. Unlike other anxiety disorders, GAD appears at any age and is prevalent in all age groups. It is also the most frequent anxiety disorder in the elderly. GAD is one of the most common anxiety disorders in the general population (Table 1), but in clinical settings it is usually not seen alone and is often diagnosed as a secondary or co-occurring condition, with the primary or main condition being depression or another anxiety disorder. Individuals with GAD often do not seek professional help; when they do so, they tend to present to general practitioners with depression or with bodily symptoms, not with pathological worry.[19] These patterns of clinical presentation may "mask" GAD and delay recognition and accurate diagnosis.

The similarities between genetic predisposition to GAD and genetic predisposition to major depressive disorder have led to a suggestion that the two conditions are genetically indistinguishable.[20] Various neurobiological mechanisms (e.g., hyperactivity of the norepinephrine system, decreased function of the GABA-A receptors) have been postulated to play a role in the pathogenesis of GAD, suggesting that it is a heterogeneous condition; this heterogeneity is also suggested by the effectiveness of the pharmacological agents with various mechanisms of action (Table 3). Psychological models of GAD have focused on pathological worry as the key feature. Thus, it has been postulated that worrying allows a person to avoid unpleasant bodily symptoms that accompany strong emotional states (i.e., that worry serves as "cognitive avoidance").[21] Beliefs about the benefits of worrying (i.e., that worrying can prevent catastrophic outcomes) may maintain worry.[22] Intolerance of uncertainty, with an interpretation of ambiguous stimuli and information as threatening, may also play a role in the development of GAD.[23] However, intolerance of uncertainty has been found to be a transdiagnostic phenomenon that characterizes several other anxiety disorders as well.[24]

ADULT SEPARATION ANXIETY DISORDER

Separation anxiety disorder in adults is characterized by excessive and persistent anxiety about separation from attachment figures. It is manifested through worries about losing these figures or about possible harm to them with the refusal to go out, travel, or participate in events or activities that entail physical separation; fear of being alone; and recurring nightmares involving separation. The onset of separation anxiety disorder typically occurs in childhood, but the DSM-5 acknowledges that the disorder can also begin in (early) adulthood. It is uncertain whether adult-onset and childhood-onset separation anxiety disorder represent the same condition.

Separation anxiety disorder in adults has much symptom overlap and close relationships with other anxiety disorders (especially agoraphobia and panic disorder), complicated grief, post-traumatic stress disorder, and dependent personality disorder. In clinical practice, other anxiety disorders and depression are diagnosed much more frequently than adult separation anxiety disorder, which has been suggested to be due to a tendency to neglect adult separation anxiety disorder.[25] However, it has also been argued that there may be a relatively low clinical need for the diagnosis of separation anxiety disorder in adults, not necessarily its poor recognition, misdiagnosis, or neglect.[26] Cultural factors may influence the conceptualization of high levels of separation anxiety as a disorder and in more collectivistic milieus prominent separation anxiety is not necessarily regarded as a disorder.

Considering the diagnostic conundrums, it is not surprising that little is known about epidemiology of adult separation anxiety disorder. While the disorder may be common in the general population (Table 1), further studies need to ascertain its prevalence more precisely, as well as its gender distribution and demographic correlates. Adult separation anxiety disorder can be associated with significant impairment in functioning.

Genetic factors may play a role in the etiology of adult separation anxiety disorder, but the heritability of the disorder has not been estimated. Attachment abnormalities, perception of parents as overprotective, and

TABLE 3 Evidence-Based Pharmacotherapy for Panic Disorder, Social Anxiety Disorder, and Generalized Anxiety Disorder

	Panic disorder	Social anxiety disorder	Generalized anxiety disorder
First line	• Selective serotonin reuptake inhibitors: sertraline, paroxetine, escitalopram, fluoxetine • Serotonin and norepinephrine reuptake inhibitors: venlafaxine	• Selective serotonin reuptake inhibitors: sertraline, paroxetine, escitalopram, fluvoxamine • Serotonin and norepinephrine reuptake inhibitors: venlafaxine	• Selective serotonin reuptake inhibitors: escitalopram, paroxetine, sertraline • Serotonin and norepinephrine reuptake inhibitors: venlafaxine, duloxetine • Pregabalin
Second line	• Tricyclic antidepressants: imipramine, clomipramine • Benzodiazepines: clonazepam, alprazolam	• Benzodiazepines: clonazepam	• Agomelatine • Benzodiazepines • Quetiapine
Third line		• Classical, irreversible monoamine oxidase (MAO) inhibitors: phenelzine	• Imipramine • Buspirone • Hydroxyzine

low self-esteem have been associated with adult separation anxiety disorder, but the specificity of these relationships is unlikely. The disorder has been linked with a preoccupied, insecure attachment style in adult relationships.[27] Pathological separation anxiety in children may be a precursor to various anxiety and depressive disorders in adults and does not necessarily continue into adulthood.

TREATMENT

The treatment of anxiety disorders aims to decrease anxiety and anxiety-related behaviors (e.g., avoidance), promote better coping with anxiety, decrease vulnerability to anxiety disorders, prevent recurrences and complications, and improve functioning and quality of life. Treatment modalities differ in terms of their ability to address these goals and how they go about reducing the negative impact of anxiety. Of all the treatments used for the anxiety disorders, pharmacotherapy[28] and cognitive-behavioral therapy (CBT)[29] have received most empirical support. Modifications of CBT, such as mindfulness-based therapy[30] and acceptance and commitment therapy,[31] have also demonstrated efficacy. Likewise, controlled studies of psychodynamic psychotherapy for various anxiety disorders have produced favorable results.[32] Pharmacological and psychological treatments are often combined in clinical practice and there is some evidence to support this approach.[33]

PHARMACOTHERAPY

Medications are used to alleviate the anxiety symptoms and distress. They do not improve coping with anxiety and are effective only for as long as they are taken. Consequently, relapses following the cessation of pharmacotherapy are not rare. Symptom relief produced by medications is often valued by the sufferers and may facilitate learning of the strategies for effective and long-term coping with anxiety. Pharmacotherapy is usually used for rapid relief of acute anxiety, when the disorder is more severe and in the presence of co-occurring conditions such as depression. The key task in pharmacological treatment is to find the right balance between effectiveness and adverse effects of the medication.

The aim of pharmacotherapy is remission, defined as minimal symptoms and return to normal functioning. Remission promotes recovery and is believed to decrease the risk of relapse after medication discontinuation. In order to achieve remission, medication should be used long enough (usually for at least 6-12 weeks) and in an adequate dose (often the highest recommended dose). If an adequate trial with one medication fails, another medication may be administered or the initial medication is augmented with another pharmacological agent. Maintaining remission entails a continuous, daily use of the medication for at least 6 months. If the remission has lasted for about 6-12 months and the person is ready for medication discontinuation, medication taper can be planned. The duration of this gradual reduction in dose until medication discontinuation depends on the type of medication and personal circumstances. As a rule, medication should not be ceased abruptly after long-term treatment.

Medications used in the treatment of panic disorder, social anxiety disorder, and GAD are listed in Table 3. Pharmacotherapy has not been studied in the treatment of adult separation anxiety disorder and it is considered to be of little value for specific phobia. The role of medication treatment for agoraphobia in the absence of panic attacks or panic disorder is unknown. Selective serotonin reuptake inhibitors and serotonin and norepinephrine reuptake inhibitors are considered by the contemporary treatment guidelines[34,35] to be the pharmacological treatments of choice. However, these agents are not always effective against the prominent anxiety symptoms, they do not work quickly, and are often associated with adverse effects. For these reasons, benzodiazepines and other medications have been used as alternative or even preferred pharmacotherapy for anxiety disorders, albeit there are some concerns about dependence with benzodiazepines.[36]

COGNITIVE-BEHAVIORAL THERAPY

The main advantage of CBT is that it brings about changes in the thinking and behavioral patterns that may reduce vulnerability to the anxiety disorders and decrease the risk of relapse following the cessation of treatment. In addition, CBT fosters an active attitude toward treatment and makes it easier to develop a sense of ownership of treatment gains. For these reasons, therapeutic effects of CBT are more likely to last longer than those of pharmacotherapy.

In clinical practice, cognitive and behavioral therapy approaches are usually combined. They both entail psychoeducation (i.e., a provision of the relevant information and explanation and correction of any misconceptions). CBT for anxiety disorders has been administered in individual and group formats and in recent years it has also been successfully delivered via the Internet.[37] Table 4 lists the specific aspects of CBT and related treatments for the specific anxiety disorders. There are no controlled studies of any psychological treatment, including CBT, in which adult separation anxiety disorder was the specific focus of treatment.

Maladaptive, anxiety-related assumptions, beliefs, misinterpretations, and appraisals, which play a role in

TABLE 4 Specific Aspects of Cognitive-Behavioral Therapy and Related Treatments for Anxiety Disorders

	Panic disorder	Agoraphobia	Social anxiety disorder	Specific phobia	Generalized anxiety disorder
Cognitive therapy techniques	• Correcting misinterpretations of bodily symptoms • Modifying beliefs about body-based threat and the dangerous nature of anxiety • Learning not to be afraid of panic and its symptoms		• Modifying assumptions and beliefs about oneself, others and social situations • Modifying appraisals of social situations as threatening and perception of the social environment as hostile		• Imagery exposure to the content of worries • Modifying beliefs about the benefit of worrying • Improving coping with uncertainty • Improving decision-making and problem-solving
Behavior therapy techniques		Exposure to agoraphobic situations: • Gradual • Self-directed • In vivo	Exposure to relevant social situations: • Gradual • First therapist-assisted (with role-play), then self-directed • First imaginal (often in-session), then in vivo	Exposure to phobic stimuli: • Gradual (rarely "flooding") • Therapist-assisted and/or self-directed • Imaginal (guided imagery) or in vivo	
Symptom control techniques	Breathing retraining				Progressive muscle relaxation
Other techniques			Social skills training		

the development and maintenance of anxiety disorders, are targeted by cognitive therapy techniques. These approaches are particularly useful in the treatment of panic disorder, GAD, and social anxiety disorder (Table 4). Behavior therapy aims to change and eliminate behaviors (e.g., avoidance) that help maintain pathological anxiety; with the disappearance of these behaviors, pathological anxiety usually diminishes as well. The key behavioral therapy technique is exposure to situations or stimuli that elicit anxiety; it is most effective for phobic disorders (Table 4) because they are usually characterized by prominent avoidance. Symptom control techniques (e.g., muscle relaxation, breathing retraining) suppress bodily symptoms of anxiety and may be used in conjunction with CBT.

CONCLUSION

Existing knowledge about anxiety disorders has allowed more precise diagnoses and improvement in treatment outcomes. However, etiological understanding of this realm of psychopathology remains insufficient and calls for further research. Greater insight into the origins and pathogenesis of anxiety disorders is expected to lead to a further refinement of treatment approaches.

References

1. Somers JM, Goldner EM, Waraich P, Hsu L. Prevalence and incidence studies of anxiety disorders: a systematic review of the literature. *Can J Psychiatry*. 2006;51:100–113.
2. Sareen J, Cox BJ, Afifi TO, et al. Anxiety disorders and risk for suicidal ideation and suicide attempts: a population-based longitudinal study of adults. *Arch Gen Psychiatry*. 2005;62:1249–1257.
3. Abrignani MG, Renda N, Abrignani V, Raffa A, Novo S, Lo Baido R. Panic disorder, anxiety, and cardiovascular diseases. *Clin Neuropsychiatry*. 2014;11:130–144.
4. Papp LA, Klein DF, Gorman JM. Carbon dioxide hypersensitivity, hyperventilation, and panic disorder. *Am J Psychiatry*. 1993;150:1149–1157.
5. Klein DF. False suffocation alarms, spontaneous panics, and related conditions: an integrative hypothesis. *Arch Gen Psychiatry*. 1993;50:306–317.
6. Beck AT, Emery G, Greenberg RI. *Anxiety Disorders and Phobias: A Cognitive Perspective.* New York, NY: Basic Books; 1985.

7. Reiss S, McNally RJ. Expectancy model of fear. In: Reiss S, Bootzin RR, eds. *Theoretical Issues in Behavioral Therapy*. San Diego, CA: Academic Press; 1985:107–121.
8. Clark DM. A cognitive approach to panic. *Behav Res Ther*. 1986;24:461–470.
9. Faravelli C, Cosci F, Rotella F, Faravelli L, Dell'Osso MC. Agoraphobia between panic and phobias: clinical epidemiology from the Sesto Fiorentino Study. *Compr Psychiatry*. 2008;49:283–287.
10. Wittchen H-U, Nocon A, Beesdo K, et al. Agoraphobia and panic: prospective-longitudinal relations suggest a rethinking of diagnostic concepts. *Psychother Psychosom*. 2008;77:147–157.
11. Ralevski E, Sanislow CA, Grilo CM, et al. Avoidant personality disorder and social phobia: distinct enough to be separate disorders? *Acta Psychiatr Scand*. 2005;112:208–214.
12. Chambless DL, Fydrich T, Rodebaugh TL. Generalized social phobia and avoidant personality disorder: meaningful distinction or useless duplication? *Depress Anxiety*. 2008;25:8–19.
13. Kagan J, Reznick JS, Clarke C, Snidman N, Garcia-Coll C. Behavioral inhibition to the unfamiliar. *Child Dev*. 1984;55:2212–2225.
14. Clark DM, Wells A. A cognitive model of social phobia. In: Heimberg R, Liebowitz M, Hope DA, Schneier FR, eds. *Social Phobia: Diagnosis, Assessment and Treatment*. New York, NY: Guilford Press; 1995:69–93.
15. Rapee RM, Heimberg RG. A cognitive-behavioral model of anxiety in social phobia. *Behav Res Ther*. 1997;35:741–756.
16. Stinson FS, Dawson DA, Chou SP, et al. The epidemiology of DSM-IV specific phobia in the USA: results from the National Epidemiologic Survey on Alcohol and Related Conditions. *Psychol Med*. 2007;37:1047–1059.
17. Menzies RG, Clarke JC. The etiology of phobias: a nonassociative account. *Clin Psychol Rev*. 1995;15:23–48.
18. Starcevic V, Portman ME, Beck AT. Generalized anxiety disorder: between neglect and an epidemic. *J Nerv Ment Dis*. 2012;200: 664–667.
19. Rickels K, Rynn MA. What is generalized anxiety disorder? *J Clin Psychiatry*. 2001;62(suppl 11):4–12.
20. Kendler KS, Neale MC, Kessler RC, Heath AC, Eaves LJ. Major depression and generalized anxiety disorder: same genes, (partly) different environments? *Arch Gen Psychiatry*. 1992;49:716–722.
21. Borkovec TD, Ray WJ, Stoeber J. Worry: a cognitive phenomenon intimately linked to affective, physiological, and interpersonal behavioral processes. *Cogn Ther Res*. 1998;22:561–576.
22. Freeston MH, Rhéaume J, Letarte H, Dugas MJ, Ladouceur R. Why do people worry? *Personal Individ Differ*. 1994;17:791–802.
23. Ladouceur R, Talbot F, Dugas MJ. Behavioral expressions of intolerance of uncertainty in worry: experimental findings. *Behav Modif*. 1997;21:355–371.
24. Starcevic V, Berle D. Cognitive specificity of anxiety disorders: a review of selected key constructs. *Depress Anxiety*. 2006;23:51–61.
25. Silove D, Manicavasagar V. Let's not abandon separation anxiety disorder in adulthood. *Aust N Z J Psychiatry*. 2013;47: 780–782.
26. Starcevic V. Looking for balance between promoting adult separation anxiety disorder and overlooking it. *Aust N Z J Psychiatry*. 2013;47:782–784.
27. Manicavasagar V, Silove D, Marnane C, Wagner R. Adult attachment styles in panic disorder with and without adult separation anxiety disorder. *Aust N Z J Psychiatry*. 2009;43:167–172.
28. Emilien G, Durlach C, Lepola U, Dinan T. *Anxiety Disorders: Pathophysiology and Pharmacological Treatment*. Birkhäuser: Basel; 2002.
29. McLean PD, Woody SR. *Anxiety Disorders in Adults: An Evidence-Based Approach to Psychological Treatment*. New York, NY: Oxford University Press; 2001.
30. Arch JJ, Ayers CR, Baker A, Almklov E, Dean DJ, Craske MG. Randomized clinical trial of adapted mindfulness-based stress reduction versus group cognitive behavioral therapy for heterogeneous anxiety disorders. *Behav Res Ther*. 2013;51:185–196.
31. Bluett EJ, Homan KJ, Morrison KL, Levin ME, Twohig MP. Acceptance and commitment therapy for anxiety and OCD spectrum disorders: an empirical review. *J Anxiety Disord*. 2014;28: 612–624.
32. Keefe JR, McCarthy KS, Dinger U, Zilcha-Mano S, Barber JP. A meta-analytic review of psychodynamic therapies for anxiety disorders. *Clin Psychol Rev*. 2014;34:309–323.
33. Stahl SM, Moore BA, eds. *Anxiety Disorders: A Guide for Integrating Psychopharmacology and Psychotherapy*. New York, NY: Routledge; 2013.
34. Katzman MA, Bleau P, Blier P, Chokka P, Kjernisted K, Van Ameringen M. Canadian clinical practice guidelines for the management of anxiety, posttraumatic stress and obsessive-compulsive disorders. *BMC Psychiatry*. 2014;14(suppl 1):S1.
35. Baldwin DS, Anderson IM, Nutt DJ, et al. Evidence-based pharmacological treatment of anxiety disorders, post-traumatic stress disorder and obsessive-compulsive disorder: a revision of the 2005 guidelines from the British Association for Psychopharmacology. *J Psychopharmacol*. 2014;28:403–439.
36. Starcevic V. The reappraisal of benzodiazepines in the treatment of anxiety and related disorders. *Expert Rev Neurother*. 2014;14:1275–1286.
37. Andersson G, Titov N. Advantages and limitations of Internet-based interventions for common mental disorders. *World Psychiatry*. 2014;13:4–11.

CHAPTER

25

The Post-Traumatic Syndromes

D.J. Castle[1,2,3,4], V. Starcevic[1,2]

[1]University of Melbourne, Melbourne, VIC, Australia
[2]St Vincent's Mental Health, Melbourne, VIC, Australia
[3]The University of Cape Town, Cape Town, South Africa
[4]Australian Catholic University, Banyo, Australia

OUTLINE

Abstract

The post-traumatic disorder syndrome cluster is defined in terms of a stressor being associated with subsequent psychological and behavioral consequences. This chapter traces the history of the post-traumatic syndromes; outlines the current nosology of this group of disorders; addresses some of the more contentious areas in the field, such as vicarious exposure to stressful events; and finally summarizes treatment approaches.

Of all disorders in psychiatry, it is the post-traumatic cluster that is defined in terms of a stressor. Other maladies such as depression can be precipitated by stressful events, but they can also occur in the absence of such events. It is unique to the post-traumatic disorders that a stressor must be identifiable.

This chapter traces the history of the post-traumatic syndromes; outlines the current nosology of this group of disorders; addresses some of the more contentious areas in the field, such as vicarious exposure to stressful events; and finally summarizes treatment approaches.

THE EVOLUTION OF THE CONSTRUCT

Arguably post-traumatic psychiatric symptoms have been recognized for millennia, but in the modern age, the responses of some soldiers to the horrors of the battlefields of World War 1 brought this to the stark attention of psychiatry. Descriptions of hysterical conversion and so-called shell shock have been the subject of plays and novels and the clinical manifestations were a major impetus to the then fledgling field of psychodynamic psychotherapy.

The Second World War and the horrors of the concentration camps and the sequelae thereof led to the inclusion of "gross stress reaction" as a distinct entity in the first edition of the US Diagnostic and Statistical Manual of Mental Disorders (DSM) in 1952. However, this condition was listed under "transient situational personality disorders," a reflection of then-current thinking about the problem. As Andreasen[1] explains, it was seen as a "severe stress reaction occurring in a normal personality as a mechanism for dealing with overwhelming fear—the disorder was by definition acute and reversible." These core definitional aspects have subsequently been challenged, with the recognition that response to trauma depends in large part on the individual's make-up pre-exposure; and also, that all too often the symptoms can be enduring and very disabling.

The issue of combat-related post-traumatic syndromes is a particularly emotive one. In a cluster analysis of post

http://dx.doi.org/10.1016/B978-0-12-800951-2.00025-X

combat syndromes from the Boer to the Gulf War, Jones et al.[2] showed that a "syndrome characterized by unexplained medical symptoms" was ubiquitous, but that the form changed over the period, from an emphasis on disability (Boer War) to somatization (First World War) to neuropsychiatric presentations such as fatigue syndromes (Second World War, Gulf War).

KEY POINTS

- Post-traumatic stress disorder (PTSD) is the most important psychopathological syndrome appearing in the aftermath of trauma; it has also been one of the most controversial psychiatric disorders.
- PTSD is an important public health problem in terms of its high prevalence and disability levels associated with it.
- Most trauma victims do not develop PTSD, many recover spontaneously from early PTSD-like symptoms and only a minority go on to develop a chronic and severe PTSD.
- With longer duration of PTSD, the likelihood of recovery decreases sharply.
- The definition of trauma associated with PTSD has shown a tendency to be broadened.
- The diagnostic criteria for PTSD reflect the heterogeneity of the concept of PTSD.
- There has been a tendency to misuse PTSD in the context of compensation claims and litigation.
- There may have been too much emphasis on vulnerability to develop PTSD and too little understanding of the factors that promote resilience in the face of adversity and trauma.
- The current treatment approaches to PTSD are largely symptom-oriented and should be replaced by treatments (both pharmacological and psychological) that target some key underlying pathophysiological and psychological mechanisms.
- Results of the treatment of chronic PTSD have generally been modest, with psychological therapies achieving somewhat better results than pharmacological treatment.

It took another war, this time the Vietnam War, to refocus US psychiatry on this topic. The fact that many veterans of the Vietnam War had to contend with a highly ambivalent populace and that they did not receive a universal heroes' welcome upon return to the United States, added to what had for many been extremely difficult combat experience in terms of jungle conditions, an elusive enemy, and ultimately defeat. Many had complex and prolonged, or even delayed stress reactions which challenged then-current clinical constructs. It was also increasingly realized that it is not only military conflicts in which people can be exposed to damaging traumas, with post-trauma psychiatric sequelae. Thus, the term "Post Traumatic Stress Disorder" (PTSD) entered the third edition of the DSM in 1980. That term has subsequently stuck, albeit its definition has been massaged as knowledge has advanced and social and other attitudes have changed. The current edition of DSM, the fifth, sees the post-traumatic syndromes in their own chapter, and including a range of disorders including acute stress disorder (ASD) (by definition less than a month of symptoms) and the now iconic PTSD.

It will be noted that PTSD in the DSM requires at least 1 month of symptoms. This leaves a hiatus in the month immediately after the traumatic event. DSM plugged this gap by introducing a separate entity, called ASD. The main reason for conceptualizing ASD as a separate condition was that ASD was assumed to be a valid predictor of the development of PTSD. This has been challenged, however, as many individuals with symptoms of ASD recover spontaneously and do not develop PTSD.[3] The World Health Organization's International Classification of Mental Disorders (ICD-10) does not have a similar construct, and the strict "1 month" rule imposed by the DSM does seem simplistic and not truly reflective of the highly diverse symptom trajectories experienced by individuals in the wake of trauma. But the DSM, it seems, wishes to distinguish brief from more prolonged stress reactions.

The features of PTSD have been organized in several symptom clusters in the DSM system. The number and composition of these clusters have been debated. According to DSM 5, there are four such clusters: intrusion symptoms, avoidance, negative alterations in cognitions and mood, and alterations in arousal and reactivity. Each cluster consists of a number of symptoms, with the total of 20 PTSD symptoms. This large number of symptoms of equal diagnostic value and the requirement for the symptoms from each cluster to be present have proven problematic for several reasons. First, it is too cumbersome for routine clinical practice, as it is complicated to memorize all the symptoms and diagnostic symptom thresholds within each cluster. Second, many PTSD symptoms are nonspecific and characterize depression or other anxiety disorders. Third, no hierarchy between PTSD symptoms is postulated by the DSM system, so that for example, symptoms of reexperiencing the trauma are not deemed more characteristic of PTSD and therefore more important than problems with concentration or irritable behavior. Fourth, the DSM 5 diagnosis of PTSD cannot be made in the absence of avoidance, even if many other prominent features of PTSD are present. Finally, this diagnostic conceptualization perpetuates the notion that PTSD is too heterogeneous; in fact, it has been estimated that there are 79,794 possible combinations of the PTSD diagnostic criteria qualifying for the DSM-IV diagnosis of PTSD and 636,120 possible combinations of the diagnostic criteria qualifying for the DSM 5 diagnosis of PTSD.[4]

CONTENTIOUS ISSUES

There are a number of areas of particular ongoing debate regarding PTSD and related disorders. These include the nature of stressors and what can be considered "legitimate" stressors for the precipitation of PTSD; the problem of vicarious exposure to trauma; the issue of PTSD as a social or political construct; the problem of memory for traumatic events being somewhat ephemeral; the role that issues of financial compensation might play in the maintenance of symptoms; and whether there are specific subtypes of PTSD. These are considered briefly, in turn.

The Nature of the Stressor

One of the major problems for the field has been defining an event that is of sufficient magnitude to result in a post-traumatic disorder (see reviews by Weathers and Keane[5] and Friedman et al.[6]). Of course, this is rather circular as it is only with the manifestation of symptoms that one can make the diagnosis, and many people exposed to really horrific events do not develop such symptoms: why they are so resilient is a fascinating issue that has hardly been addressed in scientific studies. But if an individual does exhibit a symptom set that meets the other criteria for PTSD in response to what might be objectively seen as a rather prosaic (albeit distressing) event such as an argument with a spouse or the failure of one's sporting team to win the grand finale, could one uphold the diagnosis? DSM-III (1980) defined the stressor as "a psychologically traumatic event that is generally outside the range of usual human experience," but this falls down when one considers that in some cultural and geographical settings, certain traumas would (regrettably) be much more common (and thus "within the range"). Also, as Andreasen[1] points out, the DSM-III terminology was very much about the psychological response rather than the physical. DSM-IV (1987) reversed this and defined the stressor in terms related to likely death or physical injury: as Andreasen[1] states: "The pendulum had swung fully across the divide, placing an emphasis on soma and deemphasizing psyche."

Furthermore, there is the key issue of how the individual perceives the trauma. DSM-IV divided the stressor criterion in objective (so-called A1 criterion) and subjective (A2) components. Regarding the latter, the individual was to have experienced "fear, helplessness, or horror" at the time of the trauma. Brewin et al.[7] have questioned the utility of the A2 criterion, arguing, *inter alia,* that such emotional responses at the time of the trauma are only weekly predictive of later PTSD and that some people who develop PTSD symptoms do not experience such trauma-related responses. Also, other emotional responses such as anger and shame could be important in the evolution of PTSD. Balancing these findings, DSM 5 has dropped the A2 criterion.

Vicarious Exposure

A highly contentious issue is whether vicarious exposure can count as a traumatic event and justify a PTSD diagnosis (see Friedman et al.[6]). If this were the case, PTSD could be seen to be consequent upon such exposures as watching images of war or natural disaster on television or even reading about them in the news. But not allowing it would exclude people in whom high rates of PTSD have been reported in the wake of, for example, death of a family member by homicide (71% in one study) or in the 9/11 attack (22%). Another group of individuals are vicariously exposed to trauma through their work: for example, mortuary workers, police, and troops exposed to human remains. As a seemingly sensible compromise, DSM 5 "allows" vicarious exposure as part of the definition of the traumatic experience and PTSD diagnosis, but restricts it to learning about trauma to a loved one or close friend, whereby the event must have been "violent or accidental." Exposure to "gruesome evidence" in the line of duty (e.g., police, ambulance) is accepted as traumatic experience, but witnessing of trauma through media is explicitly excluded, unless it is work-related (e.g., journalists). This broadening of the definition of trauma has already been implicated in a tendency to overdiagnose PTSD and it may lead to an increased use of PTSD for compensation and litigation purposes.

PTSD as a Social and Political Construct

PTSD has been criticized on the grounds that it "medicalizes" normal trauma- or stress-related emotions and distress.[8] Many modern Western societies obsessed with risk-free lifestyles and relentless pursuit of "happiness" do not leave much room for the naturally occurring frustrations and disappointments. However, when these do occur, the associated "unnatural" distress is transformed into a disorder, which then legitimizes distress and allows one to enjoy secondary gain and seek help. This explains the popularity of both PTSD and "disorders" modeled on it: "traumatic grief disorder," "prolonged duress stress disorder," "post-traumatic embitterment disorder," "post-traumatic relationship syndrome," and others.

The Ephemeral Nature of Traumatic Memories

PTSD relies on people's recollection of trauma. On the face of it, this seems pretty simple but in reality it is a complex and changing phenomenon. The ephemeral nature of some memories is well described, notably of early-life traumas. But recollection of adult traumas can also wax and wane. For example, Southwick et al. (1995)[9] assessed soldiers' accounts of traumatic events a month after their return from combat and again 2 years later. Up to 70% of respondents recalled events at 2 years that they had not reported at the 1-month time-point. Of particular importance was that those soldiers who developed PTSD were the most likely to "amplify" (to use the term adopted by Hales and Zatzick[10]) their traumatic memories. Hales and Zatzick[10] go on to state of this finding: "the associations of changes in memory with greater PTSD symptom level brings into question the retrospective methodology traditionally used to demonstrate severity of combat exposure and the development of PTSD." It also brings into question the whole nature of traumatic memory: it is not as though people simply "make up" memories, but it does leave the field with a major challenge in terms of establishing causal relationships between trauma and subsequent psychiatric symptomatology.

Compensation

The role of compensation and of the legal system in impacting the process whereby people deal with stress is hotly debated. Early commentators such as Miller[11] tended to take the view that many individuals simply "manufactured" their symptoms for financial gain. It is clear now that the matter is far more complex than this. Indeed, it is usually the case that people with so-called compensation neurosis do not relinquish their symptoms or experience a dramatic cure immediately after their compensation claim has been settled, albeit relatively rare cases of true malingering do exist.[12] Importantly, there is evidence that the legal process can have a negative impact on the trajectory of illness, expressly if prolonged and complicated. It has also been asserted that some members of the legal profession can be so wedded to the PTSD diagnosis as legitimating a financial claim for their client that they *sotto voce* influence the claimant to not relinquish their symptoms before a claim is settled.[13]

Putative Subtypes

There have been a number of suggested ways of subtyping PTSD. Perhaps the most compelling has been the notion that some people's response to traumatic stress involves the mechanism of dissociation. Dissociation is defined as "a disruption of the usually integrated functions of consciousness, memory, identity, or perception of the environment" (DSM-IV). Lanius et al.[14] make a good case for a putative subtype of PTSD characterized predominantly by dissociation. Such individuals appear to deal with trauma by a profound inhibition of emotions and a disengagement from the emotional pain associated with the trauma, which manifests as a persistent emotional numbing. This might particularly be the case for children exposed to sexual abuse, as children are more likely to respond to trauma by dissociation, compared to adults. Brain activation studies suggest a neurobiological substrate characterized by suppression of limbic and regions by prefrontal "over-ride" (see Chu[15]). As Chu[15] pointed out, there are implications of this model for treatments, as classic exposure paradigms (see below) would presumably fail because they rely on emotional engagement for efficacy.

There is also a subtype of PTSD with "delayed expression." This refers to PTSD manifestations that are fully expressed at least 6 months after a traumatic event, although they may be present at a diagnostically subthreshold level soon after the trauma.[16] Such a trajectory of PTSD symptoms calls for an adequate early assessment of trauma victims and their follow-up.

WHO GETS PTSD?

PTSD is a response to stress that has, by definition, to occur after a trauma. But it is clear, as stated above, that many people exposed to trauma do not develop the symptoms. Indeed, trauma is regrettably a very common occurrence in modern society but rates of PTSD do not match rates of trauma. For example, in a study from Detroit, MI, USA, lifetime exposure to a stressor that met DSM criteria A was 39%, but the rate of PTSD in such exposed individuals was 23.6%.[17] In the large-scale US National Comorbidity Survey, some 50% of women and 60% of men reported having been exposed to a traumatic event, but the lifetime prevalence of PTSD was 6.7%. Although rates of exposure to traumatic events are generally higher in men than in women, rates of PTSD tend to be higher in women.

So, what determines whether a particular individual develops PTSD? The role of the type of trauma has been discussed above, and it appears that repeated trauma at particular stages of emotional development and in particularly vulnerable contexts is more likely to lead to later psychiatric sequelae. For example, childhood sexual abuse is disturbingly common, but the likelihood of it being associated with psychiatric disorders depends upon a complex matrix including the family context.[18] This raises another issue, namely that the association of sexual abuse in particular with psychiatric disorders is not limited to PTSD: indeed, many psychiatric disorders have shown such an association, including borderline personality disorder and depression, as well as schizophrenia. It is also the case that early abuse might not result in symptoms until a later event of similar nature ("re-traumatization") unmasks them. This is seen, for example, in women who have suffered sexual abuse as a child and who are then sexually abused in adulthood and suggests some emotional and physiological "priming" from the early abuse.[19]

As Rosen et al.[13] point out, predicators of PTSD have most to do with the individual's make-up "coming into" the trauma and the response to the trauma (e.g., post-trauma social support) than the nature of the trauma itself. What constitutes the underlying genetic liability to PTSD is not clear, but "inoculation" type models of repeated traumas seem parsimonious, such that repeated traumas during specific developmental stages make later severe responses more likely, akin to Post's notion of "kindling" in mood disorders.

The differences between individuals in trauma response are particularly important in understanding why post-incident debriefing can make some people more rather than less likely to develop PTSD. McFarlane and Yehuda[20] suggest that the failure of such debriefing may be interpreted as either "interference" with a normative stress response in those who would otherwise not have developed PTSD; or exacerbating an abnormal stress response in those who would be destined to develop PTSD but whose trajectory is hastened by the intervention. These authors go on to cite a study by Resnick et al.[19] who assessed rates of PTSD in women who had been raped. Those women with a prior history of sexual assault were less likely to show a heightened cortisol response in the aftermath of the rape and were more likely to develop PTSD than those women who had not previously been raped and had a "normal" cortisol response. Thus, a diminished acute physiological stress response appears to leave one at higher risk of a later post-traumatic syndrome.

TREATMENT

Given the complexity of PTSD as a construct and noting that there may well be biologically definable subtypes (see above), it is hardly surprising that treatment is required to be individualized and multifaceted. There are often added complexities such as physical disability sustained as part of the traumatic event (e.g., accidents), chronic pain syndromes, loss of livelihood, and loss of role. Other psychological factors include so-called survivor guilt and also guilt and loss associated with loved ones or friends and colleagues who might have perished or been injured in the traumatic situation (e.g., a motor vehicle accident in which the driver survives but all passengers are killed or maimed; comrades who died in war). The use of alcohol and illicit substances can further perturb recovery, as can prolonged legal and compensation battles, as outlined above. Presence of the cooccurring depression, sleep disturbance, or anxiety disorders usually complicates treatment. "Complex PTSD"[21] refers to significant and often permanent changes in personality following the complex and/or repeated traumatic experiences. It has been argued that treatment of individuals with such a "type" of PTSD requires long-term treatment that also addresses various aspects of personality functioning.

PTSD is unique in that knowledge of one of its crucial etiological factors—the occurrence of trauma—allows prevention if the appropriate measures are implemented soon after traumatic event. However, this is easier said than done. Despite decades of research, it has been difficult to ascertain a combination of risk factors that predicts PTSD reliably, which would identify trauma victims who are likely to develop PTSD and who might benefit from prevention or early treatment strategies.

As discussed, the notion that early debriefing after a traumatic event can ameliorate risk of later PTSD has been shown to be well considered but mostly unhelpful or even damaging.

McFarlane and Yehuda[20] provide a useful overview psychological models that have been applied in PTSD. These include:

(a) Psychodynamic/reprocessing: focuses on aberrant integration of cognitions and affects and seeks to change the attitude to the trauma through addressing feelings of helplessness and shame.
(b) Behavioral therapy: applies learning theory and exposes the individual to cues related to the trauma, using a graded approach.
(c) Cognitive therapy: employs cognitive restructuring techniques such as relating to beliefs that by some different action on the part of the individual the trauma could have been avoided; this is often combined with behavioral elements.

It appears that trauma-focused psychological therapies are more effective for PTSD than supportive psychological interventions.[22] Of the former, various types of cognitive-behavioral therapy have the most studied. Prolonged exposure therapy, whether imaginal or in vivo, has been found to be particularly useful.[23] Benefits of adding cognitive therapy techniques to prolonged exposure have not been unequivocally confirmed by research.

A rather more controversial technique is so-called rapid eye-movement sensitization Eye movement desensitisation and reprocessing (EMDR).[24] Here the patient follows the therapist's finger with their eyes, while evoking first a traumatic memory and describing associated feelings; and subsequently a "safe" image that is reassuring the patient. The mechanism whereby this method works is unclear but some researchers suggest even a single session can be remarkably effective.[28]

Many medications have been investigated in the treatment of PTSD. The best studied and most widely used pharmacological agents are selective serotonin reuptake inhibitors (SSRIs) and serotonin and norepinephrine reuptake inhibitors (SNRIs). How much of the benefit from these agents is due to their general effects on mood and anxiety and how much it is in any way specific to PTSD, is not clear. Also, as Friedman[25] points out, the remission rates with such agents is only of the order of

30% and certainly less impressive than for evidence-based psychological therapies. However, if depression is a prominent part of the clinical picture, SSRIs and SNRIs would have an important role to play in treatment.

Numerous other agents, acting on mechanisms as diverse as opioid, glutamatergic, dopaminergic, and GABA-ergic pathways, have been investigated in PTSD, but the therapeutic returns have been mostly disappointing. Often clinicians are left with using medications to target specific symptoms, such as benzodiazepines for anxiety and insomnia and anticonvulsants and mood stabilizers for outbursts of anger and impulsive and aggressive behavior. The sedative antipsychotics such as quetiapine are also widely used clinically, albeit the evidence base to support their use in this context is sparse at best. These pharmacological agents are often used in combination with antidepressants and with each other, resulting in polypharmacy.

A relatively recent target has been the adrenergic dysregulation seen in PTSD. Thus, Raskind et al.[26] showed that the alpha-1-adrenergic agent prazosin reduced PTSD-related nightmares and overall symptoms in veterans of the Iraq and Afghanistan conflicts. The agent is gaining increasing use in clinical practice.

CONCLUSIONS

There is a sharp contrast between the deceptively simple concept of PTSD and the complexity with which PTSD is often encountered in clinical practice. This reflects our insufficient understanding of the pathways between traumatic events and full-blown PTSD. A greater insight into the mechanisms that connect pre-trauma vulnerabilities, the traumatic experience itself and post-trauma factors to produce PTSD will help devise more effective treatment strategies for this common, chronic, and often debilitating condition.

References

1. Andreasen NC. Posttraumatic stress disorder: psychology, biology, and the false Manichaean warfare between false dichotomies. *Am J Psychiatr*. 1995;152:963–965.
2. Jones E, Hodgins-Vermaas R, McCartney H, et al. Post-combat syndromes from the Boer war to the Gulf war: a cluster analysis of their nature and attribution. *Br Med J*. 2002;324:321–324.
3. Bryant RA. Does dissociation further our understanding of PTSD? *J Anxiety Disord*. 2007;21:183–191.
4. Galatzer-Levy IR, Bryant RA. 636,120 ways to have posttraumatic stress disorder. *Perspect Psychol Sci*. 2013;8:651–662.
5. Weathers FW, Keane TM. The criterion A problem revisited: controversies and challenges in defining and measuring psychological trauma. *J Trauma Stress*. 2007;20:107–121.
6. Friedman MJ, Resick PA, Bryant RA, Brewin CR. Considering PTSD for DSM-5. *Depress Anxiety*. 2011;28:750–769.
7. Brewin CR, Andrews B, Rose S. Fear, helplessness, and horror in posttraumatic stress disorder: investigating DSM IV criteria A2 in victims of violent crime. *J Trauma Stress*. 2000;13:499–509.
8. Summerfield D. Cross-cultural perspectives on the medicalization of human suffering. In: Rosen GM, ed. *Posttraumatic Stress Disorder: Issues and Controversies*. New York, NY: Wiley; 2004:233–244.
9. Southwick SM, Morgan III CA, Darnell A, et al. Trauma-related symptoms in veterans of operation desert storm: a 2-year follow up. *Am J Psychiatr*. 1995;152:1150–1155.
10. Hales RE, Zatzick DF. What is PTSD? *Am J Psychiatr*. 1997;154: 143–145.
11. Miller H. Accident neurosis. *Br Med J*. 1961;1:919–998.
12. Mayou R. Accident neurosis revisited. *Br J Psychiatry*. 1996;168: 399–403.
13. Rosen GM, Spitzer RL, McHugh PR. Problems with posttraumatic stress disorder and its future in DSM-IV. *Br J Psychiatry*. 2008;192:3–4.
14. Lanius RA, Vermetten E, Lowentein RJ, et al. Emotion modulation in PTSD: clinical and neurobiological evidence for a dissociative subtype. *Am J Psychiatr*. 2010;67:640–647.
15. Chu AJ. Posttraumatic stress disorder: beyond DSM-IV. *Am J Psychiatr*. 2010;167:615–617.
16. Carty J, O'Donnell ML, Creamer M. Delayed-onset PTSD: a prospective study of injury survivors. *J Affect Disord*. 2006;90:257–261.
17. Breslau N, Davis GC, Andreski P, Peterson E. Traumatic events and posttraumatic stress disorder in an urban population of young adults. *Arch Gen Psychiatry*. 1991;48:216–222.
18. Mullen PE, Martin JL, Anderson JC, Romans SE, Herbison GP. Childhood sexual abuse and mental health in adult life. *Br J Psychiatry*. 1993;163:721–732.
19. Resnick HS, Yehuda R, Pitman RK, Foy DW. Effect of previous trauma on acute plasma cortisol level following rape. *Am J Psychiatry*. 1995;152:1675–1677.
20. McFarlane AC, Yehuda R. Clinical treatment of posttraumatic stress disorder: conceptual challenges raised by recent research. *Aust N Z J Psychiatry*. 2000;34:940–953.
21. Herman JL. Sequelae of prolonged and repeated trauma: evidence for a complex posttraumatic syndrome (DESNOS). In: Davidson JRT, Foa EB, eds. *Posttraumatic Stress Disorder: DSM-IV and Beyond*. Washington, DC: American Psychiatric Press; 1993:213–228.
22. Bisson JI, Ehlers A, Matthews R, Pilling S, Richards D, Turner S. Psychological treatments for chronic post-traumatic stress disorder. Systematic review and meta-analysis. *Br J Psychiatry*. 2007;190: 97–104.
23. Powers MB, Halpern JM, Ferenschak MP, Gillihan SJ, Foa EB. A meta-analytic review of prolonged exposure for posttraumatic stress disorder. *Clin Psychol Rev*. 2010;30:635–641.
24. Shapiro F. Eye movement densitisation: a new treatment for posttraumatic stress disorder. *J Behav Ther Exp Psychiatry*. 1989;20: 211–217.
25. Friedman MJ. Toward rational pharmacotherapy for posttraumatic stress disorder: reprise. *Am J Psychiatr*. 2013;170:944–946.
26. Raskind MA, Peterson K, Williams T, et al. A trial of prazosin for combat trauma PTSD with nightmares in active-duty soldiers returned from Iraq and Afghanistan. *Am J Psychiatry*. 2013;170:1003–1010.
27. Kessler RC, Bromet E, Hughes M, Nelson CB. Posttraumatic stress disorder in the national comorbidity survey. *Arch Gen Psychiatry*. 1995;52:1048–1060.
28. McCann DL. Post-traumatic stress disorder due to devastitang burns overcome by a single session of eye movement desensitization. *J Behav Ther Exp Psychiatry*. 1992;23:319–323.

CHAPTER

26

Distress

G. Matthews

University of Central Florida, Orlando, FL, USA

OUTLINE

Abstract

This chapter reviews the assessment of distress, the roles of situational factors and personality in generating distress, and its psychological significance, including clinical implications. Distress is typically defined as unpleasant subjective stress responses such as anxiety and depression. It may be measured using a variety of general and context-linked scales. States of distress reflect psychobiological, cognitive, and social influences, but should be understood as the outcome of dynamic self-regulation as the person confronts external threats and pressures. Individuals differ considerably in their vulnerability to distress. Traits including neuroticism, dispositional anxiety, and metacognitive style increase vulnerability, whereas traits such as hardiness, grit, and emotional intelligence support resilience. Distress is associated with abnormalities in information-processing, including attentional impairment and cognitive bias. Distress is also a key symptom in a range of emotional disorders.

INTRODUCTION

"Distress" is an imprecise term that typically refers to unpleasant subjective stress responses such as anxiety and depression. It is also sometimes used to describe behaviors and medical symptoms ("somatic distress"). The concept of distress derives from H. Selye's[1] General adaptation syndrome (GAS): the generalized temporal sequence of physiological and psychological stress responses that may be elicited by noxious or threatening life events. Often, stress responses are characterized by difficulties in adapting to the external stressor (i.e., distress, although stress may sometimes have a stimulating, energizing effect ("eustress"). Distress may thus be conceptualized as the internal "strain" provoked by an external "stressor." However, this simple metaphor has been largely superseded by the transactional model of emotion in which stress responses reflect dynamic person-environment relationships.[2] From this perspective, distress signals that adaptation to environmental demands is taxing or unsuccessful. Contemporary research on distress is dominated by two issues. The first is the extent to which a global distress concept is to be preferred to more specific negative affective responses such as anxiety, depression, and anger. The second theme is that distress must be understood within a wider framework of the person's active attempts to adapt to the physical and social environment.

ASSESSMENT OF DISTRESS

Distress is a broad label for a variety of stress responses, and so there is no single, generally accepted means of assessment. The term typically refers to negative emotions, such as anxiety, depression, and anger,

Stress: Concepts, Cognition, Emotion, and Behavior
http://dx.doi.org/10.1016/B978-0-12-800951-2.00026-1

as well as somatic distress. Numerous questionnaires have been published that assess these constructs. Some scales refer to the general emotional state or disposition of the person,[3,4] whereas others are geared toward a specific context, such as distress resulting from a medical condition such as cancer.[5] In animal and child research, distress may be operationalized as behaviors such as vocalizations or facial expressions of negative emotion. Distress may be experienced as both an acute emotional state and a chronic condition,[6] so that it may be assessed over various timescales:

(i) A transient state lasting for a few minutes; i.e., the person's immediate state of mind;
(ii) An episodic condition lasting for weeks or months; e.g., the distress provoked by a life event or a short-lived clinical disorder;
(iii) A personality trait, such as trait anxiety, which may show stability over decades.

Distress may be defined as a latent construct through multivariate psychometric research. Studies of mood and basic emotions differentiate orthogonal dimensions of negative affect and positive affect, such that negative affect is a superordinate category relating to tension, unhappiness, irritability, and other negative emotions, whereas positive affect covers states such as happiness and energy. More fine-grained psychometric models of distress have also been developed. Different basic negative emotions may be discriminated. Spielberger pioneered questionnaires that assessed anxiety, depression, and anger as both transient states and stable traits.[6] Distress is defined more precisely within a three-factor model of subjective states as an affective-cognitive dimension, defined by tension, unpleasant mood and cognitions of lack of control, and low confidence.[3,7] Distress is psychometrically distinct from two further broad state factors of task engagement (e.g., energy, task motivation) and worry (e.g., self-focused attention, intrusive thoughts). That is, transient distress states may involve not only negative moods but also disturbances of cognitive function and motivational states.

Dimensional models have been applied also to understanding distress symptoms in a range of psychiatric conditions, especially anxiety and mood disorders. Psychometric evidence suggests a broad distinction between distress disorders (e.g., generalized anxiety disorder, depression, post-traumatic stress disorder, and fear disorders (e.g., panic disorder, phobias).[8] This distinction contrasts with the classification of emotion disorders in DSM-V. Watson[9] has proposed a quadripartite model of distress symptoms for multiple disorders. Symptoms are classified according to (1) the magnitude of their general distress component and (2) their level of specificity in discriminating depression and anxiety disorders. Thus, distress symptoms may either present similarly across the spectrum of disorders, or be more characteristic of specific disorders. General distress is similar to dysphoria, including symptoms such as negative mood, loss of interest, worry, and worthlessness. By contrast, symptom dimensions such as suicidality and lassitude are expressions of distress that are more prevalent in depression than in anxiety disorders.

Finally, distress may be assessed in relation to a wide range of specific stressors or threatening contexts. For example, studies of evaluative anxiety are concerned with contexts in which the person's self-esteem is under threat, including test anxiety, sports anxiety, and computer anxiety.[10] Similarly, serious medical disorders often elevate distress, and scales have been developed for distress associated with diseases including cancer, cardiovascular diseases and diabetes, as well as general health distress. In addition, conditions associated with psychological distress, such as depression, may represent a risk factor for medical illness, as shown in prospective studies.[11]

KEY POINTS

- Distress is a general term for a range of negative emotional states.
- Distress is shaped by a variety of biological, cognitive, and social processes that underpin person-environment interaction.
- People differ in their vulnerability to stress, depending on traits for resilience that promote effective self-regulation in challenging environments.
- Distress is a common symptom of a range of emotional disorders; understanding metacognitive and self-regulative processes can guide cognitive-behavioral therapies and stress management.

INFLUENCES ON DISTRESS

Distress reflects both situational influences, such as life events, and intrapersonal influences such as personality traits. This section reviews situational influences, investigated either through controlled experiments, or through correlational studies of real-life stressors. Events that threaten or damage the person's wellbeing often provoke distress. Environmental factors promoting distress include (i) traumatic events, (ii) physical factors such as loud noise, (iii) failure to accomplish personal goals or perform effectively, (iv) social factors such as criticism by others, and (v) ill health. Adverse effects of agents of these kinds are well-documented,[12] although there is considerable variation in their effects across individuals and across different occasions. However, there are competing physiological, cognitive, and social accounts of the mechanisms that link these external stressors to psychological stress responses.

Physiological Influences

Historically, the distress concept has been important to physiological research, for example, as a concomitant of Selye's GAS.[1] Physiological studies of distress focus primarily on the brain mechanisms that may output and regulate negative affect. Evidence for a biological basis for distress is provided most directly by studies of brain damage in humans and animals, as well as brain imaging studies. Distress may be regulated by both subcortical structures such as the amygdala, as well as areas in prefrontal cortex that regulate emotional response.[13,14] Pharmacological studies link negative affects to the neurotransmitters sensitive to the drug concerned. Different negative emotions may be controlled by different brain systems: anxiolytic drugs operate through benzodiazepine receptors, and antidepressants such as fluoxetine (Prozac) through serotonergic pathways.

The contemporary neuroscience of stress suggests that there are a variety of brain systems that may underpin and regulate psychological distress, over differing timespans.[13] The autonomic nervous system (ANS) supports immediate response to stressor exposure, via its sympathetic and parasympathetic branches. The hypothalamic-pituitary-adrenocortical (HPA) axis produces elevations in circulating glucocorticoids which afford stress regulation over periods of tens of minutes. Top-down cortical regulation of the limbic structures implicated in the ANS and HPA is especially important when the source of stress is social or psychogenic in nature, as opposed to physical stressors such as blood loss or pain. Chronic stress produces various changes in neurochemistry and brain anatomy that may sensitize certain stress responses, such as those of the HPA and the sympatho-adrenomedullary systems, while suppressing others. Damaging changes to stress physiology associated with repeated, poorly regulated exposure to stressors are referred to as allostatic load.[14] Sensitization of response in chronic stress is consistent with the increased vulnerability to distress of individuals experiencing chronic "burnout," characterized by exhaustion and helplessness.[15]

Cognitive Influences

Cognitive models of stress are supported by experimental studies demonstrating that both the psychological and physiological impact of stressors are moderated by the person's beliefs and expectancies.[2,16] For example, negative moods may be induced by suggestive techniques such as reflecting on negative events or making statements that one is unhappy. Distress in the workplace reflects not just work demand, but also perceived control and decision latitude. Appraisal theories of emotion seek to link affect to the person's evaluation of external stimuli and their personal significance. Anxiety may then relate to appraisals of personal threat and uncertainty, and depression to uncontrollable personal harm. Appraisal is not necessarily accessible to consciousness, and there may be several distinct appraisal processes operating in tandem. Appraisal theories tend to focus on the differing cognitive antecedents of specific emotions, rather than on undifferentiated distress. Studies of transient distress measured as a unitary state show that it relates to appraisals of threat, loss and uncontrollability of events, and to use of emotion-focused coping strategies such as self-criticism.[7]

The transactional model of stress[2] embeds appraisal within a wider matrix of person-environment interaction within which negative emotions are tied to core themes describing the person-environment relationship. Core relational themes also relate to action tendencies. Transactional theory emphasizes the person's active attempts to cope with threatening or damaging events. In general, distress is likely to develop when the people appraise themselves as failing to cope adequately, and lacking personal control over significant events. Contemporary theory also emphasizes the differing relational themes associated with different negative emotions. The dynamic nature of the stress process is supported by evidence from longitudinal studies for reciprocal relationships between distress and life events, and between distress and health perceptions. In cases of chronic distress, negative cognitions drive distress responses which in turn feed back into further negative cognitions.[16]

Consistent with the transactional perspective, people also attempt to regulate distress through a variety of strategies, varying in their efficacy. Research has delineated different strategies, their neurological concomitants, and their effectiveness in countering distress.[17] For example, constructive reappraisal of a stressful event appears to be more effective than suppression of outward emotional response. Wells[16,18] has identified dysfunctional metacognitions as perpetuating maladaptive emotion-regulation across a range of anxiety and mood disorders. Perseverative worry and rumination may be driven by metacognitive beliefs that maintain the focus of attention on disturbing thoughts and images, such as the (typically false) belief that brooding on problems is an effective means for solving them. Metacognitive processes elicit a cognitive-attentional syndrome (CAS) that perpetuates distress. The CAS is at the core of maladaptive feedback loops that maintain the focus of attention onto stress symptoms, reinforce negative self-referent beliefs, and discourage the person from confronting feared situations directly. Clinically oriented research has also addressed the concept of distress tolerance, defined as the capacity to withstand aversive experiential states.[19]

Social Influences

Disruption of social relationships associated with bereavement, marital discord, and unemployment are among the most potent factors eliciting distress.[12] Conversely, availability of social support often functions to alleviate stress responses.[20] Early studies focused on the role of objectively defined social networks in protecting ("buffering") the person from the mental and physical health impacts of adverse events. Subsequently, it has become clear that perceptions of social support may be as influential in mitigating distress as actual support is, consistent with cognitive perspectives on negative affect. In addition, perceived social support may promote well-being even in the absence of significant external stressors.

Some social psychologists characterize as inadequate attempts to locate distress solely within the individual. Instead, distress may reflect people's discourses about their problems, and interpersonal interactions within a social and cultural context. For example, depression has been attributed to loss of socially defined roles.[21] Social psychologists also emphasize the importance of the expression and display of emotion, which reflects social motivations and norms. Distress may propagate within social networks as a consequence of threats common to the group and emotional contagion.[22]

INDIVIDUAL DIFFERENCES IN VULNERABILITY TO DISTRESS

Neuroticism and Negative Affect

Some individuals are more distress-prone than others. Psychometric and experimental studies identify stable personality traits that relate to the person's predisposition to experience negative emotion.[23] The Five Factor Model (FFM) of personality includes a dimension of neuroticism (vs. emotional stability) associated with proneness to negative affects including distress. Studies of temperament in infants and children, based on observations of behavior, also identify negative affectivity or distress-proneness as a fundamental dimension.[24] Biological models link neuroticism broadly to sensitivity to punishment cues, and more specifically to sensitivity of brain systems controlling fight-flight response and behavioral inhibition in response to threat.[25] Neuroimaging studies of neuroticism and allied traits confirm their basis in brain systems for negative emotion, such as amygdala.

Neuroticism predicts negative moods such as depression and anxiety, although the strength of the relationship varies with situational stressors.[23] It also predicts vulnerability to worry and disturbances of cognition. The association between neuroticism and distress is substantiated by both controlled experimental studies, and field studies of everyday mood using diary or experience-sampling methods. Neuroticism is also elevated in patients suffering from emotional disorders. Traits for specific negative emotions, such as Spielberger's trait anxiety and trait depression constructs,[6] correlate both with neuroticism and with state measures of the emotion concerned. Longitudinal studies suggest that neuroticism is a factor predisposing clinical disorder, although there is some reciprocity between the personality trait and disorder. Elevated neuroticism may also be an outcome or a "scar" of mental illness. Other FFM traits including low extraversion and (in some circumstances) low conscientiousness have also been linked to vulnerability to distress.

Neuroticism also relates to self-reported somatic distress and medical symptoms, including various psychosomatic conditions.[23] It remains controversial whether more neurotic individuals are genuinely more prone to illness, or whether they are just prone to complain about physical symptoms. However, a causal role for neuroticism is suggested by growing evidence that various negative affects, including depression, anger, and anxiety, operate as risk factors for illnesses including coronary heart disease,[26] although effect sizes for the association between distress and illness are often small. Relationships may reflect both direct effects of stress on physiological functioning, and indirect effects of personality that are mediated by health behaviors.

Current personality models generally adopt an interactionist or diathesis-stressor perspective, such that the distress response depends on the interaction of person and situation factors.[23] Hence, although more neurotic individuals are more vulnerable to distress, the extent to which they experience greater distress than less neurotic persons on a given occasion varies with situational factors. A more subtle view of neuroticism is also suggested by life event research. Consistent with the negative affectivity hypothesis, more neurotic individuals seem to experience life events as more distressing. However, neurotic persons also seem to experience a higher frequency of life events, which may reflect behavioral difficulties in adapting to life circumstances. For example, neuroticism relates to a higher frequency of interpersonal conflicts. Thus, there are at least four models of the relationships between neuroticism and distress. First, high neurotics may be generally distress-prone, independent of external circumstances. Second, high neurotics may react to external events with elevated distress. Third, high neurotics may behave in ways that elicit a higher frequency of distressing events. Fourth, high neurotics may be poor at regulating distress following its initial elicitation. For example, consistent with models emphasizing metacognitive processes in distress,[18] neurotic individuals may be prone to ruminate on their distress and its sources, prolonging the experience. The dynamic nature of distress response is demonstrated also in studies of temperament in children. Overly distress-prone infants may be perceived as whiny or clingy, eliciting parental annoyance or neglect, which provokes further dysregulative responses from the child.

There are also multiple causal explanations for the various impacts of neuroticism on distress. Neuroticism is known to have a substantial inherited component, and some progress has been made in molecular genetics in identifying specific alleles for vulnerability to negative emotions that may be associated with the trait.[27] However, the pathways connecting genes for distress-prone personality to specific brain mechanisms are still poorly understood. From the perspective of the transactional model of stress, neuroticism and similar traits appear to possess a distinctive style of processing information related to threat in relation to the self.[16,23] High neuroticism individuals are prone to negative self-appraisals; they appraise the world as demanding and threatening, and themselves as ineffective in dealing with external demands. More neurotic individuals are prone to cope with potentially stressful events through emotion-focused strategies such as ruminating on their problems or criticizing themselves. Such strategies are frequently maladaptive and may sometimes even exacerbate the problem and contribute to clinical disorder. Neuroticism may also relate to further forms of ineffective coping, such as avoidance of the problem and to reduced task-focused coping.

Distress Vulnerability and Resilience

Research has also focused on traits that specifically predict distress response to challenging events,[28] as opposed to traits that correlate with distress irrespective of external stressors. Spielberger's pioneering work[6] showed that trait anxiety is a vulnerability factor for distress. That is, it represents a predisposition to experience transient state anxiety in threatening environments, but, consistent with interactionism, the trait anxious individual does not experience state anxiety on all occasions. Experimental studies suggest that trait anxiety correlates with state anxiety under conditions of ego-threat, such as personal criticism. Various dimensions of metacognitive style that promote excessive attention to anxious thoughts and images elicited by stressful events, leading to rumination, avoidance of distressing situations, and vulnerability to emotional disorder.[18] Other traits such as optimism-pessimism, self-focus of attention, self-efficacy, and attributional style have also been considered as possible resilience factors, with mixed results.[23]

By contrast, traits that attenuate stress response are said to confer resilience. Several traits of this kind have been identified.[28] Hardiness allows people to interpret stressful and painful experiences as a normal aspect of existence.[29] It has multiple facets referring to, respectively, a sense of life and work commitment, strong feelings of control, and openness to change and challenges. Hardiness may be especially valuable to warfighters who must cope with the stresses of combat; low hardiness may increase the risk of traumatic stress. Grit is a somewhat similar trait, defined as strong motivations to accomplish long-term goals accompanied by perseverance and determination in the face of obstacles.[30]

Another family of traits that may confer resilience is that associated with emotional intelligence, defined as a set of aptitudes and competencies for perceiving, understanding, and managing emotion.[31] A metaanalysis has confirmed that emotional intelligence is associated with better mental health.[32] Experimental studies provide mixed results, but there is some evidence that higher emotional intelligence is associated with weaker distress responses to stress manipulations, as well as with psychophysiological stress markers.[31]

Distress-related traits tend to intercorrelate with each other, and also with broader measures of neuroticism and negative affectivity.[23] One view is that the various traits predict distress because they overlap with neuroticism, and the narrower trait constructs may add little to the broad trait. Similarly, when measured by questionnaire, emotional intelligence may be highly correlated with low neuroticism. Nevertheless, there is evidence for discriminative validity for various resilience and vulnerability traits: distress is best predicted from multiple traits. As for neuroticism, both psychobiological and cognitive processes may mediate the effects of these traits.

Contextualized Traits

Neuroticism and other traits appear to act as generalized factors predisposing distress across a variety of contexts. However, in many cases people possess traits linked to specific contexts, which predict distress over and above more generalized traits. A paradigmatic case is that of test anxiety, measured with scales that ask respondents specifically about their thoughts and feelings when required to take tests and examinations.[10] People show stable individual differences in vulnerability to test anxiety, which predict how they feel and perform in examination settings. Researchers have focused on evaluative anxiety in various other settings, using measures of social anxiety, math anxiety, computer anxiety, and so forth. Such measures typically correlate with neuroticism but have better predictive validity in the appropriate context. The contexualized trait may reflect both neuroticism and the person's more specific stable cognitions of the context concerned. Much of this work has focused on anxiety as a distress symptom, but it is likely that contextualized traits predict other distress symptoms in a similar way. For example, individuals vulnerable to driver stress are prone to experience both anxiety and unhappiness during vehicle operation.[33] In health psychology, there is interest in scales geared toward the distress that patients feel as a result of their illnesses.[5,34]

PSYCHOLOGICAL CONCOMITANTS OF DISTRESS

Distress has various behavioral correlates including impairment of objective performance, bias in selective attention, and overt clinical symptoms. These findings are explained by theories of self-regulation.[16,35] In these models, self-regulation operates homeostatically to reduce the discrepancies between actual and preferred status. Distress signals self-discrepancy and the need for coping efforts to reduce self-discrepancy and maintain adaptation to the external environment. The regulatory system may have physiological as well as psychological expressions. It also operates within a social context in that interactions with others provide strong cues toward actual status, and ideal status reflects in part societal and cultural norms. Experimental studies of attention and performance identify biases in the cognitions supporting self-regulation, which may be the source of clinical disorders associated with distress.

Attention and Performance

Trait and state distress measures are associated with slower and/or less accurate performance on a variety of tasks.[16] Both anxiety and depression tend to be associated with performance impairment. Contextualized distress measures are also typically associated with performance impairments within the context concerned: for example, test anxiety predicts poorer examination performance.[10] It is the cognitive rather than the emotional aspects of the anxiety state which interfere with performance. Self-referent worries appear to be especially damaging, because attention and/or working memory may be diverted from the task at hand onto personal preoccupations. Detrimental effects of depression have been similarly ascribed to worry and rumination.[16] Loss of performance may reflect loss of attentional resources, controlled processing, or working memory, although performance effects may also be moderated by motivational factors. Thus, attentional impairment is not a fixed "cost" of distress, but a consequence of a style of self-regulation which promotes worry, self-focus of attention, and emotion-focused coping, at the expense of performance goals.

Distress is also associated with cognitive bias and enhanced processing of negative stimuli.[16,36] Such effects have been demonstrated in studies of experimentally induced moods, in studies comparing the performance of distressed patients with controls, and in studies comparing nonclinical groups selected for high and low distress. Anxious individuals are slow to name the color in which threat-related words are printed (the emotional Stroop effect), and also appear to attend preferentially to threat stimuli in other paradigms. Other distressed groups are similarly biased toward negative stimuli congruent with their source of distress. These effects may reflect either an automatic bias toward threat stimuli in vulnerable individuals, or a consequence of deliberately chosen strategies, or both voluntary and involuntary processes. Various other tasks, including judgment, decision-making, and memory, may also show cognitive bias. From the self-regulative perspective, distress may relate to the extent to which the person balances the goal of maintaining awareness of threat against focusing on the immediate task at hand.

Clinical Disorder

Distress is a central symptom of both mood and anxiety disorders.[9] A simple view of clinical distress is that it represents the upper extreme of a continuum of vulnerability to negative emotion. However, clinical disorder involves qualitative abnormalities of behavior in addition to distress such as avoidance of challenging situations, self-harm, and difficulties in social interaction. Distress in clinical depression and anxiety is bound up with problems in behavioral adaptation to everyday life. Self-regulative approaches to understanding distress contribute to understanding the maladaptive nature of these disorders.[16,18] Central to both anxiety and depression are stable but false negative beliefs about the self, which may be represented as "schemas." Anxious persons believe themselves to be especially vulnerable to threats of certain kinds, whereas the depressed patient experiences cognitions of lack of self-worth and hopelessness. Cognitions and behaviors differ across clinical conditions, but, in general, unrealistic negative self-beliefs feed into faulty appraisal of the person's place in the external world, choice of ineffective coping strategies such as worry, and consequent distress.

The view of distress as a sign of maladaptive self-regulation has implications for therapy.[16,18] In mild, subclinical distress conditions, it may be sufficient for the person to learn specific coping skills that permit them to manage the specific situations that provoke distress. For example, individuals with mild social anxiety may be helped by explicit instruction in social skills. Distress is alleviated both through the person's increased confidence and positive self-appraisal, and through the likelihood that the person's greater skill will lead to more positive outcomes in the problematic situation. In more severe cases, stress management techniques such as skills training fail to address the underlying cognitive distortions which drive both distress and behavioral problems. Cognitive-behavior therapy seeks to uncover and modify faulty cognitions through a variety of techniques. Empirical studies show that clinical improvement in behavior and affect is accompanied by a decline in cognitive bias.

Glossary

Affect A general term for emotional or feeling states, including specific emotions such as anxiety and depression.

Metacognition The process of monitoring and controlling internal thoughts, images, and subjective states.

Neuroticism A personality trait associated with a predisposition to experience negative affect.

Resilience The traits and processes that protect the person against adverse effects of stressors.

Self-regulation The motivational and cognitive processes controlling goal-directed personal adaptation to the external environment.

Stress A general term for the processes and responses associated with adaptation to demanding or challenging environments.

Trait A personal disposition or characteristic showing long-term stability and influencing behavior in a range of situations.

References

1. Selye H. *The Stress of Life.* New York, NY: McGraw-Hill; 1976.
2. Lazarus RS. *Stress and Emotion: A New Synthesis.* New York, NY: Springer Publishing; 1999.
3. Matthews G, Campbell SE, Falconer S, et al. Fundamental dimensions of subjective state in performance settings: task engagement, distress and worry. *Emotion.* 2002;2:315–340.
4. Thayer RE. *The Origin of Everyday Moods.* Oxford: Oxford University Press; 1996.
5. Faller H, Schuler M, Richard M, Heckl U, Weis J, Küffner R. Effects of psycho-oncologic interventions on emotional distress and quality of life in adult patients with cancer: systematic review and meta-analysis. *J Clin Oncol.* 2013;31:782–793.
6. Spielberger CD, Reheiser EC, Owen AE, Sydeman SJ. Measuring the psychological vital signs of anxiety, anger, depression, and curiosity in treatment planning and outcomes assessment. In: Maruish ME, ed. *The Use of Psychological Testing for Treatment Planning and Outcomes Assessment.* 3rd ed. Mahwah, NJ: Lawrence Erlbaum Associates; 2004:421–447. Instruments for Adults; vol. 3.
7. Matthews G, Szalma J, Panganiban AR, Neubauer C, Warm JS. Profiling task stress with the Dundee stress state questionnaire. In: Cavalcanti L, Azevedo S, eds. *Psychology of Stress: New Research.* Hauppage, NY: Nova Science; 2013:49–90.
8. Waters AM, Nazarian M, Mineka S, et al. Context and explicit threat cue modulation of the startle reflex: preliminary evidence of distinctions between adolescents with principal fear disorders versus distress disorders. *Psychiatry Res.* 2014;217:93–99.
9. Watson D. Differentiating the mood and anxiety disorders: a quadripartite model. *Annu Rev Clin Psychol.* 2009;5:221–247.
10. Zeidner M. *Test Anxiety: The State of the Art.* New York, NY: Plenum Press; 1998.
11. Katon WJ. Epidemiology and treatment of depression in patients with chronic medical illness. *Dialogues Clin Neurosci.* 2011;13:7–23.
12. Hammen C. Stress and depression. *Annu Rev Clin Psychol.* 2004;1:293–319.
13. Ulrich-Lai YM, Herman JP. Neural regulation of endocrine and autonomic stress responses. *Nat Rev Neurosci.* 2009;10(6):397–409.
14. McEwen BS. Stressed or stressed out: what is the difference? *J Psychiatry Neurosci.* 2005;30:315–318.
15. Seidler A, Thinschmidt M, Deckert S, et al. The role of psychosocial working conditions on burnout and its core component emotional exhaustion—a systematic review. *J Occup Med Toxicol.* 2014;9:1–13.
16. Wells A, Matthews G. *Attention and Emotion: A Clinical Perspective.* Classic Edition. Hove, Sussex: Erlbaum; 2015.
17. Sheppes G, Scheibe S, Suri G, Radu P, Blechert J, Gross JJ. Emotion regulation choice: a conceptual framework and supporting evidence. *J Exp Psychol Gen.* 2014;143:163–181.
18. Wells A. *Metacognitive Therapy for Anxiety and Depression.* New York, NY: Guilford Press; 2009.
19. Anestis MD, Lavender JM, Marshall-Berenz EC, Gratz KL, Tull MT, Joiner TE. Evaluating distress tolerance measures: interrelations and associations with impulsive behaviors. *Cogn Ther Res.* 2012;36:593–602.
20. Taylor S. Social support: a review. In: Friedman H, ed. *Oxford Handbook of Health Psychology.* New York, NY: Oxford University Press; 2011:189–214.
21. Oatley K. *Best-Laid Schemes: The Psychology of Emotions.* Cambridge: Cambridge University Press; 1992.
22. Rimé B. Emotion elicits the social sharing of emotion: theory and empirical review. *Emot Rev.* 2009;1:60–85.
23. Matthews G, Deary IJ, Whiteman MC. *Personality Traits.* 3rd ed. Cambridge: Cambridge University Press; 2009.
24. Rothbart MK, Sheese BE, Conradt ED. Childhood temperament. In: Corr PL, Matthews G, eds. *Cambridge Handbook of Personality.* Cambridge: Cambridge University Press; 2009:177–190.
25. Corr PJ, McNaughton N. Neuroscience and approach/avoidance personality traits: a two stage (valuation–motivation) approach. *Neurosci Biobehav Rev.* 2012;36:2339–2354.
26. Suls J, Bunde J. Anger, anxiety, and depression as risk factors for cardiovascular disease: the problems and implications of overlapping affective dispositions. *Psychol Bull.* 2005;131:260–300.
27. Montag C, Reuter M. Disentangling the molecular genetic basis of personality: from monoamines to neuropeptides. *Neurosci Biobehav Rev.* 2014;43:228–239.
28. Fletcher D, Sarkar M. Psychological resilience: a review and critique of definitions, concepts, and theory. *Eur Psychol.* 2013;18:12–23.
29. Maddi SR, Kobasa SC. The development of hardiness. In: Monat A, Lazarus RS, eds. *Stress and Coping: An Anthology.* 3rd ed. New York, NY: Columbia University Press; 1991:245–257.
30. Duckworth A, Gross JJ. Self-control and grit: related but separable determinants of success. *Curr Dir Psychol Sci.* 2014;23:319–325.
31. Matthews G, Zeidner M, Roberts RD. *Emotional Intelligence 101.* New York, NY: Springer Publishing Company; 2012.
32. Martins A, Ramalho N, Morin E. A comprehensive meta-analysis of the relationship between emotional intelligence and health. *Personal Individ Differ.* 2010;49:554–564.

33. Rowden P, Matthews G, Watson B, Biggs H. The relative impact of work-related stress, life stress and driving environment stress on driving outcomes. *Accid Anal Prev*. 2011;43:1332–1340.
34. Donovan KA, Grassi L, McGinty HL, Jacobsen PB. Validation of the distress thermometer worldwide: state of the science. *Psychooncology*. 2014;23:241–250.
35. Carver CS, Scheier MF. *On the Self-Regulation of Behavior*. New York, NY: Cambridge University Press; 1998.
36. Cisler JM, Koster EH. Mechanisms of attentional biases towards threat in anxiety disorders: an integrative review. *Clin Psychol Rev*. 2010;30:203–216.

CHAPTER

27

Depersonalization: Systematic Assessment

M. Steinberg

Independent Practice, Naples, FL, USA

OUTLINE

Abstract

Depersonalization is often not the presenting complaint, making familiarity with the means for detection and assessment critical to avoiding misdiagnosis and ineffective treatment. Depersonalization occurs on a spectrum, from few/transient episodes in individuals with a variety of psychiatric disorders, to recurrent or ongoing episodes experienced in those with posttraumatic and dissociative disorders. Accurate diagnosis requires assessment of depersonalization within a context of other dissociative symptoms in order to properly characterize an underlying dissociative disorder, or to rule one out. The author reviews assessments of depersonalization in adolescents and adults using the Structured Clinical Interview for Dissociative Disorders (SCID-D) (Steinberg, 1994, DSM-5/ICD version). The SCID-D evaluates depersonalization in the context of four additional dissociative symptoms: amnesia, derealization, identity confusion, and identity alteration. Many studies have documented the SCID-D's good-to-excellent reliability and validity for detection of depersonalization and its characterization within the full spectrum of dissociative symptoms.

Depersonalization is characterized by a sense of detachment from the self. The symptom itself may manifest in a variety of axis I or axis II psychiatric disorders. The *Diagnostic and Statistical Manual of Mental Disorders*, 5th edition (DSM-5)[1] describes depersonalization as involving experiences of unreality, detachment, or being an outside observer with respect to one's thoughts, feelings, sensations, body, or actions. The sense of detachment itself may be experienced in various ways. Commonly it appears as out-of-body experiences giving a sense of division into a participating and an observing self, resulting in the sense of going through life as though one were a machine or robot.[2] In some cases, there exists a feeling that one's limbs are changing in size or are separated from the body.

It is important to distinguish between recurrent to persistent depersonalization that is characteristic of both, the dissociative disorders and of a subset of individuals with posttraumatic stress disorder,[3–11] versus the occasional episodic depersonalization which occurs in patients with other nondissociative axis I or II disorders,[3–6,8,9] versus the very brief or isolated episodes experienced by persons in the nonpsychiatric population (normal controls).[2,12]

DEFINITION AND CHARACTERISTICS

Although depersonalization was first described in 1872, it was not named until 1898, when Dugas[13] contrasted the feeling of loss of the ego with a real loss. In 1954, Ackner remedied the lack of clearly defined boundaries of the symptom by describing the four salient features: (1) feeling of unreality or strangeness regarding the self, (2) retention of insight and lack of delusional elaboration, (3) affective disturbance resulting in loss of all affective responses except discomfort over the depersonalization, and (4) an unpleasant quality that varies in

http://dx.doi.org/10.1016/B978-0-12-800951-2.00027-3

intensity inversely with the patient's familiarity with the symptom.[14] For the purposes of clinical assessment, Steinberg defined depersonalization as one of five core symptoms of dissociation, the other four consisting of amnesia, derealization, identity confusion, and identity alteration.[2,15–18] Each of the five dissociative disorders has characteristic symptom profiles of these core dissociative symptoms. For this reason, it is essential that the symptom of depersonalization is evaluated within the context of the other dissociative symptoms and not as an isolated symptom.[2,15–18]

KEY POINTS

- Depersonalization is characterized by a sense of detachment or disconnection from one's self, and is commonly experienced within a variety of psychiatric conditions, as well as in nonpsychiatric populations.
- Depersonalization occurs along a spectrum, from a few transient episodes that are not associated with dysfunction or distress, to recurrent or persistent episodes which result in dysfunction or distress.
- Depersonalization as seen in both dissociative disorders and in a subset of individuals with posttraumatic stress disorder can be distinguished from depersonalization occurring in other psychiatric and nonpsychiatric samples by use of the *SCID-D Interview* to assess the frequency, nature, and context of the patient's depersonalization experiences.
- Systematic assessment of depersonalization in the context of other dissociative symptoms, as assessed in the *SCID-D Interview*, is essential for accurate diagnosis and effective treatment of an underlying dissociative disorder.

Episodes of depersonalization can accompany or even may precipitate panic attacks and/or agoraphobia; they may also be associated with dysphoria.[19] Chronic depersonalization frequently results in the patient's acceptance of the symptoms, in a manner of resignation. Patients experience difficulty putting their experience into words, but often compare their feelings to such states as being high on drugs, seeing themselves from the outside, or floating in space and watching themselves. Other descriptions of depersonalization include feelings of being unreal, or in severe cases, include the feeling of being numb or dead, or the lack of all feeling, which may be attributed to and/or misdiagnosed as depression.

Depersonalization has been reported to be a common complaint among psychiatric patients, after depression and anxiety. The assessment and prevalence of depersonalization has been impeded by (1) its multifaceted presentation and relative strangeness of the symptoms, (2) the difficulty that patients have in communicating their depersonalization experiences, and (3) the lack of widespread training in the diagnostic tools used for the systematic assessment of depersonalization.[2] Detection is further complicated by the fact that depersonalization may not be accompanied by altered observable or social behavior indicating the patient's dysfunction or distress.

ETIOLOGY

Various biological and psychodynamic theories have been advanced for the etiology of depersonalization[20–23]: (1) physiological or anatomical disturbance, with feelings of depersonalization produced by temporal lobe function and various metabolic and toxic states, (2) the result of a preformed functional response of the brain to overwhelming traumata, (3) a defense against painful and conflictual affects such as guilt, phobic anxiety, anger, rage, paranoia, primitive fusion fantasies, and exhibitionism, (4) a split between the observing and the participating self, allowing the patient to become a detached observer of the self, and (5) the result of childhood maltreatment, especially emotional abuse.

Depersonalization has been reported to be a normal reaction to life-threatening events, such as accidents, serious illnesses, and near-death experiences and was noted in 66% of 101 survivors of life-threatening experiences.[12] Depersonalization is common among victims of sexual abuse, political imprisonment, torture, and cult indoctrination. Symptoms of depersonalization are often associated with hypnosis, hypnagogic and hypnopompic states, sleep deprivation, sensory deprivation, hyperventilation, and drug or alcohol abuse.

Depersonalization, as a brief, isolated symptom, is nonspecific and not necessarily pathognomonic of any clinical disorder. Research indicates that it is the persistence, nature, and context of depersonalization that differentiates depersonalization in people without psychiatric disorders from persons with dissociative and nondissociative disorders.[2–12] Table 1 is useful for distinguishing between common mild depersonalization and pathological depersonalization. Table 2 summarizes the spectrum of depersonalization.

The differential diagnosis tree of depersonalization (Figure 1) illustrates procedures for distinguishing between depersonalization disorder and other disorders that may resemble it. The differential diagnosis of patients experiencing recurrent or persistent depersonalization should include the dissociative disorders, posttraumatic stress disorder, and possible medical disorders/organic etiology, most commonly acute head trauma, seizure disorders, and acute drug or alcohol use.

TABLE 1 Distinguishing Between Common Mild Depersonalization and Pathological Depersonalization

Common mild depersonalization	Transient depersonalization	Pathological depersonalization
Context		
Occurs as an isolated symptom	Occurs as an isolated symptom	Occurs within a constellation of other dissociative or nondissociative symptoms or with ongoing interactive dialog
Frequency		
One or few episodes	One or few episodes that are transient	Persistent or recurrent depersonalization
Duration		
Depersonalization episode is brief; lasts seconds to minutes	Depersonalization of limited duration (minutes to weeks)	Chronic and habitual depersonalization lasting up to months or years
Precipitating factors		
• Extreme fatigue • Sensory deprivation • Hypnagogic and hypnopompic states • Drug or alcohol intoxication • Sleep deprivation • Medical illness/toxic states • Severe psychosocial stress	• Life-threatening danger. This is a syndrome noted to occur in 33% of individuals immediately following exposure to life-threatening danger, such as near-death experiences and auto accidents (Noyes and Kletti[12]) • Single, severe psychological trauma	• Not associated with precipitating factors in column 1 exclusively • May be precipitated by a traumatic memory • May be precipitated by a stressful or traumatic event but occurs even when there is no identifiable stress • Results in dysfunction or distress

Reprinted with permission from Ref. 2.

TABLE 2 The Spectrum of Depersonalization on the SCID-D-R

DID and DDNOS	Nondissociative and personality disorders	No psychiatric disorder
Depersonalization questions elicit descriptions of identity confusion and alteration	No spontaneous elaboration	No spontaneous elaboration
Includes interactive dialogs between individual and depersonalized self	No interactive dialogs	No interactive dialogs
Recurrent-persistent	None-few episodes	None-few episodes

Note: *DID, dissociative identity disorder; DDNOS, dissociative disorder; not otherwise specified.*
Reprinted with permission from Ref. 16.

ASSESSMENT WITH THE STRUCTURED CLINICAL INTERVIEW FOR DSM-IV DISSOCIATIVE DISORDERS—REVISED

Several self-administered tests are available which screen for the presence of depersonalization.[4] Screening tests should not be used for the purpose of diagnoses, but rather for identification of at-risk cases to be evaluated further by clinical interview or diagnostic test.

The *Structured Clinical Interview for DSM-IV Dissociative Disorders—Revised* (*SCID-D-R*) is a diagnostic interview for the comprehensive assessment of dissociative symptoms and disorders, including the systematic identification of depersonalization. The interview can be used in adolescents and adults.[10,24] Developed in 1985 and extensively field tested, it is the only diagnostic instrument enabling a clinician to detect and assess the presence and severity of five core dissociative symptoms (amnesia, depersonalization, derealization, identity confusion, and identity alteration) as well as the dissociative disorders (dissociative amnesia, depersonalization/derealization disorder, dissociative identity disorder, other specified dissociative disorder, and unspecified dissociate disorder) as defined by DSM-5 criteria. The updated version of the SCID-D-R also allows for diagnosis based on ICD criteria.[17] The SCID-D-R is a semi-structured diagnostic interview with good-to-excellent inter-rater and test-retest reliability and discriminant validity.[3–5,10,25] Recent neuroimaging studies have validated the SCID-D-R's ability to discriminate those with complex dissociative disorders, for example, a recent study documented that SCID-D-R identified dissociative identity disorder patients had characteristic brain activity during state-switching that could not be duplicated by either fantasy-prone controls or trained feigners.[26]

Guidelines for the administration, scoring, and interpretation of the SCID-D-R are reviewed in the *Interviewer's Guide to the SCID-D-R*. Severity rating definitions were developed to allow clinicians to rate the severity of symptoms in a systematic manner and are included in the guide.[16,18]

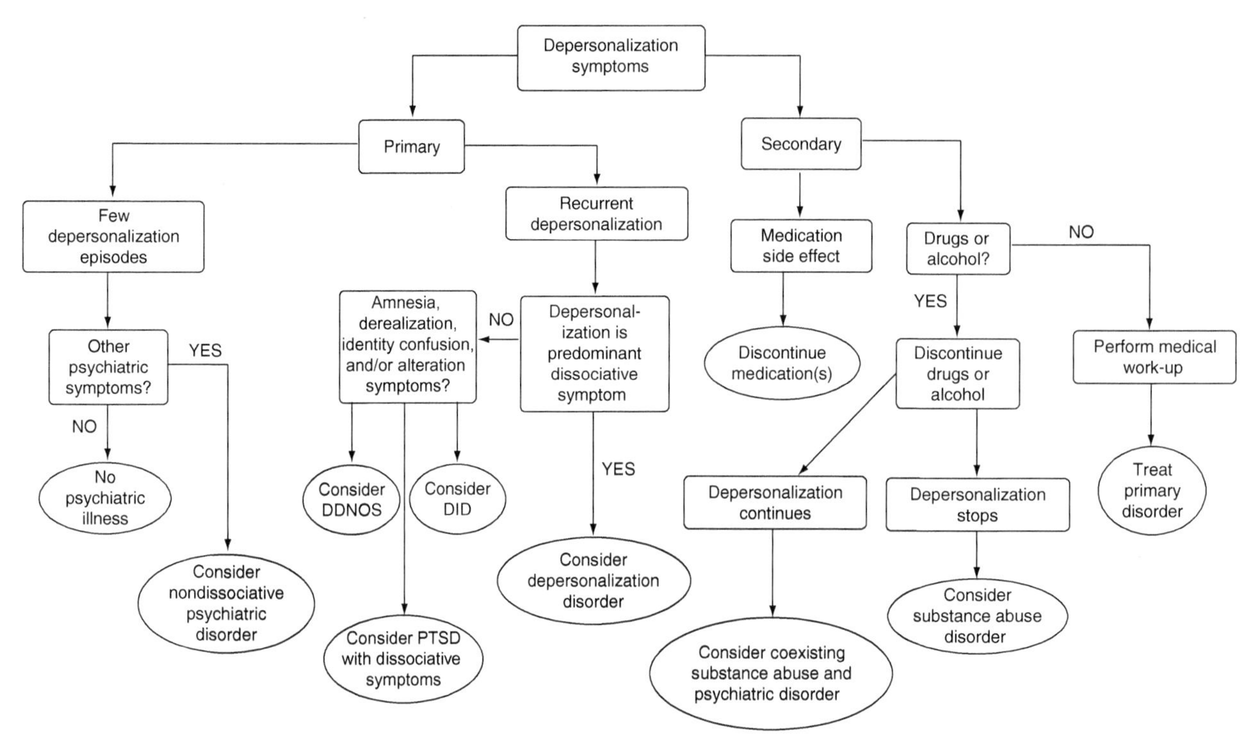

FIGURE 1 Differential diagnosis decision tree of depersonalization. Adapted with permission from Ref. 16.

The SCID-D-R can be used for symptom documentation for psychological and forensic reports.[25,27] Early detection of dissociative disorders, including depersonalization disorder, can be realized from the use of this specialized instrument, the format of which includes open-ended questions designed to elicit spontaneous descriptions of endorsed dissociative symptoms. The SCID-D-R has been demonstrated to be a valuable tool in differential diagnosis with patients of different ages (adolescents, as well as adults), backgrounds, previous psychiatric histories, and presenting complaints. It also plays a useful role in treatment planning, patient follow-up, and symptom monitoring.[28]

Correct diagnosis is vital to proper treatment of depersonalization. If the depersonalization is a symptom of a dissociative disorder, the symptoms can be alleviated by treatment of the underlying dissociative condition. The presence of depersonalization disorder itself is characterized by recurrent depersonalization. In instances in which the patient experiences only occasional episodes of depersonalization in the context of other nondissociative symptoms, the clinician should consider a diagnosis of a nondissociative psychiatric disorder.

CASE STUDY

The process of differential diagnosis of depersonalization may best be illustrated by presenting a case study. The study demonstrates the utility of the SCID-D-R in diagnostic assessment, patient education, and treatment planning. For space reasons, conventional formatting and content have been abbreviated.

Sample SCID-D-R Psychological Evaluation

Demographic information and chief complaint: Susan Walker is a 31-year-old administrative assistant at a community college who presented with the complaint of feeling detached from herself since adolescence. Past psychiatric history: Although the patient had no history of hospitalization for psychiatric disturbance, she began treatment for an episode of depression that interfered with her employment and social relationships. Although admitting to past casual use of marijuana, she had never been in treatment for substance abuse disorder.

Family history: Susan had a younger sibling; both children grew up in an intact but emotionally unsupportive family. Patient reported that both parents suffered mood swings and unpredictable temper outbursts.

Mental status exam: Susan answered questions with relevant replies; although she seemed slightly depressed, her affect appeared full range. She denied hallucinations, both auditory and visual, and evidenced no psychotic thinking. She denied acute suicidal or homicidal ideas.

SCID-D-R evaluation: The SCID-D-R was administered to systematically evaluate the patient's dissociative symptoms and was scored according to prescribed guidelines. Significant findings from the SCID-D-R interview follow. Susan denied experiencing severe episodes of amnesia, but endorsed a persistent sense of depersonalization, resulting in distress and interference with occupational and personal functioning. This feeling of depersonalization had been chronic and occurred all the time rather than episodically. Although the feeling varied in intensity with her overall stress level, the experience of depersonalization was always characterized by a general sense of detachment from life, rather than by disturbances in body image or a split between participating and observing parts of the self. Only a single isolated out-of-body experience had occurred. Susan experienced feelings of derealization that varied in intensity with the depersonalization, but she reported the depersonalization as the most distressing symptom. She described recurrent anxiety and panic episodes triggered by the depersonalization; it was the combination of depersonalization and panic attacks that led to the depression that brought her into therapy. Susan reported that the depersonalization has eroded her sense of control over her occupational functioning and other significant areas of her life, but she did not attribute feelings of loss of control to identity confusion or alteration. She denied having internal dialogs, feelings of possession, or acquiring unexplained possessions or skills. Her descriptions of internal struggle were focused on her feelings of unreality, not on conflicts between different aspects of her personality or different personalities within herself.

Assessment: Susan's symptoms are consistent with a primary diagnosis of a dissociative disorder based on DSM-5 criteria and ICD-10 criteria. Specifically, in the absence of substance abuse disorder or other organic etiology, her severe chronic feelings of unreality toward herself and the accompanying dysfunction (in the absence of other dissociative symptoms such as identity confusion and alteration) are consistent with a diagnosis of depersonalization disorder.

Recommendations: Although detailed discussion of treatment for depersonalization disorder is beyond the scope of this article, it would be standard practice to conduct a follow-up interview to review the findings of the SCID-D-R evaluation, to educate the patient regarding her symptoms, and to begin the process of individual psychotherapy.

CONCLUSIONS

Recent advances in the development of reliable diagnostic tools allow for early detection and accurate differential diagnosis of depersonalization. Research based on the SCID-D-R indicates that depersonalization occurs in individuals without psychiatric illness who experience few brief episodes following high stress, as well as in individuals with dissociative disorders who experience recurrent to ongoing episodes. In addition to the frequency of the depersonalization, the nature, severity, and context also distinguish cases of dissociative disorder from other nondissociative disorders.[1–10,15–18] To date, no double-blind trials of medication have found pharmacotherapy to be effective in the treatment of depersonalization. Further research is necessary in the form of controlled double-blind studies evaluating psychotherapeutic and/or pharmacotherapeutic agents. As the SCID-D-R allows for the assessment of the severity of depersonalization based on operationalized criteria, psychotherapy and pharmacotherapy trials can be systematically performed and can evaluate baseline and post-treatment severity levels of depersonalization. Given the frequency of misdiagnosis in patients suffering from depersonalization and other dissociative symptoms, earlier detection of dissociative symptoms and disorders using the SCID-D-R can allow for rapid implementation of effective treatment.

Glossary

Amnesia A specific and significant block of time that has passed but cannot be accounted for by memory.

Depersonalization Detachment from one's self, for example, a sense of looking at one's self as if one were an outsider.

Derealization A feeling that one's surroundings are strange or unreal. Often involves previously familiar people.

Dissociation Disruption in the usually integrated functions of conscious memory, identity, or perception of the environment. The disturbance may be sudden or gradual, transient, or chronic.

Identity alteration Objective behavior indicating the assumption of different identities or ego states, more distinct than different roles.[2,15–18]

Identity confusion Subjective feelings of uncertainty, puzzlement, or conflict about one's identity. Note: Identity confusion and identity alteration are defined as listed above by Steinberg for the purpose of Steinberg for the purpose of clinical assessment.[15–18]

References

1. American Psychiatric Association. *Diagnostic and Statistical Manual of Mental Disorders*. 5th ed. (DSM-5) Arlington, VA: American Psychiatric Association; 2013.
2. Steinberg M. *Handbook for the Assessment of Dissociation: A Clinical Guide*. Washington, DC: American Psychiatric Press; 1995.
3. Steinberg M. Advances in the clinical assessment of dissociation: the SCID-D-R. *Bull Menninger Clin*. 1995;59:221–231.
4. Mueller-Pfeiffer C, Kaspar Rufibach K, Wyss D, Perron N, Pitman R, Rufer M. Screening for dissociative disorders in psychiatric out- and day care-patients. *J Psychopathol Behav Assess*. 2013;35(4): 592–602.
5. Mueller C, Moergeli H, Assaloni H, Schneider R, Rufer M. Dissociative disorders among chronic and severely impaired psychiatric outpatients. *Psychopathology*. 2007;40(6):470–471.
6. Ginzburg K, Somer E, Tamarkin G, Kramer L. Clandestine psychopathology: unrecognized dissociative disorders in inpatient psychiatry. *J Nerv Ment Dis*. 2010;198:378–381.
7. Steinberg M, Schnall M. *The Stranger in the Mirror: Dissociation—The Hidden Epidemic*. New York, NY: Harper Collins; 2001.

8. Steinberg M, Rounsaville B, Buchanan J, Raakfeldt J, Cicchetti D. Distinguishing between multiple personality and schizophrenia using the Structured Clinical Interview for DSM-IV Dissociative Disorders. *J Nerv Ment Disord*. 1994;182:495–502.
9. Steinberg M, Siegel H. Advances in assessment: the differential diagnosis of Dissociative Identity Disorder and Schizophrenia. In: Moskowitz A, Schafer I, Dorahy M, eds. *Psychosis, Trauma and Dissociation: Emerging Perspectives on Severe Psychopathology*. 3rd ed. London, UK: John Wiley & Sons; 2008.
10. Sar V, Onder C, Kilincaslan A, Zoroglu S, Alyanak B. Dissociative identity disorder among adolescents: prevalence in a university psychiatric outpatient unit. *J Trauma Dissociation*. 2014;15 (4):402–419.
11. Bremner D, Steinberg M, Southwick S, Johnson DR, Charney DS. Use of the Structured Clinical Interview for DSM-IV Dissociative Disorders for systematic assessment of dissociative symptoms in posttraumatic stress disorder. *Am J Psychiatry*. 1993;150 (7):1011–1014.
12. Noyes Jr. R, Kletti R. Depersonalization in response to life-threatening danger. *Compr Psychiatry*. 1977;18:375–384.
13. Dugas L. A case of depersonalization [Un cas de depersonalization]. *Rev Philos*. 1898;45:500–507.
14. Ackner B. Depersonalization I: aetiology and phenomenology. *J Ment Sci*. 1954;100:838–853.
15. Steinberg M. *The Structured Clinical Interview for DSM-IV Dissociative Disorders—Revised (SCID-D)*. 2nd ed. Washington, DC: American Psychiatric Press; 1994.
16. Steinberg M. *The Interviewers' Guide to the Structured Clinical Interview for DSM-IV Dissociative Disorders—Revised*. 2nd ed. Washington, DC: American Psychiatric Press; 1994.
17. Steinberg M. *The Structured Clinical Interview for Dissociative Disorders (SCID-D) (DSM/ICD Version)*. Washington, DC: American Psychiatric Press; 1994.
18. Steinberg M. *The Interviewers' Guide to the Structured Clinical Interview for Dissociative Disorders (DSM/ICD Version)*. Washington, DC: American Psychiatric Press; 1994.
19. Simeon D, Gross S, Guralnik O, Stein DJ, Schmeidler J, Hollander E. Feeling unreal: 30 cases of DSM-III-R depersonalization disorder. *Am J Psychiatr*. 1997;154:1107–1113.
20. Somer E. Evidence-based treatment for depersonalisation-derealisation disorder (DPRD). *BMC Psychol*. 2013;1:20. http://dx.doi.org/10.1186/2050-7283-1-20.
21. Simeon D, Guralnik O, Schmeidler J, Sirof B, Knutelska M. The role of childhood interpersonal trauma in depersonalization disorder. *Am J Psychiatry*. 2001;158:1027–1033.
22. Bowman ES, Markand O. Psychodynamics and psychiatric diagnoses of pseudoseizure subjects. *Am J Psychiatr*. 1996;153(1):57–63.
23. Lambert M, Sierra M, Phillips MD. The spectrum of organic depersonalization: a review plus four new cases. *J Neuropsychiatry Clin Neurosci*. 2002;14(2):141–154.
24. Carrion V, Steiner H. Trauma and dissociation in delinquent adolescents. *J Am Acad Child Adolesc Psychiatry*. 2000;39:353–359.
25. Welburn K, Fraser G, Jordan S, Cameron C, Webb L, Raine D. Discriminating dissociative identity disorder from schizophrenia and feigned dissociation on psychological tests and structured interview. *J Trauma Dissociation*. 2003;4(2):109–130.
26. Reinders A, Willemsen A, Vos H, den Boer J, Nijenhuis E. Fact or factitious? A psychobiological study of authentic and simulated dissociative identity states. *PLoS ONE*. 2012;7(6):e39279.
27. Steinberg M, Hall P, Lareau C, Cicchetti D. Recognizing the validity of dissociative symptoms and disorders using the SCID-D-R: guidelines for clinical and forensic evaluations. *South Calif Interdiscip Law J*. 2001;10(2):225–242.
28. Steinberg M, Hall P. The SCID-D diagnostic interview and treatment planning in dissociative disorders. *Bull Menninger Clin*. 1997;61:108–120.

CHAPTER

28

Emotional Inhibition

H.C. Traue[1], H. Kessler[2], R.M. Deighton[3]

[1]The University of Ulm, Ulm, Germany
[2]The Ruhr University Bochum, Bochum, Germany
[3]The Cairnmillar Institute, Melbourne, VIC, Australia

OUTLINE

Abstract

Emotional inhibition constitutes a dysfunctional verbal and non-verbal interaction between individuals. Emotional inhibition may be classified into genetic, repressive, suppressive, and deceptive inhibition, and the extreme form, emotional implosion. Overt emotional inhibition is characterized by reduced expressiveness, unemotional language, and shyness, all of which are related to dysfunctional bodily reactions and may be adaptive in a short-term social stress situation. In the long run, emotional inhibition is likely to have a harmful effect on the individual along any of three pathways: neurobiological, social-behavioral, and cognitive. There is implicit knowledge in most societies that emotional inhibition has negative health implications. Most psychotherapeutic techniques are directed at emotional behavior and experience and focus largely on changes in intra- and inter-individual emotional regulation and the construction of meaning from emotional experience.

EMOTION AND INHIBITION

Emotions are essentially transactions between individuals and their social environment. They give personal meaning to external and internal stimuli and communicate meaning from the individual to others which is relevant to the individual's needs or concerns. Emotions are composed of interpretations of intero- and exteroceptive stimuli, intentions, physiological patterns of arousal, and motor behavior including overt emotional expressiveness. The interaction of these different components in the individual and the social and physical environment is mediated by the central nervous system. From a system regulation point of view, emotional expressiveness has two important functions: first, it serves a communicative function in that it facilitates the regulation of person-environment transactions and, second, the feedback function of behavioral expressions controls the intraindividual regulation of emotion. This means that actively responding toward an environmental trigger may influence an experience indirectly through the attenuation of a negative emotional stimulus or directly through self-regulation. In other words, we can take active (and expressive) steps to have our needs met or we can make internal adjustments (e.g., reviewing the validity of the felt need). Thus, expressive behavior can serve simultaneously as a component of emotional processes and as a coping response. Emotional inhibition may be part of dysfunctional coping.

Three prominent scientists of the turn of the century, all of whom were active at a time of major discoveries in neurophysiology, contributed to important developments in the concept of inhibition. The neurophysiology of C.S. Sherrington (born 1857), the theory of the higher nervous system of I.P. Pavlov (born 1849), and the psychoanalysis of S. Freud (born 1856) transformed the principle of inhibition into a key concept in neurophysiology

Stress: Concepts, Cognition, Emotion, and Behavior
http://dx.doi.org/10.1016/B978-0-12-800951-2.00028-5

(in the case of Sherrington) and higher mental functioning (in the cases of Pavlov and Freud).

KEY POINTS

- Emotional inhibition can be considered as part of a maladaptative interaction between an individual and his or her social environment. This chapter highlights the role of dysfunctional emotion regulation, especially emotional inhibition in relation to physiological, social-behavioral, and cognitive processes.
- Although reduced expressiveness and heightened bodily reactions may be adaptive in a short-term social stress situation, emotional inhibition is likely to have a harmful health effect on the individual in the long-term. A heuristic model featuring pathways linking emotional responses with health problems via three pathways is the core of the somatization theory of emotional inhibition. It can also be applied to understandings of traumatic stress responses and associated brain systems.
- Neurobiological pathways are being better understood and more clearly delineated, with those related to expressive suppression particularly well-investigated.
- Most societies have at least some implicit knowledge that emotional inhibition has negative health implications and there is an awareness for the need for emotional regulation as well as for disclosure, sharing, and catharsis on the other hand. Such knowledge is implicit in universal cultural rituals and for modern societies the development of psychological techniques for enhancing emotional expressiveness.
- New psychotherapy approaches are developing increasingly refined methods to explore and integrate somatic experiencing and subjective meaning systems, notably mindfulness-based approaches, emotion-focused therapy, and emotion-oriented interventions.

For many years, an inverse relationship between expressive behavior and autonomic responsivity has been documented, such that the inhibition of overt emotional expressiveness can lead to an autonomic overreaction. This has been considered to be a significant factor in the etiology and maintenance of psychosomatic disorders. A number of early researchers in the first two decades of the century reported measurements of high physiological activity in subjects suppressing emotional expression. These studies led toward the concept of internalization and externalization, wherein two behavioral coping styles for dealing with psychological tension were discerned: behaviorally, outwardly directed, or physiologically, within the individual. Following this concept, the term internalizer has been used to describe a person exhibiting a low level of overt expressiveness under stress yet a high level of physiological excitation, whereas an externalizer is characterized by high expressiveness and a low level of physiological expressiveness in social situations. Temoshok[1] proposed a model of internalizing and externalizing coping styles integrating the severity of stressors which intended to predict the occurrence of mental disorders (dependent on degree of externalizing coping) and somatic disorders (dependent on degree of internalizing coping). Another inhibition theory, put forward by Pennebaker,[2] summarizes the process by which failure to confront traumatic events results in poorer health. The principal assumption of this theory is that inhibiting ongoing behavior, thoughts, and feelings requires physiological activity. It has been suggested that the increased autonomic responses of internalizers may reflect the work of behavioral inhibition. Over time, the work of inhibition acts as a low-level cumulative stressor. As with all cumulative stressors, sustained inhibition is linked to increases in stress-related diseases and various other disorders such as cardiovascular and skin disorders, asthma, cancer, and also pain.

DOMAINS OF EMOTIONAL INHIBITION

A model (see Figure 1) of how emotional stress, in a given social situation, can trigger or modulate health disorders is described in the next two sections. Health disorders and illness behavior are considered as different, but related, processes and distinct mechanisms may make common or separate contributions to a given disorder and its behavioral consequences.

On a phenomenological level, emotional stress can occur on a severity dimension ranging from daily stressors, through more traumatic life events, to more chronic or severe traumatic stress. Each of these levels of stress challenge an individual's coping to varying degrees. Emotional stress can be seen as being processed by way of an inhibition-implosion dimension (to implode means to collapse or cause to collapse inward in a violent manner as a result of external pressure), modulated by dispositional factors (innate, personality, and socialization).

Innate and socialization factors are of particular importance. Possible individual differences in limbic opioidergic pathways associated with increased vulnerability to stress induction have been suggested. Emotional processing relating to the idea of inhibition-implosion

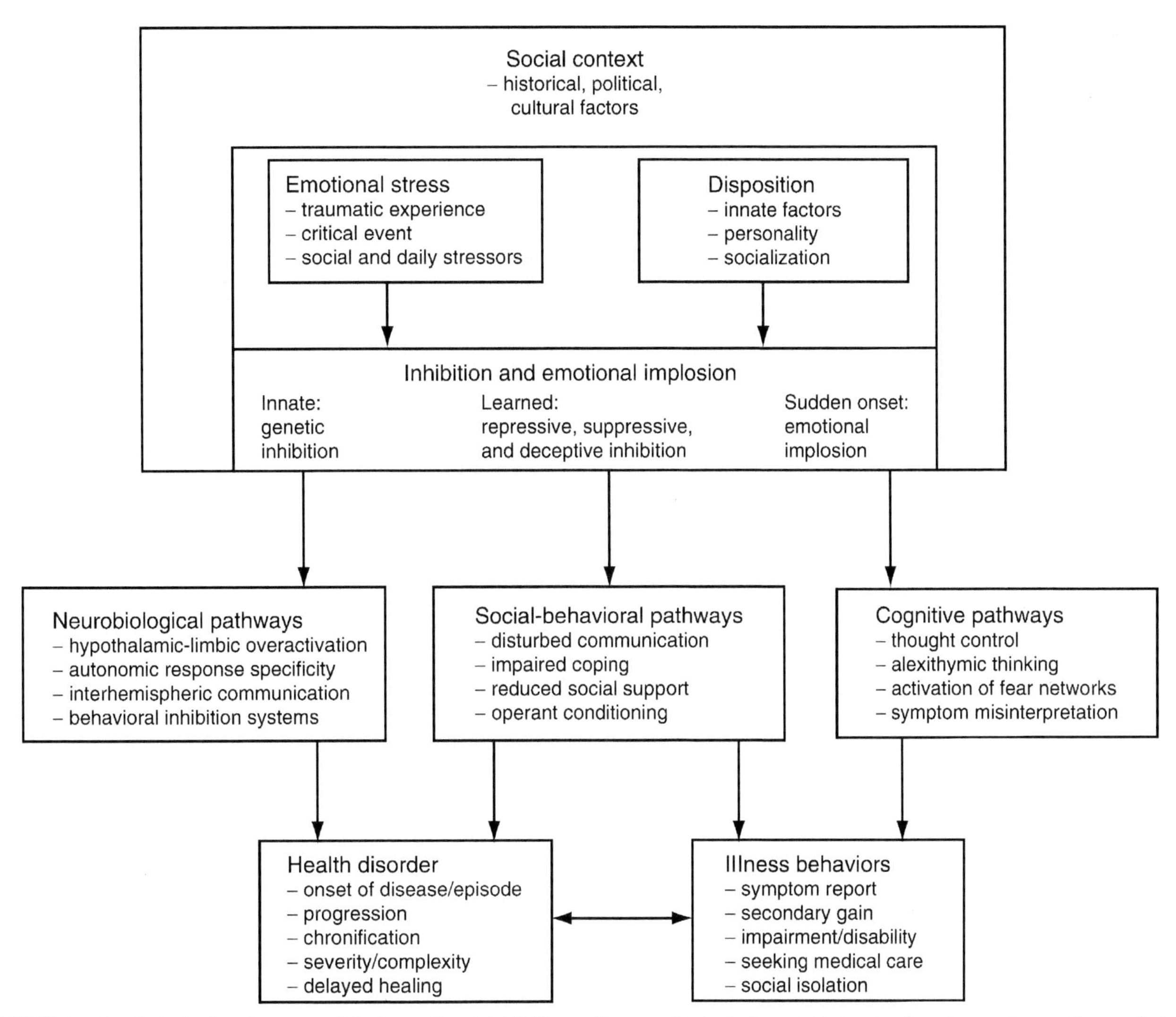

FIGURE 1 Psychological pathway model of emotional inhibition with neurobiological, social-behavioral, and cognitive pathways between stress/disposition, emotional processing, and subsequent health disorders or illness behaviors. Adapted from Traue.[3]

has been discussed in relation to several topics including control, suppression, type C personality, repression, alexithymia, ambivalence, emotional conflict, and experiential avoidance. Each of these concepts covers different aspects of overt emotional expressiveness from a personality or coping perspective.[4,5]

Inhibition is the most general term for the incomplete processing of emotional stress when these stressors induce bodily changes (physiological, endocrinological, or immunological) and the cognitive, emotional, and behavioral processing are dysfunctional such that subjective experience and spontaneous expression of emotions and action tendencies are separately or simultaneously attenuated and intra- and interpersonal regulation is disturbed. This process is seen as a product of innate and/or acquired behavior.

The mechanisms involved are classified as follows, reflecting different foci in the relevant literature: genetic inhibition, repressive inhibition, suppressive inhibition, and deceptive inhibition. All four classes of inhibition can occur in every state of emotion-induced psychophysiological arousal.

Genetic inhibition reflects the genetically determined basis of behavioral inhibition. Studies working with young children have classified those children who were least able to initiate interaction in a social situation with other children and adults as behaviorally inhibited. Most children's degree of behavioral inhibition has been shown to be stable over a period of 5 years. In inhibited children, increased levels of arousal, norepinephrine, and salivary cortisol have been found.

Repressive inhibition is defined as emotional processing with attenuated subjective experience of emotional arousal. Emotional expressive responses in repressive inhibition can be based solely on cognitive interpretation of the situation and are nonspontaneously organized. Because the individual is unable to feel his or her own arousal, insufficient response information is experienced, which in turn decreases the need to express emotions or cope with an emotional stressor. In addition, the cognitive interpretation of the situation without the emotional component may lack important needs-relevant information and be misleading. Prolonged bodily arousal and impaired coping could result. Repressive coping style is the best known model for repressive inhibition. Dissociative processes, which are often thought to have the function of diminishing awareness of otherwise unbearable (and often trauma-related) emotion, are another example.

Suppressive inhibition can be considered the habitual quashing of emotional arousal. The emotional arousal is recognized by the individual but spontaneous expressive and cognitive behaviors are involuntarily suppressed. Suppressive inhibition of emotions could result from interactions between innate factors and socialization. For example, if individuals show increased responses under stressful encounters, they are prone to socialization conditions of punishment and negative reinforcement, initiating a learning history with decreases in spontaneous expressiveness and increases in bodily reactivity.

This may be influenced in part by cultural conditions of emotion regulation. A paradoxical constellation between collectivism and high individualist goal orientation has been found to be associated with high emotional inhibition and high internalization. This pattern of emotion variables reflects inhibited emotional processing which was given the name "neoteric estrangement," and is associated with somatization.[6]

Finally, individuals under emotional stress, aware of their bodily reactions and their urge for expressiveness, can voluntarily suppress this need or try to present a false response to a receiver, called deceptive inhibition. Whether an individual is poker-faced or displays a false emotional response, such inhibition consumes cognitive capacity, reducing the individual's coping capacity and providing additional stress.

In reviewing the psychophysiological and psychosomatic data, one is led to conclude that emotional inhibition is potentially harmful. However, it should be noted that although the correlations between bodily processes and the above four forms of inhibition support such a notion, in certain circumstances inhibition can be beneficial for the individual and his/her relationship with the social environment. Inhibition becomes toxic when it is related first to physiological, endocrinological hyperarousal, or immunological dysfunction, second to longstanding disregulation of emotions within the individual on a cognitive and behavioral basis, and third if inhibition disturbs the individual's social relations.

Inhibition constitutes a risk factor for health under normal stressors. If the severity of stressors is dramatically high, the mental and physical health consequences are inevitable. In traumatic stress situations like rape, criminal bodily attacks, or torture, the individual may well lose control over strong emotional responses. The emotional responses of horror, panic, and loss of control, could literally cause a violent breakdown in the mental and bodily systems. Such an emotional implosion is visible in the symptom pattern of post-traumatic stress disorder (PTSD): cognitive and behavioral avoidance of trauma stimuli, numbing of general and emotional responsiveness (emotional anesthesia), detachment from other people, and persistent symptoms of arousal such as disturbed sleep, exaggerated startle response, and somatic complaints. In addition, persons with PTSD suffer an increased risk of social phobia and major depressive and somatizing disorders.

The psychological, physical, and social symptoms in PTSD are a form of emotional processing that describes an extreme form of inhibition. While inhibition generally develops over a long time span through interaction between innate and socialization factors, implosion can occur in a very short time as a result of a single event.

Stephen Porges[7] has recently suggested that coping with traumatic stress activates a hierarchy of brain systems which act as different "lines of defense," which can also be seen as levels of organismic emotional inhibition. In this theory, the social engagement system can be activated as a form of defense, involving social means of reducing threat (e.g., crying out to bystanders, talking down a rapist). If the social engagement system fails, sympathetically mediated mobilizing defenses are activated (e.g., fight, flight), and if these fail, the organism resorts to parasympathetically mediated immobilizing defenses (e.g., a freeze response). Others[8] have described how curtailed defenses (i.e., inhibited action tendencies of emotions), such as a failed defensive arm movement toward an attacker, or very specific patterns of arousal, can remain latent, and be re-evoked at times of traumatic re-experiencing. Hence, physically oriented mindfulness is emphasized as a central treatment element.

PATHWAYS FROM EMOTIONAL INHIBITION TO HEALTH DISORDERS AND ILLNESS BEHAVIORS

Emotional stress modulated through innate, personality, and socialization factors can trigger, maintain, or worsen health disorders and related illness behavior through neurobiological, social-behavioral, and cognitive pathways. With respect to illness behaviors, the pathways include biases in symptom reporting, secondary gain by presented symptoms, subjective feelings of being impaired, pressure to seek medical help, and social isolation.

The Neurobiology of Emotional Inhibition

Neurobiological studies in this field mainly apply functional magnetic resonance imaging as the methodological approach of choice and use the concept of emotion regulation in a broad sense. This encompasses the up- or down-regulation of emotions in order to flexibly adapt to changing environmental demands and situations. Here, the down-regulation of emotions is of key interest. Neurobiological models generally postulate prefrontally mediated inhibition of subcortical response-related brain regions. This is in accordance with the consistently reported control functions of prefrontal areas and the involvement of subcortical areas in emotional responses. Meta-analyses confirmed decreased bilateral activation the amygdala and parahippocampal gyrus as well as increased activation of the inferior, middle, and superior frontal gyrus and the left anterior cingulate cortex during emotional down-regulation. It is important to note that both groups of areas have rich functional and structural interconnections mediating the regulatory process. Conceptually closer to emotional inhibition as it is referred to here is the construct of expressive suppression introduced by James Gross.[9,10] This is the process of inhibiting already ongoing emotion-expressive behavior—a response-focused strategy after the emotional reaction itself has already been initiated. Hence, suppression decreases emotional expressive behavior but fails to change bodily reactions or negative emotional experience. During expressive suppression, activation in dorsolateral and ventrolateral prefrontal regions has been reported. Interestingly though, increased activity in the insula and amygdala has also been found, supporting psychophysiological findings of expressive suppression boosting the autonomic component of emotional responding. The response-focused nature of suppression is supported by findings of decreased activity in the medial prefrontal cortex (an active area when emotions are down-regulated in reappraisal) when subjects who habitually use suppression expect an emotional stimulus.[11] This could be interpreted as being a reflection of overwhelmed coping capacity in previous situations.

Neurobiological Pathways

It can be assumed that emotional inhibition is strongly neurobiologically based. The behavioral inhibition system and the behavioral activation system have been discussed as implicated neurobiological structures. Empirical evidence from between-subject studies shows that inhibited, repressed, or suppressed emotional expressiveness is linked to greater autonomic arousal, both under conditions of emotion induction and voluntary deception. There is rich empirical evidence for neurobiological correlates of inhibition affecting respiratory, cardiovascular, muscular, digestive, endocrine, and immune functions.[12] Immune functioning is of particular interest because it is the immune system that may be relevant in all sorts of infectious, allergic, and neoplastic illness processes. Inhibited style of processing upsetting events can compromise immune functions, resulting in higher serum antibody titers, decreased monocyte counts, and poorer natural killer cell activity. Other areas of research relevant to the neurobiological pathways of inhibition include hypothalamic-limbic overactivation, prolonged activation of physiological response specificity, hemispheric brain lateralization of emotion processing and faulty interhemispheric communication, the neuroregulation of action, and the behavioral activation and behavioral inhibition systems.

Social-Behavioral Pathways

A variety of social-behavioral pathways connect inhibition-implosion to health disorders and illness behaviors. First, neurobiologically innate factors (shyness, behavioral inhibition, hypersensitivity, introversion) are superimposed by classical and operant conditioning in the socialization of an individual. Since individuals with these characteristics in early childhood are more easily conditioned, the process of socialization involves greater vulnerability to them than it does for others. Under critical developmental conditions, the gap between emotional expressiveness and physiological hyperactivity may increase. A lack or deficit in emotional expressiveness will hinder interpersonal communication. It is implied that deficits in interpersonal communication disturb the development of emotional competence which is important for sharing experiences, maintaining psychological and physical contact, and adapting to the social environment. These are the deficiencies in healthy coping competencies that Salovey and Mayer[13] termed emotional intelligence. Other consequences of inhibited emotional expressiveness are disturbed social relations resulting in social isolation and a disrupted social support network.[14,15]

Normally, persons respond to emotion-evoking stimuli with emotional expression, and such reactive expression is realized through facial muscle activity and movements with reafferent neuronal signals in the central nervous system, which contribute to the individual's emotional experience. However, subjective emotional experience does not depend mainly on this nervous input as argued in the facial feedback hypotheses, but feedback does contribute positively to sensitivity toward the physiological aspects of emotion. If this sensitivity is disrupted, an individual will not perceive adequately increased muscle tension or other autonomic nervous system reactions caused by stress and consequently will not initiate healthy relaxing behavior. Bischoff et al. demonstrated that the hypothesis of deficient perception of muscle tension holds for myogenic pain. Patients with this kind of pain were reliably less able to judge the extent of their muscle tension than were controls.[16]

It is conceivable that, under unfavorable circumstances, expressive behavior of mainly negative emotions like anger and aggressiveness is punished socially and thus justifiably avoided. The suppression of expressive behavior can be realized by an additional increase in muscle activity. Such avoidance behavior or inhibition is very adaptable in the short term, and it helps to modify a socially stressful situation. This reduction of emotional expressiveness is conditioned by the learning mechanism of negative reinforcement (the avoidance of punishment).[17,18]

Cognitive Pathways

Memories and thoughts of emotional stressors are generally unpleasant. Increasing severity of the stressful encounter makes imagining the event painful or even unbearable in the case of traumatic experience. Most individuals attempt to suppress or inhibit the thoughts surrounding the events. As soon as the inhibition work begins, the urge to distract oneself and the mental energy put into this process fuels the images. Consequently, triggered intrusions and unwanted thoughts make life more stressful than before. In addition, thought control interferes with natural ways of coping (e.g., sharing the experience with important others and thinking through the event). Therefore inhibition may be dangerous because it hampers the individual who has suffered a critical life event from resolving the stressful experience cognitively and behaviorally.

Other facets of problematic cognitive processing include an alexithymic or low-level thinking style. As part of the inhibition process, individuals may tend to exclude the emotional content of the stressful encounter from their language representation of the event. Although this may help to avoid negative emotionality in the short term, it impairs one's own complete processing and integration of the stressful experience. The lack of integration into the self-concept makes an individual prone to activation of fear networks. Finally, impaired or unfinished cognitive processing makes an individual prone to a misinterpretation of bodily symptoms. Instead of understanding bodily reactions as part of emotional responses, the individual conceptualizes the bodily reactions as symptom patterns and seeks medical help for illnesses. Badly advised medical treatment procedures result in iatrogenic diseases, trapping the individual in a vicious cycle. Cognitive appraisal of a situation depends partly on facial feedback as a source of emotional information. When the expressive components of emotional reactions are systematically repressed by inhibition, the individual unlearns accurate assessment of stressful circumstances. This learning mechanism occurs since the estimated load of a stress situation is dependent not only on external features of the situation, but also on the subjective experience of stress-conditioned reactions. When, however, an inhibited person takes bodily reactions into account in the evaluation of a situation, his or her judgment will be impaired when the original physiological components of mainly negative emotions are interpreted as symptoms.[17]

In clinical studies, psychosomatic patients (e.g., suffering headache) have been found to report significantly lower stress levels than control groups, but showed nearly twice as much neck muscle tension as the controls. Although the arousal and muscle tension data indicated higher levels of stress, patients were unable or not willing to report those stressors. It appears that patients tend to interpret their stressors in terms of bodily symptoms rather than as underlying levels of stress.[2,3,18]

RITUALS AND THERAPEUTIC INTERVENTIONS

There is at least some implicit knowledge in most societies that emotional inhibition has negative health implications. The conflicts resulting from the need for emotional regulation on the one hand and the need for disclosure, sharing, and catharsis on the other lead to a variety of cultural phenomena to overcome these adverse consequences. These include older universal cultural rituals (such as rituals of grief or lament) or religious acts such as confessions. The Western (Wailing) Wall in Jerusalem, where Jews have been going for centuries to deliver a written prayer, is possibly an example of an ancient disclosure phenomenon. Today people can deliver their prayers to the Western Wall via the Internet, and similar services are offered in connection to Christian confession. Contemporary western societies have also introduced psychotherapy for enhancing emotional expressiveness.[3] Here, talking or writing about emotions is encouraged as well as acting out emotions, in role plays. Assertiveness training aims at effective expression of emotion, and catharsis-based techniques like confrontation are modern remedies for anxiety, PTSDs, and the like.[19,20]

The therapeutic adjustment of emotional behavior is an important part of all therapeutic schools. A particular focus on emotional behavior and associated physical experience is found in Emotion-Focused Therapy,[4,21] Exposure Therapy, Dialectical Behavior Therapy,[22] the Affect School,[23] Schema Therapy,[24] and Emotion-Oriented Interventions.[25] Therapists experienced in these forms of therapy will recognize the common elements across these approaches that are brought together in interventions with a strong focus on emotion regulation, including mindfulness, interpersonal skills training, coping skills training for strong emotional experience, stress tolerance, and self-esteem stabilization: Emotion and cognition are harmonized, the client acquires social competences in order to act assertively and deal with conflicts appropriately and effectively. The client learns to differentiate and name a variety of emotions and understand their meaning in terms of personal needs and actions. Feelings are understood as signals, and the importance of the acceptance and expression of negative feelings and stress reactions is emphasized.

All of these techniques seem to have in common that they are directed toward the construction of meaning from emotional experience. Different cultures tend to construe emotional experience (including stress) in different ways, such that many non-Western cultures tend to emphasize the somatic components of emotional suffering, whereas Western cultures focus on the psychological components.[6] Such cultural differences could influence the pathways between emotional inhibition and illness. Hence, clinicians should take into account the relevant cultural conceptions of emotional experience and behavior a client has been exposed to when choosing interventions for emotional inhibition. If a client emphasizes the somatic components of his or her reaction to a very stressful event, it may not be (dysfunctional) emotional inhibition, but rather a cultural construal of emotional suffering. Only when culture-sensitive exploration reveals lack of insight into relevant psychosocial factors should the possibility of dysfunctional emotion processing be considered.

Glossary

Alexithymia A constellation of cognitive and affective characteristics, including difficulty identifying and communicating subjective feelings, a restricted imaginative life, and a concrete and reality-oriented style of thinking.

Behavioral inhibition A concept encompassing several behaviors in about 15% of otherwise healthy children in response to unfamiliar social events, including reduced spontaneity, subdued emotional expressiveness, shyness, social avoidance, and several peripheral physiological hyperactivities. It is thought to be related to limbic-hypothalamic arousal, in response to socially stressful events.

Emotional intelligence The ability of an individual to adaptively and effectively regulate his or her emotional behavior in a social context. This encompasses the ability to recognize subjective feelings, to manage emotions, to transform emotions into expressiveness and action, to react empathetically, and to shape relationships.

Myogenic pain Pain stemming from dysfunctional muscular activity as part of motor behavior in relation to stress, posture, movement, and emotion (e.g., low back pain, tension type headache, repetitive strain injury, and myofascial pain disorder).

Socialization The process by which an individual gradually becomes integrated into the norm and value system of a social group or a society. It is based on the assumption of interaction between the biological organism and the social environment during psychological development.

Somatization The conversion or expression of mental states as bodily symptoms, which cannot be fully explained by medical conditions. Somatization is distressing by itself, can impair social and occupational functioning, and leads people to seek medical evaluation and treatment.

Torture One of the most severe stressors inflicted by governmental, police or security services agencies or by militia or criminal groups. Amnesty international[25] received between 2009 and 2013 reports of torture and other ill-treatment committed only by state officials in 141 countries, and from every world region. The adverse effects of torture stem from the manmade nature of these stressors and induce specific and nonspecific physical and mental disorders of long duration because they shatter victims' basic assumptions about human benevolence.

References

1. Temoshok L. Emotion, adaption, and disease. In: Temoshok L, Van Dyke C, Zegans LS, eds. *Emotions in Health and Illness*. New York, NY: Grune and Stratto; 1983:207–233.
2. Pennebaker JW. Emotion, disclosure, and health: an overview. In: Pennebaker JW, ed. *Emotion, Disclosure, and Health*. Washington, DC: American Psychological Association; 1995:3–10.
3. Traue HC. *Emotion und Gesundheit: Die psychobiologische Regulation durch Hemmungen*. Heidelberg: Spektrum; 1998.
4. Greenberg L. Emotion focused therapy: a clinical synthesis. *Psychotherapy*. 2010;8:32–42.
5. Hayes SC, Lillis J. Acceptance and commitment therapy processes. In: VandenBos G, Meidenbauer E, Frank-McNeil J, eds. Washington, DC: American Psychological Association; 2014:11–17. Psychotherapy Theories and Techniques: A Reader; vol. 8.
6. Deighton R, Traue HC. Emotional inhibition and somatization across cultures. *Int Rev Soc Psychol*. 2005;18:109–140.
7. Porges SW. The polyvagal perspective. *Biol Psychol*. 2007;74:116–143.
8. Ogden P, Minton K, Pain C. *Trauma and the Body. A Sensorimotor Approach to Psychotherapy*. New York, NY: Norton; 2006.
9. Gross J. Emotion regulation: affective, cognitive, and social consequences. *Psychophysiology*. 2002;39:281–291.
10. Mauss IB, Gross J. Emotional suppression and cardiovascular disease: is hiding your feelings bad for your heart? In: Nykliček I, Temoshok L, Vingerhoets A, eds. *Emotional Expression and Health. Advances in Theory, Assessment, and Clinical Applications*. Hove/New York: Brunner-Routledge; 2004.
11. Abler B, Hofer C, Walter H, et al. Habitual emotion regulation strategies and depressive symptoms in healthy subjects predict fMRI brain activation patterns related to major depression. *Psychiatry Res*. 2010;183:105–113.
12. Frank DW, Dewitt M, Hudgens-Haney M, et al. Emotion regulation: quantitative meta-analysis of functional activation and deactivation. *Neurosci Biobehav Rev*. 2014;45:202–211.
13. Salovey P, Mayer JD. Emotional intelligence. *Imagin Cogn Pers*. 1990;9:185–211.
14. Rimé B, Herbette G, Corsini S. The social sharing of emotion: illusory and real benefits of talking about emotional experiences. In: Nykliček I, Temoshok L, Vingerhoets A, eds. *Emotional Expression and Health. Advances in Theory, Assessment, and Clinical Applications*. Hove/New York: Brunner-Routledge; 2004.
15. Smith CE, Fernengel K, Holcroft C, Gerald K, Marien L. Meta-analysis of the association between social support and health outcomes. *Behav Med*. 1994;16:352–362.
16. Bischoff C, Traue HC, Zenz H, eds. *Clinical Perspectives on Headache and Low Back Pain*. Toronto/Lewiston/Göttingen/Bern: Hogrefe & Huber; 1989.
17. Traue HC. Emotional inhibition and health. In: Smelser NJ, Baltes PB, eds. *The International Encyclopaedia of the Social and Behavioural Sciences*. Oxford: Pergamon; 2001:4449–4454.

18. Traue HC, Pennebaker JW, eds. *Emotion, Inhibition and Health.* Toronto/Lewiston/Göttingen/Bern: Hogrefe & Huber; 1993.
19. Deighton RM, Gurris N, Traue HC. Factors affecting burnout and compassion fatigue in psychotherapists treating torture survivors: is the therapist's attitude to working through trauma relevant? *J Trauma Stress.* 2007;20:63–75.
20. Steffen W, Leißner N, Jerg-Bretzke L, Hrabal V, Traue HC. Pain and emotional processing in psychological trauma. *Psychiatr Danub.* 2010;22:465–470.
21. Greenberg L. *Emotion-Focused Therapy: Coaching Clients to Work Through Feelings.* Washington, DC: American Psychological Association; 2007.
22. Linehan MM. *Cognitive Behavioral Therapy of Borderline Personality Disorder.* New York, NY: Guilford Press; 1993.
23. Bergdahl J, Larsson A, Nilsson LG, Ahlstrom KR, Nyberg L. Treatment of chronic stress in employees: subjective, cognitive and neural correlates. *Scand J Psychol.* 2005;46:395–402.
24. Young JE, Klosko JS, Weishaar ME. *Schema Therapy: A Practitioner's Guide.* New York, NY: Guilford Press; 2003.
25. Traue HC, Steffen W, Kessler H. Emotionsorientierte Interventionen bei Kopfschmerzen vom Spannungstyp (KST). In: Fritsche G, Gaul C, eds. *Multimodale Schmerztherapie bei chronischen Kopfschmerzen.* Stuttgart: Thieme; 2013:151–158.
26. Amnesty international. Torture in 2014. www.amnestyusa.org/sites/default/files/act400042014en.pdf

CHAPTER

29

Chronic Stress, Regulation of Emotion, and Functional Activity of the Brain

B.J. Ragen[1,2], *A.E. Roach*[3], *C.L. Chollak*[2]
[1]New York University, New York, NY, USA
[2]NYU School of Medicine, New York, NY, USA
[3]University of South Carolina-Aiken, Aiken, SC, USA

OUTLINE

Abstract

Stress has been conceptualized as a threat to physiological or psychological homeostasis. Reappraisal, suppression, and rumination are emotion regulation techniques utilized to alter the affective components of stress. Reappraisal consists of activation of prefrontal cortical areas that reduces cortical activity in limbic areas such as the amygdala. Suppression and rumination involve similar brain areas but have different patterns of activation. Pathological chronic stress, such as posttraumatic stress disorder (PTSD), is associated with disruptions in emotion regulation, particularly suppression and rumination. Although there are no significant disruptions in reappraisal in PTSD, brain activity during reappraisal differs compared to individuals without PTSD. These findings indicate that chronic stress is associated with emotion regulation functioning and its neural underpinnings.

STRESS

The concept of stress began with Cannon and Selye. Cannon[1] introduced the concept of homeostasis and the idea that stress constituted a perturbation of homeostasis. Selye[2] continued this research and expanded the concept of stress as a direct disruption of homeostasis or anything that could be interpreted as a threat to physiological or psychological homeostasis. Furthermore, Selye introduced the concept of the General Adaptation Syndrome, which consists of three phases.[2] The first is the alarm phase, which is the initial, acute response to a stressor. The second is the resistance phase, which is the body's attempt to return to homeostasis. The third phase is the exhaustion phase in which the body is unable to return to homeostasis. Selye researched the physiological repercussions of the exhaustion phase, which can also be considered as chronic stress. Later research has found that chronic stress can also result in impairment of the immune system and inhibition of neurogenesis in the hippocampus.[3] Chronic stress can also negatively impact cognitive functioning, such as learning and memory and if the stress is extreme enough it can even result in psychological disorders.[3]

This chapter will address how chronic stress can impact emotion regulation, the neural functions involved

Stress: Concepts, Cognition, Emotion, and Behavior
http://dx.doi.org/10.1016/B978-0-12-800951-2.00029-7

in emotion regulation and how chronic stress disrupts its underlying neural functioning. We will also address psychological disorders brought on by severe and chronic stress, specifically posttraumatic stress disorder (PTSD) and its relationship to impaired emotion regulatory abilities and how these impairments correlate with brain functioning.

KEY POINTS

- How do we know that chronic stress, specifically posttraumatic stress disorder (PTSD), is associated with disrupted emotion regulation and is it correlated with neural brain functioning?
- Reappraisal, suppression, and rumination are common emotion regulation techniques and each one can have varying affective and behavioral outcomes. In healthy individuals, reappraisal usually results in a higher quality of life, while suppression and rumination usually have negative outcomes. Chronic stress can impact the use and success of these different emotion regulation techniques. PTSD is a stress-related disorder and can be considered a pathological form of chronic stress. A meta-analysis examining the relationship between PTSD and problems with emotion regulation found that there is a strong effect size for suppression and rumination but not for reappraisal. It has also been found that levels of suppression and rumination predict severity of PTSD symptomatology. Although individuals with PTSD appear not to have difficulties with reappraisal, there is evidence that there are differences in brain activity during reappraisal. Reappraisal in healthy individuals involves activation of areas in the prefrontal cortex (PFC) such as the dorsolateral PFC (dlPFC), ventrolateral PFC (vlPFC), and dorsomedial PFC (dmPFC), which usually result in an inhibition of limbic areas such as the amygdala. Although individuals with PTSD appear to not have difficulty with reappraisal, there are abnormalities in brain activity such as the dlPFC.

EMOTION REGULATION

The first thing to address when discussing emotion regulation is to conceptualize emotion. External events (e.g. presentation of a fearful stimulus) or internal events (e.g. increase in heart rate) result in an emotion generative process that indicates something of significance. When these emotion cues signify an important event, a coordinated set of response tendencies are activated and include experiential, behavioral, and central and peripheral physiological systems. Gross[4] proposed five main sets of emotion regulation processes: situation selection, situation modification, attention deployment, cognitive change, and response modulation. Situation selection involves avoiding or approaching certain locations, people, or objects. Situation modification involves altering one's situation to regulate one's emotions. Attention deployment involves altering attentional focus (e.g., distraction and rumination). Cognitive change involves cognitively altering the percept to a form that allows for an individual to manage the emotions that the percept may induce. A classic form of cognitive change includes reappraisal. Response modulation is the ability to appropriately respond to the physiological and psychological components that the emotion generated. Behaviors that are utilized for response modulation can include the use of drugs, biofeedback, and exercise.

This chapter will primarily focus on suppression, rumination, and reappraisal. These three emotion regulation techniques are associated with negative and positive repercussions for an individual. For example, suppression is a form of response modulation in which an individual attempts to inhibit an emotional response, usually in response to a negative experience. Suppression can result in poorer quality of life. In contrast, reappraisal results in higher quality of life and greater expression of positive emotion.[5] Reappraisal is a form of cognitive change, which allows an individual to alter emotional impact. This can be done by up-regulating or down-regulating an emotional response. Lastly, rumination is a form of attentional deployment in which an individual focuses on their feelings and the consequences of a negative event.[4] These three emotion regulation techniques have a particularly special relationship with chronic stress.

FUNCTIONAL ACTIVITY AND EMOTION REGULATION

The strategy of reappraisal is the most common emotion regulation strategy explored in functional neuroimaging studies. In these paradigms, participants are instructed to either reinterpret a stimulus by intentionally changing the perception of emotional elements of a visual stimulus or its attributes, or to distance themselves from the emotional stimulus.[6,7] Typical stimuli in reappraisal paradigms range from images or video clips of scenes depicting negative- or neutral-affect, disgust, as well as the evaluation of happy or sad faces. In general, regions in the frontal lobe involved in cognitive control, appraisal, and monitoring act on cortical and subcortical areas involved in emotion generation and/or emotion response.[8] This neural activity can alter the degree in which a particular emotional valence is experienced. Below we review functional magnetic resonance imaging (fMRI) studies examining neural activation during reappraisal, as well as the less effective strategies of suppression and rumination.

fMRI Studies of Reappraisal

In a comprehensive review of fMRI studies of emotion regulation using reappraisal, Ochsner and colleagues[6] determined the most common neural systems involved in the reappraisal process include areas in the prefrontal cortex (PFC) including the dorsolateral PFC (dlPFC), posterior PFC, ventrolateral PFC (vlPFC), dorsomedial PFC (dmPFC), and dorsal anterior cingulate cortex (dACC), all of which act on, in some way, areas involved in emotion processing, namely the amygdala, ventral striatum, and to a lesser degree, the ventromedial PFC (vmPFC) and insula (Figure 1). The finding that there is an inverse relationship between activation of PFC regions and inhibition of amygdala has been found in both rat and human studies.[9] Functional connectivity studies confirm this influence by providing evidence of the coupling between increased PFC activation and decreased amygdalar activation during reappraisal.[10–12] Variations in location and activation levels within specific brain regions are influenced by experimental differences such as the age or sex of the participants, the goal of reappraisal (e.g., increase or decrease a particular emotion), the valence of the stimuli (e.g., positive, negative, or neutral), and the specific reappraisal tactic used (e.g., distancing or reinterpreting).

Ochsner and colleagues[8] studied healthy women during a reappraisal task which instructed participants to down-regulate negative emotional responses to negative photos by either attending to or reappraising the emotional valence of negative stimuli. Contrasts to determine regions of greater activation in "reappraise" versus "attend" trials found that successful reappraisal of negative images resulted in greater activation in the left vlPFC and dmPFC. Interestingly, during successful reappraisal trials, the amygdala and MOFC demonstrated decreased activation which was correlated with an increase in vlPFC activation. This provides evidence that the strategy of reappraisal attenuates the experienced emotional response to negative stimuli and moderates the degree of activation in brain regions that respond to negative affect. These results suggest that cognitive control by

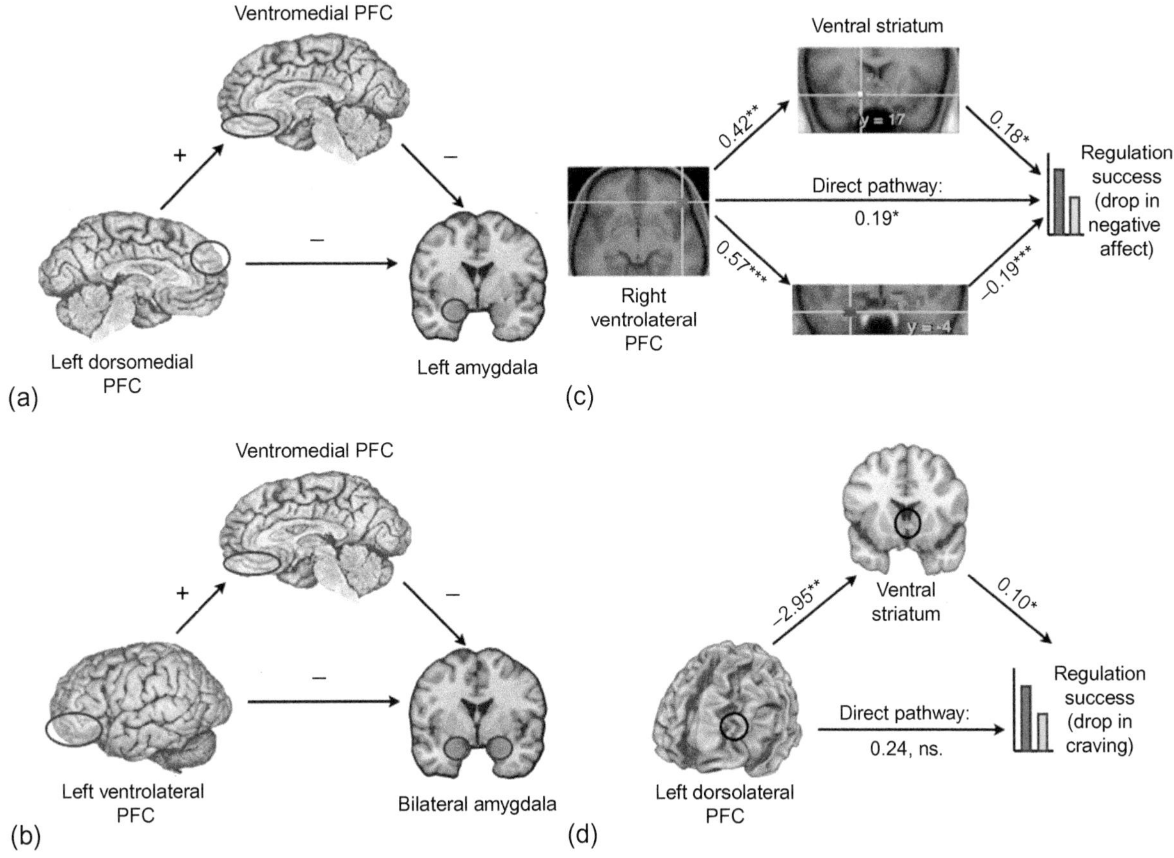

FIGURE 1 Two kinds of mediation pathways involved in reappraisal. (a) and (b) show pathways identified in two studies of the down-regulation of negative emotion whereby dorsomedial or ventrolateral prefrontal regions diminish amygdala responses via their impact on ventromedial prefrontal cortex. These studies did not report weights for the mediation paths between regions or test for full versus partial mediation. (c) and (d) show pathways identified in two studies of the down-regulation of negative or positive emotion, whereby ventrolateral or dorsolateral prefrontal regions diminish self-reports of negative affect or craving via their impact on the amygdala or ventral striatum, respectively. $*p<0.05$, $**p<0.01$, $***p<0.001$ Reproduced from Ochsner KN, Silvers JA, Buhle JT. Functional imaging studies of emotion regulation: a synthetic review and evolving model of the cognitive control of emotion. *Ann NY Acad Sci.* 2012;1251:E1–E24.

frontal regions influence the subsequent emotional response of the amygdala and MOFC.

There is also evidence that the everyday use of reappraisal is reflected in brain activity during reappraisal tasks. When undergoing specific reappraisal tasks, individuals who score high on the Emotion Regulation Questionnaire (ERQ), which can act as an index of self-reported use of daily reappraisal, have greater activation of the dmPFC, lateral OFC, and dlPFC and less amygdala activation wile assessing negative stimuli.[11] These findings are supported by Vanderhasselt and colleagues[13] in which ERQ scores were used to identify frequent users of reappraisal versus suppression. In this study, female participants were instructed to identify either the actual valence of a happy or sad face, or identify the opposite valence. High reappraisal ERQ scores were positively correlated with left dlPFC and left dACC activation during the trials that required the inhibition of negative information. These results support the idea that regular use of reappraisal recruits the dorsal frontal cognitive control network in the presence of negative information resulting in a decrease in amygdala activity and a subsequent down-regulation of negative emotional responses.

fMRI Studies of Suppression

Few functional activation studies have been conducted examining suppression. One study required female participants to view neutral and disgusting film clips either passively, or using reappraisal or suppression strategies. Goldin and colleagues[14] found that reappraisal resulted in the predicted neural responses, namely increased activation in prefrontal regions (mPFC, dlPFC, vlPFC, and lateral OFC) and a decrease in the amygdala and left insula. Interestingly, during suppression, frontal control regions did not come online early in the viewing process as they had in reappraisal, suggesting that suppression is not recruiting the cognitive control network that successfully down-regulates the amygdala. Rather, once the amygdala and insula start to respond to negative stimuli, regions in the frontal cortex are recruited to suppress the behavioral emotional response (i.e., facial expression) elicited by negative affect. Suppression did successfully attenuate the subjective negative emotional experience of disgusting film clips, but this coping strategy fails to down-regulate the underlying neurophysiological response of the amygdala and insula.

fMRI Studies of Rumination

Rumination is a maladaptive emotion regulation strategy in which an individual focuses on attributes of an event that evoke a negative emotional response. Compared to reappraisal, rumination leads to greater activation of the mPFC, subgenual ACC (sgACC), and higher subjective ratings of negative affect.[15] Activation in the sgACC has been implicated in depressive symptoms, while both the mPFC and sgACC are involved in self-referential thought and emotional salience. Additionally, during rumination of anger-induced autobiographical memories there is a correlated increase in activation of the inferior frontal gyrus (IFG), part of the lateral PFC, and the amygdala. These findings could indicate that during rumination the top-down control of IFG to decrease amygdala activity is ineffective. An alternate interpretation, but not necessarily mutually exclusive, is that a positive feedback loop between the amygdala and IFG explains why rumination increases arousal and anger because PFC regions are involved in the reflection of angry feelings.[16] Stopping rumination requires the engagement of the cognitive control network to refocus on other information. According to Vanderhasselt et al.[17] this neural network recruits the dlPFC. Healthy individuals rated high on the Rumination Response Scale (RRS) for depressive brooding had more difficulty inhibiting negative information in a go/no-go task. Furthermore, neural activation patterns revealed a positive correlation between brooding and dlPFC activity, in that high brooders had greater dlPFC activation during trials requiring inhibition of negative information. This suggests habitual brooders have impaired ability to disengage from negative information and require more neural energy in order to successfully disengage.

The results of these studies using reappraisal, suppression, and rumination provide insight into the neural substrates involved in emotion regulation (Figure 2). Reappraisal recruits prefrontal regions that enforce cognitive control in pursuit of down-regulating the emotional reactivity of the amygdala and related limbic structures, resulting in successful decreases of reported negative affect in response to negative stimuli. Suppression may successfully decrease reported negative affect, but it fails to down-regulate the neurophysiological response to negative stimuli, despite recruitment of prefrontal cognitive control regions, which may have long-term consequences. Finally, rumination does not recruit the cognitive control network effectively; rather it activates regions linked to depression and self-reflection, resulting in failure to regulate negative emotion.

CHRONIC STRESS AND EMOTION REGULATION

Both acute and chronic stress can impact emotion regulation.[18–20] These impairments have been found in individuals without psychopathology. An acute stressor,

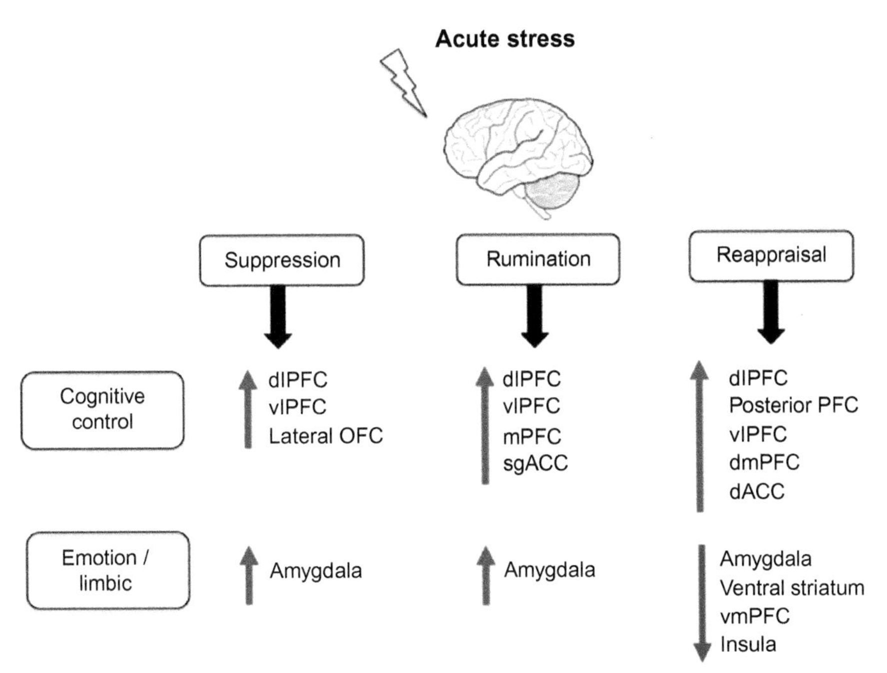

FIGURE 2 Summary of brain activity involved in suppression, rumination, and reappraisal when exposed to acute stress. During all three emotion regulation techniques there is activation in a variety of frontal cortical regions involved in cognitive control. However, only reappraisal results in a decrease in activity of brain areas involved in emotion generation and processing (i.e., amygdala). In spite of activation of frontal cortical regions in suppression and rumination, there is no subsequent decrease in amygdala activity.

such as a cold pressor task, impairs an individual's ability to utilize skills learned during a cognitive regulation training session, specifically skills involved in reappraisal.[20] Emotion regulation issues are observed in individuals suffering from work-related chronic stress, referred to as occupational "burnout." Burnout develops through a series of prolonged stressful work events and occurs in otherwise healthy individuals.[18] This type of chronic stress results in difficulty with emotion modulation particularly when attempting to down-regulate responses to negative images.[18]

Stigma and Stress

A unique population that undergoes chronic stress is that belonging to stigmatized communities. This is especially true for individuals with concealable stigmas.[21] These stigmas include mental illness, infertility, HIV status, and nonheterosexual sexual orientation. These individuals have to face the unique stress of the constant threat of discovery due to the potential repercussions of being discovered such as social rejection, social isolation, housing discrimination, and violence.[21] These chronic stressors can lead to difficulties in emotion regulation. Compared to their heterosexual counterparts, adolescents with same-sex attractions have poorer emotion regulation abilities indicated by higher levels of rumination and poorer emotion awareness both of which predict higher levels of anxiety and depression.[22]

Due to the secretive component of concealable stigma there is effort in suppressing thoughts of the stigma. Greater levels of suppression occur in individuals with nonheterosexual identities, particularly on days when there are opportunities to disclose their sexual identity.[23] However, chronic suppression can produce a rebound effect resulting in rumination. This is a potential explanation for why encountering stigma-related stressors results in higher rates of both suppression and rumination.[24]

Caretaking and Stress

Another nonclinical population that often experiences persistent chronic stress is caretakers; particularly individuals charged with the task of caring for loved ones with Alzheimer's disease. Variances of coping strategies account for the greatest differences in caretakers' perceived life satisfaction. Caretakers who use coping strategies centered around logical problem-solving have greater levels of life satisfaction, while those who rely on avoidant strategies report lower life satisfaction.[25] Emotion regulation problems also occur in people caring for individuals with schizophrenia and borderline personality disorder (BPD). For example, those who care for individuals with BPD often report distress so severe that their emotion regulation difficulties are comparable to individuals with PTSD.[19]

PTSD

In the DSM-5, PTSD is categorized as a trauma or stress-related disorder. The most current research on symptomatology of PTSD has clustered the symptoms into five factors including intrusion/reexperiencing (e.g., reliving trauma), avoidance (e.g., avoidance of reminders of the trauma), numbing (e.g., feeling detached), anxious arousal (e.g., overly alert), and dysphoric arousal (e.g., irritability).[26] PTSD is brought on after exposure to one or multiple traumatic life events. This disorder can be considered a source of chronic stress due to the reexperiencing of the trauma that includes flashbacks and nightmares. These symptoms, in essence, result in an individual repeatedly undergoing the traumatic event.

Research has consistently found that individuals with PTSD suffer from impaired emotion regulation. A meta-analysis by Seligowski and colleagues[27] found that there was a large to moderate effect size for general emotion regulation, experiential avoidance, suppression, and rumination in the relationship between emotion regulation and PTSD. Interestingly there was no effect of acceptance and reappraisal. Much of the research studying emotion regulation and PTSD has focused on the general impairments in emotion regulation. These studies frequently utilize the Difficulties in Emotion Regulation Scale (DERS). This scale focuses on the emotion regulation domains that have been described by Gratz and Roemer.[28] Some examples of these domains include: inability or unwillingness to attend to and acknowledge emotions (AWARNESS); difficulties remaining in control when upset (IMPULSE); and the extent to which individuals know the emotions they are experiencing (CLARITY).

It has consistently been found that individuals with PTSD score high on the DERS scale indicating poor emotion regulation, and DERS scores are positively correlated to PTSD severity.[29] Difficulties in emotion regulation are particularly amplified in individuals who experience chronic early-life trauma compared to other trauma groups (e.g., late life interpersonal trauma).[30] It has also been found that problems with emotion regulation indicated by DERS scores predict later PTSD symptomatology after a mass shooting.[31] DERS scores, particularly the domain of ACCEPTANCE, are predictive of the severity in the avoidance and hyperarousal symptom factors in PTSD.[32]

PTSD AND EXPERIENTIAL AVOIDANCE

Experiential avoidance has been defined by an unwillingness or inability to remain in contact with unwanted private experiences, including unwanted thoughts, emotions, memories, and physiology. This results in attempts to avoid or alter those private experiences.[33] The Acceptance and Action Questionnaire-II (AAQ) has been demonstrated to be a valid and reliable measure of experiential avoidance. This questionnaire consists of questions such as "I am afraid of my feelings" and "Emotions cause problems in my life."[34]

Studies have found that the use of experiential avoidance results in greater symptomatology and a stronger predictor of symptomatology compared to the severity of previous traumatic events.[35] For example, sexual assault in homosexual men has been found to result in PTSD, with internalized homophobia and experiential avoidance being predictors of symptom severity. Gold and colleagues[36] found that experiential avoidance was the greater predictor of severity compared to internalized homophobia. Experiential avoidance also moderates PTSD severity demonstrated by Shenk and colleagues[37] who found that experiential avoidance moderated the correlation of early life maltreatment and PTSD symptomatology severity.

PTSD AND SUPPRESSION

As discussed previously, suppression includes any mental effort to avoid specific thoughts or feelings regarding negative events or emotions. Suppression is related to experiential avoidance, in that it is a specific strategy utilized to avoid or escape emotions.[33] One scale used to measure suppression is the Cognitive Emotional Regulation Questionnaire (CERQ).[38] This questionnaire asks individuals to rate the extent to which they agree with statements such as "I keep my emotions to myself."

There are two main types of suppression, thought suppression and expressive suppression. Thought suppression is the deliberate attempt to not think about negative thoughts while expressive suppression involves attempts to not express behaviors that reflect internal negative emotions (e.g., facial expression).[5] It has consistently been observed that levels of suppression predict the probability of developing PTSD and the severity of symptomology of PTSD in a range of trauma, including sexual assault[39] and vehicle accidents.[40] The use of suppression in the face of traumatic events and intrusive memories is correlated to more severe PTSD symptoms[41] and utilization of expressive suppression predicts severity of PTSD symptoms in addition to specific symptom clusters such as reexperiencing, avoidance, and hyperarousal.[42] Suppression therefore has been found to be ineffective and maladaptive for individuals to cope with the memories of traumatic events.

PTSD AND RUMINATION

Rumination is an attentional deployment technique in which an individual repetitively focuses on negative emotions and the precipitators, symptoms of distress, and meaning of a negative event.[43] Rumination is considered a type of cognitive avoidance strategy that when utilized can perpetuate PTSD symptoms. While the idea of rumination acting as an avoidance strategy seems counterintuitive, when individuals with PTSD engage in rumination, they often consider the causes and subsequent consequences of the traumatic event instead of actively processing the traumatic event itself.[43] Rumination is often measured by the Ruminative Thought Style Questionnaire (RTSQ), which asks questions such as "I have never been able to distract myself from unwanted thoughts" and "I find that my mind goes over things again and again."

Scores on RTSQ predict the development of PTSD and the severity of symptomatology.[44] It has been found that rumination results in increased negative mood and reexperiencing symptoms.[43] Furthermore, individuals with PTSD have reported that rumination triggers intrusive thoughts and memories more frequently than individuals who have experienced trauma without developing PTSD.[45] Interestingly, specific types of rumination predict harm in PTSD. Brooding is a form of passive rumination centered on problems and their consequence. Pondering is a form of active rumination, which emphasizes understanding and problem solving.[46] This distinction is significant as it can help explain why rumination is helpful among healthy veterans who engage in pondering, while rumination is harmful in veterans with PTSD, as this population is more likely to engage in brooding.

EMOTION REGULATION, PTSD, AND BRAIN FUNCTIONING

Individuals suffering from chronic stress and PTSD demonstrate deficiencies in emotion regulation that may result from alterations in brain functioning. Research focusing on the neural substrates of emotion regulation amongst populations of individuals exposed to chronic stress, or individuals with PTSD, is far from conclusive. However, the need for translational research on this subject is critical and is rapidly gaining momentum.

PTSD and Brain Functioning

A characteristic of PTSD and other anxiety disorders is an exaggerated fear response.[47,48] In PTSD, certain brain regions are affected in ways that may lead to difficulties with emotion regulation (Figure 3). Individuals who experience pathological levels of stress and anxiety possess altered functional activation profiles. Liberzon and Martis[49] discussed how individuals with PTSD have impaired blood flow in regions such as the mPFC and ACC, and increased blood flow to the amygdala and insula. This may result in heightened levels of responsiveness to threat and negatively valenced stimuli, whereby the mPFC fails to inhibit the amygdala, as it does in healthy individuals. Additionally, hypoactivity in the vmPFC and lateral PFC has been seen in individuals with PTSD when confronted with stimuli designed to induce anxiety and negative affect.[50] Male combat veterans with and without PTSD report successful down-regulation of negative affect when using reappraisal, however PTSD veterans experience less dlPFC activation compared combat-exposed, in accordance with findings of dysfunctional frontal recruitment in PTSD. However neither group showed a decrease in amygdala activation during the reappraisal trials (Figure 3).[50]

Many neuroimaging studies have reported findings of hypoactivity of the vmPFC and hyperactivity of the amygdala in patients with PTSD,[47] and inferences have been made that the vmPFC fails to inhibit the amygdala directly. However, this causal relationship has been called into question by lesion studies from the Vietnam Head Injury Study.[48] In this cohort studied by Koenigs and Grafman, none of the veterans with amygdala lesions

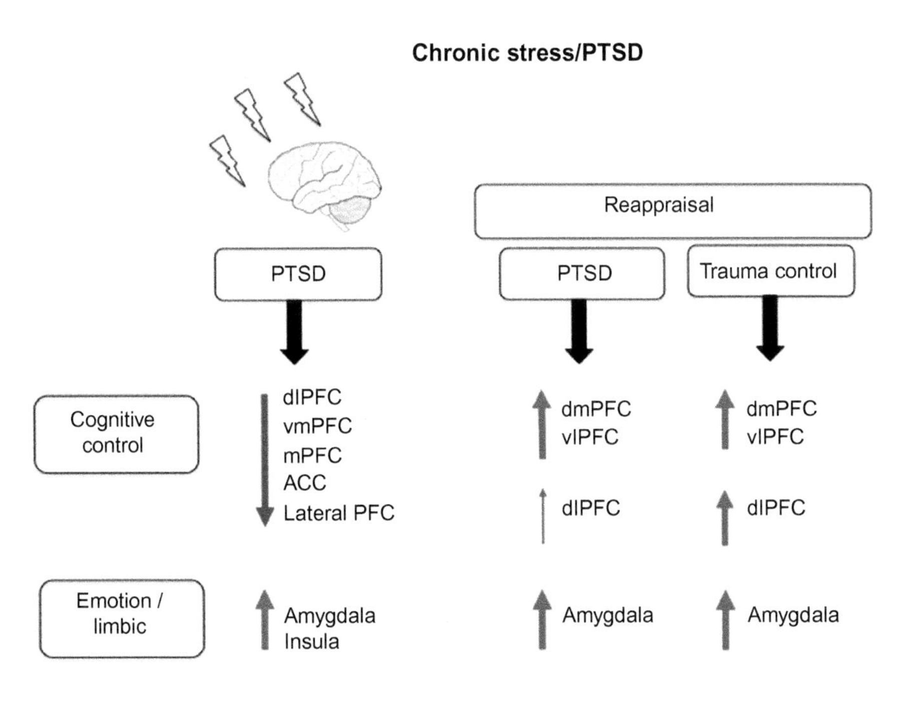

FIGURE 3 Summary of general brain activity and brain activity during reappraisal in individuals with PTSD. In general, individuals with PTSD have hypoactivation in a variety of frontal cortical regions involved in cognitive control and corresponding hyperactivation in the amygdala and insula. Reappraisal in PTSD in response to negative stimuli results in increases in dmPFC, dlPFC, and amygdala, but there is less activation in the dlPFC compared to trauma control.

went on to develop PTSD. These findings support functional imaging studies that find amygdala hyperactivation is necessary to develop PTSD. In contrast, there was no evidence of vmPFC lesions leading to the development of PTSD. Functional imaging studies predict that vmPFC hypoactivity plays a causal role in amygdala hyperactivity by failing to inhibit it. If this causal relationship were true, then veterans with vmPFC lesions would be incapable of inhibiting the amygdala which would result in higher than average rates of PTSD.

Summary

The research discussed finds that the neural activation patterns for emotion regulation in PTSD differ from healthy controls in the level of activation found in regions involved in cognitive control and emotion processing. Prefrontal regions ordinarily recruited to reappraise negative emotional valence are disrupted, and amygdala and other supporting limbic structures involved in emotional processing are hyper-responsive. Future studies should explore neural activity during emotion regulation in nonclinical populations undergoing chronic stress (i.e., caregivers, concealed stigma) and how they compare to pathological populations such as individuals with PTSD.

GENERAL CONCLUSION

As discussed in this chapter, chronic stress, especially that leading to psychopathology can result in severe impairments in psychological functioning. Emotion regulation techniques are important to modulate how emotions impact an individual. Reappraisal can be beneficial in both the short and long term, but techniques such as suppression and rumination can have negative long-term consequences. These types of emotion regulation appear to be driven by different neural patterns. Reappraisal results in activation of frontal cortical areas which in turn down-regulate emotion producing limbic areas such as the amygdala. During suppression and rumination frontal cortical areas are also activated but appear to fail in down-regulating amygdalar activity.

In individuals who have encountered extreme stress, such as those with PTSD, there is an association with disruptions in emotion regulation techniques especially suppression and rumination, which are correlated with the symptom severity. Although there is little evidence suggesting impairments of reappraisal in PTSD, it appears that there is altered brain functioning while engaging in reappraisal. This is compounded upon the fact that there is general hypoactivation in frontal cortical areas and hyperactivation in the amygdala. Future research should further explore the neural function of suppression and rumination in both healthy individuals as well as those experiencing nonpathological and pathological forms of chronic stress.

References

1. Cannon WB. Stresses and strains of homeostasis. *Am J Med Sci.* 1935;189:1–14.
2. Selye H. Evolution of the stress concept. *Am Sci.* 1973;61(6):692–699.
3. McEwen BS. Protection and damage from acute and chronic stress—allostasis and allostatic overload and relevance to the pathophysiology of psychiatric disorders. In: Yehuda R,McEwen B, eds. *Biobehavioral Stress Response: Protective and Damaging Effects.* Vol. 1032. New York: New York Academy of Sciences; 2004:1–7.
4. Gross JJ. The emerging field of emotion regulation: an integrative review. *Rev Gen Psychol.* 1998;2(3):271–299.
5. Gross JJ, John OP. Individual differences in two emotion regulation processes: implications for affect, relationships, and well-being. *J Pers Soc Psychol.* 2003;85(2):348–362.
6. Ochsner KN, Silvers JA, Buhle JT. Functional imaging studies of emotion regulation: a synthetic review and evolving model of the cognitive control of emotion. *Ann NY Acad Sci.* 2012;1251:E1–E24.
7. Ochsner KN, Gross JJ. The cognitive control of emotion. *Trends Cogn Sci.* 2005;9(5):242–249.
8. Ochsner KN, Bunge SA, Gross JJ, Gabrieli JDE. Rethinking feelings: an fMRI study of the cognitive regulation of emotion. *J Cogn Neurosci.* 2002;14(8):1215–1229.
9. Quirk GJ, Beer JS. Prefrontal involvement in the regulation of emotion: convergence of rat and human studies. *Curr Opin Neurobiol.* 2006;16(6):723–727.
10. Banks SJ, Eddy KT, Angstadt M, Nathan PJ, Phan KL. Amygdala—frontal connectivity during emotion regulation. *Soc Cogn Affect Neurosci.* 2007;2(4):303–312.
11. Drabant EM, McRae K, Manuck SB, Hariri AR, Gross JJ. Individual differences in typical reappraisal use predict amygdala and prefrontal responses. *Biol Psychiatry.* 2009;65(5):367–373.
12. Urry HL, van Reekum CM, Johnstone T, et al. Amygdala and ventromedial prefrontal cortex are inversely coupled during regulation of negative affect and predict the diurnal pattern of cortisol secretion among older adults. *J Neurosci.* 2006;26(16):4415–4425.
13. Vanderhasselt MA, Baeken C, Van Schuerbeek P, Luypaert R, De Raedt R. Inter-individual differences in the habitual use of cognitive reappraisal and expressive suppression are associated with variations in prefrontal cognitive control for emotional information: an event related fMRI study. *Biol Psychol.* 2013;92(3):433–439.
14. Goldin PR, McRae K, Ramel W, Gross JJ. The neural bases of emotion regulation: reappraisal and suppression of negative emotion. *Biol Psychiatry.* 2008;63(6):577–586.
15. Kross E, Davidson M, Weber J, Ochsner K. Coping with emotions past: the neural bases of regulating affect associated with negative autobiographical memories. *Biol Psychiatry.* 2009;65(5):361–366.
16. Fabiansson EC, Denson TF, Moulds ML, Grisham JR, Schira MM. Don't look back in anger: neural correlates of reappraisal, analytical rumination, and angry rumination during recall of an anger-inducing autobiographical memory. *Neuroimage.* 2012;59(3):2974–2981.
17. Vanderhasselt MA, Kuhn S, De Raedt R. Healthy brooders employ more attentional resources when disengaging from the negative: an event-related fMRI study. *Cogn Affect Behav Neurosci.* 2011;11(2):207–216.
18. Golkar A, Johansson E, Kasahara M, Osika W, Perski A, Savic I. The influence of work-related chronic stress on the regulation of emotion and on functional connectivity in the brain. *PLoS One.* 2014;9(9):11.

19. Bailey RC, Grenyer BFS. Supporting a person with personality disorder: a study of carer burden and well-being. *J Pers Disord*. 2014;28 (6):796–809.
20. Raio CM, Orederu TA, Palazzolo L, Shurick AA, Phelps EA. Cognitive emotion regulation fails the stress test. *Proc Natl Acad Sci U S A*. 2013;110(37):15139–15144.
21. Pachankis JE. The psychological implications of concealing a stigma: a cognitive-affective-behavioral model. *Psychol Bull*. 2007;133(2):328–345.
22. Hatzenbuehler ML, McLaughlin KA, Nolen-Hoeksema S. Emotion regulation and internalizing symptoms in a longitudinal study of sexual minority and heterosexual adolescents. *J Child Psychol Psychiatry*. 2008;49(12):1270–1278.
23. Beals KP, Peplau LA, Gable SL. Stigma management and well-being: the role of perceived social support, emotional processing, and suppression. *Pers Soc Psychol Bull*. 2009;35(7):867–879.
24. Hatzenbuehler ML, Nolen-Hoeksema S, Dovidio J. How does stigma "get under the skin"?: the mediating role of emotion regulation. *Psychol Sci*. 2009;20(10):1282–1289.
25. Wright SD, Lund DA, Caserta MS, Pratt C. Coping and caregiver well-being: the impact of maladaptive strategies. *J Gerontol Soc Work*. 1991;17(1–2):75–91.
26. Armour C, Elhai JD, Richardson D, Ractliffe K, Wang L, Elklit A. Assessing a five factor model of PTSD: is dysphoric arousal a unique PTSD construct showing differential relationships with anxiety and depression? *J Anxiety Disord*. 2012;26(2):368–376.
27. Seligowski AV, Lee DJ, Bardeen JR, Orcutt HK. Emotion regulation and posttraumatic stress symptoms: a meta-analysis. *Cogn Behav Ther*. 2015;44(2):87–102.
28. Gratz KL, Roemer L. Multidimensional assessment of emotion regulation and dysregulation: development, factor structure, and initial validation of the difficulties in emotion regulation scale. *J Psychopathol Behav Assess*. 2004;26(1):41–54.
29. Tull MT, Barrett HM, McMillan ES, Roemer L. A preliminary investigation of the relationship between emotion regulation difficulties and posttraumatic stress symptoms. *Behav Ther*. 2007;38(3):303–313.
30. Ehring T, Quack D. Emotion regulation difficulties in trauma survivors: the role of trauma type and PTSD symptom severity. *Behav Ther*. 2010;41(4):587–598.
31. Miron LR, Orcutt HK, Kumpula MJ. Differential predictors of transient stress versus posttraumatic stress disorder: evaluating risk following targeted mass violence. *Behav Ther*. 2014;45(6):791–805.
32. O'Bryan EM, McLeish AC, Kraemer KM, Fleming JB. Emotion regulation difficulties and posttraumatic stress disorder symptom cluster severity among trauma-exposed college students. *Psychol Trauma*. 2015;7(2):131–137.
33. Hayes SC, Wilson KG, Gifford EV, Follette VM, Strosahl K. Experiential avoidance and behavioral disorders: a functional dimensional approach to diagnosis and treatment. *J Consult Clin Psychol*. 1996;64 (6):1152–1168.
34. Meyer EC, Morissette SB, Kimbrel NA, Kruse MI, Gulliver SB. Acceptance and action Questionnaire-II scores as a predictor of posttraumatic stress disorder symptoms among war veterans. *Psychol Trauma*. 2013;5(6):521–528.
35. Plumb JC, Orsillo SM, Luterek JA. A preliminary test of the role of experiential avoidance in post-event functioning. *J Behav Ther Exp Psychiatry*. 2004;35(3):245–257.
36. Gold SD, Marx BP, Lexington JM. Gay male sexual assault survivors: the relations among internalized homophobia, experiential avoidance, and psychological symptom severity. *Behav Res Ther*. 2007;45(3):549–562.
37. Shenk CE, Putnam FW, Rausch JR, Peugh JL, Noll JG. A longitudinal study of several potential mediators of the relationship between child maltreatment and posttraumatic stress disorder symptoms. *Dev Psychopathol*. 2014;26(1):81–91.
38. Miklosi M, Martos T, Szabo M, Kocsis-Bogar K, Forintos DP. Cognitive emotion regulation and stress: a multiple mediation approach. *Transl Neurosci*. 2014;5(1):64–71.
39. Rosenthal MZ, Cheavens JS, Lynch TR, Follette V. Thought suppression mediates the relationship between negative mood and PTSD in sexually assaulted women. *J Trauma Stress*. 2006;19(5):741–745.
40. Ehlers A, Mayou RA, Bryant B. Psychological predictors of chronic posttraumatic stress disorder after motor vehicle accidents. *J Abnorm Psychol*. 1998;107(3):508–519.
41. Clohessy S, Ehlers A. PTSD symptoms, response to intrusive memories and coping in ambulance service workers. *Br J Clin Psychol*. 1999;38:251–265.
42. Moore SA, Zoellner LA, Mollenholt N. Are expressive suppression and cognitive reappraisal associated with stress-related symptoms? *Behav Res Ther*. 2008;46(9):993–1000.
43. Elwood LS, Hahn KS, Olatunji BO, Williams NL. Cognitive vulnerabilities to the development of PTSD: a review of four vulnerabilities and the proposal of an integrative vulnerability model. *Clin Psychol Rev*. 2009;29(1):87–100.
44. Borders A, Rothman DJ, McAndrew LM. Sleep problems may mediate associations between rumination and PTSD and depressive symptoms among OIF/OEF veterans. *Psychol Trauma*. 2015;7 (1):76–84.
45. Michael T, Halligan SL, Clark DM, Ehlers A. Rumination in posttraumatic stress disorder. *Depress Anxiety*. 2007;24(5):307–317.
46. Kashdan TB, Young KC, McKnight PE. When is rumination an adaptive mood repair strategy? Day-to-day rhythms of life in combat veterans with and without posttraumatic stress disorder. *J Anxiety Disord*. 2012;26(7):762–768.
47. Etkin A, Wager TD. Functional neuroimaging of anxiety: a meta-analysis of emotional processing in PTSD, social anxiety disorder, and specific phobia. *Am J Psych*. 2007;164(10):1476–1488.
48. Koenigs M, Grafman J. Posttraumatic stress disorder: the role of medial prefrontal cortex and amygdala. *Neuroscientist*. 2009;15 (5):540–548.
49. Liberzon I, Martis B. Neuroimaging studies of emotional responses in PTSD. In: Yehuda R, ed. *Psychobiology of Posttraumatic Stress Disorder: A Decade of Progress*. vol. 1071. Oxford: Blackwell Publishing; 2006:87–109.
50. Rabinak CA, MacNamara A, Kennedy AE, et al. Focal and aberrant prefrontal engagement during emotion regulation in veterans with posttraumatic stress disorder. *Depress Anxiety*. 2014;31 (10):851–861.

CHAPTER

30

Neuroimaging and Emotion

A.H. Brooke, N.A. Harrison
University of Sussex, Brighton, UK

OUTLINE

Abstract

Emotion forms a central part of everyday human experience. Emotional stimuli within our environment capture our attention and enhance memory encoding. Furthermore, emotion is fundamental to our social relationships, forming a foundation for empathetic interactions with others. While emotion undoubtedly influences multiple psychological functions, contemporary emotional neuroscience has tended to focus on fear processing. Throughout this chapter we follow this focus, which accords with an emphasis on responses to stress. Disorders of emotion regulation are manifest in a range of psychiatric conditions; emotional imbalance underpins much of human unhappiness, and is not confined to the anxiety disorders. The chapter reviews the contribution of neuroimaging to our neurobiological understanding of emotion. We discuss how emotion interacts with cognition, and demonstrate how understanding the mechanisms behind this can enable a deeper understanding of stress-related psychiatric morbidity.

Abbreviations

fMRI functional magnetic resonance imaging
MEG magnetoencephalography
pFC prefrontal cortex

EMOTION

Emotion is central to our everyday human experience. Emotional events or objects within our environment are assigned value, capture attention, and enhance memory encoding. Neuroimaging techniques, including modern optogenetics, have led to a growing awareness that, unlike many other psychological functions, emotion is unencapsulated, interacting with, and influencing multiple other areas of functioning. As well as effects on attention, perception, memory, and learning, emotion plays a core part in our social relationships, and forms a foundation for empathy. Emotional cues can also act as powerful reinforcers, can bias our thoughts and reasoning, and modulate subsequent behavior.

While emotions come in various forms and undoubtedly influence multiple psychological functions, contemporary emotional neuroscience has tended to focus on fear processing. For the purpose of this chapter, we have followed this focus, which also accords with an emphasis on stress responses. Disorders of emotion regulation are manifest in a wide range of psychiatric conditions, consistent with the influence of emotion across the broad spectrum of human functioning. Indeed, emotional imbalance underpins much human unhappiness, and is not confined to anxiety and stress-related disorders. This chapter reviews neuroimaging studies and their contribution to our neurobiological understanding of human emotion.

Stress: Concepts, Cognition, Emotion, and Behavior
http://dx.doi.org/10.1016/B978-0-12-800951-2.00030-3

In particular, we discuss how emotion interacts with and influences other areas of cognition, and demonstrate how an understanding of the specific mechanisms that drive these interactions can enable a deeper understanding of stress-related psychiatric morbidity.

KEY POINTS

- Emotions represent complex psychological and physiological states that index occurrences of value.
- Functional neuroimaging has considerably aided the recognition that, unlike other psychological functions, emotions are relatively unencapsulated and influence virtually all aspects of human cognition.
- Disorders of emotion and in particular, stress-related disorders, consequently lead to impairments in a wide range of human functions.

DEFINITION OF EMOTION

Differences in the conceptualization of emotion can lead to seemingly contradictory theories of emotion. One widely held view conceptualizes emotion as a transient disturbance in the function of an organism, triggered by an internal or external (emotive) stimulus. Most accounts of emotion view emotion as a multicomponential process, with physiological arousal, motor expression, and subjective feelings making up a "reaction triad." What proves more controversial is the extent to which changes in these components are necessary and sufficient to define an episode as "emotional." Scherer[1] provides an excellent discussion of distinct theoretical viewpoints with regards to emotion. For the purposes of this chapter, we use the following definition:

> Emotions are transient events, produced in response to internal or external events of significance to the individual; they are typically characterized by attention to the evoking stimulus and changes in neurophysiological arousal, motor behavior, and subjective feeling state that engender a subsequent biasing of behavior.

EMOTION PERCEPTION AND ATTENTION

This definition of emotion implies that events of significance to the individual undergo preferential perceptual processing and suggests that our attentional processes are susceptible to emotional bias. Behavioral studies have used visual search paradigms in which participants are shown target items, alongside multiple distracter items to illustrate this mechanism. Using visual search paradigms, emotionally valenced items are identified more rapidly than nonemotional items. Furthermore, presentation of an emotional cue on one side of visual space leads to faster identification of nonemotional stimuli subsequently presented on the same side, that is, attention is focused on the location of the previous emotional item. Using spatial orientation tasks, neuroimaging studies have implicated frontal and parietal cortices, as well as lateral orbitofrontal cortex as the neural substrates for this "emotional capture of attention."

Neuroimaging, electroencephalography, and patient studies have shown that even unattended emotional stimuli are more prone to entering awareness. The dominant hypothesis for this effect is that the amygdala rapidly detects the emotional salience of a stimulus after brief and superficial representational processing enabling more detailed sensory processing via projections to sensory cortices, facilitating attention and perception. Anatomical studies suggest that the amygdala mediates this via direct and indirect influence on sensory cortices. Direct reciprocal connections are present between the amygdala and sensory cortices, alongside projections from the central nucleus of the amygdala to the cholinergic nucleus basalis of Meynert, which facilitates indirect ascending neuromodulatory effects.

This hypothesis has been tested with functional neuroimaging using visual "backward masking" paradigms where visual targets are presented very briefly then rendered invisible by subsequently presented stimuli. Despite being unseen, these subliminal emotional stimuli still evoke neural responses in the amygdala supporting existence of a subcortical "low-road" pathway. Furthermore, fusiform face area (FFA: an extra-striate visual area implicated in detailed facial processing) is activated by masked fearful but not neutral facial expressions, with the degree of FFA activity predicted by the strength of amygdala activation. Patients with amygdala damage do not show this enhanced activity in FFA when tested on this paradigm[2] providing further evidence for the hypothesis that amygdala processing heightens activity in sensory cortex. Presentation of fearful face cues has also been shown to enhance early perceptual functions such as contrast sensitivity, consistent with the suggestion that the amygdala may even modulate activity within primary visual cortex.

Preattentive processing of emotional stimuli provides evidence for early representational discrimination between emotional and nonemotional events. Magnetoencephalography (MEG), a neuroimaging technique with high temporal resolution, has shown human discriminatory responses to emotional faces as early as 100-120 ms. This is supported by electrophysiological studies using intracranial electrodes that show a similarly timed (100-160 ms) response that precedes characteristic face-related responses that occur at 170 ms. Subcortical amygdala activation in humans has been shown in both normal and brain injury subjects. One study exploited two separate findings; first, that low and high spatial frequency information are processed by separate visual neural pathways, and second, that coarse emotional cues are carried in low frequency components. When participants were presented low frequency (blurred) face stimuli, fearful compared to neutral face stimuli were associated with enhanced activity in amygdala, pulvinar nucleus of the thalamus and superior colliculus, components of a proposed "low-road" subcortical circuit. In contrast, presentation of the same faces at high spatial frequency activated cortical regions including FFA. Implicit processing of emotional valence can also be demonstrated in patients with damage to primary visual cortex, who are not consciously aware of stimuli presented in their blind hemifield (i.e., blindsight). These patients show amygdala activation to fearful faces presented in their "blind" hemifield,[3] further supporting the presence of subcortical visual processing of emotion to the level of the amygdala.

Together these lesion and functional imaging studies provide mechanistic insight into the role of amygdala interactions with sensory cortices in facilitating enhanced attention and perception of visually presented emotional stimuli. Preferential early processing of emotionally valenced material via a subcortical "low-road" is proposed to establish a primary bias in subsequent information processing that regulates ongoing behavior. However, this simple concept of a quick, but dirty subcortical "low-road" and slow but precise cortical "high-road" to amygdala activation has recently been challenged by data suggesting that high spatial frequency information also plays an important role in the rapid detection of emotionally salient stimuli. Ongoing studies will ultimately determine whether a more nuanced "many roads" or "multiple waves" model that acknowledges that cortical regions may process affective stimuli adequately without the need for a faster subcortical route is adopted[4] (Figure 1).

Visual processing of emotionally valenced facial expressions has been the mainstay of neuroscientific studies of human emotion. However, facial muscles form a subset of the broader vocabulary of emotional state signaling. For example, emotional state is also signaled by a range of bodily changes, including pupillary

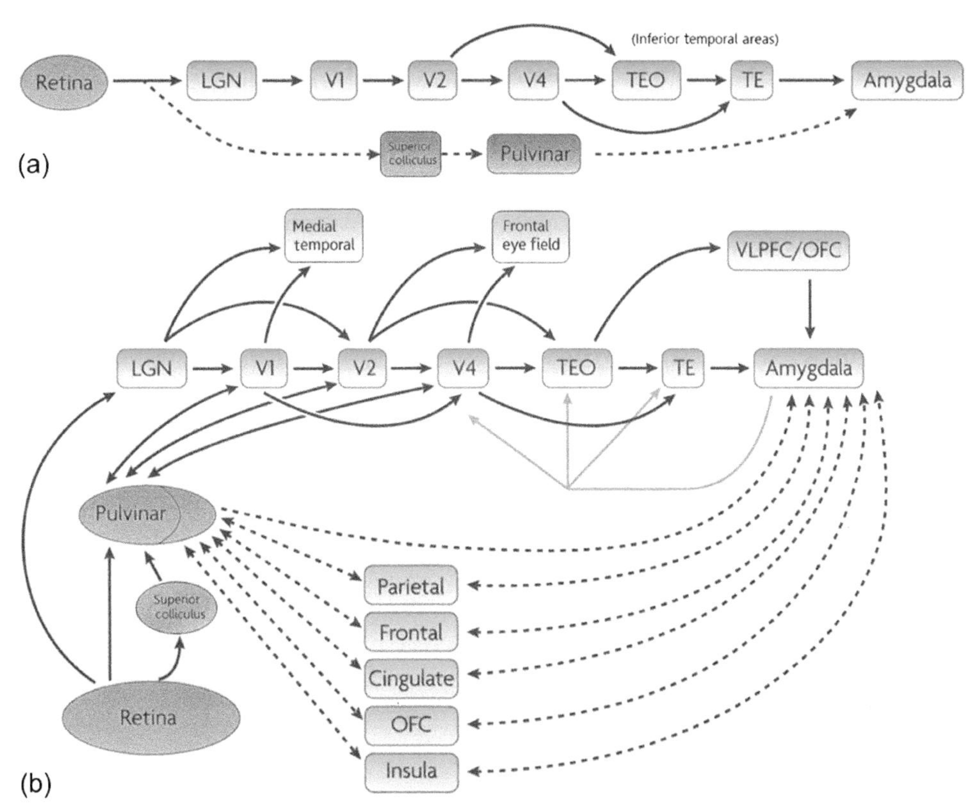

FIGURE 1 The standard hypothesis is represented by (a) a subcortical pathway comprised of the superior colliculus and the pulvinar nucleus of the thalamus leads directly to the amygdala—provoking a fast and automatic response. The alternative view is represented by (b) the idea that the flow of visual information occurs via multiple different pathways, including alternative routes and shortcuts—giving rise to the understanding of a "multiple waves of activation" approach. Reprinted with modifications from Ref. 4, Copyright (2011), with permission from Elsevier.

dilation/constriction, skin changes, body posture, gesticulations, and speech content and prosody. Prioritized processing of emotional sounds, particularly nonspeech intonations, for example, gasps, screams, are also illustrated within primary and accessory auditory cortices. As with the visual system, amygdalo-cortical interactions are implicated in enhanced auditory attention to emotive sounds. Specificity in emotional processing may also traverse sensory modalities, such that ventral insular regions activated by disgusting smells may also by activated by auditory and visual expressions of disgust.[5]

A degree of regional specialization in human emotional processing is suggested from lesion and neuroimaging studies: Acquired damage to the amygdala impairs recognition and behavioral experience of fear. A study in a patient with focal bilateral amygdala lesions (due to Urbach-Wiethe disease) showed impairment in fear perception and absence of any behavioral manifestations of fear despite being able to describe the concept of fear and subjectively feel all other basic emotions.[6,7] In contrast, the insula and striatum have been implicated in disgust processing and damage to these regions has been associated with impaired recognition of facial expressions of disgust.[8] Furthermore, insula stimulation can provoke nausea and experience of unpleasant taste.[9] These lines of evidence for specialized emotion-specific processing within different brain regions are generally circumscribed to fear/threat (amygdala), disgust (insula/striatum) and, to a lesser extent, sadness (subgenual cingulate activity). Neuroimaging studies show that the amygdala is sensitive to the emotional intensity of external emotional stimuli, with threat being perhaps the most arousing.

The insula also appears to relate perceptual information with interoceptive information about bodily state. Insula activity is observed during recognition and experience of disgust, sadness, happiness, and fear, as well as during the perception and expression of threat, experience of phobic symptoms, hunger, satiety, and even explicit categorization of facial emotion. These actions are summarized and discussed further in the section relating to feeling states.

DECISION-MAKING

Decision-making is frequently considered a highly rational process under conscious control. However, recent behavioral and neuroimaging data show that decision-making processes are far more prone to emotional or "intuitive" influences than previously appreciated. For example, Kahneman and Tversky have identified a cognitive bias, in which peoples' choice preference change based simply on the way options are presented, a concept they termed the "framing effect." De Martino and colleagues[10] investigated the neural basis for this framing effect using functional neuroimaging. In this study participants made a series of simple decisions. For example, at the start of each trial they were told that they had £50 and must choose between a "sure" or a "gamble" option presented in either a "Gain" or "Loss" frame. In the "Gain" frame, the sure option was worded as a guarantee to keep £20 of the original £50. However, in the "Loss" frame, the sure option was worded as a guarantee to lose £30 of the original £50 (i.e., they would still keep £20 of the original sum). The "gamble" option showed the probability of winning or losing the whole £50. It was designed to have the same expected value, that is, a 40% chance of keeping the £50 (expected value £20) and was identical for "Gain" and "Loss" frames.

Despite "Gain" and "Loss" frames being of identical monetary value, participants were significantly more risk-averse in the Gain frame, choosing the "sure" option more often, yet more risk-taking, choosing to "gamble" more often, in the Loss frame. Neuroimaging data showed a specific association of this framing effect with amygdala activity, suggesting a key role for the emotion system in mediating this decision-making bias. Individuals with greatest orbital and medial prefrontal cortex (pFC) activation showed least susceptibility to the framing effect raising the possibility that more "rational" individuals have better cortical representations of their emotional biases enabling them to modify behavior in circumstances where biases might lead to suboptimal decisions.

Individuals with autism spectrum disorders (ASDs) have been shown to be significantly less susceptible to framing effects, resulting in more "rational" decisions in the face of this emotional decision-making bias.[11] They also express diminished skin conductance arousal responses suggesting they may fail to incorporate emotional context into the decision-making process. This finding might help explain both enhanced analytic and impoverished social abilities in ASD, the latter reflecting an inability to deploy an affect heuristic in complex and uncertain social contexts.

MEMORY AND LEARNING

Emotion has a dramatic, and well described, impact on memory. Animal studies highlight the central role of the amygdala in fear conditioning,[12] a form of implicit memory that involves rapid learning of a potentially harmful stimulus. If an adverse event, for example, a painful shock (unconditioned stimulus (US)) occurs after a relatively benign stimulus, the latter (conditioned stimulus (CS)) comes to predict the adverse event itself and induce arousal responses associated with it. Patients with bilateral amygdala damage do not acquire conditioned fear responses (autonomic or motor) despite intact knowledge of the association between the conditioned and unconditioned stimuli. The importance of the amygdala in fear conditioning has also been confirmed by functional neuroimaging studies[13] and optogenetic studies.[14] Further, in accordance with animal data, neuroimaging studies describe amygdala engagement in more general associative reinforcement learning, including operant stimulus-reward learning, Pavlovian-instrumental transfer and appetitive behavior.

Recent optogenetic work has also begun to clarify mechanisms underlying extinction of fear responses and identified age-specific effects that challenge existing models of extinction (Figure 2). For example, in adult rodents, amygdala vmPFC interactions appear to play an important role in mediating fear extinction. However, in young rats, extinction responses are confined to the amygdala,[15] suggesting greater amygdala dependence for extinction learning at this age. It has been proposed that this absence of vmPFC involvement in young animals may contribute to reduced extinction flexibility observed in childhood, including reduced renewal, reinstatement, and spontaneous recovery of fear responses. Anxiety disorders are one of the most prevalent forms of psychopathology in children, and often persist into adulthood, with costly and debilitating consequences. Exposure therapies (based on extinction processes) are one of the most successful treatments for anxiety disorders. In adults, relapse post therapy is a common problem; however, these new findings suggest that exposure therapy in young children may be more effective and may prevent adulthood recurrence.[15]

In addition to simple associative responses, human memory is characterized by rich declarative episodic and semantic memory. Hippocampal activity supports both the encoding, and to a lesser extent, recollection of declarative memory and is strongly modulated by emotion. Emotional events and stimuli are typically remembered better than nonemotional occurrences.[17] Animal studies highlight noradrenergic mechanisms enhancing amygdala-hippocampal connectivity in the preferential encoding of emotional objects. Human brain-imaging studies have gone further, showing that amygdala activation during memory encoding actually predicts subsequent accuracy of declarative memory for emotional stimuli.

INTEROCEPTION AND SUBJECTIVE FEELING STATES

Changes in bodily state, reflected as shifts in physiological and motor behavior, are obligatory defining characteristics of emotion. James and Lange, and more recently, Damasio[7] argue that emotional feeling states must have their origin in the brain's representation of bodily physiological state.[18,19] Without changes in bodily physiological state there is no emotion. This peripheral account of emotion predicts that afferent feedback from muscles and viscera associated with physiological arousal are integral to emotional experience. These mechanisms have been illuminated in a number of neuroimaging experiments examining central representation and re-representation of autonomic states and homoeostatic sensations, such as temperature.

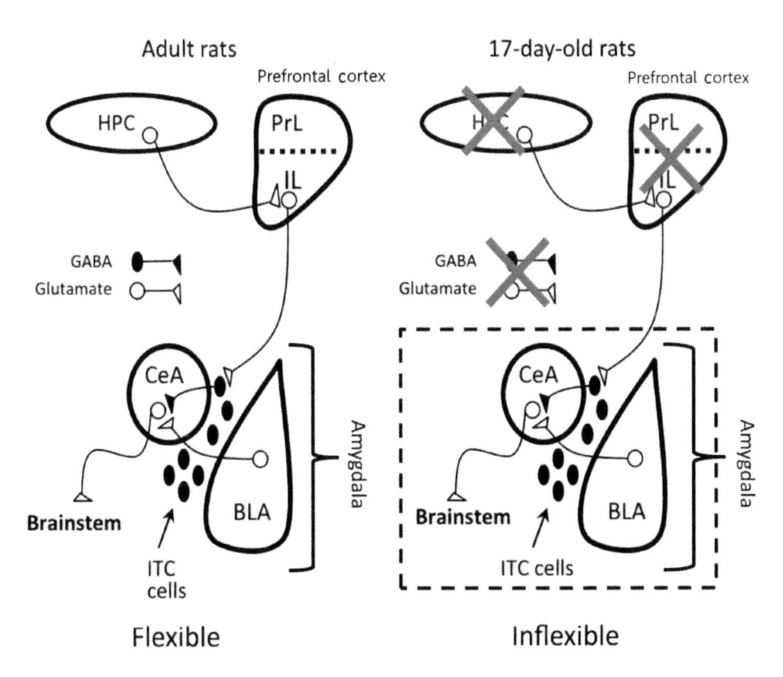

FIGURE 2 The neural structures activated in extinction of conditioned fear are similar in P24 and adult rats. The process begins with hippocampal modulation of the vmPFC, which then controls projections to the amygdala in order to modulate the fear extinction. However, in P17 rats (i.e., young rats) extinction occurs independently of the hippocampus and vmPFC, suggesting a greater reliance on only the amygdala for extinction learning at this age. Reprinted with modifications from Ref. 15, Copyright (2010), modified with permission from Ref. 16, Copyright (2004), with permission from Elsevier.

In this context, the role of insula cortex is particularly interesting: in a PET study activation within "viscerosensory" mid insula cortex correlated closely with actual stimulus temperature. However, subjective ratings of the perceived intensity or pleasantness of hot and cold stimuli correlated with activity in right anterior insula and orbitofrontal cortex. This finding suggests a dissociation between primary central representations of viscerosensory information, within mid insula, and subjective feelings of warmth or cold, in the right anterior insula and orbitofrontal cortex. A more general role for these regions in the representation of emotional feelings is also supported by their activation during generated states of sadness and anger, anticipatory anxiety and pain, panic, disgust, sexual arousal, trustworthiness, and even subjective responses to music.[20]

Behavioral and neuroimaging studies in patients with selective peripheral autonomic denervation (pure autonomic failure (PAF)) lend further support to the importance of viscerosensory feedback to emotional feelings. Patients with PAF do not have autonomic (e.g., heart rate or skin conductance) changes in response to emotional stimuli or physiological stressors, and show a subtle blunting of subjective emotional experience. In a fear conditioning study, both PAF and control participants showed amygdala activity during threat learning.[21] However, right insula cortex showed attenuated responses to threat stimuli in PAF patients compared to controls, suggesting sensitivity to the presence/absence of autonomic responses generated during emotional processing. When threat stimuli were presented unconsciously, the same insula regions were sensitive to both the conscious awareness of emotion events and induced bodily arousal suggesting a role for the right insula in the contextual integration of external emotional events with feeling states arising from bodily reactions.

If subjective emotional feelings arise from central representations of bodily arousal states, it follows that people with greater awareness of their bodily responses may also experience more intense emotional feelings. In support of this, studies investigating individual differences in "interoceptive awareness" suggest that patients with anxiety disorders may be more attuned to their own bodily reactions such as heartbeat or stomach motility. In one such interoceptive task,[22] subjects were played a series of 10 notes, presented in or out of synchrony with their own heartbeats. On half the trials subjects had to judge the timing of their heartbeat in relation to the notes, in the control condition they had to detect a rogue note. When participants focused interoceptively on the timing of their own heartbeat, there was enhanced activity in somatomotor, insula, and cingulate cortices. However, only one region, the right anterior insula cortex, reflected conscious awareness of internal bodily responses, that is, how accurately they performed the interoceptive task. Importantly, activity and even gray matter volume in this region, predicted interindividual differences in day-to-day awareness of bodily reaction and subjective experience of negative emotions, particularly anxiety.

Recent functional magnetic resonance imaging studies also show that second-by-second changes in body physiology can influence emotion processing. For example, during one study,[23] participants were shown emotional or neutrally valenced facial expressions at different points in the cardiac cycle, that is, during systole (when the heart is ejecting blood) or diastole (when the heart is refilling). During systole (when feedback from high-pressure baroreceptors is strongest), participants were better able to identify fearful faces during rapid serial visual presentation and rated them as appearing more intense. Furthermore, fearful faces presented at systole were associated with greater amygdala activity, as well as increased anterior and middle insula, and dorsal posterior cingulate cortex reactivity (Figure 3). Studies such as this provide further evidence for the influence of interoception and visceral state on responses to fear and support a role for the anterior insula as the principle neural substrate supporting experiential feeling states arising from interoceptive feedback of bodily arousal state.

SOCIAL INTERACTION

An intuitive ability to understand one another's mental and emotional states is a fundamental characteristic of social interactions. Humans show a marked tendency to mimic each other's gesticulations, emotional facial expressions, and body postures, suggesting that this mirroring of activity may facilitate emotional understanding. In monkeys, premotor cortical neurons are described that respond to observing or performing the same motor actions.[24] Correspondingly, human imaging studies of action observation illustrate premotor cortical activity when observing the actions of others, and MEG studies report resynchronization in primary motor cortex. Importantly these mirror activations are somatotopic with respect to body part performing the action.

Accumulating evidence supports extension of these action-perception mechanisms to emotions and feeling states. A common neural representation is proposed for the perception of actions and feelings in others and their experience in self. Thus viewing emotional facial expressions in others automatically activate mirrored expressions on one's own face (even when expressions are masked).[25,26] Importantly, this automatic mimicking of emotional expressions also extends to autonomic features of emotional states such as pupil size in expressions of sadness.[27] Neuroimaging studies show shared neural activation in somatosensory cortex when experiencing

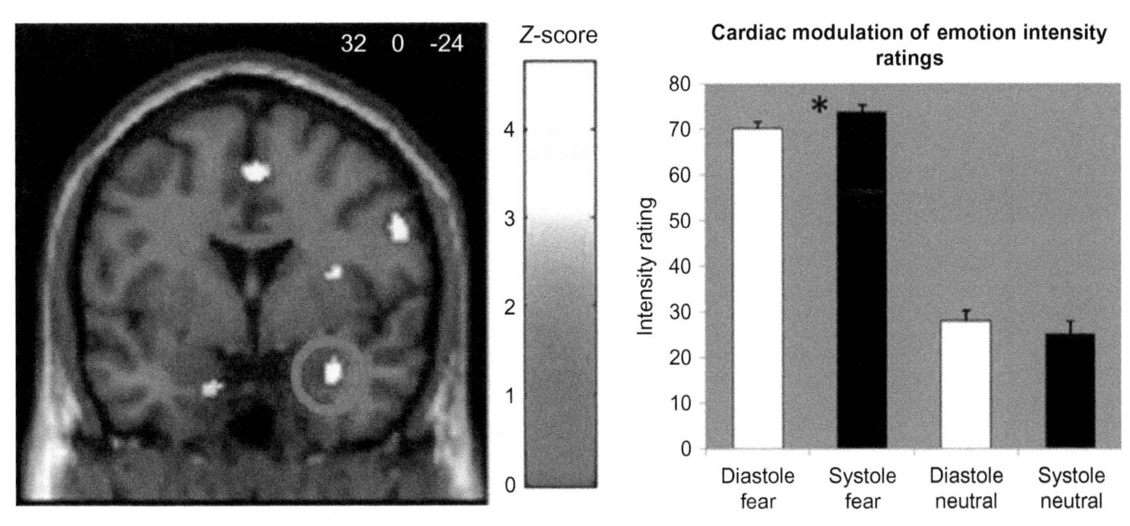

FIGURE 3 The image on the left shows increased activation in bilateral amygdala, in particular the right side (circled), during emotion processing, in relation to the presentation of emotional faces at systole and diastole. The graph depicts this interaction between perception of emotional intensity and cardiac timing; fear faces were rated as more intense during systole, while neutral faces were more intense at diastole. Reprinted with modifications from Ref. 23, Copyright (2014) with permission from *The Journal of Neuroscience*.

or observing another being touched, right insula activity to the experience of disgusting odors, and observed expressions of disgust[28] and activation of a common matrix of regions in the observation or experience of pain.[29] Imitation or observation of emotional facial expressions also activates a largely similar network of brain regions including premotor areas that are critical for action representation. There is also selectivity; compared to nonemotional facial movements, emotional expressions activate right frontal operculum and anterior insula and regions such as amygdala, striatum, and subgenual cingulate are differentially responsive when imitating happy or sad emotions. The amygdala and posterior cingulate cortex also appear to be most strongly activated when we meet someone new, allowing us to form rapid and relatively accurate first impressions after minimal exposure.[30]

EMOTION REGULATION

Understanding adaptive mechanisms that allow us to control expression of our emotions is clinically and socially important. Unlearning established emotional reactions is difficult. Many studies have examined the modulation of previously learned emotional responses with experimental extinction: Repeated presentation of a previously CS without its associated US rapidly leads to a diminished emotional response. However, the relationship between the two stimuli is not forgotten but rather inhibited and can be rapidly reinstated. Neuroimaging of extinction learning shows increased amygdala activity in early extinction, however retention of extinction beyond a day was associated with activity in the subgenual anterior cingulate and ventromedial pFCs, suggesting a role for this region in the longer term recall of extinction learning (Figure 4).

Emotional responses may also be regulated volitionally, using cognitive control or reappraisal. For example, an image of a woman crying may be interpreted as sadness, though it may be reappraised as joy if told the picture was taken at her daughter's wedding. Neuroimaging studies of emotional reappraisal demonstrate that enhanced activity in left middle frontal gyrus is associated with attenuation of amygdala activity, suggesting a role in modulating amygdala emotional reactivity, possibly via connections to medial pFC. Similar cognitive emotion-regulation techniques can powerfully diminish responses acquired through fear conditioning and neuroimaging studies show increased activity in the left middle frontal gyrus using this paradigm. Taken together, these studies suggest that cognitive emotion regulation and lower level extinction learning recruit similar overlapping neural mechanisms for the regulation of amygdala to control primary emotional responses.

Understanding interactions between emotion and cognition has important implications for clinical practice. Interindividual differences in these interactions point to individual differences in vulnerability and susceptibility to affective problems such as depression and anxiety. Indeed, it is often the inability to reconcile cognition and emotion that leads to the devastating effects of many psychiatric disorders.[17] Understanding these subtle differences may allow us to guide treatment for patients in a specific manner, based upon a better understanding

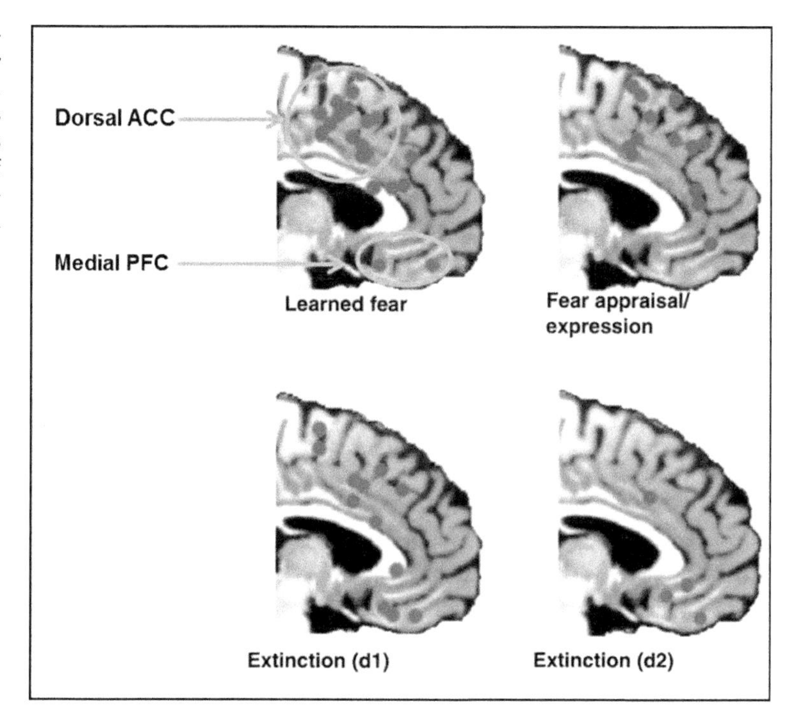

FIGURE 4 Display of areas activated during fear regulation. Dorsal ACC/mPFC activity is seen during classical fear conditioning. This is also the case during the appraisal and expression of fear. With regard to fear extinction, the focus of activation is the dorsal and ventral ACC/mPFC (d1). This remains true for subsequent delayed recall and expression of the memory of the extinction (d2). Reprinted with modifications from Ref. 31, Copyright (2011) with permission from Elsevier.

of the unique underlying brain mechanisms responsible for their illness.

EMOTION DYSREGULATION

The fact that such a variety of symptoms and presentations are encompassed by emotional disorders is demonstrative of the broad influence of emotion on psychological functions. Post-traumatic stress disorder (PTSD), arising following the experience of a highly traumatic and emotional event, is very often linked to symptoms in a number of different psychological areas. Patients display heightened arousal and startle responses to stimuli previously perceived as harmless, alongside emotional numbing, intrusive flashbacks to the traumatic event, and impairment in social function. Identification of brain regions that show increased activation associated with the enhanced perception and memory for emotional events and associated feeling states enables linkage of psychological mechanisms in conditions such as PTSD and phobias with their underlying neural substrates.

Abnormal cortisol regulation and adrenergic stress responses are also seen in PTSD as is a reduction in hippocampal volume. Interestingly, patients with endocrine disorders associated with chronically elevated cortisol levels, such as Cushing's disease, show a similar reduction in hippocampal volume, suggesting an etiological role for stress hormones. This is further supported by a recent study of healthy postmenopausal women,[32] which shows a strong correlation between hippocampal volume and lifetime levels of perceived stress suggesting a role of stress even in the normal population.

The recent development of optogenetic imaging, the combination of optics, and genetic encoding of light-gated ion channels,[33] allows real-time in vivo imaging of neural activity and the potential to develop a deeper understanding of the neurobiological processes underpinning emotional responses. To this end, recent optogenetic studies have shown that discrete hippocampal regions play specific roles in emotional processing. Dorsal areas encode context-specific learning of fear, while more ventral regions regulate anxiety-related behavior.[34] Findings such as this may have direct clinical applicability as it has been shown that ventral dentate gyrus activation can suppress innate anxiety without adversely affecting learning or memory encoding. Optogenetic stimulation of the medial pFC in mice has been shown to have antidepressant effects,[35] suggesting that this evolving technique may inform novel methods for the treatment of PTSD, affective, and anxiety disorders.

Many psychological therapies used in the treatment of emotional disorders focus on changing maladaptive cognitive interpretation of everyday events that are central to their perpetuation. A deeper understanding of the neural mechanisms underlying the modulation of emotional behavior will be essential to fully understand the neurobiology of these psychiatric disorders. In addition, improved mechanistic understanding offers the opportunity to design targeted approaches, with a specific focus on altering detrimental maladaptive emotional reactions.

Glossary

Amygdala Almond-shaped cluster of interrelated nuclei and associated cortex in the anterior medial temporal lobe.

Arousal A dimension of emotion that varies from calm to excitement and predicts behavioral activity level.

Blood oxygen level dependent (BOLD) T_2*-weighted signal in fMRI reflecting hemodynamic changes in regional perfusion evoked by neural activity.

Interoception Representation of visceral activity: the afferent limb of homoeostatic neural control of bodily organs.

Optogenetics Use of light to control neurons that have been genetically modified to be sensitive to light enabling real-time imaging of the combination of optics and genetic encoding of light-gated protein channels to monitor individual neuronal activity in vivo real-time in living tissue.

Valence A dimension of emotion that varies from unpleasant (negative) to pleasant (positive) to reflect motivational value.

References

1. Scherer KR. Psychological models of emotion. In: Borod J, ed. *The Neuropsychology of Emotion*. New York, NY: Oxford University Press; 2000.
2. Vuilleumier P, Richardson M, Armony J, Driver J, Dolan R. Distant influences of amygdala lesion on visual cortical activation during emotional face processing. *Nat Neurosci*. 2004;7(11):1271–1278.
3. Morris J. Differential extrageniculostriate and amygdala responses to presentation of emotional faces in a cortically blind field. *Brain*. 2001;124(6):1241–1252.
4. Pessoa L, Adolphs R. Emotion processing and the amygdala: from a 'low road' to 'many roads' of evaluating biological significance. *Nat Rev Neurosci*. 2010;11(11):773–783.
5. Phillips ML, Young AW, Senior C, et al. A specific neural substrate for perceiving facial expressions of disgust. *Nature*. 1997;389(6650):495–498.
6. Feinstein J, Adolphs R, Damasio A, Tranel D. The human amygdala and the induction and experience of fear. *Curr Biol*. 2011;21(1):34–38.
7. Damasio A. *The Feeling of What Happens*. New York, NY: Harcourt Brace; 1999.
8. Calder AJ, Keane J, Manes F, Antoun N, Young AW. Impaired recognition and experience of disgust following brain injury. *Nat Neurosci*. 2000;3(11):1077–1078.
9. Penfield W, Faulk M. The insula. *Brain*. 1955;78(4):445–470.
10. De Martino B. Frames, biases, and rational decision-making in the human brain. *Science*. 2006;313(5787):684–687.
11. De Martino B, Harrison N, Knafo S, Bird G, Dolan R. Explaining enhanced logical consistency during decision making in autism. *J Neurosci*. 2008;28(42):10746–10750.
12. LeDoux J. *The Emotional Brain*. New York, NY: Simon & Schuster; 1996.
13. LaBar KS, Cabeza R. Cognitive neuroscience of emotional memory. *Nat Rev Neurosci*. 2006;7(1):54–64.
14. LaLumiere R. Optogenetic dissection of amygdala functioning. *Front Behav Neurosci*. 2014;8:107.
15. Kim J, Richardson R. New findings on extinction of conditioned fear early in development: theoretical and clinical implications. *Biol Psychiatry*. 2010;67(4):297–303.
16. Sotres-Bayon F, Bush DE, LeDoux JE. Emotional perseveration: an update on prefrontal-amygdala interactions in fear extinction. *Learn Mem*. 2004;11(5):525–535.
17. Dolcos F, Iordan A, Dolcos S. Neural correlates of emotion-cognition interactions: a review of evidence from brain imaging investigations. *J Cogn Psychol*. 2011;23(6):669–694.
18. Lange C. *The Emotions*. Baltimore, MD: Williams & Wilkins; 1922 [translated by IA Haupt original work published 1885].
19. James W. The physical bases of emotion. 1894. *Psychol Rev*. 1994;101(2):205–210.
20. Craig A. How do you feel? Interoception: the sense of the physiological condition of the body. *Nat Rev Neurosci*. 2002;3(8):655–666.
21. Critchley H, Mathias C, Dolan R. Fear conditioning in humans. *Neuron*. 2002;33(4):653–663.
22. Critchley H, Wiens S, Rotshtein P, Öhman A, Dolan R. Neural systems supporting interoceptive awareness. *Nat Neurosci*. 2004;7(2):189–195.
23. Garfinkel S, Minati L, Gray M, Seth A, Dolan R, Critchley H. Fear from the heart: sensitivity to fear stimuli depends on individual heartbeats. *J Neurosci*. 2014;34(19):6573–6582.
24. Gallese V. The manifold nature of interpersonal relations: the quest for a common mechanism. *Philos Trans R Soc Lond B Biol Sci*. 2003;358(1431):517–528.
25. Dimberg U, Thunberg M, Elmehed K. Unconscious facial reactions to emotional facial expressions. *Psychol Sci*. 2000;11(1):86–89.
26. Tamietto M, de Gelder B. Neural bases of the non-conscious perception of emotional signals. *Nat Rev Neurosci*. 2010;11(10):697–709.
27. Harrison N. Pupillary contagion: central mechanisms engaged in sadness processing. *Soc Cogn Affect Neurosci*. 2006;1(1):5–17.
28. Wicker B, Keysers C, Plailly J, Royet J, Gallese V, Rizzolatti G. Both of us disgusted in my insula. *Neuron*. 2003;40(3):655–664.
29. Singer T. Empathy for pain involves the affective but not sensory components of pain. *Science*. 2004;303(5661):1157–1162.
30. Schiller D, Freeman JB, Mitchell JP, Uleman JS, Phelps EA. A neural mechanism of first impressions. *Nat Neurosci*. 2009;12(4):508–514.
31. Etkin A, Egner T, Kalisch R. Emotional processing in anterior cingulate and medial prefrontal cortex. *Trends Cogn Sci*. 2011;15(2):85–93.
32. Gianaros P, Jennings J, Sheu L, Greer P, Kuller L, Matthews K. Prospective reports of chronic life stress predict decreased grey matter volume in the hippocampus. *NeuroImage*. 2007;35(2):795–803.
33. Nieh E, Kim S, Namburi P, Tye K. Optogenetic dissection of neural circuits underlying emotional valence and motivated behaviors. *Brain Res*. 2013;1511:73–92.
34. Fournier N, Duman R. Illuminating hippocampal control of fear memory and anxiety. *Neuron*. 2013;77(5):803–806.
35. Covington H, Lobo M, Maze I, et al. Antidepressant effect of optogenetic stimulation of the medial prefrontal cortex. *J Neurosci*. 2010;30(48):16082–16090.

Further Reading

Darwin C. *The Expressions of Emotion in Man and Animals*. London: John Murray; 1872. [Ekman P, ed. 3rd ed., London: Harper Collins, 1998].

Furl N, Henson R, Friston K, Calder A. Top-down control of visual responses to fear by the amygdala. *J Neurosci*. 2013;33(44):17435–17443.

Morris JS, Ohman A, Dolan RJ. Conscious and unconscious emotional learning in the human amygdala. *Nature*. 1998;393(6684):467–470.

Ochsner K, Gross J. The cognitive control of emotion. *Trends Cogn Sci*. 2005;9(5):242–249.

Seymour B, Dolan R. Emotion, decision making, and the amygdala. *Neuron*. 2008;58(5):662–671.

CHAPTER

31

Rumination, Stress, and Emotion

D. Roger

University of Canterbury, Christchurch, New Zealand

Abstract

Psychometric instruments developed to assess moderator variables in the stress process have been hampered by a number of shortcomings, including the use of inappropriate factoring techniques and the absence of confirmatory factor analysis to endorse the obtained structures. The research has also been compromised by the absence of an unambiguous definition of stress, which has led to widespread misunderstandings about the construct. This chapter aims to resolve some of these issues by focusing on the role of personality factors that have been shown to be significantly implicated in the stress response in everyday contexts, especially emotional rumination, emotional inhibition, coping styles, and self-esteem. A key distinction is drawn between chronic as opposed to acute stress, and the evidence presented is grounded in established physiological findings based on cardiovascular and immune function.

The incidence of post-traumatic stress disorder (PTSD) following disasters ranges from as low as 3.7% to over 60%, but with most studies tending toward the mid-to-lower end of the incidence scale.[1] The variation is attributable to a variety of factors, such as whether the disaster was natural or manmade, but it is nonetheless the case that the majority of those exposed to disasters are not subsequently diagnosed with PTSD. What is it that serves either to protect these people from becoming stressed or, as in the case of the minority, to make them more vulnerable to stress? The research described in this chapter is concerned with exploring the individual differences that might account for the range in the way people respond, and will focus on the implications of the findings from four key areas of personality: emotional rumination, emotional inhibition, coping styles, and self-esteem.

Reviewing the evidence for the moderating effects of personality shows that the findings are at best equivocal. For example, it has often been argued that factors such as extraversion or locus of control act as buffers against stress, but as with many other candidate variables, the results have been inconclusive.[2,3] One of the problems is that the target personality variables may have been developed in entirely different contexts, and don't necessarily generalize to stress manipulations. The methods used to develop psychometric measures are also often inappropriate, such as biased item generation to construct the initial item pool, the use of an eigenvalue-1 criterion to decide on structure (which almost invariably leads to the extraction of too many factors), and the absence of confirmatory factor analysis to endorse obtained factor structures.[4,5]

An alternative approach is to provide a sample of participants with a series of targeted scenarios and asking them to describe how they would think, act, and feel in these situations.[6] After eliminating repetitions, the remaining responses are phrased as questionnaire items to comprise the initial item pool, but with the minimum of alterations to the grammar—rather than translating them into "academese," the vernacular is preserved. Responses from sequential independent samples are then processed using both exploratory and confirmatory factor analyses to arrive at final scales, preparatory to further concurrent and predictive validation. An illustrative case comes from research aimed at resolving the misleading implication of the neuroticism label, that high-scoring individuals are more neurotic than those with low scores. An unbiased analysis using these techniques showed that the key element is emotional sensitivity, with discriminable components for both positive other-oriented sensitivity, which correlates positively with showing empathic concern, and negative self-centered sensitivity which relates to responding with personal distress.[7]

Stress: Concepts, Cognition, Emotion, and Behavior
http://dx.doi.org/10.1016/B978-0-12-800951-2.00031-5

KEY POINTS

- A distinction is drawn between post-traumatic and everyday stress. Major traumas are unpredictable and generally lack preincident baseline measures, and the emphasis in the chapter is on studies of experimental and naturalistic everyday stress.
- The evidence for the role of personality moderators of the stress process has been inconclusive, and the psychometric shortcomings in many of the scales that have been developed have been a major obstacle to progress in stress research.
- The chapter offers an alternative procedure for scale development, and focuses on key areas where there is significant supportive evidence from appropriately constructed indices: emotional rumination, emotional inhibition, coping styles, and self-esteem.
- A distinction between pressure and stress is proposed as a less confusing description than the conventional notions of eustress and distress, and is explained in terms of the differential effects of acute as opposed to chronic stress.
- The deleterious consequences of stress are, for the most part, attributable to chronic stress, where opportunities for recovery are limited and homeostasis is compromised.
- These consequences are discussed in the context of the activation of the hypothalamic-pituitary-adrenal (h-p-a) axis, focusing on the effects of adrenocortical and adrenomedullary hormones, and chronicity is predicated on emotional rumination as the mechanism which sustains emotional responses that would otherwise dissipate.

Using appropriate psychometric procedures to develop measures that might be implicated in the stress response, Roger and his colleagues[8] developed the emotion control questionnaire (ECQ), later revised as the ECQ2.[9] Both questionnaires comprised four scales: rehearsal, emotional inhibition, benign control, and aggression control. Emotional inhibition measures the tendency to bottle up, rather than to express experienced emotion, and the emergence of an index of inhibition was not surprising in the context of stress research current at the time—the role of expressing emotion in mitigating stress responses was already well established, including making a written record of emotional experiences (see Pennebaker[10] for a review).

Rehearsal measures the tendency to dwell on emotional upsets, but the label "rehearsal" restricted the meaning to preparation for a future event. Scale items encompassed both prospective and retrospective elements, and to include the tendency to continue to churn over emotionally upsetting events that have either happened in the past or might happen in the future, the scale was renamed *rumination*. A number of subsequent rumination-related measures have been developed, but the ECQ scale was the first published psychometric index of rumination, and its role in the stress response was at that stage unclear.

Rumination is statistically orthogonal to emotional inhibition, and these two scales are in turn independent of the last two ECQ scales, but benign control and aggression control are themselves significantly positively correlated. These two scales subsequently proved to be part of the extraversion constellation—benign control, for example, is a strong inverse index of impulsiveness.[8] As mentioned earlier, the evidence linking extraversion to stress is weak, and the emotion control research program focused instead on rumination and inhibition. The emphasis of this work also shifted from stress induced by disasters to the effects of everyday stress. Disasters are unpredictable, and one of the challenges of studying their effects is the absence of pre-exposure control measures; by contrast, everyday stress can be studied in more easily controlled laboratory and naturalistic settings, where the role of individual differences can systematically be explored.

Validation studies showed that rumination in particular is associated with a range of health-related indices, including cardiovascular recovery from a laboratory stress manipulation, prolonged cortisol secretion following exposure to a naturalistic stressor, and increased postpartum analgesic demand.[11–13] Rumination has also been shown to play a role in behavioral problems: McDougall et al.[14] compared incarcerated young offenders who either did or did not display anger control problems, and rumination discriminated more powerfully between the groups than any of the other indices that were used, including measures of state and trait anger. The offenders with anger control problems scored significantly higher on rumination, indicating that rumination had served to sustain feelings of anger that might otherwise have dissipated.

Emotional inhibition has proved to be less significantly implicated in health outcomes when compared with rumination, but high inhibition scores were significantly related to delayed muscle tension recovery following an experimental speech preparation paradigm.[15] Just as there are different but interrelated retrospective and prospective aspects to the rumination construct, different elements also contribute to the general reluctance to disclose emotional feelings, including a fear of disclosure and low social intimacy needs.[6] The different strands of both rumination and inhibition were drawn together in the recent construction of a revised and expanded index of rumination and inhibition, the inhibition-rumination scale[16] (I-RS), and the two-factor structure of the new scale was unambiguously confirmed. Compared to the original ECQ measures, the revised rumination and inhibition scales have shown similar but improved relationships with indices of physical and mental well-being, and both scales were found to be significantly associated with self-harming behavior.[17]

Coping styles have featured prominently in stress research, though early scales for assessing coping suffered from a range of psychometric shortcomings, especially the extraction of an unwarranted number of factors—the Coping Orientations to Problems Experienced (COPE) questionnaire,[18] for example, claimed 14 discrete dimensions, several defined by fewer than five items each. Endler and Parker[19] proposed just three fundamental coping dimensions in their Multidimensional Coping Inventory (MCI), labeled task, emotional, and avoidance coping. Task-oriented coping, also referred to as rational coping, tends to be defined by outlining priorities and seeking information, emotional coping by self-blame, helplessness, and frustration, and avoidance by ignoring, distorting, or escaping from stimuli that are perceived to be threatening.

The same three components emerged from an independent radial parcel analysis of the COPE,[20] and the three-factor structure was again confirmed by Roger et al.[21] with their coping styles questionnaire (CSQ). However, the CSQ added a fourth coherent factor, endorsed by confirmatory factor analysis, which the authors labeled *detached* coping. Detachment assesses the ability to maintain perspective, indexed by seeing situations for what they are, responding neutrally, and not catastrophizing by becoming identified with the issues. The new factor thus provides a well-defined cognitive strategy for coping, in contrast to the direct information-seeking and active planning involved in rational coping or the emotional overinvolvement of emotional coping—indeed, emotional coping is best described as an absence of coping.

As expected, detached coping is significantly inversely correlated with I-RS rumination: the greater the degree of emotional catastrophizing, the greater the likelihood of continuing to ruminate about the emotional upset, and ruminating will in turn enhance and prolong engagement with the negative emotions associated with the issue.[21] Detached coping also significantly predicted job satisfaction and adaptational capacity amongst probation staff in a study of factors affecting resilience across four European probation services.[22]

Subsequent research has provided a better understanding of rational and emotional coping, but despite having potentially significant implications for psychological and physical health, avoidance coping is much less well understood.[23] For example, avoidance coping is seen as a defensive response which might be beneficial in the short term, allowing the opportunity to marshal effective coping resources, but since avoidance requires a significant effort, it is likely to compromise adaptation and well-being in the longer term.[24] A number of avoidance coping scales were developed in an attempt to clarify the processes involved, but the research has been hampered by psychometric shortcomings, such as the absence of confirmatory factor analyses to endorse obtained factor structures and the use of subscales with too few items to adequately sample the constructs. These issues have recently been addressed by the development of a new questionnaire, the general and specific avoidance scale (GSAQ[5]). The GSAQ was constructed using appropriate procedures, and comprises discrete components assessing general, emotional, and conflict avoidance; the new scale should provide a sound psychometric basis for future avoidance research.

Aside from psychometric inadequacies in the measures that have been developed, perhaps the most important issue that has dogged stress research generally is the absence of an agreed upon definition of stress itself. The field has been plagued by contrasting claims: that stress is either good for your health or bad for you; that it is either the property of events or a consequence of individual predispositions; that there are differences in the effects depending on whether the stress is chronic or acute.[25] Some of the ideas about stress have become so embedded in popular psychology that people unthinkingly assume that moving house or a job change is bound to be stressful. These views persist despite a history of research dating back at least three decades challenging life-events methodology, as well as more recent research showing that reporting negative life events is significantly explained by genetic predispositions moderated by individual differences.[26,27]

Attempts to clarify the issues by using new terminology, such as replacing the conventional notions of good versus bad stress with eustress versus distress, respectively, have in fact compounded rather than resolved the problem. What is required is a more precise distinction based on simple English that avoids attributing stress to both components, and a useful starting point is to distinguish instead between *pressure* and stress. Pressure is simply a demand to perform, and it can be distinguished from stress by using everyday examples: when someone is woken by their alarm clock on a working day the demand (pressure) is to get up, and although there might be some resistance to doing so it could hardly be considered stressful. If this person dozed off again and ended up rushing to get ready and leaving late for work, pressure will certainly have increased, with a corresponding increase in the degree of psychological and physiological arousal involved, but this is not necessarily stress.

Describing the sequence of events logically, the person in our story dozed off and ended up being late, but having arrived at work, explaining what had happened and apologizing for it, there is then the opportunity for the pressure to recede and for arousal levels to return to those required for attending to work. If a deadline for completing a piece of work looms, arousal will increase correspondingly to meet the greater demand—pressure varies constantly, but the periods between demands of any description, irrespective of how intense they might be, offer the opportunity for recovery. On the other hand, using our anecdote, if the individual continued to dwell on the incident—for example, churning over what they imagined their colleagues might be thinking about them, or worrying about what might happen if they arrived late for work again—the pressure has been transformed into stress, with the concomitant prolongation of emotional and physical arousal in response to imagined negative situations.

Distinguishing between pressure and stress is not just word-play, and it provides a simple resolution to the arguments about acute as opposed to chronic stress. The deleterious effects are associated primarily with chronic stress—in other words, with sustained demand—and stress needs therefore to be defined in terms of the mechanism that continues to maintain pressure in the absence of any external demand. Rumination provides this mechanism. Pressure corresponds to acute "stress," but chronicity is conferred by the preoccupation with emotional upset about events that have occurred in the past or might occur in the future. In other words, events are not inherently stressful, but they do offer themes to ruminate about in the absence of the event itself. It is rare to be exposed to constant demand, but the effect is clearly demonstrated when this does occur: as we shall see, one of the consequences of sustained physiological arousal is compromised immune function, and this is illustrated by significantly slowed wound-healing amongst individuals who have no respite from demand, such as caring for a relative suffering from Alzheimer's disease.[28]

The physiological effects of stress are complex, but the core mechanism linking the cognitive perception of a need to respond to the changes in cardiovascular function required for action is the h-p-a axis.[29] The perception of demand provokes an elevation in adrenaline (epinephrine in the United States) secreted from the adrenal medulla, leading to increased cardiovascular activation—the "fight-or-flight" response. Adrenaline causes increases in both heartrate and blood pressure, designed to facilitate action by maximizing blood flow to the musculature, but if cardiovascular strain is sustained it can lead to lesioning of the arterial wall,[30] in turn providing the opportunity for the deposition and build-up of arterial plaques (atherosclerosis).

At the same time there is an increase in the steroid hormone cortisol, secreted from the adrenal cortex, which contributes to fight-or-flight in a number of ways, such as by enhancing gluconeogenesis to provide the energy for action and by augmenting blood flow to the musculature through vasoconstriction. Cortisol also limits the degree of inflammation accompanying any injury that might be sustained in the incident (the synthetic analog, hydrocortisone, is a well-known anti-inflammatory preparation), and in addition to making energy available through gluconeogenesis, cortisol also acts to conserve energy by constraining nonessential processes. These include reducing the production of leukocytes such as natural killer (NK) cells and T-cells,[31,32] and the consequent compromise of immune function has led to the widespread use of cortisol as a biomarker for stress.

Indeed, both adrenaline and cortisol are popularly misnamed "stress hormones," when in fact they're hormones doing precisely what they're intended to do. However, the physiological system is governed by homeostatic principles and will return to resting levels as quickly as possible to ensure recovery, which will include repairing strain-related damage to the cardiovascular system and restoring normal immune functioning.[33,34] Adrenal hormones will only be implicated in stress when they are elevated for prolonged periods, hence the focus on the negative effects of chronic as opposed to acute stress. Chronic stress is conventionally operationalized as exposure to unusual ongoing demand, such as caring for relatives suffering from Alzheimer's disease in the Kiecolt-Glaser et al.[28] study referenced earlier, but in the context of everyday stress, what serves to prolong the activation of the h-p-a axis is continued rumination about emotional upset.

The link between the activation of the h-p-a axis in response to acute demand and the consequences of chronically sustaining adrenal activation is well established, but the biochemistry of stress is more complex than this would suggest. Reference to "white blood cells" in the context of immune function is a generalization across a wide range of leukocytes which have quite different functions. For example, while levels of NK and T-cells are suppressed by cortisol during chronic stress, noradrenalin (norepinephrine in the United States) secreted from the adrenal medulla during fight-or-flight stimulates a progressive increase in monocytes and neutrophils; these cells cluster at the sites of inflammation, including lesions to the arterial walls, and serve to increase the density of plaque formation at these lesions.[35] Chronic rumination thus has compounded effects: sustained elevation of adrenaline results in increased cardiovascular strain and potential arterial lesioning, sustained elevation of cortisol compromises NK and T-cell activity, and sustained elevation of noradrenalin contributes to plaque density through enhanced monocyte and neutrophil production.

Ruminating on negative emotions is thus the key cognitive mechanism to understanding stress, and especially the distinction between acute stress—more appropriately described as intermittent pressure—and chronic stress. Other than unusual circumstances where external demand is continuous, chronic stress is defined by habitual rumination. The effects of the physiological responses involving adrenaline, noradrenalin, and cortisol, combined with an appropriate definition of stress as rumination, provide an unambiguous explanation for the link between stress and physical health, and the impact on psychological well-being is equally strong owing to the negative emotional states that inevitably accompany rumination.

From an evolutionary perspective, it might be argued that repeatedly reimagining negative events facilitates learning through conditioned avoidance, but there is a difference between considering different ways in which issues might be dealt with the next time they occur and churning over the emotional upset that they provoke. Reflecting constructively to arrive at better solutions requires the perspective afforded by detached coping, a perspective which is compromised by rumination—as we've noted, detached coping and rumination are significantly inversely correlated.

Rumination is clearly maladaptive, yet it appears to be universal, manifesting across widely divergent cultural groups.[36] Rumination also occurs to varying degrees across random samples, generally conforming to a bell-curve distribution, and the fact that it can significantly be modified by practicing appropriate training strategies suggests that it is predominantly learned behavior.[37] The question which naturally follows is why do people ruminate? There will doubtless be some contribution from genetic predispositions, but the answer is more likely to do with self-perception or self-esteem.

The concept of self-esteem has a long history in psychology: as early as 1890, William James described it as the discrepancy between aspirations and actual achievements, where the greater the discrepancy, the lower self-esteem is likely to be. This suggests a process that might change across situations, but self-esteem can also be defined as a relatively permanent predisposition or trait. While it might vary according to specific criticism or praise received, these perceptions will generalize to form an enduring composite of self-attitudes. The findings from self-esteem research confirm that the construct does indeed have two facets, a relatively stable, resilient self-regard and a more changeable aspect which depends upon external circumstances, and people with genuinely high underlying self-esteem are significantly less affected by a perceived lack of competence in particular situations than those with low self-esteem.[38,39]

Ruminative "stories" tend to revolve around self-image, and the frequent replays of events in the mind will commonly be aimed at retrieving a perceived loss of face or self-esteem. This would suggest that rumination and self-esteem should be significantly inversely correlated, and this expectation is supported by research findings.[40] The evidence is correlational, and it isn't clear whether low self-esteem leads to more rumination or rumination itself engenders low self-esteem, but the link between the two constructs helps to explain the ruminative process associated with experiences such as the anecdote about arriving late for work described earlier in the chapter. More importantly, the overall evidence summarized in the chapter offers a way of conceptualizing stress in terms of individual differences in response to emotional upset, and rumination provides the key mechanism that transforms everyday pressure into stress, accounting for both the physiological and emotional sequelae that are well established in stress research.

References

1. Neria Y, Nandi A, Galea S. Post traumatic stress disorder following disasters: a systematic review. *Psychol Med*. 2008;38:467–480.
2. Jackson S, Schneider TS. Extraversion and stress. In: Haddock AD, Rutkowski AP, eds. *Psychology of Extraversion*. NY: Nova Science Publications; 2014.
3. Roger D. Emotion control, coping strategies and adaptive behavior. *Stress Emotion*. 1995;15:255–264.
4. Steed LG. A critique of coping scales. *Aust Psychol*. 1998;33:193–202.
5. Stemmet L, Roger D, Kuntz J, Borrill J. General and specific avoidance: the development and concurrent validation of a new measure of avoidance coping. *Eur J Pers Assess*. 2015;31:222–230.
6. Forbes A, Roger D. Stress, social support and fear of disclosure. *Br J Health Psychol*. 1999;4:165–179.
7. Guarino LR, Roger D, Olason DT. Reconstructing N: a new approach to measuring emotional sensitivity. *Curr Psychol*. 2007;26:37–45.
8. Roger D, Nesshoever W. The construction and preliminary validation of a scale for measuring emotional control. *Personal Individ Differ*. 1987;8:527–534.
9. Roger D, Najarian B. Construction and validation of a new scale for measuring emotion control. *Personal Individ Differ*. 1989;10:845–853.
10. Pennebaker JW. *Emotion, Disclosure, and Health*. Washington, DC: American Psychological Association; 1995.
11. Nieland M, Roger D. Emotion control and analgesia in labor. *Personal Individ Differ*. 1993;14:841–844.
12. Roger D, Jamieson J. Individual differences in delayed heart-rate recovery following stress: the role of extraversion, neuroticism and emotional control. *Personal Individ Differ*. 1988;9:721–726.
13. Roger D, Najarian B. The relationship between emotional rumination and cortisol secretion under stress. *Personal Individ Differ*. 1998;24:531–538.
14. McDougall C, Venables P, Roger D. Aggression, anger control and emotion control. *Personal Individ Differ*. 1991;12:625–629.
15. Kaiser J, Hinton JW, Krohne HW, Stewart R, Burton R. Coping dispositions and physiological recovery from a speech preparation stressor. *Personal Individ Differ*. 1995;19:1–11.
16. Roger D, Guarino de Scremin L, Borrill J, Forbes A. Rumination, inhibition and stress: the construction of a new scale for assessing emotional style. *Curr Psychol*. 2011;30:234–244.
17. Borrill J, Fox P, Flynn M, Roger D. Students with self-harm: coping style, rumination and alexithymia. *Couns Psychol Q*. 2009;22:361–372.
18. Carver CS, Scheier MF, Weintraub JK. Assessing coping strategies: a theoretically based approach. *J Pers Soc Psychol*. 1989;56:267–283.

19. Endler NS, Parker JDA. Multidimensional assessment of coping: a critical evaluation. *J Pers Soc Psychol*. 1990;58:844–854.
20. Lyne K, Roger D. A psychometric re-assessment of the COPE questionnaire. *Personal Individ Differ*. 2000;29:321–335.
21. Roger D, Jarvis G, Najarian B. Detachment and coping: the construction and validation of a new scale for measuring coping strategies. *Personal Individ Differ*. 1993;15:619–626.
22. Clarke J. *Sustaining Probation Officer Resilience in Europe (SPORE): a transnational study*. Report on Norway Grants Financial Mechanism project JUST/2010/JPEW/AG/1574; 2013.
23. Nielsen J, Shapiro S. Coping with fear through suppression and avoidance of threatening information. *J Exp Psychol Appl*. 2009;15:258–274.
24. Suls J, Fletcher B. The relative efficacy of avoidant and nonavoidant coping strategies: a meta-analysis. *Health Psychol*. 1985;4:249–288.
25. Goldberger L, Breznitz S. *Handbook of Stress: Theoretical and Clinical Aspects*. NY: Simon & Schuster, The Free Press; 1993.
26. Aagard J. Stressful life events and illness: a review with special reference to a criticism of the life-event method. In: Cullen J, Siegrist J, Wegman HM, eds. *Breakdown in Human Adaptation to 'Stress'*. Netherlands: Springer; 1986.
27. Saudino KJ, Pedersen NL, Lichtenstein P, McClearn GE, Plomin R. Can personality explain genetic influences on life events? *J Pers Soc Psychol*. 1997;72:196–206.
28. Kiecolt-Glaser JK, Marucha PT, Malarkey WB, Mercado AM, Glaser R. Slowing of wound healing by psychological stress. *Lancet*. 1995;346:1194–1196.
29. Smith SM, Vale WW. The role of the hypothalamic-pituitary-adrenal axis in neuroendocrine responses to stress. *Dialogues Clin Neurosci*. 2006;8:383–395.
30. Ahlgren AR, Cinthio M, Steen S, Persson HW, Sjöberg T, Lindstrom K. Effects of adrenaline on longitudinal arterial wall movements and resulting intramural shear strain: a first report. *Clin Physiol Funct Imaging*. 2009;29:353–359.
31. Gatti G, Cavallo R, Sartori ML, et al. Inhibition by cortisol of human natural killer (NK) cell activity. *J Steroid Biochem*. 1987;26:49–58.
32. Yang EV, Glaser R. Stress-induced immunomodulation: impact on immune defences against infectious disease. *Biomed Pharmacother*. 2000;54:245–250.
33. Geurts SAE, Sonnetag S. Recovery as an explanatory mechanism in the relationship between acute stress reactions and chronic health impairment. *Scand J Work Environ Health*. 2006;32:482–492.
34. Gleeson M. Immune function in sport and exercise. *J Appl Physiol*. 2007;103:693–699.
35. Heidt T, Sager HB, Courties G, et al. Chronic variable stress activates hematopoietic stem cells. *Nat Med*. 2014;20:754–758.
36. Roger D, Garcia de la Banda G, Lee HS, Olason D. A factor-analytic study of cross-cultural differences in emotional rumination and emotional inhibition. *Personal Individ Differ*. 2001;31:227–238.
37. Roger D, Hudson CJ. The role of emotion control and emotional rumination in stress management training. *Int J Stress Manag*. 1995;2:119–132.
38. Bracken BA, ed. *Handbook of Self-Concept: Developmental, Social, and Clinical Considerations*. New York, NY: Wiley; 1996.
39. Crocker J, Park LE. The costly pursuit of self-esteem. *Psychol Bull*. 2004;130:392–414.
40. Rector N, Roger D. Cognitive style and well-being: a prospective examination. *Personal Individ Differ*. 1996;21:663–686.

CHAPTER

32

Psychology of Suicide

D. Lester[1], J.F. Gunn III[2]

[1]The Richard Stockton College of New Jersey, Galloway, NJ, USA
[2]Montclair State University, Montclair, NJ, USA

OUTLINE

Abstract

Suicide is a major health concern that claims many lives each year. Despite the importance of the problem, the causes of suicide are mostly unknown, and current theories are somewhat speculative. A better understanding of suicidal behavior may be possible through an understanding of the linkages between suicide and stress. A number of theories, known as stress-diathesis theories, attempt to shed light on the relationship between stress and suicide. These theories postulate that suicide occurs when a diathesis (i.e., a predisposition) is acted upon by life stressors causing an individual to reach a breaking point at which suicide becomes likely. There are a number of physiological possibilities for the link between stress and suicide, including the neurotransmitter norepinephrine and the hypothalamus/pituitary gland/adrenal gland axis.

INTRODUCTION

There are a number of difficulties that have hampered the study of suicide:

(1) *Intentionality*: It is difficult to gauge what are the intentions of suicidal individuals. Is their suicidal behavior merely nonsuicidal self-injury or have they engaged in self-injury with the intention to die?

(2) *Suicide is rare*: While suicide is a tragic and devastating behavior, it is relatively rare. This very low incidence makes it impossible to predict accurately.

(3) *Loss of subjects*: Those best able to assist us in understanding suicide are those who have died by suicide. Psychological autopsy studies, while invaluable, are hampered by a number of limitations (such as whether the reports of informants are valid) and are time-consuming and difficult to undertake.

Regardless of these limitations, there is a large body of work examining the various risk and protective factors related to suicidal behavior. Table 1 lists some of the risk factors (i.e., factors that increase the risk of suicide) and protective factors (i.e., factors that decrease the risk of suicide) associated with suicidal behavior.

The most recent epidemiological data on suicide in the United States comes from 2011.[2] In 2011, there were 39,518 fatal suicides recorded, a rate of 12.7 per 100,000 per year. When broken down by male and female, men had a suicide rate of 20.2 while women had a suicide rate of 5.4. The most common methods of suicide were firearm (50.6%), suffocation/hanging (25.1%), and

Stress: Concepts, Cognition, Emotion, and Behavior
http://dx.doi.org/10.1016/B978-0-12-800951-2.00032-7

TABLE 1 Risk and Protective Factors for Suicide[1]

Risk factors	Protective factors
Mental disorders	Social support
Previous suicide attempts	Positive coping skills
Social isolation	Life satisfaction
Aggression and violence	Resiliency
Physical illness	Hopefulness
Unemployment	Self-efficacy
Family conflict	Effectiveness in obtaining helping resources
Family history of suicide	
Impulsivity	
Incarceration	
Hopelessness	
Time of year and season	
Serotonergic dysfunction	
Agitation or sleep disturbance	
Childhood abuse	
Exposure to suicide in others	
Homelessness	
Combat exposure	
Low self-esteem	

poisoning (16.6%). Suicide was the 10th leading cause of death for the general population and the 2nd leading cause of death for the young. The greatest at-risk age group was those aged 45-54 (with a rate of 19.8). The racial breakdown indicates that white males are at the greatest at risk (with a rate of 23.0).

While suicide is a rare occurrence, it is the cause of almost 40,000 lives each year in the United States, more deaths than are caused by homicide and so suicide prevention must be a vital part of the nation's public health goals. In order to prevent suicide, there must be a solid theoretical framework in which it can be better understood. The relationship between stress and suicide has been well documented, and this chapter will lay out our understanding of this relationship and then place it into a theoretical framework.

KEY POINTS

- Stress-diathesis theories of suicide propose a key role for stressors in precipitating suicide in individuals who possess the diathesis (predisposition) for suicide.
- The overall level of life stress is associated with the risk of suicide.
- The type of stressors that increase the risk of suicide varies with the characteristics of individuals, such as age and sex.
- For example, the suicides of younger people are more often precipitated by interpersonal conflicts, whereas the suicides of older people are more often precipitated by chronic stressors such as medical illnesses.

THE RELATIONSHIP BETWEEN STRESS AND SUICIDE

There have been a number of studies examining the relationship between life stress and suicidal behavior. Bonner and Rich[3] found that, among a jail population, psychosocial vulnerability factors, situational life stress, and jail stress were the best predictors of suicidal intent. Brent and colleagues[4] examined the relationship between suicide and stressful life events with a focus on adolescent suicide. Interpersonal conflict with parents and with boyfriends and girlfriends, disruption of romantic attachments, and getting into trouble with the law or in a disciplinary setting were more likely to occur in the lives of those who died by suicide as compared to living matched community controls. Some stress life events, such as disciplinary and legal problems, were statistically significant even when controlling for psychopathology.

Another examination of the relationship between stress and suicide was undertaken by Heikkinen, Aro, and Lonnqvist,[5] who reviewed some of the existing literature on life events, social support, and suicide. With regard to life events, they found that there was evidence for a relationship between suicide and bereavement, loss of employment, loss of social support, attempted suicide in a close friend or relative, and interpersonal conflict. Van Praag[6] concluded his review of stress and suicide by stating "[s]tress almost always precedes suicidality… [c]onsequently, the question of whether stress is an epiphenomenon or a decisive factor in the causation of suicidality and (certain forms of) mood disorders is crucial." Van Praag argued that, until changes in brain activity can be shown to connect stress to suicide, causation cannot be inferred. Finally, Zhang and colleagues[7] examined stress, coping styles, and suicide ideation among Chinese college students. Based on their findings, and the findings from previous work, Zhang and colleagues[7] concluded that "adverse life stress is a consistent and strong predictor in suicide behaviors."

The Adequacy of Stressors

The question next arises whether the different life stressors are equal in their impact on suicide. Older adults who die by suicide are typically viewed as having experienced more adequate stressors than younger adults and adolescents who kill themselves. Elderly suicides frequently suffer from medical problems, often severe, and have experienced a succession of losses (of loved ones and friends, of employment and work, and of physical abilities). Thus, the precipitants of their suicides appear to involve severe stress. Most of the examples of rational suicide involve such cases and, indeed, assisted-suicide laws passed in the United States and other nations are usually restricted to those who are terminally ill. In contrast, the precipitants of suicide in adolescents and young adults are sometimes seen as inadequate causes of suicide. Such suicides are, therefore, viewed as impulsive and not rational and are less predictable.

However, this may be a bias in the way we perceive suicide. Are the motivations behind elderly suicide really more justifiable than the reasons behind younger suicides or does this point to a bias in our understanding of the stressors and strains that cause suicide, regardless of the population? For example, it may be that the stress of facing down declining health and loss of family and friends (for the elderly) and the stress of being bullied and harassed because of sexual orientation (in adolescents) do not differ in their adequacy as a stressor for suicide. Judgments of the adequacy of the stressor for leading to suicide are subjective.

THE SITUATION

There is evidence that situational variables increase the likelihood of suicide. The easy accessibility of lethal methods for suicide may play a role. Research on the effects of restricting easy access to methods for suicide (such as detoxifying domestic gas, placing emission controls on cars, and restricting the number of pills prescribed for psychiatric disorders) has shown that this can be an important tactic for preventing suicide.

The publicity given to suicides, especially suicides by celebrities, has been shown to lead to an increase in the suicide rate in the following days, especially among those of the same age and sex as the celebrity. Suicide among peers can have a contagion effect, precipitating suicidal behavior in adolescents with those characteristics that predispose the individual to suicide. Suicides occurring in schools have aroused particular concern, and there are now guidelines for staff in responding to the suicide of a student.

STRESS-DIATHESIS MODELS OF SUICIDE

We know that life stressors and suicide are related, but can this be placed into a theoretical framework? Stress-diathesis models of suicidal behavior do this. Stress-diathesis refers to the relationship between a stressor (i.e., an event that causes stress) and a long-term vulnerability or predisposition to suicide (i.e., the diathesis). There are a number of stress-diathesis theories of suicide. However, before moving onto the theories, it is important that we discuss just what is meant by *diathesis*.

What is a Diathesis?

A diathesis is a long-term factor that increases the vulnerability of individuals to suicide. However, over the years what is meant by diathesis has changed considerably. For example, Mann[8] suggested that a diathesis may exist independent of a psychiatric disorder while others consider psychiatric disorders to be part of the diathesis. Schotte and Clum view the diathesis as a predisposition to have difficulties solving problems, while Wasserman listed genetic predispositions, personality traits, the presence of psychiatric disorders, and social isolation as potential diatheses.[9,10]

Therefore, although theorists differ in their definition of what is a diathesis, all agree that the diathesis confers a vulnerability that is then acted upon by life stress. The stress and the diathesis may also interact. As van Heeringen[11] points out, not only may the diathesis predispose people to the occurrence of stressful life events, but also the experience of stressful life events may have a detrimental impact on the diathesis (for example, by having a detrimental impact on the serotonergic system).

Zhang's Strain Theory of Suicide

Zhang's theory of suicide relies on the presence of multiple stressors (i.e., strains),[12] and Zhang has proposed several types of strain.

Value strains are those strains that are tied to cultural values that are competing (e.g., a religious belief that homosexuals are sinful vs. sexual attraction to someone of the same sex).
Reality versus aspirational strains are strains that are caused when the reality of a situation goes against one's aspirations. For example, wanting to be a professional basketball player is dashed by the reality that one is only 5′6″ and notathletically inclined.
Relative deprivation strain is a strain caused by comparison between your lot in life and another's. For example, if you are poor and you are constantly reminded of the wealth and success of others (e.g., by

television), this can cause you to be acutely aware of the differences in your advantages and theirs. *Deficient coping strains* are those strains linked to an inability to cope with daily problems and crises.

While Zhang does not discuss the relationship between these strains and a diathesis, it is very easy to incorporate a diathesis into this model. One would suspect that individuals deal with such strains on a regular basis, but the vast majority of them do not choose to die by suicide. By adding in a diathesis, the theory is able to account for this variation in who does and who does not die by suicide.

Schotte and Clum

For Schotte and Clum,[9,13] the diathesis is a deficit in divergent thinking which interferes with the ability to deal with life's stressors. Schotte and Clum have found that suicides are cognitively rigid, and this cognitive rigidity does not allow them to adapt to stressful situations. The pathway from cognitive rigidity to suicide is mediated through the sense of helplessness and hopelessness that occurs when the rigid cognitive style hampers an individual's ability to deal with stressors.

Williams's Cry of Pain Model

The Cry of Pain Model (also known as the defeat-entrapment model) proposed by Williams is based on evolutionary (animal) models of depression.[14,15] In this model, depression and suicide are caused by arrested flight, which fosters a sense of entrapment when the possibility of escape is blocked, or perceived to be blocked. Defeat occurs when someone perceives themselves to have fallen in social status or failed in some social endeavor. The occurrence of defeat and entrapment activate a "helplessness script" which involves giving-up behaviors.

O'Connor's Integrated Motivational-Volitional Model[16]

O'Connor's integrated motivational-volitional (IMV) model of suicidal behavior is based on Williams's Cry of Pain[14] model and the Theory of Planned Behavior.[17] The IMV is broken into three phases: (1) a premotivational phase, (2) a motivational phase, and (3) a volitional phase. The stress-diathesis component for the IMV is present in the premotivational phase. O'Connor calls this the "diathesis-environment-life events triad," which includes the background characteristics of the individual and the triggering events.[16] The motivational phase of the IMV is identical to the Cry of Pain model of suicide, while the volitional phase is made up of key risk factors for suicide (e.g., the acquired capability for suicide) which bridge the gap between suicidal ideation and suicide attempt.

Mann, Waternaux, Haas, and Malone

After interviewing a number of psychiatric inpatients for lifetime suicidal behavior, psychiatric disorder, impulsivity, traits of aggression, and other personal history variables, Mann and colleagues[18] proposed a stress-diathesis model of suicidal behavior. In this model, depression, psychosis, and life events result in feelings of hopelessness, depression, and suicidal ideation. These feelings of hopelessness, depression, and suicidal ideation lead to suicidal planning. Impulsivity, the result of several factors including low serotonergic activity, alcoholism, smoking, substance abuse, and head injury, is the key element in the transition from suicidal planning to suicidal acts. In this model, the diatheses are hopelessness (pessimism) and impulsivity. These are affected by lowered levels of noradrenaline and serotonin, respectively, in the brain. These diatheses are in turn affected by variables such as sex and childhood experiences.

PHYSIOLOGICAL CORRELATES OF STRESS AND SUICIDE

In this section, we will briefly discuss the relationship between various physiological correlates of stress and suicide.

Neurotransmitters

Norepinephrine is the neurotransmitter most likely to be associated with both suicide and stress. Norepinephrine is associated with the fight-or-flight response, along with epinephrine, and has an impact on regions of the brain, including the amygdala. However, the relationship between norepinephrine and suicidal behavior is not fully understood, and the findings of research studies often conflict. Mann[8] concluded that research indicates that there is a reduced presence of norepinephrine in the locus coeruleus (the part of the brainstem involved with physiological responses to stress and panic) in patients with major depression who die by suicide. Additionally, several studies have identified a role for tyrosine hydroxylase which is a rate-limiting enzyme responsible for the biosynthesis of norepinephrine, although the findings of the different studies are inconsistent. The role of neurotransmitters, therefore, requires further research.

The Hypothalamic-Pituitary-Adrenal Axis

The hypothalamic-pituitary-adrenal axis (HPA axis) involves complex sets of interactions between the hypothalamus, the pituitary glands, and the adrenal glands. The HPA axis is involved in functions such as body temperature, digestion, the immune system, mood, sexuality, and energy usage, in addition to its role as a component of the fight-or-flight response and the release of hormones such as cortisol. Currier and Mann[19] argued that the role of the HPA axis in suicidal behavior may be in "the context of acute stress response to life events preceding a suicidal act in which impaired stress response mechanisms contribute to risk" and the "[hypothalamus/pituitary gland/adrenal gland axis] may also be involved in suicidal behavior if increased activity of stress response to adversity during development has deleterious effects on the development[a] of other systems and brain structures implicated in suicidal behavior."

There are a number of studies that have documented a connection between the HPA axis and suicidality. Lopez, Vazquez, Chambers, and Watson[20] concluded that suicides and depressed patients generally showed evidence of over-activity of the HPA axis. Using the dexamethasone suppression test (DST), a measure of hyperactivity of the HPA axis, Coryell and Schlesser[21] found that while none of the risk factors examined (i.e., sex, age, living alone, feeling hopeless, Hamilton Depression Rating Scale scores, serious suicide attempt, bipolar affective disorder, and delusions) were predictive of completed suicide, DST results were more promising. The risk of suicide among patients with abnormal DST responses was estimated to be 26.8% while the risk was only 2.9% for those with normal DST responses. Jokinen and Nordstrom[22] found that abnormal DST responses were associated with suicidal behavior in both young and elderly suicides.

Finally, two meta-analyses, one by Lester and the other by Mann, Currier, Stanley, Oquendo, Amsel, and Ellis, both concluded that DST nonsuppressors had a greater risk of dying by suicide than were suppressors.[23,24]

PREVENTING SUICIDE

The prevention of suicide requires several strategies. Many communities now have suicide prevention centers that maintain telephone counseling services staffed by well-trained lay people who can provide crisis counseling for those who are suicidal. The American Association of Suicidology has a directory of such centers in the United States and also inspects and certifies the centers if asked to do so. Befrienders International, Life Line, and the International Federation of Telephone Emergency Services are international organizations that have established and coordinated crisis intervention and suicide prevention services in other nations.

Educational programs have also been established to teach students about suicide, especially recognizing signs of suicidal intent in their peers, and to provide them with resources to which they can refer their suicidal peers. Educational programs have also occasionally been established for general practitioners and family physicians so that they can diagnose depression in their patients more accurately and prescribe antidepressants more effectively. Efforts have been made to restrict access to lethal methods for suicide, such as fencing in the high places from which people jump to their death (such as bridges) and restricting access to medications and poisons (such as fertilizers and insecticides in rural areas).

Clinicians have developed counseling techniques designed for suicidal clients to supplement the general systems of psychotherapy that guide therapists. These new techniques have been primarily in the field of cognitive therapy, and the prominent therapy here is Dialectical Behavior Therapy proposed by Marsha Linehan.[25]

[a] To clarify, this is in reference to human development post birth.

References

1. Gunn III JF, Lester D. Risk and protective factors for male suicide. In: Lester D, Gunn III JF, Quinnett P, eds. *Suicide in Men: How Men Differ from Women in Expressing Their Distress*. Springfield: Charles C. Thomas; 2014:51–60.
2. McIntosh JL, Drapeau CW [for the American Association of Suicidology]. *U.S.A. Suicide 2011: Official Final Data*. Washington, DC: American Association of Suicidology; 2014. http://www.suicidology.org; Accessed 19.06.14.
3. Bonner RL, Rich AR. Psychosocial vulnerability, life stress, and suicide ideation in a jail population: a cross-validation study. *Suicide Life Threat Behav*. 1990;20(3):213–224.
4. Brent DA, Perper JA, Moritz G, et al. Stressful life events, psychopathology, and adolescent suicide: a case control study. *Suicide Life Threat Behav*. 1993;23(3):179–187.
5. Heikkinen M, Aro H, Lonnqvist J. Life events and social support in suicide. *Suicide Life Threat Behav*. 1993;23(4):343–358.
6. Van Praag HM. Stress and suicide: are we well-equipped to study this issue? *Crisis J Crisis Interv Suicide Prev*. 2004;25(2):80–85.
7. Zhang X, Wang H, Xia Y, Liu X, Jung E. Stress, coping, and suicide ideation in Chinese college students. *J Adolesc*. 2012;35(3):683–690.
8. Mann JJ. Neurobiology of suicidal behavior. *Nat Rev*. 2003; 4:819–828.
9. Schotte DE, Clum GA. Problem-solving skills in suicidal psychiatric patients. *J Consult Clin Psychol*. 1987;55:690–696.
10. Wasserman D. A stress-vulnerability model and the development of the suicidal process. In: Wasserman D, ed. *Understanding Suicidal Behaviour*. Chichester: Wiley; 2001:76–93.
11. Van Heeringen K. A stress-diathesis model of suicidal behavior. *Crisis*. 2000;21:192.
12. Zhang J. Conceptualizing a strain theory of suicide. *Chin Ment Health J*. 2005;19:778–782.
13. Schotte DE, Clum GA. Suicide ideation in a college population. A test of a model. *J Consult Clin Psychol*. 1982;50:690–696.

14. Williams JMG. *Cry of Pain: Understanding Suicide and Self-Harm*. New York, NY: Penguin; 1997.
15. Williams JMG, Pollock LR. Psychological aspects of the suicidal process. In: van Heeringen K, ed. *Understanding Suicidal Behaviour*. Chichester: Wiley; 2001:76–93.
16. O'Connor RC. Towards an integrated motivational-volitional model of suicidal behaviour. In: O'Connor RC, Platt S, Gordon J, eds. *International Handbook of Suicide Prevention: Research, Policy and Practice*. Hoboken, NJ: Wiley-Blackwell; 2011:181–198.
17. Ajzen I. The theory of planned behavior. *Organ Behav Hum Decis Process*. 1991;50:179–211.
18. Mann JJ, Waternaux C, Hass GL, Malone KM. Toward a clinical model of suicidal behavior in psychiatric patients. *Am J Psychiatry*. 1999;156:181–189.
19. Currier D, Mann JJ. Stress, genes and the biology of suicidal behavior. *Psychiatr Clin N Am*. 2008;31:247–269.
20. Lopez JF, Vazquez DM, Chalmers DT, Watson SJ. Regulation of 5-HT receptors and the hypothalamic-pituitary-adrenal axis: Implications for the neurobiology of suicide. *Ann N Y Acad Sci*. 1997;836:106–134.
21. Coryell W, Schlesser M. The dexamethasone suppression test and suicide prediction. *Am J Psychiatry*. 2001;158:748–753.
22. Jokinen J, Nordstrom P. HPA axis hyperactivity as suicide predictor in elderly mood disorder inpatients. *Psychoneuroendocrinology*. 2008;33:1387–1393.
23. Lester D. The dexamethasone suppression test as an indicator of suicide. *Pharmocopsychiatry*. 1992;25:265–270.
24. Mann JJ, Currier D, Stanley B, Oquendo MA, Amsel LV, Ellis SP. Can biological tests assist prediction of suicide in mood disorders? *Int J Neuropsychopharmacol*. 2006;9:465–474.
25. Linehan M. *Cognitive-Behavioral Treatment of Borderline Personality Disorder*. New York, NY: Guilford; 1993.

CHAPTER

33

Sociology of Suicide

D. Lester[1], J.F. Gunn, III[2]

[1]The Richard Stockton College of New Jersey, Galloway, NJ, USA

[2]Montclair State University, Montclair, NJ, USA

OUTLINE

Abstract

Societal stressors play an important role in the major sociological theories of suicide. In Durkheim's theory, reduced social integration and excessive social regulation both result in stress, while Henry and Short's theory focuses on whether individuals can blame others for their unhappiness, a protective factor for suicide. Anthropologists have documented the role of stress created by cultural conflict in increasing the risk for suicide. Whereas individual stressors can increase the risk of suicide in individual members of a society, societal, and cultural stressors can increase the risk of suicide in all members of the society.

INTRODUCTION

Sociological research into suicide has focused primarily on the rate of completed suicide in regions (nations or regions within nations), a rate that is relatively stable over time. Ecological research examines the suicide rates of a set of regions at one point in time, usually to see which social characteristics of the regions predict the suicide rates. Time-series research examines the suicide rates of one region every year for a period of time to see which social characteristics predict the annual suicide rate. In this chapter, three major theories of suicide are reviewed: Emile Durkheim's, Andrew Henry and James Short's, and Raoul Naroll's. The role of culture conflict in increasing the rate of suicide is also discussed. Finally, the possibility of sociological studies of nonfatal suicide is discussed.

DURKHEIM'S THEORY

The first major theory of the social suicide rate was proposed in 1897 by Emile Durkheim.[1] Durkheim argued that two broad social characteristics are responsible for determining the societal suicide rate: social integration and social regulation. Social integration is the extent to which the members of the society are bound together in social networks. Very high levels of social integration lead to high rates of *altruistic suicide*, whereas very low levels lead to high rates of *egoistic suicide*. Social regulation is the extent to which the desires and behavior of the members of the society are restricted by social norms and customs. Very high levels of social regulation lead to high rates of *fatalistic suicide*, whereas very low levels lead to high rates of *anomic suicide*.

There has been a great deal of research purporting to test this theory, for example, exploring the association between divorce rates and suicide rates. Divorce is assumed to weaken the extent of social integration and

Stress: Concepts, Cognition, Emotion, and Behavior
http://dx.doi.org/10.1016/B978-0-12-800951-2.00033-9

KEY POINTS

- Social stressors play a key role in precipitating suicide in individuals.
- These stressors include broken social relationships, oppression, and culture conflict.
- Being able to blame others for one's misery may protect against suicide in individuals.
- In addition, a society (or subculture) as a whole may have stressors that increase the risk of suicide for all (rather than only for vulnerable individuals) such as the high rate of suicide among the Austrian and German Jewish populations during their persecution prior to and during the Second World War.

reflect weak social regulation and, therefore, higher divorce rates should be positively associated with higher suicide rates. Both ecological studies and time-series studies have confirmed this predicted positive association between divorce rates and suicide rates.

However, research rarely examines the roles of both social integration and social regulation separately, partly because it is difficult to operationalize the two social characteristics independently of one another. The few studies that have attempted to do so have found that measures of social integration are stronger correlates of suicide rates than measures of social regulation.

Furthermore, no research has attempted to classify suicides in the societies under study into Durkheim's four types. Durkheim proposed, for example, that low levels of social integration result in high levels of *egoistic suicide*. Therefore, a proper test of Durkheim's theory requires the dependent variable to be the rate of egoistic suicide. The rate of egoistic suicide is not identical to the total social suicide rate, which is, instead, the combined rate of all four of Durkheim's types of suicides.

FATALISTIC SUICIDE

Stress is built into Durkheim's theory in his concept of social integration. Many factors, including divorce and the death of others, reduce social integration thereby increasing the risk of suicide. But another major source of stress comes from fatalistic suicide in which individuals choose suicide because their "future are pitilessly blocked and passions violently choked by oppressive discipline" (Durkheim, 1951,[1] p. 276).

Lester[2] has documented the role of oppression in precipitating suicide, ranging from protest suicide, such as the epidemic of suicide among Tibetan monks (and some civilians) in Tibet to protest the treatment of Tibetans by the Chinese government, suicide in the slaves brought to the United States in the 1600s and thereafter, suicide in the concentration camps (where Lester[3] calculated suicide rates as high as 25,000 per 100,000 per year), suicide in oppressed women (for example, in those forced into sati [dying on their husbands funeral pyre in India] or those forced to become suicide bombers), suicide in gay and lesbian individuals, suicide in the workplace (for example, in the spate of suicides in 2012 in the Foxconn factories in Shenzhen, China, as a result of the working conditions), to suicide in those who are bullied. Zhang and Liu[4] have documented how the sexist nature of Confucian thought (which condemns women to a subservient role) protects Chinese men against suicide but increases the risk of suicide in Chinese women.

HENRY AND SHORT'S THEORY

Henry and Short[5] proposed a theory based on both psychoanalytic theory and the frustration-aggression hypothesis. They assumed that the primary target of aggression for a frustrated individual is another person. What inhibits this other-oriented aggression and results in the aggression being directed inward onto the self?

Henry and Short proposed that the primary factor that inhibits the outward expression of aggression when people are frustrated was the strength of external restraints on behavior. When behavior is required to conform rigidly to the demands and expectations of others, the role of others in the responsibility for the self's frustration and misery is strong. As a consequence, other-oriented aggression is legitimized. When external restraints are weak, the self must bear the responsibility for the frustration and misery. Thus, other-oriented aggression is not legitimized, and self-directed aggression becomes more likely.

These proposals lead to interesting predictions, many of which have been confirmed. The oppressed in a society have clear external sources to blame for their misery—their oppressors. Therefore, other-oriented aggression is legitimized for them, and they will tend to have relatively higher rates of assault and, in the extreme, homicide. In contrast, the oppressors in the society have fewer external sources to blame for their misery because, as the dominant group, they have tremendous opportunities for advancement and gratification. Therefore, the oppressors in the society will tend to have higher rates of depression and, in the extreme, suicide. These predicted differences in the suicide and homicide rates are found for African Americans and European Americans in the United States.

Henry and Short's theory can also explain the positive association between the quality of life and suicide rates. When the quality of life in a nation is high, there are few external sources to blame for one's misery, and so suicide will tend to be more common. This association has been confirmed in ecological studies both of nations and of the American states.

NAROLL'S THEORY

Raoul Naroll,[6] an anthropologist, proposed a theory of social suicide rates that has relevance to stress. Naroll proposed that suicide was more likely in those who were socially disoriented, that is, in those who lack or lose basic social ties (such as those who are single or divorced). Naroll called these situations *thwarting disorientation contexts*. Because not all socially disoriented people commit suicide, there must be a psychological factor that makes suicide a more likely choice when a person is socially disoriented, and Naroll suggested that this factor was the person's reaction to thwarting disorientation context. Thwarting disorientation contexts involve a weakening of people's social ties as a result of the actions of other people or themselves, but not as a result of impersonal, natural, or cultural events. Being divorced by a spouse or murdering one's spouse are examples of thwarting disorientation contexts, but storm damage to property or losing a spouse to cancer are not. In thwarting disorientation contexts, some people commit protest suicide, which Naroll defined as voluntary suicide committed in such a way as to come to public notice, and this definition, therefore, excludes unconsciously motivated suicides and culturally motivated suicides such as sati.

Durkheim's theory of suicide refers more to steady-state characteristics of the society as major explanatory variables, whereas Naroll's theory refers to short-term events and is phrased in a way that makes it more usefully applied to individuals as well as to societies as a whole.

CULTURE CONFLICT

Cultures often come into conflict, a source of stress that may increase the suicide rate. For example, the conflict between traditional Native American culture and the dominant Anglo-American culture has often been viewed as playing a major role in precipitating Native American suicide. Nancy van Winkle and Philip May[7] reviewed three hypotheses that might have relevance here. In the social disorganization hypothesis, the dominance of the Anglo-American culture erodes traditional cultural systems and values. This changes the level of social regulation and social integration in the Indian subculture, important causal factors for suicide in Durkheim's theory of suicide.

A second hypothesis focuses on cultural conflict. The pressure from the educational system and mass media on Native Americans, especially on young people, to acculturate, a pressure that is opposed by their elders, leads to great stress for the youth. A third hypothesis focuses on the breakdown of the family in Native American tribes. Parents are often unemployed, substance abusers, and in trouble with the law, and divorce and desertion of the family by one or both parents are common. The suicide rates of three groups of Native Americans in New Mexico—the Apache, Pueblo, and Navajo (whose suicide rates were 43.3, 27.8, and 12.0 per 100,000 per year, respectively)—were in line with their levels of acculturation as rated by van Winkle and May (high, moderate, and low, respectively).

MEASURING REGIONAL STRESS

Linsky and coworkers[8] measured the stress level of each of the American states in three areas. For economic stress, they used business failures, unemployment claims, strikes, bankruptcies, and mortgage foreclosures. For family stress, they used divorces, abortions, illegitimate births, and fetal and infant deaths. For community stress, they used disasters, new housing starts, new welfare recipients, high school dropouts, and interstate migration. They found that states with higher levels of stress also had higher suicide rates. Interestingly, however, ratings of the stress level of each state based on responses from a national survey, which asked residents about their *perceived* level of stress, were not associated with the state suicide rates.

MULTIVARIATE STUDIES OF REGIONAL SUICIDE RATES

The accuracy of official national and regional rates of suicide has been questioned by Jack Douglas[9] and others, who have argued that these rates are biased by the values of the local coroners and medical examiners and of the resident populations. However, despite the fact that regions do have different standards for classifying a death as a suicide, Martin Voracek and Lisa Loibl[10] have shown that the suicide rates of immigrants to one country from different nations are in almost the same rank order as the suicide rates of the nations of origin.

The suicide rates of the world's nations vary greatly, and Simpson and Conklin[11] identified two clusters of variables that were associated with national suicide rates: (1) a cluster that had the highest loading from the percentage of Muslims in the population and (2) a cluster that seemed to assess economic development. Suicide rates were lower in nations with less economic development and where Islam was a more important religion.

In a similar study of the social correlates of national suicide rates, Lester[12] identified 13 orthogonal (independent) factors for the social variables studies, only one of which was associated with suicide rates, a factor that seemed to measure economic development (with high loadings from such social variables as low population growth and high gross domestic product per capita). Thus, these two studies agreed in finding that economic development is associated with higher suicide rates. Lester also found that the classical Durkheimian variables measuring social integration (such as marriage and birth rates and religiosity), as well as economic variables such as female labor-force participation and the gross domestic product per capita were strong predictors of cross-national suicide rates for both males and females and for different age groups.

For the United States, a similarly designed study conducted by Lester[13] identified a cluster of variables that seemed to measure social disintegration (high divorce and interstate migration rates, low church attendance, and high per capita alcohol consumption), and this was the strongest correlate of the suicide rates of the states. For 29 nations, Lester and Yang[14] found that the time-series suicide rates (for the period 1950-1985) were associated with marriages rates (negatively in 20 of the 29 nations) and divorce rates (positively in 22 of the 29 nations), while the association with birth rates was inconsistent (a positive association in 12 nations and a negative association in 17 nations).

NONFATAL SUICIDE BEHAVIOR

Almost all sociological research and theories have in the past been formulated for fatal (completed) suicide. Nonfatal suicidal behavior has been relatively ignored. It is important that better epidemiological studies of attempted suicide and suicidal ideation be conducted so that sociologists can develop theories of these behaviors. The World Health Organization has sponsored a comparative epidemiological study of attempted suicide in 15 sites. Unfortunately, the sites are limited to cities in European nations and do not encompass these nations as a whole. However, the study is innovative and may lay the groundwork for more comprehensive epidemiological studies in the future. Thus, research on nonfatal suicide behavior is possible, and it is to be hoped that more studies are planned along these lines and theories developed to account for the results of the research.

References

1. Durkheim E. *Le suicide.* Paris: Felix Alcan; 1897 [English edition: Suicide. New York: Free Press; 1951].
2. Lester D. Oppression and suicide. *Suicidol Online.* 2014;5:59–73.
3. Lester D. *Suicide and the Holocaust.* Hauppauge, NY: Nova Science; 2005.
4. Zhang J, Liu EY. Confucianism and youth suicide in rural China. *Rev Relig Res.* 2012;54:93–111.
5. Henry AF, Short JF. *Suicide and Homicide.* New York, NY: Free Press; 1954.
6. Naroll R. *Thwarting Disorientation and Suicide: A Cross-Cultural Survey.* Unpublished manuscript. Northwestern University; 1963.
7. Van Winkle NW, May PA. Native American suicide in New Mexico, 1959-1979. *Hum Organ.* 1986;45:296–309.
8. Linsky AS, Bachman R, Straus MA. *Stress, Culture, and Aggression.* New Haven, CT: Yale University Press; 1995.
9. Douglas JD. *The Social Meanings of Suicide.* Princeton, NJ: Princeton University Press; 1967.
10. Voracek M, Loibl LM. Consistency of immigrant and country-of-birth suicide rates. *Acta Psychiatr Scand.* 2008;118:259–271.
11. Simpson ME, Conklin GH. Socioeconomic development, suicide and religion. *Soc Forces.* 1989;67:945–964.
12. Lester D. *Patterns of Suicide and Homicide in the World.* Commack, NY: Nova Science; 1996.
13. Lester D. *Pattern of Suicide and Homicide in America.* Commack, NY: Nova Science; 1994.
14. Lester D, Yang B. *Suicide and Homicide in the 29th Century.* Commack, NY: Nova Science; 1998.

CHAPTER

34

Cortisol Awakening Response

A. Steptoe, B. Serwinski
University College London, London, UK

OUTLINE

Abstract

The cortisol awakening response (CAR) is the change in cortisol concentration that occurs in the first hour after waking from sleep. It is typically assessed using salivary cortisol samples immediately after waking and then at intervals over the next hour. The CAR has emerged as an important aspect of hypothalamic-pituitary-adrenocortical axis function because it is regulated differently from cortisol output over the rest of the diurnal cycle. It has also been related to stress, affective disorders, and physical health risk. This chapter discusses the origins and measurement of the CAR and its relationship with sleep-waking cycles, acute and chronic stress, depression, and health outcomes.

INTRODUCTION

The cortisol awakening response (CAR) is the change in cortisol concentration that occurs over the first hour after waking from sleep. It has become a topic of intense investigation following the introduction of salivary cortisol sampling, which has allowed people to collect samples noninvasively under normal life conditions at home, instead of the contrived conditions of the laboratory. Cortisol levels are low in the night but rise in the early hours before waking. After waking up, most people show a further rise, peaking 20-40 min later. This is followed by a progressive reduction of cortisol over the day.

There are several reasons why the CAR has become an important research topic. First, there is evidence that the CAR is under somewhat independent control from cortisol output over the remainder of the day. There is little association between the CAR and levels over the rest of the day or the slope of cortisol decline into the evening.[1] Second, twin studies have documented a genetic influence over the CAR that is distinct from the heritability of daytime cortisol levels.[2] Third, the CAR is associated with stress and health in distinctive and potentially significant ways, suggesting that it is a useful marker of hypothalamic-pituitary-adrenocortical (HPA) function. The fact that studies of the CAR do not require the facilities of a human psychophysiological or sleep laboratory has led to a proliferation of research ranging from intensive investigations of small samples to large-scale applications in epidemiological studies involving many thousands of participants.[3]

A number of measures of the CAR have been devised, depending on how many saliva samples are obtained over the first hour after waking. Serial sampling suggests that that peak CAR is observed around 30-40 min after waking, so the simplest method of assessment involves measures at two time points: immediately after waking and 30 min later, with the CAR being defined as a simple

http://dx.doi.org/10.1016/B978-0-12-800951-2.00034-0

difference score. More elaborate assessments involve measurements of several time points, such as immediately after waking, 15, 30, 45, and 60 min later. Under these circumstances, the area under the curve (AUC) can be computed to estimate total cortisol output. In some studies, the AUC is presented in absolute terms, but usually it is computed as an increase from the waking level (AUCi). Simple difference scores and AUC measures are consistently intercorrelated, but some studies have suggested that they have distinct correlates.

KEY POINTS

- The cortisol awakening response (CAR) is the rise in cortisol that typically takes place over the first 30-45 min after waking. It is an important component of hypothalamic-pituitary-adrenal axis function, and is regulated independently of cortisol output over the rest of the diurnal cycle.
- The CAR may be an anticipatory response, preparing the person for the demands of the upcoming day. It is greater on working than leisure days, and is related to stress exposure.
- The CAR is associated with mental health risk and with the presence of various physical health problems. But there is variation across studies about whether the CAR is heightened or diminished in different illnesses.

THE BIOLOGICAL ORIGINS OF THE CAR

The diurnal rhythm of cortisol output is well recognized, with high levels early in the day and a decrease across waking hours reaching a nadir in the hours before waking. The actual hours over which this rhythm is entrained depends on whether the species is active at night or during the day. The secretion of cortisol is pulsatile, and it has been estimated that each secretary burst induces an increase in salivary free cortisol of approximately 2.5 nmol/l. The CAR is a result of one to four secretary bursts over the period following waking. Cortisol output is under the control of a cascade of hormones initiated in the paraventricular nucleus (PVN) of the hypothalamus. The PVN itself is under strong regulation by higher structures in the brain, notably the hippocampus, amygdala, and prefrontal cortex.[4] Additionally, there are connections between the PVN and the suprachiasmatic nucleus (SCN), a small structure in the hypothalamus which is light sensitive and is involved in circadian regulation.[5] Neuroanatomical studies have shown direct axonal projections from the SCN to the PVN, complemented by indirect connections via the dorsomedial hypothalamus.[6] Vasopressin related from the SCN has a role in modifying HPA circadian activity. The SCN also links with autonomic pathways that act directly on the adrenal cortex,[7] and is probably responsible for entraining the additional cortisol release after waking.

The CAR is not simply the rising section of the diurnal profile, but is a response to the process of waking itself. Wilhelm et al.[8] demonstrated that the rise in cortisol after awakening was steeper than could be accounted for by the diurnal cycle on its own, and the CAR is not disturbed by repeat awakenings in the night.[9] Interestingly, when wake time is expected, there is a rise in adrenocorticotropic hormone in the hour beforehand which does not occur when waking was not expected, suggesting a preparatory response for the activities of the day.[4] A CAR does not take place after waking from naps in the day. The CAR, but not waking cortisol level, is sensitive to light in the environment, and can be enhanced by augmented light exposure toward the end of the sleep period. By contrast, noise in the night has little effect.

One of the striking features of the CAR is the wide variation in the magnitude of the response both across people and within the same individual on different days. Studies involving repeated assessments over several months suggest that up to 70% of the variability in the CAR is attributable to day-to-day fluctuations rather than stable characteristics.[10] This means that any theories about the biological function of the CAR need to take into account these large variations in the magnitude and stability of the response itself.

MEASUREMENT OF THE CAR

Participants in CAR studies typically collect samples at home and are instructed to take saliva samples immediately after waking and then at later defined intervals. Saliva is collected using cotton rolls or by directly spitting into test tubes, and participants are instructed not to brush their teeth, eat, drink, smoke, or physically exert themselves throughout the sampling period, as these factors can influence cortisol values. Since the CAR is dynamic, accurate timing is critical, and failure to adhere to the sampling schedule may confound studies and produce misleading results. If the "waking" sample is delayed, the CAR may already have commenced, leading to an apparent diminution in the magnitude of the rise of the next 20-30 min. Similarly, if later samples are mistimed, then the magnitude of the CAR will not be assessed correctly.

An illustration of the impact of delays in collecting the waking sample is shown in Figure 1, which compares salivary cortisol profiles over the first hours of the day in

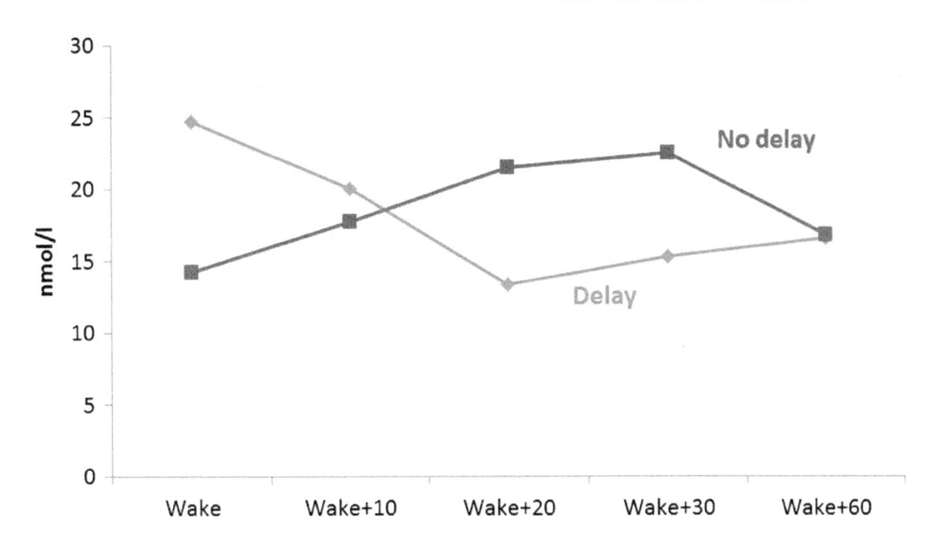

FIGURE 1 Mean salivary cortisol levels immediately after waking and at 10, 20, 30, and 60 min after waking in elderly volunteers who delayed (green line) and did not delay (red line) saliva sampling after waking in the morning. Error bars are standard errors of the mean. From Wright and Steptoe.[11]

elderly (65-80 year) volunteers who reported no delay in taking the first sample after waking, and those who admitted delaying for 10 or more minutes.[11] The no delay group showed a typical CAR with increases from waking to 30 min, while a decrease over the same period was found in the delay group. The failure to adhere to the sampling protocol is greater among people of lower socioeconomic status as defined by education or income.[12]

Several methods of improving the accuracy and reliability of CAR measurement have been proposed. Asking participants to record when they wake and when they take each sample provides a check, and limits can be set on acceptable delays, but this method depends on self-reports. Some studies involve saliva collection devices that are time stamped, and accelerometers to measure the increase in movement that indicates the end of sleep periods; in combination, these methods provide objective evidence about delays between waking and obtaining the first saliva sample. However, they are likely to remain a luxury, particularly for large-scale population studies, so self-report remains important. There is some disagreement about what reported delays are acceptable for obtaining a reliable CAR, with some researchers suggesting that delays as short as 5 min may compromise assessments while others advocate that delays up to 10-15 min in taking the waking sample may be acceptable.[13]

The issue of sampling delays is relevant to another controversy in the literature, namely whether there are genuine CAR nonresponders. It has been estimated that only 75% of adults show an increase in cortisol after waking when measures are performed at home, and nonresponders may constitute an interesting subgroup of the population. But this proportion may be distorted by noncompliance, so the prevalence of genuine nonresponders could be smaller. Positive CAR values are recorded in almost everyone who has measures taken in the sleep laboratory study under experimental conditions.[8] But the response rate under laboratory conditions might be a result of poor sleep and transient awakenings in the unfamiliar environment, and therefore not be representative.

FACTORS AFFECTING THE CAR

Gender and Age

Several investigations have shown that the CAR is somewhat greater in women than men.[4] Results have not been completely consistent, and may depend on whether measures are taken on a working or leisure day. The greater CAR in women may also be affected by the stressful demands they often encounter early in the day. Women are frequently responsible not only for preparing themselves for the day ahead, but for caregiving, and household chores as well.

Studies relating age with the CAR have generated mixed findings, with a large study of adults aged 33-84 years showing increases in the CAR with age in men, but not women,[14] while another study with a wide age range showed no association in either sex.[2] Another large study of adults aged 18-65 years found that the CAR was positively associated with shorter leukocyte telomere length after adjusting for lifestyle factors, health, and medication, suggesting links between the rise in cortisol after waking and fundamental aging processes.[15]

Time of Waking

There has been controversy about the association between time of waking and the CAR. Several researchers have shown that the CAR is greater in people who wake up early compared with those who wake later, but no

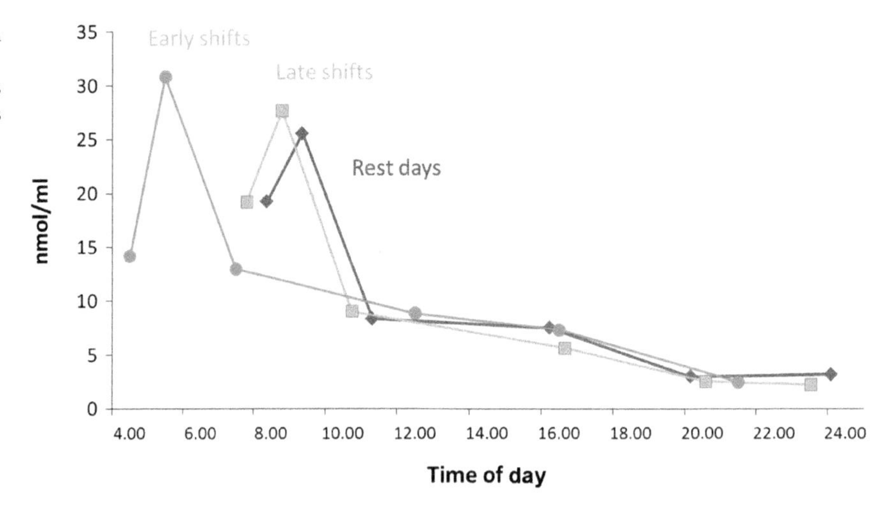

FIGURE 2 Mean salivary cortisol levels on waking, waking + 30 min, waking + 2.5 h, waking + 8 h, waking + 12 h, and bedtime for early shifts (blue line), late shifts (yellow line), and rest days (red line). From Bostock and Steptoe.[16]

association with time of waking has been reported in other large-scale investigations.[5,7] The issue is complicated by two factors. First, people who say they wake up late in the morning may in fact have woken earlier and have subsequently dozed. Their CAR may therefore have partly taken place prior to the reported waking time. Second, waking early may be a reflection of greater stress, worry, or work demands, which themselves influence the CAR.

Within-person comparisons of days on which people wake at an early or later hour have produced more consistent results, with the CAR being greater when wake time is earlier. This is illustrated in Figure 2, which shows results for a study of people tested on days when they had an early work shift, a late work shift, or a rest day.[16] The CAR was substantially larger on the early work day, due both to a lower value on waking and larger peak 30 min later. The effect remained significant after sleep duration, sleep quality, and feeling of stress, tiredness, and happiness were taken into account.

Sleep Duration and Quality

In view of the literature concerning time of waking and the CAR, it is not surprising that shorter sleep duration has been associated with a larger CAR.[17] Associations between the CAR and sleep quality are inconsistent. Some large-scale studies have failed to document changes in the CAR, while experimentally induced sleep disruption during the night is not related to the CAR.[9,17] There is very little evidence from sleep medicine concerning the CAR, so firm conclusions cannot be drawn.[18] The CAR seems to be unaffected by whether or not the person wakes spontaneously or is woken by an alarm clock.[4]

Smoking

Cortisol increases acutely when people smoke. In addition, the CAR is generally greater in smokers than nonsmokers, even when they have not smoked during the early morning period.[3] The relationship between smoking and acute cortisol is dose dependent.[19] This indicates that smoking has a sustained impact on HPA regulation that outlives the short-term effect of nicotine stimulation.

SIGNIFICANCE FOR STRESS AND HEALTH RESEARCH

An Acute Anticipation Response

One prominent theory about the function of the CAR is that it prepares the individual for the demands of the upcoming day.[20] This may be the explanation of the well-documented difference between the CAR on work and leisure days. Figure 3 shows CAR results from a sample of working men and women.[21] Cortisol levels on waking did not differ between the 2 days, but the CAR was

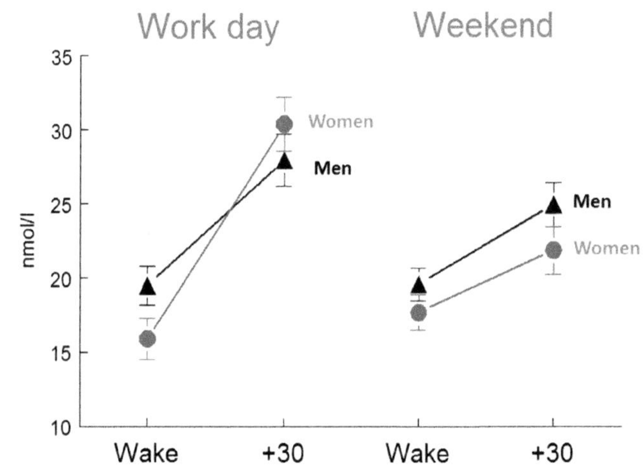

FIGURE 3 Mean salivary cortisol on waking (wake) and 30 min later (+30) on a work day and weekend day in men (black lines) and women (blue lines). Error bars are standard errors of the mean. From Kunz-Ebrecht et al.[21]

substantially greater on the work than weekend days. Additionally, this study showed a larger CAR in women than men, possibly reflecting the greater demands on women early in the day.

The anticipation hypothesis is also consistent with evidence that the CAR is greater among people reporting worry or preoccupation with work.[22] An intensive within-person study showed that feelings of threat, sadness, and lack of control on the day before predicted a larger CAR on the following day.[23] The CAR is larger on the morning of important occasions such as competitive events.[24] Another finding that illustrates the importance of anticipation is the observation that patients with severe amnesia do not show any CAR or rise in cortisol after waking.[25]

Life Stress and the CAR

The notion that the CAR prepares the person for the rigors of the day would suggest that certain forms of chronic stress would be positively associated with the CAR. A meta-analysis of more than 140 studies demonstrated that the CAR is generally larger among people experiencing job stress and work overload, and also in those reporting other forms of chronic life stress such as high levels of daily hassles.[22] But this pattern is not universal. In particular, posttraumatic stress appears to be associated with a reduced CAR. The explanation is not certain, but one possibility is that the CAR is greater under conditions in which people have to cope actively with the demands of the day, whereas severely stressful conditions that cannot be addressed by active behavioral responses may show the reverse effect.

Nearly all of these effects have been cross-sectional, so the causal significance of associations is unclear. However, one study assessed changes in financial strain over a 3-year period, and demonstrated that a reduction in financial strain was accompanied by a decrease in the CAR, indicating a parallel between changes in chronic stress and the magnitude of the CAR.[26]

Depression and Anxiety

Diurnal rhythms of cortisol secretion are disrupted in a substantial proportion of people with major depressive disorder.[27] However, the literature relating depression and anxiety with the CAR is mixed, with some studies showing associations with a larger CAR while others have reported null or even negative effects.[22] But there is an important body of evidence indicating that a high CAR is a risk factor for future depressive illness. It has now been demonstrated that high morning cortisol or a larger CAR predicts future depression, particularly among individuals at risk because of other factors such as elevated subclinical depressive symptoms or family history.[28,29] The data are not entirely consistent,[30] but do suggest that the CAR may be a significant marker of HPA dysfunction related to mental health risk in vulnerable individuals. Additionally, studies of people with major depressive disorder who are in remission have shown that a larger CAR is associated with greater risk of relapse.[31]

Fatigue and Burnout

The meta-analysis of factors associated with the CAR published in 2009 showed that fatigue, exhaustion, and burnout are associated with a lower rise in cortisol after awakening.[22] A more recent review has confirmed this association in people diagnosed with chronic fatigue syndrome, with less consistent effects for reports of fatigue that are not clinically diagnosed.[32] An issue with these studies is that people suffering from severe fatigue often report sleeping fitfully, so the CAR may already have occurred before participants regard themselves as properly waking up, leading to a diminished rise in cortisol.

Physical Health

The CAR has been studied in relation to a number of physical health conditions. Unfortunately, it is difficult to know at present whether larger or smaller CARs are more problematic. Investigations have often been carried out with small samples of patients that may not be representative of the clinical group in general. But even in well characterized studies of large population cohorts, results have been inconsistent. For example, some studies have observed associations between type 2 diabetes and a diminished CAR, while others have shown no differences.[33,34] A Swedish study showed that women with the metabolic syndrome had large CARs,[35] but the same was not true of men, and a study in Germany reported lower CARs in people with components of the metabolic syndrome.[36] Results for obesity have also been inconsistent, while hypertension has been related to a reduced CAR.[37] A large prospective study of all-cause and coronary heart disease mortality reported no association with the CAR, although other aspects of diurnal cortisol rhythm did predict mortality.[38] An interesting study detailing the rate of healing and recovery from an experimentally administered cutaneous wound showed that the rate of healing was slower in people with larger CARs, but this has yet to be replicated in clinical settings.[39] Research on multiple sclerosis suggests that individuals with progressive relapsing-remitting disease have a larger CAR than those with other forms of the disease.[40] Other work has measured the CAR in people with chronic pain conditions, various cancers, and autism, but larger scale studies with robust methodologies are needed before any general conclusions can be drawn.

CONCLUSIONS

The CAR has been increasingly studied over the last two decades, and has been shown to provide valuable information about the relationship between HPA function, psychosocial factors, and health risk.

References

1. Schmidt-Reinwald A, Pruessner JC, Hellhammer DH, et al. The cortisol response to awakening in relation to different challenge tests and a 12-hour cortisol rhythm. *Life Sci.* 1999;64(18):1653–1660.
2. Wust S, Wolf J, Hellhammer DH, Federenko I, Schommer N, Kirschbaum C. The cortisol awakening response—normal values and confounds. *Noise Health.* 2000;7:77–85.
3. Adam EK, Kumari M. Assessing salivary cortisol in large-scale, epidemiological research. *Psychoneuroendocrinology.* 2009;34 (10):1423–1436.
4. Fries E, Dettenborn L, Kirschbaum C. The cortisol awakening response (CAR): facts and future directions. *Int J Psychophysiol.* 2009;72(1):67–73.
5. Clow A, Hucklebridge F, Thorn L. The cortisol awakening response in context. *Int Rev Neurobiol.* 2010;93:153–175.
6. Spiga F, Walker JJ, Terry JR, Lightman SL. HPA axis-rhythms. *Compr Physiol.* 2014;4(3):1273–1298.
7. Kalsbeek A, Yi CX, la Fleur SE, Buijs RM, Fliers E. Suprachiasmatic nucleus and autonomic nervous system influences on awakening from sleep. *Int Rev Neurobiol.* 2010;93:91–107.
8. Wilhelm I, Born J, Kudielka BM, Schlotz W, Wust S. Is the cortisol awakening rise a response to awakening? *Psychoneuroendocrinology.* 2007;32(4):358–366.
9. Dettenborn L, Rosenloecher F, Kirschbaum C. No effects of repeated forced wakings during three consecutive nights on morning cortisol awakening responses (CAR): a preliminary study. *Psychoneuroendocrinology.* 2007;32(8–10):915–921.
10. Ross KM, Murphy ML, Adam EK, Chen E, Miller GE. How stable are diurnal cortisol activity indices in healthy individuals? Evidence from three multi-wave studies. *Psychoneuroendocrinology.* 2014;39:184–193.
11. Wright CE, Steptoe A. Subjective socioeconomic position, gender and cortisol responses to waking in an elderly population. *Psychoneuroendocrinology.* 2005;30(6):582–590.
12. Golden SH, Sanchez BN, Desantis AS, et al. Salivary cortisol protocol adherence and reliability by socio-demographic features: the Multi-Ethnic Study of Atherosclerosis. *Psychoneuroendocrinology.* 2014;43:30–40.
13. Smyth N, Clow A, Thorn L, Hucklebridge F, Evans P. Delays of 5-15 min between awakening and the start of saliva sampling matter in assessment of the cortisol awakening response. *Psychoneuroendocrinology.* 2013;38(9):1476–1483.
14. Almeida DM, Piazza JR, Stawski RS. Interindividual differences and intraindividual variability in the cortisol awakening response: an examination of age and gender. *Psychol Aging.* 2009;24(4):819–827.
15. Revesz D, Milaneschi Y, Verhoeven JE, Penninx BW. Telomere length as a marker of cellular aging is associated with prevalence and progression of metabolic syndrome. *J Clin Endocrinol Metab.* 2014;99(12):4607–4615.
16. Bostock S, Steptoe A. Influences of early shift work on the diurnal cortisol rhythm, mood and sleep: within-subject variation in male airline pilots. *Psychoneuroendocrinology.* 2013;38(4):533–541.
17. Kumari M, Badrick E, Ferrie J, Perski A, Marmot M, Chandola T. Self-reported sleep duration and sleep disturbance are independently associated with cortisol secretion in the Whitehall II study. *J Clin Endocrinol Metab.* 2009;94(12):4801–4809.
18. Elder GJ, Wetherell MA, Barclay NL, Ellis JG. The cortisol awakening response—applications and implications for sleep medicine. *Sleep Med Rev.* 2014;18(3):215–224.
19. Badrick E, Kirschbaum C, Kumari M. The relationship between smoking status and cortisol secretion. *J Clin Endocrinol Metab.* 2007;92(3):819–824.
20. Powell DJ, Schlotz W. Daily life stress and the cortisol awakening response: testing the anticipation hypothesis. *PLoS One.* 2012;7(12). e52067.
21. Kunz-Ebrecht SR, Kirschbaum C, Marmot M, Steptoe A. Differences in cortisol awakening response on work days and weekends in women and men from the Whitehall II cohort. *Psychoneuroendocrinology.* 2004;29(4):516–528.
22. Chida Y, Steptoe A. Cortisol awakening response and psychosocial factors: a systematic review and meta-analysis. *Biol Psychol.* 2009; 80(3):265–278.
23. Adam EK, Hawkley LC, Kudielka BM, Cacioppo JT. Day-to-day dynamics of experience-cortisol associations in a population-based sample of older adults. *Proc Natl Acad Sci U S A.* 2006;103(45): 17058–17063.
24. Rohleder N, Beulen SE, Chen E, Wolf JM, Kirschbaum C. Stress on the dance floor: the cortisol stress response to social-evaluative threat in competitive ballroom dancers. *Personal Soc Psychol Bull.* 2007;33(1):69–84.
25. Wolf OT, Fujiwara E, Luwinski G, Kirschbaum C, Markowitsch HJ. No morning cortisol response in patients with severe global amnesia. *Psychoneuroendocrinology.* 2005;30(1):101–105.
26. Steptoe A, Brydon L, Kunz-Ebrecht S. Changes in financial strain over three years, ambulatory blood pressure, and cortisol responses to awakening. *Psychosom Med.* 2005;67(2):281–287.
27. Herbert J. Cortisol and depression: three questions for psychiatry. *Psychol Med.* 2013;43(3):449–469.
28. Owens M, Herbert J, Jones PB, et al. Elevated morning cortisol is a stratified population-level biomarker for major depression in boys only with high depressive symptoms. *Proc Natl Acad Sci U S A.* 2014;111(9):3638–3643.
29. Vrshek-Schallhorn S, Doane LD, Mineka S, Zinbarg RE, Craske MG, Adam EK. The cortisol awakening response predicts major depression: predictive stability over a 4-year follow-up and effect of depression history. *Psychol Med.* 2013;43(3):483–493.
30. Carnegie R, Araya R, Ben-Shlomo Y, et al. Cortisol awakening response and subsequent depression: prospective longitudinal study. *Br J Psychiatry.* 2014;204(2):137–143.
31. Hardeveld F, Spijker J, Vreeburg SA, et al. Increased cortisol awakening response was associated with time to recurrence of major depressive disorder. *Psychoneuroendocrinology.* 2014;50:62–71.
32. Powell DJ, Liossi C, Moss-Morris R, Schlotz W. Unstimulated cortisol secretory activity in everyday life and its relationship with fatigue and chronic fatigue syndrome: a systematic review and subset meta-analysis. *Psychoneuroendocrinology.* 2013;38(11):2405–2422.
33. Champaneri S, Xu X, Carnethon MR, et al. Diurnal salivary cortisol and urinary catecholamines are associated with diabetes mellitus: the Multi-Ethnic Study of Atherosclerosis. *Metabolism.* 2012;61(7): 986–995.
34. Hackett RA, Steptoe A, Kumari M. Association of diurnal patterns in salivary cortisol with type 2 diabetes in the Whitehall II study. *J Clin Endocrinol Metab.* 2014;99(12):4625–4631.
35. Bengtsson I, Lissner L, Ljung T, Rosengren A, Thelle D, Wahrborg P. The cortisol awakening response and the metabolic syndrome in a population-based sample of middle-aged men and women. *Metabolism.* 2010;59(7):1012–1019.
36. Kuehl LK, Hinkelmann K, Muhtz C, et al. Hair cortisol and cortisol awakening response are associated with criteria of the metabolic

syndrome in opposite directions. *Psychoneuroendocrinology*. 2015;51:365–370.

37. Wirtz PH, von Kanel R, Emini L, et al. Evidence for altered hypothalamus-pituitary-adrenal axis functioning in systemic hypertension: blunted cortisol response to awakening and lower negative feedback sensitivity. *Psychoneuroendocrinology*. 2007;32(5):430–436.
38. Kumari M, Shipley M, Stafford M, Kivimaki M. Association of diurnal patterns in salivary cortisol with all-cause and cardiovascular mortality: findings from the Whitehall II study. *J Clin Endocrinol Metab*. 2011;96(5):1478–1485.
39. Ebrecht M, Hextall J, Kirtley LG, Taylor A, Dyson M, Weinman J. Perceived stress and cortisol levels predict speed of wound healing in healthy male adults. *Psychoneuroendocrinology*. 2004;29(6): 798–809.
40. Kern S, Krause I, Horntrich A, Thomas K, Aderhold J, Ziemssen T. Cortisol awakening response is linked to disease course and progression in multiple sclerosis. *PLoS One*. 2013;8(4). e60647.

CHAPTER

35

Anger

R.W. Novaco
University of California, Irvine, CA, USA

OUTLINE

Abstract

Anger is an affective response to survival threats or otherwise stressful experiences. It is a primary emotion having adaptive functions linked to survival mechanisms that are biological, psychological, and social in nature. Threat perception is intrinsic to its activation, and symbolic structures govern such perception. Cognitive processing of anger-provoking experiences can alternatively prolong or disengage anger. Anger is primed and demarcated by neurophysiological arousal, and, as a high arousal state, anger can constitute an internal stressor, causing wear and tear on the body when it is recurrently activated. Behaviorally, anger is associated with approach motivational systems and can activate aggressive behavior. While anger expression is governed by social rules, it can be part of an antagonistic style of coping with the stressors of daily life, particularly in responding to interpersonal conflict. The role of anger as an activator of violent behavior is interpersonally and societally problematic. Anger dysregulation produces impairment in functioning across life domains and is associated with various psychiatric disorders through transdiagnostic processes, such as selective attention, threat perception, and rumination. The efficacy of psychotherapeutic interventions for anger, principally cognitive-behavioral therapy, is well-established for a wide variety of clinical populations.

ANGER AND STRESS

Anger is a negatively toned emotion, subjectively experienced as an aroused state of antagonism toward someone or something perceived to be the source of an aversive event. Prototypically, it is triggered or provoked by events that are perceived to constitute deliberate harm-doing by an instigator toward oneself or toward those to whom one is endeared. It can also be a product of goal-blocking or frustrations, particularly when recurrent, or be a reactive response to pain, physical or psychological. Provocations usually take the form of insults, unfair treatments, or intended thwarting. Anger is prototypically experienced as a justified response to some "wrong" that has been done. While anger is situationally triggered by acute, proximal occurrences, it is shaped and facilitated contextually by conditions affecting the cognitive, physiological, and behavioral systems that comprise anger reactions and by social rules that govern anger expression. Anger activation is centrally linked to threat perceptions and to survival responding. Although it is neither necessary nor sufficient for aggression or violence, anger impels aggressive behavior, particularly when its intensity overrides regulatory control mechanisms.[1] Clinically problematic anger often has traumatic stress origins.

The experience of anger can be prolonged or revivified by cognitive processes, such as rumination, imagery, and symbolic cues, in reciprocal feedback loops with neurological and somatovisceral systems. As anger impels antagonistic action, it can amplify the noxious qualities of the circumstances that have evoked the anger activation, through escalating exchanges of anger and aggressive behavior. The cognitive, physiological, and behavioral bases of anger activation correspondingly provide portals for anger regulating interventions.

Stress: Concepts, Cognition, Emotion, and Behavior
http://dx.doi.org/10.1016/B978-0-12-800951-2.00035-2

Anger is a primary human emotion, observable in infancy.[2] Since the landmark works of Darwin[3] and Cannon,[4] anger has been understood as an adaptive response to survival threat (danger or pain). Although there are sociocultural variations in the acceptability of its expression and the form that such expression takes,[5] in the face of adversity, anger can mobilize psychological resources, energize behaviors for corrective action, and facilitate perseverance. Anger serves as a guardian to self-esteem, operates as a means of communicating negative sentiment, potentiates the ability to redress grievances, and boosts determination to overcome obstacles to our happiness and aspirations. Akin to aggressive behavior, anger has functional value for survival.[6] Anger *energizes* behavior as a high arousal state, increasing the amplitude of responding; it *focuses* attention on situational elements having threat significance; it *expresses* or communicates negative sentiment to convey displeasure and to prompt conflict resolution; it *defends* the self by social distancing and fear suppression, and it also defends self-worth by externalizing attributions of blame for misfortune; it *potentiates* a sense of personal control or empowerment among social groups as well as individuals; it *instigates* aggressive behavior due to its survival relevance, symbolic linkages, and learned connections; it *signals* information about personal state and situational significance which is relevant to self-monitoring; and it *dramatizes* a social role enactment, in the sense of anger expression being dramaturgy played out in accord with social scripts.

Despite having multiple adaptive functions, anger can have maladaptive effects on personal and social well-being. Generally, strong physiological arousal impairs the processing of information and lessens cognitive control of behavior. Because heightened physiological arousal is a core component of anger,[7] people are not cognitively proficient when they become angry. Also, because the activation of anger is accompanied by aggressive impulses, anger can motivate harm toward other people, which in turn can produce undesirable consequences for the angered person, either from direct retaliation, loss of supportive relationships, or social censure[8]. An angry person is not optimally alert, thoughtful, empathic, prudent, or physically healthy. Being a turbulent emotion ubiquitous in everyday life, anger is now known to be substantially associated with a range of physical health problems, including stress-related cardiovascular disorders.[9,10] Anger is also a symptom of posttraumatic stress disorder (PTSD), and it has high relevance to PTSD derivative of disasters, health traumas, violent crime victimization, and especially to combat or war zone exposure.[11] Among combat veterans, anger is a salient postdeployment problem affecting social relationships, job performance, physical health, and violence risk, and it is intensified when PTSD and depression are comorbid.[12,13]

KEY POINTS

- Anger is prototypically experienced as a justified response to some "wrong" that has been done. Its activation is centrally linked to threat perceptions and to survival responding. Although anger has multiple adaptive functions, when its intensity is high and/or prolonged, it can impair social relationships, work performance, and health, as well as propel harmful aggressive behavior.
- Whether or not anger has problematic status can be gauged by its frequency, latency, intensity, duration, and mode of expression. High intensity arousal overrides inhibitory controls on aggressive behavior. Rumination about provoking events extends or revivifies anger reactions.
- Aversive events or stressful circumstances activate anger through cognitive processes of attention and meaning structures. Anger arousal is marked by activation in the cardiovascular, endocrine, and central nervous systems, and by tension in the skeletal musculature. Anger has evolutionary roots in preparing the organism for attack, including signaling attack readiness so as to ward off opponents or to coerce compliance. In dealing with survival threat, anger serves to suppress fear, pain, and shame.
- There are feedback loops between anger's cognitive, physiological, and behavioral systems, along with its environmental triggers. This can involve deviation-amplifying processes, such as the escalation of anger and aggression, or deviation-counteracting processes, such as self-control strategies.
- Anger dysregulation appears in a wide range of psychopathologies, as a product of transdiagnostic processes, such as selective attention, threat perception, interpretive bias, rumination, and self-control deficiencies. Perceived malevolence is a common anger-inducing appraisal, invoking anger by externalizing blame and entraining justification.
- There is a preponderance of evidence for the efficacy of psychotherapeutic interventions for anger dyscontrol, especially cognitive-behavioral therapies.

THE EXPERIENCE AND EXPRESSION OF ANGER

There is a duality of psychosocial images associated with anger experience and anger expression. The emotional state is depicted as eruptive, destructive, unbridled, savage, venomous, burning, and consuming, but also as energizing, empowering, justifying, signifying, rectifying, and relieving. The metaphors, on the one hand, connote something pressing for expression and utilization, and, alternatively, they imply something requiring containment and control. This duality in psychosocial imagery reflects conflicting intuitions about anger, its expression, and its consequences that abound in ordinary language and are reflected in both scholarly literature and artistic works from the classical period to contemporary times. This Janus-faced character of anger foils attempts to understand it and to therapeutically intervene with recurrently angry individuals.

The facial and skeletal musculature is strongly affected by anger, mobilized by a mixture of adrenaline and noradrenaline hormonal secretions. The face becomes flushed, and the brow (corrugator) muscles move inward and downward, fixing a hard stare on the target. The eyes narrow, nostrils flare, and the jaw tends toward clenching. This is an innate pattern of facial expression that can be observed in toddlers,[2,14] and angry faces are rapidly detected even when there are distractors.[15] Tension in the skeletal musculature, including raising of the arms and adopting a squared-off stance, as well as squaring the jaw, are preparatory actions for attack and defense. The muscle tension provides a sense of strength and self-assurance. An impulse to strike out accompanies this subjective feeling of potency. From an evolutionary perspective, our perceptual system has been shaped to detect angry faces and angry postures rapidly, especially those of angry males. Correspondingly, arousal of anger engages an approach motivational system[16] that has survival value for defense or for corrective action.

When people report anger experiences, they most typically give accounts of things that have "happened to them." For the most part, they describe events physically and temporally proximate to their anger arousal. As a rule, they provide accounts of provocations ascribed to events in the immediate situation of the anger experience. This fosters the illusion that anger has a discrete external cause. The provocation sources are ordinarily identified as the aversive and deliberate behavior of others; thus, anger is portrayed in the telling as being something about which anger is quite fitting. People are very much inclined to attribute the causes of their anger to the personal, stable, and controllable aspects of another person's behavior—akin to what is called the "fundamental attribution error" in social psychology.

However, the response to the question, "What has made you angry?" hinges on self-observational proficiencies and is often based on intuitions. Precisely because getting angry involves a loss in self-monitoring capacity, people are often neither good nor objective observers when they are angry. Inspecting any particular episode, the immediate activators ("causes") of the anger are readily identifiable. People far less commonly disaggregate their anger experiences into multicausal origins, some of which may be prior, remote events and ambient circumstances, rather than acute, proximal events. Anger experiences are embedded or nested within an environmental-temporal context. Disturbances that may not have involved anger at the outset leave residues that are not readily recognized but which operate as a lingering backdrop for focal provocations.

Anger, as an approach motivation system affect, is inherently a disposition to respond aggressively, but aggressive behavior is not an automatic consequence of anger, as it is regulated by inhibitory control mechanisms, engaged by internal and external cues.[1,17] In this regard, physical constraints, expectations of punishment or retaliation, empathy, consideration of consequences, and prosocial values operate as regulatory controls on aggression. While the experience of anger creates a readiness to respond with aggression, that disposition may be otherwise directed, suppressed, or reconstituted. Thus, the expression of anger is differentiated from its experience.

One aspect of anger that influences the probability of aggression is its degree of intensity. The higher the level of arousal, the stronger the motivation for aggression, and the greater the likelihood that inhibitory controls will be overridden.[18] Strong arousal not only impels action, it impairs cognitive processing of aggression-mitigating information. A person in a state of high anger arousal is perceptually biased toward the confirmation of threat, is less able to attend to threat-discounting elements of the situation, and is not so capable of reappraising provocation cues as benign. Because anger and aggression occur in a dynamic interactional context, the occurrence of aggression will, in turn, influence the level of anger. Thus, anger reactivity can be seen as a mode of responding characterized by automaticity, high intensity, and short latency.

An important aspect of the dynamic interrelation of anger and aggression is the escalation of provocation. Escalation involves increases away from equilibrium, whereby succeeding events intensify their own precursors. In the case of anger and aggression, escalation refers to incremental change in their respective probabilities, occurring as reciprocally heightened antagonism in an interpersonal exchange. The consumption of alcohol can further amplify this process. Anger-elicited aggression may evoke intensified anger in response, thus progressively generating justification for retaliation. A model of the neural organization of the escalation of both anger and aggression has been formulated by Potegal.[19]

When physical aggression is deployed by an angry person against the anger instigator, and there is no retaliation, anger arousal and subsequent aggression are then diminished in that situation. Konecni[20] called this the "cathartic effect," and its conditions should not be confused (as they often are) with those involving aggression by nonangry people, vicarious or observed aggression, or aggression not received by the anger-instigator. Such nonqualifying conditions are often prevalent in social psychology laboratory studies of catharsis as "venting,"[21] which moreover do not ecologically map onto the clinical contexts in which anger catharsis might be encouraged (i.e., in clinical implementation, catharsis is a substitute and preventive for harm-doing and the client returns for subsequent therapy visits). However, the arousal-reducing cathartic effect of aggression carried out by angry people against those who have made them angry does reinforce aggressive behavior. This means that when anger is reinstated by a new provocation, the likelihood of aggressive behavior is increased. The cathartic expression of anger, whether through destructive aggression or through verbal communication intended to be constructive, can be understood as an organismic action to restore equilibrium.

Alternative to the deliberate expression of anger is suppression, which is largely a product of inhibitory controls, and which is often discussed in the larger context of "emotion regulation." While suppression, as a habitual coping strategy has adverse consequences,[22] anger suppression can be quite functional in promoting interpersonal or social conciliation, and diminishing the likelihood of triggering a physical assault. Whether in a domestic, occupational, or street context, anger is adaptively muffled when physical retaliation can be expected or when a cool head is needed to solve a problem. Depending on the context, suppressing even the verbalization of anger may not only be beneficial interpersonally, it may also serve to regulate physiological reactivity levels. However, recurrent deployment of anger suppression as a stress-coping style will likely have deleterious effects on cardiovascular health.[9] The suppression of anger, as a customary mode, is associated with rumination, which can lead to prolonged anger, cognitive perseveration, aggressive behavior, and heightened sympathetic nervous system activation.[21,23]

Because anger and aggression are thought to be differentially socialized for males and females, the question of gender differences in the experience and expression of anger arises. It has generally been found that the anger of women is comparable to that of men from the standpoint of experienced intensity.[5,24] An exception is that women in prison/correctional settings report higher anger than do men.[25] Style of anger expression does vary by gender, especially according to the context of anger activation and its anticipated consequences. Males are more likely to be angered in a public place or by impersonal triggers, whereas females are more likely to be angered at home or by being let down by someone close to them. Females are more likely to become angered by verbal aggression and insensitive/condescending behavior, and males more likely to be angered by behavior causing physical harm. Men, when angered, are more inclined to use physical aggression than women, who in turn are more likely to fear aggressive retaliation. On the other hand, a number of studies have found that, compared to women, anger suppression among men, especially those at risk for hypertension, is associated with higher blood pressure reactivity.[26]

At a more aggregate social level, anger has a pivotal role in the general strain theory of Agnew[27] to account for criminal conduct and delinquency, especially aggressive behavior. This sociological theory is convergent with psychological conceptions.[1] Strain occurs when people are prevented from achieving goals, lose valued possessions, and are exposed to noxious stimuli. Strain results in negative affect, which creates pressure for corrective action, but anger has superordinate value. Anger is considered the crucial emotion, as it is produced by strain when others are blamed for personal adversity, it increases the sense of being injured or wronged, and it creates a motivation for retaliation or revenge, energizes action, and lowers inhibitory control. Many large sample studies have supported Agnew's theory with multifaceted life stress measures and finding a mediation effect of anger on aggression.[28]

ANGER PHYSIOLOGY

A defining condition of anger is physiological arousal, the activation of which has evolutionary roots, as anger serves to mobilize us to response energetically and to sustain effort. The "flight-or-fight" response[4] refers to this hard-wired physiological mechanism that is triggered instantaneously to engage survival behavior, to focus attention on the survival threat, and to enable the organism to not succumb to fear, pain, or shame. Anger is the emotional complement of the organismic preparation for attack, which also entails the orchestration of signals of attack readiness so as to ward off opponents or to coerce compliance. The latter has been conceptualized in an evolutionary perspective in terms of a formidability posture to induce an opponent to recalibrate the "welfare tradeoff ratio."[29]

The arousal of anger is marked by activation in the cardiovascular, endocrine, and limbic systems, as well as other autonomic and central nervous system areas, and by tension in the skeletal musculature. The autonomic signature of anger corresponds to a mixture of adrenaline and noradrenaline. Autonomic system arousal, especially cardiovascular, has been commonly observed in conjunction with anger by scholars from the classical age (such as Seneca, Aristotle, and Plutarch) to the early behavioral scientists of the nineteenth and twentieth centuries (especially Charles Darwin, William James, G. Stanley Hall, and Walter B. Cannon). Laboratory research has reliably found anger arousal to entail increases in both systolic and diastolic blood pressure, in respiration, heart rate, and skin conductance responses. It is differentiated from fear by a stronger increase in diastolic pressure, in muscle tension (electromyogram recordings), in total peripheral resistance, and in facial temperature, the latter associated with facial flushing, often reported by people reflecting on their anger experience. The sensation of anger is highly correlated with anger's physiological profile.[7] In terms of psychosocial imagery, there is no better metaphor for anger than hot fluid in a container.

Autonomic arousal is primarily engaged through adrenomedullary and adrenocortical hormonal activity. In anger, the catecholamine activation is more strongly noradrenaline than adrenaline (the reverse being the case for fear). The adrenocortical effects, which have longer duration than the adrenomedullary ones, are mediated by secretions of the pituitary gland, which also influences testosterone levels. The pituitary-adrenocortical and pituitary-gonadal systems are thought to affect readiness or potentiation for anger responding. Testosterone provides vigor and reduces fear.

A number of central nervous system structures have been identified in anger activation, most prominently the amygdala, the almond-shaped, limbic system component located deep in the temporal lobe that is well-established for its activation in threat-detection, its association with trauma, and anger-activated aggressive behavior.[30,31] Activation in the amygdala has been found to be associated with anger and attack priming. Interconnections with the ventromedial and orbital frontal cortex, which are recruited during anger states, serve to regulate behavior and mediate how anger affects aggressive responding.[32]

The amygdala is the key site for aversive motivational system, and it has been thought that anger is derived from that system, as it is a negatively valenced emotion that is evoked by aversive experiences. Similarly, anger is conjectured to be a product of a defensive, "rage system."[33]

However, anger has also been linked with asymmetric left-prefrontal cortical activity, which has typically been associated with positive affect and approach motivation.[16,34] Although the complexities of the neuropsychological and psychophysiological processing of anger seem to be far from straightforward, particularly how it might bear on psychopathology,[35] Potegal and Stemmler[32] have put forward a neuroanatomical conjecture about the dynamics of anger involving the amygdala, temporal lobe, and the ventromedial and orbital frontal cortex.

The central nervous system neurotransmitter serotonin, which is also present in blood platelets, affects anger potentiation, as low levels of this hormone are associated with irritable mood, as are functional polymorphisms in serotonin receptor genes.[36] Serotonin imbalances are related to deficits in the modulation of emotion. While serotonin and other neurotransmitters (noradrenaline and dopamine) are involved in anger activation, the neural structures and circuitry in anger dysregulation and aggressive behavior remain to be disentangled.

These various physiological mechanisms pertain not only to the intensity of anger arousal but also its duration. Arousal activation eventually decays to baseline levels, but recovery time may be prolonged by exposure to new arousal sources or by rumination. The potency of a provocation may be heightened by the carryover of undissipated excitation from a prior arousal source, which may not have been anger-specific (i.e., an otherwise stressful circumstance, such as exposure to bad news, work pressure, or traffic congestion). This "excitation transfer" of arousal residues facilitates anger, augments its intensity, amplifies blood pressure, and raises the probability of aggression.[18] Residual arousal from unresolved anger events can transfer to future conflicts and further intensify anger reactivity to instigating events. In turn, unexpressed anger is associated with exaggerated and more prolonged cardiovascular responses to a variety of stressful stimuli. In this regard, a stress framework is highly useful. Both acute and prolonged exposure to common stressors (e.g., noise, crowding, difficult tasks, and high-pressure job environments filled with time demands, or exposure to abrasive interactions) may induce physiological activation that decays slowly. When someone experiences an event that pulls for the cognitive label "anger," and this event occurs concurrently with already elevated arousal, the anger system is then more easily engaged.

ANGER DYSCONTROL: PHYSICAL AND PSYCHOLOGICAL HEALTH PROBLEMS

Anger is a highly functional human emotion and is to be appreciated as a rich part of cultural life, but the survival value of the aggression-enabling function of anger is an archaic remnant with rare contemporary necessity. Outside of warfare, the survival challenges presented by civilized society are predominantly psychological, rather than physical. Effective coping with the demands of modern life requires understanding complex information, problem-solving, and prudent action, not energized rapid responding. Even in emergency situations, anger requires regulation. Contrary to intuitions, anger can be detrimental to survival in a physical threat crisis. It is counterproductive for energy conservation in a prolonged fight, for monitoring additional threat elements and hazards, and for effective strategy selection in circumstances where survival threat lingers and/or remains obscure. The regulation of the intensity and duration of anger arousal is pivotal to its merit or utility.

To get angry about something, one must pay attention to it. Anger is often the result of selective attention to cues having high provocation value. A principal function of cognitive systems is to guide behavior, and attention itself is guided by integrated cognitive structures, known as schemas, which incorporate rules about environment-behavior relationships. What receives attention is a product of the cognitive network that assigns meaning to events and the complex stimuli that configure them. The appraisal process is *in* the seeing and the hearing, not something tandem to the perception, and it is an ongoing process. Expectations guide attentional search for cues relevant to particular needs or goals. Once a repertoire of anger schemas has been developed, events (e.g., being asked a question by someone) and their characteristics (e.g., the *way* the question was asked, *when* it was asked, or *who* asked it) are encoded or interpreted as having meaning in accord with the preexisting schema. Rumination about provoking events extends or revivifies anger reactions. Because of their survival function, the threat-sensing aspect of anger schemas carries urgent priority and can preempt other information processing.

Perceived malevolence is one of the most common forms of anger-inducing appraisal. It pulls for anger by involving the externalization of blame and the theme of justification. That, in turn, engages social norms of retaliation and retribution. Averill's view of anger is that it is a socially constituted syndrome or a transitory social role governed by social rules.[5] While anger and physical aggression may be viewed as applying a legitimate punitive response for transgression or as ways of correcting injustice, justifications can be embellished to serve the exoneration of blame for destructive outcomes of expressed anger.

Physiological components of anger, such as increased blood flow, may be adaptive for survival in a short-term danger episode, but the byproducts of recurrent engagement of anger are hazardous in the long term. Unregulated anger is associated with physical and psychological health impairments, including detrimental effects on the cardiovascular system bearing on mortality.[9] Persons who are reactively angry are at considerable risk for coronary heart disease. An angry, hostile, and distrusting outlook necessitates high vigilance for thwarting and malevolence, resulting in prolonged neurohormonal activation conducive to atherosclerosis. In addition to these pathogenic effects for a personality style that is overly expressive of anger, the coronary system is also impaired by recurrently suppressed anger, long identified as a causal variable in the etiology of essential hypertension. People having difficulties expressing anger tend to be at risk for chronically elevated blood pressure, as mediated by high plasma renin activity and norepinephrine. The suppression of anger has been robustly correlated to elevated blood pressure and greater cardiovascular reactivity to provocation in laboratory studies and to sustained hypertension in field studies. Anger suppression also amplifies pain sensitivity.[37]

With regard to psychological well-being, anger occurs in conjunction with a wide range of psychiatrically classified disorders, emerging in conjunction with the emotional instability attributes of personality disorders, irritability and "attacks" in mood disorders, delusions and command hallucinations in psychotic disorders (especially paranoid schizophrenia), impulse control disorders, intellectual disabilities, dementia, substance abuse disorders, and exotic cultural-bound syndromes.[38,39] As anger often results from trauma, it can be salient in PTSD, affecting the severity and course of PTSD symptoms, and it is associated with major adjustment problems for military veterans.[12,13] The central quality of anger in the broad context of clinical disorders is dysregulation—its activation, expression, and experience occur without appropriate controls. Transdiagnostic processes of threat perception, selective attention, interpretative bias, confirmation bias, rumination, and self-control deficits are involved.

Among hospitalized psychiatric patients in long-term care in both civil commitment and forensic institutions, anger is a pervasive problem, as identified by both clinical staff and by the patients themselves. Studies with multiple control variables show anger to be related to the violent behavior of psychiatric patients before, during, and after hospitalization and to physical aggression within institutions by incarcerated adults.[40] Anger is not only an important clinical need among many psychiatric and custodial populations, it also bears on the therapeutic milieu and on the wellbeing of clinical and custodial staff. Anger dysregulated people often have traumatic life histories, replete with experiences of abandonment and rejection, as well as economic and psychological impoverishment. For them, anger becomes entrenched as a mode of reactance to stressful or aversive experiences. Chronically angry people are reluctant to surrender the anger-aggression system that they have found useful to engage, because they discount the costs of its engagement.

ANGER TREATMENT

The first step in the provision of therapeutic intervention for anger is facilitating client "readiness" for anger treatment, which can be very challenging due to client background adversities, multifaceted clinical comorbidities, and resource limitations in facilities where treatment might be implemented.[41] Nevertheless, there is substantial evidence for the efficacy of anger treatment. Nine meta-analyses on the effectiveness of psychotherapy for anger have been published, involving a wide range of clinical populations, which overall have found medium to strong effect sizes, indicating that approximately 75% of those receiving anger treatment improved compared to controls. Cognitive behavioral therapy (CBT) approaches have greatest efficacy in reducing anger, and now with encouraging evidence for reducing physically assaultive behavior in hospital.[42]

CBT approaches incorporate training in self-monitoring, relaxation, and social skills, but centrally seek to modify cognitive structures and the way a person processes information about social situations. They strongly emphasize self-regulation, cognitive flexibility in appraising situations, arousal control, and learning prosocial values and scripts. Making extensive use of therapist modeling and client rehearsal, anger proneness is modified by first motivating client engagement and then restructuring cognitive schemas, increasing capacity to regulate arousal and facilitating the use of constructive coping behaviors. Priority is given to self-regulatory controls of anger activation. The parameters or state markers for anger activation that receive attention in CBT anger treatment are: *reactivity* (frequency on onset and how easily anger is triggered); *latency* (how rapidly activated); *intensity* (how strongly engaged), and *duration* (persistence of arousal). Treatment aims to minimize anger reactivity, intensity, and duration and to moderate anger expression to reduce the costs of anger dyscontrol.

To facilitate anger regulation, anger treatment procedures strive to disconnect anger from the threat system. This is done first through the provision of safety, patience, and psychological space for reflection, exploration, and choice. The client's view of anger is normalized, to obviate worries about being a "bad" or unworthy person. The therapist will acknowledge the legitimacy of the client's feelings, affirming his or her self-worth. Building trust in the therapeutic relationship is pivotal. As self-regulation hinges on knowledge, education about anger and discovery of the client's personal anger patterns or "anger signature" is facilitated. Much is done to augment self-monitoring and to encourage the moderation of anger intensity. The therapist models and reinforces nonanger alternative responding so as to build replacements for the automatized angry reactions that had been the client's default coping style.

One CBT approach to anger treatment that has received significant support for its efficacy is called "stress inoculation" (SI). In this treatment approach, anger provocation is simulated by therapeutically paced progressive exposure to anger incidents created in imaginal visualization and in role play, based on a hierarchy of anger incidents produced by the collaborative work of client and therapist. This graduated, hierarchical exposure is the basis for the "inoculation" metaphor. The SI involves the following key components: (1) client education about anger, stress, and aggression; (2) self-monitoring of anger frequency, intensity, and situational triggers; (3) construction of a personal anger provocation hierarchy, created from the self-monitoring data; (4) arousal reduction techniques of progressive muscle relaxation, breathing-focused relaxation, and guided imagery training; (5) cognitive restructuring by altering attentional focus, modifying appraisals, and using self-instruction; (6) training behavioral coping in communication and respectful assertiveness as modeled and rehearsed with the therapist; and (7) practicing the cognitive, arousal regulatory, and behavioral coping skills while visualizing and role playing progressively more intense anger-arousing scenes from the personal hierarchies.

While therapeutic mechanisms underlying anger treatment gains are not clear, nor are their sustainability or generalizability, the field is fortified by the evidence base and continues to seek advances in providing remedies for anger dyscontrol.

References

1. Anderson CR, Bushman BJ. Human aggression. *Annu Rev Psychol.* 2002;53:27–51.
2. Izard CE. *Human Emotions.* New York, NY: Plenum Press; 1977.
3. Darwin C. *The Expression of Emotions in Animals and Man.* 3rd ed. London: HarperCollins; 1872/1998.
4. Cannon WB. *Bodily Changes in Pain, Hunger, Fear, and Rage.* New York, NY: Appleton; 1915.
5. Averill JR. *Anger and Aggression: An Essay on Emotion.* New York, NY: Springer Verlag; 1982.
6. Novaco RW. The functions and regulation of the arousal of anger. *Am J Psychiatry.* 1976;133:1124–1129.
7. Stemmler G. Somatovisceral activation during anger. In: Potegal M, Stemmler G, Spielberger C, eds. *International Handbook of Anger.* New York, NY: Springer; 2010:103–121.
8. Agnew R. A revised strain theory of delinquency. *Soc Forces.* 1985;64:151–167.
9. Iyer P, Korin MR, Higginbotham L, Davidson KW. Anger, anger expression, and health. In: Suls JM, Davison KW, Kaplan RM, eds. *Handbook of Health Psychology and Behavioral Medicine.* New York, NY: Guilford Press; 2010:120–132.
10. Siegman AW, Smith TW. *Anger, Hostility, and the Heart.* Hillsdale, NJ: Erlbaum; 1994.
11. Orth U, Wieland E. Anger, hostility, and posttraumatic stress disorder in trauma-exposed adults: a meta-analysis. *J Consult Clin Psychol.* 2006;74:698–706.
12. Novaco RW, Swanson RD, Gonzalez O, Gahm GA, Reger MD. Anger and post-combat mental health: validation of a brief anger measure with U.S. soldiers post-deployed from Iraq and Afghanistan. *Psychol Assess.* 2012;24:661–675.
13. Gonzalez OI, Novaco RW, Reger MA, Gahm GA. Anger intensification with combat-related PTSD and depression comorbidity. *Psychol Trauma Theory Res Pract Policy.* 2015; http://dx.doi.org/10.1037/tra0000042.
14. Feldman R, Dollberg D, Nadam R. The expression and regulation of anger in toddlers: relations to maternal behavior and mental representations. *Infant Behav Dev.* 2011;34:310–320.

15. Pinkham AE, Griffin M, Baron R, Sasson NJ, Gur RC. The face in the crowd effect: anger superiority when using real faces and multiple identities. *Emotion*. 2010;10:141–146.
16. Carver CS, Harmon-Jones E. Anger is an approach-related affect: evidence and implications. *Psychol Bull*. 2009;135:183–204.
17. Bandura A. Psychological mechanisms in aggression. In: Geen R, Donnerstein E, eds. *Aggression: Theoretical and Empirical Reviews*. New York, NY: Academic Press; 1983:1–40.
18. Zillmann D. Cognition-excitation interdependencies in aggressive behavior. *Aggress Behav*. 1988;14:51–64.
19. Potegal M. Temporal and frontal lobe initiation and regulation of the top-down escalation of anger and aggression. *Behav Brain Res*. 2012;231:386–395.
20. Konecni VJ. Annoyance, type, and duration of postannoyance activity, and aggression: "The cathartic effect" *J Exp Psychol Gen*. 1975;104:76–102.
21. Bushman B. Does venting anger feed or extinguish the flame? Catharsis, rumination, distraction, anger, and aggressive responding. *Pers Soc Psychol Bull*. 2002;28:724–731.
22. Gross JJ, John OP. Individual differences in two emotion regulation processed: implications for affect, relationships, and well-being. *J Pers Soc Psychol*. 2003;85:348–362.
23. Ray RD, Wilhelm FH, Gross JJ. All in the mind's eye? Anger rumination and reappraisal. *J Pers Soc Psychol*. 2008;94:133–145.
24. Archer J. Sex differences in aggression in real-world settings: a meta-analytic review. *Rev Gen Psychol*. 2004;8:291–322.
25. Suter JM, Byrne MK, Byrne S, Howells K, Day A. Anger in prisoners: women *are* different from men. *Pers Individ Dif*. 2002;32:1087–1100.
26. Vogele C, Jarvis A, Cheeseman K. Anger suppression, reactivity, and hypertension risk: gender makes a difference. *Ann Behav Med*. 1997;19:61–69.
27. Agnew R. Foundation for a general strain theory of crime and delinquency. *Criminology*. 1992;30:47–86.
28. Aseltine RH, Gore S, Gordon J. Life stress, anger and anxiety, and delinquency: an empirical test of general strain theory. *J Health Soc Behav*. 2000;41:256–275.
29. Sell A, Tooby J, Cosmides L. Formidability and logic of human anger. *Proc Natl Acad Sci U S A*. 2009;106:15073–15078.
30. Davidson RJ. Dysfunction in the neural circuitry of emotion regulation—a possible prelude to violence. *Science*. 2000;289:591–594.
31. Blair RJR. Considering anger from a cognitive neuroscience perspective. *Wiley Interdiscip Rev Cogn Sci*. 2012;3:65–74.
32. Potegal M, Stemmler G. Constructing a neurology of anger. In: Potegal M, Stemmler G, Spielberger C, eds. *International Handbook of Anger*. New York, NY: Springer; 2010:39–59.
33. Panksepp J, Zellner MR. Towards a neurobiologically based unified theory of aggression. *Rev Int Psychol Soc*. 2004;17:37–61.
34. Harmon-Jones E, Gable PA, Peterson CK. The role of asymmetric frontal cortical activity in emotion-related phenomena: a review and update. *Biol Psychiatry*. 2010;84:451–461.
35. Dougherty DD, Shin LM, Alpert NM, et al. Anger in healthy men: a PET study using script-driven imagery. *Biol Psychiatry*. 1999;46:466–472.
36. Conner TS, Jensen KP, Tennen H, Furneaux HM, Kranzler HR, Covault J. Functional polymorphisms in the serotonin 1B receptor gene [HTR1B] predict self-reported anger and hostility among young men. *Am J Med Genet B*. 2010;153B:67–78.
37. Quartana PJ, Burns JW. Painful consequences of anger suppression. *Emotion*. 2007;7:400–414.
38. Novaco RW. Anger and psychopathology. In: Potegal M, Stemmler G, Spielberger C, eds. *International Handbook of Anger*. New York, NY: Springer; 2010:465–497.
39. DiGuiseppe R, Tafrate RC. *Understanding Anger Disorders*. New York, NY: Oxford University Press; 2007.
40. Novaco RW. Anger dysregulation: driver of violent offending. *J Forensic Psychiatry Psychol*. 2011;22:650–668.
41. Howells K, Day A. Readiness for anger management: clinical and theoretical issues. *Clin Psychol Rev*. 2003;23:319–337.
42. Novaco RW, Taylor JL. Reduction of assaultive behavior following anger treatment of forensic hospital patients with intellectual disabilities. *Behav Res Ther*. 2015;65:52–59.

CHAPTER

36

Aggressive Behavior and Social Stress

S.F. de Boer, B. Buwalda, J.M. Koolhaas
University of Groningen, Groningen, The Netherlands

OUTLINE

Abstract

Across the animal kingdom, social interactions, ranging from affiliative cooperative exchanges to intensely aggressive contests, have a fundamental impact upon the health and wellbeing of an individual. Social life is often quite intricate, full of rewards, but also the cause of serious conflict and intense stresses. In this chapter, the sociobiology of feral rodents and their natural defense mechanisms are described as a starting point in the evaluation of the causes and consequences of aggressive behaviors and social stress in laboratory animals. The use of ecologically relevant conditions provides animal models with a high degree of face and construct validity. Considering the ecological significance of the dynamic changes in behavior and physiology of animals living in a social structure allows an answer to the fundamental question of the adaptive or maladaptive nature of the changes observed. It is important to notice that societies are by definition based on individuals, each with its own individual trait characteristics. Hence, a focus on individual differences may provide answers to the question of individual vulnerability and resilience.

INTRODUCTION

Throughout the animal kingdom, aggression is one of the most widespread and functional forms of social behavior that ultimately contributes to fitness (procreation) and survival of individuals. Clearly, aggression is the behavioral weapon of choice for essentially all animals and humans to gain and maintain access to desired resources (food, shelter, and mates), defend themselves and their offspring from rivals and predators, and establish and secure social status/hierarchical relationships. Although most individuals engage in these social conflicts with appropriate and well-controlled (functional) forms of aggressive acts and postures, a relatively small fraction of individuals can become excessively aggressive and extremely violent. This small percentage of escalated aggressive, antisocial, and violent individuals is not only a major cause of deaths worldwide, it is in particular the major source of serious health problems and disability in the surviving victims and/or witnesses of violent conflicts. These include various bodily diseases (e.g., cardiovascular, gastrointestinal, and immune dysfunctions) and mental disorders (e.g., depression, alcohol and substance abuse, eating, sleeping, and anxiety disorders). Clearly, inappropriate and/or excessive forms of violent aggression and the consequent intense stress of experiencing social defeat and/or chronic subjugation constitute one of the most significant problems for the public health. Hence, this capacity of humans to persistently and pervasively exert aggressive antisocial behaviors and violent outbursts has motivated

Stress: Concepts, Cognition, Emotion, and Behavior
http://dx.doi.org/10.1016/B978-0-12-800951-2.00036-4

much of the scientific interest in social stress and aggressive behaviors in animals. The scientific rationale of these studies is that the mechanisms underlying aggressiveness and social stress have a common biological basis in animals and humans. In particular, there is a need to understand these behaviors, both in its normal and pathological forms, in terms of their underlying causal physiological mechanisms, modulating factors, and grave detrimental consequences. In general, animal models are essential to obtain experimental support for this. As a matter of fact, a considerable part of our current knowledge on the ethology, etiology, neurobiology, genetics, and pharmacology of aggressive, defensive and social stress-induced submissive behavior is based on experimental and laboratory studies of social conflict in a wide variety of animals (i.e., ranging from fruit flies, crickets, zebra fish, songbirds, mice, rats, hamsters, voles, tupaias, dogs, cats, and monkeys).

In this chapter, the sociobiology of feral rodents and its natural defense mechanisms are described as a starting point in the evaluation and description of offensive and defensive aggressive behaviors and different social stress models in laboratory rats and mice.

KEY POINTS

- Aggressive behavior among conspecifics is a fundamental and adaptively significant component of the natural sociobiology of virtually every animal species.
- Under captive seminatural laboratory conditions, the resident-intruder paradigm allows detailed analyses of the spontaneous and natural expression of both offensive aggression and defensive behavior in laboratory rodents.
- Winning or losing aggressive confrontations is a salient life experience with enduring physiological and neurobiological consequences that profoundly impact individual's status, health, and fitness.
- Regular unpredictable and uncontrollable episodes of social defeat and/or chronic subordination induce chronic social stress that physiologically compromises individuals to a degree that leads to grave detrimental breakdown in adaptive capacity.
- Individual differences in aggressiveness reflect general stress coping styles that importantly determine the individual vulnerability and resilience to stress-related disorders.

THE SOCIOBIOLOGY OF FERAL RODENTS

Most neurobehavioral, pharmacological, and genetic studies of aggressive behavior in the laboratory setting is performed in rat and mice species that are highly adaptive in their social behavior both within their natural habitats and under captive laboratory conditions. In their classic works, Calhoun,[1] Barnett,[2] and Telle[3] gave a detailed account of the social interactions and ecology of wild brown Norway rats (*Rattus norvegicus*). Under natural environmental conditions and at low population densities, one adult male rat dominates a small group of females and young rats by vigorously marking, patrolling, and defending the region (called the territory) around their feeding and underground nest sites (burrows) from other male intruders. Outside these territories, neutral areas exist where fighting is minimal and avoidance occurs. In general, a colony comprises a number of territories and neutral areas. In case of an encounter with an intruder, the male displays fierce territorial aggressive behavior causing the intruder to flee. Although it is mainly the resident male who drives away male intruders, lactating female rats defend their nest, against both males and females. Such dispersive territorial behavior can only persist when invasions of a territory are infrequent.

With increasing population densities, rats and mice become socially more tolerant and cohesive, and adapt to a despotic social structure, with one male being socially dominant (the *alpha* rat) and prevailing most often over rival or *beta* males and over subordinate or *omega* animals in aggressive confrontations. The attacks and threats exerted by dominant individuals are responded to by defensive, evasive, and submissive postures in the beta and omega animals that decrease the probability of being attacked further. Although overt aggression may initially be necessary to obtain a dominance-subordinate relationship, once established, a stable hierarchy suppresses further aggressive conflicts and unwanted fights among the group members. The decision to initiate aggressive behavior is likely the result of a cost-benefit analysis that depends on social context, requires higher (prefrontal) cortical regulation, and is not necessarily restricted to socially dominant individuals.

When environmental conditions such as weather, food supplies, nesting opportunities, infectious, and predatory pressures are favorable, wild rats and mice crowd together in colonies that rapidly increase and eventually may number many hundreds. Nevertheless, life in the colony is not easy. Colonies may crash periodically due to increasing and intense social conflict in spite of conducive environmental conditions. It is very well known that populations of many small rodent species undergo marked fluctuations with a periodicity of 3-7 years (see Figure 1a). The causes of these cycles have interested ecologists for nearly a century,[4,5] and an old hypothesis by Chitty[6] suggests that the cyclic nature of rodent populations might be due to disruptive selection for aggressive behavior in the course of the population

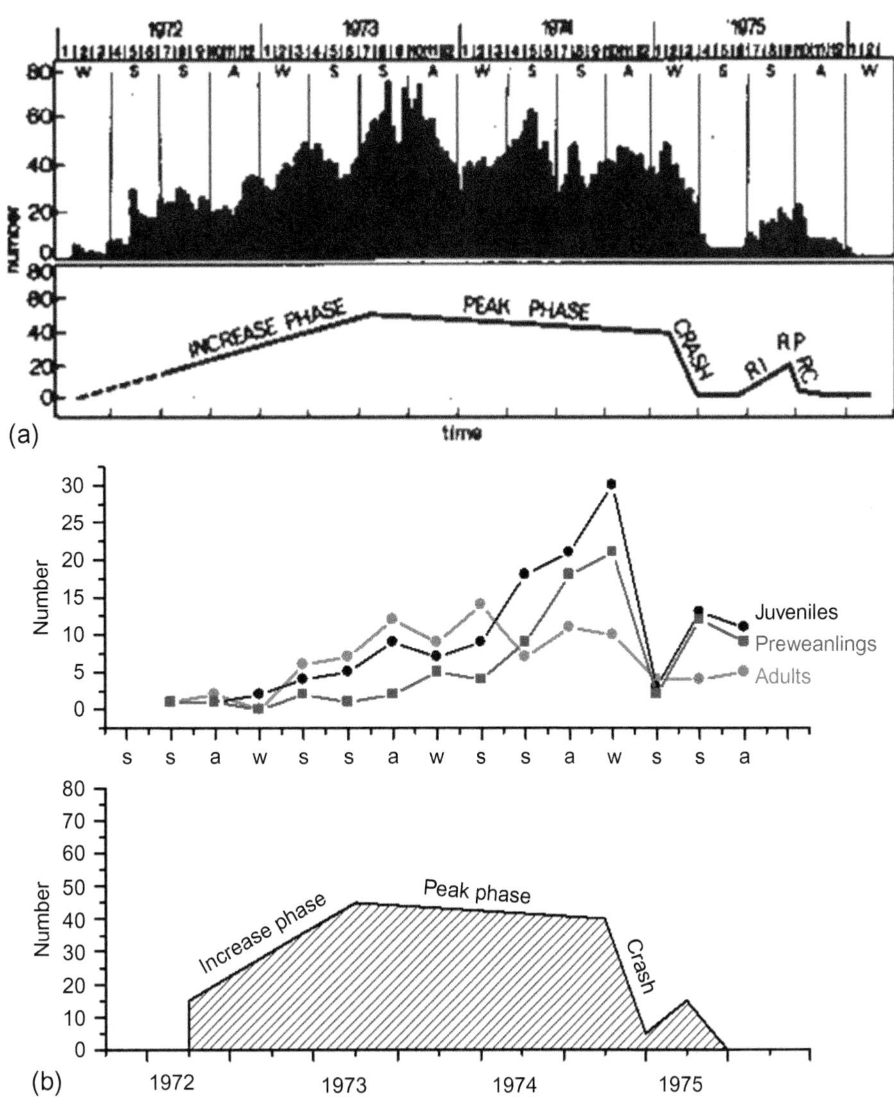

FIGURE 1 (a) Change in population density of a feral population of house mice in large outdoor enclosure. (b) Mortality of adult, juvenile, and preweanling house mice during the various phases of the population cycle. (Revised figure from Van Oortmerssen and Busser 7)

cycle. Aggressive individuals might be at a selective advantage as the population increases. This hypothesis is strongly supported by the early studies of Van Oortmerssen and Busser[7] on seminatural populations of house mice consisting of both males and females. Phenotypic characterization of individuals caught from these feral outside colonies, revealed a bimodal distribution of attack latencies as measured in a standardized resident-intruder paradigm. Subsequent selective breeding for high and low attack latencies resulted within five generations in a stable short attack latency selection line. Subsequent embryo transfer, cross-fostering and back-cross experiments showed that the phenotypic differentiation in aggressive behavior as observed in the wild has a strong genetic basis that is only marginally influenced by the maternal environment.[8] Generally, heritability estimates for aggressive behavioral traits in both animals and humans range between 30% and 60%.[9] Several additional data support the general idea that stable genetic variation for aggressive behavior may be an important factor in the population dynamics of the wild house mouse. While nonaggressive mice fare better in establishing and growing colonies, aggressive mice do better in settled stable demes. Moreover, analysis of mortality in these feral populations reveals a strong increase in dead females, juveniles, and preweanling juveniles just before the crash of the population (see Figure 1b[8]).

It is tempting to consider the possibility that this is due to high levels of intermale aggression in that phase of the population cycle (i.e., development of hyperaggressiveness within social groups could then be the direct cause of the population decline). Indeed, increasing social conflicts and frequent hyperaggressive behavior among individuals in unstable social groups of laboratory and feral rats not only increases the risk of direct fatal injuries during fighting, but induces social stress and chronic subordination that physiologically compromises the immune system, diverts energy from reproductive activities and foraging, disrupts circadian and physiological rhythms, places prolonged demands on endocrine functions resulting in gonadal atrophy and adrenal hypertrophy, and ultimately shortens lifespan.[10–12] Not surprisingly, social instability and/or chronic subordination in rats, mice, and other animals are employed as chronic social stress model in fundamental research of the mechanisms underlying stress-related pathology in humans (see "Chronic Social Stress Models" section).

EXPERIMENTAL LABORATORY MODELS OF AGGRESSION

The popularity of rats and mice in experimental studies is largely due to their high adaptive capacity and social plasticity when reared and housed in laboratory settings that generally are widely divorced from their natural habitats. Housing conditions of laboratory rats and mice rarely allow for the formation of breeding units (demes), but rather force several adolescent/adult males or females to cohabitate. Groups of laboratory rats or mice quickly develop a rather strict social hierarchy as a necessity of the restraint of their cage. Agonistic behavior by the dominant male serves to solidify social hierarchies that in turn prevent the need for frequent fighting bouts and associated risk of injury. However, certain dominant individuals under specific conditions and experiences (i.e., crowding, repetitive winnings fights, or if subordinates do not respond with appropriate submissive behavior) may develop excessive or violent-like aggressive behavior that may cause severe injury or death of cage mates.[13,14]

Given the general despotic territorial and/or gregarious hierarchical social structure of feral rodents, much of the preclinical aggression research is conducted in territorial male resident rats/mice confronting an unfamiliar intruder conspecific. This so-called resident-intruder paradigm allows the spontaneous and natural expression of both offensive aggression and defensive behavior in laboratory rodents in a seminatural laboratory setting. By recording the frequencies, durations, latencies, and temporal and sequential patterns of all the observed behavioral acts and postures in the combatants during these social intercourses, a detailed quantitative picture (ethogram) of offensive (resident) and defensive (intruder) aggression is obtained (see Figure 2).

The resident-intruder paradigm brings this natural form of behavior into the laboratory allowing controlled studies of both the resident aggressor and the intruder victim.[15] The paradigm is strongly based on the fact that under low population densities, an adult male rat or mouse will establish a territory when given sufficient living space and resources. Territoriality is significantly enhanced in the presence of females and/or sexual experiences. As a consequence of territoriality, the resident will attack unfamiliar male conspecifics intruding in its home cage. The intruder in turn will show defensive and submissive behavior in response to the offensive attacks by the resident. Although typical patterns of aggressive behavior differ between species, there are several concordances in the ethology and neurobiology of aggression among rodents, primates, and humans.

AGGRESSIVE BEHAVIOR: DIFFERENT FORMS IN BOTH ANIMALS AND HUMANS

The existence of different kinds of aggression has long been recognized mainly on the basis of laboratory animal research.[16–18] As already indicated in the previous section, there are generally two types of agonistic behaviors in both males and females related to conspecific attacks: *offense* and *defense*. These differ in motor patterns, bite/attack targets, physiological/neuroendocrine response profiles, ultimate functional consequences and proximate neurobiological control mechanisms. For offense, the motor patterns are approach to the opponent followed by anogenital sniffing, offensive upright/sideways posture, attacks (simple bites or bite and kick), chase, piloerection and tooth-chattering (mainly in rats) or tail-rattling (mostly in mice). In the minutes leading up to intense attack bites, the resident rat emits brief pulses of ultrasonic vocalizations in the 50 kHz ranges that may reflect high excitement. The bite targets are primarily the hindquarters of the flanks, back and base of the tail (less-vulnerable body regions). The function is to obtain and retain resources like space, food, and mates. The eliciting and motivating stimuli for offense in rodents are predominantly odors acting on both the main and accessory olfactory receptors.[19] These project to many core and other structures in the brain involved in offense. The lateral septum, medial amygdala, bed nucleus of the stria terminalis (BNST), medial preoptic nucleus, anterior hypothalamus, the ventrolateral aspects of the ventromedial hypothalamus, and dorsolateral periaqueductal gray (dlPAG) are core neural structures involved in offensive aggression of rodents.[20,14] Other structures involved in rodent offense are the prefrontal cortex, the hippocampus, dorsal raphe nucleus, locus coeruleus, and nucleus accumbens. Possible interconnections among some of these structures relevant to their role in offensive aggression of rodents have been reviewed by Nelson and Trainor[21] and by Takahashi and Miczek.[22]

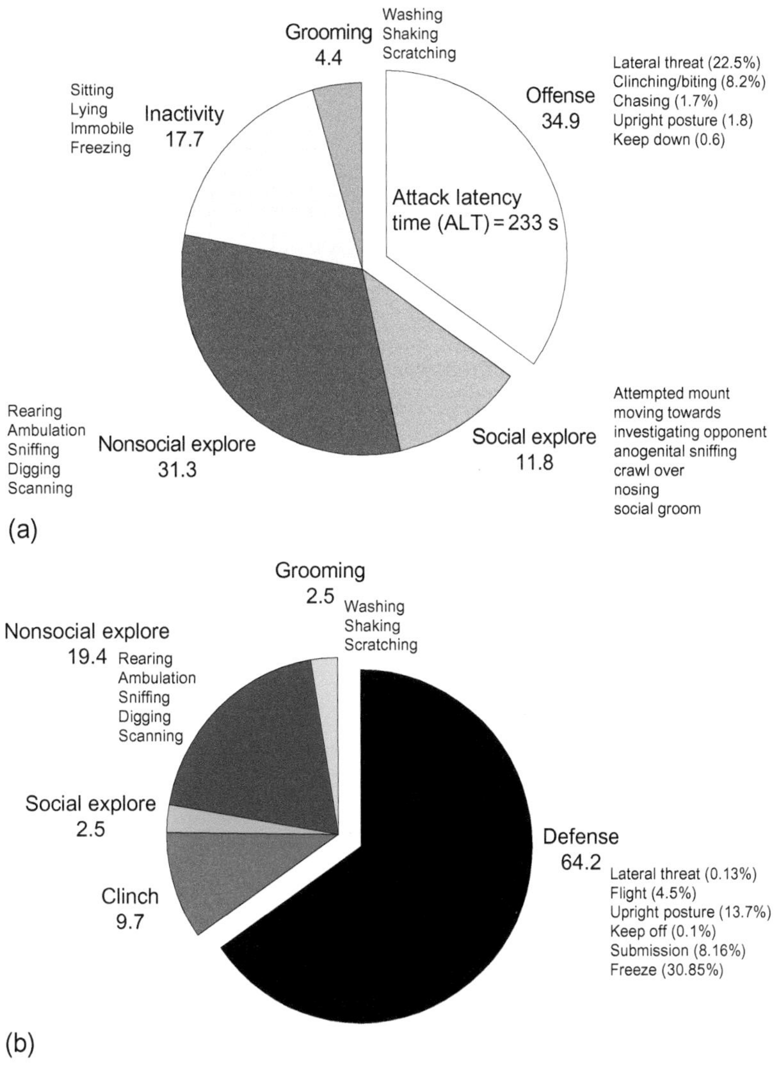

FIGURE 2 Behavioral profile of resident males WTG rats (a) and Wistar intruder rats (b) during a 10 min resident-intruder test.

For defense, the motor patterns are avoidance/freezing, defensive upright and sideways posture (keep-away), flight, and attacks (lunge and bite). These defensive motor acts are usually accompanied with urination/defecation and emittance of 22 kHz ultrasonic vocalizations. The lunge and attack-bite targets are primarily the face (snout), neck, and belly (vulnerable body regions). The function is to defend one's self, mates, and progeny from attacks of another animal of the same or different species. As with offense, conspecific odors elicit and arouse defense in rodents via the main and accessory olfactory systems.[19] These conspecific odor cues act on the posterior part of the basomedial amygdala, which is connected to the medial parts of the ventromedial nucleus of the hypothalamus (mVMH). The mVMH in turn projects to the dorsomedial premammilary nucleus which in turn projects to the dorsomedial PAG leading to the somatomotor and autonomic physiological patterns of defense.[23]

Besides offense and defense, additional forms of aggressive behavior in animal research can be distinguished as well, such as infant-directed aggression or infanticide, predatory aggression, play-fighting (in juvenile animals), and maternal aggression. The latter can be observed in females during the late stages of pregnancy and the early phases of nursing. Predatory aggression is known as quiet-biting attack observed as the swift killing of a mouse or a cricket by a rat.

The most basic acts of aggression in humans are hitting, kicking, biting, pushing, grabbing, pulling, shoving, beating, twisting, and choking. Threatening (vocal) and using objects (weapons) to aggress are also included into this definition.[24] Different forms of aggression are also recognized in humans and the offensive pattern of aggression in animals generally relates to the "hot-tempered" *hostile aggression* subtype in humans (also called *reactive*, *emotional*, *affective*, and *impulsive* aggression). This form of aggression has its strong externalizing engagement and autonomic/neuroendocrine arousal in common with offensive aggression in animals. Moreover, both in animals and humans, this form of aggressive behavior is usually initiated in response to a perceived threat, such as the intrusion of an unfamiliar conspecific into the territory or in response to fear and frustration (omission of expected rewards). In contrast, "cold-blooded" *instrumental aggression* (also called *premeditated* and *proactive* aggression) is callous-unemotional aggression that seems to resemble more the quiet-biting attack or predatory forms of aggressive behavior in rodents.

Although both male and female rodents perform offensive aggression, there is a clear gender difference in the frequency and intensity of aggression similar to what is generally observed in humans. Males may perform frequent and fierce offensive aggression in a territorial and sociosexual context. Females show defensive aggression mostly in a maternal context, but low to medium levels of offensive aggression can certainly be observed in all female groups in relation to competition within the social hierarchy.[25]

Finally, it should be noted that aggression in both animals and humans has to be conceptualized into two components: trait-like aggressiveness and state-like aggressive behavior. Whereas *trait-like aggressiveness* refers to an individual's predisposition to act persistently aggressive in various different contexts, *state-like aggression* refers to the actual execution of aggressive behaviors. This distinction appears to be of crucial importance when linking certain physiological or neurobiological parameters to aggression.

AGGRESSIVE BEHAVIOR AND STRESS

Aggressive behavior or the threat of aggression is accompanied by strong neuroendocrine and autonomic stress responses in the two participants of the social interaction. The magnitude and the nature of this response depend on the outcome of the aggressive interaction (i.e., winning or losing) as well as the individual's aggressiveness and/or coping style. Because winning and losing social interactions form the basis of the social structure in groups of animals, this differential response may underlie the relationship between the position in the social structure and the vulnerability for stress pathology. Furthermore, the formation and maintenance of social hierarchies is one of the most important functions of aggressive behavior and much of the stress of everyday life has its origin in this social structure. Several studies in free ranging social groups of animals indicate that the stability of social environment is an important factor in health and disease (i.e., fitness). Unstable social groups and failure of social adaptive capacities may lead to serious forms of stress pathologies like hypertension, cardiovascular abnormalities, stomach ulcers, and immune system-mediated diseases and increased propensity to compulsively self-administer drugs of abuse.

The relationship between aggression and stress obviously relates to the physiology of aggression. Therefore, the most important neuroendocrine and physiological mechanisms underlying aggressive behavior are briefly summarized by discussing some social stress models used in adult rodents that challenge the defense mechanisms and hence call on the natural adaptive capacity of the animal.

ACUTE SOCIAL STRESS MODELS

The activation of the so-called stress systems such as the hypothalamic pituitary axis (HPA) and the sympathetic-adrenomedullary system (SAM) is not necessarily a sign of exposure to adverse physical or psychological stress. These systems have an important adaptive function in the cardiovascular, thermogenic, and metabolic support of any overt behavioral reaction to salient environmental challenges or opportunities, irrespective of their emotional valuation. For example, the response of these systems to rewarding, but physically activating stimuli (like sexual behavior) can be just as high as to aversive situations like social defeat[26] (Figure 5). This point is also nicely illustrated by comparing the immediate neuroendocrine and autonomic responses to acute social stress as can be studied in the resident-intruder paradigm. This paradigm allows analysis of both the winner and the loser of the conflict, and while the latter situation produces dangers and adverse emotionality that must be avoided, winning a social conflict produces pleasures which can be enjoyed and are highly rewarding.[27]

Defeat

Most social stress studies concentrate on defense and defeat (i.e., the loser of a dyadic social interaction and/or the subordinate male in a social dominance structure). Social defeat by a male conspecific induces an immediate and robust increase in the sympathetic arm of the autonomic nervous system (ANS) with consequent rapid rises in the plasma catecholamines noradrenaline and adrenaline (Figure 4), that in turn promptly triggers increases in heart rate, blood pressure, core body temperature (Figure 3), as well as energy mobilization (increased plasma glucose and free fatty acids). In addition, a slower but strong neuroendocrine activation/inhibition of the anterior and posterior pituitary gland results in increases of Adrenocorticotrophic hormone (ACTH), prolactin, and reduced Luteinizing hormone (LH) that consequently induce the release of the adrenocortical glucocorticoid steroid hormone

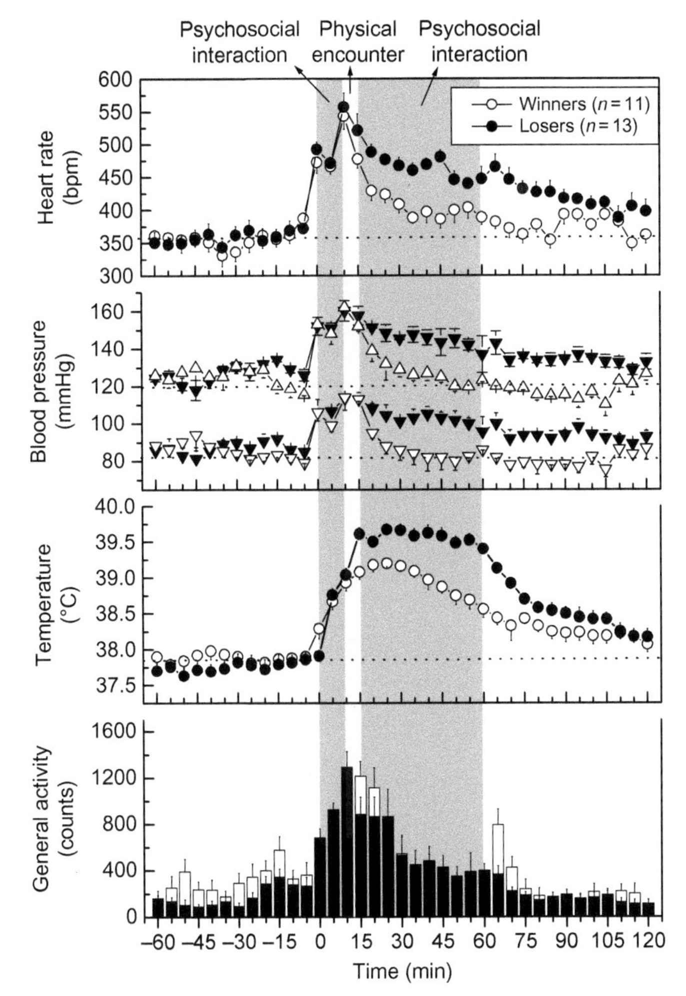

FIGURE 3 Changes in heart rate, blood pressure, and core body temperature in male rats during winning or losing a social interaction in a resident-intruder paradigm. The animals were provided with permanently implanted radio telemetry transmitters.

corticosterone and inhibition of the gonadal androgen hormone testosterone, respectively.[28,29]

Social defeat also increases a variety of central nervous system neurotransmitters including serotonin, norepinephrine, dopamine, and neuropeptides like Corticotropin releasing hormone (CRH), Arginine Vasopressin (AVP), Oxytocin (OXT), and endorphins. These responses, including the behavioral defense reaction (flight, immobility, and subjugation), can be considered as part of the classic response to an acute stressor. A comparison of a range of different laboratory stressors measured in terms of the magnitude of the corticosterone[29] reveals that social stimuli are among the many laboratory challenges the most potent "stressors" (Figure 5).

Recent evidence reveals that stressors not only quantitatively differ but may also differ qualitatively in the pattern of activated physiological systems. For example, a comparison of the electrocardiac response to restraint stress and social defeat reveals that the two types of stressors differ strongly in the balance of activation of the two branches of the ANS. Social defeat seems to exclusively activate the sympathetic branch together with a vagal withdrawal, whereas during restraint stress the parasympathetic nervous system shows a considerable activation as well. This differential autonomic balance may explain the high incidence of cardiac abnormalities (i.e., ventricular tachyarrhythmias and sudden cardiac death) observed during social defeat.[30]

More important, however, is the temporal pattern of the social stress responses.[31] Studies using more chronic recordings indicate that the various stress parameters have a different time course. The cardiovascular and catecholaminergic response to defeat diminishes within 1-2 h after the event, but the corticosterone response lasts for longer than 4 h (Figure 4). After an initial rise, plasma testosterone drops below baseline levels and remains at extremely low levels for at least 2 days.[28] A single social defeat appears to induce a reduction in the circadian variation in core body temperature, growth, sexual interest, and open field exploration that may last from 2 to 14 days after the social stress. Clearly, a single social defeat induces changes in physiology and behavior that can last from days to weeks and months. Hence, an acute stressor or major life event may have chronic consequences. These lasting alterations may be adaptive but may just as well be considered as the early signs of stress pathology. Clearly, the social defeat model allows further analyses of the changes in time of factors known to be involved in stress and adaptation.

Victory

Relatively few studies consider the physiological changes during aggressive behavior in the victor of the social intercourse. However, by using the resident male in the resident-intruder paradigm as the experimental animal, one can study the consequences of executive control and/or threat to control. Although the resident ultimately controls its social environment, this is preceded by a certain degree of unpredictability and threat to losing control over the social environment. This is clearly indicated by the fact that the stress response in terms of plasma corticosterone and catecholamines, heart rate, and blood pressure is initially almost as high as in the defeated intruder, but these stress parameters rapidly return to baseline levels as soon as the dominance relationship becomes clear and control is experienced. Typical for the winner of the social interaction are the cardiovascular abnormalities observed in the electrocardiogram immediately after the interaction. These abnormalities indicate a strong shift in autonomic balance toward high sympathetic dominance (Figure 5).[30]

Traditionally, high levels of offensive aggressive behavior have been linked to high levels of plasma

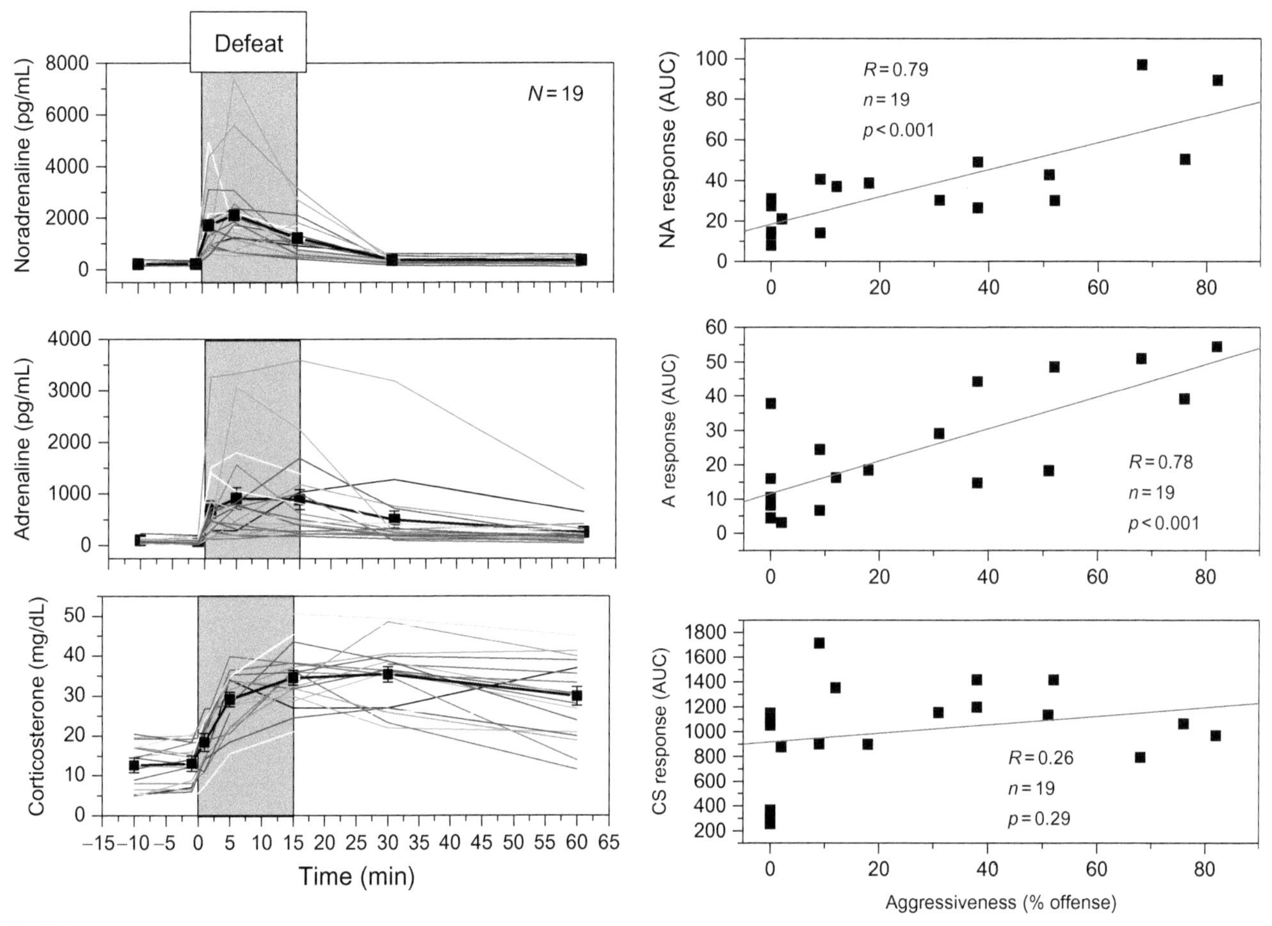

FIGURE 4 Individual time-response curves (AUCs) of plasma noradrenaline, adrenaline, and corticosterone responses to social defeat. The right-hand graphs show the relationship between the total neuroendocrine response (area under the curve, AUC) and the individual level of offensive aggressive behavior measured in the resident-intruder paradigm. Blood samples were collected using permanently implanted jugular vein cannula.

testosterone. However, the direct causal relationship between testosterone and the expression of offensive behavior is still a matter of debate. It seems that various kinds of aggressive behavior may be differentially dependent on testosterone, whereas experiential factors strongly interfere with the degree in which aggression depends on the dynamics of plasma testosterone alterations. Winning experience results in a temporary increase in plasma testosterone levels, whereas defeat results in a long-lasting decrease. To elucidate the causal role of testosterone dynamics and aggression, testosterone levels have been manipulated in mice with and without previous winning experience. While testosterone administration in mice without winning experience were generally more aggressive, they were not more likely to win future competitive interactions. In contrast, both intact and castrated male mice administered testosterone after winning a competitive interaction were more aggressive and more likely to win subsequent interactions.[32,33] In addition, winners treated with an antiandrogen drug, which prevented the normal increase in testosterone in response to aggressive interactions, were less likely to win a subsequent aggressive interaction.[34] Moreover, Fuxjager and coworkers[35] reported that the winner effect was due to an upregulation of androgen receptors in several key brain regions involved in reward processing (nucleus accumbens and ventral tegmental area) as well as social aggression (BNST). These studies provide compelling support for the idea that competition-induced testosterone dynamics may function to modulate ongoing and/or future aggressive behavior.

CHRONIC SOCIAL STRESS MODELS

Social stress can be studied in its most complex form using groups or colonies of rats or mice. It is well known that even in small breeding colonies of laboratory rats that are provisioned with clumped sources for food and water, subordinate members need to be rescued periodically in order to ensure their survival.[11] Repeated conflict in unstable social groups of laboratory rats, as for feral rats in its natural environment,[1] increases the risk

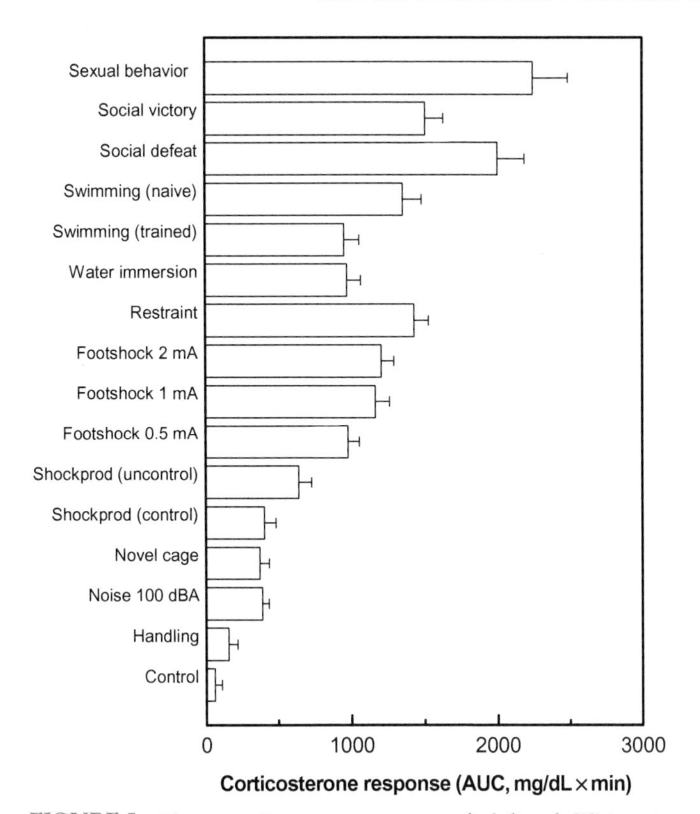

FIGURE 5 Plasma corticosterone responses of adult male Wistar rats to experimental stressor conditions, quantified as total area under the AUC. Each test consisted of a standardized series of baseline samples and a 15 min exposure to the stimulus followed by a recovery phase for the remaining hour. Blood samples were collected via a permanently implanted jugular vein cannula.

of injuries, compromises the immune system, diverts energy from reproductive activities and foraging, disrupts circadian and physiological rhythms, places prolonged demands on endocrine functions resulting in gonadal atrophy and adrenal hypertrophy, and ultimately shortens the lifespan.[10–12] Not surprisingly, social instability and/or chronic subordination in rats and other animals are used as chronic social stress models in fundamental research of the mechanisms underlying stress pathology in humans. The design of cages used for groups (colonies) of rats or mice may range from large outdoor enclosures to cages of several square meters with nest boxes or much smaller cages without any further facilities. The visible burrow system (VBS), created 30 years ago by the Blanchards, provides a unique experimental paradigm that allows for highly nuanced behavior within a laboratory setting and detailed evidence for a rodent's psychosocial state, as well as the opportunity to examine the neural and neuroendocrine cascade effects that connect behavior with potential predictive and constructive validity. Basically, the VSB habitat effectively recreates in a lab setting key environmental and behavioral features of rat colonies in the wild. Rats and mice are normally social animals; living in mixed-sex groups of up to several hundred animals in complex underground habitats consisting of interconnected tunnel and burrow systems. The VBS consists of a large central open arena and some smaller nest chambers that are connected to the central arena by means of Plexiglas tunnels. As in the wild, one of the key features of the VBS colony model is the use of mixed-sex groups, generally consisting of 4-5 adult males and 2-3 females to facilitate territoriality and aggression. The males quickly form dominance hierarchies, with the dominant guarding access to food and water and attempting to limit access to the females as well. Subordinates tend to remain largely in the "burrowed" tunnel/chamber, while the dominant male of each group utilized the open surface area. Furthermore, dominants and subordinates are readily identified by distinctive patterns of offensive or defensive wounding: the subordinates generally lose a significant amount of body weight during this period. An additional defensive behavior noted in the VSB is behavioral inhibition, involving suppression or reduction of normal activities like eating, drinking, sexual activity, and aggression in the subordinate animals of the group. The colonies are usually studied for periods of 2-3 weeks. Such a VBS setup allows one to evaluate changes in the brains and/or neuroendocrine systems of males as a function of their status as dominant, subdominant and subordinate or outcast (see Ref. 36 for review).

Living as a subordinate male in the continuous presence of a dominant male is generally considered to be a chronic stressor. However, the degree of social stress in VBS groups of rats and mice depends mainly on the stability of the social structure. In stable social groups, hardly any signs of stress pathology can be observed. Therefore, researchers tend to increase stress by reducing the social stability. This can be achieved by forming groups consisting of only aggressive males, leading to serious and regular dominance fights. However, this procedure might give a considerable bias to the experimental results due to the selection procedure. Another way to manipulate social stability is to mix groups on a regular basis. Lemaire and Mormède[37] successfully applied this method in rats. They could demonstrate that social instability triggered by daily rotations of male rats in cages housing female animals led to the development of hypertension, in particular, in a more socially active strain of rats.

INDIVIDUAL DIFFERENCES IN AGGRESSIVENESS AND COPING WITH SOCIAL STRESS

An important concept in aggression research that relates to stress concerns the relationship between aggression and coping. A number of studies in a variety of species show a strong positive correlation between the individual level of offensiveness and the tendency to actively cope with any kind of environmental stressor.[38] Aggressive males have a general tendency to actively deal with different kinds of environmental challenges. The antipode of this active problem-focused coping is a reactive or emotion-focused coping style. This is expressed as the absence of aggressive behavior in a social situation and as passivity in other challenging environmental conditions. These coping styles, as observed in several animal species, are also characterized by differential neuroendocrine and neurobiological profiles. They can be considered as important trait characteristics determining the individual adaptive capacity.

Personality factors or trait characteristics have long been recognized as playing a role in human stress psychophysiology as well. For example, the distinction between proactive and reactive coping styles in relation to individual levels of aggression is also made in human beings. The different coping styles as found in animal studies seem to be analogous to the distinction between the human type A and type B personalities used in car diovascular psychophysiology. The type A personality is described as aggressive, hostile, and competitive and is physiologically characterized by a high sympathetic reactivity. Although the A-B typology has been seriously questioned in the human psychological literature, factors related to trait-like aggressiveness such as anger and hostility have been repeatedly shown to play a role in the individual capacity of human beings to cope with environmental challenges. Recent studies in feral populations of various animal species show the ecological and evolutionary basis of individual variation in aggression and coping or personality as an (epi)genetic suite of traits.[38]

Due to its high emotional character and the high amount of physical activity, aggression is accompanied by a profound activation of the pituitary-adrenocortical system and the SAM. However, the reactivity of these neuroendocrine systems appears to differ as an important trait characteristic between high-aggressive and low-aggressive individuals (see Figure 4). Aggressive males are characterized by a high reactivity of the sympathetic nervous system and the adrenal medulla, whereas non- or low-aggressive males show a higher reactivity of the pituitary-adrenocortical axis and the parasympathetic branch of the ANS.[39] Under stress conditions, high-aggressive rats displayed a reduced vagal antagonism and an increased incidence of ventricular tachyarrhythmias compared to nonaggressive rats.[30] Recently, hypo-(re)activity of the HPA axis has been associated causally with the development of pathological forms of aggressive behavior in rats.[40]

Importantly, the long-term behavioral and physiological effects of social defeat stress appear to vary in accordance with the defensive behavior the intruder displayed during the interactive phase of the defeat episode (i.e., the social conflict). In particular, delayed submissiveness and more persistent defensive counterattacks, and guarding behavior by intruders appear to characterize social stress resilient animals.[12,41–43]

CONCLUDING REMARKS

In summary, we may conclude that the sociobiology of laboratory animals is a rich source of natural stimuli and conditions that can be exploited experimentally in the laboratory to understand the causes and consequences of aggression and social stress. The use of ecologically relevant conditions provides animal models with a high degree of face and construct validity. Considering the ecological significance of the dynamic changes in behavior and physiology of animals living in a social structure allows an answer to the fundamental question of the adaptive or maladaptive nature of the changes observed. The astonishing survival formula of rats and mice involves the fact that their exceptional social capacity is coupled with the physical resilience enjoyed by all members of the ancient and large order of rodents. It is important to note that societies are by definition based on individuals, each with its own individual trait characteristics. Hence, these animal models cannot avoid individual differences. This focus on individual differences may provide answers to the question of individual vulnerability and resilience.

References

1. Calhoun JB. *The Ecology and Sociology of the Norway Rat.* Bethesda, MD: U.S. Public Health Service; 1962.
2. Barnett SA. *The Rat: A Study in Behavior.* Chicago, IL: University of Chicago Press; 1963.
3. Telle HJ. Beitrag zur Kenntnis der Verhaltensweise von Ratten vergleichend dargestellt bei *Rattus norvegicus* und *Rattus rattus. Z Angew Zool.* 1966;53:126–196.
4. Elton CS. *Voles, Mice and Lemmings.* Oxford: Clarendon Press; 1942.
5. Christian JJ. The adreno-pituitary system and population cycles in mammals. *J Mammal.* 1950;31:247–259.
6. Chitty D. The natural selection of self-regulatory behaviour in animal populations. *Proc Ecol Soc Aust.* 1967;2:51–78.
7. Van Oortmerssen GA, Busser J. Studies in wild house mice III: disruptive selection on aggression as a possible force in evolution. In: Brain PF, Mainardi D, Parmigiani S, eds. *House Mouse Aggression: A Model for Understanding the Evolution of Social Behavior.* Chur: Harwood; 1989:87–117.
8. Sluyter F, Bult A, Lynch CB, van Oortmerssen GA, Koolhaas JM. A comparison between house mouse lines selected for attack latency or nest-building: evidence for a genetic basis for alternative behavioral strategies. *Behav Genet.* 1995;25:247–252.
9. Vassos E, Collier DA, Fazel S. Systematic meta-analyses and field synopsis of genetic association studies of violence and aggression. *Mol Psychiatry.* 2013;2013:1–7.

10. Fleshner M, Laudenslager ML, Simons L, Maier SF. Reduced serum antibodies associated with social defeat in rats. *Physiol Behav.* 1989;45:1183–1187.
11. Blanchard RJ, Blanchard DC, Flannely KJ. Social stress, mortality and aggression in colonies and burrowing habitats. *Behav Process.* 1985;11:209–213.
12. Stefanski V. Social stress in laboratory rats: behavior, immune function, and tumor metastasis. *Physiol Behav.* 2001;73:385–391.
13. De Boer SF, Caramaschi D, Natarajan D, Koolhaas JM. The vicious cycle towards violence: focus on the negative feedback mechanisms of brain serotonin neurotransmission. *Front Behav Neurosci.* 2009;3:1–6.
14. De Boer SF, Olivier B, Veening J, Koolhaas JM. The neurobiology of aggression: revealing a modular view. *Physiol Behav.* 2015;146:111–127.
15. Koolhaas JM, Coppens CM, de Boer SF, Buwalda B, Meerlo P, Timmermans PJA. The resident-intruder paradigm: a standardized test for aggression, violence and social stress. *J Vis Exp.* 2013;77: e4367.
16. Blanchard RJ, Blanchard DC. The organization and modeling of animal aggression. In: Brain PF, Benton D, eds. *The Biology of Aggression.* Alphen aan den Rijn: Sijthoff et Noordhoff; 1981:529–563.
17. Brain PF. Differentiating types of attack and defense in rodents. In: Brain PF, Benton D, eds. *Multidisciplinary Approaches to Aggression Research.* Amsterdam: Elsevier; 1979:53–77.
18. Adams DB. Brain mechanisms of aggressive behavior: an updated review. *Neurosci Biobehav Rev.* 2006;30:304–318.
19. Stowers L, Cameron P, Keller JA. Ominous odors: olfactory control of instinctive fear and aggression in mice. *Curr Opin Neurobiol.* 2013;23:339–345.
20. Haller J, Tóth M, Halasz J, De Boer SF. Patterns of violent aggression-induced brain c-fos expression in male mice selected for aggressiveness. *Physiol Behav.* 2006;88:173–182.
21. Nelson RJ, Trainor BC. Neural mechanisms of aggression. *Nat Rev Neurosci.* 2007;8:536–546.
22. Takahashi A, Miczek KA. Neurogenetics of aggressive behavior: studies in rodents. *Curr Top Behav Neurosci.* 2014;17:3–44.
23. Gross CT, Canteras NS. The many paths to fear. *Nat Rev Neurosci.* 2012;13(9):651–658.
24. Tremblay RE, Szyf M. Developmental origins of chronic physical aggression and epigenetics. *Epigenomics.* 2010;2:495–499.
25. De Jong TR, Beiderbeck DI, Neumann ID. Measuring virgin female aggression in the female intruder test (FIT): effects of oxytocin, estrous cycle, and anxiety. *PLoS ONE.* 2014;9(3):e91701.
26. Buwalda B, Scholte J, de Boer SF, Coppens CM, Koolhaas JM. The acute glucocorticoid stress response does not differentiate between rewarding and aversive social stimuli in rats. *Horm Behav.* 2012;61 (2):218–226.
27. Couppis MH, Kennedy CH. The rewarding effect of aggression is reduced by nucleus accumbens dopamine receptor antagonism in mice. *Psychopharmacology.* 2008;197:449–456.
28. Schuurman T. Hormonal correlates of agonistic behavior in adult male rats. In: McConnell PS, Boer GJ, Romijn HJ, Van de Poll NE, Corner MA, eds. *Adaptive Capabilities of the Nervous System.* Amsterdam: Elsevier; 1981:415–420.
29. Koolhaas JM, Bartolomucci A, Buwalda B, et al. Stress revisited: a critical evaluation of the stress concept. *Neurosci Biobehav Rev.* 2011;35(5):1291–1301.
30. Sgoifo A, Carnevali L, Grippo AJ. The socially stressed heart. Insights from studies in rodents. *Neurosci Biobehav Rev.* 2014;39:51–60.
31. Koolhaas JM, Meerlo P, De Boer SF, Strubbe JH, Bohus B. The temporal dynamics of the stress response. *Neurosci Biobehav Rev.* 1997;21 (6):775–782.
32. Gleason ED, Fuxjager MJ, Oyegbile TO, Marler CA. Testosterone release and social context: when it occurs and why. *Front Neuroendocrinol.* 2009;30(4):460–469.
33. Trainor BC, Bird IM, Marler CA. Opposing hormonal mechanisms of aggression revealed through short-lived testosterone manipulations and multiple winning experiences. *Horm Behav.* 2004;45(2):115–121.
34. Antunes RA, Oliveira RF. Hormonal anticipation of territorial challenges in cichlid fish. *Proc Natl Acad Sci U S A.* 2009;106 (37):15985–15989.
35. Fuxjager MJ, Forbes-Lorman RM, Coss DJ, Auger CJ, Auger AP, Marler CA. Winning territorial disputes selectively enhances androgen sensitivity in neural pathways related to motivation and social aggression. *Proc Natl Acad Sci U S A.* 2010;107 (27):12393–12398.
36. McEwen BS, McKittrick CR, Tamashiro KL, Sakai RR. The brain on stress: insight from studies using the visible burrow system. *Physiol Behav.* 2015;146:47–56.
37. Lemaire V, Mormède P. Telemetered recording of blood pressure and heart rate in different strains of rats during chronic social stress. *Physiol Behav.* 1995;58(6):1181–1188.
38. Koolhaas JM, De Boer SF, Coppens CM, Buwalda B. Neuroendocrinology of coping styles: towards understanding the biology of individual variation. *Front Neuroendocrinol.* 2010;31(3):307–321.
39. De Boer SF, van der Vegt BJ, Koolhaas JM. Individual variation in aggression of feral rodent strains: a standard for the genetics of aggression and violence? *Behav Genet.* 2003;33(5):481–497.
40. Haller J, Kruk MR. Normal and abnormal aggression: human disorders and novel laboratory models. *Neurosci Biobehav Rev.* 2006;30 (3):292–303.
41. Meerlo P, Sgoifo A, De Boer SF, Koolhaas JM. Long-lasting consequences of a social conflict in rats: behavior during the interaction predicts subsequent changes in daily rhythms of heart rate, temperature, and activity. *Behav Neurosci.* 1999;113(6):1283–1290.
42. Walker FR, Masters LM, Dielenberg RA, Day TA. Coping with defeat: acute glucocorticoid and forebrain responses to social defeat vary with defeat episode behaviour. *Neuroscience.* 2009;162 (2):244–253.
43. Wood SK, Walker HE, Valentino RJ, Bhatnagar S. Individual differences in reactivity to social stress predict susceptibility and resilience to a depressive phenotype: role of corticotropin-releasing factor. *Endocrinology.* 2010;151(4):1795–1805.
44. Lore R, Flannely K. Rat societies. *Sci Am.* 1977;236:106–116.

CHAPTER

37

The Amygdala and Fear

G.M. Goodwin[1], R. Norbury[1,2]

[1]University Department of Psychiatry, University of Oxford, Warneford Hospital, Oxford, UK

[2]Department of Psychology, University of Roehampton, London, UK

OUTLINE

Abstract

The amygdala comprises multiple subnuclei receiving input, directly or indirectly from all sensory systems. Major reciprocal connections with the insula are relevant to interoception and emotional awareness in man.

Lesion studies demonstrate that the amygdala subserves the classical paradigm of fear conditioning, in animals and man. Modern neuroscience promises a detailed understanding of mechanisms in animals. In man, rare lesions of the amygdala are associated with a loss of both the subjective experience of fear and the capacity to detect expressions of fear by others. In addition, there is impairment of social cognition and decision-making.

Brain imaging in man has confirmed that fear may be registered subliminally and independently of other nonemotional signals: thus a fearful expression is processed independently of facial recognition. Prominent fearful experience in a psychiatric disorder necessarily implicates the amygdala. Drugs that increase serotonin in the brain have early actions in reducing neuronal responses in the amygdala.

AMYGDALA ANATOMY

The amygdala is an almond-shaped structure located deep within the temporal lobe of the higher animals. It was first identified by Burdach in the early nineteenth century, who described a group of cells, or nuclei, now referred to as the basolateral complex. Subsequently, however, the amygdala has been shown to be both more complex and more extended, comprising more than 12 subnuclei. In the rat, current nomenclature divides the amygdala nuclei into three main groups: (1) the basolateral complex, which includes the lateral nucleus, the basal nucleus, and accessory basal nucleus; (2) the cortical nucleus, which includes the cortical nuclei and the lateral olfactory tract; and (3) the centromedial nucleus comprising the medial and central nuclei.

The amygdala receives input from all sensory systems: olfactory, somatosensory, gustatory and visceral, auditory, and visual. Olfactory inputs arise at the olfactory bulb and project to the lateral olfactory tract. Somatosensory inputs pass via the parietal insular cortex in the parietal lobe, and via thalamic nuclei to the lateral, basal, and central nuclei. Primary gustatory and visceral sensory areas project to the basal nucleus and central nucleus. In contrast, auditory and visual information, thought to be important in fear conditioning, arise from association areas rather than primary sensory cortex. The major reciprocal connections with the insula are of growing interest because of this structure's proposed key role in mediating the relationship between interoception and emotional awareness in man.[1]

The amygdala has widespread efferent connections to cortical, hypothalamic, and brain stem regions. The basolateral complex projects to the medial temporal lobe memory system (e.g., hippocampus and perirhinal cortex) and the basal nucleus has a major projection to

Stress: Concepts, Cognition, Emotion, and Behavior
http://dx.doi.org/10.1016/B978-0-12-800951-2.00037-6

prefrontal cortex, nucleus accumbens, and the thalamus. Thus, the anatomy of the amygdala is consistent with its role in fear processing. The amygdaloid complex receives inputs from all sensory modalities and activates brain regions important to measurable neurobehavioral correlates of fear. How it works offers an insight into the nature of emotion in man and animals.

KEY POINTS

- The amygdala is the key structure in the processing of emotional experience and emotional learning.
- Evidence for its localized role comes from lesion studies in animals and man. Its connectivity and dynamic role in behavior are shown by increasingly sophisticated physiological studies in animals and functional imaging in man.
- The amygdala is now implicated in a range of behaviors requiring the attribution of value to stimuli in the environment, so affecting social cognition and decision-making.
- The role of the amygdala in psychopathology is of emerging interest for understanding both anxiety disorders and their treatment with drugs or psychotherapy.

THE AMYGDALA AND FEAR CONDITIONING IN ANIMALS

Much of the scientific interest in the amygdala stems from its established role in fear conditioning. We define fear as an unpleasant, powerful emotion caused by anticipation or awareness of potential or actual danger.[2] Classical Pavlovian fear conditioning is a type of emotional learning originally studied in animals: an emotionally neutral conditioned stimulus (CS), often a tone, is presented in conjunction with an aversive unconditioned stimulus (US), typically a small electric shock to the foot of the animal. After one or more pairings, the emotionally neutral stimulus (CS) is able to elicit a constellation of species-specific responses that are characteristic of fear (e.g., freezing or escape behavior), autonomic responses (elevated heart rate and blood pressure), potentiated acoustic startle to aversive acoustic stimuli, and increased neuroendocrine responses (release of stress hormones). Fear conditioning therefore allows new or learned threats to activate neural programs for responding to threat that have been long established in evolution.

Numerous studies have demonstrated that lesions to the amygdala impair the acquisition and expression of conditioned fear in rats. As already described, the basolateral complex of the amygdala is a substrate for sensory convergence from both cortical and subcortical areas, and is considered a putative locus for CS-US association during fear conditioning. Thus, its cells encode this emotional learning. By contrast, the central nucleus of the amygdala projects to brain regions implicated in the generation of fear responses such as the hypothalamus: it may therefore act as a common output pathway for the generation of fear conditioned responses. Consistent with this hypothesis, lesions to either the basolateral or central nucleus of the amygdala impair both the acquisition and expression of conditioned fear.

The amygdala pathways involved in fear conditioning have also been traditionally studied using electrophysiology but understanding has been greatly expanded by the incorporation of optogenetics, pharmacogenetics, and viral-based tract tracing that take advantage of gene-targeting[3,4] (see Figure 1). Thus, the amygdala is able to both integrate and associate sensory information and influence motor and physiological responses associated with fear conditioning.[5]

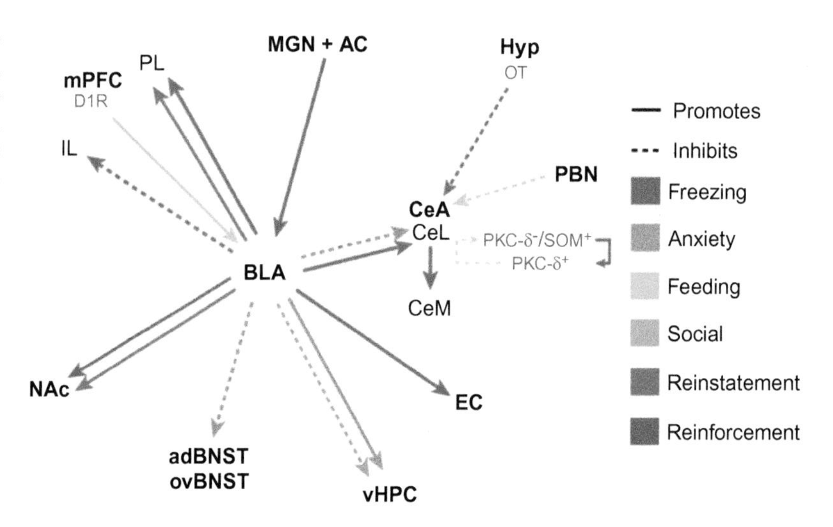

FIGURE 1 Projection-specific effects as shown by optogenetic or pharmacogenetic manipulation. The solid or dotted lines indicate the promotion or inhibition of certain behaviors. The basolateral complex of the amygdala (BLA) encompasses the lateral and basal nuclei. Specific cell types are shown in pink. For simplicity, projections that are anatomically or electrophysiologically defined but have not been shown to have a causal relationship with behavior are omitted. AC, auditory cortex; adBNST, anterodorsal bed nucleus of the stria terminalis; CeA, central nucleus of the amygdala; CeL, lateral CeA; CeM, medial CeA; D1R, dopamine 1 receptor; EC, entorhinal cortex; Hyp, hypothalamus; IL, infralimbic; MGN, medial geniculate nucleus; mPFC, medial prefrontal cortex; NAc, nucleus accumbens; OT, oxytocin; ovBNST, oval nucleus of the BNST; PBN, parabrachial nucleus; PKC, protein kinase C; PL, prelimbic; SOM, somatostatin; vHPC, ventral hippocampus.[5]

THE AMYGDALA AND FEAR IN HUMANS: EVIDENCE FROM LESION STUDIES

Damage to the amygdala, or to areas of the temporal lobe that include the amygdala also produces deficits in fear processing in humans. Thus a patient (S.M.) with a rare congenital lipoid storage disease causing bilateral degeneration of the amygdala, when exposed to live snakes and spiders, touring a haunted house, and watching emotionally disturbing films demonstrated no observable or reported anxiety experiences.[6] Compared to normal control subjects, S.M. showed no evidence of fear conditioning (as measured by galvanic skin response with either visual or auditory conditioned stimuli and a loud noise as the US). However, S.M.'s recall of the events associated with fear conditioning was fully preserved. These data support the hypothesis that the amygdaloid complex plays a key role in the acquisition of Pavlovian fear conditioning. It also implies that the hippocampus (intact in S.M.) can still mediate acquisition of declarative knowledge or memory of the conditioning contingencies.

In a further study, bilateral amygdala damage impaired recognition of fearful, sad, but not happy facial expressions.[7] Such patients have no difficulty in recognizing people by their faces or in learning the identity of new faces. Bilateral amygdala damage has also been associated with impairment in recognition of vocalized expressions of threat. Thus, impairment of the perception of expressive signs of fear in patients with amygdala damage extends beyond its facial expression. These results suggest that the human amygdala is directly involved in the experience by self and the recognition in others of negative emotion.

Finally, the startle response has been studied in patients with unilateral lesions involving the amygdala and age-matched controls. The startle response is an innate reflex observed in nearly all animals, including humans. It is manifest as an involuntary motor response to any unexpected noise, touch, or sight. In studies of emotion, it has been used to measure the aversiveness of emotive stimuli. Control subjects display a well-documented effect of aversive stimuli; potentiation of startle magnitude compared with neutral stimuli. By contrast, in patients with right or left amygdala lesions, no startle potentiation was observed in response to aversive stimuli evoking disgust and fear.[8] Thus, studies of patients with amygdala damage are consistent in suggesting a key role in the perception and production of negative emotions, particularly fear.

Further observation of patients with amygdala lesions highlight the importance of emotional signals for more complex behavior, already implied by the animal findings summarized in Figure 1. This would be expected if the amygdala helps support the function of assigning emotional value to a stimulus. Therefore, social interactions, social judgement, and decision-making have all been shown to be affected by amygdala damage. There appears to be an important functional reciprocal connection with ventromedial frontal cortex and lesions of either structure tend to have related effects, for example, leading to impulsive decision-making. There may be an interaction between sex and laterality of amygdala functioning. Unilateral damage to the right amygdala is associated with greater deficits in decision-making and social behavior in men, while unilateral damage to the left amygdala seems to be more detrimental for women.[9]

THE AMYGDALA AND FEAR IN HUMANS: EVIDENCE FROM NEUROIMAGING STUDIES

From the early 1990s, researchers have begun to explore the role of the amygdala in fear processing using noninvasive neuroimaging. These powerful techniques include positron emission tomography and functional magnetic resonance imaging (fMRI): they allow the examination of regional patterns of brain activation in the living human brain with reasonable spatial resolution, but limited temporal resolution (in the order of seconds). Both techniques estimate the dynamic vascular responses yoked with neural activity. So far, amygdala activity has been assessed predominantly using two basic paradigms: fear conditioning and presentation of emotional facial expressions.

fMRI experiments have demonstrated increased amygdala activity during both early acquisition and early extinction of fear conditioning to a visual stimulus.[10] Indeed, it has been suggested that the amygdala's role is limited to early conditioning or early extinction, when response contingencies change. That is, the amygdala is particularly important for forming new associations and relationships—emotional learning and unlearning.

The amygdala may also be responsible for generating coordinated reflexive behavioral responses to highly aversive stimuli. The animal literature shows that the amygdala, specifically the central nucleus, generates conditioned autonomic, behavioral, and endocrine responses in acute stress paradigms. Conditioning stimuli, which may be interpreted as changing the animal's emotional set, increase the response in startle paradigms. Clearly, this kind of response may be quite primitive but lesioning experiments have demonstrated definite mediation by the amygdala. It is usually assumed the amygdala may also be involved in the selection of more purposive motor/behavioral responses in response to more subtle aversive conditions.

The second major neuroimaging protocol for assaying amygdala activity is the presentation of faces expressing different emotions. It is difficult to overemphasize the importance of facial expressions in social communication. Facial expressions act at a number of levels to signal important information; expressions of disgust enable the avoidance of ingestion of harmful substances, and fearful facial expressions are able to rapidly communicate the presence of imminent threat. More subtly, cues from the faces of others are continuously informing us of our social impact and acceptability, and we in turn communicate our feelings and intentions through our own facial gestures and expressions.

It should be no surprise, therefore, that facial expressions have provided a useful experimental tool to measure fear-related amygdala activity. Many studies have demonstrated that the amygdala is preferentially activated during presentation of fearful facial expressions. Further, even when subjects are unaware of *seeing* a fearful face, the amygdala is still activated.[11] This has been shown in a backward masking paradigm where an initial fearful face is presented for a very short duration (17-33 ms) then immediately replaced by a neutral face presented for a longer duration (~200 ms). In this situation, the initial image is subliminal: subjects will report that they have seen only the neutral expression. Activation of the amygdala is also seen when attention is directed away from the fearful face. Such findings suggest that the extraction of potentially threatening information within the amygdala may not be sufficient for conscious face perception, but it may well be necessary. The amygdala may act to direct attention toward emotionally salient events that are ambiguous or require further processing. Indeed, anatomical studies in rats suggest that sensory information travels to the amygdala via two distinct pathways: a short, rapid thalamic route, and a longer cortical pathway. It is proposed that the short thalamic pathway rapidly prepares the animal for a potentially aversive encounter independent of conscious processing, which occurs later via the slower cortical route. A similar two-way route to the amygdala has yet to be anatomically defined in man, however, evidence

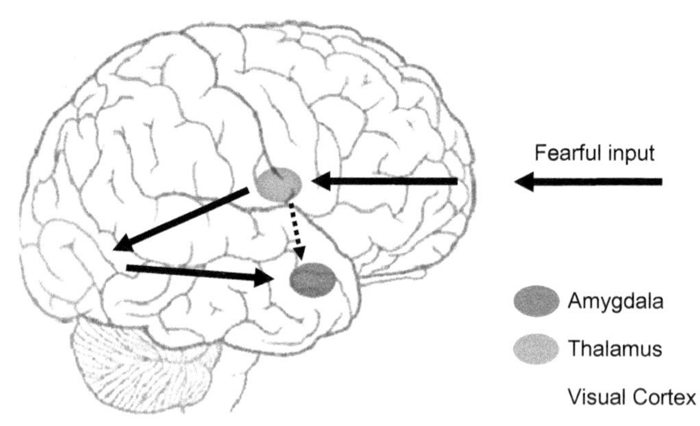

FIGURE 2 Two-pathway hypothesis for amygdala activation to fear. The short, rapid thalamic route is shown with the dashed arrow, the longer cortical pathway with bold arrows. (For clarity subcortical structures are shown overlaid on brain surface).

from a number of well-designed fMRI experiments point to the existence of these two pathways (Figure 2). First, as already described, the amygdala responds to fearful faces even when presented outside of conscious awareness. Second, in an extension of the masking paradigm, Whalen and colleagues[12] presented degraded versions of fearful and happy expressions by displaying only the eye whites immediately masked by a complete image of a neutral expression. Fearful faces are typically associated with enlarged exposure of the sclera and it was observed that the crude information provided by the "wide-eyes" of fearful expressions was sufficient to stimulate the amygdala. Analogously, Vuilleumier used fMRI to compare amygdala and cortical responses to rapidly presented (200 ms) fearful and neutral faces that had been filtered to extract either low spatial frequency information (giving a blurry image) or high spatial frequency (finely detailed, sharp image) (Figure 3)[13]. As detected by fMRI, the finely detailed images elicited a greater response in a region of the cortex called the fusiform face area—a brain region widely implicated in the conscious

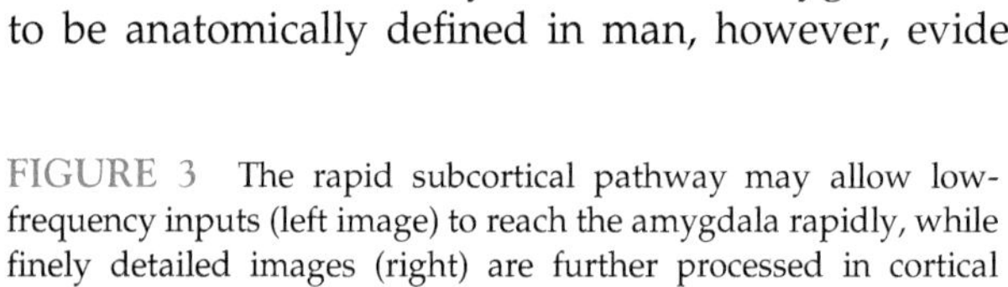
FIGURE 3 The rapid subcortical pathway may allow low-frequency inputs (left image) to reach the amygdala rapidly, while finely detailed images (right) are further processed in cortical regions (e.g., fusiform gyrus).

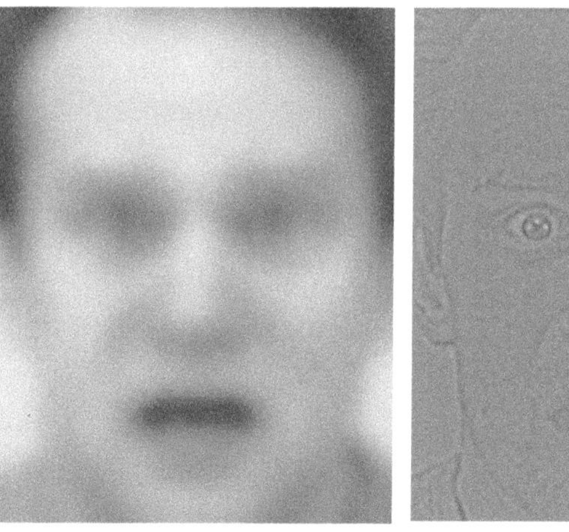
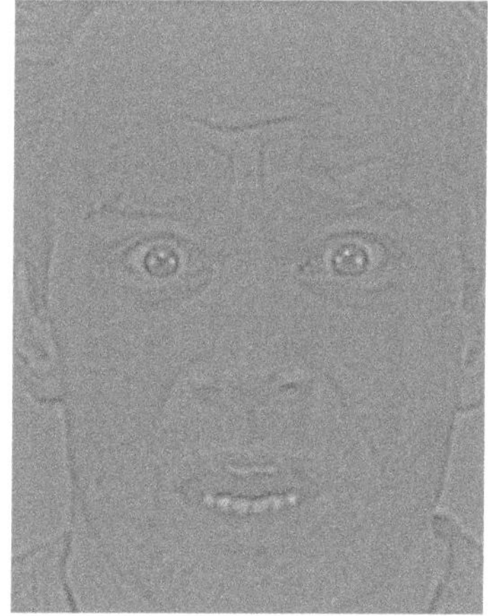

recognition of faces and facial expressions. The amygdala was relatively unresponsive to these images even if they were fearful in nature. Low spatial frequency fearful faces, however, produced a robust response in the amygdala. Although these findings could imply the existence of a two-pathway route to the amygdala in man, with dissociation between neural responses in amygdala and fusiform gyrus across different spatial frequency ranges for face stimuli, they do not prove it. However, they suggest that the amygdala does not require very complex configural stimuli, such as presented by a complete image of a face, in order to respond to emotion. It is obviously possible that crude fragmentary information related to emotionally salient signals could be rapidly communicated via a subcortical route. Extreme facial expressions of emotion may be well adapted to attract immediate attention via this mechanism.

THE AMYGDALA AND PSYCHOPATHOLOGY

Given the involvement of the amygdala in fear processing, there is increased interest in its role in anxiety disorders such as generalized anxiety disorder (GAD), panic disorder, social anxiety disorder, and posttraumatic stress disorder. Anxiety disorders are associated with increased processing of threat-related stimuli and increased attention to threats in the environment. Broadly speaking, the mechanisms to explain this behavior can be formulated either as increased sensitivity to potential threats or impaired discrimination of threatening from neutral stimuli (or both). The content specificity of threat-related attentional bias in anxiety disorders is toward disorder-congruent compared with disorder-incongruent threat stimuli. The effect appeared to be independent of age, type of anxiety disorder, and task detail but the effect size was small ($d = 0.28, p < 0.0001$).[14] The amygdala and its connections provide the obvious target for neurobiological explanation of cognitive mechanisms.

Neuroimaging can be used to dissect the mechanisms at a systems level in man. The emerging pattern is to find some effects in common across anxiety disorders in the responses to a range of experimental paradigms. However, in addition, other disorder-specific details are often apparent. For example, greater right amygdala activation when matching fearful *and* happy facial expressions was associated with greater negative affectivity across three different anxiety disorder groups (GAD, panic, and social anxiety disorder) compared with controls. However, the panic disorder group alone showed increased posterior insula activation.[15] Other data support the existence of a common abnormality in anxiety and depression in the ventral cingulate and the amygdala, but with disorder-specific compensation during implicit regulation of emotional processing apparently through engagement of cognitive control circuitry in the depressed group.[16] The complexity of the likely mechanisms involved in different disorders remains challenging.

fMRI can also be used to investigate how pharmacological treatment for anxiety and depression modulates amygdala activity.[17] A number of studies have demonstrated that the increased amygdala response to negative facial expressions seen in depression is reduced following effective treatment with serotonergic drugs used to treat anxiety and depression. While these results suggest an important role for the amygdala in recovery from depression, it remains unclear if this normalization of amygdala response to fearful faces with time is a direct action of the drug or rather reflects current symptom state (patients were scanned when depressed compared with recovered). Work from our laboratory, in healthy volunteers, has demonstrated that short-term administration (7 days) of either serotonergic (e.g., citalopram)[18] or noradrenergic (e.g., reboxetine)[19] antidepressant drugs reduces the amygdala response to facial expressions of fear as compared to placebo controls. Importantly, these effects on the amygdala were observed in the absence of changes in mood. These results suggest that antidepressants have rapid, direct effects on the amygdala. Moreover, if similar effects are seen in clinical populations, drug modulation of amygdala activity in response to negative stimuli may be a key component in recovery from depression and anxiety. Studies to explore the role of amygdala in psychological treatments have also begun.[20] Direct measurement of changes in amygdala responsiveness may provide an important assay for the development of novel drug and psychological treatments for anxiety and depression.

References

1. Critchley H, Seth A. Will studies of macaque insula reveal the neural mechanisms of self-awareness? *Neuron*. 2012;74(3):423–426.
2. LeDoux JE. Coming to terms with fear. *Proc Natl Acad Sci U S A*. 2014;111(8):2871–2878.
3. Cho J-H, Deisseroth K, Bolshakov VY. Synaptic encoding of fear extinction in mPFC-amygdala circuits. *Neuron*. 2013;80 (6):1491–1507.
4. Tye KM, Prakash R, Kim S-Y, et al. Amygdala circuitry mediating reversible and bidirectional control of anxiety. *Nature*. 2011;471 (7338):358–362.
5. Janak PH, Tye KM. From circuits to behaviour in the amygdala. *Nature*. 2015;517(7534):284–292.
6. Feinstein JS, Adolphs R, Damasio A, Tranel D. The human amygdala and the induction and experience of fear. *Curr Biol*. 2011;21 (1):34–38.
7. Adolphs R, Tranel D. Impaired judgments of sadness but not happiness following bilateral amygdala damage. *J Cogn Neurosci*. 2004;16(3):453–462.
8. Buchanan TW, Tranel D, Adolphs R. Anteromedial temporal lobe damage blocks startle modulation by fear and disgust. *Behav Neurosci*. 2004;118(2):429–437.
9. Gupta R, Koscik TR, Bechara A, Tranel D. The amygdala and decision-making. *Neuropsychologia*. 2011;49(4):760–766.

10. LaBar KS, Gatenby JC, Gore JC, LeDoux JE, Phelps EA. Human amygdala activation during conditioned fear acquisition and extinction: a mixed-trial fMRI study. *Neuron*. 1998;20(5):937–945.
11. Whalen PJ, Rauch SL, Etcoff NL, McInerney SC, Lee MB, Jenike MA. Masked presentations of emotional facial expressions modulate amygdala activity without explicit knowledge. *J Neurosci*. 1998;18 (1):411–418.
12. Whalen PJ, Kagan J, Cook RG, et al. Human amygdala responsivity to masked fearful eye whites. *Science*. 2004;306(5704):2061.
13. Vuilleumier P, Armony JL, Driver J, Dolan RJ. Effects of attention and emotion on face processing in the human brain: an event-related fMRI study. *Neuron*. 2001;30(3):829–841.
14. Pergamin-Hight L, Naim R, Bakermans-Kranenburg MJ, van IMH, Bar-Haim Y. Content specificity of attention bias to threat in anxiety disorders: a meta-analysis. *Clin Psychol Rev*. 2015;35:10–18.
15. Fonzo GA, Ramsawh HJ, Flagan TM, et al. Common and disorder-specific neural responses to emotional faces in generalised anxiety, social anxiety and panic disorders. *Br J Psychiatry*. 2015;206(3):206–215.
16. Etkin A, Schatzberg AF. Common abnormalities and disorder-specific compensation during implicit regulation of emotional processing in generalized anxiety and major depressive disorders. *Am J Psychiatry*. 2011;168(9):968–978.
17. Harmer CJ, Goodwin GM, Cowen PJ. Why do antidepressants take so long to work? A cognitive neuropsychological model of antidepressant drug action. *Br J Psychiatry*. 2009;195(2):102–108.
18. Harmer CJ, Mackay CE, Reid CB, Cowen PJ, Goodwin GM. Antidepressant drug treatment modifies the neural processing of nonconscious threat cues. *Biol Psychiatry*. 2006;59(9):816–820.
19. Norbury R, Mackay CE, Cowen PJ, Goodwin GM, Harmer CJ. Short-term antidepressant treatment and facial processing. Functional magnetic resonance imaging study. *Br J Psychiatry*. 2007;190:531–532.
20. Fonzo GA, Ramsawh HJ, Flagan TM, et al. Cognitive-behavioral therapy for generalized anxiety disorder is associated with attenuation of limbic activation to threat-related facial emotions. *J Affect Disord*. 2014;169:76–85.

CHAPTER

38

Aging and Psychological Stress

E. Zsoldos, K.P. Ebmeier

University of Oxford Department of Psychiatry, Warneford Hospital, Oxford, UK

OUTLINE

Abstract

The conceptual link between aging and psychological stress is the construct "allostatic load"—that is, the notion that a repeated and cumulative disturbance of homeostasis may lead to certain syndromes that are triggered by a dysregulation of the chronic stress-response. We summarize the markers of this hypothetical process at primary, secondary, and tertiary stages, with particular attention on brain mechanisms that are in this pathway or are even instrumental to such dysregulation and its pathological consequences. As cumulative stress and allostasis are highly correlated with chronological age, we present an example for the statistical isolation of such mechanisms from the generic effect of age.

INTRODUCTION

The repeated occurrence of stressful situations, leading to "allostatic load (AL)," are said to result in cumulative age-dependent illness.[1] As there is no strict relationship between chronological and biological age, there is no general agreement on the age at which a person becomes old.[2] Most western countries define "older age" as over 60-65 years, making workers eligible for retirement and pension benefits. It is estimated that globally the percentage of 60+-year olds is going to rise from 12% in 2014 to 21% in 2050.[3] Since the pension-paying younger generation is decreasing, it is likely that retirement ages will soon have to extend into the later 60s. The older generation will thus represent a substantial section not only of the general, but also of the working population.[4,5] While chronic diseases are generally associated with chronological age, the progressive loss of ability to deal with stress and the development of age-related distress occur at individually different rates. Psychological distress seems to promote earlier onset of age-related diseases.[6] Chronic psychosocial stress and consequent physiological dysregulations have been described as "catalysts of accelerated aging and agitators of disease trajectories."[7] The AL model of chronic stress assumes the prominent role of certain brain structures, such as the hippocampus, amygdala, and prefrontal cortex as *regulators*—and sympathetic-adrenal-medullary axis release of catecholamines and the hypothalamic-pituitary-adrenal (HPA) axis secretion of glucocorticoids as *effectors*.[7] Key elements of the model are that extraordinary (stressful) situations require temporary adaptations of the whole body ("allostasis"), while excessive adaptations to AL can lead to peripheral organ damage that may also result

Stress: Concepts, Cognition, Emotion, and Behavior
http://dx.doi.org/10.1016/B978-0-12-800951-2.00039-X

in plastic changes in certain brain structures (e.g., the hippocampus) that perpetuate maladaptive responses. Individual variability in any of these mechanisms will affect the vulnerability or resilience against stress-related diseases.[8] The model has been refined to include "Type I allostatic overload," which occurs if "energy demands exceed energy income and what can be mobilized from body stores" (maybe more common in developing countries).[9] "Type II allostatic overload" occurs if "energy demands are not exceeded and the organism continues to take in or store as much or even more energy than it needs. This may be a result of stress-related food consumption, choice of a fat-rich diet, or metabolic imbalances (prediabetic state) that favors fat deposition."[9]

KEY POINTS

- The effect of cumulative stress has been conceptualized as allostatic load.
- Primary, secondary, and tertiary markers of allostatic load have been described.
- Brain structures play a central role in the regulation of the chronic stress response.
- They are not just important effectors of the stress-response system, but are also its targets, so that chronic dysregulation can result in damage to certain brain structures.
- In order to identify "treatable" effects of aging, the mechanisms that lead to stress-related brain damage need to be identified.

REGULATION OF STRESS-RESPONSE (ALLOSTASIS) AND AGE-RELATED DISEASE (ALLOSTATIC LOAD; AL)

In everyday life and popular perception, the term "stress" has been frequently used to describe both an external environment and an internal state or feeling. This confusion is most likely due to the lack of clear academic definition of the concept. The endocrinologist János Selye, who was the first to coin the term, described *stress* as a nonspecific response of the body to a *stressor*: a real or imagined stimulus that exerts demands on the organism, thus causing stress.[10,11] Stressors can have either short- or long-term effects. The former comprise acute, time-limited events, while the latter may have long-term effects that subside over time. Examples include certain life events,[12] changes that are chronic and require the restructuring of social roles or identity (e.g., becoming a caregiver)[13] or are traumatic with lasting psychological and physiological effects (e.g., combat stress, sexual assault).[14–17] Stress can be negative (distress) or positive, the latter increasing performance ("eustress"). Both distress and eustress may pose a threat to the organism's internal steady state, its homeostasis.[9,18,19]

The amygdala evaluates stressors and relays relevant information to the hypothalamus, which delivers the stress response. This utilizes two major pathways, the *autonomic* component of the peripheral *nervous system* (*ANS*) and the *HPA axis* of the central nervous system (Figure 1).[8] The ANS supplies the internal organs of the body, hence the "visceral nervous system." It controls involuntary body functions, such as breathing, heartbeat,

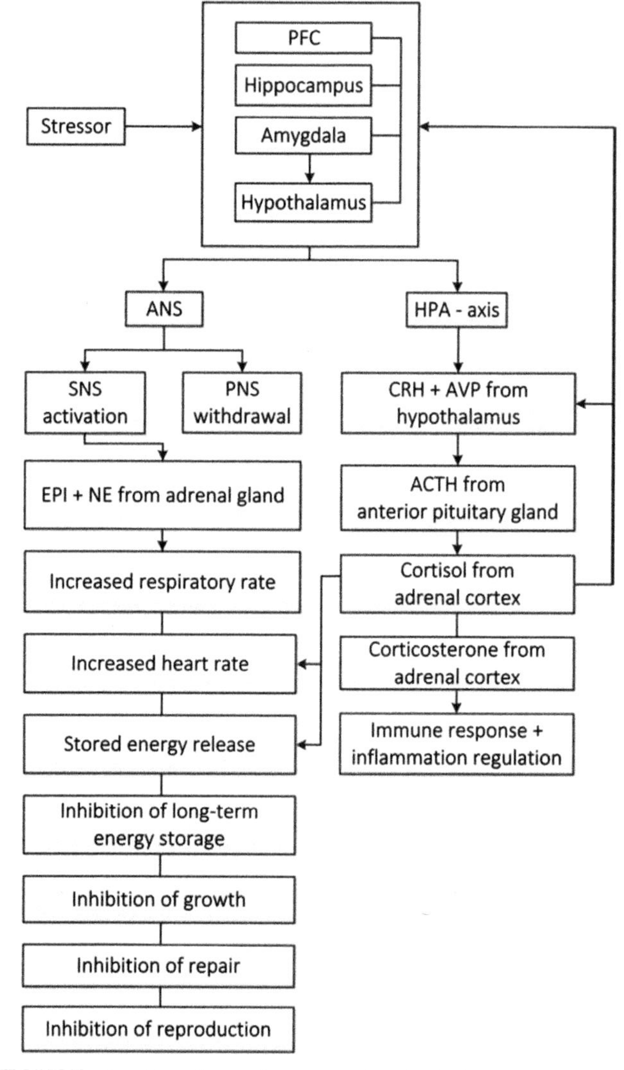

FIGURE 1 Overview of systems involved in stress response (allostasis). PFC, prefrontal cortex; ANS, autonomic nervous system; SNS, sympathetic nervous system; PNS, parasympathetic nervous system; EPI, epinephrine; NE, norepinephrine; HPA, hypothalamic-pituitary-adrenal; CRH, corticotropin-releasing hormone; AVP, arginine vasopressin; ACTH, adrenocorticotropic hormone. Adapted from Refs. 8,20–22.

blood pressure, the dilation and constriction of blood vessels and bronchioles, and certain visceral reflex actions. It responds to stressors via the catecholamine hormones and neurotransmitters epinephrine and norepinephrine, secreted by the adrenal medulla, and is associated with the fight-or-flight response.[23,24] The ANS consists of a sympathetic and parasympathetic branch. The *sympathetic ANS* is responsible for energizing the body for immediate mobilization through glucose and fat release, as well as release of proteins from nonexercising muscle. It inhibits vegetative functions, such as long-term energy storage through suppressing insulin levels, as well as growth, repair, and reproduction. The parasympathetic ANS generally counteracts sympathetic ANS activity.

The hypothalamus also activates the HPA-axis (Figure 1), which is the slower *neuroendocrine* pathway and culminates in the production of glucocorticoids by the adrenal cortex. Neurons in the medial parvocellular region of the paraventricular nucleus of the hypothalamus produce and release corticotropin-releasing hormone (CRH) and arginine vasopressin (AVP).[8] CRH and AVP in turn trigger the release of adrenocorticotropic hormone (ACTH) from the anterior pituitary gland, which acts on the adrenal cortex to release the glucocorticoids hydrocortisone (commonly known as cortisol) and corticosterone. Cortisol blocks the storage and aids the conversion of fats, proteins, and carbohydrates to energy, and it regulates cardiovascular function. Corticosterone regulates immune responses and acutely suppresses inflammation.[20,25,26] Once a perceived stressor has subsided, feedback loops are triggered at various levels of the system in order to return it to baseline homeostasis. Glucocorticoid receptors are expressed throughout the brain and HPA-axis, and have the ability to regulate gene expression; they can thus have potentially long-lasting effects on the functioning of brain regions that regulate their release.[8]

Allostasis is defined as the process of maintaining homeostasis through the adaptive change of the organism's internal environment to meet perceived and anticipated demands.[27] It mainly affects systems involved in the stress response.[27,28] Components of the ANS and HPA-axis change their physiological response within a normal range of values in response to a stressor. Mediators that participate in allostasis link together in a nonlinear regulatory network; they have the ability to regulate the activity of other mediators. Chronic allostasis, however, can be damaging, leading to AL and overload, that is, wear and tear.[9] This can result in diseases affecting one or many parts of the body that are involved in allostasis and the stress response.[1,9,29] McEwen argues that AL occurs when responses deviate from the physiological range, because of (1) repeated exposure to multiple stressors, (2) lack of response adaptation, (3) prolonged allostatic response even after the stressor subsided, or (4) inadequate response.[9,29] It is important to note that both an under- and overactive ANS and HPA-axis can be a feature of AL.[26]

PRIMARY AND SECONDARY MARKERS OF AL AND STRESS RESPONSE

Stress hormones and their antagonists, in conjunction with pro- and anti-inflammatory cytokines, such as interleukine-6 (IL-6) and tumor necrosis factor-alpha (TNF-α), form the *primary markers* ("mediators," "biomarkers") of AL (see Table 1).[9,14] While their increase is adaptive in the short run, their prolonged secretion can damage both brain and body.[29]

Cardiovascular, metabolic, and inflammatory systems shift their operational ranges in order to sustain AL, in the process reaching subclinical levels in the intermediate stage, and are referred to as *secondary markers* ("outcomes," "biomarkers," see Tables 1 and 2).[7] They serve as risk indices for stress-related illness. Their composite scores are often better predictors of physical, mental and cognitive health, and mortality risk than individual components of AL.[32] *Secondary markers* are often associated with social status and societal structure, usually defined by family's socioeconomic status (SES) and neighborhood characteristics in childhood, and by income, household income, and occupational position in adulthood, and capture the long-term effect of chronic stressors, which have a higher prevalence in low SES (see Tables 2 and 3). They thus also serve as functional biomarkers of biological aging, indicating the biological pathway from life stresses to Type II allostatic overload and ill health in epidemiological research.[34]

Disease trajectories, such as toward hypertension, left ventricular hypertrophy, irregular heartbeat, atherosclerosis, myocardial infarct, stroke, myocardial ischemia, angina pectoris, Type 2 diabetes, metabolic syndrome, autoimmune diseases, chronic fatigue syndrome, and major depression have been claimed to be consequences of AL.[20,27,35] Best evidence for the effect of stressors on health (see Table 2) exists for *cardiovascular health*, including hypertension, and disease.[35] High blood pressure reactivity to psychological stress ("white coat syndrome"), and cortisol response to experimentally induced stress, as well as depressive symptoms are associated with future increase in blood pressure. Those who have high blood pressure reactivity in response to a stressor also show increased neural activation in response to a stress task in brain areas controlling the cardiovascular system. Furthermore, lifetime hypertension predicts brain structure 28 years later. Job strain, effort-reward imbalance, job insecurity, unfairness, overtime work, self-reported adverse effects of stress on health, cortisol secretion patterns, depressive symptoms, and gray

TABLE 1 Overview of Primary and Secondary Markers of the Stress Response and Allostatic Load in the Whitehall II Study[30]

		Primary markers	Secondary markers		
			Framingham Stroke Risk Score[31]	Metabolic syndrome (ATP III)	Allostatic load index
Cardiovascular	SBP		X	X	X
	DBP			X	X
	Prior CVD		X		
	Atrial fibrillation		X		
	Left ventricular hypertrophy		X		N/A
	Hypertensive medication		X		
Metabolic	Diabetes mellitus		X		
	Diabetes medication		X		
	Fasting glucose			X	X
	Waist circumference			X	
	Weight				X
	Fat mass				X
	Percent body fat				X
	Serum triglycerides			X	X
	Low HDL cholesterol			X	X
	High LDL cholesterol				X
	Blood cholesterol				X
	BMI				X
	HbA1C				X
Immune	CRP	X			X
	IL-6	X			X
Other	Age		X		
	Cigarette smoking		X		
	Cortisol	X			X
	TNF-α	N/A			N/A

ATP III: Summary of the NCEP Adult Treatment Panel III Report (JAMA 2001, 285(19) 2486-97). SBP, systolic blood pressure; DBP, diastolic blood pressure; CVD, cardiovascular disease; HDL, high-density lipoprotein; LDL, low-density lipoprotein; BMI, body mass index; HbA1C, glycated hemoglobin; CRP, C-reactive protein; IL-6, interleukin-6, TNF-α, tumor necrosis factor-α.

matter volume affect cardiovascular health negatively. In turn, multimeasure cardiovascular and stroke risk scores, such as the Framingham Risk Scores,[31] predict subsequent cognitive decline, gray matter volume reduction, as well as white matter changes.

Metabolic syndrome[36] is a term coined to recognize the interconnection between components of the cardiovascular and neuroendocrine systems. It is defined by biomarkers of AL (elevated blood sugar levels, hyperglycemia, hypertension, insulin resistance, low levels of high-density lipoprotein (HDL) cholesterol, dyslipidemia, high levels of triglycerides, and abdominal obesity), at least three of which have to be above the respective clinical thresholds.[37] Organizational justice, job strain, low SES, and unfairness all predict metabolic syndrome. In combination, metabolic syndrome components predict cardiovascular disease, telomere shortening,[6] stroke, and mortality rates (Table 2). However, even if most are at a subclinical levels they can increase the risk of heart disease, cognitive decline, decline of

TABLE 2 Selected Literature on Secondary Markers of the Stress Response and Allostatic Load

Predictor/maker	Outcome/marker	Type	Association/ outcome	Author	Journal
Low SES	CVD	Cross	Yes	Adler NE et al., 1993	J Am Med Assoc 269, 3140-45
Low SES	Increased prevalence of stressors in the home	Review	Yes	McEwen BS and Tucker P, 2011	Am J Public Health 101, S131-139
Cortisol response to laboratory induced stress	Incident high BP	Long	Yes	Hamer M et al., 2012	J Clin Endocrinol Metab 97(1):E29-34
High BP to psychological stress	Future BP	Long	Modest	Caroll D et al., 2001	Psychosom Med 63(5): 737-43
Depression + age	Future BP	Long	Mixed, mostly due to age	Nabi H et al., 2011	Hypertension 57(4): 710-6
High BP to stress task	Functional neural reactivity to stress task in brain areas that control CV system	Cross	Yes	Gianaros PJ et al., 2007	Hypertension 49: 134-40
Lifetime hypertension	Gray matter atrophy and white matter hyperintensities	Long	Yes	Allan CL et al., 2015	Br J Psychiatry 206 (4):308-15
Job strain and low decision latitude	Risk of CHD	Long	Yes	Kuper H and Marmot M, 2003	J Epidemiol Comm Health 57(2):147-53
Job strain and SES	BP + stress ratings + happiness ratings	Cross	Yes	Steptoe A and Willemsen G, 2004	J Hypertens 22(5):915-20
Effort-reward imbalance	CHD + health	Long	Moderate	Kuper H et al., 2002	Occup Environ Med 59(11): 777-84
Effort-reward imbalance + job control	CHD	Long	Yes	Bosma H et al., 1998	Am J Pub Health 88(1): 68-74
Job insecurity	Incident CHD	Long	Yes	Ferrie JE et al., 2013	Atheroscl 227(1):178-81
Unfairness and job strain	Coronary events + mental and physical health	Long	Yes	De Vogli R et al., 2007	J Epidemiol Comm Health 61(6):513-8
Overtime work	Incident CHD	Long	Yes	Virtanen M et al., 2010	Eur Heart J 31 (14):1737-44
Self-reported adverse effects of stress on health	CHD or MI risk	Long	Yes	Nabi H et al., 2013	Eur Heart J 34 (34):2697-705
Cortisol secretion during the day	CVD mortality	Long	Yes	Kumari M et al., 2011	J Clin Endocrinol Metab 96(5):1478-85
Depression	Stroke and CVD	Long	Mixed (dose-response and reverse causation	Brunner EJ et al., 2014	Eur J Prev Cardiol 21(3): 340-6
Cardiovascular risk and structural brain changes	Framingham CHD Risk Score	Cross	Yes + gender effects	Rondina JM et al., 2014	Front Aging Neurosci 1(6):300-14
Framingham Stroke Risk and CVD Risk Scores	Cognitive decline	Long	Yes	Dregan A et al., 2012	Age and Ageing 42(3) :338-45
Framingham Stroke Risk and CVD Risk Scores + dementia risk score	Cognitive decline	Long	Framingham scores better predictors	Kaffashian S. et al., 2013	Neurology 80(14):1300-6
Framingham Stroke Risk Score	Gray matter decline + executive function decline	Long	Yes	Debette S et al., 2011	Neurology 77(5): 461-68
Framingham Stroke Risk Score in late-life depression	White matter integrity	Cross	Yes	Allan CL et al., 2012	Int Psychogeriatr 24(4): 524-31

Continued

TABLE 2 Selected Literature on Secondary Markers of the Stress Response and Allostatic Load—Cont'd

Predictor/maker	Outcome/marker	Type	Association/ outcome	Author	Journal
Organizational justice	Metabolic syndrome	Long	In men only	Gimeno D et al., 2010	Occup Environ Med 67 (4):256-62
Job strain	Metabolic syndrome	Long	Yes	Chandola T et al., 2006	BMJ 332(7540):521-5
SES (employment grade)	Metabolic syndrome	Cross	Central obesity + plasma fibrinogen	Brunner EJ et al., 1997	Diabetologia 40(11): 1341-9
SES (house hold wealth)	Metabolic syndrome	Cross	Yes	Perel P et al., 2006	Diabetes care 29 (12):2694-700
Unfairness	Metabolic syndrome	Long	Yes, but reduced after adjusting for covariates	De Vogli R. et al., 2007	J Psychosom Res 63(4):413-9
Metabolic syndrome	Cardiovascular disease	Meta-analysis	Yes	Galassi A et al., 2006	American J of Medicine 119(10):812-9
Metabolic syndrome	Cardiovascular disease mortality	Long	In men only	Lakka H-M et al., 2002	JAMA 288(21):2709-16
Metabolic syndrome	Telomere shortening	Review	Yes	Epel ES, 2009	Hormones 8(1):7-22
Metabolic syndrome	Stroke + mortality rate	Cross	Yes	Isomaa B et al., 2001	Diabetes Care 24(4): 683-9
Metabolic syndrome	Heart disease + physical functioning + mortality	Cross	Yes	Gardner AW et al., 2006	J Vasc Surg 43(6):1191-6
Cumulative metabolic syndrome	Poor cognition	Long	Yes	Akbaraly TN et al., 2010	Diabetes care 33(1):84-9
Metabolic syndrome	Onset of depressive symptoms	Long	Obesity + dyslipidemia	Akbaraly TN et al., 2009	Diabetes care 32(3): 499-504
Job strain and caring for a relative	Allostatic load	Long	Yes	Dich N et al., 2015	Psychosom Med [2015 May 15 Epub ahead of print]
Negative emotional response to stressful life events	Allostatic load	Long	Nonlinear	Dich N et al., 2014	Psychoneuroendocrinol 49:54-61
Allostatic load	Physical + cognitive decline	Long	Yes	Seeman TE et al., 1997	Arch Intern Med 157 (19): 2259-68
Allostatic load	Physical + cognitive function	Long	AL better predictor than individual components	Karlamangla AS et al., 2002	J Clin Epidemiol 55: 696-710
Allostatic load	CVD + mortality	Long	Yes	Seeman TE et al., 2001	Proc Natl Acad Sci U S A 98(8): 4770-75
Allostatic load	Physical + general health	Long	Yes	Read S and Grundy E, 2014	Psychosom Med 76(7): 490-496
Allostatic load	Cognitive function	Cross	Yes	Karlamangla AS et al., 2014	Neurobiol Aging 35(2): 387-394
Allostatic load	Mental health	Cross	AL better predictor than individual components	Bizik G et al., 2013	Harv Rev Psychiatry 21(6):296-313
Allostatic load reduction	Mortality risk decrease	Long	AL better predictor than individual components	Karlamangla AS et al., 2006	Psychosom Med 68(3): 500-7

AL, allostatic load; BP, blood pressure; CHD, coronary heart disease; CV, cardiovascular; CVD, cardiovascular disease; MI, myocardial infarct; SES, socioeconomic status.

TABLE 3 Structural and Functional Studies of the Brain in Allostatic Load and Overload

Examined effect	Author	Journal	Notes
Hippocampal/temporal lobe shrinkage			
MCI and AD	de Leon MJ et al., 1997	Neurobiol Aging 18, 1-11	
T2D	Gold SM et al., 2007 McEwen BS et al., 2007	Diabetologia 50, 711-19 Physiol Rev 87, 873-904	Besides its response to glucocorticoids, the hippocampus is an important target of metabolic hormones that have a variety of adaptive actions in the healthy brain, which is perturbed in metabolic disorders, such as T2D. The hippocampus has receptors for insulin-like growth factor-1 (IGF1), which mediates exercise-induced neurogenesis
Prolonged MDD	Sheline YI et al., 2003	Biol Psychiatry 54, 338-52	
Autopsy on MDD-suicide	Stockmeier CA et al., 2004	Biol Psychiatry 56, 640-50	Most likely, due to glial cell loss + small soma size, which suggests shrinkage of dendritic trees in MDD
Postnatal depression	Lupien SJ et al., 2011	Proc Natl Acad Sci 108, 14324-29	Hippocampal volume *not* affected
Cushing's disease	Starkman MN et al., 1999	Biol Psychiatry 46, 1595-02	
PTSD	Gurvits TV et al., 1996	Biol Psychiatry 40, 1091-99	
Chronic stress	Gianaros PJ et al., 2007	NeuroImage 35, 795-803	
Chronic inflammation	Marsland AL et al., 2008	Biol Psychiatry 64, 484-90	
Lack of physical activity	Erickson KI et al., 2009	Hippocampus 19, 1030-39	
Jet lag	Cho K, 2001	Nat Neurosci 4, 567-68	
Low SES	Hanson JL et al., 2011	PloS One 6, e18712	
Amygdala hyperactivity and enlargement			
MDD	Sheline YI et al., 2001 Frodl T et al., 2003	Biol Psychiatry 50, 651-58 Biol Psychiatry 53, 338-44	Hyperactivity Enlargement
Anxiety	Drevets WC, 2000	Biol Psychiatry 48, 813-29	Hyperactivity
PTSD	Rao RP et al., 2012 Zohar J et al., 2011	Biol Psychiatry 72, 466-75 Eur Neuropsychopharmacol 21, 796-809	Corticosteroid treatment of PTSD (characterized by low cortisol levels) alleviate symptoms in patients and increases dendritic growth and spine density in *rat* amygdala
Postnatal depression	Lupien SJ et al., 2011	Proc Natl Acad Sci 108, 14324-29	Enlargement in offspring
CVD	Gianaros PJ et al., 2009	Biol Psychiatry 65, 943-50	Increased reactivity to angry + sad faces mediated by increased sympathetic activity
Traumatic event in healthy	Ganzel BL et al., 2008	NeuroImage 40, 788-95	Increased reactivity to angry + sad faces
Sleep deprivation	Yoo S-S et al., 2007	Curr Biol 17, R877-78	Increased reactivity to angry + sad faces
Upbringing in low SES environment	Gianaros PJ et al., 2008	Soc Cogn Affect Neurosci 3, 91-96	Increased reactivity to angry + sad faces
Prefrontal cortex (PFC) impairment			
Medial PFC	Liston C et al., 2006	J Neurosci 26, 7870-74	Shrinkage of dendrites and loss of spines through excitatory amino acids + glucocorticoids Expansion of dendrites

Continued

TABLE 3 Structural and Functional Studies of the Brain in Allostatic Load and Overload—Cont'd

Examined effect	Author	Journal	Notes
Orbitofrontal cortex (OFC)			
Poor cognitive flexibility	Dias-Ferreira E et al., 2009 Liston C et al., 2006 Karatsoreos et al., 2011	Science 325, 621-25 J Neurosci 26, 7080-74 Proc Natl Acad Sci 108, 1657-62	Animal and human studies
Reduced functional connectivity	Liston C et al., 2009	PNAS 106, 912-17	
Aging	Bloss EB et al., 2010	Exp Neurol 210, 109-17	Failure to reverse shrinkage of medial PFC neurons after chronic stress in *rats*
Low subjective SES	Gianaros PJ et al., 2007	Soc Cogn Affect Neurosci 2, 161-73	Reduction in PFC gray matter

MCI, mild cognitive impairment; AD, Alzheimer's disease; T2D, Type 2 diabetes; MDD, major depressive disorder; PTSD, post-traumatic stress disorder; SES, socioeconomic status; CVD, cardiovascular disease; PFC, prefrontal cortex; OFC, orbitofrontal cortex.
Adapted from Ref. 33.

physical functioning, mortality, and the onset of depressive symptoms.[32] Organ systems, once damaged, are more sensitive to physical and psychological acute stressors, thus creating a vicious cycle of AL.

There is no consensus about the exact combination of biomarkers that define *allostatic load index*,[7] but the list tends to include a combination of primary markers, that is, direct measures of HPA and sympathetic ANS activity hormones, and secondary markers, including those that constitute metabolic syndrome (Tables 1 and 2). Recently a couple of studies of the Whitehall II cohort have found that in combination, high levels of job strain and caring for a relative as well as negative emotional response to stressful life events predict high levels of AL. The authors defined AL merely by the combination of secondary markers (cardiovascular, metabolic, and inflammatory markers) over a 20-year period. In the elderly, the combination of such measures is a significant predictor of cognitive and physical decline over a 3-year period and risk of cardiovascular disease and mortality over a 7-year period, along with mental health outcomes.[32] Lowering AL decreases mortality rates. Furthermore, combinations of such markers are better predictors of tertiary markers (see Table 3) than the metabolic syndrome or any one of the components are on their own,[32] thus proving useful indicators of risk of subsequent poor health or function. However, further studies are needed to understand the mechanisms in which primary and secondary markers come together to predict cognitive decline[38] and what role gender plays in these.[7,38,39]

While the *allostasis*, *allostatic stage*, *-load*, and *-overload* model provides a theoretical framework for how an individual adapts under stress, it has a number of shortcomings. Delineating time courses of dysregulation is difficult.[40] Quantifying AL at a biological level, let alone with respect to the multiplicity of psychosocial contributing factors has represented a significant challenge. A concerted research effort over the last decade has helped to illustrate the complexity of AL as a construct.[7] Cardiovascular outcomes have, for example, been linked to job strain and SES in the Whitehall II cohort (Table 2), but the underlying allostatic processes have not been characterized in detail.[41]

This lack of detailed mechanistic explanations has been used to argue that the AL concept does not add heuristic value to the currently used concept of stress (stressors, stress response).[42] Alternative formulations and approaches include defining stress as an event or change in environment that elicits adaptive changes in behavior and physiological systems. For example, the *adaptive calibration model* examines whether early life experiences need to be viewed as adaptive or maladaptive to the stress system.[43] While allostasis is an attractive and plausible concept, it remains difficult to test. How useful it will be, will depend on its translation into predictions within medical practice.[33,34]

TERTIARY MARKERS OF AL AND STRESS-RESPONSE AND CENTRAL ROLE OF THE BRAIN

Pathology of the brain is an example of a *tertiary marker of AL* and *overload* (Table 3). While the brain plays an essential role in the appraisal of stressors as well as the regulation of both the physiological stress response of the periphery and accompanying behavioral changes, such responses intrinsically change the appraisal of future stressors and can be deleterious to the brain in the end. Table 3 presents an overview of the key brain areas, hippocampus, amygdala, and prefrontal cortex, and their involvement in allostasis and AL.

Glucocorticoids and Vascular Risk

Amygdala, hippocampus, and prefrontal cortex regulate physiological and behavioral responses to stress, thus coordinating allostasis.[21] While adaptive in the short run, for example, facilitating memory formation through long-term potentiation, and formation of synapses and neurons,[44] in the end these responses can be deleterious to the very system that produces them. Brain changes associated with chronic stress and elevated glucocorticoid levels, such as transient dendritic remodeling, weakening of synapses, inhibition of adult neurogenesis in the hippocampus, atrophy of hippocampal neurons,[20] further impair the ability to appropriately (and cognitively) evaluate and respond to stressors.[7] This creates a vicious circle of further damage (e.g., hippocampal neuron death) through feedback loops, which also may affect memory performance. The initial surge of glucocorticoid supplies to the hippocampus and cortex decreases if the stressor is chronic, inhibiting energy storage by these neurons. Cutting back glucocorticoid energy supplies from neurons that are undergoing neurological insult such as hypoglycemia, hypoxia-ischemia, and seizure, can lead to neuronal death. "Glucocorticoid neurotoxicity" described above[45] is also a feature of healthy aging in humans,[46] and additional stress in old age accelerates hippocampal degeneration and memory decline in rats.[20] While findings in humans are mixed, a continuous rise of resting glucocorticoid levels over the years in older adults seems to be associated with severe loss of hippocampal volume and memory decline.[8] Elevated diurnal and reactive glucocorticoid levels in a group of elderly males ($N=90$, 73 years) were also associated with more white matter hyperintensities and decreased integrity (mean diffusivity), especially in anterior thalamic radiation, uncinate, arcuate, and inferior longitudinal fasciculi.[47] This is in line with previous in vitro postmortem studies that found stimulating effects of low- and depressing effects of high glucocorticoid concentrations on axonal transport in prefrontal neurons, suggesting that "glucocorticoids potentially have negative effects on prefrontal cortex neurons' survival and function."[8]

While the brain is involved in sympathetic ANS activity that leads to wear and tear of the cardiovascular system by repeated activation, the latter also affects brain regional vasodilation and vascular reactivity, which are required for tasks and clearing of waste products. Midlife-vascular risk factors predict increased rates of progression of vascular brain injury, global and hippocampal atrophy, and decline of executive function over 10 years.[48] Middle-aged and older participants of the English Longitudinal Study of Ageing followed up at 4 and 8 years showed that Framingham Stroke Risk Score at baseline predicted lower global cognition, memory, and executive function at follow-up.[49]

Major depressive disorder (MDD) is one of the most obvious examples of Type II allostatic overload associated with structural gray and white matter changes in the human brain. Early-onset MDD is characterized by prolonged HPA-axis dysregulation and elevated glucocorticoid levels, resulting in volumetric decreases in the hippocampus, disruption of hippocampal neurogenesis, and prefrontal cortex regulation, which are associated with memory and concentration problems, and an increased risk of dementia. Longer duration of MDD is associated with a greater degree of hippocampal atrophy, amygdala, anterior cingulate cortex, and dorsomedial prefrontal cortex volume decline,[50] while a later onset is associated with widespread "nonspecific brain atrophy, consistent with evidence of generalized brain changes," possibly representing vulnerability for first-time depressive episodes in late life.[51,52] White matter hyperintensities and deep white matter lesions, reflecting underlying ischemic changes, are associated with reduced cognitive function and quality of life. Depression, particularly late-life depression, is associated with changes in the integrity of white matter tracts in frontal-subcortical, temporal, and limbic regions. Vascular disease and vascular risk factors, such as the Framingham Stroke Risk Score, contribute to an increased risk of depression, mediated through the effects of white matter lesions and integrity in those that are otherwise vulnerable.[53] Recently two mechanisms were suggested to explain the relationship between white matter changes and depressive symptoms: (1) vascular damage leads to altered neural connectivity (disconnection hypothesis) and (2) vascular changes mediated by inflammatory markers cause altered brain function and subsequent clinical symptoms (inflammatory and hypoperfusion hypothesis; for a review see Ref. 52).

Cognitive impairment, especially of processing speed and executive function, is also a feature of late-life depression, and is associated with altered cognitive function in other domains, total white matter hyperintensity volumes and reduced white matter integrity in frontal tracts.[52] Furthermore, late-life depression and cognitive impairment are both associated with hippocampal atrophy and cardiovascular risk factors. It is not a surprise therefore that depression is a risk factor for dementia and dementia is a risk factor for depression.[52] It seems likely that depression thus acts as a marker of organic dysfunction, and contributes to progression toward Alzheimer's disease and vascular dementia via mild cognitive impairment.

A related clinical example is post-traumatic stress disorder (PTSD). Lower glucocorticoid levels in PTSD, but higher sensitivity to glucocorticoids may contribute to reduced hippocampal size. It is not clear if this is due to the trauma itself or the period that follows the trauma, or whether hippocampal size is smaller in patients with PTSD before symptoms appear.[54]

Traumatic Life Stress

There is ample evidence of the negative effects of traumatic life stress on brain structure, in particular the hippocampus, amygdala, anterior cingulate cortex, and medial frontal gyrus.[12] The majority of studies examine the structural brain outcomes of early life events such as childhood neglect or trauma, in participants with certain psychopathologies such as depression or PTSD.[40,55] Despite the obvious involvement of the hippocampus in the stress response, only a handful of studies have examined the effect of stressful life events on brain structure in adults *without* a clinical syndrome. While they all found deleterious effects on brain structure, some are difficult to interpret. Average chronic perceived stress, for example, predicted decreased gray matter volume in the right orbitofrontal cortex and right hippocampus in healthy postmenopausal women, even after controlling for various demographic variables.[56] In another study, the number of stressful life events appeared to be correlated with decreased gray matter volume in the left parahippocampal gyrus, the left and right anterior cingulate, and the right hippocampus.[12] In a community sample of middle-aged adults, cumulative exposure to adverse life events was associated with prefrontal and limbic gray matter volume reduction.[57]

Metabolic Syndrome

Individual metabolic syndrome components have a negative impact on the human brain.[58] For example, higher blood glucose levels in the normal range are associated with lower gray and white matter regional volumes in the frontal cortices that correlate with poorer cognitive performance in men.[59] We know little of the effect as a syndrome on the aging brain.[58,60] The association of metabolic syndrome with brain and cognitive health has not been extensively studied, and results from existing studies are mixed. Ten studies compared cognitive function in older adults with and without metabolic syndrome. Seven out of 10 studies found a negative association between metabolic syndrome and cognition, while 3 found no relationship. Four examined the effect of sex; two found a more pronounced negative effect of metabolic syndrome in men, one in women, and one did not find any sex differences. The studies found an association between metabolic syndrome and cognitive deficits in several domains, even after controlling for cardiovascular and coronary heart disease, Type 2 diabetes, silent brain infarcts, depression, education, and SES: memory, visuospatial skills, executive function, processing speed, and overall IQ.[58] Metabolic syndrome was negatively associated with memory performance and executive function in men, but not in women, and the more criteria were fulfilled, the worse the performance was in men.[61] There were no MRI differences, for example, focal ischemic lesions or brain volumes that could have accounted for the cognitive differences.

To date, only one study has evaluated cortical thickness and subcortical volumes in metabolic syndrome.[60] Comparing participants with ($N = 40$) and without metabolic syndrome ($N = 46$), and after controlling for age, gender, education, and total intracranial volume, they found significant volume reductions in the right nucleus accumbens in participants with metabolic syndrome compared with controls. Furthermore, cortical thickness was significantly lower in participants with metabolic syndrome in both hemispheres, in the left insular, superior parietal, postcentral, entorhinal, and right superior parietal cortices.

Metabolic syndrome has been associated with ischemic stroke, subclinical ischemic brain damage, intracranial arteriosclerosis, periventricular white matter hyperintensities, and subcortical white matter lesions in middle-aged individuals, and with silent brain infarction in both middle-aged and older individuals.[58] Fifty- to eighty-year olds with metabolic syndrome compared with age-matched controls had reduced white matter integrity in frontal and temporal regions.[62] There was no relationship between any of the individual vascular risk factors and white matter integrity. Others have found a positive association between the number of metabolic syndrome components present and the degree of white matter damage, driven by vascular risk factors.[58] A small-scale diffusion tensor MRI study compared seven participants with and seven without metabolic syndrome to show decreased white matter integrity in participants with metabolic syndrome in the right external capsule, entire length of the corpus callosum, and part of the deep white matter in the right frontal lobe.[63] Increased myoinositol/creatine and glutamate/creatine ratios in occipitoparietal regions of cognitively healthy middle-aged adults with metabolic syndrome suggest microglia- or neuroinflammation, often seen in Type 2 diabetes.[64] Such subclinical alterations in cerebral metabolism and cerebrovascular reactivity may represent early brain damage associated with peripheral metabolic disturbances.[58]

Allostatic Load Index

Few studies have examined the association among allostatic load indices, brain, and cognitive health.[7]

A study of the Lothian birth cohort ($N=658$) reported a negative association of AL with total brain volume and white matter volume, a positive (sic!) association with hippocampal volume, and no association with gray matter volume.[65] AL was also negatively associated with general cognitive ability, processing speed, and knowledge, but not with memory or nonverbal reasoning. This dissociation between global and hippocampal involvement may be the result of different pathways (cortisol-hippocampal cascade vs. generalized peripheral vascular illness) in this completely age-controlled cohort. Interestingly, the associations of AL with cognitive impairment appeared not to be mediated by these brain volume measures. AL did not predict cognitive change from age 11 to approximately age 73.[65]

We will need a number of longitudinal prospective cohort studies to disentangle the relationship between metabolic syndrome and components of allostatic load index and their effect on brain and cognition. Figure 2 gives an example of the identification of vascular risk

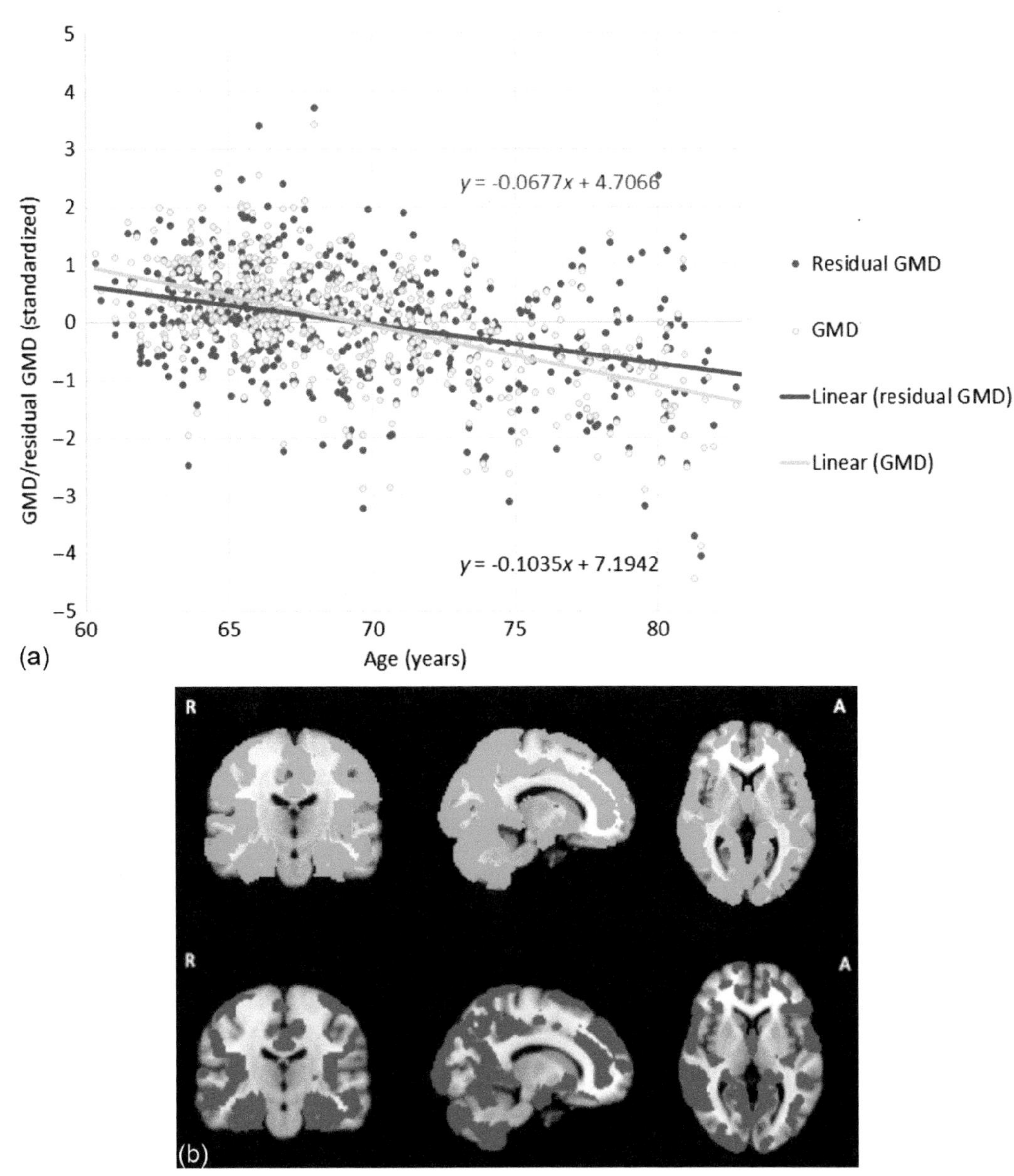

FIGURE 2 Association between age and average gray matter density (GMD). (a) Before-and-after controlling for vascular risk [measured by % Framingham Stroke Risk Score (FSRS)] in a sample of community-dwelling adults ($N=527$, age $=69.5\pm5.29$ years, 421 male).[66] The change in regression slope was significant, leaving a regression slope after controlling for % Framingham Risk Score that remained significant (i.e., Framingham Risk Score partially accounts for age effect). (b) Images were analyzed with FSL-VBM and optimized voxel-based morphometry (VBM) protocol. Using randomize and correcting for multiple comparisons, voxelwise general linear model (GLM) was applied between age and GM (top image), and age and GM with FSRS as confounder (bottom image). Significance threshold was set at $p<0.05$ using threshold-free cluster-enhancement (tfce). Extracted standardized average gray matter densities (residuals) are shown in the plot.

(Framingham Stroke Risk Score) as mediator of age effects on gray matter atrophy in 527 participants at the Whitehall II MRI substudy.[66]

In conclusion, the AL/chronic stress model is a useful platform to explore the effects of time on the brain, but the underlying mechanisms require exploration in greater detail, along with individual factors, which could include personality traits, genetic risk, and resilience-conferring factors such as education, SES, and support networks.

References

1. McEwen BS, Stellar E. Stress and the individual. Mechanisms leading to disease. *Arch Intern Med*. 1993;153(18):2093–2101.
2. Gorman M. Development and the rights of older people. In: Randel J, German T, Ewing D, eds. *The Ageing and Development Report: Poverty, Independence and the World's Older People*. London: Routledge; 1999:3–21.
3. Unite Nations Population Division. *Population Ageing and Sustainable Development*. New York, NY: United Nations; 2014.
4. Topiwala A, Patel S, Ebmeier KP. Health benefits of encore careers for baby boomers. *Maturitas*. 2014;78(1):8–10.
5. Zsoldos E, Mahmood A, Ebmeier KP. Occupational stress, bullying and resilience in old age. *Maturitas*. 2014;78(2):86–90.
6. Epel ES. Psychological and metabolic stress: a recipe for accelerated cellular aging? *Hormones (Athens)*. 2009;8(1):7–22.
7. Juster RP, McEwen BS, Lupien SJ. Allostatic load biomarkers of chronic stress and impact on health and cognition. *Neurosci Biobehav Rev*. 2010;35(1):2–16.
8. Lupien SJ, McEwen BS, Gunnar MR, Heim C. Effects of stress throughout the lifespan on the brain, behaviour and cognition. *Nat Rev Neurosci*. 2009;10(6):434–445.
9. McEwen BS, Wingfield JC. The concept of allostasis in biology and biomedicine. *Horm Behav*. 2003;43(1):2–15.
10. Selye H. What is stress? *Metabolism*. 1956;5(5):525–530.
11. Selye H. History of the stress concept. In: Goldberger L, Breznitz S, eds. *Handbook of Stress: Theoretical and Clinical Aspects*. 2nd ed. New York, NY: The Free Press; 1993.
12. Papagni SA, Benetti S, Arulanantham S, McCrory E, McGuire P, Mechelli A. Effects of stressful life events on human brain structure: a longitudinal voxel-based morphometry study. *Stress*. 2011;14(2): 227–232.
13. Graham JE, Christian LM, Kiecolt-Glaser JK. Stress, age, and immune function: toward a lifespan approach. *J Behav Med*. 2006;29(4):389–400.
14. Segerstrom SC, Miller GE. Psychological stress and the human immune system: a meta-analytic study of 30 years of inquiry. *Psychol Bull*. 2004;130(4):601–630.
15. Widom CS, Horan J, Brzustowicz L. Childhood maltreatment predicts allostatic load in adulthood. *Child Abuse Negl*. 2015;47:59–69. pii: S0145-2134(15)00040-X.
16. Friedman EM, Karlamangla AS, Gruenewald TL, Koretz B, Seeman TE. Early life adversity and adult biological risk profiles. *Psychosom Med*. 2015;77(2):176–185.
17. Barboza GE. The association between school exclusion, delinquency and subtypes of cyber- and F2F-victimizations: identifying and predicting risk profiles and subtypes using latent class analysis. *Child Abuse Negl*. 2015;39:109–122.
18. Bernard C. *An Introduction to the Study of Experimental Medicine*. New York, NY: Dover Publications; 2003.
19. Cannon WB. *The Wisdom of the Body*. New York, NY: WW Norton; 1932.
20. Sapolsky RM. *Why Zebras Don't Get Ulcers* [Revised and Updated]. New York, NY: Owl Books; 2004
21. McEwen BS, Gianaros PJ. Central role of the brain in stress and adaptation: links to socioeconomic status, health, and disease. *Ann N Y Acad Sci*. 2010;1186:190–222.
22. Klein LC, Corwin EJ. Seeing the unexpected: how sex differences in stress responses may provide a new perspective on the manifestation of psychiatric disorders. *Curr Psychiatry Rep*. 2002;4(6):441–448.
23. Sapolsky RM, Romero LM, Munck AU. How do glucocorticoids influence stress responses? Integrating permissive, suppressive, stimulatory, and preparative actions. *Endocr Rev*. 2000;21(1):55–89.
24. McEwen BS. Interacting mediators of allostasis and allostatic load: towards an understanding of resilience in aging. *Metabolism*. 2003;52 (10 suppl 2):10–16.
25. Chrousos GP. The hypothalamic-pituitary-adrenal axis and immune-mediated inflammation. *N Engl J Med*. 1995;332 (20):1351–1362.
26. Chrousos GP. Stress and disorders of the stress system. *Nat Rev Endocrinol*. 2009;5(7):374–381.
27. Sterling P, Eyer J. Allostasis: a new paradigm to explain arousal pathology. In: Fisher S, Reason J, eds. *Handbook of Life Stress, Cognition and Health*. New York, NY: Wiley; 1988:629–649.
28. McEwen BS, Lasley E. *The End of Stress as We Know It*. Washington, DC: Joseph Henry Press; 2002.
29. McEwen BS. Protective and damaging effects of stress mediators: central role of the brain. *Dialogues Clin Neurosci*. 2006;8(4):367–381.
30. Marmot M, Brunner E. Cohort profile: the Whitehall II study. *Int J Epidemiol*. 2005;34(2):251–256.
31. D'Agostino RB, Wolf PA, Belanger AJ, Kannel WB. Stroke risk profile: adjustment for antihypertensive medication. The Framingham Study. *Stroke*. 1994;25(1):40–43.
32. Seeman TE, McEwen BS, Rowe JW, Singer BH. Allostatic load as a marker of cumulative biological risk: MacArthur studies of successful aging. *Proc Natl Acad Sci U S A*. 2001;98(8):4770–4775.
33. McEwen BS, Gray J, Nasca C. Recognizing resilience: learning from the effects of stress on the brain. *Neurobiol Stress*. 2015;1:1–11.
34. Ewbank DC. Biomarkers in social science research on health and ageing: a review of theory and practice. In: Christensen K, Hankinson S, Seeman TE, et al., eds. *Biosocial Surveys*. Washington, DC: The National Academies Press; 2007:156–171.
35. Steptoe A, Kivimaki M. Stress and cardiovascular disease. *Nat Rev Cardiol*. 2012;9(6):360–370.
36. Reaven GM, Reaven EP. Effects of age on various aspects of glucose and insulin metabolism. *Mol Cell Biochem*. 1980;31(1):37–47.
37. Grundy SM, Brewer Jr. HB, Cleeman JI, et al. Definition of metabolic syndrome: report of the National Heart, Lung, and Blood Institute/ American Heart Association conference on scientific issues related to definition. *Circulation*. 2004;109(3):433–438.
38. Gruenewald TL, Seeman TE, Ryff CD, Karlamangla AS, Singer BH. Combinations of biomarkers predictive of later life mortality. *Proc Natl Acad Sci U S A*. 2006;103(38):14158–14163.
39. Seeman TE, Singer B, Wilkinson CW, McEwen B. Gender differences in age-related changes in HPA axis reactivity. *Psychoneuroendocrinology*. 2001;26(3):225–240.
40. McEwen BS. Central effects of stress hormones in health and disease: understanding the protective and damaging effects of stress and stress mediators. *Eur J Pharmacol*. 2008;583(2–3):174–185.
41. Adler NE, Boyce T, Chesney MA, et al. Socioeconomic status and health. The challenge of the gradient. *Am Psychol*. 1994;49(1):15–24.
42. Dallman MF. Stress by any other name ...? *Horm Behav*. 2003; 43(1):18–20. discussion 28–30.
43. Hostinar CE, Gunnar MR. The developmental psychobiology of stress and emotion in childhood. In: Weiner IB, Freedheim DK, Lerner RM, eds. *Handbook of Psychology*. 2nd ed. Hoboken, NJ: John Wiley & Sons; 2013:121–141.

44. Lynch MA. Long-term potentiation and memory. *Physiol Rev.* 2004;84(1):87–136.
45. Sapolsky RM, Krey LC, McEwen BS. Prolonged glucocorticoid exposure reduces hippocampal neuron number: implications for aging. *J Neurosci.* 1985;5(5):1222–1227.
46. Raskind MA, Peskind ER, Wilkinson CW. Hypothalamic-pituitary-adrenal axis regulation and human aging. *Ann N Y Acad Sci.* 1994;746:327–335.
47. Cox SR, Bastin ME, Ferguson KJ, et al. Brain white matter integrity and cortisol in older men: the Lothian Birth Cohort 1936. *Neurobiol Aging.* 2015;36(1):257–264.
48. Debette S, Seshadri S, Beiser A, et al. Midlife vascular risk factor exposure accelerates structural brain aging and cognitive decline. *Neurology.* 2011;77(5):461–468.
49. Dregan A, Stewart R, Gulliford MC. Cardiovascular risk factors and cognitive decline in adults aged 50 and over: a population-based cohort study. *Age Ageing.* 2013;42(3):338–345.
50. Frodl TS, Koutsouleris N, Bottlender R, et al. Depression-related variation in brain morphology over 3 years: effects of stress? *Arch Gen Psychiatry.* 2008;65(10):1156–1165.
51. Sexton CE, Le Masurier M, Allan CL, et al. Magnetic resonance imaging in late-life depression: vascular and glucocorticoid cascade hypotheses. *Br J Psychiatry.* 2012;201(1):46–51.
52. Allan CL, Zsoldos E, Ebmeier KP. Imaging and neurobiological changes in late-life depression. *Br J Hosp Med (Lond).* 2014;75(1):25–30.
53. Allan CL, Sexton CE, Kalu UG, et al. Does the Framingham Stroke Risk Profile predict white-matter changes in late-life depression? *Int Psychogeriatr.* 2012;24(4):524–531.
54. Deppermann S, Storchak H, Fallgatter AJ, Ehlis AC. Stress-induced neuroplasticity: (mal)adaptation to adverse life events in patients with PTSD—a critical overview. *Neuroscience.* 2014;283:166–177.
55. Cohen RA, Grieve S, Hoth KF, et al. Early life stress and morphometry of the adult anterior cingulate cortex and caudate nuclei. *Biol Psychiatry.* 2006;59(10):975–982.
56. Gianaros PJ, Jennings JR, Sheu LK, Greer PJ, Kuller LH, Matthews KA. Prospective reports of chronic life stress predict decreased grey matter volume in the hippocampus. *Neuroimage.* 2007;35(2):795–803.
57. Ansell EB, Rando K, Tuit K, Guarnaccia J, Sinha R. Cumulative adversity and smaller gray matter volume in medial prefrontal, anterior cingulate, and insula regions. *Biol Psychiatry.* 2012; 72(1):57–64.
58. Yates KF, Sweat V, Yau PL, Turchiano MM, Convit A. Impact of metabolic syndrome on cognition and brain: a selected review of the literature. *Arterioscler Thromb Vasc Biol.* 2012;32(9):2060–2067.
59. Mortby ME, Janke AL, Anstey KJ, Sachdev PS, Cherbuin N. High "normal" blood glucose is associated with decreased brain volume and cognitive performance in the 60s: the PATH through life study. *PLoS One.* 2013;8(9):e73697.
60. Song SW, Chung JH, Rho JS, et al. Regional cortical thickness and subcortical volume changes in patients with metabolic syndrome. *Brain Imaging Behav.* 2015;9(3):588–596.
61. Cavalieri M, Ropele S, Petrovic K, et al. Metabolic syndrome, brain magnetic resonance imaging, and cognition. *Diabetes Care.* 2010;33 (12):2489–2495.
62. Segura B, Jurado MA, Freixenet N, Falcon C, Junque C, Arboix A. Microstructural white matter changes in metabolic syndrome: a diffusion tensor imaging study. *Neurology.* 2009;73(6):438–444.
63. Shimoji K, Abe O, Uka T, et al. White matter alterations in metabolic syndrome—diffusion tensor analysis. *Diabetes Care.* 2013;36:696–700.
64. Haley AP, Gonzales MM, Tarumi T, Miles SC, Goudarzi K, Tanaka H. Elevated cerebral glutamate and myo-inositol levels in cognitively normal middle-aged adults with metabolic syndrome. *Metab Brain Dis.* 2010;25(4):397–405.
65. Booth T, Royle NA, Corley J, et al. Association of allostatic load with brain structure and cognitive ability in later life. *Neurobiol Aging.* 2015;36(3):1390–1399.
66. Zsoldos E, Mahmood A, Filippini N, et al. Chronological and biological ageing: predicting grey matter density by age and Framingham Stroke Risk Scores in community-dwelling older adults. *Biol Psychiatry.* 2015;77:213S.

CHAPTER

39

Childbirth and Stress

A. Horsch[1], S. Ayers[2]

[1]University Hospital Lausanne, Lausanne, Switzerland

[2]City University London, London, UK

OUTLINE

Abstract

Childbirth is an intense event that involves extreme physical stress and is of emotional, cognitive, social, and cultural significance. It is a very common and predictable event that enables us to study the interaction between pre-birth and birth factors in perceived stress and the effect of this on physical and psychological outcomes. Pregnancy and childbirth are complex physiological processes that involve many of the same processes involved in stress responses. Approximately 15-20% of women in Western countries rate their childbirth as traumatic. Postpartum posttraumatic stress disorder is a distressing and disabling condition that can also have important negative consequences for the attachment relationship with the baby and the development of the child. This chapter provides an overview of the current evidence and theoretical models of stress and trauma in childbirth for the parents and maternity staff, and discusses the economic costs and clinical implications.

Childbirth is an intense event that involves extreme physical stress and is of emotional, cognitive, social, and cultural significance. Physically, women have to cope with acute changes as the uterus contracts, the cervix dilates, and the baby and placenta are born.

General responses to stress such as release of endorphin and catecholamines have been shown in laboring women, as well as increases in cardiac output and blood pressure. Emotionally, labor and the birth of the baby usually involve intense positive and negative emotions. Cognitive demands are also placed on the woman as she copes with the events of birth. Interpersonal dynamics between the woman, her birth partner, and maternity staff can be supportive or can increase stress if birth attendants are perceived as unhelpful or dismissive. Culturally, birth and motherhood are associated with many cultural expectations and norms. In Western societies, birth is highly medicalized yet emphasis is often placed on women trying to achieve a natural birth with minimal intervention or pain relief.

Worldwide, approximately 135 million births occur each year.[1] Birth is therefore a very common and predictable event which enables us to study the interaction between prebirth and birth factors in perceived stress and the effect of this on physical and psychological outcomes. This is consistent with the diathesis-stress approach to health outcomes, in which preexisting vulnerability (physical, psychological, and social) interacts with the event to determine outcomes. The impact of childbirth must also be placed in the context of previous adjustment to pregnancy and postpartum adjustment to motherhood and a new baby, all of which are also potentially stressful and difficult.

Research into the role of psychosocial factors in birth outcomes has generally shown that psychosocial factors are important in physical outcomes, such as morbidity and mortality, and psychological outcomes, such as maternal satisfaction and mental health. The postpartum period is associated with a number of psychological problems that are seen as specific to childbirth, such as the

Stress: Concepts, Cognition, Emotion, and Behavior
http://dx.doi.org/10.1016/B978-0-12-800951-2.00040-6

baby blues, postpartum depression, and puerperal psychosis. There is also evidence that women can develop posttraumatic stress disorder (PTSD) in response to difficult birth experiences. Meta-analyses suggest PTSD affects 3.17% of women after birth at diagnostic levels and around 15% of women in high-risk groups, for example, after preterm birth or neonatal death.[2] Worldwide birth rates mean this prevalence of 3.17% equates to 4.3 million women potentially developing PTSD as a result of childbirth every year. There are also a substantial number of women who suffer with highly distressing PTSD symptoms that are below diagnostic threshold level. PTSD after birth is important to understand and address for many reasons. It is a severe stress response that arises in direct response to the events of pregnancy and/or birth, and can therefore help us understand more about severe stress responses. Theoretical frameworks of postpartum PTSD can be used to inform general theories of stress. Presumably, a large proportion of postpartum PTSD should be preventable if appropriate care and support is provided during labor and birth. Qualitative research suggests PTSD can have a wide ranging negative impact on women and their families.[3,4]

This chapter provides an overview of the current evidence and understanding of stress and trauma in childbirth by looking at: (i) childbirth and stress; (ii) theoretical models of childbirth and trauma; (iii) the impact of traumatic childbirth; and (iv) the costs and clinical implications.

KEY POINTS

- Childbirth is a common and predictable event that enables us to study stress processes and outcomes for both the mother and her child.
- Approximately 15-20% of women in Western countries rate their childbirth as traumatic.
- Postpartum posttraumatic stress disorder (PTSD) is a distressing and disabling condition that can also have important negative consequences for the attachment relationship with the baby and the development of the child.
- General theories of stress and PTSD have been applied or adapted to childbirth and new, specific theories have been developed.
- To date, there is a lack of evidence-based interventions for women after a traumatic childbirth.

CHILDBIRTH AND STRESS

Pregnancy and childbirth are complex physiological processes that involve many of the same processes involved in stress responses. Interrelated maternal and fetal hormones and hormonal changes are vital to a healthy pregnancy, birth, and the developing fetus so it is not surprising that these processes can be affected by stress. Studies of stress, anxiety, and depression in pregnancy show these are associated with altered patterns of fetal behavior and heart rate responses.[5] Postpartum follow-up shows infants of women who are stressed or anxious in pregnancy are more likely to show fearful or anxious behavior and are more at risk of poor development and adverse outcomes such as attention deficit hyperactivity disorder (ADHD).[6] Stress hormones and epigenetic mechanisms are thought to underlie the effect of women's emotional state in pregnancy on the developing baby.[7] Severe stress or anxiety in pregnancy is associated with babies being born preterm and having a lower birth weight,[8] both of which are associated with poorer infant outcomes. The evidence for anxiety in pregnancy suggests this has a dose-response effect on the fetus, so that moderate symptoms of anxiety will also have a negative impact.

Stress hormones are also a part of the normal physiological process of labor and birth. In late labor there is a natural surge in stress hormones such as epinephrine, prolactin, and cortisol which is thought to be important in preparing the mother and baby physiologically for the birth. The surge in hormones may promote contractions, facilitate the effect of oxytocin (the "bonding" hormone), and promote physiological changes in the baby that maximize their chances of surviving the transition through the vaginal canal and establishing respiration. Additional stress during labor therefore has the potential to interfere with or disrupt normal physiological processes. Animal research suggests stress in early labor may slow down or stop labor, and stress in late labor may lead to a quick birth—termed the "fetus ejection reflex." More research is needed to examine this in women but the available evidence has led to organizations and individuals highlighting the potential negative impact of common maternity care practices, such as induction of labor, on normal physiological processes.[9]

In some cases, childbirth can be experienced as traumatic by women. Approximately 15-20% of women in Western countries rate childbirth as traumatic according to the Diagnostic and Statistical Manual of Mental Disorders, 4th Edition (DSM-IV) diagnostic criteria for PTSD.[10] To meet criteria for a traumatic stressor, women should witness or experience death, threat of death, or serious injury to themselves or a significant other—in this case usually the baby (Criterion A1); and respond to this with intense fear, helplessness, or horror (Criterion A2).[11] Approximately 3% of women develop PTSD as a result of birth or postpartum events.[2] PTSD is characterized by three types of symptoms: (1) reexperiencing symptoms (e.g., flashbacks, nightmares); (2) avoidance symptoms (e.g., avoiding reminders of the stressor); and

(3) hyperarousal symptoms (e.g., insomnia, hypervigilance). In the recent revision of the DSM, a fourth symptom cluster of altered cognitions and mood was added (e.g., anhedonia, negative beliefs).[12] For a diagnosis of PTSD, symptoms should be experienced for longer than 1 month and cause significant distress, disability, and impaired functioning. There is also emerging evidence that between 33% and 44% of women report significant levels of posttraumatic growth after birth.[13]

An issue that is relevant to all studies of emotion and psychopathology in pregnancy and after birth is that factors normally associated with pregnancy and having a new baby can confound measurement and inflate estimates of low mood and PTSD. The birth of a new baby requires substantial adjustment, with physiological changes and fatigue almost universal after birth. The postpartum period may be associated with increased vigilance and parental preoccupation with the newborn which creates an environment for meeting the needs of the infant and keeping it safe. This process might be mediated by oxytocin, which has also been associated with obsessional and hypervigilant behavior, usually associated with obsessive compulsive disorder.[14] Hypervigilance may manifest itself as heightened anxiety up to the first few weeks after the birth.[15] Ayers et al.[16] examined this and found that two or more hyperarousal symptoms were reported by 50% of women who did *not* have traumatic births. They conclude that hyperarousal symptoms have poor specificity and should therefore not be relied on as a sole indicator of PTSD. Thus, measurement issues are important when examining stress responses in pregnancy and postpartum and it should not be assumed that generic measures are applicable or valid in this population.

THEORETICAL MODELS OF CHILDBIRTH AND TRAUMA

Theories are vital to ensure clarity of concepts and develop understanding of stress and childbirth. General theories of stress and PTSD have been applied or adapted to childbirth, as well as specific theories being developed. Relevant stress theories include the transactional model, which emphasizes the importance of appraisal and coping in stress responses.[17] Diathesis-stress frameworks accounting for the interaction between individual vulnerability and events have been adapted to explain postpartum PTSD. Figure 1 shows a diathesis-stress framework of the etiology of postpartum PTSD.[18,19] An alternative conceptual framework was put forward by Slade,[20] which outlined predisposing, precipitating, and maintaining factors relating to internal, external, and interactional influences. Theories of PTSD have also been applied and tested in relation to postpartum PTSD. For example, Ehlers and Clark's[21] theory proposes that PTSD occurs if individuals process the event or its sequelae in a way which produces a *sense of current threat*, with negative thoughts and cognitions about the event, and a disturbance or block in memory processing. This model has been tested prospectively in childbirth research and found to be a good predictor of postpartum PTSD symptoms.[22]

Reviews and meta-analyses of factors associated with postpartum PTSD suggest an interaction between prenatal factors, birth experience, and postpartum factors as summarized in Figure 1.[2,19] Prenatal vulnerability factors associated with PTSD include a history of psychopathology, depression in pregnancy, and fear of childbirth. Risk factors in birth associated with PTSD include a negative subjective birth experience, lack of support, dissociation, and having an operative birth (i.e., assisted vaginal or caesarean birth). Postpartum factors that contribute to the maintenance of PTSD include poor coping and stress. These reviews also indicate that postpartum PTSD is highly comorbid with depression and other measures of poor emotional health. Interestingly, one of the meta-analyses showed that, over time, the events of pregnancy and birth become less associated with PTSD whereas individual vulnerability and psychosocial circumstances become more associated with PTSD.[19] This

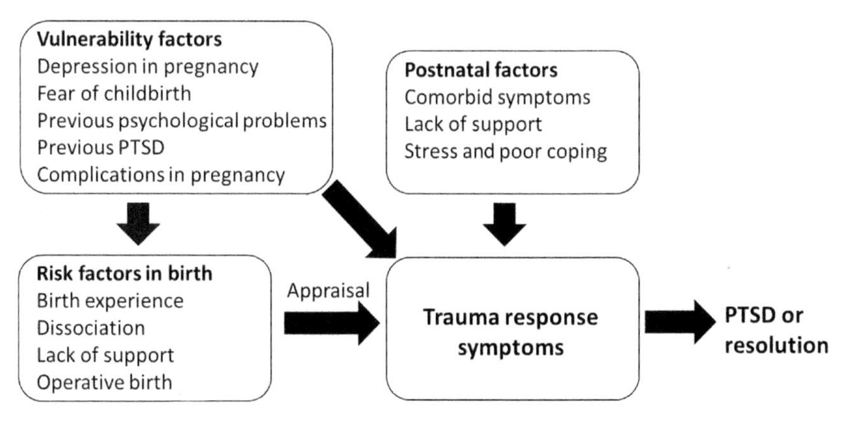

FIGURE 1 Diathesis-stress model of postpartum PTSD (Ayers et al., in press).

demonstrates how vulnerability and psychosocial circumstances may influence long-term recovery or maintenance of severe stress responses.

IMPACT OF TRAUMATIC CHILDBIRTH

Postpartum PTSD is highly distressing for women and some experience panic, anxiety, grief, anger, and tearfulness.[23] Some women report having suicidal thoughts and ideations following traumatic childbirth and one study quoted a mother who had thoughts about harming her baby; "I just thought oh, I wanna strangle, you know I didn't want to, it just came into my head, strangle" (p. 395).[3] Women may be haunted by painful memories for many years, as one woman described; "it was sort of a long black hole, just endless, endless pain" (p. 304).[24] Women talk about a loss of identity and self-esteem, particularly with regards to their competencies as a mother.[3]

This distressing and disabling condition also has important negative consequences for the attachment relationship with the baby and the development of the child.[25] Mothers may feel emotionally detached and therefore unable to show affection toward their baby, as was described by one mother; "mechanically I'd go through the motions of being a good mother. Inside I felt nothing" (p. 220).[26] Following birth, traumatized mothers have been shown to be more controlling and less sensitive toward their child. Women also report being overprotective toward their children. This may be a consequence of hypervigilance, one of the symptoms of PTSD. Furthermore, there is neuropsychological evidence that women's brain responses to the cry of their own baby are negatively affected by a cesarean section compared with women who had a vaginal delivery,[27] therefore interfering with the bonding between the mother and her baby. Finally, compared with vaginal delivery, women following cesarean delivery are less likely to breastfeed, which may have negative implications for their infants' well-being.[28]

Variations in maternal care during the early postpartum period may permanently alter systems in the infant that modulate stress or enable social adaptation.[29] The mother's role in the emotional coregulation with her infant is crucial for the infant's development of capacities for self-regulation and prosocial behavior. One study showed that traumatized mothers have difficulties remaining available for the regulation of their infants' emotions, arousal, and aggression during and immediately following a stressful interaction.[30] Furthermore, research considering other postpartum mental health disorders, such as depression, indicates potential for long-term negative effects on child development and behavior.[31] However, the longer term impact of maternal postpartum PTSD on the development (cognitive, language, motor, social-emotional) of their infant is unknown and research is needed to investigate this.

Research has also shown that mothers with PTSD are more likely to report relationship problems with their partner, with some women blaming their partners for the events that took place regarding their traumatic childbirth.[4] Sexual problems are also common, as sexual intercourse may trigger intrusive memories of the traumatic childbirth. Women with higher levels of PTSD symptoms at 6 weeks postpartum show an increased likelihood of deciding not to have further children; and women after a traumatic birth who do embark on a subsequent pregnancy report delaying this longer than those without such an experience.[32] Therefore, postpartum PTSD may have a significant negative impact on family planning.

Postpartum PTSD is linked with avoidance of medical care and other high-risk health behaviors, such as smoking, alcohol consumption, substance use, and excessive weight gain. This may be motivated by an intention to avoid triggers, such as smells, sounds, persons, or places that might remind women of their traumatic birth. Following traumatic childbirth, women may develop fear of subsequent pregnancy and childbirth (secondary tocophobia) and they are more likely to request an elective cesarean section.[33] This in turn may increase the likelihood of having complications or experiencing a traumatic childbirth again. A previous traumatic childbirth can also significantly influence the experience of subsequent pregnancies, with some women experiencing distress when they found out they were pregnant again; "I took the test and crumpled over the edge of our bed, sobbing and retching hysterically for hours" (p. 245).[34] This may have negative implications for the subsequent child, as we have already outlined in terms of stress in pregnancy and poor outcomes such as preterm birth and low birth weight.[8] During the subsequent pregnancy, higher utilization of healthcare services has been reported.

Working with women in labor and witnessing traumatic childbirth can also negatively impact on the well-being of maternity staff. A study of 464 labor and delivery nurses found that 35% reported moderate to severe levels of secondary traumatic stress,[35] which "results from helping or wanting to help a traumatized or suffering person" (p. 10).[11] Content analysis of nurses who were present at traumatic births resulted in six themes: "magnifying the exposure to traumatic births," "struggling to maintain a professional role while with the traumatized patients," "agonizing over what should have been," "mitigating the aftermath of exposure to traumatic births," "haunted by secondary traumatic stress symptoms," and "considering foregoing careers in labor and delivery suite to survive." Furthermore, 26% of this sample also met diagnostic criteria for PTSD. Maternity staff who are traumatized themselves are likely to struggle to deliver high quality care to laboring women, which in turn may make it more likely for traumatic births to reoccur.

COSTS OF TRAUMATIC CHILDBIRTH AND CLINICAL IMPLICATIONS

It is clear that stress in pregnancy and childbirth can interfere with normal physiological processes and, although this does not always result in poor mental health, it can be a triggering factor for some women—particularly those who develop birth-related PTSD. The short and long-term costs of postpartum PTSD are difficult to quantify, but a report from the London School of Economics concluded that perinatal depression, anxiety disorders (including PTSD), and psychosis carry a total long-term cost to the UK of about £8.1 billion for each 1-year cohort of births. This is equivalent to a cost of just under £10,000 for every single birth. The report also concluded that 72% of this cost relates to adverse impacts on the child rather than the mother. This shows the importance of early screening of women for mental health problems in the postpartum period and the need to develop early interventions to prevent the negative impact of maternal mental health problems on the mother-infant attachment relationship and the development of the infant.[36]

To date, there is a lack of evidence-based interventions for women after a traumatic childbirth. A recent Cochrane review concluded that there is little or no evidence to support either a positive or adverse effect of psychological debriefing for the prevention of psychological trauma in women following childbirth and no evidence to support routine debriefing for women who perceive giving birth as psychologically traumatic.[37]

A review[38] of psychological interventions for PTSD in other adult populations recommended individual trauma-focused cognitive-behavioral therapy, eye movement desensitization and reprocessing, stress management, and group trauma-focused cognitive-behavioral therapy as effective treatments. However, research into the efficacy of these approaches in the postpartum population is so far sparse.

Given that cognitive appraisals of the birth and its aftermath, as well as preexisting traumatic experiences and beliefs play an important role in the development of PTSD, managing the expectations of the labor process may be helpful. Ayers et al.[39] suggest maternity staff ensure that during the antenatal period they discuss with individual women that the process of labor and childbirth may not be as planned. Additionally, sensitive and supportive management of events during the labor process may increase the woman's and her partner's sense of control and positively influence their appraisals of these events. This includes giving the woman and her partner as much information and choice as possible and being aware that even routine procedures during labor may be stressful and make women anxious. Even in emergency situations that are not life threatening, maternity staff should provide reassurance. Health professionals need to know that women are not only at risk of developing postpartum PTSD because of their traumatic childbirth experience, but remain vulnerable in the postpartum period if they continue to experience problems with themselves and/or the baby.[39]

Finally, it is well established that perceived social support can help prevent the development of PTSD. Therefore, women should be encouraged to make use of their social network especially at this difficult time, and partners may also benefit from psychoeducation and normalization of the women's (and/or their own) symptoms to provide support.[40]

CONCLUSIONS

Childbirth is an intense event that involves extreme physical stress and is of emotional, cognitive, social, and cultural significance. It is a very common and predictable event that enables us to study the interaction between prebirth and birth factors in perceived stress and the effect of this on physical and psychological outcomes. Pregnancy and childbirth are complex physiological processes that involve many of the same processes involved in stress responses. Approximately 15-20% of women in Western countries rate their childbirth as traumatic. Postpartum PTSD is a distressing and disabling condition that can also have important negative consequences for the attachment relationship with the baby and the development of the child. General theories of stress and PTSD have been applied or adapted to childbirth and new, specific theories have been developed. Working with women in labor and witnessing traumatic childbirth can also negatively impact on the well-being of maternity staff. To date, there is a lack of evidence-based interventions for women after a traumatic childbirth. Managing the expectations and providing sensitive support during the labor process may contribute to the woman's and her partner's sense of control and positively influence their appraisals of these events.

References

1. Data UN. Annual number of births: United Nations Children's Fund; 2011 [cited 26.05.2015]. Available from: http://data.un.org/Data.aspx?d=SOWC&f=inID%3A75.
2. Grekin R, O'Hara MW. Prevalence and risk factors of postpartum posttraumatic stress disorder: a meta-analysis. *Clin Psychol Rev.* 2014;34(5):389–401.

3. Ayers S, Eagle A, Waring H. The effects of childbirth-related post-traumatic stress disorder on women and their relationships: a qualitative study. *Psychol Health Med*. 2006;11(4):389–398.
4. Nicholls K, Ayers S. Childbirth-related post-traumatic stress disorder in couples: a qualitative study. *Br J Health Psychol*. 2007;12(4): 491–509.
5. Kinsella MT, Monk C. Impact of maternal stress, depression & anxiety on fetal neurobehavioral development. *Clin Obstet Gynecol*. 2009;52(3):425.
6. Talge NM, Neal C, Glover V. Antenatal maternal stress and long-term effects on child neurodevelopment: how and why? *J Child Psychol Psychiatry*. 2007;48(3-4):245–261.
7. Wadhwa PD. Psychoneuroendocrine processes in human pregnancy influence fetal development and health. *Psychoneuroendocrinology*. 2005;30(8):724–743.
8. Rogal SS, Poschman K, Belanger K, et al. Effects of posttraumatic stress disorder on pregnancy outcomes. *J Affect Disord*. 2007;102(1): 137–143.
9. Buckley SJ. *Executive Summary. Hormonal Physiology of Childbearing: Evidence and Implications for Women, Babies, and Maternity Care*. Washington, DC: Childbirth Connection Programs, National Partnership for Women & Families; 2015.
10. Boorman RJ, Devilly GJ, Gamble J, Creedy DK, Fenwick J. Childbirth and criteria for traumatic events. *Midwifery*. 2014;30(2): 255–261.
11. Association AP. *Diagnostic and Statistical Manual of Mental Disorders*. 4th ed., Text Revision Washington, DC: American Psychiatric Publishing; 2000.
12. Association AP. *Diagnostic and Statistical Manual of Mental Disorders*. 5th ed. Arlington, VA: American Psychiatric Publishing; 2013.
13. Sawyer A, Nakić Radoš S, Ayers S, Burn E. Personal growth in UK and Croatian women following childbirth: a preliminary study. *J Reprod Infant Psychol*. 2014;33(3):294–307 [ahead-of-print].
14. Leckman JF, Goodman WK, North WG, et al. The role of central oxytocin in obsessive compulsive disorder and related normal behavior. *Psychoneuroendocrinology*. 1994;19(8):723–749.
15. Paul IM, Downs DS, Schaefer EW, Beiler JS, Weisman CS. Postpartum anxiety and maternal-infant health outcomes. *Pediatrics*. 2013;131(4):e1218–e1224.
16. Ayers S, Wright DB, Ford E. Hyperarousal symptoms after traumatic and nontraumatic births. *J Reprod Infant Psychol*. 2015;33 (3):282–293 [ahead-of-print].
17. Lazarus RS, Folkman S. *Stress, Appraisal and Coping*. New York: Springer Publishing; 1984.
18. Ayers S. Delivery as a traumatic event: prevalence, risk factors, and treatment for postnatal posttraumatic stress disorder. *Clin Obstet Gynecol*. 2004;47(3):552–567.
19. Ayers S, Bond, R, Bertullies S, Wijma K. The etiology of posttraumatic stress disorder following childbirth: a meta-analysis. Psychol Med; in press.
20. Slade P. Towards a conceptual framework for understanding post-traumatic stress symptoms following childbirth and implications for further research. *J Psychosom Obstet Gynaecol*. 2006;27(2):99–105.
21. Ehlers A, Clark DM. A cognitive model of posttraumatic stress disorder. *Behav Res Ther*. 2000;38(4):319–345.
22. Ford E, Ayers S, Bradley R. Exploration of a cognitive model to predict post-traumatic stress symptoms following childbirth. *J Anxiety Disord*. 2010;24(3):353–359.
23. Fenech G, Thomson G. Tormented by ghosts from their past': a meta-synthesis to explore the psychosocial implications of a traumatic birth on maternal well-being. *Midwifery*. 2014;30(2):185–193.
24. Nilsson C, Bondas T, Lundgren I. Previous birth experience in women with intense fear of childbirth. *J Obstet Gynecol Neonat Nurs*. 2010;39(3):298–309.
25. Muller-Nix C, Forcada-Guex M, Pierrehumbert B, Jaunin L, Borghini A, Ansermet F. Prematurity, maternal stress and mother-child interactions. *Early Hum Dev*. 2004;79(2):145–158.
26. Ayers S, Wright DB, Wells N. Symptoms of post-traumatic stress disorder in couples after birth: association with the couple's relationship and parent-baby bond. *J Reprod Infant Psychol*. 2007;25(1):40–50.
27. Swain JE, Tasgin E, Mayes LC, Feldman R, Todd Constable R, Leckman JF. Maternal brain response to own baby-cry is affected by cesarean section delivery. *J Child Psychol Psychiatry*. 2008;49 (10):1042–1052.
28. Prior E, Santhakumaran S, Gale C, Philipps LH, Modi N, Hyde MJ. Breastfeeding after cesarean delivery: a systematic review and meta-analysis of world literature. *Am J Clin Nutr*. 2012;95(5):1113–1135.
29. Meaney MJ. Epigenetics and the biological definition of gene × environment interactions. *Child Dev*. 2010;81(1):41–79.
30. Schechter DS, Willheim E, Hinojosa C, et al. Subjective and objective measures of parent-child relationship dysfunction, child separation distress, and joint attention. *Psychiatry*. 2010;73(2):130–144.
31. Grace SL, Evindar A, Stewart D. The effect of postpartum depression on child cognitive development and behavior: a review and critical analysis of the literature. *Arch Womens Ment Health*. 2003; 6(4):263–274.
32. Gottvall K, Waldenström U. Does a traumatic birth experience have an impact on future reproduction? *BJOG*. 2002;109(3):254–260.
33. Waldenström U, Hildingsson I, Ryding E-L. Antenatal fear of childbirth and its association with subsequent caesarean section and experience of childbirth. *BJOG*. 2006;113(6):638–646.
34. Beck CT, Watson S. Subsequent childbirth after a previous traumatic birth. *Nurs Res*. 2010;59(4):241–249.
35. Beck CT. Secondary traumatic stress in nurses: a systematic review. *Arch Psychiatr Nurs*. 2011;25(1):1–10.
36. Excellence NIoC. Antenatal and postnatal mental health: clinical management and service guidance. CG192; 2014.
37. Bastos MH, Furuta M, Small R, McKenzie-McHarg K, Bick D. Debriefing interventions for the prevention of psychological trauma in women following childbirth. *Cochrane Database Syst Rev*. 2015;4: CD007194.
38. Bisson J, Andrew M. Psychological treatment of post-traumatic stress disorder (PTSD). *Cochrane Database Syst Rev*. 2007;18(3): CD003388.
39. Ayers S, McKenzie-McHarg K, Eagle A. Cognitive behaviour therapy for postnatal post-traumatic stress disorder: case studies. *J Psychosom Obstet Gynecol*. 2007;28(3):177–184.
40. Horsch A. Post traumatic stress disorder following childbirth and pregnancy loss. *Clin Psychol Pract*. 2009;274–287.

CHAPTER

40

Stress Generation

K.L. Harkness, D. Washburn
Queen's University, Kingston, ON, Canada

OUTLINE

Abstract

The stress generation hypothesis posits that individuals with particular personality, cognitive, and interpersonal vulnerabilities generate stressful life events. As a result, individuals with these vulnerabilities are theorized to experience higher rates of life events that may be due to their own behavior and characteristics than nonvulnerable individuals. In contrast, these two groups are not expected to differ in rates of independent life events. The stress generation hypothesis has been very influential to theories of the role of stress in the onset and maintenance of diseases, such as depression, by proposing that individuals are not passive victims of stress, but instead actively contribute to their environments. This chapter will review the state of the literature with an emphasis on predictors and moderators of stress generation effects. The chapter will conclude with directions for future research and implications of the stress generation hypothesis for the treatment and prevention of stress-related disease.

INTRODUCTION

Major stressful life events (SLEs), such as the loss of a romantic relationship or getting fired from one's job, are strongly predictive of the onset and maintenance of various forms of psychopathology, most notably major depression.[1] Indeed, SLEs are the strongest proximal causal factor in depression, and significantly moderate the effects of biogenetic, early environmental, and trait vulnerabilities on depression onset.[2] One of the most compelling advances in the field of life events is the understanding that the reverse causal association is also evident; that is, stable vulnerability factors, including a history of depression, are associated with the generation of SLEs. Specifically, Hammen's[3,4] stress generation hypothesis proposes that individuals with a history of depression generate SLEs in their environment due to maladaptive personality patterns and disrupted social support networks that persist as scars following depressive episodes. This hypothesis is innovative in proposing that individuals are not passive victims of stress, but instead actively contribute, however implicitly, to creating their own environments. Current theoretical formulations of stress have been heavily influenced by the stress generation hypothesis, and now emphasize the bidirectional and transactional relations between SLEs and state and trait characteristics across development and over the ongoing pathology of disorders such as depression.[5]

In the 25 years since Hammen's original formulation of the stress generation hypothesis, over 100 empirical articles have been published testing and expanding on its basic premise, including two comprehensive reviews.[4,6] This chapter will review the current state of this literature with a particular emphasis on predictors and moderators of stress generation. The stress generation hypothesis was originally formulated within the context of depression,

Stress: Concepts, Cognition, Emotion, and Behavior
http://dx.doi.org/10.1016/B978-0-12-800951-2.00041-8

and has only very recently been applied to other forms of psychopathology. Therefore, given space constraints, the current review will focus primarily on stress generation in depression.

KEY POINTS

- The stress generation hypothesis proposes that individuals with a history of depression generate stressful life events due to maladaptive personality characteristics and disrupted social support networks that are a consequence of depression.
- Cross-sectional and prospective studies confirm that those with depression experience higher rates of events that are at least partly dependent on their behavior than nondepressed individuals.
- Negative cognitive style, maladaptive interpersonal behavior, and a history of stress in childhood are vulnerability factors that are prospectively associated with stress generation.
- The risk allele of the serotonin gene raises risk for stress generation, particularly in the presence of childhood maltreatment.
- Women are more likely to generate interpersonal stress than men, and this is mediated by the personality trait of interpersonal dependency.

TABLE 1 Examples of Independent, Dependent-Interpersonal, and Dependent-Noninterpersonal Stress Life Events

Independent	Earthquake causes damage to foundation, costing S $50,000 to repair S[a] is laid off from his job as a machinist due to the closure of his department S's sister, with whom she is close, is diagnosed with multiple sclerosis S's close friend attempts suicide following a romantic relationship breakup A routine mammogram discovers a lump in S's breast. S's brother and his wife have their first child. This is S's first niece.
Dependent-interpersonal	S and her mother have a heated argument, which results in not speaking for several days. S breaks up with his girlfriend of 6 months because he feels they are not compatible. S supports her best friend who is in crisis following an episode of spousal abuse. S decides to reconcile with his sister after 2 years of not speaking. S and her four siblings have a large argument regarding the distribution of their father's estate. S serves as the maid-of-honor at her sister's wedding.
Dependent-noninterpersonal	S is fired from her job as a cashier due to chronic tardiness. S causes a fender-bender. There were no injuries or damages, but her insurance will go up. S finds out he did very poorly on his graduate record examination. S receives a notice from creditors that they will begin garnisheeing his wages. S moves from her parent's house to a rented house with three friends.

[a] *S = Subject (the respondent, or actor in the event).*

DEFINITIONAL AND METHODOLOGICAL CONSIDERATIONS

An important advance provided by the stress generation hypothesis is the understanding that SLEs are not a unitary construct and that certain stressors are more likely to be generated than others. Specifically, Hammen[3] made an important distinction between SLEs that are at least in part dependent on the behavior or characteristics of the individual, and SLEs that are independent of the individual's actions. The stress generation hypothesis states specifically that individuals with a history of depression will have higher rates of *dependent* SLEs than nondepressed individuals, whereas depressed and nondepressed individuals will not differ in rates of *independent* SLEs.[3] Further, dependent SLEs in the *interpersonal* domain are theorized to be most strongly relevant to stress generation in depression.[4] Table 1 presents examples of each type of SLE. These examples are not exhaustive but simply represent examples of the sorts of SLEs that would be included in each category. Also, the independence of the SLE is not connected to its severity, or threat. Therefore, each category includes events that represent the full range of severity.

It is important to note two points of clarification with regard to the SLEs implicated in the stress generation hypothesis. First, we *all* generate life events through our intentional actions (e.g., we break up with our romantic partner; we start and quit jobs; we move house; etc.), or sometimes unintentionally (e.g., we get "dumped" by our romantic partners; we get fired; we have an unplanned pregnancy; etc.). Therefore, the stress generation hypothesis does not state that individuals with depression generate SLEs and nondepressed individuals do not. Instead, it asserts that depressed individuals will generate *higher rates* of dependent, and particularly interpersonal, SLEs than nondepressed individuals, again because maladaptive characteristics and behaviors make the occurrence of these SLEs more likely.

Second, some studies comparing samples of individuals with severe major depression to nondepressed controls have found evidence for higher prospective rates of *independent* events.[7] These researchers have suggested that while the depressed individuals may not have played a role in *causing* these events, severely depressed individuals may be more likely than nondepressed individuals to live in family and peer contexts that make these sorts of events more likely. For example, while I may not *cause* my brother's suicide attempt, or my best friend's pregnancy scare, or my parents' divorce (i.e., these are, indeed, events that are independent of my behavior), these sorts of events may be more likely to occur in my environment if I suffer from the severe syndrome of major depression than if I have always been free of psychopathology (e.g., through the mechanisms of assortative mating, intergenerational transmission of psychopathology, etc.). There is controversy regarding whether exposure to a higher context of independent events is the same as stress generation, but it does raise an important caveat regarding the strict dichotomy between independent and dependent events.

In the stress generation literature SLEs have been assessed using two main methodologies: self-report checklists or contextual interview and rating systems. The main advantage of checklist measures of major SLEs or minor daily hassles is that they are easy to administer and score. However, checklists are subject to depressive and other cognitive biases that can serve to inflate reports of stress.[8] Furthermore, checklist approaches assume that all events of a certain category are equally stressful across individuals. However, individuals can vary greatly in their experience of the "same" event. For example, "loss of job" will have very different implications for a 35-year-old single mother who is fired than for a 20-year-old woman fully supported by her parents who is laid off because the shop she was working at shut down. The checklist approach would consider these two events equal in terms of severity despite the important contextual differences.[1] Even more importantly in the present discussion, "loss of job" may be independent in one context (e.g., laid off due to shop closure) and dependent in another (e.g., fired). An inability to take context into account makes it very difficult, if not impossible, for checklist measures to fully test hypotheses germane to stress generation.

Several groups have improved the reliability and validity of life event assessment by developing semi-structured life event interviews and rating systems.[1] These interviews encourage respondents to talk at length about the context of each event. Following the interview, trained raters who are unaware of the respondent's clinical status and subjective reactions to the events apply operational criteria to determine the severity and independence of each event based on the relevant contextual information. For example, the differences in the lives of the two women in the example above would be taken into account in the severity rating, and determinations of independence would be clear. A comparison of checklist versus interview measures suggests that the interview measures have superior reliability and validity.[8] Although the interview method requires considerably more time and labor, the benefits associated with these more rigorous measures yield greater precision and validity than self-report measures.

SUMMARY

The stress generation hypothesis posits that individuals with particular personality, cognitive, and interpersonal vulnerabilities generate SLEs. As a result, these individuals experience higher rates of SLEs that are dependent on their own behavior than nonvulnerable individuals. In contrast, these two groups are not expected to differ in rates of independent SLEs. The stress generation hypothesis has been hugely influential to theories of the role of stress in the onset and maintenance of diseases such as depression, by proposing that individuals actively contribute to their environments.

Individuals with a history of depression generate higher rates of dependent, interpersonal SLEs than controls, and this generated stress mediates prospective increases in depression symptoms. In contrast, depression is not associated with higher rates of independent SLEs. The generation of dependent-interpersonal stress is predicted by a negative cognitive style focused on interpersonal dependency, maladaptive interpersonal behaviors, such as ERS, and an early context of childhood maltreatment that sets the developmental stage for these vulnerabilities. Interpersonal stress generation is stronger in women than in men and is moderated by polymorphisms in genes known to affect the stress response. Future research is focusing on interventions to prevent stress generation and, ultimately, reduce rates of stress-related disease.

EVIDENCE FOR STRESS GENERATION

Hammen[3] originally proposed that individuals with a *history* of depression should be particularly vulnerable to generating SLEs. Consistent with this early formulation, Hammen[3] reported that women with recurrent unipolar major depression reported significantly more dependent and interpersonal SLEs, as assessed by contextual interview, than matched samples of women with bipolar disorder, chronic medical illness, and healthy controls. In contrast, no differences across groups were found for independent SLEs. This result was confirmed in a study reporting that adult outpatients with recurrent unipolar depression reported higher rates of dependent, but not independent, SLEs, assessed with a contextual interview, than patients on their first episode of depression.[9]

Prospective, longitudinal studies have provided even more compelling evidence for the role of depression in generating stress. For example, in a sample of 140 young women making the transition from adolescence to young adulthood, Daley and colleagues found that women with unipolar major depression, and particularly depression comorbid with another Axis I disorder, at time 1 reported significantly higher rates of dependent and conflict SLEs, assessed by contextual interview, over a 1-year interval than nondepressed controls.[10] In contrast, no group difference was found for independent SLEs. A second study of married couples found that women's initial level of depression symptoms predicted the level of dependent, interpersonal stress reported over a 1-year follow-up interval, and this generated stress mediated the relation of time 1 to time 2 depression symptoms.[11] This latter finding is consistent with the theorized mechanism of stress generation; that is, depression leads to the generation of SLEs that contribute to ongoing psychopathology.

One of the most ambitious prospective tests of stress generation included two independent samples of over 500 adolescents covering up to 12 waves of prospective follow-up over 6 years.[12] In both samples they found that depression scores at previous waves predicted increases in general SLEs, assessed by checklist, at the next wave, and these SLEs predicted subsequent prospective increases in depression symptoms. In this study the investigators were able to model trait depression (i.e., the aspect of depression that is stable across time) and state depression (i.e., the aspect of depression that varies with time). Their depression-stress-depression links were stronger for trait depression than state depression, and they became stronger over time. Therefore, these results support the hypothesis that the tendency to generate stress is a trait disposition that emerges over development.

PREDICTORS OF STRESS GENERATION

If the tendency to generate stress is a trait, then it should be present even prior to the onset of depression in individuals with preexisting depressive vulnerabilities. As Hammen[3] originally noted, stress generation is a consequence of intrapersonal vulnerabilities (maladaptive personality patterns) and interpersonal vulnerabilities (disrupted social support). These vulnerabilities were originally conceptualized as scar consequences of depression. However, a large amount of research conducted since Hammen's original paper has confirmed that these vulnerability factors are observable even before the onset of depression and, indeed, are strong predictors of depression onset in the face of SLEs. Therefore, a compelling hypothesis emerging from stress generation theory is that preexisting personality, cognitive, and interpersonal vulnerabilities may heighten the generation of SLEs, which then predict the onset of depression and/or other forms of psychopathology.

Negative Cognitive Style

A large number of studies in diverse samples have provided support for a direct relation between maladaptive traits, and particularly a negative cognitive style, and high rates of SLEs. Negative cognitive style in these studies has been variously defined as, for example, negative cognitive schemas (i.e., negative core beliefs about the self, world, or future), a negative inferential style (i.e., a tendency to make global, stable attributions about negative events), or rumination (i.e., a tendency to cognitively elaborate upon one's distress). For example, in a large 6-month prospective study of 1187 community-sampled adolescents, both time 1 depressive symptoms and cognitive vulnerability (schemas and negative inferential style) predicted time 2 SLEs assessed by checklist, although these SLEs were not broken down by type (e.g., interpersonal, dependent, or independent).[13] Similarly, in a sample of 157 college students with a history of depression, those characterized as high in negative cognitive style reported significantly more dependent and interpersonal SLEs (but not independent or noninterpersonal SLEs), assessed by contextual interview, over a 6-month follow-up than those characterized as low in negative cognitive style.[14] This finding was replicated in a sample of 301 early adolescents. Specifically, time 1 scores on a measure of negative cognitive style predicted dependent-interpersonal SLEs and relational victimization over a 9-month follow-up.[15] Further, in this latter study, the generated stress mediated the relation of negative cognitive style to subsequent depression symptoms. A negative cognitive style may lead to the generation of stress through several mechanisms, but the one proposed most consistently is that a stable tendency toward negative thinking may result in maladaptive interpersonal behaviors that erode individuals' support networks, thereby increasing the probability of tension, conflict, and even rejection in interpersonal relationships.[15]

Maladaptive Interpersonal Behavior

Empirical study of the toxic role of maladaptive patterns of interpersonal behavior on the generation of stress has focused primarily on excessive reassurance seeking (ERS). ERS is defined as repeated requests for assurance of one's lovability and worth from close others that persist despite the provision of these assurances.[16] ERS is theorized to lead directly to a degrading of relationships, resulting in the generation of interpersonal conflict and, ultimately, rejection.[16] A number of prospective studies have confirmed this model, finding that high levels of self-reported ERS behaviors at time 1 predict the occurrence of general interpersonal SLEs reported over brief follow-up periods ranging from 4 to 5 weeks in college student samples.[17–19] A recent study replicated and extended this finding in a study of 118 college student romantic couples. High self-reported ERS behaviors in the women at time 1 significantly predicted greater likelihood of being rejected (dumped) by their male partner, and a shorter time to rejection, over an 18-month follow-up.[20]

Given the theorized pathway from negative cognitive style to maladaptive interpersonal behaviors to interpersonal stress generation, several investigators have sought to test an integrated cognitive-interpersonal model of stress generation. For example, in a sample of 66 adults with major depression, those who endorsed a negative cognitive style specifically focused on fears of interpersonal rejection (high rejection sensitivity) reported higher rates of dependent, but not independent, SLEs over a 4-month follow-up, as assessed by contextual interview.[21] Further, in a sample of 122 college students, depressive rumination prospectively predicted dependent, interpersonal SLEs over a 9-month follow-up, assessed by checklist and follow-up interview.[22] This relation was mediated by low perceptions of social support. In contrast, an earlier study of 198 college students failed to find that ERS mediated the relation of the specific negative cognitive schema of self-criticism and rates of dependent SLEs over 5 weeks.[18] However, SLEs in this study were assessed with a brief self-report checklist. Further, it is possible that a 5-week period may not be enough time to see the transactional relations between cognition and interpersonal behavior that result in the generation of SLEs. The balance of the literature, however, implicates maladaptive interpersonal behavior as a strong predictor of stress generation that may drive the relation of stable trait characteristics to the generation of SLEs particularly in the interpersonal domain.

Childhood Maltreatment

Patterns of negative thinking and maladaptive interpersonal behavior are theorized to emerge in a transactional manner over development through relationships with primary attachment figures.[23] Therefore, a history of parental maltreatment may set an early stage for stress generation. Consistent with this hypothesis, in two independent samples of undergraduates, a history of emotional abuse and parental discord in childhood were significantly associated with elevated rates of SLEs experienced over a 10-week (sample 1) and a 2-year (sample 2) prospective period.[24] Further, the generated stress mediated the relation of childhood maltreatment to time 2 depressive symptoms. Similarly, in a sample of 208 undergraduate students, a history of emotional abuse in childhood predicted increases in daily hassles reported on a weekly basis over an 8-week follow-up, and the generated hassles significantly mediated the relation of abuse to time 2 depression symptoms.[25] That is, the results of these studies suggest that childhood maltreatment may cause depression at least in part through the mechanism of stress generation.

The relation of childhood maltreatment to stress generation has also been found in samples of individuals with major depression. For example, in a sample of 58 adolescents in a current episode of major depression, those with a history of severe childhood maltreatment reported a significant increase in dependent, interpersonal SLEs from the 3-months preonset to the 3-months postonset, assessed by contextual interview, whereas this increase was not found in those with no history of maltreatment, nor was it found for independent or noninterpersonal SLEs.[26] Similarly, in a sample of 66 young adults with a history of major depression, childhood emotional abuse significantly predicted higher rates of negative dependent, but not independent, SLEs over a 4-month follow-up, assessed by contextual interview.[27] Emotional abuse may be especially and preferentially predictive of interpersonal stress generation because the content of the abuse (e.g., harsh criticism, name-calling, hostility) provides the child with the cognitive content that supports the development of schemas of worthlessness and failure.[25] Consistent with the strong role of childhood maltreatment in setting up patterns of negative thinking, Liu and colleagues[27] found that negative cognitive style fully mediated the relation of emotional abuse to stress generation.

The studies reviewed above provide preliminary evidence for an emerging developmental model of stress generation. In this model, experiences of significant stress and trauma in childhood shape the development of negative cognitive schemas and a negative cognitive style. This pervasive negative cognition underlies the maladaptive behaviors, such as ERS, that disrupt social support and lead to the generation of stress in interpersonal relationships. The SLEs that are generated may be the very ones that trigger onsets of depression, which then further perpetuate the stress generation process as part of the chronic and recurrent pathology of the illness. A very important further direction for research on stress generation is to provide a full test of this transactional model. A further important question is to determine the extent to which this model is moderated by variables that have been shown to influence exposure to stress.

MODERATORS OF STRESS GENERATION

Genetic Factors

Evidence reviewed above indicating that stable dispositional factors, such as a negative cognitive style, are associated with stress generation suggests that the tendency to generate stress may be, at least in part, genetically determined. For example, in Kendler and colleagues' pioneering twin studies, they found that approximately one third of the association between SLEs and depression could be accounted for by genetic factors that increased risk for both generating stress and having a depression onset.[2] To explain these findings, Kendler and his colleagues suggested "genes have feet" that lead individuals to select themselves into particular environments.

Molecular genetic work has focused on the serotonin-transporter-linked promoter region (5-HTTLPR) polymorphism of the serotonin transporter gene (SLC6A4) given the significant relation of this gene to stress sensitivity, rumination, and negative cognition.[28] Very recent studies have provided evidence that the 5-HTTLPR polymorphism of the SLC6A4 significantly moderates the relation of depression to stress generation. In a longitudinal study of 381 adolescents, those with the short(s) risk allele of the 5-HTTLPR, in the context of elevated levels of depression symptoms at age 15, had significantly higher levels of dependent and interpersonal SLEs at age 20, assessed by contextual interview, than those homozygous for the non-risk long (l)-allele. In contrast, no moderation relation was found for independent SLEs.[29] Consistent with the role of genetic factors in shaping cognitive and interpersonal functioning, this stress generation effect for s-allele carriers was further moderated in the same sample by cognitions about one's place in interpersonal relationships.[30]

As suggested above, the development of a stress generation trait, or "phenotype," results from the transactional relation of an individual's genotype and their environment over development. Support for this premise was recently found in a community sample of 297 young adults. Specifically, s-allele carriers of the 5-HTTLPR reported significantly higher rates of dependent and interpersonal SLEs, assessed by contextual interview, than those homozygous for the l-allele, but only if they also reported a history of maternal emotional maltreatment or sexual maltreatment.[31] This gene by early environment moderation relation was not found for independent SLEs. Therefore, genes and the early environment exert important interactive effects on stress generation, and these effects may be mediated by emerging cognitive and interpersonal vulnerabilities.

Sex Differences

Women, particularly young women, report experiencing more SLEs than young men, in both dependent and independent domains.[32,33] Evidence that women are exposed to higher rates of SLEs through the mechanism of stress generation is supported by some findings that young women report specifically higher rates of interpersonal stressors than men,[34] and that this stress exposure mediates the significantly higher rates of depression seen in women versus men.[5]

There is also evidence that sex moderates the effects of stable trait variables on stress generation. For example, in a sample of 99 college students, individuals high in the personality trait of sociotropy, defined as a trait tendency to high rejection sensitivity, reported significantly higher rates of dependent-interpersonal SLEs over a 6-week period than those low in sociotropy, but only in women.[35] Further, this sex difference in stress generation partially mediated the relation of sociotropy to depression symptoms at time 2. Similarly, in a sample of 350 adolescents followed over 5 months, high levels of anxious attachment, which characterizes individuals who fear rejection and abandonment, predicted the generation of dependent, interpersonal SLEs for girls but not for boys.[36] Interestingly, in this same study, a general negative cognitive style (i.e., not focused specifically on rejection sensitivity) was associated with the generation of dependent, interpersonal SLEs for boys but not girls. Further, heightened stress generation prospectively predicted increases in depression symptoms for both boys and girls. Therefore, interpersonal stress generation in girls and women may be specifically mediated by cognitions reflecting dependency in interpersonal relationships.

SUMMARY AND DIRECTIONS FOR FUTURE RESEARCH

The stress generation hypothesis has had an enormous impact on research and theory regarding the relation of SLEs to psychopathology, and has changed the way we think about individuals' agency in creating their own environmental contexts. In general, results from studies testing the stress generation hypothesis have confirmed that individuals with particular genetic, early environmental, cognitive, personality, and behavioral vulnerabilities prospectively generate higher rates of events that are at least in part dependent on their own behavior and characteristics than do individuals without these vulnerabilities. For depressive vulnerabilities, and particularly in women, these generated SLEs are especially interpersonal in nature. The model of stress generation that emerges is presented in Figure 1. Specifically, genetic vulnerabilities in serotonergic neurotransmission interact with a history of childhood maltreatment to produce a phenotype characterized by negative cognitive style and maladaptive personality traits focused on rejection sensitivity, anxious attachment, and interpersonal

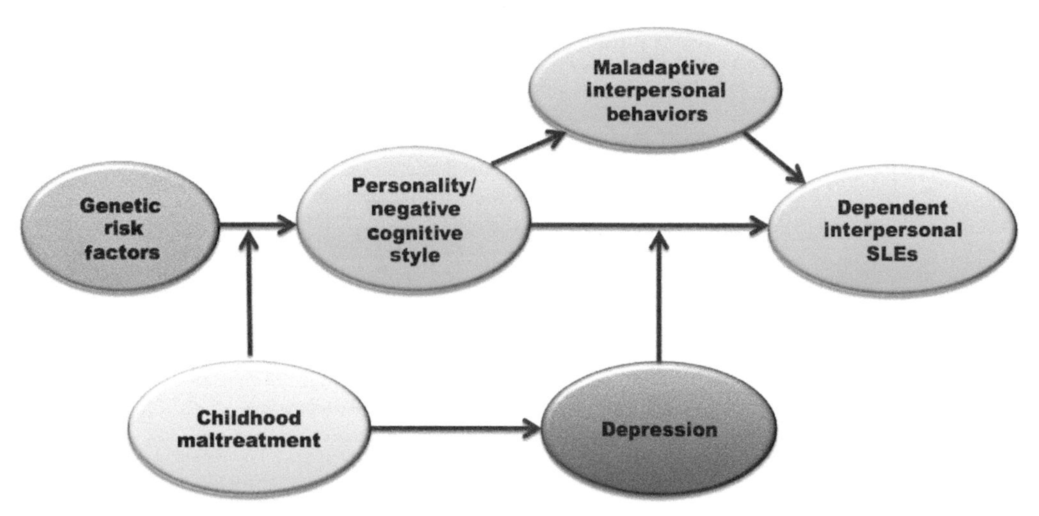

FIGURE 1 Theoretical model including predictors and moderators of stress generation.

dependency. This phenotype drives maladaptive interpersonal behaviors, such as ERS, which then generate tension, conflict, and ultimately rejection in peer, family, and romantic relationships. These dependent, interpersonal SLEs may be the very ones that then trigger onsets and recurrences of depression. It is important to note that the vast majority of studies reviewed in this chapter utilized rigorous, state-of-the-art contextual interview measures of SLEs, thereby assuring a valid and reliable assessment of the key construct in the model.

Despite the large amount of progress made in understanding the mechanism of stress generation, there are still a number of important areas of research that remain undeveloped. First, what is the mechanism of stress generation in men? Men generate SLEs, but the particular types of stressors that are generated, and the specific cognitive and behavioral mechanisms that drive stress generation, appear to be different from those of women.[34,35] Second, what is the mechanism of stress generation in psychopathologies other than unipolar depression? The basic stress generation phenomenon has been expanded to a range of other disorders, including anxiety disorders, eating disorders, and bipolar disorder, as well as chronic medical conditions. Indeed, recent work has proposed a transdiagnostic approach to understanding stress generation.[37] However, the mechanisms through which stress generation operates in these disorders have not been fully elucidated. In bipolar disorder, for example, there is evidence that hypomanic symptoms are prospectively associated with the generation of *positive* events.[38] Third, what are the implications of stress generation for treatment? The assertion that certain individuals may create their own stress as a result of their own behavioral shortcomings may sound like "blaming the victim." However, the good news about ascribing agency in creating stress is that this implies that making changes in identified aspects of cognition and behavior may result in a reduction in stress generation. Strategies for integrating interpersonal risk models, notably stress generation, into cognitive-behavioral treatment for depression were recently proposed.[39] These strategies include, for example, educating individuals about the interplay between cognitive and interpersonal factors in depression, modifying negative and maladaptive thoughts about relationships, and reducing ERS behaviors. The translation of basic research on stress generation mechanisms into evidence-based interventions has great potential for preventing the occurrence of dependent SLEs and, by extension, reducing rates of depression and other stress-related disease.

References

1. Harkness KL. Life events and hassles. In: Dozois DJA, Dobson KS, eds. *Risk Factors in Depression*. Oxford: Elsevier; 2008:317–342.
2. Kendler KS, Karkowski LM, Prescott CA. Causal relationship between stressful life events and the onset of major depression. *Am J Psychiatry*. 1999;156(June):837–841.
3. Hammen C. Generation of stress in the course of unipolar depression. *J Abnorm Psychol*. 1991;100(4):555–561. http://dx.doi.org/10.1037/0021-843X.100.4.555.
4. Hammen C. Stress generation in depression: reflections on origins, research, and future directions. *J Clin Psychol*. 2006;62(9):1065–1082. http://dx.doi.org/10.1002/jclp.
5. Hankin BL, Abramson LY. Development of gender differences in depression: an elaborated cognitive vulnerability-transactional stress theory. *Psychol Bull*. 2001;127(6):773–796.
6. Liu RT, Alloy LB. Stress generation in depression: a systematic review of the empirical literature and recommendations for future study. *Clin Psychol Rev*. 2010;30(5):582–593. http://dx.doi.org/10.1016/j.cpr.2010.04.010.Stress.
7. Harkness KL, Stewart JG. Symptom specificity and the prospective generation of life events in adolescence. *J Abnorm Psychol*. 2009;118(2):278–287. http://dx.doi.org/10.1037/a0015749.

8. Monroe SM. Modern approaches to conceptualizing and measuring human life stress. *Annu Rev Clin Psychol*. 2008;4:33–52. http://dx.doi.org/10.1146/annurev.clinpsy.4.022007.141207.
9. Harkness KL, Monroe SM, Simons AD, Thase ME. The generation of life events in recurrent and non-recurrent depression. *Psychol Med*. 1999;29:135–144.
10. Daley SE, Hammen C, Burge D, et al. Predictors of the generation of episodic stress: a longitudinal study of late adolescent women. *J Abnorm Psychol*. 1997;106(2):251–259. http://dx.doi.org/10.1037/0021-843X.106.2.251.
11. Davila J, Bradbury TN, Cohan CL, Tochluk S. Marital functioning and depressive symptoms: evidence for a stress generation model. *J Pers Soc Psychol*. 1997;73(4):849–861.
12. Cole DA, Nolen-Hoeksema S, Girgus J, Paul G. Stress exposure and stress generation in child and adolescent depression: a latent trait-state-error approach to longitudinal analyses. *J Abnorm Psychol*. 2006;115(1):40–51. http://dx.doi.org/10.1037/0021-843X.115.1.40.
13. Calvete E, Orue I, Hankin BL. Transactional relationships among cognitive vulnerabilities, stressors, and depressive symptoms in adolescence. *J Abnorm Child Psychol*. 2013;41(3):399–410. http://dx.doi.org/10.1007/s10802-012-9691-y.
14. Safford SM, Alloy LB, Abramson LY, Crossfield AG. Negative cognitive style as a predictor of negative life events in depression-prone individuals: a test of the stress generation hypothesis. *J Affect Disord*. 2007;99:147–154. http://dx.doi.org/10.1016/j.jad.2006.09.003.
15. Hamilton JL, Stange JP, Shapero BG, Connolly SL, Abramson LY, Alloy LB. Cognitive vulnerabilities as predictors of stress generation in early adolescence: pathway to depressive symptoms. *J Abnorm Child Psychol*. 2013;41(7):1027–1039. http://dx.doi.org/10.1007/s10802-013-9742-z.
16. Coyne JC. Depression and response of others. *J Abnorm Psychol*. 1976;85:186–193.
17. Eberhart NK, Hammen CL. Interpersonal predictors of stress generation. *Pers Soc Psychol Bull*. 2009;35(5):544–556. http://dx.doi.org/10.1177/0146167208329857.
18. Shahar G, Joiner Jr. TE, Zuroff CD, Blatt SJ. Personality, interpersonal behavior, and depression: co-existence of stress-specific moderating and mediating effects. *Pers Individ Dif*. 2004;36:1583–1596.
19. Potthoff JG, Holahan CJ, Joiner TE. Reassurance seeking, stress generation, and depressive symptoms: an integrative model. *J Pers Soc Psychol*. 1995;68(4):664–670. http://dx.doi.org/10.1037/0022-3514.68.4.664.
20. Stewart JG, Harkness KL. The interpersonal toxicity of excessive reassurance-seeking: evidence from a longitudinal study of romantic relationships. *J Soc Clin Psychol*. 2015;34:392–410.
21. Liu RT, Kraines MA, Massing-Schaffer M, Alloy LB. Rejection sensitivity and depression: mediation by stress generation. *Psychiatry*. 2014;77(1):86–97. http://dx.doi.org/10.1521/psyc.2014.77.1.86.
22. Flynn M, Kecmanovic J, Alloy LB. An examination of integrated cognitive-interpersonal vulnerability to depression: the role of rumination, perceived social support, and interpersonal stress generation. *Cognit Ther Res*. 2010;34(5):456–466. http://dx.doi.org/10.1007/s10608-010-9300-8.
23. Beck AT. The evolution of the cognitive model of depression and its neurobiological correlates. *Am J Psychiatry*. 2008;165:969–977.
24. Hankin BL. Childhood maltreatment and psychopathology: prospective tests of attachment, cognitive vulnerability, and stress as mediating processes. *Cognit Ther Res*. 2005;29(6):645–671. http://dx.doi.org/10.1007/s10608-005-9631-z.
25. Uhrlass DJ, Gibb BE. Childhood emotional maltreatment and the stress generation model of depression. *J Soc Clin Psychol*. 2007;26(1):119–130.
26. Harkness KL, Lumley MN, Truss AE. Stress generation in adolescent depression: the moderating role of child abuse and neglect. *J Abnorm Child Psychol*. 2008;36:421–432. http://dx.doi.org/10.1007/s10802-007-9188-2.
27. Liu RT, Choi JY, Boland EM, Mastin BM, Alloy LB. Childhood abuse and stress generation: the mediational effect of depressogenic cognitive styles. *Psychiatry Res*. 2013;206(2–3):217–222. http://dx.doi.org/10.1016/j.psychres.2012.12.001.
28. Pezawas L, Meyer-lindenberg A, Drabant EM, et al. 5-HTTLPR polymorphism impacts human cingulate-amygdala interactions: a genetic susceptibility mechanism for, depression. *Nat Neurosci*. 2005;8(6):828–834.
29. Starr LR, Hammen C, Brennan PA, Najman JM. Serotonin transporter gene as a predictor of stress generation in depression. *J Abnorm Psychol*. 2012;121(4):810–818. http://dx.doi.org/10.1037/a0027952.
30. Starr LR, Hammen C, Brennan PA, Najman JM. Relational security moderates the effect of serotonin transporter gene polymorphism (5-HTTLPR) on stress generation and depression among adolescents. *J Abnorm Child Psychol*. 2013;41(3):379–388. http://dx.doi.org/10.1007/s10802-012-9682-z.
31. Harkness KL, Bagby RM, Stewart JG, et al. Childhood emotional and sexual maltreatment moderate the relation of the serotonin transporter gene to stress generation. *J Abnorm Psychol*. 2015;124:275–287.
32. Hankin BL, Mermelstein R, Roesch L. Sex differences in adolescent depression: stress exposure and reactivity models. *Child Dev*. 2007;78(1):279–295. http://dx.doi.org/10.1111/j.1467-8624.2007.00997.x.
33. Harkness KL, Alavi N, Monroe SM, Slavich GM, Gotlib IH, Bagby RM. Gender differences in life events prior to onset of major depressive disorder: the moderating effect of age. *J Abnorm Psychol*. 2010;119(4):791–803. http://dx.doi.org/10.1037/a0020629.
34. Shih JH. Sex differences in stress generation: an examination of sociotropy/autonomy, stress, and depressive symptoms. *Pers Soc Psychol Bull*. 2006;32(4):434–446. http://dx.doi.org/10.1177/0146167205282739.
35. Shih JH, Eberhart NK, Hammen CL, Brennan PA. Differential exposure and reactivity to interpersonal stress predict sex differences in adolescent depression. *J Clin Child Adolesc Psychol*. 2006;35:103–115.
36. Shapero BG, Hankin BL, Barrocas AL. Stress generation and exposure in a multi-wave study of adolescents: transactional processes and sex differences. *J Soc Clin Psychol*. 2013;32(9):989–1012. http://dx.doi.org/10.1521/jscp.2013.32.9.989.
37. Bender RE, Alloy LB, Sylvia LG. Generation of life events in bipolar spectrum disorders: a re-examination and extension of the stress generation theory. *J Clin Psychol*. 2010;66(9):907–927. http://dx.doi.org/10.1002/jclp.
38. Conway CC, Hammen C, Brennan PA. Expanding stress generation theory: test of a transdiagnostic model. *J Abnorm Psychol*. 2012;121(3):754–766. http://dx.doi.org/10.1037/a0027457.
39. Dobson KS, Quigley L, Dozois DJ. Toward an integration of interpersonal risk models of depression and cognitive-behaviour therapy. *Aust Psychol*. 2014;49(6):328–336. http://dx.doi.org/10.1111/ap.12079.

CHAPTER

41

Caregivers and Stress

S.H. Zarit[1], J. Savla[2]

[1]Penn State University, State College, PA, USA

[2]Virginia Tech University, Blacksburg, VA, USA

OUTLINE

Abstract

Family care of a person with chronic physical or mental health problems is an increasingly common, and often stressful situation. Caregivers in intensive care situations, such as assisting a relative with dementia, often experience high levels of emotional distress and are at increased risk of illness. Stress models describe an unfolding process whereby illness-related stressors and caregivers' subjective experience of them can spill over to affect work, family relationships, and other areas of the caregiver's life. These stressors in turn affect critical physiological processes, health, and well-being. Skill-building interventions that improve caregivers' abilities in managing stressors show considerable promise in lowering subjective stress and improving well-being. Respite interventions, such as adult day services lower stressor exposure and lead to better regulation of stress responses at biological and psychological levels.

Caring for a family member who has chronic physical or mental health problems has become an increasingly common and often stressful activity. According to surveys conducted by the National Alliance for Caregiving, 29% of the adult population provides regular care to another person for a medical or behavior problem or disability.[1] Much of this care (70%) goes to people 50 years of age or older. As life expectancy has increased, more people live to advanced ages where disability is common. Among older people with long-term care needs, over three-quarters of care is given by family and friends, while another 14% receive a mix of paid help and help from family. Only a small proportion of the older population receives care entirely from paid sources or in institutions. The amount of care ranges from occasional help to a parent who is largely independent to around-the-clock assistance to an older person suffering from severe dementia. These chronic care situations are characterized by high levels of stressors directly related to the time and effort of providing assistance, and by the spillover of stressors into other areas of the caregiver's life such as employment, finances, and leisure activities. In this chapter, we emphasize care of older people, but we also note that there are other types of high stress care situations, for example, parents assisting an adult offspring with a severe disabling condition, such as autism, and parents or spouses assisting war veterans with traumatic brain injury and/or post-traumatic stress disorder.

Although families have always cared for their parents and grandparents, changes in contemporary society have added to the challenges they face. With women increasingly employed outside the home, the role of caregiver often falls on a spouse, who may have his or her own age-related limitations, or on a daughter or daughter-in-law, who must balance the multiple roles of worker, homemaker, caregiver, and, in some cases, mother to her own young children. Higher divorce rates as well as low growth in earnings and wealth during the last 30 years for a sizable portion of the population have reduced the resources that families can draw upon in caring for an older person or other special needs individual. It is not surprising, then, that much of the emphasis in

Stress: Concepts, Cognition, Emotion, and Behavior
http://dx.doi.org/10.1016/B978-0-12-800951-2.00042-X

research on caregiving has focused on the stressors and emotional strains of informal care. Collectively, these studies have found that caregivers are more likely to be depressed and angry than age-matched controls. Studies have also reported that caregivers in high stress situations have higher rates of illness and mortality than people the same age who are not caregivers.[2]

STRESS PROCESS MODEL AS A GUIDE

Caregiving stress has multiple components that interact over time and lead to adverse physical and health outcomes. An understanding of caregivers depends on recognizing the different types of stressors that can affect them. We draw on the widely used stress process model of caregiving[3] (Figure 1) to illustrate the types of stressors that can affect caregivers and the processes leading to adverse physical and mental health outcomes. Stress begins with the care responsibilities and demands placed on the caregiver, which are called *primary objective stressors*. Primary objective stressors include the type and frequency of care provided, such as a physically demanding task like lifting a care recipient from the bed or an emotionally demanding task such as responding to the care recipient's behavioral problems. Objective stressors have an immediate subjective impact on caregivers, called *primary subjective stressors*. Examples include feelings of overload for all the tasks one has to do or feelings of loss of valued aspects of the relationship with the other person. This distinction between objective and subjective stressors is central to understanding caregivers. People respond differently to different challenges, and so while it is important to know which tasks or challenges a caregiver may be facing, it is also necessary to identify which of those tasks are stressful or difficult for them. Furthermore, subjective stressors and the related construct of subjective burden have been found to account better for caregivers' health and emotional well-being than measures of objective stressors.[3]

The accumulation of primary stressors can lead to a proliferation of stress into other areas of life, or what are called *secondary role strains*. For instance, providing care may interfere with work responsibilities, with the time that caregivers have for activities with one's spouse and children or the caregiver's leisure and social activities. Role strains also have a subjective component. Some caregivers can manage competing roles fairly well, but others experience strain or problems from the multiple demands of these roles. Some caregivers, for example, have difficulty balancing work and care, while others find relief from the pressures of caregiving at work.[3,4] The build-up of primary and secondary stressors over time undermines emotional well-being and stimulates physiologic responses that result in subclinical and clinical illness.[5] This process, however, can be altered by resources that lessen the impact of stressors on caregivers. Income, social support, coping skills, self-efficacy, and

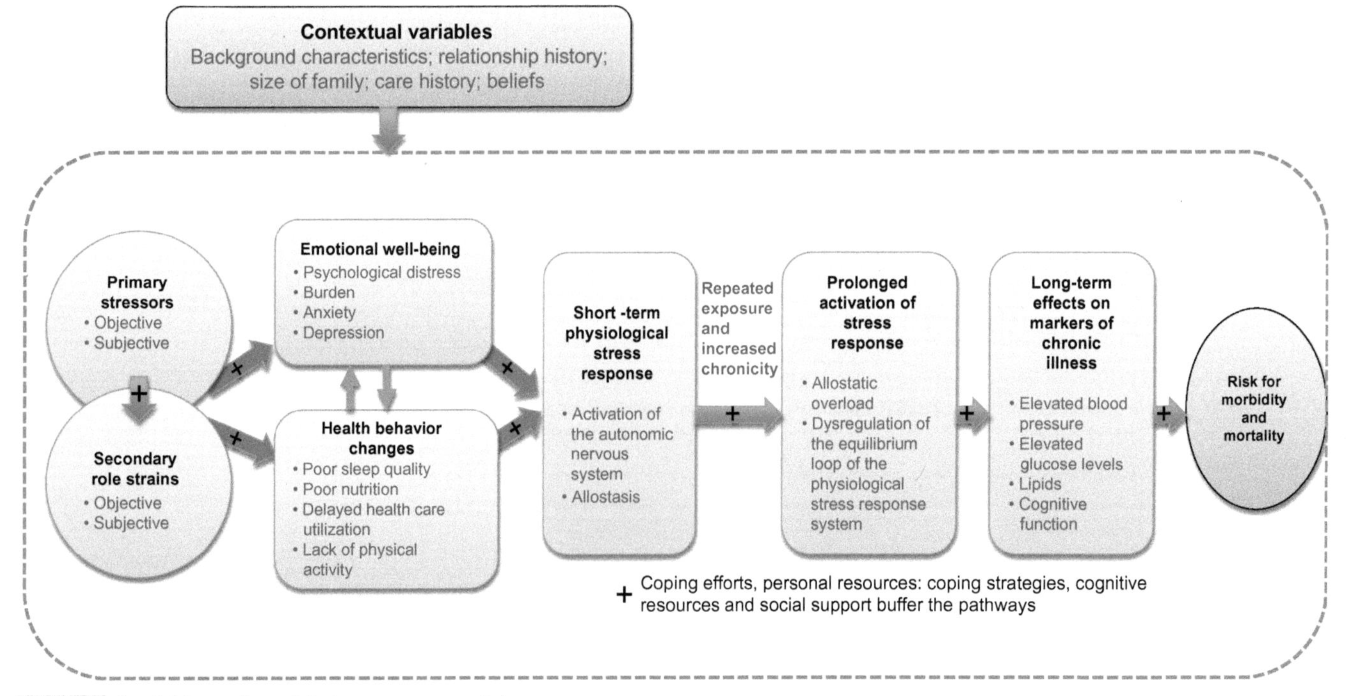

FIGURE 1 A biopsychosocial stress process model.

other resources can moderate the impact of stressors on caregivers' health and well-being.[6] This dynamic relationship between stressors and resources is the source of substantial individual variation in the extent of health consequences among caregivers.

HEALTH EFFECTS OF CARE-RELATED STRESS

Many types of health outcomes have been examined in the last two decades of caregiving research, ranging from psychological and physical health effects to cellular and biological measures. Early studies on the consequences of caregiving focused mainly on psychological well-being of caregivers. These studies consistently found that caregivers experienced high rates of symptoms of depression, anger, grief, anxiety, and a reduction in positive experiences. Caregivers also reported poor quality of life and rated their health to be worse than before.[3] More recently, caregiving research has broadened its focus to include both subjective reports of well-being and health and more objective indicators of health outcomes. These studies, for example, found a direct relationship between caregiving stress and hypertension, cardiovascular diseases, metabolic syndrome, and even mortality.[5,7]

The focus of many studies has been on identifying the mechanism by which caregiving stressors disrupt the physiologic processes and proliferate into chronic health conditions. These studies are built on the premise that the two major physiological stress pathways of the autonomic nervous system (i.e., the sympathetic-adrenal-medullary axis and the hypothalamic-pituitary-adrenal axis) are activated when we perceive a stressor, threat, or challenge. The response to this stressor instigates increased levels of circulating hormones such as α-amylase, cortisol, and dehydroepiandrosterone-sulfate (DHEA-S). Each time the stress response system is aroused, physiological adjustments are made to adapt to the stressors and bring the body back to equilibrium.[8] Over time and with repeated and prolonged activation response to stressors, the allostatic processes wear down and fail at achieving homeostasis—this condition is termed *allostatic overload*.[9–11] This state of allostatic overload results in changes such as tissue damage, dysregulation of the immune system, and cardiovascular and metabolic changes, all of which increases the risks of clinical and subclinical diseases. Studies have confirmed that with chronic stress, caregivers show diminished cortisol and DHEA-S production, and elevated levels of interleukin (IL)-6, C-reactive protein, and D-dimer, which are indicative of cognitive decline, depression, inflammation, vascular diseases, and frailty.[12,13] Other studies demonstrated changes in immune system functioning.[14] Moreover, coping and resiliency factors such as self-efficacy have been found to have a protective influence on various physiologic processes. For instance, caregivers with high self-efficacy were found to have lower levels of IL-6, in comparison to caregivers with low self-efficacy.[15]

Another contributing factor is the bidirectional role of health behaviors. As an example, low levels of physical activity and poor sleep quality have been found to increase stress reactivity and contribute to poor health outcomes.[12,16] A caregiver, however, who provides several hours of hands-on care a day may not have enough time for exercise, thereby contributing to poor health outcomes. Studies suggest that interventions aimed at encouraging healthy lifestyle behaviors and opportunities for respite may be beneficial for distressed caregivers with health risks.[17]

INTERVENTION FOR CAREGIVERS

A growing body of research has identified interventions that help lower subjective burden and emotional distress among caregivers, and improve health and biological markers of stress. Most studies have focused on caregivers of persons suffering from neurocognitive disorders, particularly dementia, but recent literature looks at other types of problems where both caregiver and care receiver can be actively involved in a treatment program.[18] The goal of intervention programs is to reduce the impact of stressors on caregivers through three primary strategies: (1) providing and helping caregivers understand information about their relative's illness; (2) building caregivers' skills in managing primary stressors, particularly behavioral and emotional problems, as well as their own emotional reactions to these problems; and (3) helping caregivers identify and use support from family or paid sources that provides the primary caregiver with temporary relief from giving care.

Helping caregivers understand their relative's illness and what can be done for management of symptoms and disability provides the foundation for intervention. Caregivers need first to understand the prospects for medical treatment of their relative's illness and the extent to which recovery is possible before they are willing to make changes needed to adapt to the challenges that their relative's disabilities and other problems impose on daily life. In neurocognitive disorders in particular, caregivers often assume there must be some simple fix that will restore memory or they may assume that merely explaining to persons with dementia that they have memory loss will create the awareness needed to reduce recurring problems, such as asking the same question over and over again, or denying that anything is wrong.[19] Information about legal contingencies such as the importance of obtaining power of attorney for someone with reduced cognitive functioning is also of critical importance.

Interventions designed to improve skills in managing primary stressors have demonstrated that caregivers can learn to apply behavioral strategies that fall under the broad umbrella of "problem-solving." Problem-solving typically involves teaching caregivers how to utilize behavioral principles to develop new approaches to daily problems.[20,21] These approaches involved one-to-one training, but there are also examples of successful time-limited group interventions.[22] Behavior problems, such as agitation and restlessness, are particularly upsetting to caregivers but are responsive to behavioral control strategies. These approaches utilize several inter-related steps, including training caregivers to identify events that trigger or reinforce the care recipient's behavior problems and brain-storming to generate new strategies that modify these contingencies, for example, reducing exposure to triggering events or providing positive reinforcement for adaptive behaviors. Programs may also explicitly teach caregivers how to implement these behavioral changes and to evaluate the effects on their relative and themselves. Caregivers can use problem-solving to address other aspects of their lives, such as lack of support from relatives or difficulties managing work and caregiving responsibilities.

A limitation of skills-based training programs is that they often follow a fixed training protocol, but caregivers vary in terms of what types of problems they are experiencing, as well as what problems they want help with. One innovative approach, tailoring the treatment to focus on the issues that most concerned caregivers, was used in the second phase of the Resources for Enhancing Alzheimer's Caregiver Health (REACH II) project.[23,24] Based on an initial assessment, counselor and caregiver identified up to three problems that they would focus on in treatment. This approach has shown positive outcomes in a racially and ethnically diverse population for reducing depression and improving quality of life. It has also been adapted for implementation in community settings.[25]

Another way of addressing the variability of stressors experienced by caregivers is to focus treatment on one specific type of problem or challenge and then recruit caregivers who have that problem. That way, treatment can be designed to maximize the impact on that one problem. Three promising studies by Teri and her colleagues[26–29] illustrate the advantages of this approach. In the first study, they trained family caregivers to manage depressive symptoms among persons with dementia. They found that two treatments, problem-solving and increasing the care receiver's engagement in pleasant activities decreased depression for both the person with dementia and the caregiver compared to control groups.[26] In a second study, caregivers implemented a program of regular exercise with the care receiver, which, over time, resulted in less decline in functional ability compared to a control group.[27] A third intervention addressed sleep disturbance among persons with dementia, which is one of the more common and stressful problems faced by caregivers. Caregivers who implemented sleep hygiene strategies or a program of increased walking and bright light exposure reduced difficulties falling asleep and nighttime awakening among persons with dementia compared to control groups.[28,29]

One other recent approach has been to help caregivers maintain their emotional well-being through strategies such as mindfulness training and yoga. These studies have shown promising results in lowering depression among caregivers and other positive changes.[30] Since mindfulness trains caregivers to be thoughtful, rather than reactive to daily problems, combining this type of training with problem-solving approaches could help caregivers become better at implementing new skills.

Most interventions are short-term, but the course of caregiving can span many years. New skills may be needed as the care recipient's condition and caregiver's needs change. Evidence of long-term benefits of caregiver interventions, however, is limited. One exception is the NYU Caregiver Intervention.[31] This program provides training in skills for managing behavioral and other problems and uses treatment sessions that involve other family members as a way of increasing support from the caregiver's family. At the end of the formal training (about 16 weeks), caregivers are encouraged to join support groups for ongoing assistance and also can seek occasional help from the counselors who initially worked with them. Not surprisingly, findings show that initial benefits are sustained over a 3-year follow-up period (Figure 2).

Another way to address the long-term course of caregiving is the use of respite programs, such as adult day services (ADS), in-home respite, and overnight respite. In contrast to skills-based programs, which emphasize management of stressors, respite has the effect of decreasing exposure to stressors by giving caregivers regular time away from the care receiver.[32] A recent study compared family caregivers of persons with dementia on days they received ADS for their relative and days they provided all the care. Lower exposure to care-related stressors on ADS days resulted in reduced anger and depressive symptoms.[4] Further, ADS use led to improved regulation of two stress hormones, cortisol and DHEA-S.[33,34] Taken together, these effects of ADS use on emotions and stress hormones may help reduce the risk of illness. In other words, reducing exposure to stressors interrupts the stress process described earlier in the chapter. Caregivers who receive regular relief may also be able to care for their relative at home for a longer period, while levels of stress remain at a more manageable level. ADS programs that specifically address caregivers' psychological needs may further

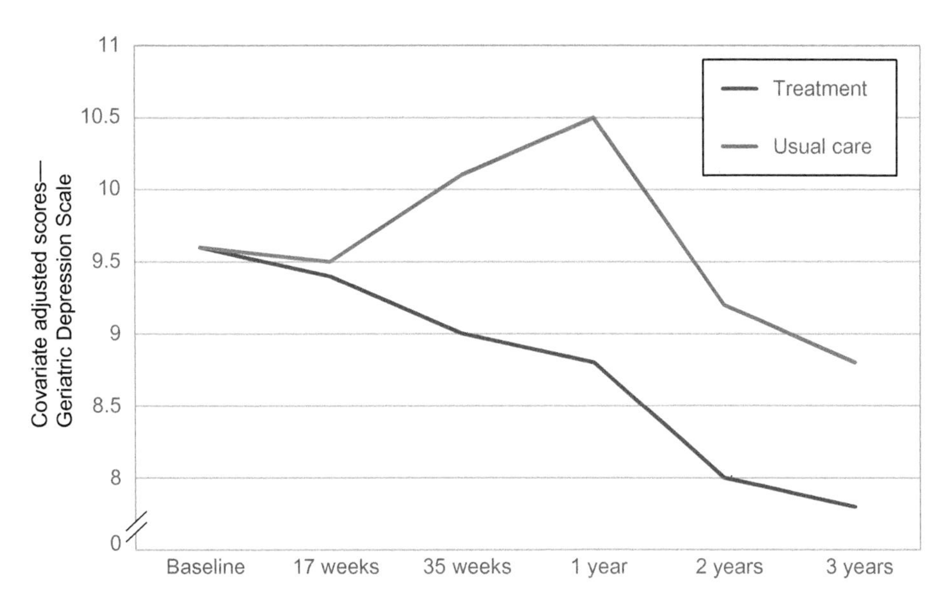

FIGURE 2 Long-term effects on depression of the New York University Caregiver Intervention. Data from Mittelman et al.[31]

increase the positive effects of respite.[35] Moreover, unlike time-limited interventions, ADS and other respite services can be sustained over the course of caregiving.

INSTITUTIONAL CARE

For many caregivers, the physical and emotional strain of providing care becomes too high, and they make the decision to place their relative in a care facility. Institutional care relieves the pressure on caregivers of providing around-the-clock care, but the placement decision and its aftermath are associated with a new set of stressors.[36] The decision to place can be very difficult, particularly for caregivers who are spouses, and can be accompanied by feelings of guilt and depression. Caregivers may also have problems finding an institutional setting that provides the quality of care they want and that they can afford. Their decision to place a parent or spouse may also become a source of conflict within the family. After placement, visiting the nursing home becomes a challenge for some caregivers, both in terms of the distance they have to travel to the facility and how they spend their time with their relative. Spouse caregivers, in particular, may visit the facility frequently and may also continue providing some daily care. Caregivers may experience frustration and conflict with staff around the care provided to their relative. Spouse caregivers may also experience feelings of loneliness. These stressors often go unaddressed, because everyone involved—health care professionals, family members, and caregivers themselves, assume that placement solves the caregiver's problems. Thus, the time leading up to and following placement can be particularly stressful for many caregivers. It is a time when they especially need the ongoing support of family members and friends.

SUMMARY

Family care has become a frequent and often stressful experience for many families, and the numbers of families providing assistance will only increase in the coming years as the number of older people in the population continues to grow. Caregiving stress is multidimensional, and encompasses the direct support that caregivers provide to the care recipient as well as the spillover of these stressors into other areas of the caregiver's life, such as work, family life, and leisure activities. Caregivers have increased risk of physical and mental health problems and mortality compared to age-matched persons who are not caregivers. Interventions can lower stress by training caregivers in skills that improve how they manage the daily challenges they face and that provide respite from care-related activities. Research suggests that sustained interventions have the potential to reduce the risks to caregivers' physical and mental health. Unfortunately, it is still the case that most family caregivers get little or no help.

References

1. National Alliance for Caregiving. Caregiving in the U.S. 2009. Retrieved from http://www.caregiving.org/pdf/research/CaregivingUSAllAgesExecSum.pdf.
2. Lazzarino AI, Hamer M, Stamatakis E, Steptoe A. The combined association of psychological distress and socioeconomic status with all-cause mortality: a national cohort study. *JAMA Intern Med*. 2013;173:22–27.

3. Aneshensel C, Pearlin L, Mullan J, Zarit S, Whitlatch C. *Profiles in Caregiving: the Unexpected Career*. San Diego, CA: Academic Press; 1995.
4. Zarit SH, Kim K, Femia EE, Almeida DM, Klein LC. The effects of adult day services on family caregivers' daily stress, affect and health: outcomes from the DaSH study. *Gerontologist*. 2014;54:570–579. http://dx.doi.org/10.1093/geront/gnt045.
5. Vitaliano PP, Zhang J, Scanlan JM. Is caregiving hazardous to one's physical health? A meta-analysis. *Psychol Bull*. 2003;129:946–972. http://dx.doi.org/10.1037/0033-2909.129.6.946.
6. Mausbach BT, Patterson TL, Känel RV, et al. The attenuating effect of personal mastery on the relations between stress and Alzheimer caregiver health: a five-year longitudinal analysis. *Aging Ment Health*. 2007;11:637–644. http://dx.doi.org/10.1080/13607860701787043.
7. von Känel R, Mills PJ, Mausbach BT, et al. Effect of Alzheimer caregiving on circulating levels of C-reactive protein and other biomarkers relevant to cardiovascular disease risk: a longitudinal study. *Gerontology*. 2012;58:354–365. http://dx.doi.org/10.1159/000334219.
8. Savla J, Granger DA, Roberto KA, Davey A, Blieszner R, Gwazdauskas F. Cortisol and alpha-amylase reactivity to care-related stressors in spouses of persons with mild cognitive impairment. *Psychol Aging*. 2013;28:666–679. http://dx.doi.org/10.1037/a0032654.
9. McEwen BS. Stress, adaptation, and disease: allostasis and allostatic load. *Ann N Y Acad Sci*. 1998;840:33–44.
10. McEwen BS, Gianaros PJ. Stress- and allostasis-induced brain plasticity. *Annu Rev Med*. 2011;62:431–445. http://dx.doi.org/10.1146/annurev-med-052209-100430.
11. Selye H. The story of the adaptation syndrome. *JAMA*. 1952;224 (6):711. http://dx.doi.org/10.1097/00000441-195212000-00039.
12. Mills PJ, Ancoli-Israel S, Känel RV, et al. Effects of gender and dementia severity on Alzheimer's disease caregivers' sleep and biomarkers of coagulation and inflammation. *Brain Behav Immun*. 2009;23(5):605–610. http://dx.doi.org/10.1016/j.bbi.2008.09.014.
13. von Känel R, Dimsdale JE, Mills PJ, et al. Effect of Alzheimer caregiving stress and age on frailty markers interleukin-6, C-reactive protein, and D-dimer. *J Gerontol A Biol Sci Med Sci*. 2006;61 (9):963–969. http://dx.doi.org/10.1159/000264654.
14. Kiecolt-Glaser JK, McGuire L, Robles TF, Glaser R. Psychoneuroimmunology: psychological influences on immune function and health. *J Consult Clin Psychol*. 2002;70:537–547. http://dx.doi.org/10.1037//0022-006X.70.3.537.
15. Harmell AL, Chattillion EA, Roepke SK, Mausbach BT. A review of the psychobiology of dementia caregiving: a focus on resilience factors. *Curr Psychiatry Rep*. 2011;13:219–224. http://dx.doi.org/10.1007/s11920-011-0187-1.
16. Puterman E, O'Donovan A, Adler NE, et al. Physical activity moderates effects of stressor-induced rumination on cortisol reactivity. *Psychosom Med*. 2011;73:604–611. http://dx.doi.org/10.1097/PSY.0b013e318229e1e0.
17. von Känel R, Mausbach BT, Dimsdale JE, et al. Regular physical activity moderates cardiometabolic risk in Alzheimer's caregivers. *Med Sci Sports Exerc*. 2011;43:181–189. http://dx.doi.org/10.1249/MSS.0b013e3181e6d478.
18. Martire LM, Schulz R, Helgeson VS, Small BJ, Saghafi EM. Review and meta-analysis of couple-oriented interventions for chronic illness. *Ann Behav Med*. 2010;40:325–342. http://dx.doi.org/10.1007/s12160-010-9216-2.
19. Zarit SH, Zarit JM. *Mental Disorders in Older Adults*. 2nd ed. New York, NY: Guilford Press; 2007.
20. Coon DW, Thompson L, Steffen A, Sorocco K, Gallagher-Thompson D. Anger and depression management: psychoeducational skill training interventions for women caregivers of a relative with dementia. *Gerontologist*. 2003;43:678–689. http://dx.doi.org/10.1093/geront/43.5.678.
21. Livingston G, Barber J, Rapaport P, et al. Clinical effectiveness of a manual based coping strategy programme (START, STrAtegies for RelaTives) in promoting the mental health of carers of family members with dementia: pragmatic randomized controlled trial. *BMJ*. 2013;347:f6276. http://dx.doi.org/10.1136/bmj.f6276.
22. Hepburn KW, Tornatore J, Center B, Ostwald SW. Dementia family caregiver training: affecting beliefs about caregiving and caregiver outcomes. *J Am Geriatr Soc*. 2001;49:450–457.
23. Belle SH, Burgio L, Burns R, et al. Enhancing the quality of life of dementia caregivers from different ethnic or racial groups. *Ann Intern Med*. 2006;145:727–738. http://dx.doi.org/10.7326/0003-4819-145-10-200611210-00005.
24. Elliott AF, Burgio LD, DeCoster J. Enhancing caregiver health: findings from the resources for enhancing Alzheimer's caregiver health II intervention. *J Am Geriatr Soc*. 2010;58:30–37. http://dx.doi.org/10.1111/j.1532-5415.2009.02631.x.
25. Burgio LD, Collins IB, Schmid B, Wharton T, McCallum D, DeCoster J. Translating the REACH caregiver intervention for use by area agency on aging personnel: the REACH OUT program. *Gerontologist*. 2009;49:103–116. http://dx.doi.org/10.1093/geront/gnp012.
26. Teri L, Logsdon RG, Uomoto J, McCurry SM. Behavioral treatment of depression in dementia patients: a controlled clinical trial. *J Gerontol B Psychol Sci Soc Sci*. 1997;52:159–166. http://dx.doi.org/10.1093/geronb/52B.4.P159.
27. Teri L, Gibbons LE, McCurry SM, et al. Exercise plus behavioral management in patients with Alzheimer's disease: a randomized control trial. *JAMA*. 2003;290:2015–2022. http://dx.doi.org/10.1001/jama.290.15.2015.
28. McCurry SM, Gibbons LE, Logsdon RG, Vitiello MV, Teri L. Nighttime insomnia treatment and education for Alzheimer's disease: a randomized controlled trial. *J Am Geriatr Soc*. 2005;53:793–802. http://dx.doi.org/10.1111/j.1532-5415.2005.53252.x.
29. McCurry SM, Pike KC, Vitiello MV, Logsdon RG, Larson EV, Teri L. Increasing walking and bright light exposure to improve sleep in community-dwelling persons with Alzheimer's disease: results of a randomized, controlled trial. *J Am Geriatr Soc*. 2011;59:1393–1402. http://dx.doi.org/10.1111/j.1532-5415.2011.03519.x.
30. Hou RJ, Wong SY, Yip BH, et al. The effects of mindfulness-based stress reduction on the mental health of family caregivers: a randomized controlled trial. *Psychother Psychosom*. 2014;83:45–53. http://dx.doi.org/10.1159/000353278.
31. Mittelman MS, Roth DL, Coon DW, Haley WE. Sustained benefit of supportive intervention for depressive symptoms in caregivers of patients with Alzheimer's disease. *Am J Psychiatry*. 2004;161:850–856. http://dx.doi.org/10.1176/appi.ajp.161.5.850.
32. Zarit SH, Kim K, Femia EE, Almeida DM, Savla J, Molenaar PCM. Effects of adult day care on daily stress of caregivers: a within person approach. *J Gerontol B Psychol Sci Soc Sci*. 2011;66:538–547.
33. Klein LC, Kim K, Almeida DM, Femia EE, Rovine MJ, Zarit SH. Anticipating an easier day: effects of adult day services on daily cortisol and stress. *Gerontologist*. 2014; http://dx.doi.org/10.1083/geront/gne060. Advance online publication.
34. Zarit SH, Whetzel CA, Kim K, et al. Daily stressors and adult day service use by family caregivers: effects on depressive symptoms, positive mood and DHEA-S. *Am J Geriatr Psychiatry*. 2014;http://dx.doi.org/10.1016/j.jagp.2014.01.013. Advance online publication.
35. Gitlin LN, Reever K, Dennis MP, Mathieu E, Hauck WW. Enhancing quality of life of families who use adult day services: short- and long-term effects of the adult day services plus program. *Gerontologist*. 2006;46:630–639. http://dx.doi.org/10.1093/geront/46.5.630.
36. Gaugler JE, Zarit SH, Pearlin LI. Family involvement following institutionalization: modeling nursing home visits over time. *Int J Aging Hum Dev*. 2003;57:91–117. http://dx.doi.org/10.2190/8MNF-QMA3-A5TX-6QQ3.

CHAPTER

42

Fatigue and Stress☆

W.J. Kop, H.M. Kupper
Tilburg University, Tilburg, the Netherlands

OUTLINE

Abstract

High levels of fatigue that interfere with daily life activities are common in the general population with prevalence estimates ranging from 5% to 20%. Fatigue is a multidimensional construct with mental and physical aspects. Fatigue-related disorders (e.g., chronic fatigue syndrome) are often associated with elevated levels of psychological distress. Evidence suggests a bidirectional association between fatigue and stress. Fatigue and related constructs such as exhaustion can result from prolonged stressful demands. Biological correlates of fatigue include dysregulation of the autonomic, neurohormonal, and/or immune systems, but no specific biomarkers of fatigue exist. Stress-related psychological disorders (e.g., posttraumatic stress disorder and depression) have fatigue as a core characteristic. Fatigue is also a common symptom in various medical diseases including cardiovascular disease and cancer. The relationship between stress and fatigue may be important in the development of multidisciplinary interventions targeted at improving health-related outcomes.

A substantial number of individuals (5-20% of the general population) suffers from persistent fatigue that interferes with routine daily life activities.[1] Fatigue is also one of the most common complaints in primary care. Prevalence estimates of fatigue depend on the measurement instruments, cut-off points for presence versus absence of fatigue, and the persistence of the complaint.[1] Patients generally regard fatigue as important because it is disabling, whereas physicians often do not place a strong emphasis on fatigue because it is diagnostically nonspecific for most diseases. Episodes of extreme fatigue are often precipitated and maintained by psychological distress. In addition, multiple other factors contribute to the onset and maintenance of fatigue, including immune system-related processes, poor sleep, and functional limitations associated with general medical diseases. This chapter provides a selective review of the interrelationship between stress and fatigue and provides evidence for a bidirectional association between stress and fatigue.

Several theories have been developed to conceptualize and investigate fatigue. These theories can be broadly divided into five general perspectives: (a) fatigue as a diagnosable disorder; (b) fatigue as an energy-preserving condition in response to stress and other environmental demands; (c) fatigue as a consequence of dysregulated central nervous system, autonomic nervous system (ANS), neurohormonal, and/or immunological functioning; (d) fatigue as related to exogenous (work-related) demands; and/or (e) fatigue as a correlate of general medical diseases. These perspectives each have their strengths and limitations, and reflect the way in which scientists and clinicians operationalize and implement their approaches to investigate and treat this complex symptom. In the following sections, we will review these perspectives and their implications for the interrelationships between stress and fatigue.

☆ This chapter is based on a previous version published in the 2nd edition of this encyclopedia: Appels A, Kop WJ. Fatigue and stress. In: Fink G, ed. *Encyclopedia of Stress*, vol. 2, 2nd ed. Oxford: Academic Press; 2007:11–15.

http://dx.doi.org/10.1016/B978-0-12-800951-2.00043-1

KEY POINTS

- Fatigue is common in the general population (prevalence = 5-20%).
- Chronic uncontrollable stress is a major factor in the etiology of fatigue.
- Fatigue is associated with multiple indicators of dysregulated autonomic, neurohormonal, and immune system measures, but there is no specific biomarker of fatigue.
- There is similarity between interventions that are designed to reduce fatigue and those designed to alleviate stress, including cognitive behavioral therapy and physical exercise.

CLINICAL DIAGNOSTIC PERSPECTIVE

Fatigue can be conceptualized as a disorder or as a normally distributed phenomenon. The challenges associated with defining and assessing fatigue have been long recognized.[2] In general, fatigue can be considered as a multidimensional construct with physical and mental components. The definitional issues related to the concept of stress are discussed elsewhere in this volume and will not be repeated in this chapter. Fatigue as a continuous construct can be measured using questionnaires and interviews. Examples of such inventories are the Multidimensional Fatigue Inventory and the Fatigue Assessment Scale, among many others.[3] Current efforts to systematize the assessment of fatigue are promising,[4] but implementing uniform assessment tools remains challenging and may not always be desirable. In addition, the correlation between objective performance-based fatigue indices with subjectively experienced fatigue is very low, suggesting that a multimethod approach may be needed to optimally quantify fatigue.

In addition to the continuous conceptualization of fatigue, diagnostic classification of fatigue-related conditions into specific syndromes can be very useful, particularly for individuals who are severely disabled by their symptoms. Fatigue-related diagnostic criteria have been published in the psychiatric and internal medicine literature and have proven to be beneficial from a clinical as well as a research perspective.

Psychiatric classifications and diagnoses have originally been based on the presumed pathophysiological origins of a complaint. In 1869, George M. Beard introduced the concept of neurasthenia to describe an organically based disorder characterized by fatigability of the body and mind, largely resulting from environmental factors. Neurasthenia was primarily treated by rest, and reportedly common among well-educated and professional individuals. Other concepts that have been developed to describe sustained fatigue complaints include "effort syndrome," "neurocirculatory asthenia," or "postinfectious fatigue syndrome."

Contemporary diagnostic classifications, such as the "Diagnostic and Statistical Manual of Mental Disorders" (DSM), are predominantly based on description and classification of symptoms. The latest edition of the DSM (DSM-5) does not include any specific fatigue-related constructs or syndromes. Some clinicians and investigators consider fatigue-related disorders (e.g., chronic fatigue syndrome [CFS]) as "functional somatic syndromes" or "medically unexplained syndromes."[5] However, the DSM-5 has moved away from emphasizing the importance of an underlying medical disorder in the definition of somatic symptom disorders and rather focuses on the impact of the symptoms on an individual's thoughts, feelings, and behaviors.

In the last few decades, several new constructs have been introduced outside the field of psychiatry in order to characterize conditions in which fatigue plays a primary role, including the CFS and "myalgic encephalomyelitis" (ME). CFS is an operationally defined syndrome characterized by a minimum of 6 months of severe physical and mental fatigue and fatigability made worse by minor exertion. This postexercise fatigue characteristic of CFS has played a role in the recent proposal to relabel this condition as systemic exercise intolerance disease (SEID). Because of the similarities between the prior neurasthenia construct, CFS, ME, and SEID, it has been questioned whether these more recent concepts are in fact "old wine in new bottles."[6]

Multiple psychiatric disorders are associated with fatigue. Posttraumatic stress disorder (PTSD) is characterized by a common cooccurrence of fatigue as well as psychological distress. The concept of PTSD describes the consequences of exposure to severe psychological stress, for example following major disasters, war, and other life-threatening events. Increased fatigue and fatigability belong to the core characteristics of PTSD.[7] Several symptoms are relatively specific for PTSD versus CFS, justifying their heuristic value as different diagnostic entities. However, there is nonetheless substantial overlap between the symptoms that constitute these syndromes,[8] which may suggest that stress is a common factor in both CFS (and other fatigue-related conditions) and PTSD.

The DSM lists decreased energy, tiredness, and fatigue among the symptoms of depressive mood disorders. The question of whether fatigue states should be distinguished from depressive mood disorders has been a topic of debate since the late 1800s. There is little doubt that depressed individuals often feel tired, and fatigue is one of the diagnostic symptoms for depression.[9] In contrast to the DSM, the International Classification of Diseases (ICD-10) considers fatigue as one of the three "key symptoms" of depression. However, not all individuals who feel fatigued and/or depleted of energy suffer from mood disturbances, lack of self-esteem, or guilt feelings.[10] Therefore, it is useful to distinguish common disorders in psychiatry, such as PTSD and depression, from conditions that are mainly characterized by profound fatigability such as CFS, ME, and SEID.[11]

Given the overlap between psychiatric disorders such as PTSD and depression with fatigue, it is important to realize that fatigue and stress (i.e., psychological distress) are common symptoms of negative affectivity. Thus, conditions associated with high levels of negative affect (e.g., depression) are likely to be associated with increased levels of both stress and fatigue and such conditions may account for the high correlation between self-reported stress and fatigue.

FATIGUE AS AN ADAPTIVE RESPONSE TO STRESS

One of the early theories of fatigue postulates a disequilibrium of the "milieu interieur" as a primary etiological factor, in which stress plays a potentially critical role. In later models, fatigue has been conceptualized as the result of an imbalance between "demand and supply" or as a breakdown in adaptation to stressful challenges.[12] From this perspective, the aforementioned "diagnostic" approach falls short in identifying the "normal" fatigue response to environmental demands.

The association between stress and fatigue becomes clear when examining Selye's General Adaptation Syndrome model. In this model, the individual response to stress has three phases: (1) acute/alarm (similar to the fight-or-flight response); (2) resistance (associated with sustained stressors and concomitant elevated activity of a wide range of biological systems, including hypothalamic-pituitary-adrenal [HPA]-axis activation); and (3) recovery/exhaustion (recovery when the stressor is successfully "removed," and exhaustion when the stressor remains present and biological resources are gradually depleted). The third "exhaustion" stage is of particular relevance to the stress-fatigue relationships. From this perspective, fatigue can be conceptualized as an adaptive response to prolonged stress such that the individual will at some point discontinue (ineffective) efforts to alleviate a chronic stressor. In other words, the behavioral consequences of exhaustion may be adaptive in the sense that they could attenuate potentially adverse effects of long-term depletion of biological resources.

PSYCHOPHYSIOLOGICAL AND BIOLOGICAL PERSPECTIVE

Psychophysiological approaches typically address behaviors, emotions, and cognitions that correspond with physiological reactions to normal and abnormal internal or external stimuli. For example, psychophysiologists and other behavioral scientists investigate the association between fatigue with the ANS and neuroimmune measures. Most psychophysiological theories conceptualize fatigue as a consequence of a breakdown in adaptation to environmental challenges, and focus on the physiological and biological responses to stress as related to fatigue.

Several brain areas are likely to be involved in fatigue following prolonged stress exposure, including parts of the central autonomic network (e.g., the anterior cingulate cortex, insula, ventromedial prefrontal cortex, amygdala, and hypothalamus, among other areas).[13] Evidence based on animal models indicates that the suprachiasmatic nucleus, areas involved in sleep regulation, the ascending arousal system, and reward-related structures may also be associated with fatigue.[14] However, there is no specific brain area or neurohormonal profile that is specific to the stress-fatigue association.

The ANS is one of the two primary outflow pathways of the central nervous system. The primary role of the ANS is to support "vegetative" functions including bodily processes that are directly relevant to life maintenance: appetite, sleep, menstruation, and bowel, bladder, and sexual function. The other outflow pathway involves the neurohormonal system and includes the HPA and sympathetic-adreno-medullary (SAM) axes. Similar to the lack of a well-defined brain structure or network that is specific to fatigue, there are also no ANS or neurohormonal "biomarkers" that are specific to acute or chronic fatigue. Nonetheless, the central and autonomic stress-related systems receive much attention in psychophysiological research relevant to the stress-fatigue relationship as ANS and neurohormonal processes play a crucial role in the human stress response.

The ANS has two main branches: the sympathetic nervous system (involved in the acute stress response) and the parasympathetic nervous system (which is commonly more active during periods of rest and recovery, and in general counterbalances the sympathetic nervous system). The neurohormonal stress response is complex and involves the SAM system and the HPA system. Stimulation of the SAM system is characterized by increased secretion of catecholamines (epinephrine and norepinephrine) as well as increased heart rate and blood pressure; typical vegetative symptoms include sweating, palpitations, dizziness, and other vegetative symptoms (i.e., related to the "vegetative" = ANS activation). Vegetative symptoms such as dizziness, palpitations, and dyspnea are often observed in individuals with CFS and other fatigue-related conditions. Most of the ANS and neurohormonal responses to stress exposure are normal, enabling individuals to adapt to environmental changes.

Chronic stress may induce sustained activation of the ANS and HPA-axis (see "Fatigue as an Adaptive Response to Stress" section). As a result, the ANS capacity to adjust may become diminished such that ANS dysregulation develops, characterized by reduced parasympathetic activity and sympathetic dominance. Both branches of the ANS have immunomodulatory properties, thereby possibly adding to the onset and maintenance of fatigue. Heart rate variability (HRV) is a measure of autonomic control of the heart, and quantifies the variability of time between successive heart beats. Decreased HRV is indicative of reduced parasympathetic activity and associated with relatively poor health. Fatigue-related conditions are also associated with measures of ANS dysregulation, including increased heart rate, altered adrenergic receptor function, and HRV-based indices of higher sympathetic and lower parasympathetic ANS activity.[15,16]

The HPA-axis correlates of fatigue display a complex pattern of increased and decreased activity levels. Increased HPA-axis activity is observed in response to acute stress. Increased HPA-axis activity as evidenced by elevated secretion of cortisol is also well documented among individuals with melancholic depression, panic disorder, central obesity, and Cushing's syndrome. These elevated cortisol levels may coincide with suppression of inflammation. Emotional exhaustion has been related to hyperreactivity of the HPA-axis.[17] However, most evidence is suggestive of *decreased* HPA-axis activity in chronic stress as well as persistent fatigue and "atypical" depression (i.e., hyperphagia and hypersomnia). These conditions are associated with a decreased production of cortisol,[18] and purportedly with activation of immune mediated inflammation.[19,20]

During inflammation, the body produces cytokines that play a pivotal signaling role in immune system activation. Some of these cytokines reach the brain or elicit release of cytokines in the brain, resulting in what has been labeled as "sickness behavior." Sickness behavior is characterized by locomotor retardation, general malaise, anorexia, and inhibition of sexual behavior. Sickness behavior is part of the defensive response to infection or inflammation. Evidence consistently demonstrates that fatigue and inflammation are interrelated, although there is not a 1:1 association between the two phenomena.[20] Thus, there is probably a bidirectional or circular association between inflammation and fatigue, and this association is mediated in part by the central and ANSs. The psychophysiological approach makes a unique contribution to the study of fatigue by investigating which symptomatic and physiological reactions belong to normal adaptive, health-protecting behaviors, and which reactions are markers of a breakdown in adaptation to environmental challenges.

EXTERNAL DEMANDS IN THE WORK ENVIRONMENT

Fatigue-related endpoints of long-lasting (work-related) stress have also been investigated in various occupational health settings, with a focus on "burnout." Burnout is a negative state of physical, emotional, and mental exhaustion resulting from a gradual process of sustained overburdening or disillusionment. In the mid-1970s, Freudenberger was the first to describe "burnout" as a common psychological characteristic observed in healthcare professionals and volunteers, many of whom experienced a gradual emotional depletion combined with a loss of motivation and commitment. Burnout has three main components: emotional exhaustion, depersonalization, and reduced personal accomplishment.[21]

Several instruments have been developed to assess burnout, despite the substantial controversies regarding the optimal definition of this construct. The scale developed by Maslach, for example, assesses the three dimensions of burnout (emotional exhaustion, depersonalization, and reduced personal accomplishment) and has become the primary tool for evaluating burnout.[21] Empirical research has shown that the emotional exhaustion component is related to depression, whereas relationships between depersonalization and personal accomplishment with depression are less strong. Fatigue in this context can be considered to result from an individual's evaluative motivational assessment in relation to work demands and resources to meet these demands. The burnout literature tends to focus on situational precursors of fatigue, which may partly reflect the origins of this construct in occupational health. The extension of the burnout concept to nonoccupational domains (e.g., feelings of emotional exhaustion caused by prolonged caregiving or marital conflicts) has been controversial, especially because the concepts of depersonalization and reduced personal accomplishment are less applicable outside the realm of occupational settings.[22] Therefore, the contributions of research on burnout to fatigue lie primarily in the identification of social and occupational "exposures." Regarding physiological correlates, a meta-analysis showed that burnout was not association with altered HPA-axis measures and that insufficient information was available for other autonomic, neurohormonal, or immune system-related measures.[23]

Another construct relevant to the relationship between stress and fatigue in the workplace is "effort-reward imbalance," developed by Siegrist and colleagues. Overcommitment to work, one of the dimensions of the effort-reward imbalance model has been associated with attenuated HPA-axis reactivity (i.e., reduced ACTH and cortisol responsiveness).[12,17] These studies show that stress-related fatigue in the workplace can express itself in multiple ways including burnout or effort-reward imbalance, and that the latter is associated with biological changes such as HPA-axis dysregulation.

BIOBEHAVIORAL AND PSYCHOSOMATIC PERSPECTIVE

In biobehavioral and psychosomatic medicine, and related disciplines, the purported etiology of fatigue depends on the nature of coinciding medical disorder. For example, among patients with cancer, the evidence linking inflammation and fatigue is relatively strong. However, the instigating factors are unclear as the tumor itself, but also radiation and/or chemotherapy may be responsible for the increased proinflammatory state as well as fatigue.[24,25] There is promising research examining the effects of cytokine antagonists as a treatment for fatigue, showing that reducing inflammation may result in improvements in fatigue.[25]

In cardiovascular disease, the association of stress with fatigue and (vital) exhaustion is of particular interest. Exhaustion (i.e., feelings of unusual fatigue/loss of energy, increased irritability, and demoralization) is an important precursor of myocardial infarction and sudden cardiac death.[26] Exhaustion and depression also predict recurrent cardiac events among high-risk populations and more research is needed to further document the divergent validity of these two partially overlapping constructs.[11] In heart failure, fatigue is one of the core symptoms that may reflect energy-related problems resulting from poor cardiac pump function or consequences of heart failure-related functional limitations.

Cancer and cardiac diseases are examples of medical conditions in which fatigue plays an important role; the stress-fatigue interrelationship is also a crucial factor in most other disorders, particularly stroke,[27] multiple sclerosis[28] as well as Parkinson's disease, rheumatic diseases, and infectious diseases.

Regarding interventions, many scientists in cardiovascular behavioral medicine interpret the premonitory symptoms of extreme fatigue as manifestations of a clinical or subsyndromal depression. Consequently, cognitive behavioral therapy approaches have been used to treat depressive symptoms in patients with coronary artery disease.[29] Other investigators have focused on reducing exhaustion resulting from long-lasting exposure to uncontrollable stressors and the accompanying experiences of psychological distress. The interventions targeting exhaustion used group therapy, including relaxation and decreasing exposure to and negative appraisal of stressors that can lead to exhaustion.[30] The effects of both types of intervention on the risk of new cardiac events have been tested in well-designed randomized controlled trials[29,30] and neither of the two strategies resulted in a reduction of new cardiac events. However, the interventions were successful in (a posteriori determined) subsamples. Other studies targeting psychological distress in patients with coronary artery disease revealed positive results regarding clinical outcomes (for review, see Ref. 31). In addition to cognitive behavioral and other psychological therapies, both fatigue and stress can be improved by physical exercise interventions. Future clinical trials are needed to demonstrate that a behavioral treatment of fatigue and depressive symptoms will not only improve quality of life but also reduce the risk of recurrent adverse medical outcomes.

CONCLUSIONS

Fatigue is a ubiquitous phenomenon. However, less than 10% of patients presenting with fatigue in primary care have a disease that plays a direct causal role in this symptom. It has been debated whether concepts such as neurasthenia, burnout, or (vital) exhaustion are useful in elucidating the stress-fatigue-disease relationships, and whether the "procrustean bed" of DSM-5 and other diagnostic classification systems do more harm than good in stress research. We expect that the law of parsimony will also result in more focused research in this area. Each of the aforementioned theoretical perspectives has contributed to the accumulating knowledge base of the origins of fatigue, its biological and physiological correlates, and the possibilities and limitations of helping individuals who suffer from unusual and/or debilitating levels of fatigue. The domain of fatigue research has unique epistemological problems that require detailed scientific attention. More research is needed on the distinction between fatigue as a health-protecting factor versus fatigue as a marker of overtaxing of the body or an indicator of underlying (subclinical) disease processes. The precipitating and maintaining factors of chronic fatigue conditions are complicated to understand because some of the original etiological factors may have become undetectable (for example long-term consequences of viral infections). In addition, multiple confounding factors may account for the association between stress and fatigue, among which age, sleep disorders, physical inactivity, overweight, disease-related functional limitations, and pain. It will be a challenge to answer the important question of whether chronic fatigue can be best approached as a form of adaptation to prolonged and uncontrollable environmental challenges, or whether fatigue is a manifestation of depression and/or dysregulation of the autonomic, neurohormonal, or immune systems. Future multidisciplinary research will continue to add to our understanding of the role of stress and other psychological factors in the origins, the mental and physical consequences, and the treatment of fatigue, particularly in conditions where fatigue is a known predictor of adverse health outcomes.

Glossary

Burnout A state characterized by emotional exhaustion that may occur after exposure to prolonged stress.

Hypothalamic-pituitary-adrenal (HPA) axis Part of the neurohormonal stress system. Increased as well as decreased activity of the HPA-axis can be observed depending on the nature of the disorder.

Neurasthenia Psychiatric term formerly used to describe a state characterized by fatigue, impaired cognitive functioning, and increased emotional sensitivity.

Vital exhaustion A state characterized by unusual fatigue and loss of energy, increased irritability, and feelings of demoralization. Often precedes cardiac events. In the original conceptualization of the exhaustion construct, the term "vital" was included to reflect the far-reaching consequences of this condition on daily life function (similar to vital depression).

References

1. Sharpe M, Wilks D. Fatigue. *BMJ*. 2002;325(7362):480–483.
2. Muscio B. Is a fatigue test possible? *Br J Psychol*. 1921;12:31–46.
3. Hjollund NH, Andersen JH, Bech P. Assessment of fatigue in chronic disease: a bibliographic study of fatigue measurement scales. *Health Qual Life Outcomes*. 2007;5:12.
4. Barsevick AM, Cleeland CS, Manning DC, et al. ASCPRO recommendations for the assessment of fatigue as an outcome in clinical trials. *J Pain Symptom Manage*. 2010;39(6):1086–1099.
5. Tak LM, Riese H, de Bock GH, Manoharan A, Kok IC, Rosmalen JG. As good as it gets? A meta-analysis and systematic review of methodological quality of heart rate variability studies in functional somatic disorders. *Biol Psychol*. 2009;82(2):101–110.
6. Wessely S. Old wine in new bottles: neurasthenia and 'ME'. *Psychol Med*. 1990;20(1):35–53.
7. Afari N, Ahumada SM, Wright LJ, et al. Psychological trauma and functional somatic syndromes: a systematic review and meta-analysis. *Psychosom Med*. 2014;76(1):2–11.
8. Skapinakis P, Lewis G, Mavreas V. Unexplained fatigue syndromes in a multinational primary care sample: specificity of definition and prevalence and distinctiveness from depression and generalized anxiety. *Am J Psychiatry*. 2003;160(4):785–787.
9. American Psychiatric Association. *Diagnostic and Statistical Manual of Mental Disorders*. 4th ed. Washington, DC: American Psychiatric Association; 1994.
10. Appels A. Depression and coronary heart disease: observations and questions. *J Psychosom Res*. 1997;43(5):443–452.
11. Kop WJ. Somatic depressive symptoms, vital exhaustion, and fatigue: divergent validity of overlapping constructs. *Psychosom Med*. 2012;74(5):442–445.
12. Siegrist J. Contributions of sociology to the prediction of heart disease and their implications for public health. *Eur J Publ Health*. 1991;1(1):10–29.
13. de Morree HM, Szabo BM, Rutten GJ, Kop WJ. Central nervous system involvement in the autonomic responses to psychological distress. *Neth Heart J*. 2013;21(2):64–69.
14. Harrington ME. Neurobiological studies of fatigue. *Prog Neurobiol*. 2012;99(2):93–105.
15. Klimas NG, Broderick G, Fletcher MA. Biomarkers for chronic fatigue. *Brain Behav Immun*. 2012;26(8):1202–1210.
16. Tanaka M, Mizuno K, Yamaguti K, et al. Autonomic nervous alterations associated with daily level of fatigue. *Behav Brain Funct*. 2011;7:46.
17. Wolfram M, Bellingrath S, Feuerhahn N, Kudielka BM. Emotional exhaustion and overcommitment to work are differentially associated with hypothalamus-pituitary-adrenal (HPA) axis responses to a low-dose ACTH1-24 (Synacthen) and dexamethasone-CRH test in healthy school teachers. *Stress*. 2013;16(1):54–64.
18. Powell DJ, Liossi C, Moss-Morris R, Schlotz W. Unstimulated cortisol secretory activity in everyday life and its relationship with fatigue and chronic fatigue syndrome: a systematic review and subset meta-analysis. *Psychoneuroendocrinology*. 2013;38(11):2405–2422.
19. Chrousos GP. The hypothalamic-pituitary-adrenal axis and immune-mediated inflammation. *N Engl J Med*. 1995;332(20):1351–1362.
20. Kop WJ. The integration of cardiovascular behavioral medicine and psychoneuroimmunology: new developments based on converging research fields. *Brain Behav Immun*. 2003;17(4):233–237.
21. Maslach C, Jackson SE. *Maslach Burnout Inventory*. Palo Alto, CA: Consulting Psychologists Press; 1986.
22. Leone SS, Wessely S, Huibers MJ, Knottnerus JA, Kant I. Two sides of the same coin? On the history and phenomenology of chronic fatigue and burnout. *Psychol Health*. 2011;26(4):449–464.
23. Danhof-Pont MB, van Veen T, Zitman FG. Biomarkers in burnout: a systematic review. *J Psychosom Res*. 2011;70(6):505–524.
24. Wagner LI, Cella D. Fatigue and cancer: causes, prevalence and treatment approaches. *Br J Cancer*. 2004;91(5):822–828.
25. Bower JE, Lamkin DM. Inflammation and cancer-related fatigue: mechanisms, contributing factors, and treatment implications. *Brain Behav Immun*. 2013;30(Suppl):S48–S57.
26. Appels A. Mental precursors of myocardial infarction. *Br J Psychiatry*. 1990;156:465–471.
27. Wu S, Barugh A, Macleod M, Mead G. Psychological associations of poststroke fatigue: a systematic review and meta-analysis. *Stroke*. 2014;45(6):1778–1783.
28. Krupp LB. Fatigue in multiple sclerosis: definition, pathophysiology and treatment. *CNS Drugs*. 2003;17(4):225–234.
29. Berkman LF, Blumenthal J, Burg M, et al. Effects of treating depression and low perceived social support on clinical events after myocardial infarction: the enhancing recovery in coronary heart disease patients (ENRICHD) randomized trial. *JAMA*. 2003;289(23):3106–3116.
30. Appels A, Bar F, van der Pol G, et al. Effects of treating exhaustion in angioplasty patients on new coronary events: results of the randomized exhaustion intervention trial (EXIT). *Psychosom Med*. 2005;67(2):217–223.
31. Rutledge T, Redwine LS, Linke SE, Mills PJ. A meta-analysis of mental health treatments and cardiac rehabilitation for improving clinical outcomes and depression among patients with coronary heart disease. *Psychosom Med*. 2013;75(4):335–349.

CHAPTER

43

Burnout

C. Maslach[1], M.P. Leiter[2]

[1]University of California, Berkeley, CA, USA

[2]Acadia University, Wolfville, NS, Canada

OUTLINE

Abstract

Burnout is a prolonged response to chronic emotional and interpersonal stressors on the job. It is defined by the three dimensions of exhaustion, cynicism, and professional inefficacy. As a reliably identifiable job stress syndrome, burnout clearly places the individual stress experience within a larger organizational context of people's relation to their work. Burnout impairs both personal and social functioning. This decline in the quality of work and in both physical and psychological health can be costly—not just for the individual worker, but for everyone affected by that person. Interventions to alleviate burnout and to promote its opposite, engagement with work can occur at both organizational and personal levels. The social focus of burnout, the solid research basis concerning the syndrome, and its specific ties to the work domain make a distinct and valuable contribution to people's health and well-being.

DEFINITION AND ASSESSMENT

The relationship that people have with their work, and the difficulties that can arise when that relationship goes awry, have long been recognized as a significant phenomenon in people's lives. However, the identification of the name, "burnout" occurred in the 1970s, first in a volume of articles about free clinics,[1] and then in an article about workers in health care and human service occupations.[2] In both instances, burnout was rooted within caregiving and service occupations, in which the core of the job is the relationship between provider and recipient. This interpersonal context of the job meant that, from the beginning, burnout was studied not simply as an individual stress response, but in terms of an individual's relational transactions in the workplace. Moreover, this interpersonal context focused attention on the individual's emotions, and on the motives and values underlying his or her work with other people. Job burnout has remained an active research focus across a wide range of academic and professional disciplines in the decades since its introduction in the 1970s, inspiring the development of theoretical constructs and practical methods for alleviating the syndrome.

Burnout is predominantly defined by its three main components: exhaustion, cynicism, and professional inefficacy. Exhaustion refers to feelings of being overextended and depleted of one's emotional and physical resources. Workers feel drained and used up, without any source of replenishment. They lack enough energy to face another day or another person in need. The exhaustion component represents the basic individual stress dimension of burnout. Cynicism refers to a negative, hostile, or an excessively detached response to the job, which often includes a loss of idealism. It usually develops in response to the overload of emotional exhaustion, and is self-protective at first—an emotional

Stress: Concepts, Cognition, Emotion, and Behavior
http://dx.doi.org/10.1016/B978-0-12-800951-2.00044-3

buffer of "detached concern." But the risk is that the detachment can turn into dehumanization. The cynicism component represents the interpersonal dimension of burnout. Professional inefficacy refers to a decline in feelings of competence and productivity at work. People experience a growing sense of inadequacy about their ability to do the job well, and this may result in a self-imposed verdict of failure. The inefficacy component represents the self-evaluation dimension of burnout.

What has been distinctive about burnout is the interpersonal framework of the phenomenon. The centrality of relationships at work—whether it's relationships with clients, colleagues, or supervisors—has always been at the heart of descriptions of burnout. These relationships are the source of both emotional strains and rewards, they can be a resource for coping with job stress, and they often bear the brunt of the negative effects of burnout. Thus, if one were to look at burnout out of context, and simply focus on the individual exhaustion component, one would lose sight of the phenomenon entirely.

KEY POINTS

- Burnout, as an enduring response to poor person/job or person/organization fit, is experienced as a syndrome of exhaustion, cynicism, and professional inefficacy.
- Since its introduction to psychological research in the 1970s, people have worked to understand the dynamics of the syndrome, its causes, and its consequences.
- Burnout has also inspired conceptual work, considering its place in a broad range of psychological connections with work, such as work engagement and chronic fatigue.
- Although some connections have been made of burnout with personality or demographic characteristics, the evidence consistently points toward workplace variables as the primary drivers of burnout.
- Important workplace factors influencing burnout include workload, control, reward, community, fairness, and value congruence.
- Burnout has close associations with workplace outcomes, including health problems, and performance deficits.
- Interventions have explored various forms of programs to ameliorate or prevent burnout. Initiatives to enhance the quality of workplace social environments have shown encouraging results.

Assessment

The three dimensions of burnout are assessed by the Maslach Burnout Inventory (MBI), which has been translated and validated in many languages.[3] Recently, an alternative burnout questionnaire has been proposed—the Oldenburg Burnout Inventory—that assesses the two dimensions of exhaustion and disengagement from work, and has shown high convergent validity with the MBI.[4] Other conceptualizations of burnout have considered exhaustion to be the sole defining criterion.[5,6] Although more recent theories make distinctions between various aspects of exhaustion—e.g., physical fatigue, emotional exhaustion, and cognitive weariness[7] or physical and psychological exhaustion[8]—their measures inevitably produce a single overriding factor of exhaustion. However, the fact that exhaustion is a necessary criterion for burnout does not mean it is sufficient. Exhaustion is not something that is simply experienced—rather, it prompts actions to distance oneself emotionally and cognitively from one's work, presumably as a way to cope with the work overload. Distancing is such an immediate reaction to exhaustion that a strong relationship from exhaustion to cynicism is found consistently in burnout research, across a wide range of organizational and occupational settings. The inclusion of distancing (cynicism) and discouragement (inefficacy) draws an important distinction between burnout and chronic exhaustion. People experiencing burnout are not simply fatigued or overwhelmed by their workload. They also have lost a psychological connection with their work that has implications for their motivation and their identity. If the only issue were exhaustion, the term "burnout" would add nothing beyond what is already and more straightforwardly captured by the term chronic fatigue. Instead, the concept and operationalization of burnout through the MBI captures a disaffection with work, as well as a crisis in work-based efficacy expectations.

Research has established that, without definitive changes in work settings, burnout remains fairly constant for long periods.[9] Recently, longitudinal analyses of burnout have developed approaches that concentrate on subgroups with a greater probability of change. In one study, employees who had inconsistent scores on exhaustion and cynicism (high scores on one; low scores on the other) were more likely to change over time than were their colleagues with consistent scores. The direction of the subsequent change was associated with employees' sense of fairness: those who found their workplace to be fair tended to change in the positive direction of low scores on both exhaustion and cynicism.[10] This same pattern of inconsistent scores and change was replicated among another group of employees, but the positive change toward lower burnout scores was associated with their access to work-relevant information control in their job.[11]

Engagement

An important development has been that researchers have turned their attention to the positive antithesis of burnout, which has been defined as "engagement." Although there is general agreement that engagement with work represents a productive and fulfilling state within the occupational domain, there are some differences in its definition. From the point of view of burnout researchers, engagement is considered to be the opposite of burnout and is defined in terms of the same three dimensions as burnout, but the positive end of those dimensions rather than the negative. Thus, engagement consists of a state of high energy, strong involvement, and a sense of efficacy.[12] An alternative view is that work engagement is an independent, distinct concept, which is not the opposite of burnout, although it is negatively related to it. In this model, engagement is defined as a positive state of mind characterized by vigor, dedication, and absorption.[13] Recent theory and research is bringing new understanding of both the work engagement and burnout concepts.[14,15]

One important implication of the burnout-engagement continuum is that strategies to promote engagement may be just as important for burnout prevention as strategies to reduce the risk of burnout. A workplace that is designed to support the positive development of the three core qualities of energy, involvement, and effectiveness should be successful in promoting the well-being and productivity of its employees, and thus the health of the entire organization.

PSYCHOSOCIAL FACTORS

Once burnout had been defined, and measures had been created to capture its complexity, it became a topic for theory development and empirical research. Although initially identified within North America, burnout was soon being studied in many different countries, and it is now recognized as a global construct. From the beginning, the burnout syndrome has been primarily attributed to work environments, where employees experience a breadth of emotional, physical, and psychological demands. Researchers have now begun to expand their scope of what leads to burnout, and what are its effects. In particular, two main developmental models have emerged: the job demands-resources model (JD-R) and the conservation of resources model (COR). The JD-R model focuses on the notion that burnout arises when individuals experience incessant job demands and have inadequate resources available to address and to reduce those demands.[16,17] The COR model follows a basic motivational theory following the idea that burnout arises as a result of persistent threats to available resources.[18] When individuals perceive that the resources they value are threatened, they strive to maintain those resources. The loss of resources or even the impending loss of resources may aggravate burnout. Both the JD-R and the COR theory of burnout development have received confirmation in research studies.

Unlike acute stress reactions, which develop in response to specific critical incidents, burnout is a cumulative stress reaction to ongoing occupational stressors. With burnout, the emphasis has been more on the process of psychological erosion, and the psychological and social outcomes of this chronic exposure, rather than just the physical ones. Because burnout is a prolonged response to chronic interpersonal stressors on the job, it tends to be fairly stable over time.

Of the three burnout dimensions, exhaustion is the closest to an orthodox stress variable, and therefore is more predictive of stress-related health outcomes than the other two dimensions. Exhaustion is typically correlated with such stress symptoms as: headaches, chronic fatigue, gastrointestinal disorders, muscle tension, hypertension, cold/flu episodes, and sleep disturbances. These physiological correlates mirror those found with other indices of prolonged stress. Similarly parallel findings have been found for the link between burnout and various forms of substance abuse. A 10-year follow-up study of Finnish industrial workers found burnout to predict subsequent hospital admissions for cardiovascular problems.[19] Other Finnish research found that a one-unit increase in burnout score related to a 1.4 unit increase in risk for hospital admission for mental health problems, as well as a one-unit increase in risk for hospital admissions for cardiovascular problems. Both the exhaustion and cynicism dimensions of burnout were closely related to mental and cardiovascular problems, with little connection to the efficacy dimension of burnout.[20] Other research has provided a more detailed examination of the link between burnout and cardiovascular disease, noting the role of high-sensitivity C-reactive protein and fibrinogen concentrations in the link.[21]

In terms of mental, as opposed to physical, health, the link with burnout is more complex. It has been assumed that burnout may result in subsequent mental disabilities, and there is some evidence to link burnout with greater anxiety, irritability, and depression. However, an alternative argument is that burnout is itself a form of mental illness, rather than a cause of it. Much of this discussion has focused on depression, and whether or not burnout is a different phenomenon. Research has demonstrated that the two constructs are indeed distinct: burnout is job-related and situation-specific, as opposed to depression, which is general and context-free. A recent study found a reciprocal relationship between burnout and depression with each predicting subsequent developments in the other. It was noteworthy that burnout fully mediated the relationship of workplace strains with depression:

when problems at work contribute to depression, experiencing burnout is a step in the process.[22]

Burnout has also been associated with various forms of job withdrawal—absenteeism, intention to leave the job, and actual turnover. For example, cynicism has been found to be the pivotal aspect of burnout to predict turnover,[23] and burnout mediates the relationship between being bullied in the workplace and the intention to quit the job.[24] However, for people who stay on the job, burnout leads to lower productivity and effectiveness at work. To the extent that burnout diminishes opportunities for satisfying experiences at work, it is associated with decreased job satisfaction and a reduced commitment to the job or the organization.

People who are experiencing burnout can also have a negative impact on their colleagues, both by causing greater personal conflict and by disrupting job tasks. Thus, burnout can be "contagious" and perpetuate itself through informal interactions on the job.[25,26] There is also some evidence that burnout has a negative "spillover" effect on people's home life. The critical importance of social relationships for burnout is underscored by studies that show that burnout increases in work environments characterized by interpersonal aggression.[27,28] In contrast, when there is greater social support and trust among coworkers, burnout is lessened.[29,30]

A MEDIATION MODEL OF BURNOUT

Inherent to the fundamental concept of stress is the problematic relationship between the individual and the situation. Thus, prior research has tried to identify both the key personal and job characteristics that put individuals at risk for burnout. In general, far more evidence has been found for the impact of job variables than for personal ones. These job factors fall into six key domains within the workplace: workload, control, reward, community, fairness, and values.

However, more recent theorizing has argued that personal and job characteristics need to be considered jointly within the context of the organizational environment. The degree of fit, or match, between the person and the job within the six areas of worklife will determine the extent to which the person experiences engagement or burnout, which in turn will determine various outcomes, such as personal health, work behaviors, and organizational measures.[31] In other words, the burnout-engagement continuum (with its three dimensions) mediates the impact of the six areas of worklife on important personal and organizational outcomes (see Figure 1).

Job Characteristics: Six Areas of Worklife

An analysis of the research literature on organizational risk factors for burnout has led to the identification of six major domains. Both workload and control are reflected in the demand-control model of job stress, and reward refers to the power of reinforcements to shape behavior. Community captures all of the work on social support and interpersonal conflict, while fairness emerges from the literature on equity and social justice. Finally, the area of values picks up the cognitive-emotional power of job goals and expectations.

Workload. Both qualitative and quantitative work overload contribute to burnout by depleting the capacity of people to meet the demands of the job. When this kind of overload is a chronic job condition, there is little

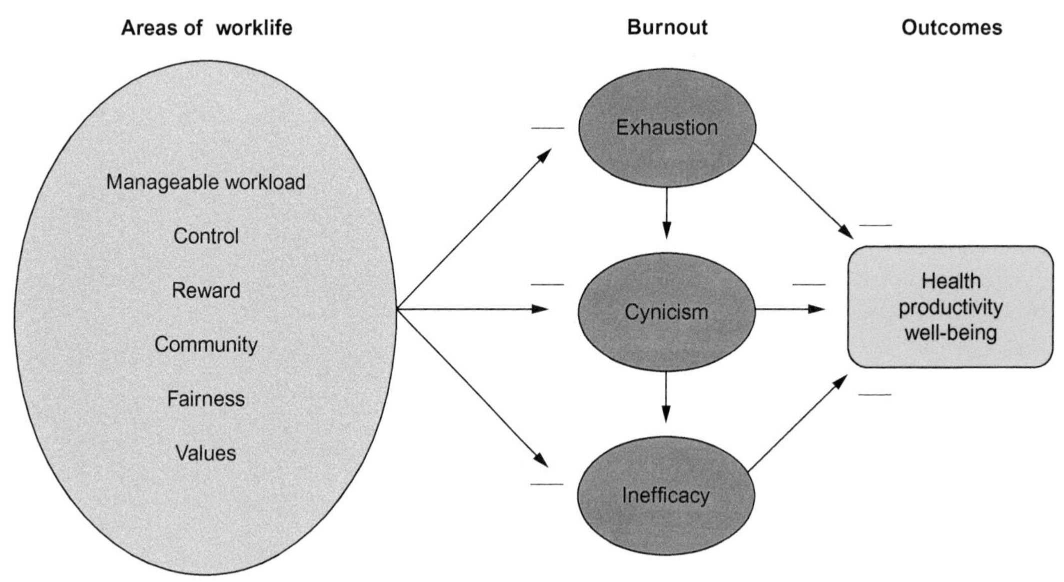

FIGURE 1 The mediation model of burnout.

opportunity to rest, recover, and restore balance. A sustainable and manageable workload, in contrast, provides opportunities to use and refine existing skills as well as to become effective in new areas of activity.

Control. Research has identified a clear link between a lack of control and high levels of stress and burnout. However, when employees have the perceived capacity to influence decisions that affect their work, to exercise professional autonomy, and to gain access to the resources necessary to do an effective job, they are more likely to experience job engagement.

Reward. Insufficient recognition and reward (whether financial, institutional, or social) increases people's vulnerability to burnout, because it devalues both the work and the workers, and is closely associated with feelings of inefficacy. In contrast, consistency in the reward dimension between the person and the job means that there are both material rewards and opportunities for intrinsic satisfaction.

Community. Community has to do with the ongoing relationships that employees have with other people on the job. When these relationships are characterized by a lack of support and trust, and by unresolved conflict, then there is a greater risk of burnout. However, when these job-related relationships are working well, there is a great deal of social support, employees have effective means of working out disagreements, and they are more likely to experience job engagement.

Fairness. Fairness is the extent to which decisions at work are perceived as being fair and equitable. People use the quality of the procedures, and their own treatment during the decision-making process, as an index of their place in the community. Cynicism, anger, and hostility are likely to arise when people feel they are not being treated with the respect that comes from being treated fairly.

Values. Values are the ideals and motivations that originally attracted people to their job, and thus they are the motivating connection between the worker and the workplace, which goes beyond the utilitarian exchange of time for money or advancement. When there is a values conflict on the job, and thus a gap between individual and organizational values, employees will find themselves making a trade-off between work they want to do and work they have to do, and this can lead to greater burnout.

Personal Characteristics

Although job variables and the organizational context are the prime predictors of burnout and engagement, a few personality variables have shown some consistent correlational patterns. In general, burnout scores are higher for people who have a less "hardy" personality, who have a more external locus of control, and who score as "neurotic" on the five-factor model of personality. There is also some evidence that people who exhibit type-A behavior (which tends to predict coronary heart disease) are more prone to the exhaustion dimension of burnout.

There are few consistent relationships of burnout with demographic characteristics. Although higher age seems to be associated with lower burnout, it is confounded with both years of experience and with survival bias (i.e., those who "survive" early job stressors and do not quit). Thus it is difficult to derive a clear explanation for this age pattern. The only consistent gender difference is a tendency for men to score slightly higher on cynicism. These weak demographic relationships are congruent with the view that the work environment is of greater significance than personal characteristics in the development of burnout.

IMPLICATIONS FOR INTERVENTIONS

The personal and organizational costs of burnout have led to proposals for various intervention strategies. Some try to treat burnout after it has occurred, while others focus on how to prevent burnout by promoting engagement. Intervention may occur on the level of the individual, workgroup, or an entire organization. In general, the primary emphasis has been on individual strategies, rather than social or organizational ones, despite the research evidence for the primary role of situational factors. Many of these individual strategies have been adapted from other work done on stress, coping, and health, rather than on studies of burnout.

However, the greater challenge is that actual evaluation of the effectiveness of any of these interventions has been relatively rare. Especially rare are studies modeled even loosely on randomized control trials. More common are studies with a single intervention group of volunteer participants for whom there are rarely follow-up assessments after treatment has ended.[32]

With these caveats, research indicates that burnout is responsive to a diverse range of methods. For example a metaanalysis confirmed evidence for a beneficial effect for burnout prevention programs that were designed to improve employees' coping abilities, social support, and confidence in their job performance, as well as job redesign initiatives focusing on streamlined procedures and increased job control.[33] It is not yet clear whether burnout is generally susceptible to a range of strategies or whether it is crucial to fit the strategy to the specific context of a workplace to be effective.

Using the Mediation Model in Interventions

From the perspective of the mediation model, burnout reflects a mismatch: workplace design is incompatible

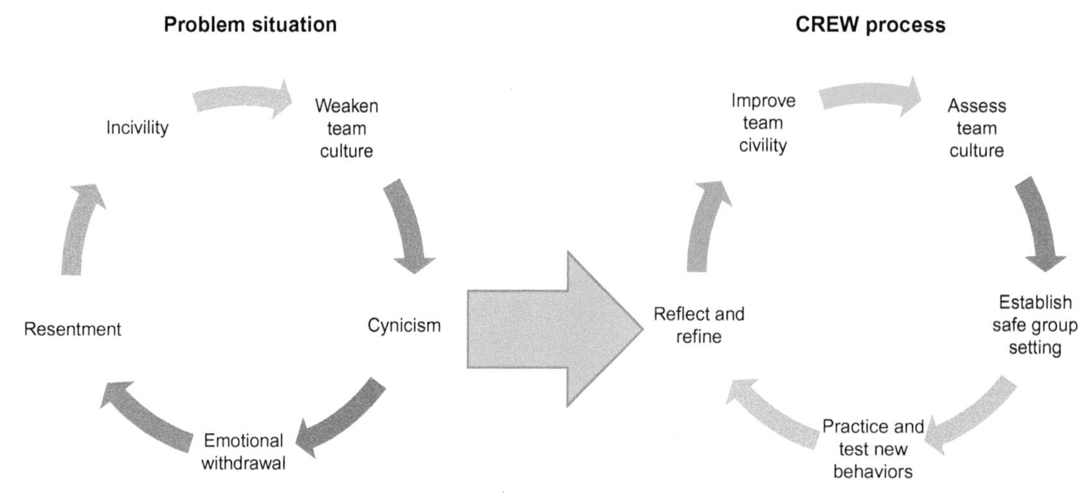

FIGURE 2 The CREW process as a response to community mismatches in a workgroup.

with employees' preferred work patterns. Experiencing that conflict is exhausting, prompting cynical withdrawal and discouragement. Addressing burnout could take two general pathways. One path is to design more employee-friendly workplaces; the other path is to increase employees' reliance to tolerate workplace mismatches. Reflecting the emphasis on the exhaustion aspect of burnout, much of the attention for interventions have been increasing employees' energy through healthy lifestyles with less focus on increasing their involvement by improving the alignment of personal and corporate values.

An intervention with a demonstrated impact on the cynicism aspect of burnout is CREW (*c*ivility, *r*espect, and *e*ngagement at *w*ork). In contrast with control groups, workgroups participating in CREW showed an increase in civility as well as decreases in incivility, burnout, and absenteeism. The decreases in cynicism underscored the importance of social relationships to this aspect of burnout. Although the exhaustion dimension of burnout is highly sensitive to work overload, the cynicism and efficacy dimensions relate closely to other areas of worklife. These studies also demonstrated the potential for workgroup-level interventions to reduce burnout.[34–36]

As outlined in Figure 2, the CREW process begins and ends with an assessment of the social dynamics of a workgroup to determine its specific challenges and the impact of the process. The CREW process provides extensive training and mentoring to local facilitators who lead the CREW meetings over 6 months. The essence of the CREW meetings is developing new behavior repertories through role-playing, followed by assignments in which participants implement these new behaviors in their work settings. It is a team-based enhancement in the community area of worklife that has its most direct impact on the cynicism dimension of burnout, increasing employees' involvement in their work and with their coworkers.

Future Work on Intervention

Burnout has been an important social issue for many years, and there have been continuing calls for solutions to the problems that burnout poses. Increasingly, these calls are coming from government agencies and organizations in both the public and private sectors. Many researchers from many countries around the world have been doing research to understand the burnout phenomenon, but so far this work has not included many rigorous tests of actual interventions to ameliorate or prevent burnout. Clearly, a serious focus on designing, implementing, and evaluating the effectiveness of such intervention programs is the next big challenge for researchers and practitioners alike, working in collaboration with each other.

References

1. Freudenberger HJ. Staff burn-out. *J Soc Issues*. 1974;30:159–165.
2. Maslach C. Burned-out. *Hum Behav*. 1976;9(5):16–22.
3. Maslach C, Jackson SE, Leiter MP. *Maslach Burnout Inventory Manual*. 3rd ed. Palo Alto, CA: Consulting Psychologists Press; 1996. Now published and distributed by Mind Garden (mindgarden.com).
4. Halbesleben JBR, Demerouti E. The construct validity of an alternative measure of burnout: investigation of the English translation of the Oldenburg Burnout Inventory. *Work Stress*. 2005;19:208–220.
5. Freudenberger HJ. Burnout: contemporary issues, trends and concerns. In: Farber BA, ed. *Stress and Burnout in the Human Service Professions*. New York, NY: Pergamon; 1983:23–28.
6. Pines A, Aronson E, Kafry D. *Burnout: From Tedium to Personal Growth*. New York, NY: Free Press; 1981.
7. Shirom A, Melamed S. Does burnout affect physical health? A review of the evidence. In: Antoniou ASG, Cooper CL, eds. *Research Companion to Organizational Health Psychology*. Cheltenham: Edward Elgar; 2005:599–622.
8. Kristensen TS, Borritz M, Villadsen E, Christensen KB. The Copenhagen Burnout Inventory: a new tool for the assessment of burnout. *Work Stress*. 2005;19:192–207.

9. Maslach C, Schaufeli WB, Leiter MP. Job burnout. *Annu Rev Psychol.* 2001;52:397–422.
10. Maslach C, Leiter MP. Early predictors of job burnout and engagement. *J Appl Psychol.* 2008;93:498–512.
11. Leiter MP, Hakanen J, Toppinen-Tanner S, Ahola K, Koskinen A, Väänänen A. Changes in burnout: a 12-year cohort study on organizational predictors and health outcomes. *J Organ Behav.* 2013;34:959–973.
12. Maslach C, Leiter MP. *The Truth About Burnout.* San Francisco: Jossey-Bass; 1997.
13. Schaufeli WB, Salanova M, Gonzalez-Roma V, Bakker AB. The measurement of engagement and burnout: a confirmation analytic approach. *J Happiness Stud.* 2002;3:71–92.
14. Innanen H, Tolvanen A, Salmela-Aro K. Burnout, work engagement and workaholism among highly educated employees: profiles, antecedents and outcomes. *Burnout Res.* 2014;1:38–49.
15. Schaufeli WB, Bakker AB. Defining and measuring work engagement: bringing clarity to the concept. In: Bakker AG, Leiter MP, eds. *Work Engagement: A Handbook of Essential Theory and Research.* New York, NY: Psychology Press; 2010:10–24.
16. Bakker AB, Costa PL. Chronic job burnout and daily functioning: a theoretical analysis. *Burnout Res.* 2014;1:112–119.
17. Bakker AB, Demerouti E. The job demands-resources model: state of the art. *J Manag Psychol.* 2007;22:309–328.
18. Hobfoll SE, Freedy J. Conservation of resources: a general stress theory applied to burnout. In: Schaufeli WB, Maslach C, Marek T, eds. *Professional Burnout: Recent Developments in Theory and Research.* New York, NY: Taylor & Francis; 1993:115–129.
19. Toppinen-Tanner S, Ahola K, Koskinen A, Väänänen A. Burnout predicts hospitalization for mental and cardiovascular disorders: 10-year prospective results from industrial sector. *Stress Health.* 2009;25:287–296.
20. Ahola K, Hakanen J. Burnout and health. In: Leiter MP, Bakker AB, Maslach C, eds. *Burnout at Work: A Psychological Perspective.* London: Psychology Press; 2014:10–31.
21. Toker S, Shirom A, Shapira I, Berliner S, Melamed S. The association between burnout, depression, anxiety, and inflammation biomarkers: C-reactive protein and fibrinogen in men and women. *J Occup Health Psychol.* 2005;10:344–362.
22. Ahola K, Hakanen J. Job strain, burnout, and depressive symptoms: a prospective study among dentists. *J Affect Disord.* 2007;104:103–110.
23. Leiter MP, Maslach C. Nurse turnover: the mediating role of burnout. *J Nurs Manag.* 2009;17:331–339.
24. Laschinger H, Wong CA, Grau AL. The influence of authentic leadership on newly graduated nurses' experiences of workplace bullying, burnout and retention outcomes: a cross-sectional study. *Int J Nurs Stud.* 2012;49:1266–1276.
25. Bakker AB, Le Blanc PM, Schaufeli WB. Burnout contagion among intensive care nurses. *J Adv Nurs.* 2005;51:276–287.
26. González-Morales M, Peiró JM, Rodríguez I, Bliese PD. Perceived collective burnout: a multilevel explanation of burnout. *Anxiety Stress Coping.* 2012;25:43–61.
27. Gascon S, Leiter MP, Andrés E, et al. The role of aggression suffered by healthcare workers as predictors of burnout. *J Clin Nurs.* 2013;22: 3120–3129.
28. Savicki V, Cooley E, Gjesvold J. Harassment as a predictor of job burnout in correctional officers. *Crim Justice Behav.* 2003;30:602–619.
29. Kalimo R, Pahkin K, Mutanen P, Toppinen-Tanner S. Staying well or burning out at work: work characteristics and personal resources as long-term predictors. *Work Stress.* 2003;17:109–122.
30. Lambert EG, Hogan NL, Barton-Bellessa SM, Jiang S. Examining the relationship between supervisor and management trust and job burnout among correctional staff. *Crim Justice Behav.* 2012;30:938–957.
31. Leiter MP, Maslach C. A mediation model of job burnout. In: Antoniou ASG, Cooper CL, eds. *Research Companion to Organizational Health Psychology.* Cheltenham: Edward Elgar; 2005: 544–564.
32. Leiter MP, Maslach C. Interventions to prevent and alleviate burnout. In: Leiter MP, Bakker AB, Maslach C, eds. *Burnout at Work: A Psychological Perspective.* London: Psychology Press; 2014: 145–167.
33. Awa WL, Plaumann M, Walter U. Burnout prevention: a review of intervention programs. *Patient Educ Couns.* 2010;78:184–190.
34. Osatuke K, Mohr D, Ward C, Moore SC, Dyrenforth S, Belton L. Civility, respect, engagement in the workforce (CREW): nationwide organization development intervention at Veterans Health Administration. *J Appl Behav Sci.* 2009;45:384–410.
35. Leiter MP, Laschinger H, Day A, Oore D. The impact of civility interventions on employee social behavior, distress, and attitudes. *J Appl Psychol.* 2011;96:1258–1274.
36. Leiter MP, Day A, Oore D, Laschinger HK. Getting better and staying better: assessing civility, incivility, distress, and job attitudes one year after a civility intervention. *J Occup Health Psychol.* 2012;17:425–434.

CHAPTER

44

Coping Process

E. Stephenson, D.B. King, A. DeLongis
University of British Columbia, Vancouver, BC, Canada

OUTLINE

Abstract

Situations that are appraised as threatening or challenging and that tax available resources are experienced as stressful. Coping encompasses the cognitive and behavioral responses to these situations. Coping includes direct efforts to solve the problem, attempts to manage one's emotions, and attempts to manage social relationships in times of stress. The effectiveness of any given coping response depends on the context in which it occurs. Coping effectiveness depends on the nature of the stressful situation, the personality characteristics of the individual, the responses of involved others, and the social and cultural context in which the coping process occurs. Since there is no coping strategy that is ideally suited to all kinds of stress, flexibility in coping is critically important. The key to successful coping may involve implementing coping responses that match the demands of a given stressful situation.

STRESS AND COPING

We all experience stress at times during our lives. Certain situations are likely to be experienced as stressful by most people. However, there are also considerable individual differences in the types of situations found to be stressful and in responses to these situations. What is extremely stressful to one individual may not be at all stressful to another. According to the transactional model of stress and coping[1] the experience of stress is a product of both the person and the situation. How a situation is evaluated or appraised is an important determinant of the degree of stress experienced by the individual. Situations perceived as threatening, harmful, or challenging, and that tax available resources for coping are experienced as stressful. Coping describes the cognitions and behaviors used to manage the demands of stressful situations.[2] If a situation is not appraised as taxing one's resources or ability to cope, it may not be experienced as stress at all. The resources one brings to a stressful situation include one's personality, age, financial assets, education, previous experiences, social support, and physical and mental health. Available resources and subjective appraisals together contribute to determine, first, the coping responses employed, and in turn, whether these responses are successful or effective in resolving the stressful situation. Whether or not a particular coping strategy was used is measured independently from whether that strategy is effective. Coping is defined as including both successful and unsuccessful attempts to manage stress.

http://dx.doi.org/10.1016/B978-0-12-800951-2.00045-5

KEY POINTS

- The experience of stress is a product of both the person and the situation. A situation is experienced as stressful if the individual perceives that situation as threatening, harmful, or challenging, and as taxing available resources for coping.
- Coping is a dynamic process that encompasses the cognitive and behavioral responses to stressful situations. This process includes efforts to solve the problem (problem-focused coping), manage emotions (emotion-focused coping), and maintain close-relationships (relationship-focused coping).
- The effectiveness of any given coping strategy depends upon personality, and the social and cultural context in which it occurs. There is not one coping strategy that is ideally suited to all contexts.

WAYS OF COPING

Many attempts have been made to categorize the ways in which people cope with stress.[3] One challenge for clinicians and researchers trying to understand how people cope is that most people vary their coping efforts and choices to match the demands of a given stressor.[4] What works well in one situation may not generalize to other contexts. General coping styles aggregated over time tend to be poorly correlated with coping in a given situation.[5] This means that researchers or clinicians cannot accurately predict how an individual will cope with any one specific stressor based on the average way in which that same individual copes across a variety of situations. The strategies that best differentiate between people (i.e., personal coping style across situations) are not necessarily the strategies that best differentiate between situations (i.e., within person fluctuations in coping responses).[6] Coping is a dynamic process, and this presents an important measurement challenge. Effective copers tend to vary their coping strategies during different phases of a stressor.[2] For example, the initial coping response to receiving a cancer diagnosis may be denial, allowing the person to more gradually adapt to a life-threatening diagnosis. However, this may change as the individual and his or her family accept the diagnosis and begin to adapt to the demands of cancer treatment. Although the denial may be effective initially, acceptance of the diagnosis may facilitate engagement in treatment in the long run.

An additional measurement issue that must be addressed is the importance of clearly identifying the stressful situation that requires a coping response. Coping with cancer can be broken down into responses to a cascade of stressful situations, starting with the perception of symptoms that require diagnosis. The treatment itself is a related but separate stressor. There are many additional stressors typically experienced by cancer patients as well, including changes in family relationships, and these may all involve very different coping responses. Stressful situations may present many competing demands that require coping responses. Here, we describe three forms of coping that address different functions of the coping process.

TRIPARTITE MODEL OF COPING

Problem-Focused Coping

Problem-focused coping describes direct efforts to solve the problem at hand.[1] Problem-focused strategies often include attempts to change the situation. This includes defining the problem, generating alternative solutions, comparing these alternatives in terms of their likely costs and benefits, selecting a likely solution, coming up with a plan, and then acting on it. Other problem-focused coping strategies may be geared toward changing ourselves, such as learning new skills and procedures, thereby increasing one's coping resources. Many problem-focused coping strategies are applicable in specific situations and do not generalize broadly across all types of stressors. For example, problem-focused strategies used to cope with chronic pain may not be useful for coping with a pressing deadline at work. Although the specific solution each situation requires may differ, both situations may be amenable to problem solving attempts.

Several factors influence the use of problem-focused coping strategies. Perceiving high levels of threat has been found to interfere with the successful use of problem-focused strategies.[1,7] A common example of this occurs when physicians must deliver bad news to their patients. The patient's feelings of threat or harm may reduce his or her capacity for information processing. At that moment, it can be more difficult for patients to engage in problem-focused strategies, such as information-seeking and treatment planning. They may have difficulty comprehending any additional information about treatment and prognosis that the physician may wish to convey. The perception that the situation is amenable to change also influences the use of problem-focused coping.[1,2] People are more likely to use problem-focused coping strategies when they feel the situation can be changed and that this change is within their control. For stressful situations that cannot be solved, such as with the death of a loved one, persistently engaging in problem-focused coping efforts may actually be maladaptive. In situations that cannot be changed, individuals may need to direct their efforts to emotion-focused coping.

Emotion-Focused Coping

The primary goal of emotion-focused coping is to lessen emotional distress.[1] Some of the ways this may be achieved are through avoidance, distancing, and wishful thinking. Early coping theories (see Parker and Endler[8] for a review) often characterized these types of strategies, such as denial, as maladaptive. However, under certain circumstances denial or denial-like strategies have been found to be quite effective.[9] For example, cognitive reappraisals involve changing the meaning of the situation without changing it objectively, yet this can be one of the most effective forms of coping and is relied upon heavily in cognitive-behavioral therapies. However, when something can be done to solve the problem, using cognitive reappraisal alone is associated with maladjustment.[10]

Relationship-Focused Coping

In addition to attempting to solve the problem and manage emotions, there is a third important function of coping. Individuals must engage in relationship-focused coping to maintain their important social relationships during periods of stress.[11,12] In the context of enduring relationships, people must concern themselves with not only their own wellbeing, but also the effects their coping efforts have on close others and on their relationships with them. Relationship-focused coping includes support provision, empathic responding, and attempting to resolve differences.[13,14] Attempts to manage relationships during stressful periods can have both positive effects and negative effects. In studies of couples coping with stress, relationship-focused coping strategies involving empathic responding have been associated with less marital tension[13] and greater marital satisfaction and stability.[15] Although relationship-focused strategies like protective buffering and overprotection may involve well-intentioned attempts to help one's partner, these strategies have been associated with increased psychological distress and decreased relationship satisfaction.[11,16]

Combined Coping Functions

Some coping strategies can have mixed functions. For example, social support seeking could be used to express emotions (emotion-focused coping), to gather information (problem-focused coping), and to maintain relationships with others (relationship-focused coping). One form of coping may facilitate the use of other strategies. For example, an individual may need to first engage in emotion-focused coping to manage his or her emotions before he or she can effectively engage in problem-focused coping efforts. As described earlier, a patient's heightened threat perception can interfere with his or her ability to comprehend the information his or her physician is trying to communicate. This patient may need to first engage in emotion-focused coping to reduce the perceived threat, before being able to effectively engage in problem-focused coping efforts. Another example could involve relationship-focused coping facilitating other forms of coping. Caregiving spouses of patients with Alzheimer's disease reported that by engaging in perspective-taking and empathy for their spouse, they were able to come up with better solutions to problems posed by the disease.[12]

COPING EFFECTIVENESS

Despite research efforts to identify the ideal coping response, there is no one strategy that is universally adaptive for all people across all situations. When we ask ourselves whether a particular coping response was effective, we need to consider what constitutes successful coping. If successful coping is defined in terms of eliminating the stressful situation, problem-focused strategies will likely be most effective. If successful coping is defined in terms of reducing psychological distress, emotion-focused strategies may be effective. If successful coping is defined in terms of maintaining relationships with close others, relationship-focused coping efforts may be necessary. We might be inclined to define successful coping in terms of all three types of goals, but sometimes these goals conflict or favor different coping responses. In a sample of individuals caring for a spouse with Alzheimer's disease, Kramer[17] found that emotion-focused coping was associated with caregiver depression whereas problem-focused and positive relationship-focused coping were both associated with caregiver satisfaction. Coping strategies that maximize individual wellbeing may come at a cost to the people involved. For example, Badr and colleagues[18] found that among couples facing metastatic breast cancer, some forms of relationship-focused coping were associated with lower distress for partners, but heightened distress for patients.

Whether a particular strategy appears to be adaptive or maladaptive also depends on the timeframe used to evaluate the effects. The same strategy can have different short-term and long-term effects. For example, interpersonal withdrawal has been associated with reduced relationship tension in families the following day, but increased tension over the next 2 years.[19] Similarly, husbands' use of empathic responding is associated with increased marital tension on the same day, but reduced tension the following day.[13] This suggests that what is adaptive in the short-term may be maladaptive in the long-term, and vice versa, and that we need to consider both short- and long-term outcomes in order to fully understand the coping process.

A CONTEXTUAL MODEL OF COPING

The effectiveness of a particular coping strategy depends on the context in which it occurs. The nature of the stressful situation, the personality of the individual, and social context can all influence whether or not a particular coping strategy is employed and whether or not it is effective.[20] The key to successful coping may involve matching or tailoring the coping response to the specific context in which in occurs.[20]

Features of the Situation

Characteristics of the stressful situation influence both which coping strategies are used and the effectiveness of these strategies. How a situation is appraised influences coping choice. Appraisals of the controllability of a given stressful situation influence the coping response.[2] Problem-focused coping is more likely to be employed when the situation is perceived to be within the coper's control; however, the accuracy of this perception influences the effectiveness of this coping response. In situations where control is low, emotion-focused coping efforts are likely to be more effective than are problem-focused coping.[21] Appraisals of threat, loss, and challenge are associated with both coping choice[22] and coping effectiveness.[23] The nature of the perceived threat has also been found to influence the coping response. Situations that threaten agency are associated with increased problem-solving; whereas, relationship-focused coping strategies such as empathic responding are more common in situations that threaten communion.[14]

Features of the Person

Although much of the variance in coping responses is explained by situational factors,[4,6] personality factors have been implicated throughout the stress and coping process. Personality has been associated with exposure to stressful events, how these events are appraised, the coping strategies used, and the effectiveness of these coping strategies.[14,20,24] Personality factors have also been related to the ability to choose an appropriate coping response for a given situation. Individuals high in neuroticism appear to be less able to tailor their coping response to the demands of the stressful situation and to select the strategy that is most likely to be effective.[25] Some individuals may be especially skilled at using particular coping strategies. For example, individuals high in extraversion or agreeableness have been found to benefit more from seeking social support, suggesting that these individuals may be especially skilled at obtaining effective social support.[26]

Social Context

Interpersonal conflicts are one of the most common sources of daily stress,[27] but social relationships are also an important part of the coping process.[20] Close relationships can serve as a resource that individuals use to help them cope. Several studies have demonstrated the benefits of social support in times of stress.[28] Support from close others may influence which coping strategies are employed and the effectiveness of these strategies. In general, individuals who perceive that support is available to them and who are more satisfied with the support they receive tend to engage in more adaptive forms of coping and are less vulnerable to the negative effects of stress. For example, patients with arthritis who felt satisfied with the support they received engaged in more adaptive forms of coping (cognitive reframing) and less maladaptive coping responses (pain catastrophizing), which in turn were associated with lower pain severity over the course of the day.[29,30] Although support attempts are usually well intentioned, they can sometimes hinder adaptive coping efforts. In the context of chronic pain, solicitous support responses can reinforce pain behaviors and are associated with greater pain severity and activity limitations.[31]

The effects of social support on stress may depend on the relationship context in which the support is provided. Both the type of relationship (e.g., spouse, friend, healthcare provider) and the quality of that relationship have been found to influence the effects of social support. For example, relationship satisfaction has been found to moderate the relationship between spousal support and psychological distress.[32]

In addition to being a source of social support, close others can cope in ways that provide a good or bad fit with the coping strategies used by others involved in the situation. There is growing recognition that coping is a social process in which people can cope in ways that facilitate or hinder adaptive coping in one another.[33–35] For example, Hagedoorn and colleagues[36] found that the effects of using self-disclosure as a coping strategy depended on whether one's spouse also engaged in self-disclosure. A similar pattern of results has been observed when spouses use different coping strategies. For example, when spouses engaged in withdrawal, the negative effects of their partners' use of confrontation and rumination in response to stress were amplified.[37,38] Rather than requiring congruent coping responses, research suggests that dyadic coping responses may be best when they are complementary.[39] Although existing research has focused primarily on dyads, this perspective could be applied to wider social networks such as provider-patient-spouse triads or larger family systems.

Cultural Context

Coping responses must also be considered within their cultural context. Some studies have found evidence for cultural differences in coping responses and their effectiveness. For example, the ways in which individuals seek social support and the stress buffering effects of social support have been found to differ across cultures.[40]

SUMMARY

Many factors contribute to the coping process. The context in which the stressful situation occurs, how the situation is appraised, the other people involved, and the personality characteristics of the individual are only some of the factors that affect how to best deal with a given situation. Further, two people might use identical strategies with different degrees of success, depending on how skillfully the strategy is implemented. For this reason, there is no single way of coping that is ideally suited to all situations. However, research suggests that successful coping involves resolving the problem, managing emotions associated with the stressful situation, and efforts to manage relationships with involved others. The ability to be flexible and adapt the coping response to the context in which it occurs may be especially useful.

Acknowledgments

Ellen Stephenson's research is supported by doctoral fellowships from the Social Science and Humanities Research Council of Canada and the Killam Trusts. David King's research is supported by a postdoctoral fellowship from the Social Science and Humanities Research Council of Canada. Anita DeLongis' research is supported by research grants from the Social Science and Humanities Research Council of Canada and the Canadian Institutes of Health Research.

References

1. Lazarus RS, Folkman S. *Stress, Appraisal, and Coping*. New York, NY: Springer Publishing Company; 1984.
2. Folkman S, Lazarus RS, Dunkel-Schetter C, DeLongis A, Gruen RJ. Dynamic of a stressful encounter: cognitive appraisal, coping, and encounter outcomes. *J Pers Soc Psychol*. 1986;50(5):992–1003.
3. Skinner EA, Edge K, Altman J, Sherwood H. Searching for the structure of coping: a review and critique of category systems for classifying ways of coping. *Psychol Bull*. 2003;129(2):216–269. http://dx.doi.org/10.1037/0033-2909.129.2.216.
4. Compas BE, Forsythe CJ, Wagner BM. Consistency and variability in causal attributions and coping with stress. *Cognit Ther Res*. 1988;12(3):305–320. http://dx.doi.org/10.1007/BF01176192.
5. Coyne JC, Racioppo MW. Never the twain shall meet? Closing the gap between coping research and clinical intervention research. *Am Psychol*. 2000;55(6):655–664. http://dx.doi.org/10.1037/0003-066X.55.6.655.
6. Roesch SC, Aldridge AA, Stocking SN, et al. Multilevel factor analysis and structural equation modeling of daily diary coping data: modeling trait and state variation. *Multivariate Behav Res*. 2010;45(5):767–789. http://dx.doi.org/10.1080/00273171.2010.519276.
7. Carver CS, Scheier MF. Situational coping and coping dispositions in a stressful transaction. *J Pers Soc Psychol*. 1994;66(1):184–195. http://dx.doi.org/10.1037/0022-3514.66.1.184.
8. Parker JDA, Endler NS. Coping and defense: an historical overview. In: Zeidner M, Endler NS, eds. *Handbook of Coping: Theory, Research, Application*. New York, NY: Wiley; 1996:3–23.
9. Taylor SE, Brown JD. Positive illusions and well-being revisited: separating fact from fiction. *Psychol Bull*. 1994;116(1):21–27. http://dx.doi.org/10.1037/0033-2909.116.1.28.
10. Wethington E, Kessler R. Situations and processes of coping. In: Eckenrode J, ed. *The Social Context of Coping*. New York, NY: Plenum Press; 1991:13–29. http://dx.doi.org/10.1007/978-1-4899-3740-7_2.
11. Coyne JC, Smith DA. Couples coping with a myocardial infarction: a contextual perspective on wives' distress. *J Pers Soc Psychol*. 1991;61(3):404–412. http://dx.doi.org/10.1037/0893-3200.8.1.43.
12. DeLongis A, O'Brien TB. An interpersonal framework for stress and coping: an application to the families of Alzheimer's patients. In: Stephens MAP, Crowther JH, Hobfoll SE, Tennenbaum DL, eds. *Stress and Coping in Later Life Families*. Washington, DC: Hemisphere Publishers; 1990:221–239.
13. O'Brien TB, DeLongis A, Pomaki G, Puterman E, Zwicker A. Couples coping with stress. *Eur Psychol*. 2009;14(1):18–28. http://dx.doi.org/10.1027/1016-9040.14.1.18.
14. O'Brien TB, DeLongis A. The interactional context of problem-, emotion-, and relationship-focused coping: the role of the big five personality factors. *J Pers*. 1996;64(4):775–813.
15. Bodenmann G, Cina A. Stress and coping among stable-satisfied, stable-distressed and separated/divorced Swiss couples: a 5-year prospective longitudinal study. *J Divorce Remarriage*. 2006;44(1–2):37–41. http://dx.doi.org/10.1300/J087v44n01.
16. Hagedoorn M, Kuijer RG, Buunk BP, DeJong GM, Wobbes T, Sanderman R. Marital satisfaction in patients with cancer: does support from intimate partners benefit those who need it the most? *Health Psychol*. 2000;19(3):274–282. http://dx.doi.org/10.1037//0278-6133.19.3.274.
17. Kramer B. Expanding the conceptualization of caregiver coping: the importance of relationship-focused coping strategies. *Fam Relat*. 1993;42(4):383–391.
18. Badr H, Carmack CL, Kashy DA, Cristofanilli M, Revenson TA. Dyadic coping in metastatic breast cancer. *Health Psychol*. 2010;29(2):169–180. http://dx.doi.org/10.1037/a0018165.
19. DeLongis A, Preece M. Emotional and relational consequences of coping in stepfamilies. *Marriage Fam Rev*. 2002;34(1–2):37–41.
20. DeLongis A, Holtzman S. Coping in context: the role of stress, social support, and personality in coping. *J Pers*. 2005;73(6):1633–1656. http://dx.doi.org/10.1111/j.1467-6494.2005.00361.x.
21. Park CL, Armeli S, Tennen H. Appraisal-coping goodness of fit: a daily internet study. *Pers Soc Psychol Bull*. 2004;30(5):558–569. http://dx.doi.org/10.1177/0146167203262855.
22. McCrae RR. Situational determinants of coping responses: loss, threat, and challenge. *J Pers Soc Psychol*. 1984;46(4):919–928. http://dx.doi.org/10.1037/0022-3514.46.4.919.
23. Mattlin JA, Wethington E, Kessler RC. Situational determinants of coping and coping effectiveness. *J Health Soc Behav*. 1990;31(1):103–122. http://dx.doi.org/10.2307/2137048.
24. Carver CS, Connor-Smith J. Personality and coping. *Annu Rev Psychol*. 2010;61:679–704. http://dx.doi.org/10.1146/annurev.psych.093008.100352.
25. Lee-Baggley D, Preece M, DeLongis A. Coping with interpersonal stress: role of big five traits. *J Pers*. 2005;73(5):1141–1180. http://dx.doi.org/10.1111/j.1467-6494.2005.00345.x.

26. Vollrath M. Personality and stress. *Scand J Psychol*. 2001;42:335–347.
27. Bolger N, DeLongis A, Kessler RC, Schilling EA. Effects of daily stress on negative mood. *J Pers Soc Psychol*. 1989;57(5):808–818.
28. Uchino BN. Understanding the links between social support and physical health: a life-span perspective with emphasis on the separability of perceived and received support. *Perspect Psychol Sci*. 2009;4 (3):236–255. http://dx.doi.org/10.1111/j.1745-6924.2009.01122.x.
29. Holtzman S, DeLongis A. One day at a time: the impact of daily satisfaction with spouse responses on pain, negative affect and catastrophizing among individuals with rheumatoid arthritis. *Pain*. 2007;131 (1–2):202–213. http://dx.doi.org/10.1016/j.pain.2007.04.005.
30. Holtzman S, Newth S, DeLongis A. The role of social support in coping with daily pain among patients with rheumatoid arthritis. *J Health Psychol*. 2004;9(5):677–695. http://dx.doi.org/10.1177/1359105304045381.
31. Leonard M, Cano A, Johansen A. Chronic pain in a couples context: a review and integration of theoretical models and empirical evidence. *J Pain*. 2006;7(6):377–390.
32. DeLongis A, Capreol M, Holtzman S, O'Brien T, Campbell J. Social support and social strain among husbands and wives: a multilevel analysis. *J Fam Psychol*. 2004;18(3):470–479. http://dx.doi.org/10.1037/0893-3200.18.3.470.
33. Revenson TA, Kayser K, Bodenmann G. *Couples Coping with Stress: Emerging Perspectives on Dyadic Coping*. Washington, DC: American Psychological Association; 2005.
34. Berg CA, Upchurch R. A developmental-contextual model of couples coping with chronic illness across the adult life span. *Psychol Bull*. 2007;133(6):920–954. http://dx.doi.org/10.1037/0033-2909.133.6.920.
35. DeLongis A, Holtzman S, Puterman E, Lam M. Spousal support and dyadic coping in times of stress. In: Sullivan K, Davila J, eds. *Support Processes in Intimate Relationships*. New York, NY: Oxford Press; 2010:153–174.
36. Hagedoorn M, Puterman E, Sanderman R, et al. Is self-disclosure in couples coping with cancer associated with improvement in depressive symptoms? *Health Psychol*. 2011;30(6):753–762. http://dx.doi.org/10.1037/a0024374.
37. King DB, DeLongis A. Dyadic coping with stepfamily conflict: demand and withdraw responses between husbands and wives. *J Soc Pers Relat*. 2013;30:198–206. http://dx.doi.org/10.1177/0265407512454524.
38. King DB, DeLongis A. When couples disconnect: rumination and withdrawal as maladaptive responses to everyday stress. *J Fam Psychol*. 2014;28(4):460–469. http://dx.doi.org/10.1037/a0037160.
39. Bodenmann G, Meuwly N, Kayser K. Two conceptualizations of dyadic coping and their potential for predicting relationship quality and individual well-being. *Eur Psychol*. 2011;16(4):255–266. http://dx.doi.org/10.1027/1016-9040/a000068.
40. Kim HS, Sherman DK, Taylor SE. Culture and social support. *Am Psychol*. 2008;63(6):518–526. http://dx.doi.org/10.1037/0003-066X.

CHAPTER

45

Combat Stress

K.E. Porter[1,2], H.M. Cochran[1,2], S.K.H. Richards[1], M.B. Sexton[1,2]

[1]Ann Arbor Veterans Healthcare Administration, Ann Arbor, MI, USA

[2]University of Michigan, Ann Arbor, MI, USA

OUTLINE

Abstract

U. S. military personnel engaged in modern combat operations face a multitude of unique stressors that vary in complexity, intensity, course, and duration. This chapter aims to provide an orientation to physical, environmental, and psychological challenges prior to deployments, within warzones, and after returning home. Despite repeated trauma exposure, most service members exhibit resilience. However, a significant minority of service members do experience chronic difficulties, such as posttraumatic stress disorder (PTSD) and other mental health concerns. The text reviews current protocols for triaging care for service members with stress reactions and outlines the biopsychosocial theories associated with PTSD.

Historians, social scientists, artists, and writers have extensively documented the relationship between warfare and the human condition. Yet, collective knowledge and understanding of individual responses to combat, including the factors and the comprehensive impact, continue to evolve. This chapter explores the modern conceptualization of combat stress, etiological theories of mental health symptoms during and after combat, and factors associated with recovery and distress during postdeployment reintegration.

THE UNIQUE CONTEXT OF COMBAT STRESS

To fully understand the unique nature of combat as a stressor, it is important to examine the context and complexities of a combat deployment. Stress related to a combat deployment begins long before a service member enters the combat theater, and challenges can continue well after the service member returns home, in a process known as the deployment cycle.

The U.S. military defines the deployment process as a four-phase cycle: predeployment, deployment, postdeployment, and reintegration (see Figure 1). In the predeployment phase, the service member receives orders for an upcoming deployment. This phase initiates multiple practical and emotional demands, including increased military training activities and myriad individual and family-impacting preparation for separation (e.g., childcare arrangements, wills, financial planning, and alteration of normal routines). Predeployment may evoke anticipatory anxiety in service members regarding perception of personal and unit readiness and capability for what lies ahead. Members of the National Guard and Reserve units are called to active duty, usually requiring a temporary

Stress: Concepts, Cognition, Emotion, and Behavior
http://dx.doi.org/10.1016/B978-0-12-800951-2.00046-7

end to current civilian employment and reporting to a military installation that may be hundreds of miles from home and support networks.

KEY POINTS

- Combat deployments involve numerous stressors that begin prior to entering the combat zone and continue after the service member returns home.
- The majority of cases of combat stress resolve, but a significant minority of people develop a mental health disorder such as PTSD.
- There is evidence for cognitive, behavioral, social, and biological factors that may contribute to the development of PTSD.
- Factors such as unit cohesion and broader social support can be protective factors and facilitate reintegration following a combat deployment.

Deployment, the second phase of the cycle, refers to the combat zone experience. As later discussed, this period involves marked and repeated stress exposure during which several trauma-related response trajectories may occur. Following their tour, service members return to primary duty stations. This postdeployment phase comprises screenings, evaluations, and educational sessions, which is followed by resuming noncombat routines. For members of National Guard or Reserve units, this often involves returning to nonactive duty status.

Finally, the reintegration phase involves reconnecting with life outside of deployment. While the stressors are considered less physically threatening and acute, several unique demands occur: adapting to changes in family roles, acclimating to expectations of civilian life, economic and occupational stressors, and managing lasting emotional and physical deployment sequelae. During the most recent conflicts of Operation Iraqi Freedom/Operation New Dawn and Operation Enduring Freedom, centrally in Iraq and Afghanistan, respectively, many service members contend with serial deployments, introducing a novel stressor rarely seen in past U.S. military conflicts. Service members remaining on active duty military, National Guard, or reserve units are aware of the potential for redeployment, which may impede full reintegration.

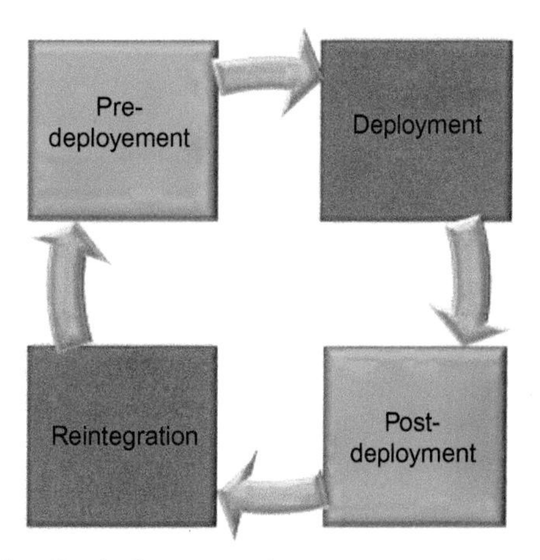

FIGURE 1 The deployment cycle.

Combat Zone Stressors

Contextual factors often compound the experience of combat. The most evident stressors in a warzone revolve around physical threat to oneself or others, unpredictability of danger, and coping with potential or actual use of lethal force. Direct exposure to traumatic combat experiences varies in frequency and severity influenced by service branch, assigned duties, location, and events in the area at that time.

Environmental and physical stressors such as extreme temperatures, hazardous terrain, noise, exposure to toxins, sleep deprivation, dehydration, and poor hygienic conditions may further service members' stress loads.[1] For example, a soldier deployed to Iraq may engage in demanding physical activity for 16 h, suffer temperatures exceeding 100 °F, and reside near power generators and mortar fire. The negative psychological impact of such environmental stressors in noncombat environments has been well established,[2,3] is likely amplified within the context of war, and is expected to further tax existing coping resources.

Combat-deployed service members must also endure innumerable emotional and psychological pressures. The nature of war decreases the predictability of risk and, when events do occur, they must make quick decisions based on ambiguous information with potentially dire consequences. Service members commonly wrestle with doubts about the efficacy of self and others, the value of portraying strength, and moral and spiritual questions. Social support from family and friends is limited by separation, sporadic contact, and restrictions on disclosures (both mandated by military and personal preference). Given this physical and emotional distance from others, within-unit conflict and cohesiveness are more salient during combat operations.

Self-confidence, trust in command, unit cohesion, and aspects of battles moderate responses to combat events.[4,5] Specifically, short- and long-term adverse combat stress reactions are more prevalent when unit cohesion is reduced, concern about direct leadership exists, and doubts about personal combat abilities arise.[4] Conversely, when strong unit bonds are developed and a person has confidence in personal and command abilities, service members are apt to respond resiliently. Initial combat exposure, enduring heavy combat for extended periods, and holding positions rather than making forward or retreat movements substantially increases risk of negative outcomes.[5] Similarly, conflicts resulting in increased casualty rates are further associated with combat stress responses.

COMBAT STRESS REACTIONS AND CLINICAL INTERVENTION

Individual response trajectories following combat traumas are varied. While deployed, a portion of service members exhibit emotional, mental, and physical symptoms that range from mild disturbances to impairing conditions. A Marine Corps report on combat stress described mild symptoms as changes that are subjectively experienced and may not be overtly apparent.[6] Examples include jumpiness, fatigue, insomnia, indecisiveness, anxiety, and tearfulness. Severe symptoms were described as debilitating and with risks of impeding job performance or failing to maintain safety. These presentations may involve indifference to danger, social withdrawal, hallucinations, aggression, recklessness, trembling, panicking under fire, and psychomotor agitation.

Clinical recommendations vary, with nonclinical observation and supports in milder circumstances and significant intervention and/or removal from the combat zone when impairments are pronounced. As a policy, the military aims to provide care for combat stress rapidly with minimal separation from the unit or roles. Specifically, the military relies on the principles of Brevity, Immediacy, Centrality/Contact, Expectancy, Proximity, and Simplicity (BICEPS)[6,7] when treating combat stress. Under this model, Brevity and Immediacy signify initial intervention should occur quickly, be of limited duration, and result in most service members returning to their units and full range of duties. When symptoms remain unresolved, a higher level of care may be necessary. The principles of Contact or Centrality (utilized by the Marine Corps and Army, respectively), Expectancy, and Proximity center on maintaining connection to the service member's unit and their identity as a fighter.[6,7] The service member's command is generally involved in care. The individual is treated in proximity to their unit, ideally, outside of a medical facility. Further, the principle of Expectancy emphasizes normalizing the responses to combat stressors and instilling an expectation of recovery and resilience rather than adoption of a sick role. These conventions are intended to allow the service member to receive the benefits associated with unit cohesion and to foster restoration. Finally, the principle of Simplicity indicates preferred use of the least complicated or intensive level of intervention necessary (in many cases rest). While implementation of BICEPS varies across branches of service, by unit, and for individuals, these principles form guidelines for decision-making regarding treating combat stress reactions in theater. Figure 2 illustrates one such combat stress triage model.

When Combat Stress Reactions Become Chronic: The Development of PTSD

In the majority of cases, symptoms of combat stress resolve and individuals make full recovery. Unfortunately, 11-37% of combat veterans develop mental health symptoms that persist long after returning home.[8,9] Combat exposure can be associated with a variety of mental health outcomes (e.g., depression, anxiety, substance use disorders), one of which is the development of posttraumatic stress disorder (PTSD).

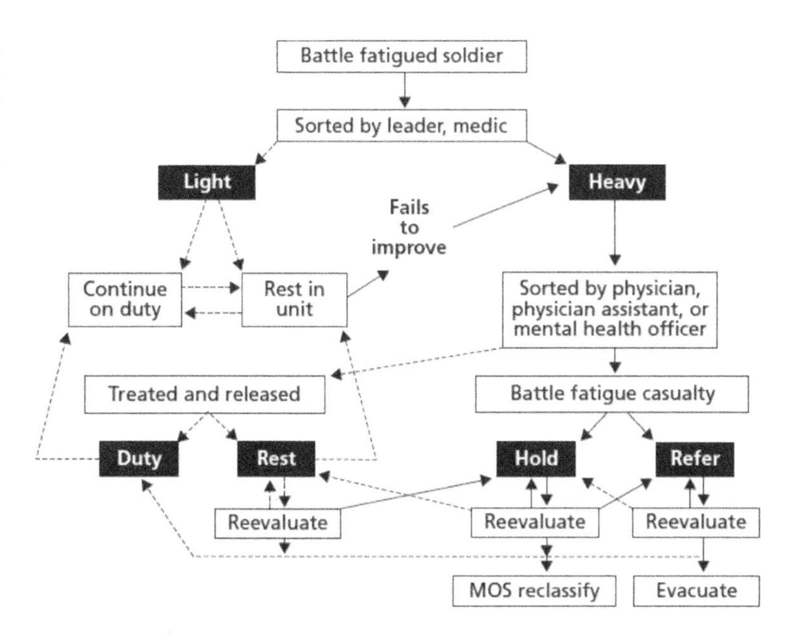

FIGURE 2 Combat stress triage model. Reproduced from Helmus TC, Glen RW. Steeling the mind: combat stress reactions and their implications for urban warfare. RAND Center for Military Health Policy Research, http://www.rand.org/pubs/monographs/MG191.html; Published 2005 Accessed 26.04.15.

PTSD is characterized by the Diagnostic and Statistical Manual of Mental Disorders, fifth edition as a condition with four clusters of symptoms: intrusions, avoidance, negative affect and cognitions, and arousal accompanied by impairment in occupational, social, self-care, or other areas of functioning.[10] Symptoms of intrusions include spontaneous mental images and recollections of specific traumatic events, distressing dreams, or upsetting emotional and physical reactions to reminders of the trauma. Avoidance symptoms encompass attempts to distract, suppress, or neutralize thoughts and emotions connected to traumatic memories. This includes avoiding places, people, situations, or sensory stimuli that cue the trauma memories. Other avoidance may be more subtle such as excessive work, substance abuse, or safety behaviors (e.g., weapon carrying, body positioning, scanning). For example, a marine who experienced a firefight in which children were injured might avoid places where she must hear them screaming such as parks, playgrounds, or malls. Negative affect and cognitive symptoms encompass overall reduction in positive emotions, such as enjoyment of activities, emotional connection to others, pervasive negative mood (e.g., irritability, guilt, fear), self-blame, and pessimistic views about oneself and the world. Finally, hyperarousal symptoms reflect elevated physiological reactivity (e.g., insomnia, poor concentration, excessive startle response, angry outbursts).

Biological Theories

Biological and neurocognitive research offer a rich body of evidence for understanding the physical and neurological mechanisms implicated in the etiology of PTSD symptoms. Researchers have used animal and human models to explore biological predispositions to developing PTSD, as well as physical and neurological changes that occur as a result of exposure to trauma and chronic stress.[11–13] Several variables of interest include structural and physiological changes in regions of the brain responsible for emotion, memory, and sensitization of the central nervous system, as well as gene activity associated with those processes.[11,12,14] Much of this research focuses on fear conditioning as the primary theory, with evidence for distinct roles of neurocognitive systems involved in the acquisition of the conditioned fear response and in extinction learning (or the ability to naturally unlearn the fear response to objectively safe stimuli).[11–13,15]

Fear conditioning occurs when an individual is exposed to a dangerous or life-threatening situation that activates the "fight-flight-or-freeze" response. Physiologically, this involves activation of the hypothalamic-pituitary-adrenal (HPA) axis, brain stem, noradrenergic systems, and amygdala.[13] These fast-acting systems assist individuals responding to trauma (e.g., explosion from a roadside bomb in a combat zone) with immediate increase in heart rate, blood pressure, and respiration. Rapid activation of this system in the face of acute danger is adaptive and protective, allowing the individual to recognize the threat, respond with necessary speed and intensity, and learn from a single instance that the situation is dangerous, potentially preventing, or minimizing future risk.

The initial fear reaction is paired with a wide range of stimuli. In an example of a roadside bomb, the immediate threat is experienced simultaneous to the sound of the explosion, the sight of debris, the smell of diesel fuel, and high ambient temperatures. Based on fear conditioning, stimuli associated with the event may subsequently elicit the fear response.[13,15] Extinction learning gradually follows fear conditioning in most cases. In this process, the neutral stimuli are encountered in absence of the threatening stimuli, and the fear response diminishes and eventually extinguishes to the nonthreatening sights, sounds, and smells. The individual retains a fear response to explosions (threatening stimulus), but does not continue to fear fuel smells, summer temperatures, or roadside litter.

In some instances, this adaptive evolutionary response can become overgeneralized or overconsolidated and result in PTSD symptoms.[12,13] Stress hormones, particularly catecholamines and cortisol, are repeatedly implicated in the development and maintenance of PTSD.[12–14] In studies of Vietnam veterans, those with PTSD (compared to combat veterans without PTSD) had higher catecholamine levels at both baseline and when exposed to trauma cues.[13] Elevated baseline catecholamine levels increase fear sensitivity, while higher levels after a cue escalate the intensity and duration of the fear response. Genetics research has begun to identify specific genotypes associated with the onset and severity of PTSD symptoms. For example, certain variants of the *Met158* gene, which is known to be responsible for breaking down catecholamines, are associated with poor concentration, hypervigilance, and sleep disruption.[14] Related, combat veterans with PTSD have demonstrated lower levels of cortisol, which provides negative feedback within the HPA axis to reduce arousal. This may explain, in part, difficulties with extinction learning in PTSD.[13,14]

Extinction learning involves the amygdala and specific regions of the medial prefrontal cortex and ventromedial prefrontal cortex (vmPFC).[11,13] In combat veterans with PTSD, the vmPFC is not deactivated in the presence of emotional stimuli, a difference from combat veterans without PTSD and noncombat controls.[11] Even in studies of aversive stimuli that are not trauma-related, individuals with PTSD show impaired or delayed ability to extinguish physiological stress responses.[12]

The roles of memory and attention in developing and maintaining PTSD symptoms have also received significant attention. Imaging studies reveal smaller volume of the hippocampus, an area of the brain essential for formation and consolidation of new memories, in combat veterans, civilians, and children with PTSD.[11,12] It is not yet known whether smaller hippocampal size represents a predisposing factor for PTSD or rather size is reduced as the result of trauma exposure and subsequent neurophysiological changes.[11,12]

Researchers hypothesize elevated catecholamines during and following trauma exposure may contribute to incorporation of the memory into the fear response.[14] The fear response can be elicited in individuals with PTSD even when the conditioned stimulus is not consciously perceptible.[11,13] This may explain frequent intrusive memories, flashbacks, and emotional reactivity to trauma reminders that are seminal components of PTSD.[12,13] Raised catecholamines are also implicated in attention to pleasurable and aversive stimuli. Thus, changes to the noradrenergic system are related to emotional and physiological reactivity to trauma reminders, panic attacks, and overall elevated anxiety.[12–14]

Cognitive Behavioral Theories

Cognitive behavioral theories of trauma responses, including combat, have garnered empirical attention and are largely based on populations with adverse mental health trajectories.[16,17] Two prominent theories are emotional processing theory (EPT)[18] and social cognitive theory.[19] These theories form the foundation of the empirically supported interventions for PTSD (Prolonged Exposure therapy and Cognitive Processing therapy, respectively) most often employed by mental health providers treating service members and veterans with PTSD through the Department of Defense and the Department of Veterans Affairs.[20]

Emotional Processing Theory

EPT is grounded in the fear conditioning model, positing that during a trauma a person creates connections between stimuli and the fear response. EPT further emphasizes the cognitive and emotional meanings that the trauma survivor assigns to those elements, as well as associated negative mood states (e.g., anxiety, anger, guilt, shame). Such appraisals subsequently influence whether the individual experiences recovery or prolonged distress. Furthermore, incorporating meaning into EPT offers a possible explanation of why individuals who survive similar traumas may have different long-term responses. For instance, in the example above in which a soldier experiences a roadside bomb, the resulting learned association network in EPT would include the stimuli (e.g., loud noise, debris, hot temperatures), the emotional reaction (e.g., fear), and also the person's interpretation of the significance of those elements. A soldier who assigns to the fear reaction the meaning that he is weak or incompetent may experience emotions such as shame. This in turn may lead to avoidance of the memory and associated emotions. Service members may use a wide range of avoidance strategies that yield short-term reduction in anxiety and other negative emotions, but, when sustained, are involved in the development and maintenance of PTSD. Once an EPT network is established (with or without the maladaptive elements), it must be activated in order to modify it with new information. Thus, avoiding thoughts, emotions, and reminders of the trauma further prohibits recovery (Figure 3).

Social Cognitive Theory

Resick et al.[19] have similarly focused on the salience of meaning and emotional expression in recovery from combat trauma in social cognitive theory. Compared to EPT, social cognitive theory focuses more on the content of specific cognitions about the traumatic event, and how those cognitions influence emotions and behavior. For instance, a combat medic attending to lethal injuries on the battlefield is likely to experience fear, anger, or sadness, due to the inherent danger and suffering. However, if the medic subsequently and repeatedly concluded that she was responsible for the death of her fellow soldiers, then she is likely to experience persistent guilt or shame. This negative interpretation and associated guilt may prompt her to avoid the memory, thus preventing natural dissipation of distressing emotions.

This paradigm is likewise concerned with how individuals reconcile events into existing beliefs about themselves, others, and the world. Trauma often violates long-standing worldviews, and people will attempt to resolve this resulting dissonance by either integrating the event into an existing belief system or by altering their beliefs to accommodate the new information from the trauma. For example, a commonly held belief among service members and civilians alike is the "just-world hypothesis": the belief that good things happen to good people, and bad things happen to bad people. When traumatic events occur and conflict with centrally-held beliefs (e.g., someone viewed as "good" is killed in action), maintaining belief in the just-world hypothesis may lead a surviving

FIGURE 3 United States Army, 2007. Sgt. Auralie Suarez and Private Brett Mansink take cover during a firefight with guerrilla forces in the Al Doura section of Baghdad on the 7th of March. Image by Staff Sgt. Sean A Foley, United States Army. http://www.army.mil/images/2007/03/21/3462.

service member to blame themselves for the catastrophe. In the case of the medic, attempting to assimilate the experience into a just-world belief may require her to ignore some information in the memory (e.g., discounting the severity of injuries), or to distort other aspects of the memory (e.g., perceiving that she had more time, resources, or prior knowledge than were actually available).

Problems incorporating the trauma into existing beliefs also occur when individuals dramatically shift beliefs, or over accommodate, after the event. If the medic concludes, "I can never trust my judgment again," she is likely to experience anxiety, shame, and other negative emotions. Here, healthy resolution of trauma-related symptoms may require her to adjust her broad view of the world to incorporate the reality of the level of risk in a combat zone irrespective of her merit. Improved emotional responses and behaviors arise from satisfactory resolution of threatening or challenging cognitive content. For the medic, this might mean acknowledging, "I did the best I could, but it was impossible to save everyone," or "I made some mistakes in a high-pressure situation, but I also saved many lives, and I can usually trust my judgment." Military members may experience prolonged difficulties if maladaptive beliefs regarding the causes of the event, safety, trust, control, esteem, and intimacy are recurrently activated and unresolved.[19]

REINTEGRATION: HOMECOMING AFTER COMBAT

As service members return home, their mission and the stressors they encounter change drastically. While this eagerly anticipated time is generally positive and marked by a reduction in risk, difficulties associated with transition back into noncombat roles are common. Many frequently report stress redefining their role within the family.[21] While the service member is deployed, family members adapt and expand roles to fill the voids created by the person's absence. Many returning service members describe feeling unneeded and out of place. Veterans with persisting physical or psychological wounds may not be able to reengage with previous activities or fulfill former obligations. They and their families often strive to accommodate changes in attitudes, emotions, and behaviors altered by military experiences. Similarly, veterans frequently report believing civilians cannot understand their sacrifices or combat experiences, resulting in feeling disconnected from their family, civilian peers, and their larger community.[22] This can be particularly alienating in contrast with the closeness felt with the comrades with whom they served.

Employment can be a significant source of stress during the reintegration period and unemployment is disproportionately high in veteran populations.[23] After returning from deployments involving leadership roles, making life-and-death decisions, or managing valuable equipment, civilian jobs may seem mundane. Cultivated talents may go unrecognized and underutilized, and as a result, they may struggle to find work, accept positions in which they are underemployed, and feel discouraged and devalued.

Protective Factors

Despite the detrimental impact of combat stress that may occur, most service members return from deployment and readjust without long-standing concerns.[8,9] As such, efforts have been made to identify protective factors that may mitigate the deleterious effects of trauma and combat. Social support and resilience are critical buffers. Research with returning veterans finds that support from family and friends is related to less distress and symptoms.[24,25] Unit support also has a positive impact on postdeployment readjustment and may protect against developing mental health symptoms.[25]

Psychological resilience, broadly viewed as a person's ability to adapt and recover from difficulties, similarly promotes successful reintegration. It protects against the development of mental health concerns, reduces suicidality, and is associated with improved physical health outcomes.[25,26] Individuals with greater resilience tend to report sense of belonging in one or more groups (e.g., family, military, community), perceive self-control over their roles, and believe their activities have meaningful purpose.[27,28] Currently, interventions are being developed and tested to foster resilience within military personnel.[29]

CONCLUSIONS

The stresses associated with combat are tremendous and multifaceted. To more fully understand the nature of combat stress, we must collectively consider the tensions and changes that begin with deployment notices and may persist long after returning home. The demands of a wartime deployment can result in myriad trajectories, including those associated with combat stress reactions and PTSD. However, factors such as strong social support and unit cohesion appear to be significant buffers for this stress, and many service members adapt and demonstrate strong resilience in the face of combat.

References

1. *Department of the Army*. Combat and operational stress control. http://armypubs.army.mil/doctrine/DR_pubs/dr_a/pdf/fm4_02x51.pdf; Published July 2006 Accessed 12.04.15.
2. Evans GW, Cohen S. Environmental stress. In: *New York*. NY: John Wiley & Sons; 1987:571–610. Stokols D, Altman I, eds. Handbooks of Environmental Psychology; vol. 1.
3. Sundstro E, Bell P, Busby P, Asmus C. Environmental psychology 1989–1994. *Annu Rev Psychol*. 1996;47:485–512.

4. Kellett A. *Combat Motivation: The Behavior of Soldiers in Battle*. Boston, MA: Kluwer Nijhoff Publishing; 1982.
5. Helmus TC, Glen RW. Steeling the mind: combat stress reactions and their implications for urban warfare. RAND Center for Military Health Policy Research, http://www.rand.org/pubs/monographs/MG191.html; Published 2005 Accessed 26.04.15.
6. *U.S. Marine Corps*. Combat stress. http://www.au.af.mil/au/awc/awcgate/usmc/mcrp611c.pdf; Published June 2000 Accessed 26.04.15.
7. *Department of the Army*. Combat and operational stress control manual for leaders and soldiers. http://armypubs.army.mil/doctrine/DR_pubs/dr_a/pdf/fm6_22x5.pdf; Published July 2009 Accessed 05.05.15.
8. Hoge C, Auchterloine J, Milliken C. Mental health problems, use of mental health services, and attrition from military service after returning from deployment to Iraq and Afghanistan. *J Am Med Assoc*. 2006;295:1023–1032.
9. Seal K, Metzler T, Gima K, Bertenthal D, Maguen S, Marmar C. Trends and risk factors for mental health diagnoses among Iraq and Afghanistan veterans using Department of Veterans Affairs health care, 2002–2008. *Am J Public Health*. 2009;99:1651–1658.
10. American Psychiatric Association. *Diagnostic and Statistical Manual of Mental Disorders*. 5th ed. Washington, DC: American Psychiatric Association; 2013.
11. Garfinkel SN, Liberzon I. Neurobiology of PTSD: a review of neuroimaging findings. *Psychiatr Ann*. 2009;39:370–381.
12. Pitman R, Rasmussen AM, Koenen KC, et al. Biological studies of post-traumatic stress disorder. *Nat Rev Neurosci*. 2012;13:679–787.
13. Yehuda R, Ledoux J. Response variation following trauma: a translational neuroscience approach to understanding PTSD. *Neuron*. 2007;56:19–32.
14. Almli LM, Fani N, Smith AK, Ressler KJ. Genetic approaches to understanding post-traumatic stress disorder. *Int J Neuropsychopharmacol*. 2014;17:355–370.
15. Amstadter AB, Nugent NR, Koenen KC. Genetics of PTSD: fear conditioning as a model for future research. *Psychiatr Ann*. 2009;39: 358–367.
16. Brewin CR, Holmes EA. Psychological theories of posttraumatic stress disorder. *Clin Psychol Rev*. 2003;23:339–376.
17. Dalgleish T. Cognitive approaches to posttraumatic stress disorder: the evolution of multirepresentational theorizing. *Psychol Bull*. 2004;130:228–260.
18. Foa EB, Steketee G, Rothbaum BO. Behavioral/cognitive conceptualizations of post-traumatic stress disorder. *Behav Ther*. 1989;20: 155–176.
19. Resick PA, Monson CM, Chard KM. *Cognitive Processing Therapy Manual: Veteran/Military Version*. Washington, DC: Department of Veterans Affairs; 2006.
20. U.S. Department of Veterans Affairs and U.S. Department of Defense. VA/DoD clinical practice guidelines for management of post-traumatic stress disorder. http://www.healthquality.va.gov/Post_Traumatic_Stress_Disorder_PTSD.asp; October 2010 Accessed 29.05.15.
21. Bowling UB, Sherman MD. Welcoming them home: supporting service members and their families in navigating the tasks of reintegration. *Prof Psychol Res Pr*. 2008;39:451–458.
22. Demers A. When veterans return: the role of community in reintegration. *J Loss Trauma*. 2011;16:160–179.
23. Sayer NA, Carlson KF, Frazier PA. Reintegration challenges in U.S. service members and veterans following combat deployment. *Soc Issues Policy Rev*. 2014;8:33–73.
24. Pietrzak RH, Johnson DC, Goldstein MB, Malley JC, Southwick SM. Psychological resilience and postdeployment social support protects against traumatic stress and depressive symptoms in soldiers returning from operations Enduring Freedom and Iraqi Freedom. *Depress Anxiety*. 2009;26:745–751.
25. Pietrzak RH, Johnson DC, Goldstein MB, et al. Psychosocial buffers of traumatic stress, depressive symptoms, and psychosocial difficulties in veterans of Operations Enduring Freedom and Iraqi Freedom: the role of resilience unit support, and postdeployment social support. *J Affect Disord*. 2010;120:188–192.
26. Green KT, Calhoun PS, Dennis MF, The Mid-Atlantic Mental Illness Research, Education and Clinical Center Workgroup, Beckham JC. Exploration of the resilience construct in posttraumatic stress disorder severity and functional correlates in military combat veterans who have served since September 11, 2001. *J Clin Psychiatry*. 2010;71:823–830.
27. Gibbons SW, Shafer M, Aramanda L, Hickling EJ, Benedek DM. Combat health care providers and resiliency: adaptive coping mechanisms during and after deployment. *Psychol Serv*. 2014;11:192–199.
28. Pietrzak RH, Southwick SM. Psychological resilience in OEF-OIF Veterans: application of a novel classification approach and examination of demographic and psychosocial correlates. *J Affect Disord*. 2011;133:560–568.
29. Meredith LS, Sherbourne CD, Gaillot S, et al. Promoting psychological Resilience in the U.S. Military. RAND Center for Military Health Policy Research, http://www.rand.org/pubs/monographs/MG996.html; Published 2011 Accessed 20.05.15.

CHAPTER

46

Survivor Guilt

P. Valent

Melbourne, VIC, Australia

OUTLINE

Abstract

Guilt is an internal moral judgement that aims to modify instinctive survival drives and pleasures in a prosocial direction. Guilt is felt as a specific emotional pain that is part of a bad conscience. Shame and not being fair are other moral modifiers. In traumatic situations, survival drives (survival strategies) are insufficient to prevent suffering of others. Unappeased moral judgements, including survivor guilt, are then relived and avoided like other posttraumatic symptoms.

This chapter examines the history and purpose of survivor guilt. Subsets of survivor guilt are examined in terms of the survival strategies that have not fulfilled their purpose. Lastly, treatment of survivor guilt is examined.

INTRODUCTION

Guilt is a pervasive feature of civilization. It is a judgement that is applied from early childhood (good boy, good girl) to cosmic sin such as Eve eating the forbidden apple. Guilt has been conceived as an inner badness that evokes traumatic consequences.[1] Thus, akin to human travail being caused by Eve's badness, disasters have been blamed on previous sins.

A scientific view of guilt is relatively recent. It involves three interrelated concepts.

1. Guilt is one of three moral judgements. The other two are shame and justice.[2] Shame judges a person's worth. Justice judges an action's fairness. Guilt judges a person's actions as good or bad, and in a more cosmic sense, as virtuous or sinful.
2. Moral judgements are normative for specific circumstances. For instance, following a bush fire people's guilt for their houses not having burnt down induced them to provide shelter to those who were made homeless.[3]
3. Guilt is a consequence of traumatic situations, not its cause. The history of survivor guilt has contributed significantly to this concept.

HISTORY OF THE CONCEPT OF SURVIVOR GUILT

Survivor guilt as an unwarranted clinical consequence of trauma was named and emphasized for the first time in Holocaust literature in the 1960s. Psychiatrists such as Krystal and Niederland[4] were struck with the ubiquity of intense, unabated guilt for having survived, while others had died, among their Holocaust survivor patients.

Survivor guilt was quickly recognized in other traumatic situations. They included the Hiroshima atom bomb, combat, disasters, and civilian bereavement. It was noted after 9/11[5]; indeed, it could occur in all traumatic situations, whether natural or manmade.

Stress: Concepts, Cognition, Emotion, and Behavior
http://dx.doi.org/10.1016/B978-0-12-800951-2.00047-9

SUMMARY

Survivor Guilt

1. Is a traumatic moral judgement that is relived and avoided.
2. Can arise from all traumatic situations. It doesn't cause them.
3. Specific survivor guilt arise from specific insufficient survival strategies.
4. Attempts to retrospectively modify traumatic events.
5. May be the central knot that maintains traumatic disorders.
6. Treatment requires recognition, processing, and grief.
7. Grief can lead to new hope and includes retrieval of morality.

DIFFERENT TYPES OF SURVIVOR GUILT

Until now survivor guilt has been seen as a generic concept. However, as Valent[2] indicated, a number of survival instincts (survival strategies) are active or potentially active in traumas. Each one has specific survivor guilt.

1. *Rescue/caretaking*. Guilt is an agony for having been neglectful and self-concerned, and as a result caused harm, suffering, or death to others, especially for whom they bore responsibility. Holocaust survivors blamed themselves for the deaths of their children and spouses. Rescuers and helpers blame themselves for not having saved those for whom they viewed themselves to be responsible.
2. *Attachment*. Guilt feels as having been bad, sinful, disobedient. One has caused the catastrophe, or it, such as abandonment, must be punishment for one's badness.
3. *Goal achievement*. Guilt here is associated with failure. One feels inadequate, incompetent, stupid, clumsy, or lazy.
4. *Goal surrender*. People may feel guilty for having given in or given up, for despairing, for attempting suicide.
5. *Fight*. While it is a virtue to fight in self-defense or in the defense of others, mistaken causing of damage and death evokes horror and dismay. One feels like a wicked murderer.
6. *Flight*. Guilt here may be for having left others behind, shirked responsibility, being a coward, lacking moral fiber, having turned away, being a bystander.
7. *Competition*. Priority guilt implies that people survived undeservedly at the expense of others, that really they should have died instead of the ones who perished. Survivors often say that the dead should be alive instead of themselves.
8. *Cooperation*. Sense of cheating, lying, exploitation, and hurt of partners evokes betrayal guilt. Sexual desire has been branded sinful and has evoked intense guilt over many centuries.

POSTTRAUMATIC MANIFESTATIONS OF SURVIVOR GUILT

Traumatic events generally are relived in feelings, thoughts, images, dreams, and flashbacks. Incorporated in these images is intense survivor guilt. This includes general anguish, and specific subguilt ruminations about the different survival possibilities that the survivor should have done but did not do. Specific cognitive schemas and meanings develop out of these ruminations, such as the survivor is an irresponsible person, a betrayer, a failure, a coward, even destroyer of life (see above). Survivor guilt is often central in maintenance of posttraumatic conditions.

Objectively the self-blame is seldom justified. The contrast between objective innocence and the torment of self-blame is frequent and conspicuous. The person merely acted instinctively according to the calculus of maximum survival, or without any objective choice being involved. For instance, a man blamed himself for killing his mother in childbirth. A veteran blamed himself for his comrade having been killed. A surgeon blamed himself for not having saved an unsaveable patient. Holocaust survivors often blamed themselves for not following or saving their loved ones who were murdered, when actually they were separated at gunpoint through physical force. The Holocaust demonstrated the frequent paradox that victims are prone to survivor guilt while perpetrators rationalize theirs.

Alongside reliving guilt, survivors may withdraw. They avoid enjoyment and vitality, as they can exacerbate their feelings of guilt. Socially, they want to avoid the imagined blame in the question "Why did you survive?" For this reason survivors may play down their survivorship.

A variety of psychological defenses may mitigate survivor guilt. Psychic numbing, dissociation, and repression fragment coherent awareness of survivor guilt. The person may displace guilt and blame and be angry with others. Blaming a spouse for a child's death can lead to breakup of the marriage. Alternately, the person may punish him- or herself by identifying with the dead person, for instance by assuming symptoms similar to those the dead person had suffered. The guilty person may seek redemption through devotion to saving others, for instance in rescuer and helper professions.

PURPOSE OF SURVIVOR GUILT

In acute phases of traumatic situations, guilt and its external provocateur, anger, modify pure instinctive responses. "Don't just leave me here!" may modify an impulse to flee and change it to one of helping the other person. We saw that priority guilt for having a house while others didn't led to sharing one's house.

Survivor guilt thus helps to preserve as many people (or genes) in the community as possible.

Posttraumatic survivor guilt is less useful. It may prime adaptive reparative action for subsequent traumatic situations. It may also preserve the survivor's retrospective hope that the "if onlys" may yet be able to alter the situation. As Danieli[6] suggests, the guilt can act as a buffer against moral chaos and existential helplessness, as the guilt preserves assumption of moral order and one's capacity to adhere to it. Last, the guilt perpetuates close emotional links to the dead.[7]

TREATMENT OF SURVIVOR GUILT

Prevention of survivor guilt. Adherence to clear demarcations of responsibility and to protocols regarding prioritization and procedures provides comfort in the knowledge that one had done the best one could in the circumstances. Prevention of compassion fatigue and early debriefing can also ameliorate survivor guilt.

Treatment. Once established, the first principle of treatment of survivor guilt is its recognition both in its overt reliving and in its veiled defended forms.

Next, thorough investigation of the facts of the circumstances usually reveals that the survivor had done his or her best. It was the circumstances, not the survivor that were irrational and bad.

Next, the evolutionary sense of objectively untenable self-blame is explained and alternative, hopeful views are explored.

In the context of an empathic, therapeutic relationship the emotional pains and meanings of survivor guilt are identified, processed, and placed in perspective. At last the survivor can grieve, establish new hopeful meanings, retrieve a sense of morality and reclaim a purposeful life.

Glossary

Guilt An internal emotional judgment (an aspect of conscience) of having been bad or having done wrong to another.

Survivor A person who has lived through a trauma.

Survival strategies Instinctual drives used to survive traumatic situations.

Trauma An experience in which a person's life has been grossly threatened and out of which a variety of biological, psychological social wounds, and scars results.

References

1. Jordan MR. A spiritual perspective on trauma and treatment. *NCP Clin Q*. 1995;5:9–10. http://www.ncptsd.va.gov/publications/cq/v5/n1/jordan.html.
2. Valent P. *From Survival to Fulfilment: A Framework for the Life-Trauma Dialectic.* Philadelphia, PA: Brunner/Mazel; 1998.
3. Valent P. The Ash Wednesday bushfires in Victoria. *Med J Aust.* 1984;141:300.
4. Krystal H, Niederland WC, eds. *Psychic Traumatization: After-Effects in Individuals and Communities.* Boston, MA: Little, Brown and Company; 1971.
5. Ochs R. Decade later, trauma haunts 9/11 survivors. *Newsday.* Sept 7th, 2011; http://www.newsday.com/911-anniversary/decade-later-trauma-haunts-9-11-survivors-1.3152854.
6. Danieli Y. Treating survivors and children of the Nazi Holocaust. In: Ochberg FM, ed. *Post-Traumatic Therapy and Victims of Violence.* New York: Brunner/Mazel; 1988.
7. Nader K. *Guilt Following Traumatic Events. Gift from Within; PTSD Resources for Survivors and Caregivers; 2015.* http://giftfromwithin.org/html/Guilt-Following-Traumatic-Events.html.
8. Raphael B. *When Disaster Strikes.* London: Hutchinson; 1986.

CHAPTER

47

Refugees: Stress in Trauma

J.D. Kinzie

Department of Psychiatry, Oregon Health and Science University, Portland, OR, USA

OUTLINE

Abstract

Armed conflicts throughout the world have created millions of refugees who in their tragic lives may have experienced death of family members, starvation, torture, and long periods in refugee camps. In the country of resettlement, many refugees will have major psychiatric problems including post-traumatic stress disorder, depression, traumatic brain injury, and psychosis. The evaluation of refugees is difficult because of language, culture, and the effect of trauma itself. Many forms of technical treatment have been suggested, but probably the personal characteristics of the therapist and providing safety and continuity over time is the most helpful. Much symptomatic relief can be given with medicine targeting insomnia, nightmares, and irritability. More research is needed to understand why some traumatized refugees get symptoms and others do not; what treatments are effective for patients from different cultures, and how to treat the high rate of cardiovascular disease among refugees.

INTRODUCTION

One of the consequences of armed conflicts, civil war, and genocide has been the creation of refugees. Refugees are defined as crossing a national border to seek safety from conflicts in their own country. This is distinguished from internally displaced persons who had forced migration within their own borders. Another group is the asylum seekers, who often enter the country legally or illegally and request asylum from possible persecution, or even death, if forced to return to their own country.

Refugees, once they are legally accepted into a country, have certain benefits and rights. Asylum seekers have no such benefits until their asylum case is approved, a process which can take years. Overall, there were 11.7 million refugees recognized by the United Nations High Commission on Refugees (UNHCR) in 2013.

Table 1 shows the top five countries of origin of refugees and the top recipients of refugees in 2013. The large disparity, the number of refugees and those being received legally indicates that the vast majority of refugees are in refugee camps in a second country awaiting acceptance to a third country.

There were 1.2 million asylum seekers in 2013. Until they receive approved status which permits them to stay in the country legally, their condition is in limbo; not obtaining legal status in their current and at any time they be judged not to be in danger of returning and be deported.

A large population of concern for UNHCR is the internally displaced persons sometimes protected and assisted by UNHCR. As of 2013, there were 23.9 million internally displaced persons.

Stress: Concepts, Cognition, Emotion, and Behavior
http://dx.doi.org/10.1016/B978-0-12-800951-2.00048-0

TABLE 1 Major Countries of Origin of Refugees & Recipient Countries

Top countries of origin of refugees		Top recipients	
2013			
Afghanistan	2.56 million	Germany	109,600
Syria	2.47 million	United States	84,400
Somalia	1.12 million	South Africa	70,000
Sudan	648,900	France	62,000
Congo	499,600	Sweden	54,300

Since 1975, the United States has admitted 3 million refugees and in 2013 it admitted 84,400 persons. In 2012, there were 7400 asylum seekers waiting for their cases to be judged and hopefully granted legal status. In 2012, there were 25,300 positive acceptances for asylum seekers in the US, representing many who have waited many years for a decision.

KEY POINTS

- Refugees who have fled their own country to find safety in another often have endured severe trauma. Many suffer from PTSD, depression, traumatic brain injury, and even psychosis. Long-term following of refugees with PTSD indicates a high rate of diabetes, hypertension, and cardiovascular disease.
- Multiple psychotherapies are advocated, but often long-term follow-up is lacking.
- Medicine can give symptomatic relief and may reduce the cardiovascular morbidity.
- More research is needed, both on the efficiency of treatment and the prevention and treatment of the resultant medical problems.

REFUGEE TRAUMA

The life of a refugee consists usually of catastrophic stress and trauma. Many often has witnessed murders, sometimes of their family, extreme cruelty, often torture, periods of starvation, and death of children and relatives, traveling on a dangerous path to a refugee camp. Refugee camps are often poorly run, sometimes invaded by criminals who rape, brutalize, and steal food. They wait while their application is being processed, usually several years, until its acceptance in a third country.[1]

Many refugees and asylum seekers have been victims of torture. Torture is defined by the UN Declaration against torture in 1975 "torture means any means by which severe pain or suffering, whether physical or mental, is intentionally inflicted by or at the instigation of a public official... to obtain information, confession or punishment." Torture has been commonly experienced, often multiple times, by refugees and asylum seekers. The estimate of torture ranged from 10% to 30%.[2] A more formal evaluation of Iraqi refugees indicated that 56% experienced torture before arrival in the United States and torture was the most significant predictor of mental illness.[3] Sexual violence among refugees is a horrific occurrence. A systematic review of female refugees found a prevalence rate of sexual violence of 21.4%.[4] This is likely an underestimation.

In addition to trauma and dislocation, refugees represent a very diverse group of people and cultures. From hill tribes in Burma and illiterate farmers in Somalia to highly educated professionals from Iran and Iraq. They also practiced various forms of Islam, Buddhism, Christianity, and local belief systems. These refugees are a complex mixture of people, cultures, and traumatic experiences. Their issues need to be recognized in the treatment process.

The following case examples come from our Intercultural Psychiatric Program (IPP) at Oregon Health & Science University in Portland, Oregon. The IPP has existed since 1977 and currently has about 1300 active patients from 15 different language groups. The patients are treated by a faculty psychiatrist and aided by counselors from each corresponding ethnic group. The counselors act as interpreters during the psychiatric visits and as case managers between visits.[5]

Case I: This refugee patient is a 43-year-old separated Somali female refugee referred by community members. Her primary symptom is very poor sleep (2-3 h per night) with frequent nightmares about real traumas. She is always fearful and startles at unexpected noises. She describes much irritability and feels sad all the time, and has no pleasure in life. In 1977, thieves dressed in military uniforms broke into her house and her father, two brothers, a step-mother and two cousins were killed. She was beaten unconscious and was raped by five men. She and her children made their way to a refugee camp in Uganda, where they stayed for about 15 years before coming to the United States.

Case II, Asylum seeker: This patient is a 40-year-old Ethiopian male asylum seeker. While in the United States, he has extremely poor sleep, frequent nightmares, and depression with a 20-pound weight loss. There is poor concentration associated with irritability. He was well-educated and worked for private companies. He is married with two children. He loosely associated with a political party which lost an election and subsequently he was taken prisoner for several months. He described being tied with beatings on his head and feet. He lost consciousness and was thought to be dead, but was taken to the hospital and recovered, only to be sent back to prison. At one time, he remained in a dark room for days and was told he would be executed at some point. When released, he obtained a visa and came to the United States. He has had no direct contact with his family for 3 years while awaiting the outcome of his asylum hearing.

POSTMIGRATION

In addition to traumatic experiences, refugees have many adjustment problems. In addition to learning English, there are financial problems and difficulties raising children in America. In a study of immigrant survivors of political violence in the United States, postmigration factors such as financial and legal insecurity explained post-traumatic stress disorder (PTSD) outcomes more than premigration factors such as traumas alone.[6] Also, increased length of time, over 1 year, between arrival in the United States and clinical services was associated with increased PTSD and depression symptoms.[7] It was found among Iraqi refugees in Michigan after 1 year in the United States, there was significant increase in body mass index (BMI) and hypertension.[8] It was suggested that a new lifestyle, including dietary patterns and physical activity level, contributed to these changes.

Although premigration trauma among refugees contributes heavily to their symptoms, postmigration stresses and lifestyle changes also contribute. It is important to assess stress exposure throughout the entire refugee experience.[9]

PSYCHIATRIC DISORDERS AMONG REFUGEES

PTSD. PTSD was recognized among Cambodian refugees soon after PTSD was first described in DSM-III in 1980.[10] This was followed by a report of PTSD among other Southeast Asian refugees whose diagnosis was missed in the original evaluation.[11] Since that time, there have been many studies of diagnosis, usually concentrating on PTSD among refugees.

Many more studies of psychiatric diagnosis among refugees have followed. In a systematic review of refugees exposed to torture and traumatic events involving 161 articles, the author found a weighted prevalence of PTSD of 30.6% and depression of 30.8%.[12] A more recent review of PTSD and depression among Iraqi refugees in Western countries found a range of PTSD between 8% and 37.2% and depression 28.3% to 75%.[13] A study of psychiatric disorders among asylum seekers in Switzerland found major depression (31.4%) and PTSD (23.3%) were diagnosed most frequently.[14] Psychiatric disorders are more frequently diagnosed in those attending clinics. A sample of 61 refugees had 82% diagnoses of PTSD and while disorders of extreme stress (NOS, not otherwise specified) was diagnosed in 16%, 649 had both PTSD and depression.[15]

In a recent report on Syrian refugees in Turkey, the frequency of PTSD was 33.5%.[16] They calculated that the probability of PTSD was 71% if they were female gender being diagnosed with a psychiatric disorder in the past, having a family history of psychiatric disorder, and experiencing two or more traumas.

The evidence indicates that PTSD and major depression are highly comorbid and are highly prevalent in refugees, but at least in community samples not present in the majority of refugees. The complete emphasis on PTSD and depression probably simplifies a complex clinical pattern. Clearly there are other diagnoses but perhaps are not so easily identified by clinical scales.

Traumatic Brain Injury. Clinical experience indicates that a large number of refugees have had head injuries. In a study of torture survivors ($N=488$), 69% reported a blow to the head and of these, 55% reported loss of consciousness.[17] Those with head injuries reported more sleep disturbances and headache. A community sample of Vietnamese political detainees ($N=337$) of whom 78% had experienced a traumatic head injury, and those with traumatic head injuries had high rates of PTSD and depression.[18]

Psychosis. Psychosis, especially schizophrenia, was described in 7 of the first 100 Cambodian refugees.[19] A series of six case reports of psychosis with PTSD was described in 2011 when Kroll[20] reported that 80% of Somali males under the age of 30 presented with psychosis compared to 13.7 in no Somali same age group.

Other Symptoms. Sleep disturbances are common among all refugees but rarely systematically studied. A study from Tbilisi, Georgia of internally displaced persons found an incidence of insomnia of 41.4%. The majority of insomniacs believed that was related to remembered stress.[21]

Suicide is not often reported among refugees, most of whom are Buddhist or Muslims, who have a strong prohibition against suicide. However, between 2009 and 2012, there were 16 cases of Bhutanese refugee suicides.[22] Suicidal ideation was found to be related to psychiatric disorders and postmigration difficulties (family conflicts and inability to find work).

TABLE 2 Diagnoses of Refugee Patients

PTSD with depression	34
PTSD only	3
Depression only	13
Schizophrenia	8
Depression with psychosis	4
Traumatic brain disorder	4
Alzheimer's probable	5
Dementia unknown	1
Alcohol and substance abuse	0

Unlike American veterans with trauma experiences, alcohol and substance abuse are relatively rare, but in our clinic experience are increasing, especially in the second generation.

The overview really does not reflect the everyday clinical realities of psychiatrists evaluating refugees. In Table 2 is a summary of 50 consecutive evaluations by the author of Somali and Ethiopian patients indicating the complex clinical issues of refugees.

THE COURSE OF TRAUMA SYNDROME AMONG REFUGEES

The outcome of trauma symptoms among refugees is a difficult issue. First, PTSD has a fluctuating course with delayed onset occurring in approximately 25% of PTSD cases, thus many patients will have PTSD long after trauma exposure.[23] Second, even apparently recovered PTSD patients will show a complete exacerbation of symptoms under current stress or trauma.[24]

Nevertheless, Cambodian refugees resettled in the United States and studied two decades later were found to still have high rates of PTSD (62%), major depression (51%), and low rates of alcohol abuse (4%).[25] In a 10-year evaluation of 23 Cambodians in continuous treatment, 13 were rated as having a good outcome, while 10 had a relatively poor outcome.[26] Similarly, mental health changes in tortured refugees ($N=45$) in a multidisciplinary treatment program found no substantial changes at 9 and 23 months.[27]

In a qualitative study of recovered survivors from former Yugoslavia ($N=26$) and to those with ongoing symptoms found that participants attributed their improvement to social support, material support, mental health treatment, and used future-oriented coping.[28] Clearly, the outcome of the asylum process has a large impact on symptoms. Usually, asylum seekers who have their application accepted have a reduction in symptoms while those who are rejected do not.

With or without treatment, the long-term follow-up of traumatized refugees indicate that a substantial number will continue to have symptoms and a fluctuating course depending on ongoing stress.

DIABETES, HYPERTENSION, AND CARDIOVASCULAR DISEASE

Very early after the arrival of refugees, hypertension and other risk factors for cardiovascular disease were reported.[29] A later report with Vietnamese, Cambodian, Somali, and Bosnian refugees ($N=459$) found a prevalence of hypertension of 42% and of diabetes of 15.5% (much more than a match to the US population). Asylum seekers with PTSD have a higher prevalence of type II diabetes (T2D) than those without PTSD, and indicated that PTSD is related to T2D among asylum seekers, independent of comorbid depression in the Netherlands.[30] In a recent study, the prevalence of diabetes, hypertension, and hyperlipidemia in Cambodian refugees greatly exceeded that of a comparable US population.[31] These factors point to a high risk for cardiovascular disease among refugees and indeed, a higher mortality from cardiovascular disease in refugees was found in Sweden.[32]

In nonrefugees, PTSD has been found to be related to cardiovascular disease and myocardial ischemia. Clearly the disorders of trauma are not just psychiatric disorders, but have major implication in cardiac disease.

EVALUATION OF THE REFUGEE PATIENT

There are many difficulties in evaluating refugee patients: language, religion, and culture may be quite different and unfamiliar to the interviewer. These make treatment evaluation difficult. The process is long, complicated, and requires patience. Another very important factor is a competent interpreter, as mentioned, our program has full-time counselors from the patients' culture groups who act as interpreters for our psychiatrists.

Rating scales have been used in most PTSD and refugee research. It provides a quantitative evaluation which shows changes after treatment. There are many problems with self-rating scales. They only give information about the symptoms they are designed to measure is PTSD and depression. They only give cross-sectional data and no longitudinal course. Even traumatic scales do not indicate how often the traumas occurred and how the patient reacts to each trauma. What is necessary is a thorough psychiatric evaluation which includes several parts.

1. An evaluation of current symptoms, especially somatic symptoms, which are important to the patient (i.e., insomnia, headache, back pain, nightmares).
2. A review of the patient's psychosocial development: family life, education, marriage, children, and community relationships.
3. The traumatic experiences, usually multiple over long periods of time. The patient reactions to these which often are surprising and complicated.
4. The escape process and refugee camp experience, which are often traumatic as well.
5. The adjustment to the United States (or other Western country), with emphasis on ongoing financial or legal difficulties, which contribute to further symptoms.
6. A mental status exam, which includes cognitive functioning, orientation, memory, and psychotic symptoms which provide diagnostic information regarding traumatic brain injury and psychosis.
7. A negotiated and agreed upon treatment plan. Usually, the doctor and patient can agree on which should be targeted first for treatment.

It is important not to open the patient to a life of trauma without treatment. If the interviewer is not the treating clinician, it is best to only concentrate on preventing complaints and to leave the full evaluation to a treating physician. It is unfair to expose a refugee patient to their trauma and then abandon them. Obviously, blood pressure and fasting glucose, or A1C needs to be measured and appropriate treatment given.

TREATMENT

Multiple psychotherapies have been reported on with traumatized refugees. These have included cognitive behavioral therapy, group therapy, trauma-focused therapy, the use of lay counselors and exposure therapy. There was also a report of a controlled trial of psychotherapy for Congolese survivors of sexual violence. Most all studies report a decrease in PTSD symptoms and sometimes depression symptoms. There are now several reviews of psychological treatment for traumatized refugees. The conclusions vary. Nickerson[33] found that trauma-focused therapy may efficiently treat PTSD. McFarlane and Kaplan[34] concluded that improvement was found in a range of interventions. All described the need for better designed studies. These reports often do not take into account the real world multiple (comorbid) diagnoses and usually have short-term follow-up.[35] In short, trauma patients have multiple diagnoses, a natural fluctuating course, and in general, a poor prognosis. It is somewhat difficult to reconcile the optimistic treatment outcome with the more pessimistic natural history studies.

There are marked variations in cultures as patients respond to psychotherapy, especially with much verbal interaction,[36] which are usually not addressed in the research literature. Another factor that strikes the author is the large emphasis on technical approaches. Research on therapy outcomes indicates that common factors in therapy (warmth, genuineness, trustworthiness) accounts for more variance than therapy techniques.[37] The author experience that the therapist be able to listen, the ability to stay and provide constancy and to tolerate stories of severe trauma are essential for successful treatment.[38]

MEDICINE

Little has been written about the use of medicine for symptomatic relief among refugees. The problem may be due to trying to treat PTSD as a complete syndrome rather than addressing individuals. For PTSD, paroxetine, fluoxetine, and venlafaxine have shown some effectiveness.[39] A pragmatic approach is to address the primary concerns of patients. Table 3 is not a representative sample; it reflects our clinical experience with patients from multiple ethnic groups. Insomnia and poor sleep, often only 2-4 h per night, has persisted for many years and is also associated with nightmares. Early literature and current suggestions that tricyclic antidepressants (TCAs) such as imipramine, doxepin, and desipramine have advantage in being sedatives promoting sleep and affecting several neurotransmitters involved in PTSD.[40] We have used TCA, particularly imipramine, for over 20 years with very good response.

Nightmares are a major problem, often occurring two to three times a night. Fortunately, the central adrenergic medicines clonidine and prazosin are very effective at reducing nightmares, usually rapidly. Agitation and irritability are very disruptive to family and social life. The antipsychotic risperidone has been found to be helpful, but is limited by the increased risk of diabetes in a vulnerable population. Aripiprazole has been effective with fewer side effects.

For young patients (below 65), it is usual to start patients on imipramine and clonidine. Aripiprazole is

TABLE 3 Major Symptoms of all Refugee Patients and PTSD Patients (Subgroup)

	$N = 50$	PTSD patients ($N = 34$)
Poor sleep	33 (66%)	27 (79%)
Nightmares	25 (50%)	25 (73%)
Irritability	20 (40%)	16 (47%)
Depressed mood	12 (29%)	11 (32%)

added if irritability is still present. Hypertension needs to be treated and there is evidence that the angiotensin II AT (1) receptor blockers can reduce PTSD symptoms as a treatment for inflammatory brain disorder. In summary, trauma among refugees is a complicated, chronic disorder characterized by exacerbation and remission. Therapy with cultural understanding and humility addressing ongoing current stress is helpful. Appropriate medicine for symptoms and medical problems can provide great relief and prevent future cardiovascular disease. Our prospective study of supported psychotherapy and medicine showed good response in 20 of 22 patients.[41] Clearly, a longer term follow-up is needed to prove lasting results.

FUTURE RESEARCH APPROACHES

Much remains to learn about disorders of severely traumatized refugees. In many ways, they represent a unique group for study, since mostly the patients are not complicated by alcohol and substance abuse and do not have issues of compensation or entitlement. A major question is that despite massive trauma, usually less than 50% develop PTSD. It is known that 40-50% of the variance of PTSD is genetic but the exact mechanisms are unknown. Future work is progressing in understanding the gene-environment in PTSD. The consistent finding of high rates of hypertension and diabetes needs biological explanation. There is strong evidence for an inflammatory process in PTSD which leads to low-grade systemic inflammation. If it is confirmed it could lead to biomarkers and novel treatment. Finally, the issue of confirming effective psychological and biological treatments needs to be addressed. This is a difficult issue since there is little finding, it is difficult to find control groups and a waiting list for suffering and frightened patients seems unethical.

References

1. Boehnlein JK, Kinzie JD. Refugee trauma. *Transcult Psychiatr Res Rev*. 1995;32:223–252.
2. Modvig J, Jaranson J. A global perspective of torture, political violence, and health. In: Wilson JP, Drozdek B, eds. *Broken Spirits: The Treatment of Traumatized Asylum Seekers, Refugees, War and Torture Victims*. New York, NY: Brunner-Routledge Press; 2004:33–52.
3. Willard CL, Rabin M, Lawless M. The prevalence of torture and associated symptoms in United States Iraqi refugees. *J Immigr Minor Health*. 2014;16(6):1069–1076.
4. Vu A, Adam A, Wirtz A, et al. The prevalence of sexual violence among female refugees in complex humanitarian emergencies: a systematic review and meta-analysis. *PLoS Curr*. 2014;6;6:ecurrents.dis.835f10778fd80ae031aac12d3b533ca7. http://dx.doi.org/:10.1371/currents.dis.835f10778fd80ae031aac12d3b533ca7.
5. Kinzie JD, Tran KA, Breckenridge A, Bloom JD. An Indochinese refugee psychiatric clinic: culturally accepted treatment approaches. *Am J Psychiatry*. 1980;137(11):1429–1432.
6. Chu T, Keller AS, Rasmussen A. Effects of post-migration factors on PTSD outcomes among immigrant survivors of political violence. *J Immigr Minor Health*. 2013;15(5):890–897.
7. Song SJ, Kaplan C, Tol WA, Subica A, deJong J. Psychological distress in torture survivors: pre- and post-migration risk factors in a US sample. *Soc Psychiatry Psychiatr Epidemiol*. 2015;50(4):549–560.
8. Jen KL, Zhou K, Arnetz B, Jamil H. Pre- and post-displacement stressors and body weight development in Iraqi refugees in Michigan. *J Immigr Minor Health*. 2014;17:1468–1475.
9. Perera S, Gavian M, Frazier P, et al. Longitudinal study of demographic factors associated with stressors and symptoms in African refugees. *Am J Orthopsychiatry*. 2013;83(4):472–482.
10. Kinzie JD, Fredrickson RH, Ben R, Fleck J, Karls W. Posttraumatic stress disorder among survivors of Cambodian concentration camps. *Am J Psychiatry*. 1984;141(5):645–650.
11. Kinzie JD, Boehnlein JK, Leung PK, Moore LJ, Riley C, Smith D. The prevalence of posttraumatic stress disorder and its clinical significance among southeast Asian refugees. *Am J Psychiatry*. 1990;147(7):913–917.
12. Steel Z, Chey T, Silove D, Marnane C, Bryant RA, van Ommeren M. Association of torture and other potentially traumatic events with mental health outcomes among populations exposed to mass conflicts and displacement: a systematic review and meta-analyses. *JAMA*. 2009;302(5):537–549.
13. Slewa-Younan S, Uribe Guajardo MG, Heriseanu A, Hasan T. A systematic review of post-traumatic stress disorder and depression amongst Iraqi refugees located in western countries. *J Immigr Minor Health*. 2014;17:1231–1239.
14. Heeren M, Mueller J, Ehlert U, Schnyder U, Copiery N, Maier T. Mental health of asylum seekers: a cross-sectional study of psychiatric disorders. *BMC Psychiatry*. 2012;12:114.
15. Teodorescu DS, Heir T, Hauff E, Wentze-Larsen T, Lien L. Mental health problems and post-migration stress among multi-traumatized refugees attending outpatient clinics upon resettlement to Norway. *Scand J Psychol*. 2012;53(4):316–332.
16. Alpak G, Unal A, Bulbul F, et al. Post-traumatic stress disorder among Syrian refugees in Turkey: a cross-sectional study. *Int J Psychiatry Clin Pract*. 2015;19(1):45–50.
17. Keatley E, Ashman T, Im B, Rasmussen A. Self-reported head injury among refugee survivors of torture. *J Head Trauma Rehabil*. 2013;28(6):E8–E13.
18. Mollica RF, Chernoff MC, Megan Berhold S, Lavelle J, Lyoo IK, Renshaw P. The mental health sequelae of traumatic head injury in South Vietnamese ex-political detainees who survived torture. *Compr Psychiatry*. 2014;55(7):1626–1638.
19. Kinzie JD, Boehnlein JK. Post-traumatic psychosis among Cambodian refugees. *J Trauma Stress*. 1989;2(2):185–198.
20. Kroll J, Yusuf AL, Fujiwara K. Psychoses, PTSD, and depression in Somali refugees in Minnesota. *Soc Psychiatry Psychiatr Epidemiol*. 2011;46(6):481–493.
21. Basishvili T, Eliozishvili M, Maisuradze L, et al. Insomnia in a displaced population is related to war-associated remembered stress. *Stress Health*. 2012;28(3):186–192.
22. Centers for Disease Control and Prevention (CDC). Suicide and suicidal ideation among Bhutanese refugees - United States, 2009–2012. *MMWR. Morb Mortal Wkly Rep*. 2013, July 5. Retrieved from http://www.cdc.gov/mmwr/preview/mmwrhtml/mm6226a2.htm.
23. Bryant RA, O'Donnell ML, Creamer M, McFarlance AC, Silove D. A multisite analysis of the fluctuating course of posttraumatic stress disorder. *JAMA Psychiatry*. 2013;70(8):839–846.

24. Kinzie JD, Boehnlein JK, Riley C, Sparr L. The effects of September 11 on traumatized refugees: reactivation of posttraumatic stress disorder. *J Nerv Ment Dis*. 2002;190(7):437–441.
25. Marshall GN, Schell TL, Elliott MN, Berhold SM, Chun CA. Mental health of Cambodian refugees 2 decades after resettlement in the United States. *JAMA*. 2005;294(5):571–579.
26. Boehnlein JK, Kinzie JD, Sekiya U, Riley C, Pou K, Rosborough B. A ten-year treatment outcome study of traumatized Cambodian refugees. *J Nerv Ment Dis*. 2004;192(10):658–663.
27. Carlsson JM, Olsen DR, Kastrup M, Mortensen EL. Late mental health changes in tortured refugees in multidisciplinary treatment. *J Nerv Ment Dis*. 2010;198(11):824–828.
28. Ajdukovic D, Ajdukovic D, Bogic M, et al. Recovery from posttraumatic stress symptoms: a qualitative study of attributions in survivors of war. *PLoS One*. 2013;8(8):e70579.
29. Dodson DJ, Hooton TM, Buchwald D. Prevalence of hypercholesterolaemia and coronary heart disease risk factors among southeast Asian refugees in a primary care clinic. *J Clin Pharm Ther*. 1995;20 (2):83–89.
30. Agyemang C, Goosen S, Anujuo K, Ogedegbe G. Relationship between post-traumatic stress disorder and diabetes among 105,180 asylum seekers in the Netherlands. *Eur J Public Health*. 2012;22(5):658–662.
31. Marshall GN, Schell TL, Wong EC, et al. Diabetes and cardiovascular disease risk in Cambodian refugees. *J Immigr Minor Health*. 2015; Feb 5. Epub ahead of print.
32. Hollander AC, Bruce D, Ekberg J, Burstrom B, Borrell C, Ekblad S. Longitudinal study of mortality among refugees in Sweden. *Int J Epidemiol*. 2012;41(4):1153–1161.
33. Nickerson A, Bryant RA, Silove D, Steel Z. A critical review of psychological treatments of posttraumatic stress disorder in refugees. *Clin Psychol Rev*. 2011;31(3):399–417.
34. McFarlane CA, Kaplan I. Evidence-based psychological interventions for adult survivors of torture and trauma: a 30-year review. *Transcult Psychiatry*. 2012;49(3–4):539–567.
35. Buhmann CB. Traumatized refugees: morbidity, treatment and predictors of outcome. *Dan Med J*. 2014;61(8):B4871.
36. Morris P, Silove D. Cultural influences in psychotherapy with refugee survivors of torture and trauma. *Hosp Community Psychiatry*. 1992;43(8):820–824.
37. Dalenberg CJ. On building a science of common factors in trauma therapy. *J Trauma Dissociation*. 2014;15(4):373–383.
38. Kinzie JD. Psychotherapy for massively traumatized refugees: the therapist variable. *Am J Psychother*. 2001;190(7):437–441.
39. Hoskins M, Pearce J, Bethell A, et al. Pharmacotherapy for posttraumatic stress disorder: systematic review and meta-analysis. *Br J Psychiatry*. 2015;206(2):93–100.
40. Davidson J. Vintage treatments for PTSD: a reconsideration of tricyclic drugs. *J Psychopharmacol*. 2015;29(3):264–269.
41. Kinzie JD, Kinzie JM, Sedighi B, Woticha A, Mohamed H, Riley C. Prospective one-year treatment outcomes of tortured refugees: a psychiatric approach. *Torture*. 2012;22:1–10.

CHAPTER

48

Stress in Emergency Personnel

J.T. Mitchell

University of Maryland Baltimore County, Baltimore, MD, USA
International Critical Incident Stress Foundation, Ellicott City, MD, USA

OUTLINE

Abstract

Police, firefighters, emergency medical technicians, paramedics, emergency communications personnel, emergency department and critical care workers, and military personnel involved in life-saving missions are in high intensity and high-risk professions.

From the alert tones to the direct efforts to serve and save others, emergency operations personnel contend with a great deal of stress.

The rates of posttraumatic stress disorder in emergency personnel rival, and sometimes, surpass those of combat veterans. Stress impacts their personal and professional lives. Frequent exposure to traumatic stress contributes to marital and relationship discord, premature retirements, and excessive use of sick time. The psychological impacts of emergency services stress can also produce alterations in personality, chronic emotional distress, panic attacks, substance abuse and stress-related disorders.

There is obviously a need to provide a comprehensive, integrative, systematic, and multicomponent staff support program to alleviate distress in these populations.

INTRODUCTION

Every police call, every ambulance response, and every alarm-of-fire has its own stressors and psychological elements. Besides the everyday stressors that most people deal with, such as administrative hassles, traffic, and family life, emergency services personnel face frequent exposure to traumatic events. As a result, many develop chronic symptoms of distress.[1] Prevalence of posttraumatic stress disorder (PTSD) in the general population may be as low as 1-6%, but emergency services personnel are at an elevated risk.[1–4]

Although stress in emergency services has been present in warfare and disasters and in less intense events such as shootings, vehicle accidents, and fatality fires throughout the history of organized emergency response organizations, it was not until the American Psychiatric Association briefly mentioned stress in disaster workers in a 1964 guidebook on psychological first aid for American Red Cross personnel that attention was placed on stress, especially traumatic stress, as a potential problem for emergency personnel.[5] Before then, stress in emergency response personnel was simply viewed as something that was part of the job. The common belief was that if people did not want to be exposed to traumatic stress they would not select emergency work as a profession.

Stress: Concepts, Cognition, Emotion, and Behavior
http://dx.doi.org/10.1016/B978-0-12-800951-2.00049-2

KEY POINTS

- *Critical incidents* are traumatic, turning point events that can overwhelm normal coping and disrupt an emergency person or group's ability to function on-the-job or even at home.
- *Critical incident stress* is an expected and usual, although painful and disruptive, psychological reaction to a critical incident. It has a wide range of cognitive, physical, emotional, behavioral, and spiritual symptoms.
- *Crisis intervention* is a temporary, active, and supportive entry into a person or group's life during a period of extreme distress. It is not psychotherapy, but a supportive process to assist shocked and overwhelmed people until chaos subsides and they can manage on their own.
- *The seven principles of crisis intervention* are the guiding precepts of all crisis intervention services. Crisis intervention should be (1) simple, (2) brief, (3) innovative, (4) practical, (5) close to emergency services operational areas, or in someone's comfort zone, (6) immediate, and (7) should set up reasonable expectations of a positive outcome.
- *Critical incident stress debriefing (CISD)* is an interactive, seven-step group crisis intervention procedure which is one of the common critical incident stress management (CISM) techniques.
- *CISM* is a package of crisis intervention techniques that form a comprehensive, integrative, systematic, and multicomponent traumatic stress management program.
- *CISM team* is a group of mental health professionals, peer support personnel, and clergy who are trained in CISM and provide those services to emergency personnel who have encountered a significant critical incident.
- *Peer support personnel* are CISM trained personnel from the same profession who provide CISM services to their distressed colleagues. Peer support personnel have demonstrated their effectiveness in a wide range of situations.
- *Strategic CISM services* is an organized approach to applying the techniques of crisis intervention. It includes selecting the right *targets, and types of techniques.* It takes into consideration the proper *timing* of the techniques and the *themes* that influence decisions. A CISM strategy also assures that the right *team or resources* are deployed to manage the crisis.
- *CISM techniques* are the tools or tactics utilized in the provision of CISM services. The core techniques are (1) assessment and triage, (2) strategic planning, (3) individual support, (4) informational groups, (5) interactive groups, and (6) follow-up and referral services.

CRITICAL INCIDENTS AND EMERGENCY PERSONNEL

Critical incidents are significant and emotionally powerful events that may temporarily overwhelm an individual emergency worker or an entire emergency unit's ability to function at their maximum capacity. Critical incidents are pivotal events that can either stimulate psychological growth or they may cause considerable psychological damage. In some cases, the psychological distress associated with a critical incident may be the source of long-term problems.

Examples of critical incidents for emergency personnel include line-of-duty deaths, suicide of a colleague, life-threatening on-the-job injuries, injuries or deaths involving children, disasters, situations in which loved ones are injured or killed, acts of violence, accidents, combat, complicated and dangerous rescues, and events with grave personal threat. Emergency personnel who experience critical incidents describe them as horrible, terrible, overwhelming, frightening, awful, grotesque, threatening, or disgusting.

CRITICAL INCIDENT STRESS

Critical incidents often generate a wide range of physical and psychological symptoms. The reaction to a critical incident is called critical incident stress. The symptoms of critical incident stress appear in five main categories. They are *cognitive, physical, emotional, behavioral, and spiritual.* Some examples are shown below.

- Cognitive
  - Mental confusion and disorganization
  - Difficulties in decision-making and problem-solving
  - Disruption to memory functions
  - Inability to manage abstract thinking
  - Inability to name familiar objects
- Physical
  - Physical reactions such as nausea, vomiting
  - Shakes

- Headaches
- Intestinal disturbance
- Chest pain or difficulty breathing
- Muscle aches and pains

- Emotional
  - Intense anxiety, shock, denial and disbelief
  - Anger, agitation and rage
  - Helplessness
  - Lowered self-esteem
  - Loss of self confidence
  - Fear
  - Feeling emotionally subdued or depressed
  - Feelings of intense grief
  - Emotional numbness
  - Apathy
- Behavioral
  - Withdrawal from others
  - Increase use of alcohol or other substances
  - Periods of excessive activity to suppress memories of the tragic event
- Spiritual
  - Loss of faith
  - Cessation of the practice of religion
  - Crisis in faith

The term, *critical incident stress,* was developed in the mid-1970s to describe an "expected and typical response of normal, psychologically healthy people after an exposure to one or more extraordinary traumatic events." It originated from a need to have nonclinical terminology that could be easily understood by and appropriately applied to emergency services, military, and disaster response personnel. Critical incident stress is a term that is routinely used to describe the cognitive, physical, emotional, behavioral, and spiritual reactions of people who experience psychologically disturbing events as a result of their jobs. The key factor in critical incident stress is the exposure to the critical incident. Without exposure to the traumatic event, the critical incident stress reaction would not occur. Some mental health professionals use the phrase critical incident stress as a synonym for terms like "posttraumatic stress," "traumatic stress," and "post-trauma syndrome." Those terms, however, are generally viewed as being more rooted in clinical psychology and, thus, are less acceptable to hardy people, such as military and emergency services personnel, who periodically encounter severe psychological threats, unusually distressing circumstances, gory sights and sounds, and other unsettling critical incidents.

Most people do recover reasonably quickly after an exposure to a critical incident and do not require formal psychological intervention. If it is not carefully managed and resolved, it is possible that critical incident stress can become the source of long-lasting conditions such as substance abuse, phobic reactions, depression, panic disorders or PTSD.[6]

CRISIS INTERVENTION

Before a program to intervene in emergency services stress could be developed, several enlightened mental health professionals set the foundation for what would eventually became a comprehensive and systematic approach to emergency services stress. Their pioneering work was in the newly formulated field of crisis intervention that began in World War I and grew in World War II.

The theories of crisis intervention were developed, discussed, researched, and put into action most especially through the work of Erik Lindermann and Gerald Caplan in the 1940s, 1950s, and 1960s. The concepts and precepts of Crisis Intervention that they developed remain in use today. Lindermann and Caplan are credited with setting the foundation of Crisis Intervention. Developments and the addition of new advances in the field are established on the strong theoretical, historical, and practical foundations established by these two giants.

Their theories and practical applications were put to the test in the horrible and tragic Coconut Grove fire in Boston on November 28, 1942 in which 492 people lost their lives and 166 were injured. Lindermann used Crisis Intervention principles to support the survivors, their families, and the families of the deceased. Lindermann and his colleagues were also able to identify specific stages of grieving during the time he worked with those involved in the Coconut Grove fire.

By the 1950s Crisis Intervention was commonly applied in disaster work by the American Red Cross and other disaster response organizations. Crisis Intervention services consisted of a broad spectrum of support services. It was very flexible and could be adjusted to the needs of the people involved in a particular disaster. Crisis Intervention as a group process was also being successfully applied to distressed families in the 1960s.[7–9]

PRINCIPLES OF CRISIS INTERVENTION

There are seven core principles of crisis intervention. They are:

- Simplicity—People respond to simple, not complex things, during a crisis.
- Brevity—Most crisis intervention contacts are short in duration, some lasting only a few minutes. It is typical

to have three to five contacts to complete crisis intervention work with an individual.
- Innovation—Crisis Intervention providers must be creative to manage unique and emotionally painful situations.
- Pragmatism—Suggestions must be practical if they are to work in resolving a crisis.
- Proximity—Most effective crisis intervention contacts occur closer to the operational zone or in someone's comfort zone.
- Immediacy—A crisis reaction demands rapid intervention. Delays cause more pain and complications.
- Expectancy—When possible, the crisis intervener works to set up expectations of a reasonable positive outcome.[10]

GOALS OF CRISIS INTERVENTION

There are three primary goals of crisis intervention. They are: (1) to reduce emotional tension in the immediate aftermath of a critical incident, (2) to facilitate the restoration of adaptive functions, and (3) to identify people who need additional help and to provide appropriate referrals.[10]

CRITICAL INCIDENT STRESS MANAGEMENT

The advances in the field of crisis intervention in the 1940s, 1950s, and 1960s eventually led to the development of the current field of critical incident stress management (CISM). CISM is a comprehensive, integrated, systematic, and multicomponent "package" of crisis intervention techniques or "tools." Comprehensive accurately suggests that there are techniques in CISM that are applied before a critical incident, during a critical incident, and after the critical incident concludes.

The techniques are blended and integrated to achieve the best possible effects. Combining or blending CISM intervention techniques adds to their strength. In other words, no single intervention, used as a stand-alone technique, is as effective as a combination of crisis intervention procedures.

A systematic approach in CISM means that many of the applications of CISM techniques come in a logical, sequential order to achieve maximum effects. Multicomponent means that CISM has many techniques and no one technique is the singular focus of the system.

It is interesting to note that the initials C.I.S.M. represent both the name of the program, "Critical Incident Stress Management" and a description of the contents of the program (*comprehensive, integrated interventions, systematic sequence of interventions, and a multicomponent approach*).

CISM is a *subset* of the field of crisis intervention. A subset is defined as a collection of elements within a certain category that are clearly related to each other and which can be found within a larger "umbrella" category. All of the elements contained within a subset, for example, CISM, can be found within the main category, in this case, crisis intervention. There are six core elements in a CISM program. They are: (1) surveillance, assessment, and triage, (2) strategic planning, (3) individual support, (4) informational groups, (5) interactive groups, and (6) follow-up services.[11]

From its inception in the mid-1970s, CISM was deeply rooted in the field of crisis intervention. CISM, as a subset of crisis intervention, shares in the history, theories, guiding principles, goals and objectives, strategies, procedures, methods, techniques, and practices of crisis intervention. The CISM program was developed with emergency personnel in mind and it is well suited for that population. Some of the crisis intervention tools are used to support individuals undergoing critical incident stress. Other tools are used to assist groups, and some are used to assist families and organizations. Different crisis interventions tools are used for individuals than those that are used for groups.[12]

STRATEGIC CISM SERVICES

A best practices approach to the provision of crisis intervention support services for emergency personnel is the provision of these services within the context of a well-organized and well-trained CISM program. All services within a CISM program must be strategic in design and sensible in their application. Too much is at stake for a haphazard, careless approach. A strategic approach to CISM considers who receives help, the types of assistance they receive, the timing of the interventions, all of the themes or issues, concerns, questions, and problems that might be encountered. A strategic CISM approach must also take into account the makeup of the teams or resources that are providing the crisis intervention services. A straightforward, easy to remember, planning formula has been established to assure a strategic approach to CISM. The formula is presented in the section below.

STRATEGIC CISM PLANNING FORMULA

Emergency personnel should always use a simple prescription to develop the strategic plan when intervening in the aftermath of any critical incident. It incorporates the following elements: *target, type, timing, theme,* and *team.*

The parts of the strategic planning formula are detailed in the paragraphs below.

1. *Target*. That is, "who" should receive assistance and who might not need assistance? Answering the who question helps the CISM team members focus their support services on-the-right "targets." This aspect of the planning formula also helps the team to determine if it is an individual or a group seeking support? How many individuals are there? If it is a group, what is the nature of the group? Are their groups outside of emergency personnel who would benefit from crisis support services? Is the group a homogeneous group or a heterogeneous group?
2. *Type(s)*. This aspect of the formula focuses on what type, or, more typically, types, of interventions are necessary. In other words, what interventions will be selected to help others? Will the services be individual or group support or both? Are families or organizations involved and do they need information or intervention? Is there a plan for the necessary follow-up and referral services? It is very important to match the most appropriate interventions to address the specific needs of those suffering through critical incident stress.
3. *Timing*. Choosing the right timing is one of the most important considerations in developing a strategic CISM plan. It has to do with psychological readiness to receive help. It answers the question of when to apply specific interventions or support services. When interventions are not carefully planned, they will be seen as interfering, intrusive, or unnecessary. If they are appropriately timed, they can make a very positive impact on the person or people involved in the critical incident stress experience.
4. *Themes*. This portion of the planning formula will influence every other aspect of the strategic formula. CISM team members need to think of many issues and concerns. Where is the assistance to be provided? Are there any threats to the people involved or to the helpers? What has been achieved so far? Are there any special concerns? Are the people involved heterogeneous or homogeneous groups? Are there any special needs (elderly, children, and handicapped people) or considerations (e.g., continuing danger) to be considered? CISM team members must make sure that they have a full understanding of the issues, concerns, questions, and facts about the incident itself and the services that may have already been provided. Themes are anything that influences decision-making or the choice of interventions. They must be considered throughout the entire crisis intervention process. CISM team members should review the target, type, timing aspects of the strategic plan before instituting the slate of interventions.
5. *Team*. Has the right team been selected to provide the CISM services? Do they have the appropriate training and skills? Are they familiar with the individuals or emergency services organizations? Are they the most appropriate resources? Do they have the time to intervene? Should there be a team approach or an individual helper approach? CISM support personnel should carefully choose the best resources available to provide the services that were selected while looking at the target, type, timing, and themes of the situation. The selection of the appropriate team to provide CISM services rounds out the strategic planning process.[13]

CISM TECHNIQUES

CISM services are best delivered by means of an organized program. CISM teams serving emergency personnel should be able to provide a variety of services before, during, and after a critical incident. Before a critical incident, precrisis education, policy and protocol development, resistance training, and strategic planning should be instituted to prepare to manage a critical incident and the resulting critical incident stress reactions.

Assessment, incident-specific strategic planning, informational group interventions, individual crisis intervention and advice to command personnel are typical during an event.

Once a critical incident concludes the application of CISM techniques include, but are not limited to, interactive group interventions such as defusing and Critical Incident Stress Debriefing (CISD), individual support, significant other support services, follow-up services and postincident education.

The most common crisis intervention tools in a CISM program are

- Preincident planning, policy development, education, training
- Crisis assessment
- Strategic planning
- Individual crisis intervention
- Informational group interventions (demobilization, crisis management briefing)
- Interactive crisis interventions (defusing, CISD)
- Pastoral crisis intervention
- Family support services
- Significant other support services
- Follow-up services
- Referral services
- Follow-up meetings with communities impacted by a critical incident to determine additional needs
- Postincident education
- Links to preincident planning and preparation for the next critical incident[14]

THE STATUS OF CISM

Today, there are approximately 1500 CISM teams operating in about 30 nations. They often use different names (e.g., Critical Incident Response Team, Staff Outreach Support, Critical Incident Stress Team, Critical Incident Response Program) and each team lists a wide range of crisis intervention functions and services (business staff support, fire service crisis intervention teams, hospital staff support teams, law enforcement teams, crisis response teams for communities, etc.). In 2007, the United Nations adopted the name Critical Incident Stress Management Unit for its internal staff support program.[15–17]

The use of peer support personnel is one feature of the CISM system that makes CISM unique. Peer support personnel, under the guidance of mental health professionals, apply the crisis intervention tools within the CISM field. Emergency services peer support personnel enjoy immediate credibility among their fellow workers. The various crisis intervention models within CISM provide structure for the peer support personnel who do not have university degrees and certifications in mental health. Peers are highly motivated to assist distressed colleagues and they are very serious about their crisis work. Specially trained peer support personnel have proven to be extremely effective in delivering excellent support service to their emergency services colleagues.[6]

POSITIVE EFFECTS OF CISM

Critical incident stress concepts have developed over the last 40 years into a sensible staff support program that is widely accepted by, among others, educational institutions, businesses, fire services, law enforcement agencies, emergency medical systems, hospitals, school systems, community groups, the military, and the United Nations. CISM teams have demonstrated their worth for a wide range of personnel, who have worked at scenes of violence, medical emergencies, threatening circumstances, disasters, and in combat situations.[18–26]

THE FUTURE OF CISM

The critical incident stress field remains a dynamic entity in which improvements and refinements are made as experience in the field is gained. A high degree of flexibility is incorporated into the CISM training protocols so that the emergency personnel who provide CISM support services are able to respond quickly and efficiently to new demands that may arise in the midst of actual traumatic events.

The last four decades have been fruitful in developing critical incident stress concepts, policies, training programs, protocols, procedures, strategies, and effective tactics. Additionally a considerable number of research projects have offered insights into what is working and why certain procedures are viewed as helpful for people struggling through critical incident stress reactions.[18–26]

The future of the critical incident stress field lies in additional research and continued developments and refinements within the field. Each new research project opens an additional window that allows crisis support personnel to see their interventions more clearly and to determine what new directions may be necessary to serve their constituents. It is likely that future research will demonstrate that the very same principles that have guided the critical incident stress field to date will serve as a foundation for future progress.

Emergency personnel are the beneficiaries of the widespread support and assistance that is currently provided through the organized CISM programs that serve them in many communities nationally and internationally. CISM services reduce sick time utilization, disabilities, and premature retirements among emergency personnel. These programs enhance personal resistance, resiliency, and the capacity to recover fairly quickly from overwhelming traumatic events.

References

1. Corneil W, Beaton R, Murphy S, Johnson C, Pike K. Exposure to traumatic incidents and relevance of posttraumatic stress symptomatology in urban firefighters in two countries. *J Occup Health Psychol*. 1999;1999(4):131–141.
2. Helzer JE, Robins LN, McEvoy L. Post-traumatic stress disorder in the general population. *N Engl J Med*. 1987;317:1630–1634.
3. Kessler RC, Berglund P, Demler O, Jin R, Merikangas KR, Walters EE. Lifetime prevalence and age-of-onset distributions of DSM-IV disorders in the national comorbidity survey replication. *Arch Gen Psychiatry*. 2005;62:593–602.
4. McFarlane AC. The longitudinal course of posttraumatic morbidity: the range of outcomes and their predictors. *J Nerv Ment Dis*. 1988;176:30–39.
5. American Psychiatric Association. *First Aid for Psychological Reactions in Disasters*. Washington, DC: American Psychiatric Association; 1964.
6. Mitchell JT. *Group Crisis Support: Why It Works, When and How to Provide It*. Ellicott City, MD: Chevron Publishing; 2007.
7. Lindemann E. Symptomatology and management of acute grief. *Am J Psychiatr*. 1944;101:141–148.
8. Caplan G. *An Approach to Community Mental Health*. New York: Grune and Stratton; 1961.
9. Caplan G. *Principles of Preventive Psychiatry*. New York: Basic Books; 1964.
10. Mitchell JT, Visnovske WL. *Crucial Moments: Stories of Support in Times of Crisis*. North Charleston, SC: CreateSpace; 2015.
11. Mitchell JT. *Group Crisis Intervention*. 5th ed. Ellicott City, MD: International Critical Incident Stress Foundation; 2015.
12. Everly Jr GS, Mitchell JT. *Critical Incident Stress Management: A New Era and Standard of Care in Crisis Intervention*. Ellicott City, MD: Chevron Publishing Corp; 1997.

13. Mitchell JT, Clark D, Everly Jr GS. *Strategic Response to Crisis: Student Course Manual.* Ellicott City, MD: International Critical Incident Stress Foundation; 2006.
14. Everly Jr GS, Mitchell JT. *Integrative Crisis Intervention and Disaster Mental Health.* Ellicott City, MD: Chevron Publishing Corporation; 2008.
15. United Nations Department of Safety and Security CISMU Staff. *UNDSS CISMU certification training for counselors. DSS Newsletter.* New York: United Nations Secretariat, Department of Safety and Security; 2007.
16. United Nations Department of Safety and Security CISMU Staff. *New Crisis and Stress Management Training Programme Launched. I Seek (May 2, 2007).* New York: United Nations Secretariat; 2007.
17. United Nations Department of Safety and Security CISMU Staff. *Certification Training in Crisis and Stress Management.* New York: UN Department of Safety and Security, Consultative Working Group on Stress in Collaboration with the International Critical Incident Stress Foundation, the American Academy of Experts in Traumatic Stress and the Comite National de L'urgence Medico-Psychologigue; 2007.
18. Adler A, Litz BT, Castro CA, Suvak M, Thomas JL, Burrell L. Group randomized trial of critical incident stress debriefing provided to US peacekeepers. *J Trauma Stress.* 2008;21:253–263.
19. Adler A, Bliese PD, McGurk D, Hoge CW, Castro CA. Battlemind debriefing and battlemind training as early intervention with soldiers returning from Iraq: randomized by platoon. *J Consult Clin Psychol.* 2009;77:928–940.
20. Boscarino JA, Adams RE, Figley CR. A prospective cohort study of the effectiveness of employer-sponsored crisis interventions after a major disaster. *Int J Emerg Ment Health.* 2005;7:9–22.
21. Boscarino JA, Adams RE, Figley CR. Mental health service use after the world trade center disaster: utilization trends and comparative effectiveness. *J Nerv Ment Dis.* 2011;199:91–99.
22. Castro CA, Adler AB. Re-conceptualizing combat-related posttraumatic stress disorder as an occupational hazard. In: Adler AB, Bliese PB, Castro CA, eds. *Deployment Psychology: Evidence-Based Strategies to Promote Mental Health in the Military.* Washington, DC: American Psychological Association; 2011:217–242.
23. Everly Jr GS. *Fostering Human Resilience.* Ellicott City, MD: Chevron Publishing; 2013.
24. Everly GS, Mitchell JT. *Critical Incident Stress Management CISM: Key Papers and Core Concepts.* Ellicott City, MD: Chevron Publishing; 2013.
25. Tuckey MR. Issues in the debriefing debate for the emergency services. *Clin Psychol Sci Pract.* 2007;14:106–116.
26. Tuckey MR, Scott JE. Group critical incident stress debriefing with emergency services personnel: a randomized controlled trial. *Anxiety Stress Coping.* 2013;27:38–54.

CHAPTER

49

Stress in Policing

S.M. Conn

University of British Columbia, Vancouver, BC, Canada

OUTLINE

Abstract

This chapter begins with a discussion of the scope of the stress problem with police officers. Three categories of police stress—critical incidents, secondary traumatic stress (STS), and organizational stress are discussed next. Critical incident stress refers to events where police officers experience a threat to their safety. STS, also known as vicarious traumatization, cumulative career stress, and operational stress injury, stems from the exposure to the suffering of others. Lastly, the most impactful source of stress, organizational stress, includes workplace hassles such as shiftwork and paperwork. The impact of stress on the psychological, physiological, behavioral, and spiritual health of police officers is also discussed. A discussion of the impact on police families follows. The chapter concludes with a brief treatise of initiatives for preventing or mediating the impact of stress on police.

SCOPE OF THE STRESS PROBLEM IN POLICING

Being a police officer is ranked one of the top 10 most stressful jobs. Each day in policing presents a new challenge and is both unpredictable and exciting. Police stress research has burgeoned in the last three decades as researchers have examined the sources, scope, consequences, and management of police officers' daily exposure to high levels of stress.

One recent study was undertaken to measure the stress of work-life conflict in the policing profession with Canadian police officers.[1] Half the sample (50%) reported high levels of stress while close to the remaining half (46%) reported moderate levels of stress. Alarmingly, only 4% of the sample reported low levels of stress.[1] Another extensive study of the effects of psychosocial work stress on police officers conducted in the United States found that approximately 25% of the officers who responded would be categorized as high stress.[2] Although the incident rate of high levels of stress is variable, it is clear that police work is stressful. When you consider that, like other jobs, most police officers will be required to work for 20-25 years before retiring, it makes sense to investigate and attempt to mediate the impact of stress.

SOURCES OF STRESS IN POLICING

Police work exposes police officers to various forms of stress. Waters and Ussery[3] suggest that stressful life events for police can be placed into three categories: explosive events, implosive events, and corrosive events. Explosive events include crimes in progress, natural disasters, and terrorist situations. In order to respond effectively to these volatile situations, police officers must learn to repress human tendencies such as emotional

Stress: Concepts, Cognition, Emotion, and Behavior
http://dx.doi.org/10.1016/B978-0-12-800951-2.00050-9

reactions. Explosive events may result in critical incident stress. Implosive events are construed as internal conflicts. Waters and Ussery[3] suggest a sense of not being able to make a difference, the conflicts among family and personal responsibilities, and job-related considerations as implosive events that may take a toll on the health of police officers. There are many terms that refer to the effects of implosive events such as secondary traumatic stress (STS), cumulative career stress, compassion fatigue, and complex or chronic stress. Lastly, corrosive events are daily conflicts that erode police officers' levels of hardiness and resiliency. Corrosive events include organizational stress and organizational hassles. Each of these forms of stress will be discussed in turn beginning with critical incident stress.

KEY POINTS

- Research shows that between 25% and 50% of police officers report high levels of stress in their work.
- Research shows that police officers have higher post-traumatic stress disorder (PTSD) and suicide rates than the general population.
- There are three primary sources of stress in policing: (1) critical incident stress, (2) secondary traumatic stress, and (3) organizational stress.
- Organizational stress is a stronger predictor of psychological distress than other forms of stress.
- The impact of stress on police officers is psychological, physiological, behavioral, and spiritual.
- The psychological impact of stress includes post-traumatic stress symptoms of avoidance, re-experiencing, and heightened startle response, as well as mood disorders and cognitive changes.
- Physiological changes in police officers who are chronically exposed to stress include cortisol dysregulation, chronic pain, heart disease, and high blood pressure.
- Behavioral impacts of stress on police officers includes substance abuse, suicidal behavior, absenteeism, and narrowing of social roles.
- Stress has also been found to erode police officers' spiritual beliefs.
- The impact of stress extends to the police officers' family members.
- Preventative and responsive initiatives are being developed and provided to address the impact of stress in police work.

Critical Incident Stress

Critical incident stress is most widely known due to its depictions in the media. Critical incidents include events such as officer-involved shootings, assaults, accidents, disasters, and crimes in progress. They involve feelings of shock, horror, and helplessness. The body of research on critical incident stress is more comprehensive than research on STS or organizational stress. This is likely because the last two types of stress have only recently come to be understood as having a stronger impact than critical incident stress. Critical incidents have shown to have lasting effects on police officers with reports of having vivid visual, tactile, and olfactory memories lasting more than 20 years.[4] Critical incident stress may culminate in post-traumatic stress disorder (PTSD).

PTSD rates in first responders are variable but seem to support that police officers have higher rates of PTSD than the general population.[5] Rates for PTSD in police range from 7% to 20%[6] compared to the rate of the general population, which is 6.8%.[7] Although most, if not all, police officers are exposed to traumatic events, the majority of police offices exposed to critical incidents do not develop PTSD. Conflicting reports of PTSD rates can be confusing for some trying to make sense of the data. There are many reasons for this variability. First, police officers typically conceal their struggles with PTSD for fear of reprisals from their employers and even their coworkers. Second, the rates of PTSD in this population are often mitigated by (1) the hardy constitutions of persons who apply to be police officers, (2) preemployment psychological screening that eliminates psychologically vulnerable persons from gaining employment, and (3) training offered by police organizations that assist employees in their ability to manage their exposure to trauma. Therefore, the fact that PTSD rates rise to the level of the general population is troubling. When it exceeds this rate, which it does in many reports, it is very compelling evidence of the negative impact of the work.

However, PTSD rates are also a very small part of the bigger picture of police stress. The prevalence of PTSD symptoms that do not meet the threshold for a diagnosis of PTSD appear to be significantly higher than instances where the full criteria is met. A large-scale study of officers indicated 33% reported intrusive thoughts, memories, or dreams about the work event, 24% felt detached from people and activities related to the stressful event, and 23% avoided anything related to the event.[8] Some experts suggest that officers who do not meet the full criteria for PTSD should be included in morbidity rates because they are indistinguishable from those who meet full criteria.[9] Making matters worse, PTSD contributes to increased levels of guilt, depression, illness, job turnover, and reduced decision-making ability.[10]

Secondary Traumatic Stress

Beyond the traumatic stress of violent encounters, police officers experience STS in the course of assisting crime and accident victims. The concept of STS is less well known despite its well-documented occurrence in the helping professions. Unlike traumatic stress, which occurs occasionally, STS may be experienced on a daily basis. The term STS is sometimes used interchangeably with terms such as *compassion fatigue*,[11] *vicarious traumatization*,[12] *burnout*,[13] *cumulative career traumatic stress*,[14] and *occupational stress injury*.[15] Whereas a singular traumatic event or series of events have happened personally to the person suffering from PTSD, STS occurs when a person in a helping profession is repeatedly exposed to the traumatization of others and develops symptoms that parallel those of PTSD. This has been documented in a variety of helping professions such as emergency room doctors and nurses, psychologists, therapists, social workers, and first responders such as firefighters, ambulance personnel, and police officers.[16,17] Police officers who repeatedly respond to trauma victims are at risk of developing STS, if not PTSD.[14,17] The diagnosis for PTSD was changed in the most recent edition of the Diagnostic and Statistical Manual of Mental Disorders (DSM-V) in recognition of the impact of the work on police officers' health by adding:

> Experiencing repeated or extreme exposure to aversive details of the traumatic event(s) (e.g., first responders collecting human remains; police officers repeatedly exposed to details of child abuse). Criterion A4 does not apply to exposure through electronic media, television, movies, or pictures unless this exposure is work related.[18]

Inclusion of police officers' repeated exposure in the diagnosis criteria for PTSD demonstrates official recognition by the psychiatric community of the insidious impact of STS on police while assisting victims.

STS in police officers is exacerbated by a number of factors such as the perceived inability to make a difference, the officer's ability to relate to victims, their exposure to the dark side of human nature, and the heightened vulnerability of the victim.[19] Many people enter the policing profession to help other people. Their sense of being able to help others can feel stunted by the inherent limitations of the police role. Oftentimes, police officers become involved in problems at the point when they have become completely unmanageable by others. Police officers may believe that it is their duty to provide a solution to a problem that took days, weeks, months, or even years to develop. The regularity of ongoing exposure to the unfixable suffering of others can also incrementally exact a toll on police officers.

Organizational Stress

Another form of police stress that tends to be overshadowed by critical incident stress is organizational stress. Exposure to routine occupational stress has been found to be a stronger predictor of psychological distress, including post-traumatic stress symptoms, than the cumulative exposure to critical incidents or danger.[1,20] Organizational stressors include unsupportive policies, procedures, colleagues, and supervisors, as well as workplace hassles such as rotating shifts, staffing shortages, court appearances on days off, and the promotional and assignment processes. Organizational stress is more problematic because police officers do not anticipate these stressors and therefore are not prepared to manage them.

Linda Duxbury and Christopher Higgins conducted a large-scale study of a common organizational stressor, work-life conflict, with 4500 police officers across Canada.[21] Fifty-five percent of respondents indicated that they felt "overloaded" by the demands of their work. Approximately a quarter of the respondents reported high levels of interference of their home life with work demands. Between 40% and 50% of respondents indicated that work had a negative impact on their personal time for leisure activities, including spending time with family.

IMPACT OF STRESS ON POLICE OFFICERS

The literature is replete with documentation of the effects of stress on police officers. The impact of critical incident stress, STS, and organizational stress are strikingly similar, lending credence to the notion that concepts such as trauma and stress are relative to the person experiencing them. Its effects are psychological, physiological, behavioral, and spiritual. The effects are not limited to police officer but extend to police families as well. A discussion of each category follows.

Psychological

The psychological impact of stress can manifest both emotionally and cognitively. Figley provides a comprehensive list of psychological indicators of STS, drawn from a large pool of literature by experts in the field of traumatic stress.[11] Two common psychological indicators are distressing emotions and intrusive imagery.[12] An investigation into the impact of cumulative stress on police officers found that 74% reported experiencing recurring memories of an incident, 62% experienced recurring thoughts or images, 54% avoided reminders of an incident, and 47% experienced flashbacks of an incident.[14]

Secondary traumatic stress disorder is believed to be a much larger threat to the psychological well-being of police officers than PTSD. While critical incidents are oftentimes addressed by police agencies, the majority of STS events would not result in the provision of a formal supportive response for the officers involved. Inadvertently, this may send the message that officers should not be expected to be impacted by the event(s). However, chronic exposure to stress can have a detrimental impact on police officers' emotions. van der Kolk[22] found that "Chronic physiological arousal, and the resulting failure to regulate autonomic reactions to internal or external stimuli, affect people's capacity to utilize emotions as signals" (p. 218). van der Kolk elaborates, stating

> In PTSD, emotional arousal and goal-directed action are often disconnected from each other. As a result, people who suffer from PTSD no longer use arousal as a cue to pay attention to incoming information. Instead, they tend to go immediately from stimulus to response without first being able to figure out the meaning of what is going on; they respond with fight-or-flight reactions (Ref. 22, p. 219).

This can have dire consequences for police officers. If they do not interpret meaning from a stimulus, there may be a tendency to overreact to the presenting situation. This overreaction is exacerbated by the tendency for police officers to be hyperaroused and hypervigilant to perceived threats in their work.

There is also evidence that stress can lead to depression in police officers. A study of work-life balance in Canadian police officers revealed that 30% of respondents reported having high levels of depressed mood, and an additional 40% reported having moderate levels of depressed mood.[1] Taken together, this study indicates that 70% of respondents reported moderate to high levels of depressed mood.

In addition to the emotional toll, stress can impact cognitive functioning as well.

Seasoned police officers demonstrate multiple cognitive shifts such as a heightened sense of vulnerability, cynicism, suspiciousness of others, and a sense of powerlessness.[16] Salston and Figley[17] stated that exposure to traumatic material "begins to affect one's worldview, emotional and psychological needs, the belief system, and cognitions, which develop over time" (p. 169). Marshall[14] found evidence of worldview changes (vicarious traumatization) with 96% of the sample of police officers reporting that their opinion of others had changed, 92% reporting they no longer trusted others, 82% believed the world was an unsafe place, and 88% experienced prejudices they did not hold prior to being on the job.

Police officers' worldview changes can be explained by Just World Theory.[23] According to this theory, individuals believe that they have influence over matters in a world that is predictable. It is predictable due to the belief that good things happen to good people and bad things happen to bad people. All one needs to do is be a good person and he or she will not be befallen with negative consequences. These assumptions are shattered in police work. Police officers routinely witness the victimization of "good" people. Common examples include families who are killed by impaired drivers and children who are sexually abused or exploited.

Physiological

The physiological impact of police stress is profound. John Violanti and colleagues have conducted studies of cortisol levels in police officers exposed to stress and traumatic stress. Police officers with PTSD symptoms show higher waking levels of cortisol which rises throughout the day with a sharp decline at the end of the day.[24] The higher the rates of symptomatology, the higher the cortisol levels were. Chronic stress and acute stress have been found to lead to a condition called *allostatic load*, where the hypothalamus-pituitary-adrenal (HPA) axis is dysregulated.[25] Furthermore, traumatic stress also leads to HPA overreactivity and an elevated startle response[24] and can also lead to a constant state of physiological arousal.[26] The police culture oftentimes stresses that officers maintain a constant state of vigilance, which also contributes to disease from the constant surge of stress hormones. Stress has been found to impact older police officers' health in a number of ways such as lower back pain (45%), high blood pressure (42%), heart disease (16%), migraines (14%), and insomnia (13%).[27] Another study reflected that officers reported more injuries and higher levels of chronic health problems, including approximately one-third of the respondents reported being burned out from the job.[2] Burnout entails physical and emotional exhaustion.

Behavioral

Stress in policing results in various behavioral changes in police officers. Some of the behavioral changes are exacerbated by the police culture, which influences how police officers cope with stressful situations. The most commonly documented behavioral manifestations relate to substance use and abuse,[28,29] absenteeism,[1] narrowing of social roles,[30,31] distancing from families,[32] and suicidal behavior.[29]

One of the most common maladaptive means of coping with stress in policing is alcohol or substance abuse. In a study of alcohol abuse by police officers nearly one-quarter of the respondents were found to abuse alcohol due to the stress from policing.[29] This estimate was believed to be low due to the reluctance of police officers to report alcohol use or dependence. Coping with stress in policing by abusing alcohol contributes to a host of other behavioral issues such as high absenteeism, citizen and supervisor complaints of misconduct, intoxication on duty, traffic accidents, and overall poor performance.[29]

Research indicates that alcohol use oftentimes begins as a social interaction with fellow police officers and evolves into a coping mechanism for those struggling to deal with stress and exposure to trauma on the job.[28] Police officers have easy access to drugs by virtue of their position, making it easier to turn to drug use to numb the pain.[28] According to Cross and Ashley, who have researched substance abuse in police officers, the negative impact that drugs and alcohol have on the police officers' performance damage the officers' perceptions of themselves as officers, creating an emotional downward spiral. Cross and Ashley assert that the amount of substance use is proportionate to the amount of stress, adding to the vicious cycle.

Stress also contributes to absenteeism from the job. In a large-scale study of work-life balance and stress, three primary factors were identified as contributing to absenteeism in the police organization: health problems (51%), mental or emotional fatigue (28%), and childcare issues (27%).[1] Respondents indicated they missed approximately 8 days per year due to emotional and mental fatigue from the stress of the job. Higher work demands were found to be linked to higher levels of stress and, consequently, increased absenteeism.[1] Stress also contributes to what is referred to as "presenteeism" on the job. Presenteeism refers to physically attending the job but, due to mental or emotional fatigue, mentally withdrawing from engagement in the work.

The relationship between stress and life roles is complex. Stress from police work reduces participation in other life roles such as friend, volunteer, and parent. Participation in other life roles sometimes reduces police officers' stress levels but, at other times, compounds the stress due to role overload. Participating in other life roles has been reported to help police officers connect with other aspects of their lives where they feel they have more control and satisfaction.[19] Police culture promotes a reduction of nonpolice life roles.[30,31] As police officers assume their professional roles as police officers, nonpolice roles are enacted less and less. This can be problematic for a number of reasons. As police officers experience stress, their ability to manage it will be compromised by the limited personal and social resources they have with the singular police role. The singular police role constricts police officers' perceived options for problem solving, leading them to solve nonpolice problems with the mindset used to solve police matters.

Traumatic stress oftentimes leads police officers to emotionally distance themselves from trauma victims. This coping mechanism allows police officers to minimize feelings of being able to relate to victims leading to a sense of vulnerability and emotional overwhelm while on the job. Unfortunately, this distancing behavior does not automatically stop when police officers leave work to go home to their families. As a result, they may demonstrate emotional detachment with their families, leading to marital problems, including divorce. High levels of stress have also been found to correlate with police officers committing domestic abuse.[2]

The most detrimental behavioral impact of stress in policing is suicide. Police suicide rates are found to be higher than rates of suicide in the general population.[33] Male police officers have shown to have a suicide rate 8.3 times that of homicide and 3.1 times higher than work accidents.[34] Increased rates of exposure to certain types of traumatic events significantly increase the risk for PTSD symptomatology and, consequently, heighten the risk of alcohol use and suicidal ideation.[29] Suicide also stems from organizational and STS. Easy access to firearms and a police culture that shuns help seeking contribute to suicide risk in police officers. In a study relating to cumulative stress in policing, 11% of the police respondents reported that they experience suicidal ideation as a result of the occupation.[14] This last statistic is compelling evidence of the harmful impact of police stress.

Spiritual

The impact of the police stress is believed to change the "souls" of police officers.[35] Regular exposure to violence, toxicity, and suffering has been found to erode police officers; spirituality.[36] Being exposed to the abuse and suffering of vulnerable persons such as children can result in police officers questioning their faith and understanding of mankind. Making matters worse, the conventional positive coping mechanisms for stress that involve aspects of spirituality such as connectedness to others and emotional expression are also discouraged by the traditional police culture.

Impact on Police Families

Aside of the impact on the family through the changes in police officers, themselves, the psychological impact of police stress can also extend to police families. Research has shown that the spouses and partners of police officers who experienced PTSD symptoms, but were not exposed to a line-of-duty critical incident, experienced secondary stress symptoms that also mirrored PTSD symptoms.[37] The secondary stress reactions develop from both the knowledge of the traumatic event as well as a form of symptom contagion from one partner to the other.[37] STS of police wives has been strongly correlated with psychological distress, depression, anxiety, and increased alcohol consumption.[38]

Limited research also addresses the impact of police stress on the children of police officers. Police officers routinely face stressful events that challenge their sense of control and safety. These experiences tend to spill over into their home environment. Police officers may enact their police role in the place of their parental roles. This role spillover could result in police officers being overly critical and authoritative in directing their children's lives. Children who do not submit to the authoritative stance of the police officer parent may experience even more discipline and attempts to control their behavior. Children are also impacted by the organizational hassles their police parents struggle within their work-shift work, overtime, and working holidays.

CONCLUSION

Multiple sources of stress exact their toll on police officers psychologically, physiologically, behaviorally, and spiritually. Recognition of the deleterious effects of stress have produced several initiatives to prevent or at least mediate its effects on police officers and their families. Some of the initiatives include police recruit training, ongoing in-service training, acute incident support, peer support teams, and procedural changes to facilitate police officers' access to mental health support. The police culture is also changing, resulting in a shift to positive coping in lieu of some of the maladaptive coping methods that have worsened police officers' management of stress reactions. Despite the seemingly dark picture of the impact of stress on police, it should be noted that many police officers practice proactive strategies as well as healthy coping mechanisms that shield them from the detrimental effects of the job and allow them to do the job they love.

References

1. Duxbury L, Higgins C. *Caring for and about those who serve: work-life conflict and employee well being within Canada's police departments.* Commission for Public Complaints against the RCMP. http://www.publicsafety.gc.ca/cnt/rsrcs/lbrr/ctlg/shwttls-eng.aspx?d=CPC&i=8151; Published March, 2012. Accessed 07.06.14.
2. Gershon R. *National Institute of Justice final report "Project Shields"* (Document No. 185892). www.ncjrs.gov/pdffiles1/nijgrants/185892.pdf; Published December, 2000. Accessed 19.02.10.
3. Waters JA, Ussery W. Police stress: history, contributing factors, symptoms, and interventions. *Policing*. 2007;30(2):169–188.
4. Karlsson I, Christianson S. The phenomenology of traumatic experiences in police work. *Policing*. 2003;26(3):419–438.
5. Canadian Mental Health Association. *Post-traumatic stress disorder.* www.cmha.ca/bins/content_page.asp?cid=3-94-97; Published 2014. Accessed 22.02.15.
6. West C, Bernard B, Mueller C, Kitt M, Driscoll R, Tak S. Mental health outcomes in police personnel after hurricane Katrina. *J Occup Environ Med.* 2008;50(6):689–695. http://dx.doi.org/10.1097/JOM.0b013e3181638685.
7. Kessler RC, Berglund P, Delmer O, Jin R, Merikangas KR, Walters EE. Lifetime prevalence and age-of-onset distributions of DSM-IV disorders in the National Comorbidity Survey Replication. *Arch Gen Psychiatry*. 2005;62(6):593–602.
8. Gershon R, Barocas B, Canton AN, Li X, Vlahov D. Mental, physical, and behavioral outcomes associated with perceived work stress in police officers. *Crim Justice Behav.* 2009;36:275–289.
9. Weiss DS, Marmar CR, Schlenger WE, et al. The prevalence of lifetime and partial post-traumatic stress disorder in Vietnam theater veterans. *J Trauma Stress*. 1992;5:365–376.
10. Drewitz-Chesney C. Posttraumatic stress disorder among paramedics: exploring a new solution with occupational health nurses using the Ottawa Charter as a framework. *Workplace Health Saf.* 2012;60(6):257–263.
11. Figley CR, ed. *Compassion Fatigue: Coping with Secondary Traumatic Stress Disorder in Those Who Treat the Traumatized.* Levittown, PA: Brunner/Mazel; 1995.
12. McCann IL, Pearlman LA. Vicarious traumatization: a framework for understanding the psychological effects of working with victims. *J Trauma Stress*. 1990;3(1):131–149.
13. Freudenberger HJ. Staff burnout. *J Soc Issues*. 1974;30(1):159–165.
14. Marshall EK. *Occupational stress and trauma in law enforcement: a preliminary study in cumulative career traumatic stress.* Unpublished doctoral dissertation, Cincinnati, OH: Union Institute and University; 2003 [UMI No. 3098255].
15. The Subcommittee on Veterans Affairs of the Standing Senate Committee on National Security and Defence. *Occupational stress injuries: The need for understanding.* Report of the subcommittee on veterans affairs of the standing senate committee on national security and defence. Ottawa, Ontario. http://www.parl.gc.ca/Content/SEN/Committee/372/vete/rep/rep14jun03-e.htm; Published June 19, 2003. Accessed 19.02.15.
16. Collins S, Long A. Working the psychological effects of trauma: consequences for mental health workers—a literature review. *J Psychiatr Ment Health Nurs.* 2003;10:417–424.
17. Salston MD, Figley CR. Secondary traumatic stress effects of working with survivors of criminal victimization. *J Trauma Stress*. 2003;16(2):167–174.
18. American Psychiatric Association. *Diagnostic and Statistical Manual of Mental Disorders.* 5th ed. Washington, DC: American Psychiatric Association; 2013.
19. Conn SM, Butterfield LD. Coping with secondary traumatic stress by general duty police officers: practical implications. *Can J Couns Psychother*. 2013;47(2):272–298.
20. Liberman AM, Best SR, Metzler TJ, Fagan JA, Weiss DS, Marmar CR. Routine occupational stress and psychological distress in police. *Policing*. 2002;25(2):421–439.
21. Duxbury L, Higgins C. *Work-life balance in the new millennium: where are we? Where do we need to go?* Canadian Policy Research Networks discussion paper no. W/12. www.cprn.org; Published October 22, 2001. Accessed 01.06.14.
22. van der Kolk BA. The body keeps the score: approaches to the psychobiology of posttraumatic stress disorder. In: van der Kolk BA, MacFarlane AC, Weisaeth L, eds. *Traumatic Stress: The Effects of Overwhelming Experience on Mind, Body, and Society.* New York, NY: Guilford Press; 1996.
23. Lerner MJ, Miller DT. Just world research and the attribution process: looking back and ahead. *Psychol Bull.* 1978;85(5):1030–1051.
24. Violanti JM, Andrew ME, Burchfiel CM, Hartley TA, Charles LE, Miller DB. Post-traumatic stress symptoms and cortisol patterns among police officers. *Policing*. 2007;30(2):189–202.

25. McEwen BS. Allostasis and allostatic overload and relevance to the pathophysiology of psychiatric disorder. *Ann N Y Acad Sci.* 2004;1032:1–7.
26. Yehuda R. Risk and resilience in post-traumatic stress disorder. *J Clin Psychiatry*. 2004;65(1):29–36.
27. Gershon RRM, Lin S, Li X. Work stress in aging police officers. *J Occup Environ Med*. 2002;44(2):160–167.
28. Cross CL, Ashley L. Police trauma and addiction: coping with the dangers on the job. *FBI Law Enforc Bull*. 2004;73(10):24–32.
29. Violanti JM. Suicide and the police culture. In: Hackett D, Violanti JM, eds. *Police Suicide: Tactics for Prevention*. Springfield, IL: Charles C. Thomas; 2003:66–75.
30. Paton D, Violanti JM, Burke K, Gehrke A. Conceptualization: the police career course and traumatic stress. In: *Traumatic Stress in Police Officers: A Career-Length Assessment from Recruitment to Retirement*. Springfield, IL: Charles C. Thomas Publisher; 2009:3–29.
31. Gilmartin KM. *Emotional Survival for Law Enforcement: A Guide for Officers and Families*. Tucson, AZ: E-S Press; 2002.
32. Thompson AJ. *Operational Stress and the Police Marriage: A Narrative Study of Police Spouses*. Master's thesis, Vancouver, BC: University of British Columbia; 2012.
33. O'Hara AF, Violanti JM. Police suicide: a comprehensive study of 2008 national data. *Int J Emerg Ment Health*. 2009;11(1):17–23.
34. Violanti JM, Vena JE, Marshall JR. Suicides, homicides, and accidental deaths: a comparative risk assessment of police officers and municipal workers. *Am J Ind Med*. 1996;30:99–104.
35. Powers W. Managing the problem employee. In: *Poster presented: Northwestern University Police Staff and Command School, Centennial, CO, October*; 2004.
36. Smith J, Charles G. The relevance of spirituality in policing: a dual analysis. *Int J Police Sci Manage*. 2010;12(3):320–338.
37. Hirshfeld A. *Secondary Effects of Traumatization Among Spouses and Partners of Newly Recruited Police Officers*. Doctoral dissertation, San Francisco, CA: The California School of Professional Psychology; 2005.
38. Dwyer LA. *An Investigation of Secondary Trauma in Police Wives*. Doctoral dissertation, Hempstead, NY: Hofstra University; 2005.

C H A P T E R

50

Peacekeeping

B. Litz[1], S. Maguen[2], A. Tankersley[3], C. Hundert[3]

[1]VA Boston Healthcare System, Boston, MA, USA; Boston University School of Medicine, Boston, MA, USA
[2]UCSF Medical School, San Francisco, CA, USA; San Francisco VA Medical Center, San Francisco, CA, USA
[3]VA Boston Healthcare System, Boston, MA, USA

O U T L I N E

Abstract

The role of peacekeeping in a military context has shifted over time since the inception of the United Nations in 1945. Over the last few decades, peacekeepers have increasingly been deployed to areas with active hostilities. As a result of this shift, peacekeepers are at high risk for exposure to typical warzone stressors such as life threat, as well as stressors unique to peacekeeping missions, such as role conflicts. Despite these risks, most service members deployed on peace enforcement missions tend to adapt well and report positive outcomes. However, some peacekeepers develop mental health problems such as posttraumatic stress disorder. Moving forward, it is important to appreciate and attend to the mental health needs of service members deployed to peacekeeping missions.

TYPES OF PEACEKEEPING MISSIONS

Peacekeeping has changed considerably following the inception of the United Nations (UNs) in 1945. There have been 69 peacekeeping missions since 1948, and currently there are 16 active operations throughout the world.[1] Historically, the role of the military in peacekeeping operations has been that of maintaining a strictly neutral presence by overseeing peace accords between formerly warring parties. However, post-Cold War, peacekeepers have shifted to enforcing peace in the midst of active hostilities. These modern missions have placed peacekeepers into more dangerous and conflict-laden environments. Modern missions require service members to be combat trained, yet able to model restraint and maintain neutrality. As a result, peace enforcement is one of the most stressful duties a service member can be exposed to in the modern military, and the possibility of exposure to potentially traumatic events and death has become even more likely in recent missions. For example, according to data released by the United Nations, there have been 857 fatalities due to "malicious acts" on peacekeeping missions since 1948, and 25% of those have occurred in the last decade.[2]

This chapter provides brief summaries of the stressors associated with peacekeeping, briefly reviews recent peacekeeping missions, summarizes the unique stressors of peacekeeping, and highlights the psychological impact of these missions.

Operation Restore Hope and Operation Continue Hope in Somalia

As a result of the civil war that erupted in Somalia in 1991, humanitarian relief efforts aimed at curbing famine and the spread of disease were interrupted and sabotaged. The United Nations, with extensive support from the United States, decided to guarantee the provision of humanitarian aid as well as to enforce the peace in

http://dx.doi.org/10.1016/B978-0-12-800951-2.00051-0

KEY POINTS

- Peacekeeping refers to a military presence that increases the likelihood of peace and reduces the likelihood of fighting among two or more formerly warring parties. However, modern peacekeeping missions often take place in regions where hostilities are still ongoing, thereby increasing physical and psychological risk for peacekeepers.
- Peacekeeping involves duties such as guarding and patrolling vulnerable areas, helping maintain crowd control, operating checkpoints, providing humanitarian assistance, and facilitating with the rebuilding of infrastructure, all of which can expose peacekeepers to potentially traumatic events.
- Some peacekeepers may experience role conflict (i.e., exposure to competing demands of neutrality and restraint in the face of life threat or danger), which may lead to feelings of helplessness or frustration. Peacekeeping rules of engagement prohibit the use of weapons except in extreme circumstances, which differs from how service members are traditionally trained to handle life threat.

Somalia. During this mission, peacekeepers became peace enforcers, and the risk of exposure to potentially traumatic events grew exponentially. Although the first phase of the peacekeeping mission, Operation Reserve Hope (ORH), was a great success with regard to the provision of medical and food supplies, the mission is often considered to be a failure because Somalia continues to be at risk for the devastating effects of famine, political instability, and violence. Operation Continue Hope was initiated when ORH concluded, in May 1993.

While in Somalia, US military personnel were assigned to a variety of tasks, ranging from policing duty to combat-like duty (e.g., patrols, disarming civilians). Peacekeepers were exposed to a fairly well-armed civilian population who were actively engaged in interclan war. Peacekeepers in Somalia were also subjected to acts of aggression by some Somalis, yet the strict rules of engagement sharply restricted options for protection or retaliation, creating a general sense of threat.

Operation Joint Endeavor and Operation Joint Guard in Bosnia-Herzegovina

In 1991, following the declaration of independence of Slovenia and Croatia, a civil war erupted in the former Yugoslavia. The disillusion of the state gave rise to power struggles over governing control of the remaining states and to old race-based hatreds. Horrible atrocities were perpetrated, including genocidal acts. In response, approximately 60,000 North Atlantic Treaty Organization (NATO) and US military personnel were deployed as peacekeepers. Peacekeepers in Bosnia-Herzegovina were subjected to shelling and sniper attacks and a minority of peacekeepers were also taken hostage. Some UN peacekeepers had to stand by helplessly while atrocities were taking place.

Recently Established Peacekeeping Missions

Locations of recent UN peacekeeping missions include: (1) the Central African Republic (MINUSCA) since April 2014, (2) Mali (MINUSMA) since April 2013, (3) Abyei, Sudan (UNISFA) since June 2011, (4) South Sudan (UNMISS) since July 2011, (5) the Democratic Republic of the Congo (MONUSCO) since July 2010, and (6) Darfur (UNAMID) since July 2007.[1]

UNIQUE STRESSORS OF PEACEKEEPING

Peacekeepers are at risk for exposure to stressors that are typical of a war zone (e.g., artillery fire, land mines, small arms fire, bearing witness to the aftermath of malicious violence). Even when peacekeepers are tasked with maintaining a firm peace, they are faced with the possibility of life threat. They may be fired upon as a result of a misunderstanding, accidentally in cross fire between two armed feuding parties, or during firing close, which occurs when the opponent wishes to intimidate the peacekeepers in order to keep them away from a certain area. Peacekeepers may also witness violence and atrocities committed against fellow peacekeepers and civilians as well as the malicious destruction of property. Other types of stressful experiences for peacekeepers are boredom, isolation, family separation, exhaustion, unfavorable climatic conditions, and demoralization about the mission's efficacy.

Role conflicts may also provoke additional stress in peacekeepers. For combat-trained service members from larger nations, peacekeeping duty may feel incongruent with their training. For example, peacekeepers, unlike traditional service members, are often restrained from taking offensive action in conditions of life threat. In addition, the types of defensive military structures that are commonplace in war are often not as available on peacekeeping missions due to the proximity that is required in order to provide humanitarian assistance and protection. The emphasis on proximity rather than protection creates considerable hypervigilance and arousal in peacekeepers and contributes to a general sense of fear. Peacekeepers who are unclear about how to respond to threats and/or who experience repeated threats of injury, with little or no opportunity for recourse, are likely to experience great anxiety, helplessness, and anguish. Furthermore, peacekeepers who are forced to suppress their frustration, fear, and anger are at risk for acting out their feelings during a mission and/or upon their return home.

Roméo Dallaire, Force Commander of the UN peacekeeping force for Rwanda between 1993 and 1994, has poignantly described his unique experience of witnessing the Rwandan genocide as someone sent there to establish and maintain peace:

> Rwanda will never ever leave me. It's in the pores of my body. My soul is in those hills, my spirit is with the spirits of all those people who were slaughtered and killed that I know of, and many that I didn't know. ... Fifty to sixty thousand people walking in the rain and the mud to escape being killed, and seeing a person there beside the road dying. We saw lots of them dying. And lots of those eyes still haunt me, angry eyes or innocent eyes, no laughing eyes. But the worst eyes that haunt me are the eyes of those people who were totally bewildered. They're looking at me with my blue beret and they're saying, 'What in the hell happened? We were moving towards peace. You were there as the guarantor' – their interpretation – 'of the mandate. How come I'm dying here?' Those eyes dominated and they're absolutely right. How come I failed? How come my mission failed? How come as the commander who has the total responsibility – We learn that, it's ingrained in us, because when we take responsibility it means the responsibility of life and death, of humans that we love.[3]

PSYCHOLOGICAL AND BEHAVIORAL IMPACT OF PEACEKEEPING

Although there are many potentially traumatizing experiences and psychological conflicts associated with being a peacekeeper, the great majority of service members appear to adapt well. However, a sizeable number of peacekeepers are at risk for the development of psychopathology related to their deployment. The most frequent mental health problems associated with peacekeeping are posttraumatic stress disorder (PTSD), depression, anger and hostility problems, and alcohol abuse. Mehlum and Weisaeth reported that 7 years following service, 16% of prematurely repatriated service members from southern Lebanon met criteria for PTSD.[4] Passey and Crocket found that more than 20% of the service members deployed as peacekeepers to Bosnia endorsed symptoms of PTSD and depression.[5] In Litz and colleagues' examination of US military personnel who deployed to Somalia, 25% of Somalia veterans reported clinically significant psychological distress, particularly hostility and anger problems, and 8% reported clinically significant PTSD[6]; Dirkzwager and colleagues found similar rates of PTSD among peacekeepers who served in Yugoslavia.[7] Among Kosovo peacekeepers, Maguen and colleagues found that 4% endorsed clinically significant PTSD symptoms[8]; Dirkzwager and colleagues found similar rates of PTSD among peacekeepers who served in Cambodia.[7] Gray and colleagues found that although most follow a standard pattern of PTSD development, some peacekeepers exhibit delayed-onset PTSD after a period of minimal distress.[9]

Adler and colleagues examined factors that predicted PTSD in US military personnel deployed to the Bosnia peacekeeping mission and found that longer deployments and first-time deployments were associated with an increase in distress scores.[10] Interestingly, the association between deployment length and distress was significant only for male service members. In a prospective study of Kosovo peacekeepers, Maguen and Litz found that preexisting stress symptoms and exposure to stressful war zone events (e.g., going on dangerous patrols) were the most robust predictors of PTSD symptoms.[11] Similarly, in Litz and colleagues' study of Somalia peacekeepers, the extent of exposure to stressful war zone events and frustrations with aspects of the peace enforcement mission (e.g., restrictive rules of engagement) predicted the severity of PTSD.[12] The relationship between war zone exposure and PTSD was strongest for Somalia veterans who had high levels of frustration with the negative aspects of peacekeeping duty. Additional variables associated with PTSD symptoms reported by Dirkzwager and colleagues were feelings of powerlessness and threat, belief that the mission had become meaningless, and lack of perceived control during deployment.[7] Litz and colleagues also found that cohesion and morale during deployment were negatively associated with symptoms of psychological distress and PTSD.[12] However, Barnes and colleagues found that higher stress in a peacekeeping context was associated with worsening subsequent perceptions of military support.[13] Thus, although organizational support may protect peacekeepers against the development of PTSD, stress symptoms can also affect the manner in which individuals perceive said support. On the other hand, Michel and colleagues found that postdeployment adversities most strongly contributed to poor mental health among Swedish peacekeepers that deployed to Bosnia.[14]

Peacekeeping duty can also promote positive outcomes. Dirkzwager and colleagues found that the majority of peacekeepers reported positive consequences of their deployment, with 82% reporting a broadening of their horizon and 52% reporting increased self-confidence.[7] Britt, Adler, and Bartone documented that following deployment to Bosnia, many peacekeepers reported an increase in political understanding, stress tolerance, and professional qualifications.[15] They also found that service members who identified more closely with the role of peacekeeper and who believed their assignment to be important were more likely to report a perceived benefit from their peacekeeping experience and were less likely to report burnout and adverse psychological consequences. Finally, the researchers noted that service members who reported benefits as a result of the deployment had at least some exposure to the physical damage caused to the local civilians and service members from other nations.

CONCLUSIONS AND RECOMMENDATIONS

Even in the context of conflicts that arise as a result of counterterrorism, peacekeeping and peace enforcement missions will thankfully outnumber wars internationally. Although the psychological casualties resulting from peacekeeping stress do not compare to combat, it is important to appreciate the mental health needs of service members who return from peacekeeping missions, particularly ones that entail unforeseen escalations in hostilities. Peacekeepers may experience a wide range of potentially traumatic events and numerous uniquely stressful experiences. Peacekeeping, and especially peace enforcement, missions should include provisions for secondary prevention interventions for those service members most at risk for chronic mental health problems following homecoming.

References

1. United Nations. *Current peacekeeping operations.* http://www.un.org/en/peacekeeping/operations/current.shtml; Accessed 10.02.15.
2. United Nations. *Fatalities by year and incident type.* http://www.un.org/en/peacekeeping/fatalities/documents/stats_5.pdf. Published January 31, 2015. Updated February 9, 2015. Accessed 10.02.15.
3. *Interview: General Romeo Dallaire.* PBS Frontline website. http://www.pbs.org/wgbh/pages/frontline/shows/ghosts/interviews/dallaire.html. Published April 1, 2004. Accessed 18.02.15.
4. Mehlum L, Weisæth L. Predictors of posttraumatic stress reactions in Norwegian U.N. peacekeepers 7 years after service. *J Trauma Stress.* 2002;15(1):17–26.
5. Passey G, Crocket D. Psychological consequences of Canadian UN peacekeeping in Croatia and Bosnia. In: *Paper Presented at 7th Annual Meeting of the International Society for Traumatic Stress Studies; 1995; Boston, MA*; 1995.
6. Litz B, Orsillo S, Friedman M, Ehlich P, Batres A. Posttraumatic stress disorder associated with peacekeeping duty in Somalia for U.S. military personnel. *Am J Psychiatry.* 1997;154(2):178–184.
7. Dirkzwager A, Bramsen I, van der Ploeg H. Factors associated with posttraumatic stress among peacekeeping soldiers. *Anxiety Stress Coping.* 2005;18(1):37–51.
8. Maguen S, Litz B, Wang J, Cook M. The stressors and demands of peacekeeping in Kosovo: predictors of mental health response. *Mil Med.* 2004;169(3):198–206.
9. Gray M, Bolton E, Litz B. A longitudinal analysis of PTSD symptom course: delayed-onset PTSD in Somalia peacekeepers. *J Consult Clin Psychol.* 2004;72(5):909–913.
10. Adler A, Huffman A, Bliese P, Castro C. The impact of deployment length and experience on the well-being of male and female soldiers. *J Occup Health Psychol.* 2005;10(2):121–137.
11. Maguen S, Litz B. Predictors of morale in U.S. peacekeepers. *J Appl Soc Psychol.* 2006;36(4):820–836.
12. Litz B, King L, King D, Orsillo S, Friedman M. Warriors as peacekeepers: features of the Somalia experience and PTSD. *J Consult Clin Psychol.* 1997;65(6):1001–1010.
13. Barnes J, Nickerson A, Adler A, Litz B. Perceived military organizational support and peacekeeper distress: a longitudinal investigation. *Psychol Serv.* 2013;10(2):177–185.
14. Michel P, Lundin T, Larsson G. Stress reactions among Swedish peacekeeping soldiers serving in Bosnia: a longitudinal study. *J Trauma Stress.* 2003;16(6):589–593.
15. Britt T, Adler A, Bartone P. Deriving benefits from stressful events: the role of engagement in meaningful work and hardiness. *J Occup Health Psychol.* 2001;6(1):53–63.

CHAPTER

51

Optimism, Pessimism, and Stress

L. Solberg Nes

Oslo University Hospital, Oslo, Norway

OUTLINE

Abstract

Dispositional optimists expect more good things to happen to them than bad. Optimism is linked with goal engagement and persistence, and optimists tend to adjust better to stressor exposure than pessimists. Dispositional optimism is also positively associated with approach coping strategies seeking to solve or manage stressors, and negatively associated with avoidance coping strategies. There is flexibility in this concept as optimists tend to choose coping strategies depending on the stressor and stressor controllability. In the context of stressor exposure, a solid link exists between dispositional optimism and psychological and physiological well-being. Some findings indicate short-term physiological costs for optimists in this process, potentially due to persistent goal engagement and approach coping. However, the short-term costs are expected to be outweighed by long-term benefits involving goal achievement and associated positive psychological and physiological well-being.

People differ in how they approach challenges, experiences, and expectations. Some are optimistic in their outlook and tend to expect more positive than bad things to happen to them, while others are pessimistic and expect more negative outcomes. Even though the thought that "optimism is good" is rooted in folk wisdom, scientific approaches have linked the concepts of optimism and pessimism to expectancy models of motivation. The optimism construct is grounded in decades of theory and research, and a large body of research has shown optimists, compared with pessimists, to adjust better to challenges. In particular, optimists tend to adjust better to stress and stressor exposure than pessimists, experiencing less psychological distress and less negative impact on long-term physical well-being.[1–3]

DISPOSITIONAL OPTIMISM

Optimism has been conceptualized in several ways. This chapter focuses on dispositional optimism, generalized positive outcome expectancies, one of the most recognized contemporary theories of optimism.[1] Other approaches to optimism include attributional or explanatory style,[4] which assumes that expectancies are based on individual interpretations of previous experiences. Optimism can also be situational or state related, referring to positive outlook about a specific situation such as a sports task or an academic test.

Dispositional optimism is considered an individual difference or a trait, with people high in dispositional optimism displaying a generalized positive outlook on their future that is stable and applies more or less across a person's entire life span.[1] Dispositional optimism is measured by the Life Orientation Test-Revised (LOT-R),[5] a revised version of the Life Orientation Test.[6] The LOT-R consists of six items worded positively as well as negatively (e.g., "In uncertain times, I usually expect the best" and "If something can go wrong for me, it will"), and also four filler items. The LOT-R has acceptable internal consistency (0.78) and construct validity with regard to related constructs.[5]

http://dx.doi.org/10.1016/B978-0-12-800951-2.00052-2

KEY POINTS

- Dispositional optimism entails generalized positive outcome expectancies and is associated with adaptive adjustment during stressor exposure.
- Likely based on the generalized belief in positive outcomes, dispositional optimists tend to engage in approach coping strategies seeking to overcome or manage challenges, while their more pessimistic counterparts tend to engage in avoidance coping strategies seeking to avoid or disengage from challenges.
- The goal engagement and persistence displayed by optimists is linked to better chance of goal achievement, which again relates to positive well-being.
- Dispositional optimism has been linked to better psychological as well as physiological well-being during and after stressor exposure.
- There are, however, some indications that dispositional optimists may pay a short-term physiological cost during stressor exposure. This is likely due to their tendency to engage and persist despite stress, however, and in the long-term their persistence is expected to pay off, leading to goal achievement and positive psychological as well as physiological well-being.

Despite the verbal convenience of being able to categorize people as either optimistic or pessimistic, using a cut-off point between optimism and pessimism may not be completely representative and measuring optimism along a continuum, that is, from high to low dispositional optimism, rather than either optimism or pessimism, may be most accurate. This is also how the LOT-R is mainly used, measuring optimism along a 1-5 score continuum (from "strongly agree" to "strongly disagree"). In this chapter, optimism is subsequently discussed along a continuum, from high optimism (i.e., dispositional optimist) to low optimism (i.e., dispositional pessimist).

Generally being considered a fairly stable personality trait, there are indications that the degree of dispositional optimism may vary some across time.[7] There is also flexibility in the outlook of optimists and despite generally expecting more good than bad things to happen to them, degree of optimism may vary depending on the situation at hand.

OPTIMISM AND STRESS

When someone describes "feeling stressed," it usually means that the person meets demands that are perceived to exceed his or her available resources. The term stress is used to describe overwhelming or threatening situations, or the pressure that people encounter when experiencing such circumstances. Stress can elicit a complex array of psychological and physiological reactions, and the sources of stress can be infinite. Stress can be physical or psychological, controllable or uncontrollable, acute or chronic, and the word stress is often used imprecisely referring either to a stressor or a stress response.

It is possible that stress and the consequences of stress may arise from how people appraise the events rather than from the events themselves. As dispositional optimists are defined by a generalized positive outlook on the future, this may impact how they appraise and approach stressors. In fact, dispositional optimists generally report experiencing less distress during stressor exposure compared to their more pessimistic counterparts, and a large body of research suggests that dispositional optimism can have a protective role during stressor exposure, potentially "buffering" from the adverse impact of stressful events. Dispositional optimism has been linked to better psychological adjustment to stressors ranging from academic stressors to health-related challenges and even extreme trauma.[8–10] Similarly, dispositional optimism has been associated with better physiological well-being as reflected in cardiovascular and immune functioning.[11,12] Seeking to identify underlying and contributing components to this phenomenon, a number of optimism-related factors have been explored. As it appears, goal engagement and choice of coping strategies are essential factors in this concept.

GOAL ENGAGEMENT

The construct of dispositional optimism arose from a general self-regulatory framework.[1] According to this framework, people have two options when encountering challenges; engage to overcome challenges and achieve goals, or disengage to avoid challenges and give up on their goals. The decision may depend on whether or not they see their desired outcome as attainable. Because optimists see positive outcomes as attainable, they are more likely to invest continued effort in order to achieve their goals, instead of disengaging and giving up, as pessimists might do.[13,14] So, if a person's outcome expectancy is positive (i.e., optimism) he/she may decide to engage and persist in order to overcome the challenge and attain the goal, but if a positive outcome is not expected (i.e., pessimism), he/she may decide to disengage from the challenge.

A number of studies have shown how dispositional optimists persist longer on tasks compared with pessimists, in some cases particularly when self-awareness was high, as awareness tends to highlight own goals.[1,13–15] For example, in two studies examining optimism's effect during a brief mental effort stressor, both undergraduate students and adults with or without chronic pain conditions displayed longer persistence/goal engagement when high in dispositional optimism.[14,15] The tendency for optimists to expect positive outcomes and remain engaged in challenges creates a self-fulfilling prophecy in which positive outcomes and success can be actualized. For pessimists on the other hand, the tendency to expect negative outcomes and give up on challenges creates a self-fulfilling prophesy of failure.

Although persistent goal pursuit can be associated with goal attainment, which again is linked to well-being, the tendency of dispositional optimists to expect positive outcomes and persist in their efforts is also linked to a greater likelihood of experiencing goal conflict.[16] When experiencing goals that conflict (e.g., "work more" and "spend more time with family and friends"), the obvious solution may be to give up on one of the goals in order to achieve the other. However, dispositional optimists are less likely to give up on goals than their pessimistic counterparts,[1,14] and hence are more likely to experience goal conflict. For example, in one cross-sectional and one longitudinal study, optimism associated with higher goal conflict.[16] However, the goal conflict experienced did not impact adjustment. Other studies have supported the notion of higher likelihood of goal conflict for optimists, but have also shown a physiological cost (i.e., lower in vivo cellular immunity[17]). The issue of potential short-term physiological costs related to goal engagement will be discussed in the Physical Well-being section.

COPING

Coping is central to how people seek to manage internal or external demands, making choice of coping strategies essential in stress management. The concept of coping is well researched, and even though different types of coping models have been proposed, the most recognized and used conceptualizations are the *problem focused* versus *emotion focused* (i.e., addressing external vs. internal demands of stressors)[18] and the *approach* versus *avoidance* (approaching vs. avoiding or disengaging from the demands presented by a stressor)[19] coping strategy distinctions.

Coping also plays a major role in the link between dispositional optimism and better adjustment to stressor exposure. Because of their positive outlook, *optimists* are more likely to appraise goals as achievable and hence more likely to *approach* challenges and strive to achieve their goals. *Pessimists* on the other hand, who are more likely to appraise goals as unachievable, are more likely to *avoid* or disengage from demanding challenges and give up. A large body of research supports this notion, showing dispositional optimism to be associated with more *approach* (i.e., aiming to reduce, eliminate, or manage the demands of the stressors) and *problem-focused* (i.e., aiming to change or eliminate the specific stressor) coping, and negatively associated with *avoidance* (i.e., aiming to ignore, avoid, or withdraw from stressor and related consequences) and *emotion-focused* (aiming to reduce or manage the emotional consequences related to the stressor) coping.[20]

Stressor controllability may also play a significant role in the effectiveness and choice of coping strategies.[21] If facing an academic test, for example, the stressor is controllable and utilizing an approach problem-focused coping strategy such as preparing well for the test will be a proper choice of action. If surviving a traumatic event, or experiencing fear of cancer recurrence, on the other hand, the stressor is less controllable and engaging in a problem-focused approach will likely be less beneficial. This suggests that flexibility in choice of coping strategies, depending on stressor type and controllability of the stressor at hand, is the most adaptive coping approach. As optimists appear to engage more in problem-focused rather than emotion-focused coping strategies, could this mean that dispositional optimists cope effectively with controllable, but inappropriately with uncontrollable, stressors?

Examining this issue, meta-analytic findings showed that there is considerable flexibility involved and that optimists likely adjust their choice of coping strategy depending on the stressor at hand.[20] Rather than consistently choosing approach problem-focused coping strategies, optimists tend to choose approach problem-focused coping strategies when the stressors are *controllable*, yet approach emotion-focused coping strategies when the stressors are *less controllable*.[20]

This suggests that a combination of the well known coping conceptualizations (i.e., approach problem/emotion and avoidance problem/emotion) may capture finer nuances and be more informative than the conceptualizations used so far.[20] Following this approach, high dispositional optimism is, for example, positively correlated with problem approach (e.g., planning) and emotion approach (e.g., acceptance) coping, but negatively correlated with problem avoidance (e.g., behavioral disengagement) and emotion avoidance (e.g., denial).[20] Considering this, dispositional optimists appear to cope well with controllable as well as uncontrollable stressors as they display flexibility in their response to the demands of the stressor, and in their approach of either seeking to overcome or modify their response to the stressor.

One note of caution: In the meta-analysis noted above,[20] studies conducted in the United States and in English-speaking countries had significantly larger effect sizes in terms of coping categories compared with studies conducted in non-USA and non-English-speaking countries. As most research examining the relationship between dispositional optimism, coping, and adjustment has been conducted with U.S. and English-speaking participants so far, one should be careful about generalizing findings to other populations without further research.

SELF-REGULATION

Given the way optimism seems to impact how people approach tasks, goals, and challenges,[1] it has been hypothesized that dispositional optimism also might play a role in the mechanisms of self-regulation. Self-regulation involves any effort to control internal or external, mental or physical activities.[1] The capacity to self-regulate appears to vary, however, and self-regulatory efforts, such as having to control thoughts, feelings, and behavior seem to depend on a limited source that can be depleted or fatigued.[22] Identifying underlying mechanisms in this, exploring potential components and finding ways to improve or maintain self-regulatory capacity and prevent self-regulatory fatigue could contribute to a better understanding of adjustment, and adjustment to stressor exposure in particular.

In a study examining potential impact of dispositional optimism on self-regulatory capacity,[15] it was hypothesized that optimism would buffer the depleting effects from self-regulatory effort. Supporting previous research,[1,14] higher dispositional optimism did predict goal engagement in the study. However, contrary to predictions, optimists persisted more than pessimists only when self-regulatory effort was *not* required, and in fact persisted *less* than pessimists when self-regulatory effort was required.[15] Contrary to expectations, this indicates that the positive relationship between dispositional optimism and engagement may be decreased and perhaps even reversed in the presence of self-regulatory fatigue. It is possible that these results may be indicative of optimist's focus on conserving effort rather than overcoming self-regulatory fatigue,[15] but this needs to be determined by future research. The study results also suggest that people with chronic multisymptom illnesses may benefit less from dispositional optimism in stressful situations compared with healthy controls. The dispositional optimism/self-regulation area is clearly where further research is warranted.

MENTAL WELL-BEING

Dispositional optimism has consistently been associated with higher levels of psychological well-being, while pessimism has been associated with lower psychological well-being.[1] Optimism has shown significant links to better mood and emotional adjustment, better life satisfaction and social support,[1] and optimists also seem less likely to experience mental health problems, especially related to stressor exposure.[6,23]

The link between dispositional optimism and psychological well-being appears robust, with significant findings in a wide variety of settings. For example, dispositional optimists in their first year of college described experiencing less stress, depression, and loneliness as well as feeling more socially supported than their more pessimistic counterparts.[1,8,24] Dispositional optimism has also been associated with better adjustment to major life event stressors, such as life threatening illness and trauma. In early-stage breast cancer patients, for example, higher optimism at diagnosis was indicative of lower distress at 3, 6, and 12 months post diagnosis,[25] regardless of medical variables and history of distress. Also, in patients 6 months or more, post bone marrow transplantation, higher dispositional optimism was linked with higher life satisfaction and lower negative mood.[26]

For people having experienced trauma or life threatening situations, pessimistic disaster victims appear more at risk for severe depression and obsessive-compulsive symptoms than their more optimistic counterparts,[27] and among healthy adult men in the Veterans Administration, dispositional optimism has been linked with less depression and overall better mental health and vitality.[28] Considering the large body of research supporting the notion, it should not be controversial at this point to claim a buffering role for dispositional optimism in the stress-distress relationship.

PHYSICAL WELL-BEING

A number of studies have suggested that dispositional optimism has a positive link not only to psychological, but also physiological well-being. Optimism has, for example, been associated with better physiological well-being in terms of cardiovascular and immune functioning.[11,12,29] Compared with pessimists, optimists have also been seen to report less pain,[30,31] better physical functioning, and also to experience fewer physical symptoms.[32] A meta-analysis examining results from 84 studies testing the relationship between optimism and physical health supported this notion, revealing a small but significant effect size pointing to optimism as a significant predictor of physical health.[3]

Despite evidence of a link between dispositional optimism and better physiological adjustment to stressors, some contradictory findings exist, suggesting that the engagement and persistence derived from positive expectancies may have, at least temporary, physiological costs. For example, optimists exposed to uncontrollable noise experienced reduction in natural killer cell cytoxicity,[33] and optimists engaging in increasingly difficult mental arithmetic tasks have experienced smaller delayed-type hypersensitivity reactions compared with pessimists, indicating worse cellular immunity.[34] In fact, several studies have found dispositional optimism in combination with high challenge to correlate with lower cellular immunity.[35] Similarly, in a study exposing participants to a stressful mental effort task, optimists were found to display goal engagement, persisting longer than pessimists on the task, but were also found to experience short-term physiological costs in the form of increased salivary cortisol and skin conductance level post task.[14] There are also indications that the positive connection between dispositional optimists and goal engagement may entail higher likelihood of goal conflict for optimists, which again has been linked to physiological cost through lower in vivo cellular immunity.[17]

These results indicate that the engagement displayed by optimists in the face of stressors may be taxing, and the engagement model[13] proposes that although the goal engagement displayed by optimists is likely to be beneficial in the long run, there may be short-term physiological costs.[13,14,29,36] Despite potential short-term physiological costs, the persistence demonstrated by dispositional optimists is likely to be beneficial in the long run, resulting in goal achievement and related positive physical and psychological well-being (see Figure 1).

HOW TO IMPROVE OPTIMISM

Given the many positive links between optimism and goal engagement, coping, adjustment, and well-being, establishing ways in which optimism can be increased would be of benefit. As a personality trait and an individual difference, however, dispositional optimism is mainly considered stable, with test-retest correlations ranging from $r=0.58$ to $r=0.79$ for up to 3 years.[2] Also, a 25% heritability is estimated for optimism,[37] and financial security as well as warmth and attention from parents in childhood may also predict adult degree of optimism.[38]

Despite being considered a fairly stable personality trait, there are some indications that degree of optimism may change over time. In a study examining links among dispositional optimism, resources and health, test-retest correlations for optimism was only $r=0.35$ after 10 years.[7] Also, some interventions have found, although unintentional, increases in dispositional optimism over time. For example, following a 10-week cognitive-behavioral stress management intervention for women recently treated for breast cancer, dispositional optimism levels were significantly increased immediately and 3 and 9 months post intervention.[39] In the study, the women who were initially the least optimistic appeared to experience most

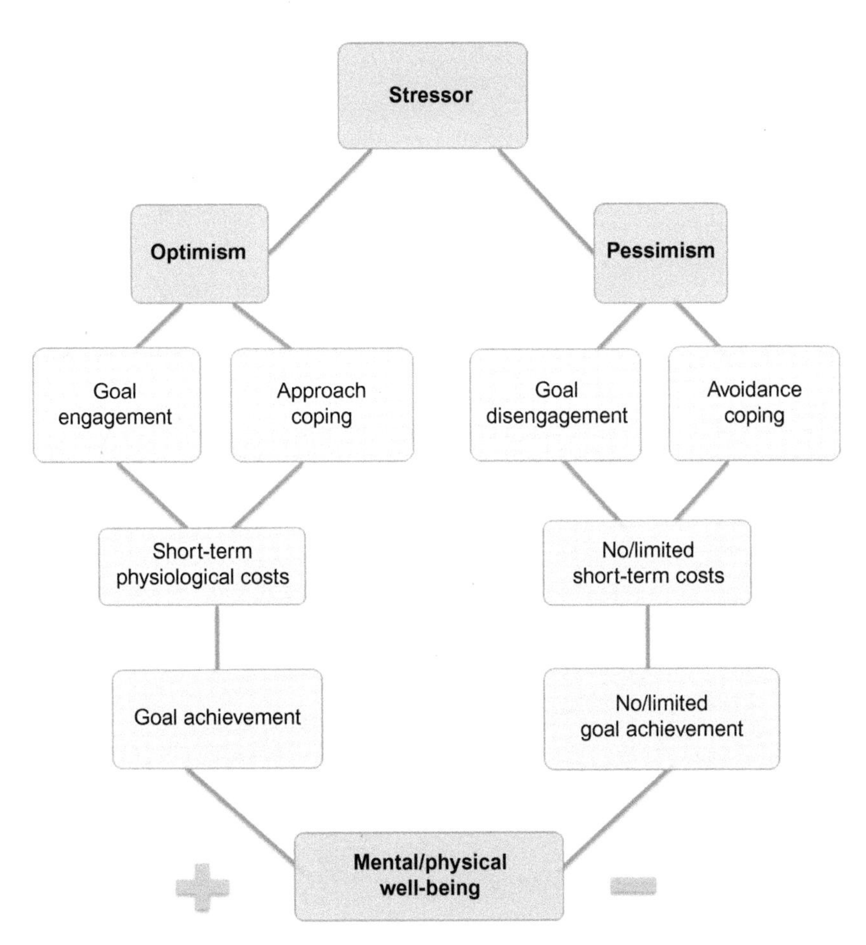

FIGURE 1 Optimism, pessimism, and stress. The stress management model displayed by dispositional optimists is based on positive outcome expectancies, goal engagement, and approach coping strategies, may entail short-term physiological costs, but is associated with goal achievement and positive mental and/or physical well-being. Illustrated by Trude Nordby Bøe.

improvement in optimism, suggesting that providing people with coping skills/stress management skills might have potential in helping them improve their outcome expectancies for the future.

As stated, optimists engage and persist in goal pursuit based on their belief in positive outcomes,[20] but the positive association between optimism, adjustment, and well-being is likely based on how optimists approach and cope with stressors. Rather than pushing people to "pull themselves together and become more optimistic," it seems giving people the knowledge and tools to better cope with specific challenges, as in the stress management study above,[39] may be the best solution. That way, coping skills may in fact contribute to a more positive outlook, and subsequently a better approach in how to cope with stress and stressor exposure, or vise versa.

CONCLUSION

Dispositional optimism plays an important role in how people respond in the face of stressful situations. While pessimists tend to expect more bad things to happen to them than good, dispositional optimists expect more good things to happen to them and when exposed to stressors, they believe in positive outcomes, persist at goal engagement, and use approach coping strategies to deal with the stressor at hand. Optimism is positively linked with use of approach coping strategies seeking to eliminate, reduce or manage stressors, or the emotional impact of stressors, and negatively linked to use of avoidance coping strategies seeking to avoid, ignore, or withdraw from stressors or the emotional impact of such. The persistent goal engagement and approach coping strategies used by optimists may help them adjust better to stressful situations.

Dispositional optimism has been associated with better psychological and physiological well-being, including less distress, better life satisfaction, and social support, as well as better cardiovascular and immune functioning. The persistent engagement and approach coping strategies displayed by optimists may entail short-term physiological costs, however, including lower cellular immunity and increased salivary cortisol and sympathetic activity during stressor exposure. Nevertheless, these short-term costs are expected to be outweighed by long-term benefits involving goal achievement and associated positive psychological and physiological well-being.

References

1. Carver CS, Scheier MF. *On the Self-Regulation of Behavior*. New York: Cambridge University Press; 1998.
2. Carver S, Scheier MF, Segerstrom SC. Optimism. *Clin Psychol Rev*. 2010;30:879–889.
3. Rasmussen HN, Scheier MF, Greenhouse JB. Optimism and physical health: a meta-analytic review. *Ann Behav Med*. 2009;37: 239–256.
4. Peterson C, Seligman MEP. Causal explanations as a risk factor for depression: theory and evidence. *Psychol Rev*. 1984;91:347–374.
5. Scheier MF, Carver CS, Bridges MW. Distinguishing optimism from neuroticism (and trait anxiety, self-mastery, and self-esteem): a reevaluation of the Life Orientation Test. *J Pers Soc Psychol*. 1994;67:1063–1078.
6. Scheier MF, Carver CS. Optimism, coping and health: assessment and implications of generalized outcome expectancies. *Health Psychol*. 1985;4:219–247.
7. Segerstrom SC. Optimism and resources: effects on each other and on health over 10 years. *J Res Pers*. 2007;41:772–786.
8. Aspinwall LG, Taylor SE. Modeling cognitive adaptation: a longitudinal investigation of the impact of individual differences and coping on college adjustment and performance. *J Pers Soc Psychol*. 1992;63:989–1003.
9. Dougall AL, Hyman KB, Hayward MC, McFeeley S, Baum A. Optimism and traumatic stress: the importance of social support and coping. *J Appl Soc Psychol*. 2001;31:223–245.
10. Stanton AL, Snider PR. Coping with breast cancer diagnosis: a prospective study. *Health Psychol*. 1993;12:16–23.
11. Räikkönen K, Matthews KA, Flory JD, Owens JF, Gump BB. Effects of optimism, pessimism, and trait anxiety on ambulatory blood pressure and mood during everyday life. *J Pers Soc Psychol*. 1999;76:104–113.
12. Segerstrom SC, Taylor SE, Kemeny ME, Fahey JL. Optimism is associated with mood, coping, and immune change in response to stress. *J Pers Soc Psychol*. 1998;74:1646–1655.
13. Segerstrom SC. Optimism, goal conflict, and stressor-related immune change. *J Behav Med*. 2001;24:441–467.
14. Solberg Nes L, Segerstrom SC, Sephton SE. Engagement and arousal: optimism's effects during a brief stressor. *Personal Soc Psychol Bull*. 2005;31:111–120.
15. Solberg Nes L, Carlson CR, Crofford LJ, de Leeuw R, Segerstrom SC. Individual differences and self-regulatory fatigue: optimism, conscientiousness, and self-consciousness. *Personal Individ Differ*. 2011;50:475–480.
16. Segerstrom SC, Solberg Nes L. When goals conflict but people prosper: the case of dispositional optimism. *J Res Pers*. 2006;40: 675–693.
17. Segerstrom SC. How does optimism suppress immunity? Evaluation of three affective pathways. *Health Psychol*. 2006;25:653–657.
18. Lazarus RS, Folkman S. *Stress, Appraisal, and Coping*. New York: Springer; 1984.
19. Suls J, Fletcher B. The relative efficacy of avoidant and non-avoidant coping strategies: a meta-analysis. *Health Psychol*. 1985;4:249–288.
20. Solberg Nes L, Segerstrom SC. Dispositional optimism and coping: a meta-analytic review. *Personal Soc Psychol Rev*. 2006;3:235–251.
21. Conway VJ, Terry DJ. Appraised controllability as a moderator of the effectiveness of different coping strategies: a test of the goodness-of-fit hypothesis. *Aust J Psychol*. 1992;44:1–7.
22. Baumeister RF, Bratslavsky E, Muraven M, Tice DM. Ego depletion: is the active self a limited resource? *J Pers Soc Psychol*. 1998;74:1252–1265.
23. Fotiadou M, Barlow JH, Powell LA, Langton H. Optimism and psychological well-being among parents of children with cancer: an exploratory study. *Psychooncology*. 2008;17:401–409.
24. Brissette L, Scheier MF, Carver CS. The role of optimism in social network development, coping, and psychological adjustment during a life transition. *J Pers Soc Psychol*. 2002;82:102–111.
25. Carver CS, Pozo C, Harris SD, et al. How coping mediates the effect of optimism on distress: a study of women with early stage breast cancer. *J Pers Soc Psychol*. 1993;65:375–390.

26. Curbow B, Somerfield MR, Baker F, Wingard JR, Legro MW. Personal changes, dispositional optimism, and psychological adjustment to bone marrow transplantation. *J Behav Med*. 1993;16:423–443.
27. Van der Velden PG, Kleber RJ, Fournier M, Grievink L, Drogendijk A, Gersons BPR. The association between dispositional optimism and mental health problems among disaster victims and a comparison group: a prospective study. *J Affect Disord*. 2007;102:35–45.
28. Achat H, Kawachi I, Spiro III A, DeMolles DA, Sparrow D. Optimism and depression as predictors of physical and mental health functioning: the Normative Aging Study. *Ann Behav Med*. 2000;22:127–130.
29. Cohen D, Hamrick N, Rodriguez MS, Feldman PJ, Rabin BS, Manuck SB. The stability of and intercorrelations among cardiovascular, immune, endocrine, and psychological reactivity. *Ann Behav Med*. 2000;22:171–179.
30. Costello NL, Bragdon EE, Light KC, et al. Temporomandibular disorder and optimism: relationships to ischemic pain sensitivity and interleukin-6. *Pain*. 2002;100:99–110.
31. Mahler HIM, Kulik JA. Optimism, pessimism and recovery from coronary bypass surgery: prediction of affect, pain and functional status. *Psychol Health Med*. 2000;5:347–358.
32. Fournier MA, de Ridder D, Bensing J. How optimism contributes to the adaptation of chronic illness. A prospective study into the enduring effects of optimism on adaptation moderated by the controllability of chronic illness. *Personal Individ Differ*. 2002;33:1163–1183.
33. Sieber WJ, Rodin J, Larson L, Ortega S, Cummings N. Modulation of human natural killer cell activity by exposure to uncontrollable stress. *Brain Behav Immun*. 1992;6:141–156.
34. Segerstrom SC, Castaneda JO, Spencer TE. Optimism effects on cellular immunity: testing the affective and persistence models. *Personal Individ Differ*. 2003;35:1615–1624.
35. Segerstrom SC. Optimism and immunity: do positive thoughts always lead to positive effects? *Brain Behav Immun*. 2005;19:195–200.
36. Maier KJ, Waldstein SR, Synowski SJ. Relation of cognitive appraisal to cardiovascular reactivity, affect, and task engagement. *Ann Behav Med*. 2003;26:32–41.
37. Plomin R, Scheier MF, Bergeman CS, Pedersen NL, Nesselroade JR, McClearn GE. Optimism, pessimism, and mental health: a twin/adoption analysis. *Personal Individ Differ*. 1992;13:1217–1223.
38. Heinonen K, Räikkönen K, Keltikangas-Järvinen L. Dispositional optimism: development over 21 years from the perspectives of perceived temperament and mothering. *Personal Individ Differ*. 2005;38:425–435.
39. Antoni MH, Lehman JM, Kilbourn KM, et al. Cognitive-behavioral stress management intervention decreases the prevalence of depression and enhances benefit finding among women under treatment for early-stage breast cancer. *Health Psychol*. 2001;20:20–32.

CHAPTER

52

Chronic Pain and Perceived Stress

C.D. King[1], A. Keil[2], K.T. Sibille[2]

[1]Cincinnati Children's Hospital Medical Center, Cincinnati, OH, USA

[2]University of Florida, Gainesville, FL, USA

OUTLINE

Abstract

Pain is a complex, multidimensional experience that varies by individual and condition. Living with chronic pain is stressful and is associated with increased morbidity and mortality. Chronic pain is difficult to treat and has developed into a significant public health concern. Comprised of sensory and affective processes, the experience of pain is often perceived as a threat. Understanding the biological interface of chronic pain and associated psychosocial stressors will be important as we work to prevent, reduce, and eliminate the associated physiological toll and health-related consequences. Pain theories identify the importance of psychological and stress-related factors in the perception and persistence of pain. An array of neurobiological and biological evidence demonstrates altered and/or dysregulated functioning with chronic pain. Further investigations are needed to improve the characterization of chronic pain and perceived stress, to better understand the biological burden, and to identify targets for chronic pain prevention and treatment.

INTRODUCTION

Pain represents an enormous public health issue with expenses as high as $635 billion annually.[1] Approximately 30% of adults report experiencing "chronic, recurrent, or long-lasting pain for at least 6 months."[2] Making matters worse, current treatments are minimally effective and there is consistent evidence that chronic pain is associated with increased morbidity and mortality.[3–5]

Pain is a complex, multidimensional phenomenon, complicated by significant individual differences. The generally accepted definition of pain is "an unpleasant sensory and emotional experience associated with actual or potential tissue damage, or described in terms of such

Stress: Concepts, Cognition, Emotion, and Behavior
http://dx.doi.org/10.1016/B978-0-12-800951-2.00053-4

damage."[6] Pain is typically classified as either acute or chronic, with an intermittent category becoming more frequently included. Acute pain is an ecologically conserved, physiological response to a situation typically associated with an identifiable injury or inciting stimulus within a 3-month period. Chronic pain is pain experienced for a minimum of 3 months, though some definitions require the presence of pain for at least 6 months.[7] Mechanistically, acute pain is associated with adaptive biological changes in central and peripheral processing resulting in hyperalgesia (e.g., a heightened sensitivity to a painful stimulus), spontaneous pain, and allodynia (e.g., perception of pain in response to normally nonpainful stimuli). The transient augmentation of pain sensitivity helps protect an injured area from further harm and typically subsides with recovery.[8,9] However, if the peripheral and/or central nociceptive processes do not reset or become altered even after evidence of tissue recovery, hyperalgesia, spontaneous pain, and allodynia can persist and the transition to chronic pain will occur.

KEY POINTS

- Pain is a complex, multidimensional, dynamic, and individualized experience comprised of sensory and affective processes.
- The experience of pain is often perceived as a threat, activating stress-regulatory systems.
- Chronic pain is a central nervous system disorder which is exacerbated by psychological and stress-related factors.
- Living with chronic pain is stressful and is associated with increased morbidity and mortality.
- Altered and/or dysregulated neurobiological and biological functioning is indicated in chronic pain.
- Understanding the biological interface of chronic pain and the associated psychosocial stressors will be necessary in order to prevent, reduce, and eliminate the physiological burden and health-related consequences.
- There are a number of potential behavioral and environmental clinical targets that may help reduce the allostatic load (AL) of chronic pain.

Understanding the biological burden of chronic pain and associated psychosocial stress will be important as we work to prevent, reduce, and eliminate the physiological toll and health-related consequences. Topics addressed in this chapter include (1) the experience of pain and related mechanisms (i.e., nociception, sensory and affective processes, threat perception, and associated psychosocial stress); (2) theories and models of pain; (3) the neurobiological and biological interface of chronic pain and associated stress; and (4) future directions including areas for further investigation to better understand the relationship between chronic pain and perceived stress on stress-systems functioning and to identify targets for prevention and treatment.

THE EXPERIENCE OF PAIN

Pain is often perceived as a threat. Sensory and affective components of the pain experience (e.g., pain intensity, unpleasantness, and suffering) can activate stress response systems to prepare the body to deal with the pain-evoking situation.[9,10] An overview of the sensory and affective dimensions of pain perception and associated mechanisms will be presented. Additionally, factors contributing to chronic pain onset and increased pain sensitivity will be reviewed. Combined, the dimensions of pain perception and the contributing factors provide the framework to better understand the experience of pain and the associated biological interface.

Pain perception is a dynamic and personal experience. Price proposed a definition of pain from a phenomenological perspective *"a somatic perception containing 1) a bodily sensation with qualities like those reported during tissue-damaging stimulation, 2) an experienced threat associated with this sensation, and 3) a feeling of unpleasantness or other negative emotion based on this experienced threat"* (pp. 1–2).[9] The sensory dimension of pain reflects the processing of the sensory discriminative attributes of the stimuli (e.g., intensity, location, sensation qualities). The affective dimensions entail an immediate perception of threat and unpleasantness and then a secondary experience of suffering. It is the combination of the sensory perception and the associated evaluation of unpleasantness that result in the experience of pain.[9]

Sensory Dimension

Nociception is the process by which noxious stimulation is communicated through the peripheral and central nervous system. Nociceptors are specific receptors within the skin, muscle, skeletal structures, and viscera that detect potentially damaging stimuli. Nociceptive afferent neurons extend through the spinal cord via the dorsal horn. In addition to receiving incoming nociceptive information, the dorsal horn integrates information from descending pathways from the brainstem and brain, which modulates the flow of nociceptive information back to the higher structures (Figure 1). Dorsal horn neurons proceed through the anterolateral system to a complex array of subcortical and cortical areas including but not limited to the brainstem, thalamus, forebrain, and somatosensory cortex.[8]

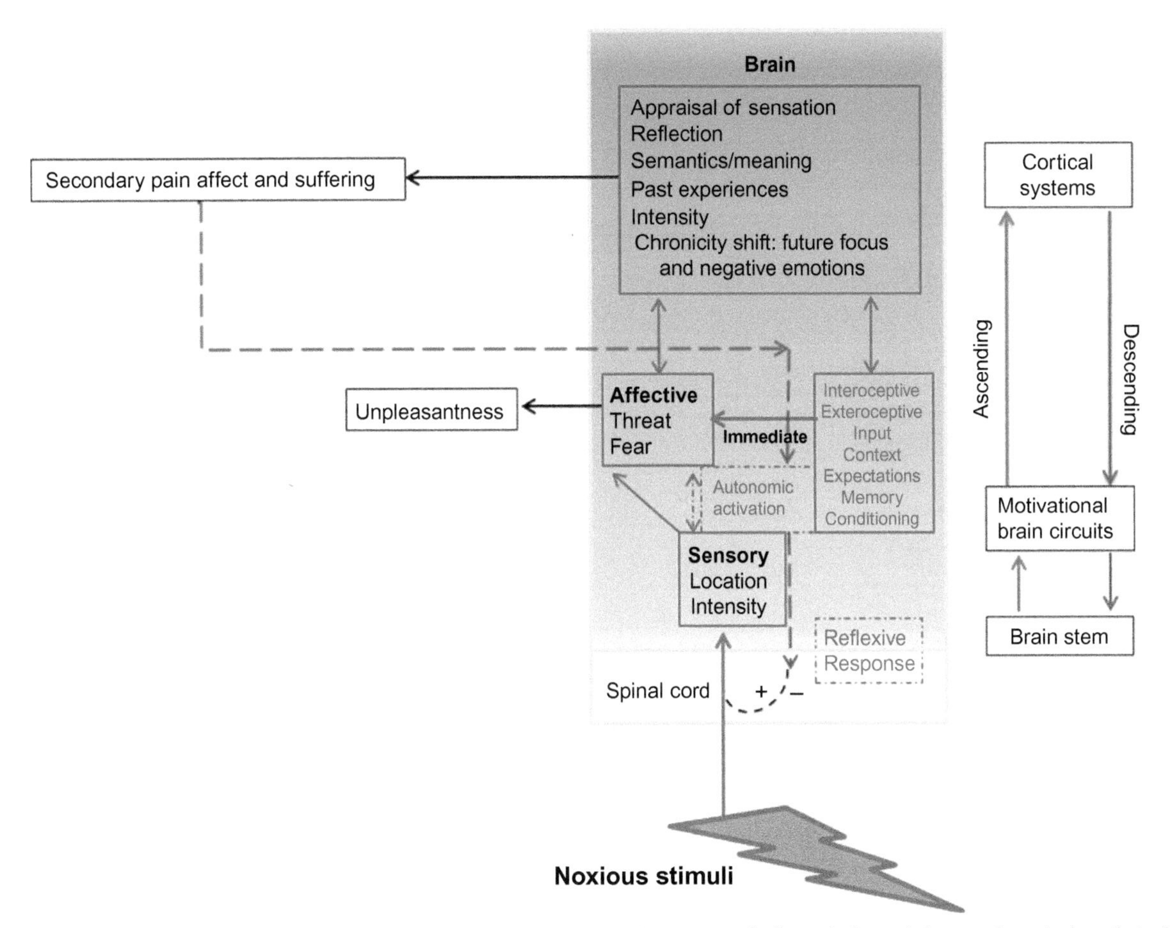

FIGURE 1 Integrated conceptual model of the pain experience. Nociceptive input extends through the periphery to the spinal cord via the dorsal horn through the anterolateral system to a complex array of subcortical and cortical areas including but not limited to the brainstem, thalamus, forebrain, and somatosensory cortex (red lines). Pain affective and motivational processing also involves an extensive network of brain structures that act in unison and sequentially, communicating through multiple ascending (red lines) and descending (blue lines) pathways involving brainstem, subcortical and cortical regions. The dorsal horn also integrates information from descending pathways, which modulates the flow of nociceptive information back to the higher structures. (Components in the model were informed and influenced by the work and teachings of Donald D. Price, PhD.)

The areas of the brain involved and the extent of activation vary by the sensory features of the nociceptive stimuli.

Affective Dimension

Pain affective and motivational processing occurs in stages and involves an extensive network of brain structures that act in unison and sequentially, communicating through multiple ascending and descending pathways involving brainstem, subcortical and cortical regions.[8,9] Two distinct stages are indicated in affective processing. The first stage is experienced as "unpleasantness," which occurs immediately and encompasses a limited time frame.[9] A number of factors contribute toward the immediate experience of unpleasantness, including the nociceptive sensation, context, expectations, autonomic activation, temporal factors, prior experiences, and somatic/motor responses (Figure 1).[9] The second stage is identified as "secondary pain affect," which involves an overall evaluation of the pain experience, including reflections, future projections, and personal meaning and is frequently referred to as suffering.[9] Some acute pain experiences may be limited to the immediate sensory and affective dimensions. Both the immediate and secondary affective aspects of pain provide feedback that either further increases or decreases the pain modulation and stress response through descending pathways as indicated in Figure 1.[8,9]

Persistent pain also contributes to a cascade of psychosocial and behavioral consequences associated with physical functioning, activities, employment, relationships, and self-identity. Changes and losses in these domains further facilitate negative secondary pain affect appraisal (suffering), and propel a vicious cycle of pain, distress, and altered psychological and physiological functioning. Repetitive experiences of pain unpleasantness, perceived threat, and suffering continue to activate the stress response, reinforce associated neurobiological and biological pathways, and condition patterns of thinking, feeling, and behaving.

Responding to Threat

The physiological, cognitive, and behavioral responses to threat are intricately linked to the activity of the brain's defensive circuitry, which is organized around evolutionarily old structures such as the amygdaloid complex and insula. These structures are also tightly integrated with the body's stress-related systems. Having evolved under the pressure to meet the challenges posed by an ever-changing human habitat, motivational brain circuits such as the amygdala and insula possess dense afferent and efferent connectivity, which enables both the efficient identification of potential danger and the implementation of appropriate, adaptive responses.[11] In line with this latter notion, recent research has consistently demonstrated that brain structures traditionally associated with perceptual, motor, or memory processes are crucially involved in mobilizing the organism when detecting threat or danger. Exposure to threat triggers activity in multiple cortical and subcortical regions that coalesce into large-scale functional assemblies (Figure 1).[12] Additionally, defensive responses are not limited to brain circuits, but include peripheral and endocrine systems. For instance, the amygdala projects to the lateral hypothalamus, mediating autonomic nervous system arousal when the defense system is activated.

Current neurobiological theories of threat processing hold that *adaptive* threat processing involves flexible and dynamic responses of the entire organism, to maximize behavioral outcomes and minimize harm.[13] This is accomplished by widespread activation of structures at all levels of the nervous system, including the integrative communication between brain regions mediating perception, attention, motivation, and action. Given this mechanism, one may hypothesize that *maladaptive* threat processing in the context of chronic pain involves exaggerated activation of defensive circuits in response to focal pain sensation. Alternatively, diffuse chronic engagement of defensive circuitry may mediate long-lasting changes relevant to pathological mood symptoms often accompanying chronic pain. Recent research in the psychophysiology of mood and anxiety supports a model as shown in Figure 2, where the initial experience of pain is associated with strong, focal defensive/aversive system activation.[14] With increasing chronicity, maladaptive plasticity in defensive/stress systems is hypothesized to accompany a pathological trajectory toward diffuse (indiscriminate) and chronic defensive engagement, associated with diminished reactivity to specific events.

The diffuse nonreactive aspect of defensive engagement in the maladaptive threat-processing model is readily illustrated by the features of chronic pain. Since many chronic pain conditions are the result of dysregulated central nervous system activation, an adaptive defensive response is not attainable—as there is no identifiable injury or inciting stimulus to address in order to "maximize behavioral outcomes and minimize harm." As an obvious example, in phantom limb pain, the perceived source of the painful sensation does not physically exist. Likewise, a common symptom of neuropathy, a "burning sensation in the foot," is not the result of noxious thermal stimuli on the exterior of the foot. Thus, chronic pain is often characterized by the absence of adaptive response options to minimize harm or reduce discomfort, and threat response behaviors are thus ineffective in this condition.

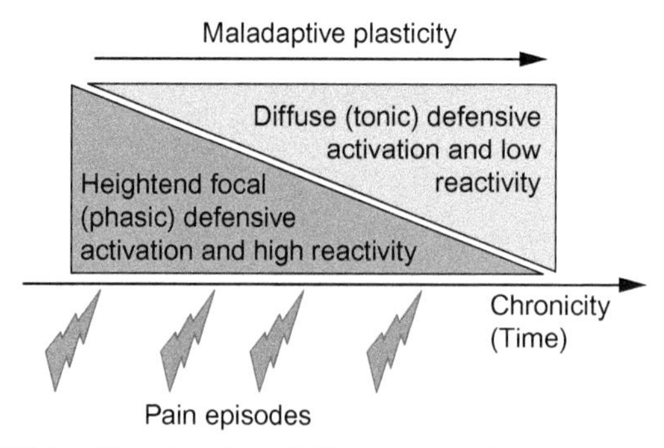

FIGURE 2 Chronic pain and discrimination between aversive and nonaversive states. The initial experience of pain is associated with strong, focal defensive/aversive system activation. With increasing chronicity, maladaptive plasticity in defensive/stress systems is hypothesized to accompany a pathological trajectory toward diffuse (indiscriminate) and chronic defensive engagement, associated with diminished reactivity to specific events.

Factors Contributing to Chronic Pain and Pain-Related Stress

In addition to underlying genetic and biological contributions, key factors associated with the onset and persistence of chronic pain include early life experiences, acute pain intensity, psychosocial stress, and psychological functioning.[7] Life experiences involving physical and psychological trauma, particularly in early childhood, but also into adulthood, have been linked with high pain sensitivity and increased risk of chronic pain. Acute pain intensity is also associated with increased risk of transition to chronic pain. As previously described, living with chronic pain contributes to an increasing cascade of psychosocial stressors. Importantly, the relationship between chronic pain and psychosocial stress is bidirectional. High levels of psychosocial stress also predict the onset of chronic pain.[7]

Multiple psychological factors have been consistently linked to the onset and severity of chronic pain as well as to heightened pain sensitivity.[7,15] Examples of pain-related psychological factors include negative affect, fear avoidance, pain catastrophizing, pessimism, anxiety, depression, passive coping, and pain hypervigilance. The various individual psychological factors may reflect the influence of psychological characteristics associated with temperament and neurobiological-based dispositional traits.[9,16] Additionally, across a number of studies, subgroups of chronic pain patients classified by psychosocial and behavioral characteristics responded differently to treatment interventions.[15] Thus, consideration of dispositional or trait-related characteristics may help account for and improve our understanding of the array of biopsychosocial factors associated with chronic pain. In total, an individual's experience of pain and perceived stress is comprised of a unique and complex web of nociceptive, affective, and predisposing factors. Theories to capture the dynamic, multidimensional, and complex interaction of peripheral and central nervous system processes and physiological systems associated with the pain experience and perception have been developed and continue to evolve.

THEORIES OF PAIN AND THE RELATIONSHIP TO STRESS

The gate control theory by Melzack and Wall in 1965 was the first theory of pain that combined evidence for physiological specificity with central integration. [7] Not only was the facilitatory and inhibitory role of the dorsal column addressed, but also the significance of psychological factors and their role in influencing pain transmission and perception. In 2001, the neuromatrix theory of pain was proposed to account for the features of chronic pain that extended beyond the concepts proposed in the gate control theory which were predominately related to acute or short-term pain experiences.[17]

This theory explicitly recognized that in the case of chronic pain, pain is often experienced based on activation from neural networks rather than solely resulting from afferent nociceptive transmission from the periphery. The neuromatrix framework addressed the importance of considering stress, both biologically and psychologically.[17] Importantly, the theory described the bidirectional, dynamic, mutually contributing relationship between the stress-regulatory system responses and chronic pain. Following this proposed theoretical relationship between pain and stress, other models have evolved and expanded on the underlying concepts.

Chapman and colleagues extended the understanding of the reciprocal relationship between pain and the stress response by proposing a "systems model of pain" which included in addition to sensory signaling systems, the role of the nervous, endocrine, and immune systems.[10] The three systems are described as complex, interactive, and adaptive; working together as a collective whole and identified as the "supersystem." A key feature in the model is the bidirectional relationship between the stress systems and pain: chronic pain contributes to a dysregulated supersystem and a dysregulated supersystem can contribute to the onset of chronic pain.[10] Within this model, the contributing interactive influence of genetics, epigenetics, environment, and past experiences on pain conditions is emphasized. The proposed relationship between pain and the stress-system response align with the concepts of allostasis and AL (see Chapter 5). A developing array of neurobiological and biological evidence supports the pain and stress relationship.[18]

PAIN-RELATED NEUROBIOLOGICAL AND BIOLOGICAL FINDINGS

Neurobiological Findings

Neuroimaging studies have provided a valuable noninvasive tool to examine the neurological mechanisms underlying acute and chronic pain. While a detailed description of imaging studies is beyond the scope of the current chapter, several important techniques to examine pain-induced changes in the brain will be highlighted.

Acute Pain

Functional MRI (fMRI) is a technique that measures changes in brain activity as indicated by changes in blood oxygenation. Findings from fMRI studies indicate that a number of cortical and subcortical areas are responsive to nociceptive input including the areas involved in the sensory (primary and secondary somatosensory, S1 and S2 respectively) and affective (anterior cingulate cortex, ACC; prefrontal cortex, PFC) dimensions of pain.[19] Additional structures often associated with affective (including defensive) engagement (i.e., insula, amygdala) tend also to be activated by painful stimulation. Collectively, these areas coordinate the perceptual experience of pain and represent a "pain matrix." The magnitude of brain activity (i.e., blood flow) is often correlated with the perceived intensity of an experimental pain stimulus and is sensitive to modulation by psychological and cognitive factors like expectations and placebo.[20,21] Finally, while noxious stimulation can engage a number of brain regions within the "pain matrix," the pattern of brain activation is not exclusively pain-specific.

Chronic Pain

Imaging studies have provided insight into the dysfunction of the central nervous system as it relates to the development and persistence of chronic pain. Differences in functional and structural variables have been reported in a variety of chronic pain conditions. Functionally, pain patients exhibit augmented pain-evoked activation of brain areas compared to pain-free controls, which indicates not only increases in nociceptive drive (e.g., S1, S2) but augmented emotional responses (e.g., potentially mediated by the ACC), negative affective states (amygdala), and negative appraisals (insula) to pain in addition to a decrease in pain inhibitory drive (periaqueductal gray).[22]

Another method to evaluate neurobiological dysfunction in chronic pain is the functional (e.g., synchronous activity between regions, resting state) connectivity between brain regions considering that these regions rarely operate in isolation.[23] Chronic pain is typically associated with altered connectivity with individual variations in connectivity related to the frequency of pain and psychological factors like depression.[5,24] The mechanisms behind these observations are still being examined, but may arise from either preexisting vulnerability factors (e.g., biological, psychological, social), pain-induced changes in neuronal plasticity, or both, which lead to characteristics commonly observed in chronic pain including spontaneous pain, deficits in cognition/attention, and altered pain sensitivity due to an enhancement or reduction in descending pain facilitation or inhibition, respectively.[23]

In addition to alterations in functional connectivity, chronic pain patients are more likely to exhibit structural abnormalities, including reductions in gray matter volume in PFC, ACC, S1, and insula. Chronic pain seems to be associated with a greater age-related gray matter loss due to changes in neuronal structure and loss of neuronal or glia cells through pathological processes like neuroinflammation.[23] For example, studies have highlighted an abnormal immunological state within the central nervous system. Activation of glia (e.g., microglia and astroglia) is associated with an enhanced nociceptive processing, development of central sensitization, and pain chronification.[23] Ultimately, these studies support the conceptualization that changes in central nervous system structure and function are involved in the development and persistence of chronic pain.

Acute Stress

In the context of acute stress, brain imaging has provided insight to the interaction of acute stress and pain-related activation. Acute stress has been shown to affect areas of the brain related to affective dimensions of pain processing.[9,25] Alternatively, pain engages a number of "stress-related" areas like the hippocampus, which commonly shows augmented activation in chronic pain due to memories related to the pain experience,[22] and the hypothalamus, which is involved in the release of cortisol following a noxious stimulus.[25] The interaction of stress and pain is adaptive in acute situations in which stress inhibits the transmission of noxious information to the brain via descending inhibition or reorganization of brain regions and networks related to pain processing.[26] However, chronic stress resulting from persistent pain may produce negative effects that could contribute to reductions in functional connectivity, which would feedback to maintain abnormal pain processing.[25]

Biological

As indicated in the various theories and models of pain and in biological evidence, the experience of pain extends beyond the central nervous system. Complex, interactive, nonlinear, and bidirectional pathways extend between the central nervous system (brain) and other stress-regulatory systems.[27,28] Just as there is clear evidence of activation in the brain with acute pain and changes in the brain's structure and function with chronic pain, the acute experience of pain is measurable across biological indices and there are indications of greater cellular aging and biological dysregulation in individuals with chronic pain.

Acute Pain

Pain-related studies evaluating biological measures from the neuroendocrine and immune system are prevalent. Specific to the neuroendocrine system, engagement of the hypothalamic-pituitary-adrenal (HPA) axis serves to ready the body to deal with the potential or actual painful stimulus. One of the classical markers of HPA activation, cortisol, is often detected following an experimental pain stimulus. Cortisol is gradually released and peaks around 30 min following exposure to the noxious stimulus.[29,30] Increases in cytokines (e.g., interleukin-6, tumor necrosis factor-alpha) are evident following acute thermal pain.[29,30]

As part of the stress response to an actual or potential injury, the immune system can be activated to survey tissue for damage-associated molecular pattern or pathogen-associated molecular pattern molecules to limit infection and promote healing. In addition to peripheral effects, the release of inflammatory cytokines can stimulate areas in the brain to induce "sickness-behaviors" (e.g., negative mood, cognitive impairment, increased pain), which are commonly observed in chronic pain patients.

Activation of the immune system may vary based on the type of nociceptive stimulus. For example, cold pain was a potent stimulator of immune activity compared to heat, suggesting that noxious cold may be more stressful than heat.[29,30] Cold-pain activation of the immune system could be driven by biological (e.g., greater adrenergic tone) and psychological (e.g., greater catastrophizing) factors. Psychological stress can activate the stress response and since the experience of pain results from a perception of unpleasantness and potential threat, the overlap between pain-related stress and psychological stress is intricately intertwined.[10]

Chronic Pain

Dysregulated biological measures have been indicated in the onset of chronic pain and identified in chronic pain conditions.[31,32] Following a low-grade stress, individuals with chronic pain compared to those with episodic pain demonstrated dysregulated stress-system responses specific to blood pressure, blood flow, and heart rate.[33] Activation of brain areas is associated with the release of stress-related markers including cortisol.[25] Immune system dysregulation may also contribute to chronic pain via the development of neuronal hyper-excitability in the peripheral and central nervous system, which in turn, increases pain sensitivity.[10]

Another biological measure demonstrating potential relevance in chronic pain is telomere length.[34–36] Telomeres are DNA protein complexes that protect the ends of chromosomes. Telomeres decrease in length with replication, aging, and chronic stress, and mechanistically are influenced by inflammation and oxidative stress. Telomere length is considered a downstream, cumulative marker of AL.[28] Telomere length was found to be shorter in individuals with chronic pain reporting high perceived stress compared to those individuals without chronic pain and reporting low perceived stress. [35] Similarly, in a study of women with fibromyalgia, individuals with high pain and high depression had shorter telomeres than individuals with low pain and low depression. Additionally, associations between telomere length and pain-related gray matter volume and evoked pain were found.[36]

Perceived Stress

Perceived stress outside of the chronic pain experience can contribute to stress-system dysregulation. Perceived stress or psychological stress is the personal interpretation that experiences are extending beyond one's coping capacity.[27] Chronic elevated levels of perceived stress and negative emotions are associated with changes in stress-related biological measures, accelerated cellular aging, poor health outcomes, and increased disease risk.[27] Chronic perceived stress, as measured by a 12-month period across 10 life domains, was predictive of osteoarthritis onset in women.[37] High levels of perceived stress can contribute to maladaptive coping responses which in turn further contribute to stress-system dysregulation.[27,28] Importantly, decreasing levels of perceived stress and increasing adaptive coping can improve stress-system functioning and is addressed further in the next section.

FUTURE DIRECTIONS

Based on the current understanding of pain and its associations with stress-related system responses, further investigations are needed to (1) improve the characterization of chronic pain and associated perceived stress, (2) to better understand the associated biological burden, and (3) to identify targets for chronic pain prevention and treatment.

Quantification of Pain

The importance of considering frequency, intensity, and duration of the stressor is evident in AL research[38]; indicated when evaluating the relationship between stress and pain[10]; and demonstrated in pain-related neurobiological findings and pain-related morbidity and mortality.[3–5] Preliminary evidence suggests that low or intermittent pain experiences may evoke adaptive responses while persistent and prolonged chronic pain experiences are associated with altered functioning and dysregulation.[24,25,33] Frequently, chronic pain studies categorize chronic pain based on a definition of >3 or 6 months. However, quantification of the chronic pain experience should incorporate the frequency (e.g., intermittent or persistent), intensity, duration, and bodily extent of pain in order to improve our understanding of the biological burden of pain and inform our clinical interventions.

Composite of Biological Measures

The majority of studies thus far have targeted specific biological indices or a few biomarkers from a specific stress-related system. It is important to recognize biological measures do not work in isolation and are adaptive in nature.[10,39] Consistent with the AL conceptualization, dysregulation across multiple systems is more reflective of overall system functioning compared to elevated dysregulation in one system.[28,39] Second, it is not known if stress-related systems and associated biological measures are equally predictive across disease states or whether system dysregulation varies by condition.[18] In order to better understand the biological burden of chronic pain, investigations of a number of biological measures from stress-related systems (primary and secondary mediators) would be beneficial, a conceptual model is depicted in Figure 3.[28] The identification of a pain-related AL composite would have significant clinical and research utility. Future studies evaluating the correspondences of changes in the brain with other primary and secondary stress-regulatory systems would be informative.

Interventions

There is a developing body of evidence that environmental and behavioral factors might reduce the consequences of chronic pain on the brain.[5] Activities such as yoga, meditation, and social support may buffer the

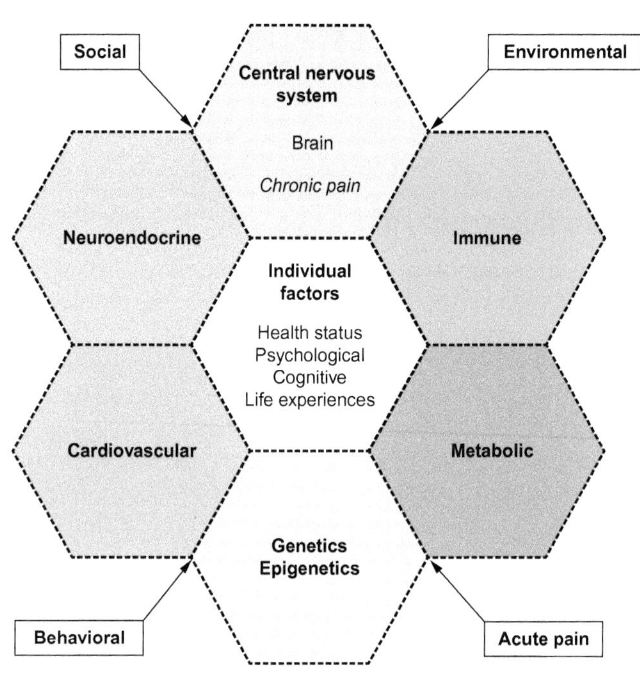

FIGURE 3 Pain, individual factors, and biological systems. Conceptual model illustrating the biological interface of the individual experience of pain. Individual factors (center) include but are not limited to internal (e.g., health status, psychological, cognitive, life experiences) and external (e.g., social, behavioral, environmental, and acute pain events). Surrounding shapes, the primary and secondary biological systems associated with the stress response.

negative neurobiological consequences of chronic pain.[5] Consistent with the AL model, reducing the number and/or intensity of stressors on the system may help reduce the biological burden and promote physiological balance thus reducing chronic pain suffering and improving quality of life.[27,28] There are a number of potential behavioral and environmental clinical targets that may help reduce the AL of chronic pain (Figure 4).

CONCLUSION

Pain is a complex, multidimensional, dynamic, and individualized experience sculpted by sensory and affective processes. Recognized as a central nervous system disorder, the perception and experience of chronic pain is influenced by psychological and stress-related factors. Altered and/or dysregulated neurobiological and biological functioning is consistently indicated in chronic pain. Understanding the biological translation of chronic pain and the associated psychosocial factors will be necessary in order to prevent, reduce, and eliminate the physiological burden and health-related consequences. Further investigations are needed to improve the characterization of chronic pain and associated psychosocial stress, to better understand the associated biological consequences, and to identify targets for chronic pain prevention and treatment.

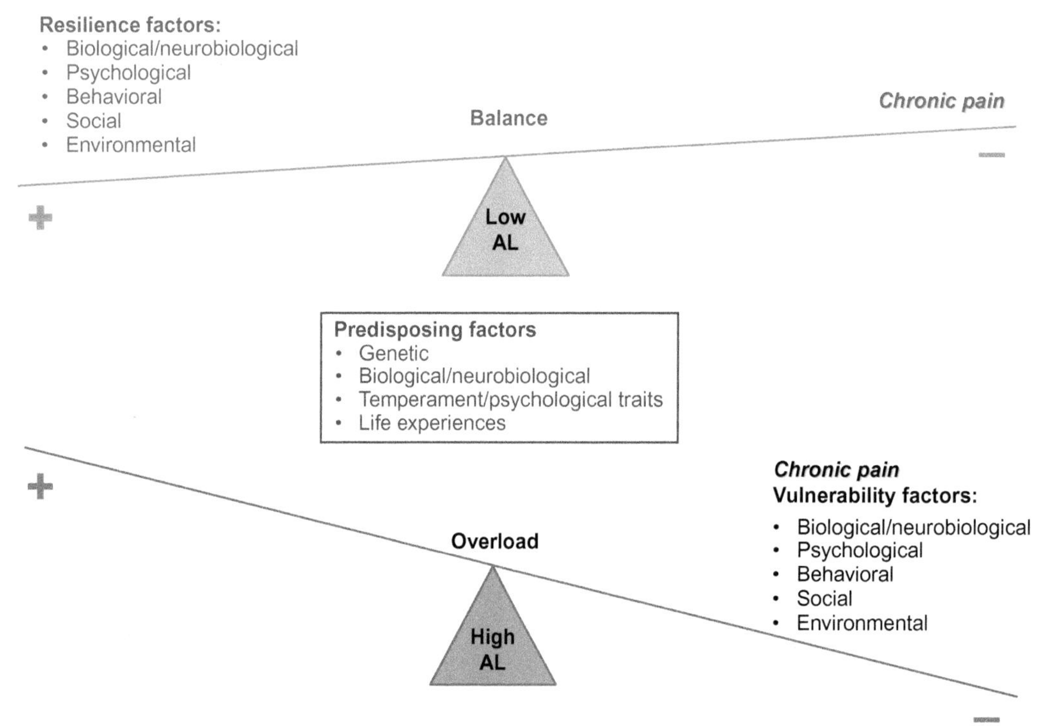

FIGURE 4 Chronic pain, allostatic load (AL), and potential clinical targets. There are a number of potential interventions that may help reduce the AL of chronic pain by reducing vulnerability factors (e.g., inadequate sleep, lack of exercise, obesity, smoking, poor quality diet, social isolation, maladaptive pain coping strategies) while enhancing resiliency factors (e.g., healthy diet, exercise, yoga, adaptive pain coping strategies, sound sleep, social support).

Acknowledgments

Many thanks to Roger B. Fillingim, PhD and Bruce S. McEwen, PhD for their mentoring in the areas of pain and stress/AL respectively and for their contributions to this chapter.

Drs. King and Sibille greatly appreciate the funding from the National Institutes of Health (NIH) that has contributed toward their training, development, and research efforts. Content in the chapter does not necessarily reflect the views of the NIH.

References

1. IOM Committee on Advancing Pain Research CaE. *Relieving Pain in America: A Blueprint for Transforming Prevention, Care, Education, and Research.* Washington, DC: National Academies Press; 2011.
2. Johannes CB, Kim LT, Zhou X, Johnston JA, Dworkin RH. The prevalence of chronic pain in United States adults: results of an internet-based survey. *J Pain.* 2010;11:1230–1239.
3. Torrance N, Elliott AM, Lee AJ, Smith BH. Severe chronic pain is associated with increased 10 year mortality. A cohort record linkage study. *Eur J Pain.* 2010;14:380–386.
4. McBeth J, Symmons DP, Silman AJ, et al. Musculoskeletal pain is associated with a long-term increased risk of cancer and cardiovascular-related mortality. *Rheumatology.* 2009;48:74–77.
5. Bushnell MC, Case LK, Ceko M, et al. Effect of environment on the long-term consequences of chronic pain. *Pain.* 2015;156:S42–S49.
6. IASP Task Force on Taxonomy. Pain terms, a current list with definitions and notes on usage. In: Merskey H, Bogduk N, eds. *Classification of Chronic Pain.* 2nd ed. Seattle, WA: IASP Press; 1994:209–214.
7. Fillingim RB. *Concise Encyclopedia of Pain Psychology.* Binghamton, NY: Haworth Press; 2005.
8. Purves D, Augustine GJ, Fitzpatrick D, et al. eds. Pain. In: *Neuroscience.* 4th ed. Sunderland, MA: Sinauer Associates, Inc.; 2008: 231–251.
9. Price DP. In: *Psychological Mechanisms of Pain and Analgesia.* Seattle: International Association for the Study of Pain; 1999: Progress in Pain Research and Management; vol. 15.
10. Chapman CR, Tuckett RP, Song CW. Pain and stress in a systems perspective: reciprocal neural, endocrine, and immune interactions. *J Pain.* 2008;9:122–145.
11. Bradley MM, Keil A, Lang PL. Orienting and emotional perception: facilitation, attenuation, and interference. *Front Psychol.* 2012;3:493.
12. Pessoa L, Adolphs R. Emotion processing and the amygdala: from a 'low road' to 'many roads' evaluating biological significance. *Nat Rev Neurosci.* 2010;11:773–783.
13. Miskovic V, Keil A. Escape from harm: linking affective vision and motor responses during active avoidance. *Soc Cogn Affect Neurosci.* 2014;9:1993–2000.
14. McTeague LM, Lang PJ. The anxiety spectrum and the reflex physiology of defense: from circumscribed fear to broad distress. *Depress Anxiety.* 2012;29:264–281.
15. Turk DC, Okifuji A. Psychological factors in chronic pain: evolution and revolution. *J Consult Clin Psychol.* 2002;70:678–690.
16. Sibille KT, Kindler LL, Glover TL, Staud R, Riley III JL, Fillingim RB. Affect balance style, experimental pain sensitivity, and pain-related responses. *Clin J Pain.* 2012;28:410–417.
17. Melzack R. Pain and the neuromatrix in the brain. *J Dent Educ.* 2001;65:1378–1382.
18. Sibille KT, Riley III JL, McEwen B. Authors build an important foundation for further research. *J Pain.* 2012;13:1269–1270.
19. Bushnell MC, Apkarian AV. Representation of pain in the brain. In: McMahon SB, Koltzenburg M (Eds.) Wall and Melzack's Textbook of Pain, Philadelphia, PA: Elsevier Churchill Livingstone; 2006:107–124.
20. Coghill RC, Sang CN, Maisog JM, Iadarola MJ. Pain intensity processing within the human brain: a bilateral, distributed mechanism. *J Neurophysiol.* 1999;82:1934–1943.
21. Zubieta JK, Stohler CS. Neurobiological mechanisms of placebo responses. *Ann N Y Acad Sci.* 2009;1156:198–210.
22. Borsook D, Becerra L, Hargreaves R. Biomarkers for chronic pain and analgesia. Part 2: How, where, and what to look for using functional imaging. *Discov Med.* 2011;11:209–219.
23. Davis KD, Moayedi M. Central mechanisms of pain revealed through functional and structural MRI. *J Neuroimmune Pharmacol.* 2013;8:518–534.
24. Maleki N, Becerra L, Brawn J, McEwen B, Burstein R, Borsook D. Common hippocampal structure and functional changes in migraine. *Brain Struct Funct.* 2013;218:903–912.
25. Vachon-Presseau E, Roy M, Martel M-O, et al. The stress model of chronic pain: evidence from the basal cortisol and hippocampal structure and function in humans. *Brain.* 2013;136:815–827.
26. Vachon-Presseau E, Martel MO, Roy M, et al. Acute stress contributes to individual differences in pain and pain-related brain activity in healthy and chronic pain patients. *J Neurosci.* 2013;33:6826–6833.
27. Glaser R, Kiecolt-Glaser J. Stress-induced immune dysfunction: implications for health. *Nat Rev Immunol.* 2005;5:243–251.
28. McEwen BS. Biomarkers for assessing population and individual health and disease related to stress and adaptation. *Metabolism.* 2015;64:S2–S10.
29. Cruz-Almeida Y, King CD, Wallet SM, Riley 3rd JL. Immune biomarker response depends on choice of experimental pain stimulus in healthy adults: a preliminary study. *Pain Res Treat.* 2012;2012:538739.
30. Goodin BR, Quinn NB, Kronfli T, et al. Experimental pain ratings and reactivity of cortisol and soluble tumor necrosis factor-alpha receptor II following a trial of hypnosis: results of a randomized controlled pilot study. *Pain Med.* 2012;13:29–44.
31. Hasselhorn HM, Theorell T, Vingård E, MUSIC-Norrtäje Study Group. Endocrine and immunological parameters indicative of 6-month prognosis after the onset of low back pain or neck/shoulder pain. *Spine.* 2001;26:D1–D6.
32. Generaal E, Vogelzangs N, Macfarlane G, et al. Reduced hypothalamic-pituitary-adrenal axis activity in chronic multi-site musculoskeletal pain: partly masked by depressive and anxiety disorders. *BMC Musculoskelet Disord.* 2014;15:227.
33. Leistad RB, Nilsen KB, Stovner LJ, Westgaard RH, Rø M, Sand T. Similarities in stress physiology among patients with chronic pain and headache disorders: evidence for a common pathophysiological mechanism? *J Headache Pain.* 2008;9:165–175.
34. Sibille KT, Witek-Janusek L, Mathews HL, Fillingim RB. Telomeres and epigenetics: potential relevance to chronic pain. *Pain.* 2012;153:1789–1793.
35. Sibille KT, Langaee T, Burkley B, et al. Chronic pain, perceived stress, and cellular aging: an exploratory study. *Mol Pain.* 2012;8:12. http://dx.doi.org/10.1186/1744-8069-8-12.
36. Hassett A, Epel E, Clauw DJ, et al. Pain is associated with short leukocyte telomere length in women with fibromyalgia. *J Pain.* 2012;13:959–969.
37. Harris ML, Loxton D, Sibbritt DW, Byles JE. The influence of perceived stress on the onset of arthritis in women: findings from the Australian Longitudinal Study on women's health. *Ann Behav Med.* 2013;46:9–18.
38. Karlamangla AS, Singer BH, McEwen BS, Rowe JW, Seeman TE. Allostatic load as a predictor of functional decline. MacArthur studies on successful aging. *J Clin Epidemiol.* 2002;55:696–710.
39. Karlamangla A, Tinetti M, Guralnik J, Studenski S, Wetle T, Reuben D. Comorbidity in older adults: nosology of impairment, diseases, and conditions. *J Gerontol A Biol Sci Med Sci.* 2007;62:296–300.

CHAPTER

53

Industrialized Societies

J. Siegrist

University of Duesseldorf, Duesseldorf, Germany

OUTLINE

Abstract

The industrial and political revolution of the late eighteenth century initiated a process of profound economic and social transformation of European countries and, successively, of other parts of the world. Demographic aging and the transition from infectious diseases to chronic diseases as major drivers of population health were some of the relevant long-term outcomes. The spread of chronic diseases was closely linked to the expansion of health-adverse lifestyles resulting from societal modernization. Additional large-scale societal changes became apparent that go along with chronic stressful experience, in particular with the processes of social disintegration and social isolation. Moreover, far-reaching changes in the nature of work and employment occurred, with significant improvements for working people, but equally so with new challenges of stressful work environments. While industrialized societies benefit from substantial human progress, substantial threats remain to be addressed, not the least, man-made disasters and persisting social inequalities in health.

PROCESS OF INDUSTRIALIZATION AND EPIDEMIOLOGICAL TRANSITION

The industrial revolution started in Great Britain during the second half of the eighteenth century and has subsequently spread across a number of economically advanced countries. This revolution, together with the political revolution initiated in France, must be considered one of the most radical changes in human history. Within a very short time period, fundamental ways of living and working were transformed from a basically agricultural society to a society that is driven increasingly by technology and market economy. With the invention of engines and machines, productivity was increased in an unprecedented way.

A large part of the workforce, employed formerly as peasants and craftsmen, moved to urban areas to engage in unskilled or semi-skilled work. The rapid growth of cities and the formation of an industrial workforce contributed to two subsequent, most significant demographic trends: (1) a take-off period of population growth with declining infant mortality rates and (2) a decline in fertility with slow population growth in combination with increased life expectancy. At the same time, the social institutions of marriage and family underwent marked transformations, and a new division of labor between partners started to develop.

After severe poverty and economic exploitation in the early stages of industrial capitalism, economic progress, and the development of a welfare state were experienced by a growing proportion of industrial populations. This progress included the availability of healthier food, increased opportunities of energy consumption, better housing, transport, education, and general hygiene. In terms of population health, the process of industrialization was associated with a marked increase in life

Stress: Concepts, Cognition, Emotion, and Behavior
http://dx.doi.org/10.1016/B978-0-12-800951-2.00054-6

expectancy and a change in the pattern of prevailing diseases. Overall, 2-3 years are added to life expectancy at birth with each decade passing in advanced societies. In the first period, this increase in life expectancy was mainly due to a substantial decline in infant mortality. A dramatic reduction in mortality from infectious diseases played an important role in this process. In a more recent second period, the observed increase in life expectancy is additionally attributed to an extension of old age, where a growing proportion of the population survives into the seventh and eighth decade of the lifespan. In advanced industrial societies, the typical pattern of leading causes of disability and death is characterized by a high prevalence of chronic degenerative physical and mental disorders, in particular, coronary heart disease, cancer, stroke, type 2 diabetes, accidents, depression, and dementia. This change from infectious to chronic multifactorial diseases is referred to as the epidemiological transition.

KEY POINTS

- Industrialized societies represent one of the most radical economic and social transformations in history, initiated by the industrial and political revolutions of the late eighteenth and early nineteenth century in Europe.
- Economic growth, technological progress, improved living and working conditions, societal modernization, and welfare state development are key features.
- A marked increase in life expectancy contributes to population aging, where a growing burden of chronic, noninfectious diseases is observed.
- Health-adverse lifestyles triggered by technological and economic progress contribute to the spread of the "diseases of affluence" which progressively became more prevalent among socially disadvantaged groups.
- Social inequalities in population health persist in modern societies, aggravated by higher exposure of poorer people to health-adverse socioenvironmental stressors; these stressors include forced migration, social disintegration, and stressful conditions of work and employment.
- Substantial gains of modernization in terms of quality of life and individual freedom are challenged by man-made disasters of environmental deterioration and collective violence, thus calling for sustained efforts to preserve a sustainable, humane, and healthier future.

While the development of modern medicine and related improvements of healthcare are a direct outflow of the broader processes of industrialization and modernization, it has been demonstrated that the impact of medicine on this major change in the pattern of morbidity and mortality has been limited. In fact, the bulk of the decline in mortality from infectious diseases occurred in a period before medicine had developed effective forms of treatment and prevention, thus documenting the impact of sustainable public health measures on population health.[1] However, during the twentieth century, modern medicine has made very significant advances in treating and preventing a broad spectrum of diseases and in improving quality of life. Despite this progress, new challenges of maintaining and improving population health emerged, such as urbanization and environmental pollution, political and economic crises, and significant transformations of people's everyday life in major societal institutions. The next section discusses some of these challenges in more detail.

STRESSFUL SOCIAL ENVIRONMENTS AND HEALTH

Stressful Experience and Health-Related Lifestyles

Many of the aforementioned degenerative diseases are considered diseases of affluence because their development is strongly influenced by a more comfortable lifestyle. For instance, with technological advances and with the expansion of commercially available food and drugs, several behavioral health risks were programmed (e.g., access to a car reduced physical activity significantly), and it increased the number of injuries and deaths from accidents. Frequent consumption of fat and meat, at first available to the wealthier groups, contributed to elevated blood lipids, overweight, and associated health risks. Spending money on drugs such as cigarettes or alcohol increased the risk of cancer and heart disease, among others. Consequently, during the first stage of the epidemiological transition, these behaviorally induced risks associated with a wealthier lifestyle were more prevalent among socially and economically privileged groups. At a later stage of the process of industrialization and modernization, however, this social pattern changed: diseases of affluence increasingly became the diseases of the poor. In contemporary societies, this is clearly the case in almost all northern, western, and central European countries as well as in North America.[2,3] The same pattern may be experienced in the near future in countries that are currently exposed to a process of rapid industrialization. Hence, substantial social inequalities in health are observed, mainly due to higher rates of the former affluent diseases among lower socioeconomic groups.[4]

To explain this shift in the social distribution of a broad range of degenerative diseases and their behavioral risk factors (e.g., smoking, alcohol consumption, overweight, hypertension, metabolic disorders, cardiovascular diseases, some manifestations of cancer, AIDS), the concept of health-related lifestyle needs to be described more precisely.

Health-related lifestyles are defined as collective patterns of health-promoting or health-damaging behavior based on routinized choices people make about food, exercise, hygiene, safety, and related matters. In many cases, these choices form a coherent pattern that is structured by specific needs, attitudes, and social norms and by the constraints of a group's socioeconomic condition. Once established among a sociocultural group, such behavioral patterns may be transferred from one generation to the next through socialization processes, or they may be adopted by model learning from peer groups during adolescence. The latter is often the case with respect to health-damaging behaviors (smoking, alcohol and drug consumption, unsafe sex practices, etc.). As these behaviors are experienced as relief and reward in stressful psychosocial circumstances, they may be reinforced easily and, thus, trigger addiction later in life.[5]

Alternatively, with higher levels of education and knowledge, and with the availability of alternative means of coping with stressful psychosocial circumstances, health-damaging lifestyles may be abandoned more easily. This is most probably what happened to the wealthier populations in advanced societies in the recent past and what contributed, and still contributes, to the widening gap in life expectancy between socioeconomic groups.[6]

HEALTHY AGING IN INDUSTRIALIZED SOCIETIES?

The continuous extension of human life expectancy is one of the major long-term outcomes of industrialized modern societies. Many older men and women reach their "third age" in good health and continue to be free from severe disability for often up to 20 more years, thus postponing senescence.[7] Yet, a limited number of widely prevalent chronic disorders account for a main part of the burden of disease in aging societies. Their distribution follows a social gradient across society, leaving those in lower socioeconomic positions at higher risk. While genetic and early life influences cannot be neglected, health-adverse lifestyles play a major role in the development of these diseases. These lifestyles are reinforced by socioenvironmental adversities, which in addition, expose people to chronic stressors in terms of social disorganization and unhealthy work and employment conditions.

To the extent that these factors can be modified, investments into the prevention of health-adverse lifestyles and into reducing socioenvironmental adversities might result in improvements of healthy aging and in a compression of morbidity in older ages. Such investments are promising given the fact that continued productivity of elderly people enriches a society's economic and social capital.

To date, there are still powerful constraints toward maintaining health-damaging lifestyles among less privileged social groups. A major constraint concerns the exposure to adverse social environments. A large number of epidemiologic studies document associations between the amount of exposure to these unfavorable circumstances and the prevalence of unhealthy behavior. For instance, with respect to cigarette smoking, poor housing, low education, unemployment, social isolation, and exposure to stressful work and home environments were identified as social determinants.[8] When health-related lifestyles are transferred from one generation to the next, more subtle processes occur. It is well known that practices of primary socialization are strongly patterned according to the parents' socioeconomic status. In particular, children born in families with a low educational level face more difficulties in deferring their gratifications and in pursuing long-lasting goals in future life. Their sense of self-efficacy and self-esteem is low, and they are more likely to experience their environment as uncontrollable and fateful. Thus, when facing frustrations, they may be less capable of coping in an active, effective way. Rather, they seek relief in health-adverse modes of consumption, including drug consumption and unhealthy food.

This knowledge, derived from sociological and psychological research, has direct implications for health promotion activities and the design of preventive programs. It is evident that comprehensive approaches tackling structural and individual conditions are needed to change health lifestyles in a successful way.[9] Unfortunately, influential economic interests operate in almost all advanced societies to counteract these measures, most evidently the tobacco and alcohol industry. Yet, recent reports indicate that impressive progress in smoking cessation has been achieved.[10]

Impact of Social Disintegration on Health

Industrialized societies are characterized by a high level of mobility. A substantial part of this mobility is due to migration. Economic constraints and wars are the major causes of large migration waves that have shaped the life of many generations throughout the processes of industrialization and modernization. Social mobility is a second component of these secular trends.

As mentioned briefly, the onset of early industrialization in Europe and the United States has been facilitated by a broader sociocultural development of modernization that promoted formal rationality and individualism as the dominant modes of thinking and behaving in western societies. Formal rationality is the purposeful calculation of the most efficient means to achieve a goal. Rational choices are made to maximize individual benefit. Importantly, successful achievement in paid work and in other domains of productivity became the leading criterion of social status and social recognition. Emphasizing individual achievement and related motivations, strengthening individual rights and widening individual life chances acted as substantial drivers of modern, liberal democratic societies. These positive consequences have to be acknowledged and recognized without reservation. However, continued social mobility and the expansion of individual freedom contributed to some less favorable developments as well. Traditional ways of maintaining group solidarity, social norms, and obligations were weakened and social disintegration or even social anomie—a lack of rules and orientations guiding interpersonal exchange—became more prevalent. More recent epidemiological research documented adverse health effects resulting from social disintegration, lack of social support, and exposure to social isolation or even social discrimination.[11,12] Among the many documents of elevated health risks among people with low social integration, a meta-analysis of 148 epidemiologic studies exploring an increased mortality in association with social isolation deserves special attention. Depending on the available measure of social isolation mortality risks were elevated by 50-90% compared to those of socially more integrated people.[13]

At the aggregate level, social disintegration has been linked to widening income inequalities and their detrimental effects on social cohesion which in turn affect a population's health and wellbeing.[14] In countries with large socioeconomic inequalities, a growing proportion of the population is experiencing injustice and relative deprivation. These experiences in the long run undermine a society's stable functioning and the legitimacy of its main institutions.[15]

Stress at Work and Health

One of the most pervasive changes resulting from the process of industrialization concerns the nature of human work and employment. For a large part of the workforce, work and home became two different places and two different types of social organization. As mentioned, in all economically advanced societies, work and occupation in adult life were increasingly accorded primacy over and above other domains of life. At least two reasons account for this fact. First, having a job is a principal prerequisite for continuous income and, thus, for independence from traditional support systems (family, community welfare, etc.). Increasingly, the level of income determines a wide range of life chances. Second, training for a job and achievement of occupational status are important goals of socialization. It is through education, job training, and status acquisition that personal growth and development are realized, that a core social identity outside the family is acquired, and that goal-directed activity in human life is shaped.

Yet, the types and quality of work and employment underwent significant changes and varied strongly between and within countries during the process of industrialization. Above all, today industrial mass production no longer dominates the labor market. This is due, in part, to technological progress, and in part to a growing number of jobs that are available in the service sector and in sectors dealing with information and data management. Person-based service occupations and professions are growing rapidly, and these latter jobs may be less susceptible to downsizing and job loss than more traditional employment sectors. Major changes also occurred in the composition of the workforce and in the provision of employment contracts. Concerning workforce composition, the proportion of women working part-time or full-time has steadily grown in the recent past, and with increasing immigration, working populations are becoming more diverse. Demographic aging additionally contributes to these changes, as well as collective improvements in the level of education and qualifications. With regard to employment contracts, flexible work arrangements are becoming more common, but growing flexibility is only partly based on voluntary decision. Many working people are exposed to temporary contracts and job insecurity. With the advent of economic globalization, mergers, downsizing, and outsourcing contribute to a rise in job instability.[16] Overall, a segmentation of the labor market is apparent, where one segment is defined by precarious, unstable, low-paid jobs and where a distinct second segment consists of well-trained, well-paid employees in rather safe positions.

Variations in the quality of work and employment result in different susceptibility to occupational health hazards. Accordingly, workers with low socioeconomic status are at higher risk of suffering from occupational diseases, accidents, and impaired health due to shift work, long work hours, and stressful psychosocial work environments, compared to those in higher socioeconomic positions.[17] In contrast to a popular belief, health-adverse psychosocial stress at work, defined by either low control or high effort and low reward, is more frequent among lower skilled workers and, thus, contributes to social inequalities in health.[18] Therefore, more investments in health-promoting working conditions are required, and these investments should be prioritized according to need, providing stronger support to more deprived groups.

CONCLUDING REMARKS

In discussing some links between stress and health in the context of industrialized societies, three topics were highlighted. First, the epidemiologic transition resulting in a high prevalence of chronic diseases was mentioned. To some extent, this burden of disease limits the substantial progress in human welfare achieved through the process of industrialization and modernization. A health-adverse lifestyle was found to contribute largely to this burden. Second, large-scale societal changes became apparent that go along with chronic stressful experience, in particular with processes of social disintegration and isolation. Third, a wide prevalence of stressful experience in modern occupational life was pointed out, with potential impact on the health and productivity of working populations.

Despite continuous economic and societal progress during the past two centuries modern industrialized countries are still facing substantial social inequalities in the health of their populations. Together with man-made disasters of environmental deterioration and of collective violence these inequalities must be considered global challenges of high priority.[4]

Finally, in view of rapid and profound economic and technological transformations that are under way one must ask whether current and near-future societies can still be labeled "industrialized," or whether new terms need to be applied to delineate their core features more appropriately. Several such new labels were proposed (e.g., "post-industrial societies," "information societies," "globalized societies.") Yet, it is almost certain that the challenges mentioned will prevail and that sustained efforts are required to preserve a sustainable humane and healthier future.

References

1. McKeown T. *The Modern Rise of Population.* New York, NY: Academic Press; 1976.
2. Evans RG, Barer ML, Marmor TR, eds. *Why Are Some People Healthy and Others Not?* New York, NY: Aldine de Gruyter; 1994.
3. WHO. *Review of social determinants and the health divide in the WHO European Region: final report.* Copenhagen; www.euro.who.int.
4. WHO. *Closing the gap in a generation: health equity through action on the social determinants of health.* Final report of the Commission on Social Determinants of Health. Geneva; 2008. www.who.int. Accessed 04.03.14.
5. Cockerham WC. Healthy lifestyles: bringing structure back. In: Cockerham WC, ed. *The New Blackwell Companion to Medical Sociology.* Chichester: Wiley-Blackwell; 2010.
6. Marmot M. *The Status Syndrome: How Your Social Standing Affects Our Health and Longevity.* 1st ed. New York, NY: Holt; 2004.
7. Vaupel JW. Biodemography of human aging. *Nature.* 2010;464:536–542.
8. Jarvis MJ, Wardle J. Social patterning of individual health behaviours: the case of cigarette smoking. In: Marmot M, Wilkinson R, eds. *Social Determinants of Health.* Oxford: Oxford University Press; 2006:224–237.
9. Puska P. *The North Karelia Project.* Helsinki: National Institute for Health and Welfare; 2010.
10. Clancy L. Reducing lung cancer and other tobacco-related cancers in Europe: smoking cessation is the key. *Oncologist.* 2014;19(1):16–20. http://dx.doi.org/10.1634/theoncologist.2013-0085.
11. Berkman LF, Krishna A. Social network epidemiology. In: Berkman LF, Kawachi I, eds. *Social Epidemiology.* Oxford: Oxford University Press; 2014:234–289.
12. Krieger N. Discrimination and health inequalities. In: Berkman LF, Kawachi I, eds. *Social Epidemiology.* Oxford: Oxford University Press; 2014:63–125.
13. Holt-Lunstad J, Smith TB, Layton JB. Social relationships and mortality risk: a meta-analytic review. *PLoS Med.* 2010;7(7):e1000316. http://dx.doi.org/10.1371/journal.pmed.1000316.
14. Kawachi I, Subramanian SV. Income inequality. In: Berkman LF, Kawachi I, eds. *Social Epidemiology.* Oxford: Oxford University Press; 2014:126–152.
15. Wilkinson RG, Pickett K. *The Spirit Level: Why Greater Equality Makes Societies Stronger.* New York, NY: Bloomsbury Press; 2011.
16. Cooper C, Pandey A, Quick J, eds. *Downsizing: Is Less Still More?* Cambridge: Cambridge University Press; 2012.
17. Schnall PL, Dobson M, Rosskam E, eds. *Unhealthy Work: Causes, Consequences, Cures.* Amityville, NY: Baywood Publishing Company; 2009.
18. Wahrendorf M, Dragano N, Siegrist J. Social position, work stress, and retirement intentions: a study with older employees from 11 European countries. *Eur Sociol Rev.* 2013;29:792–802. http://dx.doi.org/10.1093/esr/jcs058.

CHAPTER

54

Indigenous Societies

W.W. Dressler

The University of Alabama, Tuscaloosa, AL, USA

OUTLINE

Abstract

Indigenous societies exist outside the major influences of the global economic system, or within a larger society but separate in a social and cultural sense. In any society there is variation in the degree to which individuals are able to achieve the ideals of that society, and difficulty in doing so can be stressful. This can be manifest in the form of culture-bound syndromes, or local idioms of distress. Major sources of stress involve the processes of acculturation and modernization, or the degree to which traditional societies' values, beliefs, and, especially, economic systems are impacted. At the same time, there are traditional forms of stress resistance that are configured specifically within traditional systems of social relationships that serve as buffers against the stressful effects of modernization and migration. The study of stress within indigenous and changing societies can help to illuminate fundamental processes in health.

INTRODUCTION

An indigenous or traditional society is characterized as native to a specific region, with a distinctive language and way of life. Also, the way of life of the people is peripheral to global capitalist market systems. This does not mean that such a society is unaffected by global market systems; in fact, one major source of stress in indigenous societies is the impact of economic change emanating from those larger systems. Nor does this mean that an indigenous society cannot be embedded within a modern, industrial society (e.g., Native Americans). Rather, in an indigenous society, culture, and related systems of social organization are structured more in terms of local context and local systems of meaning, and less in terms of the middle-class values of industrial society. Understanding stress and its effects in indigenous societies requires an examination both of social arrangements that generate stresses within the indigenous social structure, and the way in which traditional culture and social structure interact with outside influences to generate stresses.

STRESS IN UNACCULTURATED INDIGENOUS SOCIETIES

Anthropologists conventionally describe the relative degree of external influence on a society or community along a continuum of "acculturation." A society that is relatively unacculturated or traditional has been minimally influenced by processes of modernization. Usually this means that households tend to practice a mix of economic pursuits for subsistence (i.e., raising food directly for consumption within the household) and for exchange in local markets. With respect to material lifestyles, although there is some access to imported consumer goods, these often are primarily related to subsistence activities (e.g., agricultural implements, outboard motors). Most goods consumed by a household are produced locally. What wage-labor exists is usually within the community, and there is little formal education. Social relationships tend to be dominated by kinship. Systems of kinship can range from large groups formed around descent from a common ancestor (or unilineal descent

Stress: Concepts, Cognition, Emotion, and Behavior
http://dx.doi.org/10.1016/B978-0-12-800951-2.00055-8

groups) to somewhat more loosely structured kindreds that are like large ego-centered social networks. Finally, belief systems reflect local meanings and understandings, even when there have been modernizing influences (e.g., missionaries).

KEY POINTS

- Introduction
  - Societies that exist outside the major influence of global economic systems
  - Emphasis on mixed subsistence-wage labor economies; social relationships defined by extended kinship systems; belief systems relatively unaffected by major world ideologies
  - Major sources of morbidity and mortality include infectious and parasitic disease and trauma
- Stress in unacculturated indigenous societies
  - Social stresses generated by variation in cultural consonance, or the degree to which individuals are able to achieve cultural ideals
  - Social stress often manifests in the form of culture-bound syndromes, or local idioms of distress
- Stress and acculturation in indigenous societies
  - Acculturation refers to the impact of one societies' beliefs and values on those of another society
  - Modernization or economic development of an indigenous society includes a shift in subsistence and economic pursuits as well as changes in beliefs and values
  - Increasing modernization and acculturation in a community is associated with an increase in rates of obesity, hypertension, cardiovascular disease, and psychiatric disorders
  - Stressors—especially status incongruence—are generated as a part of the modernization process, as individuals aspire to new lifestyles but in the context of slowly expanding economic resources
  - These modernization stressors tend to be consistent across different societies
  - Social factors buffering the impact of stressors tend to be more culturally specific, embedded in traditional social relationships
- Stress and migration
  - Migration from traditional societies to urban areas is a consistent concomitant of the modernization process
  - Migrants face a world in which their traditional cultural models do not prepare them for adapting to the new world they face
  - This incongruence in cultural models can lead to stress and poor health outcomes
- Conclusion
  - The study of stress in indigenous societies, both in their traditional settings and under conditions of culture change, provides an important perspective on the stress process.

Patterns of morbidity and mortality within traditional societies provide one clue to patterns of stress in those societies. Generally speaking, rates of high blood pressure, coronary artery disease, stroke, and cancer tend to be very low.[1] In traditional societies, patterns of morbidity and mortality tend to be dominated by infectious and parasitic disease, especially in childhood, and by trauma in adulthood.[2] Although any generalizations must be tempered by reference to local ecological conditions, in many traditional societies, life expectancy beyond 5 years of age is comparable to life expectancy in industrial societies, and in the aged in these societies there is little evidence of the kind of pathologies (e.g., atherosclerosis) associated with aging in industrial societies.

What has been most illuminating in the study of stress in unacculturated societies is the study of "culture-bound syndromes." Culture-bound syndromes are local idioms of distress.[3] They can be thought of as culturally appropriate ways of experiencing and expressing distress arising from stressful social relationships. Some attempts have been made to equate the culture-bound syndromes with Western psychiatric diagnoses, but recent evidence indicates that there is not a direct correspondence between culture-bound syndromes and biomedical psychiatric diagnoses; it is probably more useful to think of culture-bound syndromes and Western psychiatric diagnoses as comorbid.[4]

A classic example of a culture-bound syndrome is *susto* in Latino societies of Central and South America. The individual suffering from *susto* experiences a loss of energy, difficulty in maintaining customary activities, frequent spells of crying, and diffuse somatic symptoms, such as loss of appetite and sleep disturbance. As the name in Spanish implies, *susto* is attributed to a sudden fright (e.g., seeing a snake) at which time the soul of the individual leaves the body and wanders freely.[5]

Research has shown that the distribution of *susto* is socially patterned. The prevalence is higher in females, tends to increase with age, and tends to be higher in relatively poorer communities. Furthermore, the greatest risk of *susto* has been found among persons experiencing difficulty in enacting common social role expectations. This usually arises from a lack of specific kinds of social resources (e.g., not having a large kinship network) that can be called on in carrying out expected role obligations (such as contributions to community work groups).

Similar findings have been obtained in research on other culture-bound syndromes.[6,7] Culture-bound syndromes occur when individuals are low in cultural consonance; that is, they are unable to approximate in their own behaviors the prototypes for behaviors that are encoded in widely shared cultural models.[8] The social production of stress as it manifests in a culture-bound syndrome has profound longer-term effects; persons experiencing *susto*, for example, have an increased risk of early mortality.[5]

The diagnosis of a culture-bound syndrome makes the experience of social stress meaningful and intelligible both to the person suffering the stress and to his or her social network. Furthermore, in many societies, there are cultural practices, including healing rituals and participation in religious organizations that deal directly with the syndrome and the underlying difficulties in social relationships. The aim is to mend the tear in the fabric of social relationships, and hence end the individual's suffering.

The existence of these beliefs and practices that are helpful in ameliorating cultural stresses may in part account for patterns of morbidity and mortality in indigenous societies. As noted above, the diseases conventionally associated with stress in industrial societies are relatively less important in indigenous societies. Also, in most cases, an increase of blood pressure with age is not observed in indigenous societies. Other patterns of disease distribution that are taken for granted in industrial societies are not observed in traditional societies. One of the more striking of these is the association of blood pressure with African-descent ethnicity. While it is assumed that persons of African-descent have higher blood pressures, in fact this is true primarily for African-descent persons in the Western Hemisphere, and more specifically in societies in which Africans had formerly been enslaved. So, for example, when communities of African-descent are compared, communities in Africa have the lowest average blood pressures; communities in the West Indies have intermediate average blood pressures; and, African American and African Brazilian communities have the highest average blood pressures.[9]

These patterns suggest two things. First, there may be a relatively higher level of social integration in indigenous societies, along with practices that help to moderate the impact of social stressors that account for the lower prevalence of conditions and diseases associated with stress in industrial societies. Second, the process of social change leading to the modern industrial state may itself generate profound social stresses that contribute to the distinctive pattern of morbidity and mortality in those societies.

At the same time, it is important not to romanticize life in indigenous societies, in the sense of overemphasizing social integration and cohesion, because stresses will be generated within any system of social relationships. What is important to specify across different cultural contexts is the process by which social stresses are generated, and the ability of indigenous support systems to deal with those stresses.

Promising results in this regard are emerging from research on hormones and neurotransmitters associated with the stress process. Newer techniques of data collection under difficult field conditions, along with techniques for the analysis of those data, are beginning to show how variation in social behavior within traditional societies is associated with inter- and intraindividual variation in stress hormones, which in turn is associated with acute illness. For example, research in a peasant village in the West Indies has shown that men who are perceived by their peers to emulate the ideals of manhood in this community have lower circulating levels of cortisol. Similarly, children growing up in families that are closer to the cultural ideal of the family have lower circulating cortisol levels and experience fewer acute illnesses.[10,11] This research, coupled with research on local idioms of distress, suggests that stresses in unacculturated societies are deeply embedded in the system of social relationships that organizes everyday life.

STRESS AND ACCULTURATION IN INDIGENOUS SOCIETIES

Societies that are undergoing acculturation are those that are being influenced by other social and cultural systems. In the study of stress and disease, the effect of modern industrial societies on local sociocultural systems has been of particular interest. There is considerable imbalance in this type of acculturation, because of the unequal power and influence that modern industrial states exert on traditional societies. The terms "modernization" or "development" have been used to describe this kind of influence.

Modernization in traditional societies was initiated by colonial expansion and has been particularly prominent since the Second World War and related processes of globalization. This influence has not been inadvertent. The aim has been to take advantage of both physical and social resources in developing societies.

These changes have had large effects on local social systems, including: a transition from subsistence occupations to wage-labor occupations; the replacement of indigenous languages by European languages; increased urbanization; increased emphasis on formal education; decreased emphasis on traditional social relationships, especially kinship; and, substantial changes in indigenous belief systems. Everyday life can change at a rapid pace in modernizing contexts, the result being a stressful lack of consonance between traditional culture and the demands of modern life.

Specific and general aspects of this modernization process have been found to be associated with increasing rates of chronic diseases. Urbanization has been found to be associated with increasing rates of hypertension, independent of changes in diet and physical activity. Some investigators have used summary measures of acculturation both for individuals and for communities. When communities are ranked along a continuum of traditional, intermediate, and modern (depending on aggregate characteristics of the population by the variables noted above), rates of hypertension, obesity, diabetes, and coronary artery disease consistently increase in communities with higher levels of acculturation or modernization. Also, daily circulating levels of hormones such as cortisol appear to increase in association with acculturative stress.[1,12]

The results have been somewhat less consistent using measures of acculturation operationalized at the level of the individual. That is, an individual's adoption of new economic pursuits, shifting patterns of social relationships, or changing beliefs is not as consistently related to health outcomes. The pattern can still be observed in many studies, but the strength and the replicability of the associations are not as great. This has led some researchers to speculate that the linear model of stress and acculturation is not specified well-enough to describe the process at the individual level.[1]

Because the general model of stress and acculturation has not worked very well at the level of the individual, researchers have adapted the stress model as it has been developed for European and American populations and applied it to communities experiencing change and development. A major challenge in this research has been to identify factors that generalize across different cultural contexts and to distinguish those from factors that are culturally specific. There is emerging evidence to suggest that at least one set of social stressors generalizes across modernizing societies, primarily because these stressors are generated by the modernization process itself. In most societies undergoing modernization, there is an increased availability of Western consumer goods. Frequently the ownership of these goods becomes highly valued as symbols of status or prestige, often supplanting traditional indicators of higher status. This by itself contributes to the climate of change in a modernizing community. In addition, however, aspirations for the lifestyles of the Western middle class can quickly outstrip the ability of a developing economy to provide the kinds of jobs and salaries necessary to maintain such a lifestyle. Therefore, a kind of "status incongruence" can occur, in which the desire to attain and maintain a Western middle-class lifestyle exceeds an individual's economic resources for such a lifestyle. This kind of status incongruence has been found to be related to psychiatric symptoms, high blood pressure, elevated serum lipids, the risk of diabetes, and immunological status in developing societies in Latin America, the Caribbean, and Polynesia.[13–15]

Resources for coping with stressors have been found to be more culturally variable. Social support systems are a good case in point. Social support can be defined as the emotional and practical assistance an individual believes is available to him or her during times of felt need; the social network in which this assistance is available is the social support system. In research in Europe and North America, generally speaking, emphasis has been placed on the nature of the assistance or social support transactions, rather than on who might provide that assistance. There is a growing body of cross-cultural research to indicate that, in many societies, who provides the assistance is critical in determining the relationship between social support and health. This is probably a reflection of a continuing importance of kinship in defining who is and who is not an appropriate individual with whom to enter into a social relationship.[16]

For example, in Latin American societies, people have traditionally lived in large extended families organized around a father and his married sons (known as a patrilineal extended family). In addition to these extended family relationships, there is a social practice known as *compadrazgo*, through which individuals, especially men, establish formal, kinship-like relationships with unrelated persons (known as fictive kinship). The term *compadrazgo* literally means "coparenthood," and this carries the expectation of mutual support. These ties of fictive kinship are used to establish economically and politically important alliances. Research has shown that men who perceive greater amounts of support from both their extended and their fictive kin have lower blood pressures. Women are expected to restrict themselves to the household and domestic duties. Not surprisingly, with respect to health status, women benefit primarily from support available within the household. These studies show that the definition and effects of social supports are closely related to cultural and social structural factors, and can only be understood within that context.[17]

Other forms of stress resistance show similar differences cross-culturally, although the evidence is not quite as consistent as with social support. For example, in European and North American studies, a direct-action coping style has been found to be helpful in moderating the effects of stressful events or circumstances. In this style of coping, attempting to directly confront and alter stressful circumstances contributes to better health status. In some research in traditional societies, this same relationship has been observed. In others, the opposite effect has been observed (i.e., a direct-action coping style actually exacerbates distress and poor outcomes). In the specific case of this coping style, this probably has to do with the actual resources available to individuals and families in coping with stressors. Where social and economic resources are meager at best, the belief that one can truly change the circumstances that are often thrust upon oneself may, in the long run, be deleterious.[18]

A major source of stress in societies undergoing modernization is the increase in socioeconomic inequity that accompanies modernization. Generally speaking, the distribution of wealth becomes more unequal, resulting in marked social stratification where previously such stratification was, if not absent, at least muted. Recent studies point to the effects of such stratification on cultural consonance as a potent source of stress. The capacity of an individual to achieve higher cultural consonance is severely compromised by lower socioeconomic status. Lower cultural consonance has been found to be associated with higher blood pressure and psychological distress, and to mediate the effects of socioeconomic status in contexts of modernization. The inability to act on these widely shared cultural models is a potent source of stress in developing societies.[19,20]

STRESS AND MIGRATION

The other way in which social and cultural change can influence indigenous communities is through migration. The past century has seen remarkable movements of people from traditional societies to North America and Western Europe, along with internal migration within developing societies that takes migrants into cosmopolitan urban centers in their own societies. Rarely can migration be considered an individual matter. Rather, it is much more common for entire communities of migrants to become established in their host country. This usually occurs because migration follows patterns established through social networks, especially kinship networks. Individuals and households take advantage of kin-based social support systems established in host countries in order to establish themselves there.

Research suggests that a similar pattern of stress and response develops in cases of migration as that observed in developing societies. That is, migrants come to the new setting with aspirations for a new life, aspirations that are reinforced in host countries through the depiction of middle-class lifestyles in advertising and other media forms. At the same time, migrants typically occupy the lowest levels of socioeconomic status in their host countries. Therefore, the ability of the migrant to amass the economic resources necessary to achieve a middle-class lifestyle is severely compromised. This status incongruence again has been found to be associated with chronic disease risk factors such as blood pressure and glucose levels.[21]

Patterns of social support that can help an individual to cope with this kind of social stressor will often again be found in the kin support system. There are, however, several additional complications. First, often kin support systems will be fragmented. Only some people will migrate, not entire support systems. This can mean that some households and individuals are socially isolated, lacking even the most basic social supports. This can also mean that a large burden can be placed on a support system that is fragmented, and that the resulting demands for support are simply too much for the system to bear. Second, the migrant, especially to North America, is entering a highly competitive and individualistic society. The kinds of mutual rights and obligations entailed by kinship have ceased to be recognized as strongly, for example, in the United States compared to traditional societies in Latin America or Southeast Asia. Therefore, in some specific situations, the kin support system can come to be a source of stress and tension, as opposed to a resource for resisting stress.

The complications entailed here are illustrated by migrants from Samoa to northern California in the United States. Traditionally organized into large extended kin groups represented by chiefs, Samoans have transplanted their social organization to some US urban centers. Traditionally, the chief (or *matai*) controls economic decision-making for the entire extended family. In the American urban setting, however, the economic demands of daily life for an individual household can conflict with the decisions made for the extended family, causing this traditional form of social organization to become a source of stress. At the same time, within the larger extended family, core systems of kin-based social support have emerged, especially involving networks of adult siblings. Individuals and households with a strong support system of this kind have better health status in spite of social stressors such as status incongruence. This specific case illustrates the importance of understanding the adaptation of migrants in a host society in relation to their traditional cultural context, as well as the demands of the new social setting.[22]

Some research on migrants from the developing world has shown the long-term effects of major life events experienced by families in the society of origin. This has been observed in immigrants from Latin America and Southeast Asia who have been exposed to protracted civil war and related conflicts. For example, it was found that persons who had had relatives kidnapped or murdered by nonmilitary death squads in Guatemala had continuing high levels of anxiety and depression years after the event. Similarly, anxiety and depression levels in migrants from Southeast Asia to the United States were associated with time spent in refugee camps, independently from other stressors and demographic control variables. These major crises associated with large-scale political events and circumstances can have effects that continue well after the initial, acute stages.[23]

CONCLUSION

The study of stress and indigenous societies has helped to illuminate various aspects of the stress process. Perhaps the most important has been the clear demonstration of the link between the stress process and the social and cultural context in which individuals and families live. When stressors and resistance resources are examined only within a single cultural context, it can appear as if individual differences in exposure to stressors or in access to resistance resources are the only key to understanding the process. What is lost is the recognition that what counts as a stressor or a resistance resource is itself a function of the cultural context and related social influences. Furthermore, the relationship between stressors, resources, and outcomes can also be modified by social and cultural context. Comparing the stress process in different cultural contexts has been integral to revealing this aspect of the process.[24]

Future research must be explicitly comparative in scope in order to expand on findings produced thus far. For example, research in developing societies suggests that the most important stressors in those societies are actually a function of the development process itself, such as status incongruence. Put differently, this aspect of the stress process is comparable across different settings. On the other hand, the most important resources for resisting stress appear to be specific to the local setting, as in the way in which systems of social support are structured by the existing systems of social organization. Continuing to refine these studies, including better measurements of the physiologic dimensions of stress, will increase our understanding of human adaptation.

References

1. Dressler W. Modernization, stress, and blood pressure: new directions in research. *Hum Biol*. 1999;71:583–605.
2. Armelagos GJ, Brown PJ, Turner B. Evolutionary, historical and political economic perspectives on health and disease. *Soc Sci Med*. 2005;61:755–765.
3. Nichter M. Idioms of distress revisited. *Cult Med Psychiatry*. 2010;34:401–416.
4. Guarnaccia P, Canino G, Rubiostipec M, Bravo M. The prevalence of ataques-de-nervios in the Puerto Rico disaster study. *J Nerv Ment Dis*. 1993;181:157–165.
5. Rubel AJ, O'Nell CW, Collado-Ardon R. *Susto: A Folk Illness*. Berkeley, CA: University of California Press; 1991.
6. Brooks BB. Chucaque and social stress among Peruvian highlanders. *Med Anthropol Q*. 2014;28:419–439.
7. Oths KS. Debilidad: a biocultural assessment of an embodied Andean illness. *Med Anthropol Q*. 1999;13:286–315.
8. Dressler WW. Cultural consonance. In: Bhugra D, Bhui K, eds. *Textbook of Cultural Psychiatry*. Cambridge: Cambridge University Press; 2007:179–190.
9. Madrigal L, Blell M, Ruiz E, Otarola-Duran F. The slavery hypothesis: an evaluation of a genetic-deterministic explanation for hypertension prevalence rate inequalities. In: Panter-Brick C, Fuentes A, eds. *Health, Risk, and Adversity*. New York, NY: Berghahn Books; 2009:236–265.
10. Decker S, Flinn M, England BG, Worthman CM. Cultural congruity and the cortisol stress response among Dominican men. In: Wilce Jr JM, ed. *Social and Cultural Lives of Immune Systems*. London/New York: Routledge; 2003:147–169.
11. Flinn MV, England BG. Social economics of childhood glucocorticoid stress response and health. *Am J Phys Anthropol*. 1997;102:33–53.
12. Hanna JM, James GD, Martz JM. Hormonal measures of stress. In: Baker PT, Hanna JM, Baker TS, eds. *The Changing Samoans: Behavior and Health in Transition*. New York, NY: Oxford University Press; 1986:203–221.
13. Dressler W, Mata A, Chavez A, Viteri F. Arterial blood pressure and individual modernization in a Mexican community. *Soc Sci Med*. 1987;24:679–687.
14. Dressler WW. Psychosomatic symptoms, stress and modernization: a model. *Cult Med Psychiatry*. 1985;9:257–286.
15. McDade T. Status incongruity in Samoan youth: a biocultural analysis of culture change, stress, and immune function. *Med Anthropol Q*. 2002;16:123–150.
16. Dressler WW. Cross-cultural differences and social influences in social support and cardiovascular disease. In: Shumaker SA, Czajkowski S, eds. *Social Support and Cardiovascular Disease*. New York, NY: Plenum Publishing; 1994:167–192.
17. Dressler W, Mata A, Chavez A, Viteri F, Gallagher P. Social support and arterial pressure in a central Mexican community. *Psychosom Med*. 1986;48:338–350.
18. Au EWM, Chiu C, Zhang Z-X, et al. Negotiable fate: social ecological foundation and psychological functions. *J Cross-Cult Psychol*. 2012;43:931–942.
19. Godoy R, Reyes-Garcia V, Gravlee C, et al. Moving beyond a snapshot to understand changes in the well-being of native Amazonians. *Curr Anthropol*. 2009;50:560–570.
20. Dressler WW, Balieiro MC, Ribeiro RP, dos Santos JE. Culture as a mediator of health disparities: cultural consonance, social class, and health. *Ann Anthropol Pract*. 2015;38:214–231.
21. Zimmerman C, Kiss L, Hossain M. Migration and health: a framework for 21st century policy-making. *PLoS Med*. 2011;8:e1001034.
22. Janes C. *Migration, Social Change, and Health: A Samoan Community in Urban California*. 1st ed. Stanford, CA: Stanford University Press; 1990.
23. Sabian M, Cardozo BL, Nackerud L, Kaiser R, Varese L. Factors associated with poor mental health among Guatemalan refugees living in Mexico 20 years after civil conflict. *JAMA*. 2003;290:635–642.
24. Dressler W. Modeling biocultural interactions: examples from studies of stress and cardiovascular disease. *Yearb Phys Anthropol*. 1995;38:27–56.

CHAPTER

55

Diet and Stress: Interactions with Emotions and Behavior

J. Wardle[1,†], E.L. Gibson[2]

[1]University College London, London, UK

[2]University of Roehampton, London, UK

OUTLINE

Abstract

There is no clear consensus on whether stress reliably leads to increased or decreased caloric intake, although shifts in food choice seem likely. Research suggests the relationship between stress and changes in food intake is moderated by individual differences in psychological and physiological pathways as well as the type of stressor and the foods available. These differences make some individuals particularly susceptible to eating more energy-dense foods in response to stress. Increased intake, or changes in food choice, may be part of an individual's stress-coping response, although an alternative account is that stress shifts motivational and attentional resources to fundamental habitual levels leading to selection of salient and "safe" foods. In either case, this stress response could place individuals at increased risk of excess weight gain. Real-life stressors are diverse and their relationships with eating behaviors are similarly dynamic, creating a challenging but fascinating area for future study.

INTRODUCTION

There is a widespread belief that stress influences eating behavior, but considerable uncertainty about the direction of the effect (i.e., whether stress increases or decreases food intake). An analysis of the physiology of stress would lead us to expect decreased eating because stress slows gastric emptying and increases energy substrate levels, which should reduce appetite.[1] Stress also promotes behaviors designed to cope with or escape from the source of stress, and under these circumstances, eating might be accorded a lower priority. In contrast, the literature on human reactions to stress suggests that it can be associated with increased food intake or a shift toward a higher fat and sugar diet.[2]

Our knowledge of stress and eating comes from diverse areas of research, including clinical studies on the etiology of obesity and eating disorders, laboratory studies on dieting and the regulation of food intake, surveys of stress and health-related behaviors, and animal work on stress responses.[1] The diversity of this literature means that there is great variation in the sources and intensity of the stressors that have been examined, the circumstances in which food is consumed, the quality of

[†]Deceased

http://dx.doi.org/10.1016/B978-0-12-800951-2.00058-3

dietary information that can be gathered, and the populations that have been studied. On this basis, it may not be surprising to find that the links between stress and eating appear to be complex.

KEY POINTS

- The impact of stress on eating and diet varies depending on individual variation in psychological and genetic characteristics, as well as the nature of the stress
- Personality traits, developmental history, and attitudes toward food can all influence a person's likely dietary response to stress
- Approximately equal proportions of people report eating more during stress as report eating less
- Emotional eaters (i.e., those who reporting eating more when stressed) tend to be poor at coping with stress
- Emotional eaters tend to choose energy-dense, particularly sweet fatty foods when stressed
- Such foods produce only transient relief of mood and distraction from stress
- Emotional eaters are at greater risk of weight gain and obesity
- People who struggle to restrain their food intake, and are easily tempted by energy-rich foods, may also overeat during stress due to weakening of cognitive control
- Increased abdominal obesity is more common in emotional eaters but can also help to down-regulate the hypothalamic-pituitary-adrenal stress axis
- Pathways to stress eating also involve the major neurohormonal systems, including dopamine, serotonin, opioids, and orexigenic hormones
- There is little support for the notion that stress-induced eating provides any more than very transient "affect reduction"
- Instead, when stressed, our habitual and long-established food preferences may be evoked, particularly for highly salient and pleasurable foods: "nursery foods" become "comfort foods" through familiarity more than mood regulation

ANIMAL RESEARCH INTO STRESS AND EATING BEHAVIOR

Animal research appears to offer the ideal opportunity to examine the general, physiologically mediated effect of stress on food intake. Animals can be exposed to stress or a control condition with stressors varying in quality, intensity, or duration; the animal's deprivation state can be manipulated; and the type of food supplied can be varied. However, to date, this research has produced an inconsistent set of results. Much of the early work in rats used tail pinch as the stressor and typically found that food intake was increased by this procedure. However, tail pinch is now thought to represent an atypical stressor, if it is even stressful at all, and increased eating may be part of a general increase in oral behaviors, rather than a specific behavioral response.[3] Other stressors have included electric shock, noise, immobilization (restraint stress), isolation, housing changes, cold water swims, and defeat in fights, with varied effects on food intake from study to study, ranging from substantial increases in food intake in some studies to substantial decreases in other studies. On balance, the evidence suggests that chronic stress and social stressors appear to be more likely to have a hyperphagic effect, compared to acute stress and physical stressors.[1]

Potentially more recent fruitful approaches include development of animal models that reflect human disorders related to stress. For example, stress-induced "binge eating" can be seen in rats given access to highly palatable food as well as laboratory chow, provided that both limited access to the food and intermittent stress are present.[4] Intriguingly, it may be the ability to choose from separate sources of energy-dense palatable foods that confers stress resilience, rather than an increase in energy-dense food intake per se. Even so, consumption of either or both sugar and fat have been shown to reduce the extent of stress-induced activation of the hypothalamic-pituitary-adrenocortical (HPA) axis in rats, although with differences in effects on brain limbic pathways and central mechanisms of HPA axis regulation.[5] This supports the suggestion that selection of fatty and/or sugary foods during stress may be a form of "self-medication."

HUMAN RESEARCH INTO STRESS AND FOOD INTAKE

Human research on stress and food intake has come from several research traditions. Naturalistic studies usually investigate community samples and attempt to gather information on food intake at high- and low-stress periods of life.[8,10] The focus in these studies, as in most of the animal research, has been on the general effect of stress on eating over the short or medium term. Laboratory studies usually administer a stressful procedure in parallel with an eating task, then covertly assess food intake.[6] The emphasis therefore is on the acute effects of stress. Most of the laboratory studies have taken an individual differences perspective, examining stress-related eating in relation either to weight, dietary restraint, or binge eating tendencies.

Naturalistic Studies

These observational studies make use of naturally occurring, but predictable, stressful circumstances with which to study stress-related variations in diet. School or university examinations have been used as the stressor in several studies, and periods of high work stress in others.[1] Food intake is recorded either in a diary record kept by the study participants or with a 24-h recall procedure. The food records are then analyzed to obtain estimates of intake of energy and nutrients. Biological measures such as weight or blood lipids have been included in some studies to provide another indicator of dietary change.[1,7,8]

Among the published studies, there is one element of consistency, namely, none of them has found lower food intake, on average, during the higher stress time. Likewise, there is little evidence for lower weight or lower serum cholesterol under stress.[1] Otherwise, the results have been divided: in some of the studies, both overall energy intake and intake of fat are higher at the high-stress time[9]; in other studies, there is enormous individual variability, but no average difference in energy intake between low- and high-stress periods. In one study, which unlike many naturalistic studies was not confounded by the passage of time, high workload in department store workers was associated with higher fat, sugar, and total energy intake, but only in people who habitually restrained their food intake.[10]

In a Finnish population-based study of adults, stress-driven eaters ate more energy-dense high-fat foods and had higher body mass indices.[11] Similarly, mothers caring for chronically ill children were more likely to have higher abdominal fat and to be emotional eaters.[12] Other studies have suggested a link between stressful life events and higher body weight.[13]

Several studies have examined predictors of weight gain in different occupations; stress-induced (emotional) eating was found to be the strongest predictor of weight gain in groups as diverse as paramedics and bankers.[14] The balance of evidence so far has to be that stress in everyday life rarely appears to have a hypophagic effect, and it may, in some people and some circumstances, have a hyperphagic effect.

One explanation for the variability in results in observational studies is the poor validity and reliability of measures of food intake. Keeping a food diary is an onerous and intrusive task, so compliance is poor, and even among those who return a diary, there are likely to be errors: people forget foods, are not able to assess quantities accurately, and often complete the diary well after the eating event. Twenty-four-hour dietary recall interviews avoid some of those problems, but are still susceptible to forgetting and misreporting, as well as providing a much more restricted snapshot of food intake.

A second explanation for variability in the results is the different kinds of stressors. The stress of surgery, for example, is qualitatively and quantitatively different from the stress of an examination or a period of long working hours. Some stressors could occur together with changes in circumstances that enforce a change in eating behavior even if there is no direct effect of the stress on appetite.

A third explanation that is rarely addressed in naturalistic studies is that responses to stress may vary across individuals, even if they are consistent within themselves. Thus, diary studies of daily stressors and eating, and surveys that ask respondents about how stress affects their eating typically report some effect of stress, with approximately equal proportions saying that they eat more and eat less while under stress.[8,15]

Laboratory Studies

Most laboratory studies have been based on the idea that the biologically natural response to stress is hypophagia, but that individuals who are either overweight or highly restrained eaters are unresponsive to their internal signals. Participants are therefore characterized according to these features and hypothesized to respond differently to stress.

Laboratory studies have used a range of stressors, including unpleasant films, false heart rate feedback, and threat of public speaking. In the typical design, participants are exposed either to the stressor or to a control procedure, and food intake is assessed covertly, often disguised as a taste test in which participants are asked to taste and rate some flavors of a palatable food such as ice cream.

Obesity and Stress-Induced Eating

The clinical interest in stress-related eating stemmed from the psychosomatic theory of obesity, which suggested that for the obese, eating met emotional rather than nutritional needs and that the tendency toward emotional eating explained why they had become obese in the first place.[16] Eating was hypothesized to provide reassurance, and hence stress was predicted to trigger higher-than-usual food intake, so-called emotional eating. There is no doubt that many obese people report that they eat more under stress, but these clinical reports need to be examined in controlled studies, both to establish their validity and to see whether stress-related eating is specific to obesity.[16,17]

In practice, laboratory studies on stress-related eating in the obese have produced mixed results, with some finding higher intake under stress in the obese and others not. There has also been variability in naturalistic studies, but on balance, there is probably enough evidence to conclude that obesity is an indicator of a higher risk of stress-induced hyperphagia, or at least a lower likelihood of stress hypophagia.[2] This is supported by an extensive systematic review of prospective studies examining the impact of stress on adiposity: overall, there was a significant but weak positive association.[13] The other side of the psychosomatic theory was that eating would have an anxiolytic effect, and this has received less support. Clinically, many obese people admit that any solace they derive from eating is transient and rapidly followed by shame and regret at not having shown more self-control. In laboratory studies there has been no evidence that eating successfully reduces emotional arousal, beyond a few minutes of temporary relief.[18] These observations have largely discredited the basic idea of the psychosomatic theory, the idea that obesity represents a disorder of emotional reactions. However, there is still interest in the idea that the obese have a reduced sensitivity to internal satiety cues, and the fact that they do not show stress hypophagia may be a consequence of this lack of internal responsiveness.[2]

Stress, Restraint, and Emotional Eating

In 1972, a radical alternative to the psychosomatic theory of obesity was proposed, namely, that any abnormalities of eating observed in the obese were not preexisting tendencies, but a consequence of the steps that they were taking to reduce their weight.[16] At first the emphasis was on the effects of maintaining a body size below the hypothesized set-point. This was superseded by restraint theory, which proposed that one of the important determinants of food intake regulation was the tendency toward restrained eating—the combination of concern about eating and weight fluctuation. Most obese people and a significant proportion of normal-weight adults, particularly women, are constantly trying to restrict their food intake in order to reduce their weight. The habit of trying to restrict food intake could have the effect of changing people's relationship with food such that the usual cues to hunger and fullness become less effective in regulating their eating behavior, as eating comes more strongly under cognitive control. Restrained eaters were found to limit their food intake at times when external or emotional pressures were low, but at other times, they would abandon restraint and eat to capacity, so-called disinhibition.[2,16]

One of the early observations on restrained eating was that, among individuals who experienced anxiety in response to a stressor, food intake was increased among the restrained eaters and decreased among unrestrained eaters. Across many different studies, with a wide range of stressors, restrained eaters, but not unrestrained eaters, almost always ate more in the stressed than in the unstressed condition. Several studies have also examined the role of restraint in predicting individual differences in responses to stress in real-life studies, and the results are consistent with the laboratory studies in showing that specifically restrained eaters are more likely to report stress-induced hyperphagia.

Subsequently there has been a resurgence of interest in the role of negative affect and emotional eating as predictors of problematic eating and poor control of weight.[18] One reason for this has been the realization that earlier psychometric measures to some extent conflated restrained and emotional eating. In one laboratory study, when separate scales were used to measure restraint and emotional eating (Dutch Eating Behavior Questionnaire), the latter was the better predictor of eating induced by public speaking stress.[19] A variety of foods were presented as a buffet-style meal during preparation of the speech task. Stress did not alter overall intake; however, stressed emotional eaters ate more sweet, high-fat foods (chocolate and cake) than either unstressed emotional eaters or nonemotional eaters in either condition. This supports the survey findings that sweet, fatty foods such as chocolate may be preferentially sought during stress or negative effect, at least in a subgroup of susceptible individuals.

Emotional eating may underlie previous reports that dietary restraint or female gender predicts stress-induced eating. Moreover, emotional eaters may be more susceptible to effects of stress: women who ate more from a selection of snack foods after a stressful task also showed the greatest release of the stress-sensitive hormone, cortisol, and more stress-induced negative affect.[20] These high reactors also showed a preference for sweet foods. So, it seems that emotional eaters may be more likely to experience mood disturbance when challenged.

Emotional and disinhibited eating and dietary restraint could interact[18]: the latter seems to contribute to stress-induced eating of sweet, fatty foods by women classified as disinhibited eaters. Restrained eating may increase the salience of energy-rich food cues, but emotional or disinhibited eating tendencies may be necessary to elicit overeating of such foods in the face of ego-challenging or upsetting stressors.

Stress and Food Choice

Studies of stress and eating usually focus on the amount of food consumed, but stress might also affect food choices or meal patterns. In the naturalistic studies that showed increased energy intake, there was also an increase in the proportion of energy from fat. This could reflect an increased preference for fatty foods, or alternatively a different meal pattern. In most Western countries, the proportion of fat in meal-type foods tends to be lower than in snack-type foods. If stress induced a shift from meals to snacks, this would be reflected in a higher fat intake and might also increase the total energy intake since higher fat foods have a higher energy density.

In most naturalistic studies, it is not possible to distinguish meal pattern changes from food choice changes, because the data are presented in terms of daily intake of nutrients, without any information either on foods eaten or timing of consumption. Most laboratory studies have also failed to address the food choice issue because they present only a single type of food.[6] However, there is some evidence that sweet, high-fat foods are the ones most likely to show an increase, whereas fruit and vegetables appear to be avoided during stress.[9,19] In a study of health behaviors in 11- to 13-year-old schoolchildren in London, greater perceived stress was associated with more fatty food intake, less fruit and vegetable intake, more snacking, and a reduced likelihood of daily breakfast consumption.[21]

A survey of possible changes in food choice during stress among 212 students revealed an interesting pattern of effects of stress, which was partly independent of whether participants were grouped as reporting eating more, the same, or less overall when stressed.[15] That is, sweets and chocolate were reported to be eaten more under stress by all groups, even those eating less overall; conversely, intake of fruit and vegetables, and meat and fish, were reported as less or unchanged under stress in all groups. The changes for the staple food, bread, matched the overall group self-perceptions of changes in eating due to stress. These data imply that mechanisms governing effects of stress on food choice may be somewhat separate from those influencing overall appetite under stress and that foods such as sweets and chocolate may be particularly selected in stressful circumstances.

Chocolate: A Unique Mood-Enhancing Food?

Chocolate does seem to have a special place in the relationship between food choice and mood: it is a widely liked food, and people often report choosing to eat chocolate during stress, the more so in those with poor stress-coping personalities.[2] There are popular notions, perpetuated both in the lay and scientific media, that chocolate contains psychoactive chemicals, such as phenylethylamine and cannabinoids that are responsible for its effect on mood. However, for reasons of quantity and availability to the brain, these compounds are not likely to alter mood when obtained from chocolate. Nevertheless, it has been shown experimentally that the methylxanthines in chocolate, caffeine, and theobromine can indeed improve mood (hedonic and energetic dimensions) as well as mental function.[22]

Chocolate has other characteristics that could be mood enhancing: it is high in fat and sugar and so would be likely to activate opioidergic hedonic and dopaminergic reward pathways.[2] It is also low in protein (3-6% of energy), so if eaten in sufficient amounts on an empty stomach might conceivably enhance mood via increased synthesis of the brain neurotransmitter serotonin, as outlined in a later section. Finally, chocolate could also acquire secondary reinforcing or mood-enhancing properties by association with its use as a treat or gift and convivial social interaction. However, it should be remembered that in those who may consider chocolate to be a dangerous temptation that threatens their weight control, eating it can have negative consequences for mood. In a survey of about a thousand Californians, frequency of eating chocolate was strongly associated with self-reported levels of (untreated) depression, independently of overall energy intake.[23] If not a spurious association, this finding suggests that eating chocolate does not provide any long lasting benefit to mood. Indeed, an earlier study found that self-identified chocolate addicts felt guiltier after eating chocolate than a control group.[24] The chocolate addicts also reported worse mood before eating the chocolate. Moreover, another study showed that eating just 5 g of energy-dense foods could increase negative mood in overweight women, especially those reporting emotional eating tendencies.[2] By contrast, in healthy men, experimental induction of sadness decreased appetite, whereas when the men were cheerful, chocolate tasted more pleasant and stimulating and more of it was eaten.[18] This gender difference is likely to be confounded by dispositional and attitudinal differences. In any event, such findings are not easily explained by an "affect reduction" model of stress eating, and an alternative model is discussed below.

MECHANISMS RELATING STRESS TO EATING

Neurohormonal Pathways

Eating food activates brain pathways involved in reward, and so might be expected to have a positive effect on mood.[18] Furthermore, some of the ingested nutrients could alter the function of neurohormonal systems involved in coping with stress. We now consider the main pathways that appear to underlie such effects.

Opioids

Endogenous opioid neuropeptides are released during stress and are known to be important for adaptive effects such as tolerance of pain. They are also involved in motivational and reward processes in eating behavior, such as stimulation of appetite by palatable foods.[25] This would tend to suggest that there would be a link between opioid action, stress, and food choice. Perhaps the best evidence for opioid involvement in an interaction between stress and eating is the finding that in animals and human infants, ingestion of sweet and fatty foods, including milk and even a nonnutritive sweetener, alleviates crying and other behavioral signs of distress.[26] This stress-reducing effect of sweet taste has been shown to depend on opioid transmitter systems.

However, perceived palatability (liking) may be important, since in one study it was found that only highly liked foods (chocolate chip cookies), and not bland or disliked foods, were able to increase pain tolerance in female students when eaten beforehand.[27] It is tempting, but speculative, to conclude that adults may select sweet, fatty, palatable foods for opioid-mediated relief of stress. More support for this hypothesis arises from animal studies of "comfort eating": rats that are exposed to the combination of stress and restricted food access show enhanced consumption specifically of highly palatable sweet fatty food[4]; this effect is blocked by opioid antagonists, and these drugs also reduce intake of preferred foods in humans.[28]

Dopamine

The neurotransmitter dopamine is thought to be the major chemical messenger for reward in the brain, and its motivational consequences.[25] The availability of dopamine receptors (D2) in part of the brain reward pathway, the striatum, has been shown to be inversely correlated to body mass index.[29] This finding has led to the suggestion of a neurochemical trait that elicits overeating of palatable foods so as to enhance dopamine release—a neural predisposition for comfort eating. However, a more plausible explanation is that dopamine receptors are downregulated after years of overconsumption of palatable energy-dense foods: this might in part result from neuroendocrine feedback arising from excess adiposity, perhaps interacting with stress and mood, since dopamine also mediates stress sensitivity, depression, and reinforcement of drug-taking habits.[2] A functional magnetic resonance imaging study of emotional eaters found that, compared to nonemotional eaters, these women showed enhanced activity in brain reward pathways when in a negative mood and either anticipating or actually consuming a chocolate milkshake.[30] Studies of the influence of genetic polymorphisms of the dopamine system, such as the TAQ1A A1 allele, suggest that such genetic differences may interact with developmental experience to determine how stress may affect the drive to eat energy-dense foods.[31]

Serotonin

The neurotransmitter serotonin has been implicated in mood disorders, anxiety, and stress coping, as well as in eating behavior. Synthesis of serotonin (or 5-hydroxytryptamine [5-HT]) depends on dietary availability of the precursor essential amino acid, tryptophan (TRP), due to a lack of saturation of the rate-limiting enzyme, tryptophan hydroxylase, which converts TRP to the intermediate compound 5-hydroxytryptophan. In effect, brain 5-HT can thus potentially be affected by how much TRP is eaten versus the amount of large, neutral, primarily branched chain amino acids (LNAA) which compete for the same transport system from blood to brain.[32]

The possibility that a carbohydrate-rich, low-protein meal could raise 5-HT function gave rise to the proposal that some depressed people may self-medicate by eating high proportions of carbohydrate, in a manner reminiscent of antidepressant drugs.[2] There is some support for this: for example, when participants were divided into high or low stress-prone groups, as defined by a questionnaire measure of neuroticism, carbohydrate-rich, protein-poor meals (which raised plasma TRP/LNAA ratios) prior to a stressful task were found to block task-induced depressive feelings and release of the glucocorticoid stress hormone cortisol, but only in the high stress-prone group. This finding was replicated using high- vs. low-TRP-containing proteins (alpha-lactalbumin and casein, respectively).[33] It was argued that because stress increases 5-HT activity, the poor stress coping of this sensitive group might indicate a deficit in 5-HT synthesis that is improved by this dietary intervention.[32]

Further evidence for a critical role for 5-HT in stress eating is seen in the interaction between stress, snacking, and genotypes of the 5-HT transporter-linked polymorphic region (5-HTTLPR): students carrying the susceptible short allele (linked to deficits in 5-HT signaling) were more stressed by exams and reported greater appetite for sweet snacks.[34] It has been proposed that emotional eating may arise from a 5-HT-mediated interaction between cognitive vulnerability to stress and the ability of certain foods to improve 5-HT function and mood.[34] This might also involve serotoninergic modulation of brain dopamine activity. Interestingly, twin studies have revealed an inherited link between emotional eating and liking for sweet fatty foods; however, estimates of the heritability of emotional eating from such studies vary between 9% and 60%.[35]

Neuroendocrine Systems

A critical neurohormonal system mediating the effects of stress on energy flux and appetite is the HPA axis. Neural stimulation of selected nuclei in the hypothalamus of the brain leads to corticotropin-releasing hormone (CRH) acting on the pituitary gland to release adrenocorticotropic hormone, which in turn stimulates release of the steroid glucocorticoid hormone cortisol from the adrenal glands. In animals, CRH is thought to mediate the suppression of food intake by severe stressors. On the other hand, exogenous glucocorticoids, and endogenous cortisol release following less severe stress, appears to stimulate appetite (particularly for sweet or fatty foods) by acting on the hypothalamus to inhibit CRH release.[36,37]

The involvement of the HPA axis in the influence of stress on eating behavior seems likely for several reasons, including (1) glucocorticoid hormones substantially affect energy substrate mobilization and metabolism, increasing lipolysis, proteolysis, and gluconeogenesis while protecting hepatic glycogen stores; (2) the circadian rhythm of HPA axis responsivity is dependent on patterns of food ingestion; (3) normal meals containing at least 10% protein as energy activate the HPA axis and release cortisol in humans; and (4) acute food deprivation suppresses the HPA axis response to stress.[36]

Furthermore, most animal models of obesity, genetic or otherwise, depend on an intact HPA axis, and disruption of its function may underlie at least some human obesity, in particular increased visceral adiposity, especially among obese patients suffering from anxiety and depression.[37,38] The eating disorders bulimia and anorexia nervosa are also associated with hypercortisolemia and disrupted HPA axis activity even in the absence of undernutrition. An inability to cope with stress, coupled with chronic activation of appetite-suppressing CRH, is one psychobiological model proposed for anorexia nervosa.[17]

Another hormone that has been implicated in emotional eating is the appetite hormone ghrelin. Ghrelin normally rises together with hunger during the postabsorptive phase between meals, and then declines on eating. However, ghrelin also often increases during stress, in association with rises in glucocorticoids such as cortisol, and has been proposed to have anxiety-reducing properties.[2] Moreover, there is some evidence that poststress eating has less effect in reducing ghrelin in emotional compared to nonemotional eaters.[39] Thus, in a throwback to the early psychosomatic theory, an imbalance in the possible dual role of ghrelin (anxiolytic and appetitive) might contribute to stress-induced eating in emotional eaters. However, causality has yet to be determined, and it may be that the greater (learned) appeal of eating during stress helps maintain ghrelin levels in habitual emotional eaters.

Biobehavioral Pathways

Theoretical considerations and empirical results indicate that the mechanisms linking stress to changes in food intake are far from straightforward. Figure 1 illustrates the three principal pathways that have been implicated in the work in this area. In the center is the simple biological pathway whereby the physiological effects of stress inhibit the physiological features of hunger (or perhaps mimic the effects of satiety) and hence modify appetite and food intake. Animal research has generally been used to test this pathway since it is assumed that animals' food intake is more directly controlled by the basic drives of hunger and satiety. From present evidence it would seem that some stressors probably induce hypophagia and others hyperphagia, implicating a more complex pathway from stress to food intake even in animals. In human laboratory studies, unrestrained eaters either are unaffected by stress or eat slightly less, which is somewhat supportive of a reduction in appetite, at least in some situations, and for some foods.

A second pathway (shown at the bottom of Figure 1) results from consideration of the broader effects that stressors have across a range of aspects of life. In real life, stress motivates people to deploy resources toward dealing with the source of stress (so-called problem-focused coping). This may mean that longer term goals like dieting or healthy eating are temporarily set aside, with consequent effects on food choices. This is supported by evidence that higher energy intake under stress occurs most often in female restrained eaters, who normally have a low energy intake. This mechanism is similar to a recent theoretical model for stress eating, which, in contrast to the "affect reduction" account, proposes that stress may actually reduce the pleasure of eating highly palatable foods, but instead amplifies learned motivational and attentional responses to the presence of such foods (possibly dopamine-mediated), at the expense of more cognitively demanding goal-dependent control on eating.[40] In other words, when stressed, our habitual and long-established food preferences are evoked. Particularly in the presence of energy need, this would result in selection of energy-rich sweet and/or fatty foods, and suppression of more cognitively demanding healthy eating.

The third pathway, shown at the top of Figure 1, relates to the fact that stress also comes with an emotional coloring, so coping efforts are deployed to deal not only with the source of stress but also with the emotional state (emotion-focused coping). For some people, especially those who are usually self-denying with food, food may have a significant reward value and may be used to provide emotional comfort. Women, in particular, often describe food in terms of treats and rewards, and the potential comfort to be derived from sweet foods is deeply enshrined in contemporary culture. Clearly this idea has echoes of the discredited psychosomatic theory of obesity, but it should be remembered that it was rejected not because of the absence of emotional eating, but because eating did not appear to be a successful anxiolytic. Other emotional coping strategies are equally unsuccessful in the longer term (e.g., alcohol, smoking), but that does not prevent the smokers, drinkers, and stress eaters from feeling that their habits serve a comforting role. In the field of eating

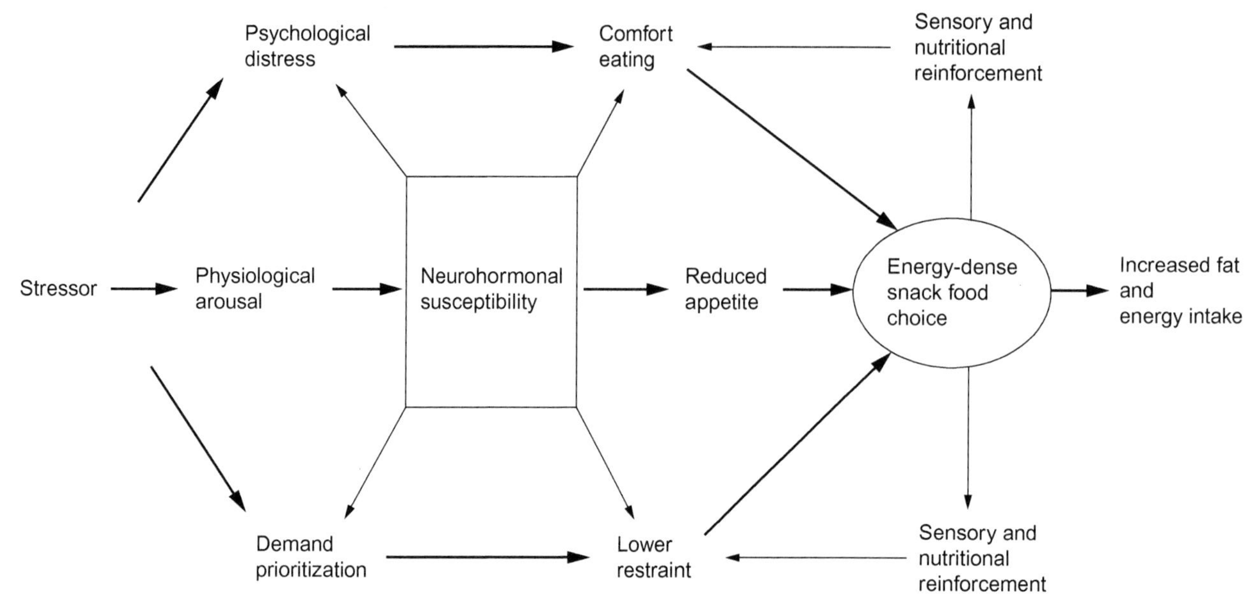

FIGURE 1 Biobehavioral pathways linking stress to changes in food choice and intake.

disorders, excessive eating has been hypothesized to provide escape from self-awareness,[17] which is especially valued in response to ego-threatening stressors, so the idea that eating has a role, functional or dysfunctional, as an emotional coping strategy may have been rejected prematurely.[41]

Figure 1 also depicts a likely final outcome of stress on diet, as discussed in the previous sections. That is, for multiple reasons, including sensory pleasure, modulation of neurotransmitter control of mood, and stress coping, susceptible stressed people may end up choosing energy-dense, sweet, fatty foods that will lead to a less healthy diet and possibly weight gain.

SUMMARY

In summary, the relationship between stress and eating is a complex one, moderated in humans not only by the type of person, the type of stress, and the types of foods that are available, but also by the dynamic relationship between the internal milieu, brain neurochemistry and behavior over time. There is increasing evidence to support the view that some people are particularly susceptible to eating more sweet, fatty, energy-dense foods during stress. Possible explanations can be found at both the psychological and physiological level, but in either case it is probable that such a change in eating behavior helps the individual to cope with stress, albeit transiently. Unfortunately, this change in diet may also result in weight gain, particularly abdominal obesity, and increased risk of cardiovascular disease.

References

1. Wardle J, Gibson EL. Impact of stress on diet: processes and implications. In: Stansfeld SA, Marmot MG, eds. *Stress and the Heart: Psychosocial Pathways to Coronary Heart Disease*. London: BMJ Books; 2002:124–149.
2. Gibson EL. The psychobiology of comfort eating: implications for neuropharmacological interventions. *Behav Pharmacol*. 2012;23(5–6):442–460.
3. Robbins TW, Fray PJ. Stress-induced eating: fact, fiction or misunderstanding? *Appetite*. 1980;1:103.
4. Boggiano MM, Chandler PC, Viana JB, Oswald KD, Maldonado CR, Wauford PK. Combined dieting and stress evoke exaggerated responses to opioids in binge-eating rats. *Behav Neurosci*. 2005;119(5):1207–1214.
5. Foster MT, Warne JP, Ginsberg AB, et al. Palatable foods, stress, and energy stores sculpt corticotropin-releasing factor, adrenocorticotropin, and corticosterone concentrations after restraint. *Endocrinology*. 2009;150(5):2325–2333.
6. Greeno CG, Wing RR. Stress-induced eating. *Psychol Bull*. 1994;115(3):444–464.
7. McCann BS, Warnick GR, Knopp RH. Changes in plasma lipids and dietary intake accompanying shifts in perceived workload and stress. *Psychosom Med*. 1990;52:97–108.
8. Stone A, Brownell KD. The stress-eating paradox: multiple daily measurements in adult males and females. *Psychol Health*. 1994;9:425.
9. O'Connor DB, Jones F, Conner M, McMillan B, Ferguson E. Effects of daily hassles and eating style on eating behavior. *Health Psychol*. 2008;27(suppl 1):S20–S31.
10. Wardle J, Steptoe A, Oliver G, Lipsey Z. Stress, dietary restraint and food intake. *J Psychosom Res*. 2000;48(2):195–202.
11. Laitinen J, Ek E, Sovio U. Stress-related eating and drinking behavior and body mass index and predictors of this behavior. *Prev Med*. 2002;34(1):29–39.
12. Tomiyama AJ, Dallman MF, Epel ES. Comfort food is comforting to those most stressed: evidence of the chronic stress response network in high stress women. *Psychoneuroendocrinology*. 2011;36(10):1513–1519.
13. Wardle J, Chida Y, Gibson EL, Whitaker KL, Steptoe A. Stress and adiposity: a meta-analysis of longitudinal studies. *Obesity*. 2011;19(4):771–778.

14. van Strien T, Koenders PG. How do life style factors relate to general health and overweight? *Appetite*. 2012;58(1):265–270.
15. Oliver G, Wardle J. Perceived effects of stress on food choice. *Physiol Behav*. 1999;66(3):511–515.
16. Ganley RM. Emotion and eating in obesity—a review of the literature. *Int J Eat Disord*. 1989;8(3):343–361.
17. Vögele C, Gibson EL. Mood, emotions and eating disorders. In: Agras WS, ed. *The Oxford Handbook of Eating Disorders*. Oxford: Oxford University Press; 2010:180–205.
18. Macht M. How emotions affect eating: a five-way model. *Appetite*. 2008;50(1):1–11.
19. Oliver G, Wardle J, Gibson EL. Stress and food choice: a laboratory study. *Psychosom Med*. 2000;62(6):853–865.
20. Epel ES, Lapidus R, McEwen B, Brownell K. Stress may add bite to appetite in women: a laboratory study of stress-induced cortisol and eating behavior. *Psychoneuroendocrinology*. 2001;26(1):37–49.
21. Cartwright M, Wardle J, Steggles N, Simon AE, Croker H, Jarvis MJ. Stress and dietary practices in adolescents. *Health Psychol*. 2003;22(4):362–369.
22. Smit HJ, Gaffan EA, Rogers PJ. Methylxanthines are the psychopharmacologically active constituents of chocolate. *Psychopharmacology*. 2004;176(3–4):412–419.
23. Rose N, Koperski S, Golomb BA. Mood food: chocolate and depressive symptoms in a cross-sectional analysis. *Arch Intern Med*. 2010;170(8):699–703.
24. Macdiarmid JI, Hetherington MM. Mood modulation by food—an exploration of affect and cravings in chocolate addicts. *Br J Clin Psychol*. 1995;34:129–138.
25. Berridge KC. 'Liking' and 'wanting' food rewards: brain substrates and roles in eating disorders. *Physiol Behav*. 2009;97:537–550.
26. Blass EM, Shide DJ, Weller A. Stress-reducing effects of ingesting milk, sugars, and fats—a developmental perspective. *Ann N Y Acad Sci*. 1989;575:292–306.
27. Mercer ME, Holder MD. Antinociceptive effects of palatable sweet ingesta on human responsivity to pressure pain. *Physiol Behav*. 1997;61(2):311–318.
28. Yeomans MR, Wright P. Lower pleasantness of palatable foods in nalmefene-treated human volunteers. *Appetite*. 1991;16(3):249.
29. Wang GJ, Volkow ND, Logan J, et al. Brain dopamine and obesity. *Lancet*. 2001;357(9253):354.
30. Bohon C, Stice E, Spoor S. Female emotional eaters show abnormalities in consummatory and anticipatory food reward: a functional magnetic resonance imaging study. *Int J Eat Disord*. 2009;42(3):210–221.
31. van Strien T, Snoek HM, van der Zwaluw CS, Engels RC. Parental control and the dopamine D2 receptor gene (DRD2) interaction on emotional eating in adolescence. *Appetite*. 2010;54(2):255–261.
32. Markus CR. Dietary amino acids and brain serotonin function; implications for stress-related affective changes. *Neuromolecular Med*. 2008;10(4):247–258.
33. Markus CR, Olivier B, Panhuysen GE, et al. The bovine protein alpha-lactalbumin increases the plasma ratio of tryptophan to the other large neutral amino acids, and in vulnerable subjects raises brain serotonin activity, reduces cortisol concentration, and improves mood under stress. *Am J Clin Nutr*. 2000;71(6):1536–1544.
34. Capello AE, Markus CR. Differential influence of the 5-HTTLPR genotype, neuroticism and real-life acute stress exposure on appetite and energy intake. *Appetite*. 2014;77:83–93.
35. Keskitalo K, Tuorila H, Spector TD, et al. The three-factor eating questionnaire, body mass index, and responses to sweet and salty fatty foods: a twin study of genetic and environmental associations. *Am J Clin Nutr*. 2008;88(2):263–271.
36. Pecoraro N, Dallman MF, Warne JP, et al. From Malthus to motive: how the HPA axis engineers the phenotype, yoking needs to wants. *Prog Neurobiol*. 2006;79(5–6):247–340.
37. Peters A, Pellerin L, Dallman MF, et al. Causes of obesity: looking beyond the hypothalamus. *Prog Neurobiol*. 2007;81(2):61–88.
38. Dallman MF. Stress-induced obesity and the emotional nervous system. *Trends Endocrinol Metab*. 2010;21(3):159–165.
39. Raspopow K, Abizaid A, Matheson K, Anisman H. Psychosocial stressor effects on cortisol and ghrelin in emotional and non-emotional eaters: influence of anger and shame. *Horm Behav*. 2010;58(4):677–684.
40. Pool E, Delplanque S, Coppin G, Sander D. Is comfort food really comforting? Mechanisms underlying stress-induced eating. *Food Res Int*. 2015;76(2):207–215.
41. van Strien T, Herman CP, Anschutz DJ, Engels RC, de Weerth C. Moderation of distress-induced eating by emotional eating scores. *Appetite*. 2012;58(1):277–284.

CHAPTER

56

Stretched Thin: Stress, In-Role, and Extra-Role Behavior of Educators

K.C. Ryan[1], L.M. Dunn-Jensen[2]

[1]Indiana University, Bloomington, IN, USA

[2]San Jose State University, San Jose, CA, USA

OUTLINE

Abstract

Ample anecdotal and empirical evidence attests to the considerable stress that teachers at all educational levels experience at work. While this stress, and the burnout that often follows, has typically been considered through the lens of the school environment or the teaching task itself, we suggest that educators' engagement in extra-role behaviors (that is, "above and beyond" their job duties) contributes substantially to feelings of stress and overload on the job. This chapter examines possible explanations for engaging in extra-role behavior in educational settings and suggests preliminary strategies for managing such behavior in order to reduce the overall stress experienced by educators in a wide variety of settings.

It's good to take control of your valuable time and realize it's all right to say no. ***Craig Williams, Author.***

INTRODUCTION

In recent years, significant attention has been focused on the broad sources and effects of stress among teachers at all levels of education. In K-12 environments, research has consistently addressed stress experienced by teachers as a result of students' behavior, discipline issues, and overall workload.[1–3] In addition, it has been shown that the negative emotional and psychological effects of stress typically result in burnout,[4,5] reduced job satisfaction, and a decreased sense of effectiveness on the job.[2,6,7] Stress related to school violence and the threat of school violence, in particular, has been shown to negatively impact both work functioning and mental health among teachers.[8] A recent study of elementary and secondary teachers by Collie et al.[9] examined the relationships among social-emotional learning, perceptions of school climate, teachers' sense of stress, teaching efficacy, and job satisfaction. Results indicated that teachers' perceived stress with respect to workload and efficacy directly impacted job satisfaction, lending additional evidence to the assumption that work-related stress may have a profound impact, not just on individual teachers, but also on students and school systems. In short, there is ample evidence to agree with the conclusions of both Brown and Nagal[10] and Kyriacou[3] that teaching is a highly stressful profession. Moreover, this conclusion is true not just for K-12 teachers, but for college educators, as well.

Empirical studies of stress and burnout among university faculty have increased as the higher education landscape evolves to include even greater numbers of part-time, adjunct, nontenure track, and online educators. As noted by Watts and Robertson,[11] "academia is no longer a comparatively low-stress working environment…" (p. 34). Similarly, Lackritz[12] concluded that "burnout clearly affects a significant percentage of

http://dx.doi.org/10.1016/B978-0-12-800951-2.00061-3

KEY POINTS

- Educators at all levels experience occupational stress.
- Engaging in extra-role or organizational citizenship behavior (OCB) can contribute significantly to feelings of stress or overload in academic settings.
- Educators can manage stress by understanding their motivation to perform OCB and thinking strategically about the OCB in which they engage.

faculty" (p. 725), noting that course load, grading, and time spent on service activities, among other variables, significantly contribute to feelings of emotional exhaustion among faculty members. Similar stressors may be felt even more acutely by those faculty members who do not have access to the benefits of tenured or tenure-track positions; the "invisible" faculty who typically work for far lower pay and without health benefits, retirement benefits, or long-term contracts.[13] For example, a recent study of online instructors revealed a significant tendency to experience burnout, indicating the need to proactively address stress and its related effects among this growing faculty population.[14] As with the K-12 community, burnout resulting from stress can have negative implications not just for university faculty, but for students and academic institutions, as well.

Clearly, there is no shortage of evidence to indicate that teaching, in virtually every educational context, is a stressful and demanding profession for a wide range of reasons. Stress in educational settings can emerge from many sources, including interactions with students, course load demands, and reduced levels of job security, as noted above. The stress emerging from these interpersonal and organizational demands deeply affects the quality of work life for instructors across educational settings. Yet, it is not just the defined teaching tasks or teaching environment that produces a significant amount of stress for educators. Perhaps there is an unexplored area of activities that instructors engage in that has not been considered.[15,16] One possibility is that educators experience considerable stress as a result of their engagement and, often, overengagement in extra-role or organizational citizenship behaviors (OCBs).

EXTRA-ROLE BEHAVIOR

Decades of research have shown that discretionary behavior, alternatively classified as extra-role or OCB captures an individual engaging in work behavior beyond defined job obligations and impacts both an organization's and an individual's success. Behavior such as helping a coworker meet a deadline, protecting the organization from potential harm, or mentoring new organizational members both increase organizational effectiveness and result in positive outcomes for individuals, such as better overall performance evaluations (see Refs. 17 for a review, 18–20).

In recent years, however, attention has turned to the negative consequences associated with such extra-role behavior. Bolino and his colleagues[21] systematically reviewed the growing evidence that individuals engaging in OCB may do so at significant personal and professional cost. They point to several studies indicating that role overload, stress, and work-family conflict may result from the decision to engage in citizenship behavior. While the positive outcomes of this behavior are not in dispute, it is increasingly clear that extra-role behavior can be associated with adverse consequences for individuals. Bolino and Turnley,[22] for example, found a positive relationship between employees displaying a high level of individual initiative to engage in OCB and job stress, while Bergeron et al.[23] argue that, particularly in outcome-based control systems in which productivity is central to job performance, time spent on OCB at the expense of time spent on task performance results in reduced career outcomes for individuals, as reflected in performance evaluations, salary increases, and promotions.

This growing body of research reframes extra-role or citizenship behavior in all organizational contexts, including education. Historically, extra-role behaviors have been considered positive, voluntary behaviors, either directed toward helping or supporting other individuals in the workplace or to facilitate organizational functioning, as a whole. Yet even while educators may expect positive outcomes from engaging in extra-role behavior, including enhanced performance evaluations and personal satisfaction, they must be prepared to experience the negative outcomes, as well. More recent research has exposed the complexity of citizenship behavior, both in terms of why individuals choose to engage in such behavior, and the outcomes associated with it, and these complexities play out markedly in the daily lives of instructors across educational settings.

EXTRA-ROLE BEHAVIOR IN EDUCATIONAL SETTINGS

It is easy to wonder why educators, already burdened with so many demands on their time, so many constituents to please, and so much uncertainty in their environment, would choose to take on more responsibility. In fact, instructors may display extra-role behaviors for a number of reasons and, in doing so, increase the stress associated with juggling multiple demands with finite resources.

Consider the situation faced by faculty members in higher education. One reason professors may choose to add to the demands of their jobs by engaging in extra-role behavior is that it can be difficult to clearly distinguish between their in-role and extra-role obligations. The lines between what is contractually required, what is implicitly or culturally expected, and what an instructor innately feels the need to give, are often blurred. This is true especially given the ambiguous nature of what is considered to be service to a university. For example, when a student asks for a letter of recommendation for a future job or to apply for a study abroad opportunity is this task considered in-role or extra-role? Moreover, how many letters of recommendation should a professor commit to in a semester? Five, ten, twenty?

Another factor that may contribute to a faculty member engaging in extra-role behaviors is what that individual might perceive to be included within the job requirements. Research suggests that individuals may differ with respect to their beliefs about what they should be contributing to the job.[17,24] Those who perceive and define their job responsibilities and expectations more broadly may be more likely to engage in extra-role behavior, even with the knowledge that such behavior is not part of their formal employment contract. Furthermore, the structure of the job itself uses a behavior-based control system which focuses on the behaviors, knowledge, skills, and abilities that teachers bring to the job, rather than focusing solely on measurable outputs.[25] Therefore, the wide range of behaviors that instructors display on the job, and how those behaviors are interpreted and classified by others, is an important factor in performance evaluations. Research in behavior-based systems, most of which reveals that both task performance and extra-role or citizenship behavior are significantly related to assessments of overall performance, also suggests that it is not always easy to distinguish where in-role performance ends and extra-role performance begins.[19,26,27]

In higher education, for example, a professor's work-role (beyond research requirements) is comprised of the two additional elements of teaching and service. The teaching obligation is generally easy to measure in terms of course load or credit hours. Service, however, is more difficult to define. Faculty may perform service in support of the department, school, or university for which they work. Service to the larger community may also be considered during performance evaluations. Examples of service include sponsoring student organizations, serving on committees, promoting the department at recruiting events, representing the department at university functions, meeting with prospective students or faculty, and any number of other activities unique to a particular institution. Equating the quantity and quality of these service activities is difficult, and it is equally hard to define the point where an acceptable level of service has been achieved. As a result, faculty members anxious to ensure that they meet their contractual obligations, may find themselves doing far more than the expected level of service, not knowing exactly the point when they have met either the contractual threshold, or the perceived cultural norm for their department.

For nontenure track faculty, the ambiguity surrounding service requirements may be even more acute. It is interesting that such decisions may also have the effect of moving the school or department norm increasingly higher as nontenure track faculty unwittingly engage in an "arms race" of service activities with respect to their peer group, exacerbating the problem of not knowing when required service ends, and discretionary extra-role behavior begins. As a result, the overall stress in trying to gauge acceptable levels of performance as well as managing the increasing number of activities with finite time, energy, and emotional investment increases. Empirical results lend support for such scenarios. In a study of contingent workers with short-term contracts, Feather and Rauter[28] found that teachers with this job status reported both more job insecurity and more citizenship behaviors with respect to their colleagues with permanent employment contracts. The authors suggest that their less secure employment status compels these teachers to take on more extra-role responsibilities in order to enhance their status within the organization and strengthen the perception of others that they are adding more value. As in the nontenure track scenario, in an uncertain and competitive environment, it seems likely that individuals will engage in extra-role behaviors to look better relative to their peer group and to increase their chances of experiencing positive consequences, such as receiving a contract renewal.

Similar challenges play out in K-12 teaching environments. A recent study by Brown and Roloff[29] investigated the relationship between extra-role activities, such as coaching and sponsoring student organizations, and burnout among teachers. Measuring the time invested in work activities beyond the contractual obligations of the work week, the authors found that teachers who voluntarily engaged in extra-role investments were more likely to experience both burnout and decreased commitment to their teaching role. They suggest that fulfillment of the psychological contract between teachers and administrators may offset these negative effects, as past research has found relationships between psychological contracts and citizenship behavior.[30] However, such explicit or implicit promises may also serve to obscure the distinction between task and extra-role behavior as interpersonal support or extra resources may be offered by administrators in exchange for teachers' willingness to engage in such extra-role behavior. Teachers, for their part, may then experience increasing stress as they agree to take on additional work in order to be perceived as "team players" or valued organizational members.

Expanding on this need to fit in and be accepted by an organization, the extent to which educators define themselves (i.e., take on the social identity) as "good citizens" or "good soldiers" may also increase their commitment to engaging in extra-role behaviors. According to social identity theory, individuals identify themselves with different social groups based on interactions with other people.[31] Creed and Scully[32] (p. 392) suggest that when claiming an identity, individuals will use "their identity to claim membership in a social category," and Bartel and Dutton[33] (p. 121), define membership claiming "as a strategic use of words and acts to assert that one is a legitimate organizational member." This identification with a group, in turn, will lead to activities that are congruent with the salient social identity.[34]

One social identity that educators may find salient in their work environment is a citizenship role or "good soldier" identity. It has been suggested that a citizenship role identity increases engagement in OCB[35] and, similarly, Mayfield and Taber[36] found a positive relationship between prosocial self-concept and OCB. Therefore, in academic settings, educators who identify as good soldiers may be more likely to volunteer for extra-role tasks to demonstrate and assert that particular social concept, regardless of the difficulties that may result in doing so.

Finally, and regardless of the potential for negative outcomes, teachers may also continue to invest in extra-role behaviors because, once they have established such a pattern, pulling back from or discontinuing their extra-role behavior altogether gives the impression that they are no longer meeting the expectations of their jobs. As Van Dyne and Ellis[37] and, relatedly, Vigota-Gadot[38] explain, others in the organization are so accustomed to seeing them perform such activities or fill such roles, that they are no longer perceived as voluntary and, moreover, supervisors or coworkers may begin to demand that individuals continue to perform such extra-role behaviors in order to meet performance expectations. For example, some teachers may work in settings in which a high level of extra-role behavior is the informal norm, and they are then covertly coerced to engage in what should be voluntary behaviors because of the external demands they experience. Such seemingly compulsory OCB has been shown to contribute to job stress and other negative outcomes.[38,39]

While teachers' decisions to engage in extra-role behavior may result from a variety of motivating factors, it is clear that simply taking on these added responsibilities contributes to the overall stress educators experience at work. This makes it imperative to develop strategies to manage extra-role commitments in a way that both reduces stress and allows individuals to continue to receive positive benefits associated with citizenship behavior.

MANAGING EXTRA-ROLE BEHAVIOR AND STRESS

Despite the challenges inherent in education, such as demanding students, unpredictable environments, and financial constraints that make it imperative for every teacher to "do more with less," it is possible for teachers to affect their overall level of job stress by paying close attention to the discretionary behaviors in which they engage. Specifically, they must first be conscious about the extra-role tasks they take on, clearly assessing their motivation and expected benefits for doing so.

While universities need professors to engage in service activities to ensure the smooth functioning of the academic institution, a faculty member must say "yes" to the right activities. Faculty members need to carefully weigh the opportunity cost of prescribed task versus extra-task behaviors and analyze the cost-benefit to both themselves and the organization. For example, willingness to be a committee member on a curriculum program committee can give the faculty member a voice in program development, the ability to interact with colleagues from other departments, and visibility with university administration. On the other hand, being a faculty member that reads 125 applications for an honors program during winter break with no true visibility to others at the university may not return as much benefit to that faculty member. While we don't advocate that faculty members should only engage in visible extra-role activities, we would encourage faculty members to consider a balance between engaging in both visible and less visible activities. Additionally, we believe that faculty members need to find a balance between engaging in not enough and too much OCB.

What then, is an appropriate level of engagement in OCB and what are the potential outcomes for an educator and his or her academic institution? Table 1 captures the challenges that an individual is faced with concerning the level of OCB in which to engage.

If the instructor engages in either not enough or too much OCB, ultimately both the individual and his or her organization may experience negative outcomes. For example, in the long run, individuals who, for whatever reason, commit to performing consistently high levels of OCB are likely to experience increasing stress, overload, conflict, and burnout. Organizations, in turn, may suffer from reduced attention and resources devoted to task behavior, as well as the direct and indirect costs of higher turnover. Thus, determining an appropriate or more balanced level of engaging in OCB should be a goal for every educator. Individuals need to become more aware of and more strategic about the extra-role behaviors in which they are engaging as well as the motivations influencing that behavior.

TABLE 1 Relationship Between Level of OCB Engagement and Outcomes

	Outcome for individual	Outcome for the organization
Too little OCB	Positive (+) short-term • Attentive to contractual job obligations • Low need to manage time and other resources to meet demands	Positive (+) short-term • Focus on completing contractual job obligations
	Negative (−) long-term • Negative impact on performance evaluations • Poor position relative to peers, cultural norms, identity • Susceptible to coercion by supervisor and others • Resentment from those who do engage in OCB	Negative (−) long-term • Lower overall performance output • Extra resources needed to complete tasks outside defined job responsibilities • Less cohesive culture; increased turnover
Balanced OCB	Positive (+)	Positive (+)
Too much OCB	Positive (+) short-term • Increased visibility, perceptions of being a "team-player" • Improved performance evaluations • Improved position relative to peers • Support of OCB identity	Positive (+) short-term • Higher overall performance output • More cohesive culture
	Negative (−) long-term • Stress, frustration, negative spillover • Inability to decrease or withdraw behaviors • Burnout	Negative (−) long-term • Time, attention, resources diverted from contractual job obligations • Direct and indirect costs of turnover

CONCLUSION

Strategically investing their time, attention, energy, and other assets across activities with the highest rate of return, educators are far more likely to find the balance of behaviors where both they and their academic institutions enjoy optimal outcomes. Skillfully managing not only their overall investment in OCB, but also engaging in OCB that will provide the outcomes individuals desire can significantly reduce the amount of stress experienced in the teaching role. There is no question that OCB contribute to positive workplace functioning or that they are an essential part of academic organizations. The opportunity to contribute to the overall quality of life in an educational institution is something many instructors value and appreciate.

However, it is equally important to recognize the stress that can accompany these choices. For long-term health and productivity, individuals must critically consider and manage their extra-role behaviors. We are certainly not advocating a wholesale boycott of OCB. It is true, however, that teaching, across contexts, is a stressful profession for many reasons outside an individual educator's control. This makes it even more critical to minimize stress where there is discretion to do so. By purposefully balancing in-role and extra-role behaviors, both individuals and organizations can experience long-term benefits.

References

1. Chaplain RP. Stress and psychological distress among trainee secondary teachers in England. *Educ Psychol*. 2008;28:195–209.
2. Klassen RM, Chiu MM. Effects on teachers' self-efficacy and job satisfaction: teacher gender, years of experience, and job stress. *J Educ Psychol*. 2010;102:741–757.
3. Kyriacou C. Teacher stress: directions for future research. *Educ Rev*. 2001;51:27–35.
4. Chang M. An appraisal perspective of teacher burnout: examining the emotional work of teachers. *Educ Psychol Rev*. 2009;21:193–218.
5. McCarthy CJ, Lambert RG, O'Donnell M, Melendres LT. The relation of elementary teachers' experience, stress, and coping resources to burnout symptoms. *Elem Sch J*. 2009;109:282–301.
6. Caprara GV, Barbaranelli C, Steca P, Malone PS. Teachers' self-efficacy beliefs as determinants of job satisfaction and students' academic achievement: a study at the school level. *J Sch Psychol*. 2006;44:473–490.
7. Klassen RM, Chiu MM. The occupational commitment and intention to quit of practicing and pre-service teachers: influence of self-efficacy, job stress, and teaching context. *Contemp Educ Psychol*. 2011;36:114–129.
8. Currier JM, Holland JM, Rozalski V, Thompson KL, Rojas-Flores L, Herrera S. Teaching in violent communities: the contribution of meaning made of stress on psychiatric distress and burnout. *Int J Stress Manag*. 2013;20:254–277.
9. Collie RJ, Shapka JD, Perry NE. School climate and social-emotional learning: predicting teacher stress, job satisfaction, and teaching efficacy. *J Educ Psychol*. 2012;104:1189–1204.
10. Brown S, Nagal L. Preparing future teachers to respond to stress: sources and solutions. *Action Teach Educ*. 2004;26:34–42.
11. Watts J, Robertson N. Burnout in university teaching staff: a systematic literature review. *Educ Res*. 2011;53:33–50.
12. Lackritz JR. Exploring burnout among university faculty: incidence, performance, and demographic issues. *Teach Teach Educ*. 2004;20:713–729.
13. Hamilton K. Getting off the burnout track. *Diverse Issues High Educ*. 2005;22:26–31.
14. Hogan RL, McKnight MA. Exploring burnout among university online instructors: an initial investigation. *Internet High Educ*. 2007;10:117–124.
15. Bynum LA, Bentley JP, Holmes ER, Bouldin AS. Organizational citizenship behaviors of pharmacy faculty: modeling influences of equity sensitivity, psychological contract breach, and professional identity. *J Leadersh Account Ethics*. 2012;9:99–111.
16. Erturk A. Increasing organizational citizenship behaviors of Turkish academicians: mediating role of trust in supervisor on the relationship between organizational justice and citizenship behaviors. *J Manag Psychol*. 2007;22:257–270.

17. Organ DW, Podsakoff PM, MacKenzie SB. *Organizational Citizenship Behavior: Its Nature, Antecedents, and Consequences*. Thousand Oaks, CA: Sage; 2006.
18. Dunlop PD, Lee K. Workplace deviance, organizational citizenship behavior, and business unit performance: the bad apples do spoil the whole barrel. *J Organ Behav*. 2004;25:67–146.
19. Podsakoff NP, Whiting SW, Podsakoff PM, Blume BD. Individual- and organizational-level consequences of organizational citizenship behaviors: a meta-analysis. *J Appl Psychol*. 2009;94:122–141.
20. Walz SM, Niehoff BP. Organizational citizenship behaviors: their relationship to organizational effectiveness. *J Hosp Tour Res*. 2000;24:301–319.
21. Bolino MC, Klotz AC, Turnley WH, Harvey J. Exploring the dark side of organizational citizenship behavior. *J Organ Behav*. 2012;34:542–559.
22. Bolino MC, Turnley WH. The personal costs of citizenship behavior: the relationship between individual initiative and role overload, job stress, and work-family conflict. *J Appl Psychol*. 2005;90:740–748.
23. Bergeron DM, Shipp AJ, Rosen B, Furst SA. Organizational citizenship behavior and career outcomes: the cost of being a good citizen. *J Manag*. 2013;39:958–985.
24. Morrison EW. Role definitions and organizational citizenship behavior: the importance of an employee's perspective. *Acad Manage J*. 1994;37:1543–1567.
25. Oliver RL, Anderson E. Behavior- and outcome-based sales control systems: evidence and consequences of pure-form and hybrid governance. *J Pers Sell Sales Manag*. 1995;15:1–15.
26. Van Dyne L, Cummings LL, Parks JM. Extra-role behaviors: in pursuit of construct and definitional clarity. *Res Organ Behav*. 1995;17:215–285.
27. Whiting SW, Podsakoff PM, Pierce JR. Effects of task performance, helping, voice, and organizational loyalty on performance appraisal ratings. *J Appl Psychol*. 2008;93:125–139.
28. Feather NT, Rauter KA. Organizational citizenship behaviours in relation to job status, job insecurity, organizational commitment and identification, job satisfaction and work values. *J Occup Organ Psychol*. 2004;77:81–94.
29. Brown LA, Roloff ME. Extra-role time, burnout, and commitment: the power of promises kept. *Bus Commun Q*. 2011;74:450–474.
30. Turnley WH, Feldman DC. Re-examining the effects of psychological contract violations: unmet expectations and job dissatisfaction as mediators. *J Occup Behav*. 2000;21:25–42.
31. Tajfel H, Turner JC. The social identity theory of intergroup behavior. In: Worchel S, Austin WG, eds. *Psychology of Intergroup Relations*. 2nd ed. Chicago, IL: Nelson-Hall; 1985:7–24.
32. Creed WED, Scully MA. Songs of ourselves: employees' deployment of social identity in workplace encounters. *J Manag Inq*. 2000;9:391–412.
33. Bartel CA, Dutton JE. Ambiguous organizational memberships: constructing organizational identities in interactions with others. In: Hogg MA, Terry DJ, eds. *Social Identity Processes in Organizational Contexts*. Philadelphia, PA: Psychology Press; 2001:115–130.
34. Ashforth BE, Mael F. Social identity theory and the organization. *Acad Manage Rev*. 1989;14:20–39.
35. Stoner J, Perrewe PL, Munyon TP. The role of identity in extra-role behaviors: development of a conceptual model. *J Manag Psychol*. 2011;26:94–107.
36. Mayfield CO, Taber TD. A prosocial self-concept approach to understanding organizational citizenship behavior. *J Manag Psychol*. 2010;25:741–763.
37. Van Dyne L, Ellis JB. Job creep: a reactance theory perspective on organizational citizenship behavior as over-fulfillment of obligations. In: Coyle-Shapiro JAM, Shore LM, Taylor MS, Tetrick LE, eds. *The Employment Relationship: Examining Psychological and Contextual Perspectives*. Oxford: Oxford University Press; 2004:181–205.
38. Vigota-Gadot E. Compulsory citizenship behavior: theorizing some dark sides of the good soldier syndrome in organizations. *J Theory Soc Behav*. 2006;36:77–93.
39. Bolino MC, Turnley WH, Gilstrap JB, Suazo MM. Citizenship under pressure: what's a "good soldier" to do? *J Organ Behav*. 2010;31:835–855.

CHAPTER

57

Stress and Coping in the Menopause

E.E.A. Simpson
Ulster University, Coleraine, UK

OUTLINE

Abstract

The menopausal transition (MT) is characterized by higher stress levels in some women. It is unclear if the MT leads to stress, as a result of increased symptoms associated with this time, or a potential vulnerability to fluctuations in hormone levels. Also, women may experience an increase in stressful life events that coincide with the MT and contribute to increased psychological distress. Few studies have looked at what women find stressful during the MT and the strategies they engage in to help them cope with this. The current chapter will provide an overview of the research that has been carried out on stress, coping, and menopause and the implications of this for menopause management.

Menopause literally means last period and is defined as "the permanent cessation of menstruation following loss of ovarian follicular activity."[1] It is regarded as a natural transition[2] from the reproductive to the nonreproductive years in a woman's life, emerging gradually from 47 to 55 years.[3] Ovarian production of estrogens begins to decline in the fourth or fifth decade of life, marking the beginning of the menopausal transition (MT) with the perimenopausal phase, and leading to menopausal symptoms such as hot flashes. Postmenopause is diagnosed retrospectively as 12 months of spontaneous amenorrhoea, usually around 51 years of age[4] and lasting for 4-5 years.[5] With increased longevity women will be postmenopausal for one third of their lives, and promotion of a healthy MT is crucial for healthy aging.[6]

PSYCHOLOGICAL STRESS

Psychological stress is defined in this chapter using the Transactional Theory of Stress,[7] which regards stress as a process comprising an interaction between the person and their environment. It has been used previously to look at chronic conditions with prolonged physiological changes. The process begins with the psychological appraisal of a stressful event, considering the suddenness, predictability, and control over the event,[8] and the emotional reactions to it, all of which will influence the amount of stress experienced and the coping styles employed.[7] This is accompanied by physiological reactions to stress that occur with the activation of the hypothalamic-pituitary-adrenal (HPA) axis. Secondary appraisal focuses on coping, and the cognitive and behavioral efforts needed to manage specific aspects of the stressful event.[7]

Physiological Reaction to Stress

Psychological stress produces physiological changes in the autonomic, neuroendocrine, and immune systems.[9] From a recent review of the literature, HPA dysregulation occurs during MT.[10] Some studies report increased physiological reactivity to stress in menopausal women with evidence of elevated levels of cortisol (stress

Stress: Concepts, Cognition, Emotion, and Behavior
http://dx.doi.org/10.1016/B978-0-12-800951-2.00062-5

KEY POINTS

- Some women experience increased psychological distress during the MT.
- Stress during the MT may be a result of the menopause itself or stressful life events occurring alongside it.
- Coping with the MT needs further research in order to establish coping styles that promote well-being at this time.
- Clinicians should be aware that stress may predict HT and symptom reporting.
- Treatment during MT should include stress management.

hormone) after controlling for social factors thought to induce stress.[11] Women with higher levels of vasomotor symptoms (hot flashes, night sweats), which are directly related to changes in estrogen, were found to have higher central sympathetic activity.[12] This suggests changes in reproductive hormones might have a detrimental effect on how women respond to stress.

Estrogen was found to regulate corticotrophin-releasing-hormone gene expression, which may explain the increase in cortisol levels found during MT. Higher levels of circulating estradiol were negatively associated with physiological reactivity to stress in one study,[13] but not in another.[14] These studies suggest estrogen may have a protective effect against stress, and additional support for this comes from studies looking at hormone therapy (HT). Long term HT was associated with enhanced parasympathetic responsiveness to stress.[15] However, progestogen, also given to some women as part of their HT, was found to antagonize the protective effects of estrogen. Dysregulation of the HPA during MT has important implications for health and well-being in postmenopause, especially in the development of chronic conditions such as coronary heart disease and changes in immune function. A recent scoping review concluded that there is very little research on how and why the HPA changes during MT.[16]

STRESS AND THE MENOPAUSE

The menopause itself may be a source of stress due to increased or persistent symptoms that occur during this time.[17] Stress during the MT may be associated with increased stressful life events such as death of a parent or children leaving home. These will now be considered.

Menopause as a Source of Stress

During the MT, women may experience biological, psychological, and social changes that influence symptom reporting during this time. A range of biological changes occur such as hormonal fluctuations. More specifically, a decline in estradiol (the main endogenous estrogen) and an increase in follicular stimulating hormone, and luteinizing hormone, the latter controlling ovulation and the production of estradiol.[5] From peri- to postmenopause, estradiol declines drastically leading to increased symptoms such as night sweats and hot flashes.[18] Psychological changes (chronic fatigue, sleep problems, changes in appetite, mood, and cognition) may lead to increased psychological distress, depression, and anxiety in some women.[18,19] Social changes during this time have been associated with losses, such as loss of maternal role, and will be discussed later.

Few studies have looked at stress, psychological distress, and menopausal symptoms.[20] Those that have tend to report increased stress is associated with greater severity of symptom reporting,[21] especially somatic and vasomotor symptoms. This may be related to estrogen decline and the physical and psychological impact of aging.[19] Severity of menopausal symptoms may be higher in late perimenopause and early postmenopause,[17] and these may also coincide with greater fluctuations in estrogen levels.

Psychological distress for some women during this time may be due to a vulnerability to changes in hormone levels, or a previous history of depression or anxiety disorders. There is a suggestion that depressive symptoms may be exacerbated by the MT, especially if vasomotor symptoms are severe.[22] Others suggest that depression that occurs during this time may be due to social factors, such as educational level, ethnicity, socioeconomic status, and partner status.[23] Both animal and human studies have provided evidence that progesterone, estrogen, and serotonin are related to changes in mood, they share pathways and receptor sites in the brain,[24] and may account for vulnerability to mood disorders in some women during MT.

Attitude to MT may mediate symptom reporting and stress. A negative attitude to MT was associated with more severe menopausal symptoms and depression, with a positive attitude being related to fewer symptoms. In a prospective study of women in early postmenopause, levels of perceived stress and attitudes to the menopause and aging were predictive of severity of reported menopausal symptoms at 12-month follow up.[25] Women who have lower emotional intelligence report more severe menopausal symptoms, have higher stress, anxiety and depression levels, a more negative attitude to the menopause and poorer physical health during MT.[2] More recently, a number of models have emerged that suggest attitudes, beliefs, and appraisal of the menopause are related to symptom reporting[26] and seeking HT or treatment during MT.[27]

Not all research has found evidence that menopause itself was particularly stressful.[28] This may be due to differences in participant samples, in a clinical sample of recently postmenopausal women attending a HT clinic, they had higher perceived stress scale scores[29] in comparison to other studies of women taken from community samples. Seeking help may be indicative of higher symptomatology and stress levels, as this was also found to predict intentions to use HT.[30] The majority of women in the MT will report hot flashes and night sweats, but only 20% of women will have severe enough symptoms that require further investigation.

Stressful Life Events During the Menopause

Stress encountered at menopause may also be due to psychosocial factors that coincide with menopause, such as death of a parent, children leaving home, divorce, or becoming a grandparent, all of which are major sources of stress[31] and are associated with increased reporting of depression and anxiety and more severe menopausal symptoms.[32] There is some evidence that stressful life events may contribute to the onset of medical and psychological problems and that the stress experienced may contribute to the development and course of chronic conditions in postmenopause. Therefore it is important to understand what causes stress during menopause and how women deal with this.

Early research tended to focus on the empty-nest syndrome, when all children had left home, a time associated with increased psychological distress. However, with considerable social change over the last few decades, women may view children leaving in a more positive light, for example, as a reduction in daily hassles.[31] Women are also having children later, so in some cases children are still quite young when their mothers are going through MT and not ready to leave home, plus there is some evidence that older children are not leaving the parental home until they are much older.[33]

Also, today women are more likely to work outside the home, so are no longer defined by their role as caregiver. However with employment can come work-related stressors. During the MT, stress can be related to work overload, role burden (employment and work outside of and within the home), and lack of support, which has been associated to functional decline if not addressed. One study identified symptoms such as tiredness, sleep disturbance, memory and attention deficits, and depressed mood as having a detrimental impact on productivity in relation to work.[34] Such symptoms were found to not only influence work but also personal and family relationships and quality of life. This may also be accompanied by equality issues, similar working conditions, opportunities and promotion, and pay in comparison to their male counterparts. Some research suggests women may have a disproportionate amount of time spent working compared to men.[35] Other research suggests that it is not the workload but how it is interpreted (positively or negatively) that affects stress levels. The workload and lack of support has implications for health and well-being in women at midlife and higher menopausal symptoms have been associated with money worries, lower education, and lack of full-time employment.

Other sources of stress during menopause are death of a parent or spouse[17] and the distress may also be related to the menopausal women facing their own mortality and aging. Another source of stress is the breakdown of a relationship. Divorce rates peak during the menopausal years, in the majority of cases (65%) women seek the divorce, occurring usually when the children leave the home and the couple discover they have nothing in common. This may impact on psychological distress at MT as McKinlay et al.[36] found that women who were depressed during menopause were more likely to be divorced, widowed, or separated. Single women and women who never had children were less likely to be depressed during MT.[28]

Some of the research on stress has looked at daily stressors or "hassles" and their effect on health. These include stressful and demanding aspects of day-to-day life, such as being stuck in rush hour traffic or being late for an appointment. This deals with general stress levels rather than focusing on specific major stressful life events as a source of stress. More recent research has tried to establish what menopausal women themselves report as stressful in everyday life. Simpson and Thompson[29] found that a clinical sample of postmenopausal women reported family problems (44%), menopause (20%), work problems (19%), daily hassles (12%), and other health problems (5%) as the most stressful events experienced in early postmenopause. In a Canadian study of midlife, women reported children's health, parenting, relationships with partner and other family members, lack of time for oneself, financial issues, unemployment, and lack of confidence and sleep as the main sources of stress.[37] These reported stressors are similar to those reported in nonmenopause studies.

COPING AND MENOPAUSE

Menopausal symptoms may act as a cue to bodily changes and a need to manage or cope with these. Coping is regarded as secondary appraisal and its function is to reduce stress. Reporting of menopausal symptoms, attitudes, beliefs, understanding, and expectations of the

MT, will have a direct effect on how women cope.[38] Improving coping skills may reduce stress levels and promote better menopause management in some women.

Coping Styles

Coping responses are infinite, but can be categorized as problem and emotion-focused coping and dysfunctional coping.[39] Emotion-focused coping regulates stressful emotions by changing the way the relationships in the environment are attended to, and reappraisal of these if necessary. It may involve emotional distancing or seeking social support, and may be used if the situation can't be changed, such as the MT. Problem-focused coping alters the environment to reduce stress and is generally employed when the situation is perceived as controllable. Dysfunctional coping is defined as those coping styles that employ avoidance, denial, self-blame, or detrimental ways of coping such as alcohol or drug abuse.[39] Greater psychological distress and symptom reporting during the MT has been associated with less effective coping styles.[21] This requires greater education and management to promote effective coping and better well-being.

Coping styles are seldom used alone, and may change over time and adapt to the demands of the situation.[7] It is suggested that people may employ preferred coping strategies that can be applied over a variety of situations. There has been very little research looking at stress, coping, and MT.

Coping with the MT

Coping styles employed during menopause have been virtually overlooked, with few studies looking at this in conjunction with stress. Menopausal women face a number of stressors and physical and social changes at this time in their lives, and how they deal with these is not well understood. The most reported coping styles displayed by a clinical sample of postmenopausal women were catharsis, taking direct action, and seeking social support.[29] Social support is a well-established stress buffer, and during MT, higher levels, has been related to fewer menopausal symptoms and better well-being. Menopause provides a unique opportunity for women to actively seek out social relationships beyond the family.[36] It is a time of uncertainty for many women, seeking advice and support from other women may be an attempt to understand the experience and to be reassured that what they are going through is normal. At menopause, having a friend or group of friends was related to better psychological well-being and adjustment to MT.

Another study looked at coping in menopause and identified three coping styles used by women: inventive, troubled, and reactive coping, and how these individual differences in coping styles impacted on the menopausal experience.[40] Effective coping in menopause could mediate the negative impact of the MT reported by some women. Women who view themselves as having control over menopausal symptoms will engage in more problem-focused coping to combat these. MT is a natural event that all women go through and for this reason, it is largely out of the women's control. HT use may be a form of participatory control and an attempt to manage symptoms more effectively. In one study, attending an HT clinic was viewed as a form of problem-focused coping to combat menopause and HT use, by coming to a clinic for advice about MT and starting HT in an attempt to cope with the symptoms associated with estrogen deficiency.[29] More research is needed to determine how women cope with the biological, psychological, and social factors that occur during the MT and how these mediate stress levels and well-being during this time.

IMPLICATIONS FOR MENOPAUSE MANAGEMENT

This chapter has highlighted the need for a better understanding of how stress and coping influence the MT and symptom reporting. The MT is complex with interactions between biological, psychological, and social factors which may impact on health and well-being. This is complicated further by individual differences in how people respond to the MT.[16] The results of these studies suggest that there may be an additive element involved in the relationship between psychological symptoms and life stress, implying that women who have greater life stresses during menopause may be more susceptible to psychological distress.

Women going through the MT should be offered education and support for all aspects of this. A stress management program could be employed as a useful alternative to HT, which is not the most effective way to treat psychological symptoms or social challenges. There are some cognitive and behavioral interventions that are showing some success in the management of menopausal symptoms and the promotion of more effective coping strategies to deal with the MT. There is a need to understand psychological stress and coping styles that occur during the MT and how these interact with underlying biological processes in order for it to be managed more effectively to promote health and well-being in menopausal women.

References

1. WHO. Research on the menopause in the 1990. Report of a WHO Scientific Group. *World Health Organ Tech Rep Ser*. 1990;866:1–107.

2. Bauld R, Brown RF. Stress, psychological distress, psychosocial factors, menopause symptoms and physical health. *Maturitas.* 2009;62:160–165.
3. Greendale G, Lee NP, Arriola ER. The menopause. *Lancet.* 1999;353:571–580.
4. Dravta J, Real F, Schindler C, et al. Is age at menopause increasing across Europe? Results on the age at menopause and determinants from two population based studies. *Menopause.* 2009;20:1–11.
5. Prior J, Hitchcock C. The endocrinology of perimenopause: need for paradigm shift. *Front Biosci.* 2011;3:474–486.
6. Jaspers L, Daanb NMP, van Dijk GM, et al. Health in middle-aged and elderly women: a conceptual framework for healthy menopause. *Maturitas.* 2015;81:93–98.
7. Folkman S, Lazarus RS. *Stress, Appraisal and Coping.* New York: Springer; 1984.
8. Wong PTP, Weiner B. When people ask "Why" questions and the heuristics of attributional search. *J Pers Soc Psychol.* 1981;40:650–653.
9. Cacioppo JT, Malarkey WB, Kiecolt-Glaser JK, et al. Heterogeneity in neuroendocrineand immune responses to brief psychological stressors as a function of autonomic cardiac activation. *Psychosom Med.* 1995;57:154–164.
10. Young E, Korszun A. Sex, trauma, stress hormones and depression. *Mol Psychiatry.* 2010;15:23–28.
11. Woods NF, Mitchell ES, Smith-DiJulio K. Cortisol levels during the menopausal transition and early postmenopause: observations from the Seattle Midlife Women's Health Study. *Menopause.* 2009;16:708–718.
12. Nedstrand E, Wijma K, Lindgren M, Hammar M. The relationship between stress-coping and vasomotor symptoms in postmenopausal women. *Maturitas.* 1998;31:29–34.
13. Komesaroff PA, Esler MD, Sudhir K. Estrogen supplementation attenuates glucocorticoid and catecholamine responses to mental stress in perimenopausal women. *J Clin Endocrinol Metab.* 1999;84:606–610.
14. Kudielka BM, Schmidt-Reinwald AK, Hellhammer DH, Kirschbaum C. Psychological and endocrine responses to psychosocial stress and dexamethasone/corticotropin-releasing hormone in healthy postmenopausal women and young controls: the impact of age and a two-week estradiol treatment. *Neuroendocrine.* 1999;70:422–430.
15. Burleson MH, Malarkey WB, Cacioppo JT, et al. Postmenopausal hormone replacement: effects on autonomic, neuroendocrine, and immune reactivity to brief psychological stressors. *Psychosom Med.* 1998;60:17–25.
16. Hoyt LT, Falconi AM. Puberty and perimenopause: reproductive transitions and their implications for women's health. *Soc Sci Med.* 2015;132:103–112.
17. Woods NF, Mitchell ES, Percival DB, Smith-DiJulio K. Is the menopausal transition stressful? Observations of perceived stress from the Seattle Midlife Women's Health Study. *Menopause.* 2009;16. http://dx.doi.org/10.1097/gme. 0b013e31817ed261.
18. Harlow SD, Gass M, Hall JE, et al. Executive summary of the Stages of Reproductive Aging Workshop + 10: addressing the unfinished agenda of staging reproductive aging. *J Clin Endocrinol Metab.* 2012;97:1159–1168.
19. Potdar N, Shinde M. Psychological problems and coping strategies adopted by post menopausal women. *Int J Sci Res.* 2014;3:293–300.
20. Freeman E, Sammel M, Lin H, et al. Symptoms associated with menopausal transition and reproductive hormones in midlife women. *Obstet Gynaecol.* 2007;110(2 part 1):230–240.
21. Igarashi M, Saito H, Morioka Y, et al. Stress vulnerability and climacteric symptoms: life events, coping behavior and severity of symptoms. *Gynecol Obstet Invest.* 2000;49:170–178.
22. Reed SD, Ludman EJ, Newton KM, et al. Depressive symptoms and menopausal burden in the midlife. *Maturitas.* 2009;62:306–310.
23. Llaneza P, Garcia-Portilla MP, Llaneza-Suarez D, Armott B, Peres-Lopez FR. Depressive disorders and the menopause transition. *Maturitas.* 2012;71:120–130.
24. Schmidt PJ, Nieman LK, Danaceau MA, Adams LF, Rubinow DR. Differential behavioral effects of gonadal steroids in women with and in those without premenstrual syndrome. *N Engl J Med.* 1998;338(4):209–216.
25. Nosek M, Kennedy HP, Beyene Y, Taylor D, Gilliss C, Lee K. The effects of perceived stress and attitudes toward menopause and ageing on symptoms of menopause. *J Midwifery Womens Health.* 2010;55:328–334.
26. Hunter M, Mann E. A cognitive model of menopausal hot flushes and night sweats. *J Psychosom Res.* 2010;69:491–501.
27. Hanisch L, Hantsoo L, Freeman E, Sullivan G, Coyne J. Hot flashes and panic attacks: a comparison of symptomatology, neurobiology, treatment and a role for cognition. *Psychol Bull.* 2008; 134:247–269.
28. Hunter MS. Somatic experiences of the menopause: a prospective study. *Psychosom Med.* 1990;52:357–367.
29. Simpson EEA, Thompson W. Stressful life events, psychological appraisal and coping style in postmenopausal women. *Maturitas.* 2009;63:357–364.
30. Simpson EEA. Predictors of intentions to use hormone replacement therapy in clinical postmenopausal women. *Climacteric.* 2012;15: 173–180.
31. Dennerstein L, Dudley E, Guthrie J. Empty nest or resolving door? A prospective study of women's quality of life in midlife during phase of children leaving and re-entering the home. *Psychol Med.* 2002;32:545–550.
32. Binfa L, Castelo-Branco C, Blumel JE, et al. Relationships between psychological complaints and vasomotor symptoms during climacteric. *Maturitas.* 2004;49:205–210.
33. ABS. Home and away: the living arrangements of young people. *Aust Social Trends.* 2009;4102:25–30.
34. Hamman RAM, Abbasa RA, Hunterb MS. Menopause and work—the experience of middle-aged female teaching staff in a Egyptian governmental facility of medicine. *Maturitas.* 2012;71:294–300.
35. Hochshild AR. *The Second Shift: Working Parents and the Revolution at Home.* New York: Viking; 1989.
36. McKinlay JB, McKinlay SM, Brambilla DA. The relative contributions of endocrine changes and social circumstances to depression in mid-aged women. *J Health Soc Behav.* 1987;28:345–363.
37. Walters V. Stress, anxiety, and depression; women's accounts of their health problems. *Soc Sci Med.* 1993;36:393–402.
38. Pimenta F, Leal I, Maroco J, Ramos C. Menopausal symptoms: do life events predict severity of symptoms in peri and post menopause? *Maturitas.* 2012;72:324–331.
39. Cooper C, Katona C, Livingston G. Validity and reliability of the Brief COPE in carers of people with dementia. *J Nerv Mental Dis.* 2008;196:838–843.
40. Kafanelis BV, Kostanski M, Komesaroff PA, Stojanovska L. Being in the script of menopause: mapping the complexities of coping strategies. *Qual Health Res.* 2009;19:30–41.

CHAPTER

58

Psychosomatic Medicine

G.A. Fava

University of Bologna, Bologna, Italy

State University of New York at Buffalo, Buffalo, NY, USA

OUTLINE

Abstract

Psychosomatic medicine may be considered as a comprehensive, interdisciplinary framework for assessment of psychological factors affecting individual vulnerability, as well as course and outcome of illness; biopsychosocial consideration of patient care in clinical practice; specialist interventions to integrate psychological therapies in the prevention, treatment, and rehabilitation of medical disease. Current advances in the field have practical implications for medical research and practice, with particular reference to the role of lifestyle, the challenge of medically unexplained symptoms, the psychosocial needs entailed by chronic illness, the function of the patient as a health producer.

INTRODUCTION

The ongoing progress in scientific medicine and technology in recent years has led to further splitting of knowledge and consequent fragmentation in the response to health issues. This is in contrast with research evidence that points to the importance of incorporating psychosocial aspects and holistic view in the approach to the person presenting with any health problem.

The interdisciplinary field of psychosomatic medicine has the potential to fill this gap, providing both the cognitive frame of reference and the practical tools that can be employed in everyday clinical practice.

KEY POINTS

- Medical assessment and treatment can be improved by using the methods that have been developed in psychosomatic medicine, with particular reference to medical unexplained symptoms, chronic illness and lifestyle modification.
- Psychosomatic medicine fosters a personalized approach to patient care, with emphasis on shared decision and self-management.
- Psychosocial therapies need to be incorporated in a multidisciplinary approach.

Stress: Concepts, Cognition, Emotion, and Behavior
http://dx.doi.org/10.1016/B978-0-12-800951-2.00064-9

HISTORY AND CURRENT DEVELOPMENTS

The term "psychosomatic" entails different meanings and connotations, which may explain its varying degrees of popularity. The concept was introduced by Heinroth in 1818, but modern psychosomatic medicine developed in the first half of the past century. It resulted from the interaction of different concepts having an ancient tradition in Western medicine: psychogenesis of disease and holism.[1] The idea of psychogenesis characterized the first phase of development of psychosomatic medicine (1930-1960), and evolved toward the definition of "psychosomatic disease" for a number of conditions in which psychological aspects seemed to have a preponderant role. Despite early criticism, the psychogenic postulate continued to exert explanatory power in the search for pathogenetic mechanisms. For example, a physical illness, such as peptic ulcer, was believed to be caused by psychological factors.

The term "psychosomatic disorder" was strongly criticized by several psychosomatic researchers in the 1960s, including Kissen,[2] Engel,[3] and Lipowski.[1] Kissen[2] provided a better specification of the term "psychosomatic." He clarified that the relative weight of psychosocial factors may vary from one individual to another within the same illness and underscored the basic conceptual flaw of considering diseases as homogeneous entities. Engel[3] wrote that the term "psychosomatic disorder" was misleading, since it implied a special class of disorders of psychogenic etiology and, by inference, the absence of a psychosomatic interface in other diseases. On the other hand, he viewed reductionism that overlooked the impact of nonbiological circumstances upon biological processes, as a major cause of mistreatment. Lipowski[1] criticized the concept of psychosomatic disorder since it tended to perpetuate the obsolete notion of psychogenesis, which is incompatible with the doctrine of multicausality, a core postulate of current psychosomatic medicine. With his work, he gave an invaluable contribution in setting the scope, mission, and methods of psychosomatic medicine.

Among major innovations, stands the introduction of the biopsychosocial model of illness by Engel.[4] It allows illness to be viewed as a result of interacting mechanisms at the cellular, tissue, organismic, interpersonal, and environmental levels. Accordingly, the study of every disease must include the individual, his body, and his surrounding environment, as essential components of the total system. This model retains today its full impact in clinical research and practice.[5]

Psychosomatic medicine is an interdisciplinary field that is concerned with the interaction between biological, psychological, and social factors in regulating the balance between health and disease.[1,6] It provides a conceptual and methodological framework for

1. scientific investigations on the role of psychosocial factors affecting individual vulnerability, course, and outcome of any type of medical disease;
2. personalized and holistic approach to patient care, which emphasizes patient-doctor interaction and aims to shared decision making;
3. multidisciplinary organization of health care that overcomes the artificial boundaries of traditional medical specialties; and
4. integration of psychological and psychiatric therapies in the prevention, treatment, and rehabilitation of medical disease.

In the United States, psychosomatic medicine is a subspecialty of psychiatry recognized by the American Board of Medical Specialties. This may lead to identifying psychosomatic medicine with consultation-liaison psychiatry.[7] Consultation-liaison psychiatry is clearly within the field of psychiatry; its setting is the medical or surgical clinic or ward, and its focus is the comorbid states of patients with medical disorders. This narrow definition pertains to the United States, but not to other countries. Psychosomatic medicine is, by definition, multidisciplinary. It is not confined to psychiatry, but concerns any field of medicine.

Until the 1970s, psychosomatic medicine was the only site for research at the interface between medicine and behavioral sciences. In those years, however, behavioral medicine developed[8] as an interdisciplinary field that integrates behavioral and biomedical knowledge relevant to health and disease. It provided a room for an increasing number of psychologists dealing with basic laboratory research on the neural and humoral systems controlled by the brain, on visceral learning, and on other aspects of behavior which lead to practical applications of medical significance.[8] This latter field was subsequently often subsumed under the rubric of health psychology. Both behavioral medicine and health psychology have a much narrower connotation than psychosomatic medicine.

In the past three decades there has been an upsurge of interest in another related discipline (mind-body medicine), sharing the same holistic, biopsychosocial connotation with psychosomatic medicine.[9] However, the concept of mind-body medicine is inextricably linked to that of alternative or complementary medicine.[10] An increasing number of people, particularly in the United States, use alternative medical therapies, such as acupuncture and herbal medicine.[10] Such highly heterogeneous practices may depart from the scientific methods of medicine that constitute a major aspect of psychosomatic research.

Psychosocial Factors and Individual Vulnerability

A number of factors have been implied to modulate individual vulnerability to disease.

Stressful Life Events

The role of early developmental factors in susceptibility to disease has been a frequent object of psychosomatic investigation.[11] Using animal models, events such as premature separation from the mother have consistently resulted in pathophysiological modifications, such as increased hypothalamic-pituitary-adrenal axis activation.[12] They may render the human individual more vulnerable to the effects of stress later in life. There has also been considerable interest in the association of childhood physical and sexual abuse with medical disorders later in life. This link has been postulated to affect several adverse health outcomes (functional disability and risk behaviors), yet the evidence currently available does not allow any firm conclusion.[11]

The notion that events and situations in a person's life which are meaningful to him or her may be followed by ill health has been a common clinical observation. The introduction of structured methods of data collection and control groups has allowed researchers to substantiate the link between life events in the year preceding the onset of symptoms and a number of medical disorders, encompassing endocrine, cardiovascular, respiratory, gastrointestinal, autoimmune, skin, and neoplastic disease.[5]

Chronic Stress and Allostatic Load

Subtle and long-standing life situations should not be dismissed too readily as minor and negligible, since chronic, daily life stresses may be experienced by the individual as taxing or exceeding his or her coping skills. McEwen[12] proposed a formulation of the relationship between stress and the processes leading to disease based on the concept of allostasis, the ability of the organism to achieve stability through change. Allostatic load reflects the cumulative effects of stressful experiences in daily life. When the cost of chronic exposure to fluctuating and heightened neural or neuroendocrine responses exceeds the coping resources of an individual, allostatic overload ensues. Allostatic overload can be assessed on clinical grounds.[13] Biological parameters of allostatic load, such as glycosylated proteins, coagulation/fibrinolysis, and hormonal markers, have been linked to cognitive and physical functioning and mortality.[12]

Personality and Psychological Well-Being

The notion that personality variables can affect vulnerability to specific diseases was prevalent in the first phase of development of psychosomatic medicine (1930-1960), and was particularly influenced by psychoanalytic investigators, who believed that specific personality profiles underlay specific "psychosomatic diseases." This hypothesis was not supported by subsequent research.[1] Two personality constructs that can potentially affect general vulnerability to disease, type A behavior and alexithymia (i.e., the inability to express emotion), have attracted considerable attention, but their relationship with health is still controversial.[14] However, personality variables (e.g., obsessive-compulsive, paranoid, impulsive) may deeply affect how a patient views illness, what it means to him/her and his/her interactions with others, including medical staff.

Positive health is often regarded as the absence of illness, despite the fact that, half a century ago, the World Health Organization defined health as a "state of complete physical, mental, and social well-being and not merely the absence of disease or infirmity." Several studies have suggested that psychological well-being plays a buffering role in coping with stress and has a favorable impact on disease course.[15]

Social Support

Prospective population studies have substantiated the role of social support in relation to mortality, psychiatric and physical morbidity, recovery, and adjustment to chronic disease.[6] An area that is now called "social neuroscience" is beginning to address the effects of the environment and social network on the brain and the physiology it regulates.[12]

Spirituality

Religiosity and spirituality (broadly defined as any feelings, thoughts, experiences, and behaviors that arise from the search for the "sacred") have been a matter of growing interest in epidemiological research.[16] Religiosity appeared to have a favorable effect on survival that is independent from behavioral factors (smoking, drinking, etc.), negative affect, and degree of social support.

Psychosomatic Assessment and Individualized Care

Psychosocial and biological factors interact in a number of ways in the course of medical disease. Their varying influence, together with each individual response determines the unique quality of the experience and attitude of every patient in any given episode of illness.[17] The psychosomatic approach thus requires a comprehensive assessment, appropriate patient-doctor interaction, and application of individualized care.

Inclusion of Psychosocial Variables in Medical Assessment

Psychosomatic evaluation is not limited to the presence of psychiatric disease, but also includes other important psychological variables.

Psychiatric illness appears to be strongly associated with medical diseases: mental disorders increase the risk for communicable and noncommunicable diseases; many health conditions increase the risk for mental disturbances; comorbidity complicates recognition and treatment of medical disorders.[18] There is evidence that psychiatric disturbances in the course of medical disease are substantially different from those which can be found in psychiatric settings in terms of clinical characteristics, prognosis, and response to treatment.[18,19] At times, mood and anxiety disturbances precede the onset of other symptoms of medical disease.[19]

Major depression has emerged as an extremely important source of comorbidity in medical disorders.[20] It was found to affect quality of life and social functioning and lead to increased health care utilization, to be associated with higher mortality, particularly in the elderly, to have an impact on compliance, and to increase susceptibility to medical illness or its complications, as was found to be particularly the case in cardiovascular disease.

There is also emerging awareness that psychological symptoms which do not reach the threshold of a psychiatric disorder may affect quality of life and entail pathophysiological and therapeutic implications. This has led to the development of Diagnostic Criteria for Psychosomatic Research (DCPR) (Table 1). The DCPR were introduced in 1995 and tested in various clinical settings.[21] The advantage of this classification is that it departs from the dichotomy between organic and functional and from the misleading and dangerous assumption that if organic factors cannot be identified, there should be psychiatric reasons which may be able to fully explain the somatic symptomatology. The psychosomatic literature provides an endless series of examples where psychological factors could only account for part of the unexplained medical disorder.[6] Similarly, the presence of an established organic cause for a medical disorder does not exclude, but indeed increases the likelihood of psychological distress.[6]

TABLE 1 The Diagnostic Criteria for Psychosomatic Research (DCPR)

1. Health anxiety
2. Thanatophobia
3. Disease phobia
4. Illness denial
5. Persistent somatization
6. Conversion symptoms
7. Functional somatic symptoms secondary to a psychiatric disorder
8. Anniversary reaction
9. Demoralization
10. Irritable mood
11. Type A behavior
12. Alexithymia

An additional target of psychosomatic investigation is the study of illness behavior, defined as the ways in which individuals experience, perceive, evaluate, and respond to their health status, that has yielded important information in medical patients.[22] In the past decades, important lines of research have been concerned with illness perception, frequent attendance at medical facilities, health care seeking behavior, delay in seeking treatment, and treatment adherence. Assessing illness behavior and devising appropriate responses by health care providers may contribute to improvement of final outcomes.[22]

Quality of life is also an object of psychosomatic assessment. While there is neither a precise nor an agreed definition of quality of life, research in this area seeks essentially two kinds of information: the functional status of the individual and the patient's appraisal of health. Indeed, the subjective perception of health status (e.g., lack of well-being, demoralization, difficulties fulfilling personal and family responsibilities, etc.) is as valid as that of the clinician in evaluating outcome.[17]

Personalized Care

Even though personalized medicine, described as genomics-based knowledge, has promised to approach each patient as the biological individual he/she is, the practical applications still have a long way to go and neglect of social and behavioral features may actually lead to "depersonalized" medicine.[23]

A basic psychosomatic assumption is the consideration of patients as partners in managing disease. The partnership paradigm includes both collaborative care, a patient-physician relationship in which physicians and patients make health decisions together,[24] and self-management, a plan that provides patients with problem-solving skills to enhance their self-efficacy.[25]

Multidisciplinary Care

There have been major transformations in health care needs in the past decades. Chronic disease is now the principal cause of disability, and use of health services consumes almost 80% of health expenditures.[25] Current health care is still conceptualized in terms of acute care perceived as a product processing, where the patients is a customer, who can, at best, select among the services that are offered. Yet, as Hart[26] has pointed out, in health care, the product is clearly health and the patient is one of the producers, not just a customer. As a result "optimally efficient health production depends on a general shift of patients from their traditional roles as passive or adversarial consumers, to become producers of health jointly with their health professionals" (Ref. 26, p. 383).

The need to include consideration of function in daily life, productivity, performance of social roles, intellectual capacity, emotional stability, and well-being, has emerged as a crucial part of clinical investigation and patient care. Patients have become increasingly aware of these issues. The commercial success of books on complementary medicine and positive practices as well as the upsurge of mind-body medicine exemplify the receptivity of the general public to messages of well-being pursuit.

Medically unexplained symptoms are common in medical patients and increase medical utilization and costs.[27] For some of the most common symptoms in primary care, such as chest pain, fatigue, dizziness, headache, and dyspnea, a medical diagnosis is not found in up to half of the cases. Such patients often spend more days in bed than patients with severe major medical disorders.[27]

The traditional medical specialties, based mostly on organ systems (e.g., cardiology, gastroenterology), appear to be more and more inadequate in dealing with symptoms and problems which cut across organ system subdivisions and require a comprehensive approach. The interdisciplinary dimension that characterizes most rehabilitation units and pain clinics exemplifies this concept. Other examples may involve clinical services concerned with psychooncology, psychonephrology, or psychoneuroendocrinology.

As Kroenke[28] argued, neither chronic medical nor psychiatric disorders can be managed adequately in the current environment of general practice, where the typical patient must be seen in 10-15 min or less. It is not that certain disorders lack an explanation; it is our assessment that is mostly inadequate, since it does not incorporate a global psychosomatic approach.[17] Further, it is idealistic to pursue shared decision and self-management, when time for interaction is so minimal.

There are several examples of multidisciplinary care guided by psychosomatic principles around the world.

Specific Inpatient Units

There is a long tradition of psychosomatic inpatient units in Germany and Japan, as well as in other countries. The characteristics of the units vary according to the type of health system. Their aim is to provide joint medical, psychiatric, and psychological care that would not be possible in traditional facilities. Recently, in Germany, psychosomatic rehabilitation hospitals have been established.[29] They aim to provide the prevention, treatment, and compensation of chronic illness and job-related disturbances (such as burnout) according to a biopsychosocial approach.

Inpatient Consultation Services

Consultation-liaison psychiatry is the most recognized and widespread modality for providing psychiatric consultation in the general hospital.[7] It is mostly geared to treat the psychiatric complications of medical illness in adults during hospitalization.

At times, specific geriatric psychiatric consultation services are available. Pediatric psychosomatic medicine services are also present around the world, in countries such as the United States, the United Kingdom, Australia, China, and Brazil.[30] The models are, of course, very different according to their geographic location, but share a need for consultation and multidisciplinary handling of clinical problems in children and adolescents during their hospitalization.[30]

There have been consultation services that have been developed within specialties and subspecialties, such as oncology, nephrology, gastroenterology, organ transplantation, and cardiology.

Much less common, but not less important, is also the location of specific medical consultation services (e.g., internal medicine, endocrinology) within the mental health system.[31] Medical comorbidity in psychiatric patients tends to be undetected in a large proportion of cases. Psychiatrists may miss the correct medical diagnosis because they may fail to think of nonpsychiatric reasons for their patients' complaints or because they may not have adequate instruments for detecting medical disorders.[19]

Outpatient Services

Pain and rehabilitation clinics have for many years provided a model for outpatient clinics guided by multidisciplinary principles. These types of services may be operated by various specialists (group approaches) or by a single specialist with a multidisciplinary background, as was reported in the field of psychoneuroendocrinology.[32] These services address complaints that fall between disciplines and require a psychosomatic approach.

Integration of Psychological Care

Unhealthy lifestyle is a recognized risk factor for most prevalent diseases, such as diabetes, obesity, and cardiovascular illness.[33] Geoffrey Rose[34] showed that the risk factors for health are almost always normally distributed and supported a general population approach to prevention, instead of targeting those at the highest risk. Accordingly, switching the general population to healthy lifestyles would be a major source of prevention. Similarly, within a biopsychosocial model, addressing the origins of disparities in physical and mental health care early in life may produce greater effects than attempting to modify health-related behaviors later or to improve access to health care in adulthood.[35]

Psychological treatments may be effective in health-damaging behaviors, such as smoking.[33] The benefits of modifying lifestyle have been particularly demonstrated in coronary heart disease[36] and type 2 diabetes.[37] Yet, at present, almost all of health care spending is directed at biomedically oriented care.

Specific psychiatric and psychological interventions may involve a wide range of health professionals and may range from reassurance and effective communication by the physician to specific psychotherapeutic and psychopharmacological treatments.

Psychotherapeutic Interventions

The use of psychotherapeutic strategies (cognitive-behavioral therapy, stress management procedures, brief dynamic therapy) in controlled investigations has yielded substantial improvements in a number of medical disorders.[38,39] Examples are interventions that increase social support and enhance coping capacity in patients with breast cancer and malignant melanoma, or employ writing about personal stressful experiences in asthma and rheumatoid arthritis.

Research on psychotherapy[6] has disclosed common therapeutic ingredients relevant to any physician-patient relationship (Table 2).

Abnormal illness behavior may greatly benefit from this type of intervention. For many years abnormal illness behavior has been viewed mainly as an expression of personality predisposition and considered to be refractory to psychotherapy. There is now evidence to challenge such pessimistic stance. For instance, several controlled studies indicate that hypochondriasis is a treatable condition by the use of simple psychological strategies.[6]

TABLE 2 Nonspecific Therapeutic Ingredients

The therapist full availability for specific times = *attention*
The patient opportunity to ventilate thoughts and feelings = *disclosure*
An emotionally charged, confiding relationship with a helping person = *high arousal*
A plausible explanation of the symptoms = *interpretation*
The active participation of patient and therapist in a ritual or procedure that is believed by both to be the means of restoring patient health = *rituals*

Another emerging area of intervention is concerned with strategies increasing psychological well-being in the setting of medical disease.[40]

Treatment of Psychiatric Comorbidity

Psychiatric disorders are frequently unrecognized and untreated in medical settings, with widespread harmful consequences for the individual and the society.[18] Treatment of psychiatric comorbidity such as depression, with either pharmacological or psychotherapeutic interventions, markedly improves health-related functioning and the quality of life, although an effect on medical outcome has not been demonstrated.[20]

CONCLUSIONS

Appraisal of the multifactorial determinants of the cure process may restore a trusting patient-doctor interaction, improving final outcomes. Indeed, the psychosomatic research background has consolidated over the past decades in dealing with complex biopsychosocial phenomena and may now provide new effective modalities of patient care.

Current advances in the field have practical implications for medical research and practice, with particular reference to the role of lifestyle, the challenge of medically unexplained symptoms, the psychosocial needs entailed by chronic illness, and the function of the patient as a health producer.

References

1. Lipowski ZJ. Psychosomatic medicine: past and present. *Can J Psychiatry*. 1986;31:2–21.
2. Kissen DM. The significance of syndrome shift and late syndrome association in psychosomatic medicine. *J Nerv Ment Dis*. 1963;136:34–42.
3. Engel GL. The concept of psychosomatic disorder. *J Psychosom Res*. 1967;11:3–9.
4. Engel GL. The need for a new medical model: a challenge for biomedicine. *Science*. 1977;196:129–136.
5. Novack DH, Cameron O, Epel E, et al. Psychosomatic medicine: the scientific foundation of the biopsychosocial model. *Acad Psychiatry*. 2007;31:388–401.
6. Fava GA, Sonino N. Psychosomatic medicine. *Int J Clin Pract*. 2010;64:1155–1161.
7. Levenson JL, ed. *Psychosomatic Medicine*. Washington, DC: American Psychiatric Publishing; 2005.
8. Miller NE. Behavioral medicine: symbiosis between laboratory and clinic. *Annu Rev Psychol*. 1983;34:1–31.
9. Pert CB, Drehert HE, Ruff MR. The psychosomatic network: foundations of mind-body medicine. *Altern Ther Health Med*. 1998;4:30–41.
10. Rakel D. *Integrative Medicine*. 2nd ed. Philadelphia, PA: Saunders; 2007.
11. Thabrew H, de Sylva S, Romans SE. Evaluating childhood adversity. *Adv Psychosom Med*. 2012;32:35–57.

12. McEwen BS. Physiology and neurobiology of stress and adaptation: central role of the brain. *Physiol Rev*. 2007;87:873–904.
13. Fava GA, Guidi J, Semprini F, Tomba E, Sonino N. Clinical assessment of allostatic load and clinimetric criteria. *Psychother Psychosom*. 2010;79:280–284.
14. Cosci F. Assessment of personality in psychosomatic medicine: current concepts. *Adv Psychosom Med*. 2012;32:133–159.
15. Pressman SD, Cohen S. Does positive affect influence health? *Psychol Bull*. 2005;131:925–971.
16. Chida Y, Steptoe A, Powell LH. Religiosity/spirituality and mortality. *Psychother Psychosom*. 2009;78:81–90.
17. Fava GA, Sonino N, Wise TN, eds. *The Psychosomatic Assessment. Strategies to Improve Clinical Practice*. Basel: Karger; 2012.
18. Sartorius N, Holt RIG, Maj M, eds. *Comorbidity of Mental and Physical Disorders*. Basel: Karger; 2015.
19. Cosci F, Fava GA, Sonino N. Mood and anxiety disorders as early manifestations of medical illness. *Psychother Psychosom*. 2015;84:22–29.
20. Katon WJ. Clinical and health services relationships between major depression, depressive symptoms and general medical illness. *Biol Psychiatry*. 2003;54:216–226.
21. Porcelli P, Sonino N, eds. *Psychological Factors Affecting Medical Conditions*. Basel: Karger; 2007.
22. Sirri L, Fava GA, Sonino N. The unifying concept of illness behavior. *Psychother Psychosom*. 2013;82:74–81.
23. Horwitz RI, Cullen MR, Abell J, Christian JB. (De)personalized medicine. *Science*. 2013;339:1155–1156.
24. Joosten EA, DeFuentes-Merillas L, de Weert GH, Sensky T, van der Staak CP, de Jong CA. Systematic review of the effects of shared decision-making on patient satisfaction, treatment adherence and health status. *Psychother Psychosom*. 2008;77:219–226.
25. Bodenheimer T, Lorig K, Holman H, Grumbach K. Patient self-management of chronic disease in primary care. *JAMA*. 2002;288:2469–2475.
26. Hart JT. Clinical and economic consequences of patients as producers. *J Public Health Med*. 1995;17:383–386.
27. Croicu C, Chwastiak L, Katon W. Approach to the patient with multiple somatic complaints. *Med Clin N Am*. 2014;98:1079–1095.
28. Kroenke K. Psychological medicine. *BMJ*. 2002;324:1536–1537.
29. Linden M. Psychosomatic inpatient rehabilitation: the German model. *Psychother Psychosom*. 2014;83:205–212.
30. Shaw RJ, Demaso DR, eds. *Textbook of Pediatric Psychosomatic Medicine*. Washington, DC: American Psychiatric Publishing; 2010.
31. Schiffer RB, Klein RF, Sider RC. *The Medical Evaluation of Psychiatric Patients*. New York, NY: Plenum Press; 1988.
32. Sonino N, Peruzzi P. A psychoneuroendocrinology service. *Psychother Psychosom*. 2009;78:346–351.
33. Tomba E. Assessment of lifestyle in relation to health. *Adv Psychosom Med*. 2012;32:72–96.
34. Rose G. Sick individuals and sick populations. *Int J Epidemiol*. 1985;14:32–38.
35. Shonkoff JP, Boyce WT, McEwen BS. Neuroscience, molecular biology and the childhood roots of health disparities. *JAMA*. 2009;301:2252–2259.
36. Rozanski A, Blumenthal JA, Kaplan J. Impact of psychological factors on the pathogenesis of cardiovascular disease and implications for therapy. *Circulation*. 1999;99:2192–2217.
37. Sonino N, Tomba E, Fava GA. Psychosocial approach to endocrine disease. *Adv Psychosom Med*. 2007;28:21–33.
38. Kaupp JW, Rapaport-Hubschman N, Spiegel D. Psychosocial treatments. In: Levenson JL, ed. *Textbook of Psychosomatic Medicine*. Washington, DC: American Psychiatric Press; 2005:923–956.
39. Smith TW, Williams PG. Behavioral medicine and clinical health psychology. In: Lambert MJ, ed. *Bergin and Garfield's Handbook of Psychotherapy and Behavior Change*. 6th ed. Hoboken, NJ: Wiley; 2013:690–734.
40. Fava GA, Tomba E. Increasing psychological well-being and resilience by psychotherapeutic methods. *J Pers*. 2009;77:1903–1934.

CHAPTER

59

Religion, Stress, and Superheroes

S. Packer

Icahn School of Medicine at Mt. Sinai, New York, NY, USA

OUTLINE

Abstract

Religion can be a cause of stress or a cure for stress. Religion's ability to relieve the stress that it induces partly accounts for its tenacious hold on its truest or newest believers. Religious sentiment and sectarianism rises during times of increased personal or societal stress. Since change and uncertainty are potent causes of stress, on a psychological, sociological, and physiological level, it stands to reason that some of history's most unusual religious movements occurred at times of rapid social change and uncertainty. Individuals undergoing personal stress and lifestyle shifts are statistically more likely to get involved with unusual or innovative religious movements. Thus, adolescents who recently left home, those who are newly divorced, bereaved or relocated, and inmates who "relocate" to prison are especially susceptible to the appeal of "radical religions." "Garden-variety" religions and also cults grow in response to the stresses of poverty, old age, war, illness, isolation, or impending death.

Religion can be a cause of stress or a cure for stress. Some say that religion's ability to relieve the very stress that it induces is also partly responsible for its tenacious hold on its truest or newest believers. It is known that religious sentiment and sectarianism rises during times of increased personal or societal stress. Since change and uncertainty are potent causes of stress, on a psychological, sociological, and physiological level, it stands to reason that some of history's most unusual religious movements occurred at times of rapid social change and uncertainty. Individuals undergoing personal stress and lifestyle shifts are statistically more likely to get involved with unusual or innovative religious movements. Thus, adolescents who recently left home, persons who are newly divorced, bereaved or relocated, and inmates who "relocate" to prison are especially susceptible to the appeal of "new religions" or "radical religions." The appeal of "garden-variety" religions also grows in response to stresses of poverty, old age, war, illness, isolation, or impending death.

RELIGION AS A CONCEPT

Our title refers to "religion" in the singular, as if religion is a single, all-embracing phenomenon. Yet it is anything but. Over 100,000 religions are registered in the United States alone. Those numbers are increasing, having been on the rise since the 1960s. Even more people

Stress: Concepts, Cognition, Emotion, and Behavior
http://dx.doi.org/10.1016/B978-0-12-800951-2.00065-0

embrace a personal "spirituality" rather than a recognized religion.

Rather than arguing about the meaning of the word religion, we shall define religion as a "system of beliefs centered around a supernatural being, power, or force." Notably, some religion-like belief systems do not revolve around the supernatural, with Confucianism being the best known.

With so many different recognized religions in existence, it makes sense that religions differ enough from one another to make each distinctive. Furthermore, some sects and subsects of recognized religions bear little resemblance to the parent religion that begot them. Some established religions tolerate a wide spectrum of views, while others are more dogmatic and unwavering. Because of this variability, it is more helpful to assess the impact of specific aspects of religion on stress, rather than to make generalizations about all religions.

KEY POINTS

- Religion relieves stress for some people but its demands and its prophecies can also cause stress.
- Paradoxically, religion may relieve the very stress it induces, so as to create a never-ending loop of religious adherence.
- In the post-9/11 era, secular superheroes with deity-like superpowers soared in popularity and seemingly relieved stress in a "nonreligious" way.

RELIGION AS CURE FOR STRESS

Both its fiercest critics and its most fervent advocates agree that most (but not all) religions dampen psychological stress. Personal testimonials on this subject are too numerous to recount. We can conjure up "screen memories" of religion's role in abating extreme stress: foot-stomping spirituals of the segregation-era South; solitary footsteps of death row inmates listening to psalms, or chaplains' prayers for troops that face the front line.

Not everyone finds relief in religion, yet we find compelling population-based statistics supporting religion's success in dissipating stress. Persons with strong internalized religious beliefs who attend religious services are less likely to complete suicide, which is arguably the ultimate response to stress. There are many interesting explanations as to why this is.

Philosophical Aspects

By positing the existence of a supernatural being who actively intervenes in the natural world, and who set a plan for each individual, religion provides an automatic antidote to the stress of being alone and awash in an unresponsive and uncaring universe. Philosophers and theologians recognize the role of existential angst on twentieth century Western society. Some equate this loss of faith in the supernatural with the stress of impending nuclear holocaust. Pastoral pop psychologists emphasize the opposite approach to reverse the tide. To counter Nietzsche's declaration that "G-d is dead," theologian Abraham Joshua Heschel assures readers that *Man is Not Alone*.[1]

Popular Culture and Mass Media

Mass media affirms religion's increased appeal during times of crisis. Consider the box-office success of wartime films like *G-d is My Co-Pilot* (Robert Florey, 1945), or the syncretic religious themes of *Lost Horizon*,[2] Frank Capra's beloved Depression-era film that deviated from James Hilton's novel and pictured Christian missionaries in isolated Tibetan Buddhist strongholds.

As the twentieth century ended, and as the uncertainty of the millennial year 2000 neared, reassuring religious imagery surged in store displays, greeting cards, even postage stamps. Angels were everywhere, but not just at Christmastime—but they were secularized and stripped of sectarian associations, to conform to New Age standards and to obscure religious origins.

In the aftermath of 9/11 (2001), preoccupation with religion was the rule. The religious motives of terrorists who toppled the Twin Towers forced everyone to reconsider religion's role in world peace (or destruction). Yet some strange reactions occurred. Yoga became more and more popular for stress relief and general exercise. Gyms that offered yoga classes deemphasized its origin's in Hindu worship, which we describe below. Within a few years, psychotherapy visits declined by a third, possibly because of economic and insurance pressures, but also because of the efficacy of this religion-based stress reliever, coupled with psychopharmacological advances. Some psychologists added religious elements to their therapies; John Cabot-Zinn appropriated "mindfulness" techniques from Zen Buddhism.[3] Martin Seligman, Ph.D. experimentally proved the utility of techniques based on Tibetan Buddhism (and Chasidic Judaism, which went unacknowledged). Hippie-era religious approaches went mainstream, their origins forgotten.

During the decade after 9/11, Americans became infatuated with superheroes, whose imaginary powers rival the deities of religion.[4] Comic book characters begot big screen blockbusters. One book about superheroes was entitled, *Our Gods Wear Spandex*.[5] Superman, comics' first superhero, began as a Moses surrogate who offered help with the same "outstretched arm and clenched fist" described in the Passover Seder.[6] *Man of Steel* (Zack Snyder, 2012), the latest film version, Christianized Superman, depicting him rising, arms extended, robes flowing, like visions of the Ascension. Stress skyrocketed among Americans, but merging recognizable religious iconography with pop culture heroes could bypass religious strife that was ready to boil over.

Psychological Aspects

Antidote to Aloneness, Abandonment, and Dependency

Psychoanalysts have different explanations for why supernatural beliefs relieve stress. They emphasize the personal, rather than societal. Some, such as the consistently controversial Freud, say that a belief in G-d substitutes for the presence of an all-protective, albeit sometimes punitive, parent. Such theories assert that religion recreates the infant's reassuring world of dependency. Freud compares the mystic's "oceanic feeling" of connectedness with the universe to the peacefulness and protectiveness of the womb, where an umbilical cord delivers nonstop nourishment while surrounding amniotic fluid buffers baby from forceful blows of the outside world. Concepts about salvation, and messianic redemption parallel rescue fantasies by returning parents.

Freudians tend to view reliance on religious reassurance as a sign of psychological immaturity, and as an obstacle to psychological growth and self-actualization. Later psychoanalytic thinkers, such as Karl Winnicot, were less critical. Winnicot saw "separation-individuation" as the critical step in psychological growth, with religious belief functioning as a transitional object, like a baby blanket or a teddy bear.[7,8] Both remind children of their parent's perpetual presence, and assuage the stress of possible abandonment during parental absence. There are many more psychoanalytic theories about ways that religion relieves stress,[9] each unique and interesting in their own right, but each as untestable by scientific standards as the tenets of religion itself.[10]

We find a more pragmatic, and less theoretical, take on religion as a surrogate dependency in the principles behind "Twelve Step" "Recovery Programs" that follow in the footsteps of Alcoholics Anonymous. These programs recommended reliance upon a "Higher Power," along with a strong group support system, to relieve immediate psychological and physiological stresses of alcohol withdrawal and to cope with general stresses that were ameliorated by addictive chemicals. Rather than admonishing their members for depending upon this spiritual "Higher Power," as many therapists did in the past, Twelve Step groups prefer spiritual dependency over chemical dependency or sensation-seeking living. Their pro-spiritual approaches garnered increasing public and professional acceptance, and have been adapted to other treatments.

Decreased Sense of Randomness and Uncertainty

Religion relieves personal stress in other ways. Religion decreases the sense of randomness and uncertainty in the world, as it details the logic—or illogic—behind a "world plan." It predicts, sometimes prophesizes, or at the very least, attempts to explain events that seem inexplicable. It creates a concept of stability that anthropologist Melvin Klass dubs "ordered universes."[11] Even religions that foretell adverse events in the future, such as an impending apocalypse, reincarnation, predestination, "bad karma," or other pessimistic prophecies, have the potential to offer partial relief from stress, simply because they help their members prepare psychologically for adverse events. Providing explanations for events, even without providing a means for altering those events, is referred to as "heuristics," and is one of the most powerful psychological tools that religion possesses. Some critics of psychoanalysis say that psychoanalysis' value also rests on heuristics because unprovable psychoanalytic explanations provide reasons for behavior rather than remedies, in the same way that religion "illuminates" cosmic events.

Religions further relieve the stress of uncertainty by providing blueprints for behavior that can theoretically change the future. By prescribing prayer, penance, codes of charity, dietary laws, sacrificial rituals, "right thought," or what not, those religions instill a sense of personal control. More optimistic prophesies, and promises of paradise, confer hope, and escape.

Cognitive Coping Techniques

Above and beyond theoretical parallels with psychoanalysis, many religions provide specific cognitive techniques that are useful in coping with stress. For instance, the act of acknowledging and articulating the experience of stress can quell some distress. Long before Freud discovered his "talking cure," religious prayers and petitions provided collective voices for stress and distress. Hebrew psalms that begin with words like, "From the depths I called out unto Thee," poeticize this experience of anguish. Christian recreations of the crucifixion dramatize Christ's ordeal on the Cross. These words and images are especially appealing to people who cannot access other avenues of expression or who prefer to deflect their own subjective sense of distress by focusing on more universal, cosmic, or collective stresses.

Role Models for Stress Endurance

Religious lore offers role models of persons who withstood extreme stress, surmounted that stress, were valued because they could endure stress, or went on to fulfill higher purposes because of those stresses. The

Christ figure is the consummate example of this sort, but is hardly the only one. Biblical stories about Jonah's ordeal inside the whale, Noah's fortitude in the face of a flood, Daniel's survival in the fiery lion's den, Job's endurance of the loss of his family, his health, and his wealth, all reassure believers that there is a relief for stress, or, if there is no immediate relief, then there is a reason, or perhaps even a reward in another life. The martyrdom of Christian saints, the Buddhist *jataka* stories about Prince Shakyamuni's wandering as a mendicant monk prior to achieving enlightenment under the Bodhi tree, the image of the Israelite tribes traversing the desert for 40 years before fulfilling their destiny as the "chosen people," are other examples of the pivotal role that stress-endurance plays in religious themes.

Consolation, Devaluation, and Dissociation

Some religions provide such a sense of consolation that they have been dubbed the "religions of consolation." Christianity's contention that the meek will inherit the earth reassures people that their "this-worldly" stress and suffering will be relieved by "other-worldly" rewards. Promises of future paradise made by prophetic religions dull the pain of the present and appeal to the impoverished, the downtrodden, and the physically or psychologically stressed. Some religions offer so much consolation for contemporary stress and distress that they devalue the material world by undermining its importance or teaching that the "real world" is illusory and no different from a dream. Buddhism's contention that the world of the senses is but a delusion (*maya*) is an extreme example of such "transcendent" thinking. The Hindu yogi's aspiration to a waking state of "dreamless sleep" is another example. Some psychotherapeutic techniques train patients to use similar techniques of detachment or "mindfulness" when confronted with stress-producing stimuli, although those people who enter similar dissociative states spontaneously and non-volitionally experience serious difficulties in life.

Temporal and Physical Escapes from Stress

Even "world-affirming" religions, which affirm the importance of the material world, often offer temporary escape routes from daily stress. Religious holidays and religious services and other "sacred times" carve out stress-free time during the ordinary workweek, and create an opportunity for rejuvenation. For those times when real-world stresses require even more relief than routine religious beliefs or rituals can confer, some religions offer physical as well as temporal "retreats," providing food, clothing, and shelter, along with social support, structure, and spiritual exercises. Ashrams, yeshivas, monasteries, convents, or any number of other religious communities provide parallels to the "retreats" popular among certain Christian denominations. Such religious retreats legitimize the need for relief from worldly stress, and even elevate the social and spiritual value of choosing to retreat from stress. In doing this, such religious retreats have the potential to provide people with a positive sense of self-worth, and a sense of connectedness with others who share their belief system, and even retraining in new vocations or avocations. Some retreats encourage members to contribute to society by doing good works or charity. The modern hospital movement evolved out of the monastic retreats in the Middle Ages, where persons who originally sought relief for spiritual, physical, or psychological ailments eventually provided care for others upon their own recovery. In contrast, psychiatric "rest cures" and "funny farms," which also offered retreats from the "real world," stigmatize the participant, pathologize the process, and artificially end social productivity. It's no wonder why religious retreats are often preferred over psychiatric treatment, and why religion is likely to be the first line of defense against stress and why more people consult clergy over psychiatry.

Social Aspects

For some people, and for some religions, religion is a solitary matter, and nothing but. For them, personal "spirituality" matters most, with respect to stress relief and everything else. This view is expressed in William James' repeatedly republished volume on the *Varieties of Religious Experience*, where James focuses on religion that is experienced in solitude.[12] Yet religion exists on a social as well as a personal level, and acts to relieve social stress—or reproduce stress—on that level as well. By providing a social support network through their communal services and activities, coupled with a sense of collective identity and purpose, religious organizations can directly combat the stress of loneliness, displacement, and "anomie." Material benefits such as charity, lodging, employment, social services, and subsidized medical care provided by some religions can counteract stresses of economic hardship and even ill-health, and act in conjunction with the psychological consolation offered by the religious theory. Religious bureaucracies can provide alternative channels for personal and political expression, and become particularly important outlets when access to legitimate channels is closed, and may even become powerful enough to challenge the existing political and economic powers.

Studies of the conversion process are especially illustrative of the importance of social forces and religion. It often comes as a surprise to persons who are ideologically committed to religion to learn that it is the social sway of religious groups that influences individuals to adopt new religious ideas, rather than the other way around. In other words, rather than experiencing a life-changing "epiphany" before converting, similar to the epiphany that the Gospels attribute to Saul of Tarsus on the road to Damascus, converts to new religions are more likely to follow a socially paved path to new religious insights. They accept more and more of the ideas of that group as they gain greater and greater acceptance into that group and as they become more reliant upon group members for social support. The process of conversion is more likely to be gradual than sudden, and, in most people, behavior changes incrementally rather than dramatically.

Not everyone who gets involved with a new religious group remains committed to that group. Several factors predict the likelihood of long-term involvement with a new religion, and one of those factors revolves around the role of stress: the more stress relief that an individual experiences at the time of joining that new religious group, the greater is the likelihood that that person will remain a member of that group. Furthermore, it is the stress that individuals experience when distancing themselves from such groups that often compels them to return to such groups, even after repeated attempts to disengage from them.[13]

People who are already in a state of stress at the time that they encounter persuasive members of a new religion stand to achieve the greatest degree of stress relief. Some religions make it a point to stress their members, through sleep deprivation, diet restriction, social isolation, overwork, enforced silence, sexual abstinence, physical or psychological threats, or other techniques, making them more amenable to the stress-relieving effects of the religion. Religions that rely upon such stress-inducing techniques run the risk of being identified as "cults" by nonmembers, who view such techniques as coercive. Such "cultish" religions often recruit through institutional settings where people are already in high stress states, and so are primed for proselytization. Colleges, prisons, and retirement communities are all places of comparatively high stress, and are also places where new religions have made major inroads.

It is the intense stress of prison that is credited with pressuring some criminals to seek relief through religious conversion. Many ministries recognize this potential, and maintain active presences in prisons. One rarely hears of public opposition to vigorous religious recruitment efforts in prisons, because most religions strive to instill more socially acceptable values that help converts avoid future criminal behavior and subsequent incarceration. (Obviously, there are notable exceptions to this truism, as we have seen in our post-9/11 world.) It is generally believed that religion helps people become rehabilitated enough to reenter society, and that recruitment to a new religion is certainly preferable to recruitment by antisocial prison gangs. It is noteworthy that religions such as the Black Muslims—made famous by Malcolm X—teach their proselytes to respect peace and avoid violence, in contrast to some newer fringe Muslim groups that recruit alienated, even affluent, youths and consecrate terrorism, suicide bombings, and abduction.

In contrast, there has been a great public outcry about "cult recruitment" on college campuses, partly because some cultish religions do the opposite of cult religions in prisons. Rather than helping integrate persons into a higher level of society, these cults isolate students from larger society, and abort attempts to achieve more in mainstream society. Some cults encourage their members to cut off contact with families and friends, and, in doing so cause great distress to relatives. Although many legitimate religions were considered to be cults in their early stages—with Methodism being the most notable American example—the mere fact that a small percentage of cults have been associated with mass suicide or homicide is enough to cause public concern and to stimulate psychiatric task forces to investigate these issues.[14] The tragic memories of 800 deaths in Jonestown, or the loss of 19 through California's Heaven's Gate, subway poisonings by the Aum Shiriko sect, or the murder of a Hare Krishna defector in North Carolina, often obscure objective evaluation of possible positive benefits of conversion, and will no doubt stimulate further studies of the interaction between religion, stress, and related factors.[15]

At present, our understanding of recruitment by religious terrorist cults is still evolving, but investigations into the Boston marathon bombings of 2013 may shed light on changing trends in this important arena. Films such as the *The Attack* (Zaid Doueiri, 2013), however fictional, posit personal and psychological reasons behind a wife's choice to become a suicide bomber on the eve that her Palestinian-Arab surgeon-husband receives the highest honors from Israeli medical societies. In this film, she steals the limelight from her high-achieving husband and achieves renown in the only way that is available to her. In death, she is hailed as a martyr, while he, the hero, loses face.

Somatic Aspects

Stress is as much a physiological response as a psychological one. Some religions use physical techniques to achieve higher "spiritual" states also alleviate both psychological and physiological stress in the process. The muscle stretching systems of *yoga* and the breathing exercises of *Zen* and the controlled movements of *Tai Chi* are examples of such "mind-body" methods.

While it is impossible to verify the existence of "higher spiritual states," it is possible to use Western scientific techniques to measure changes in heart rate, respiration, galvanic skin response (sweating), pupil size, secretions of stress hormones such as cortisol, brain waves (through an EEG or electroencephalogram), or muscle tension (via an EMG or electromyogram) in people performing those spiritual exercises. Internist Harold Benson's studies of cardiovascular effects of Transcendental Meditation and Tibetan Buddhist chanting are found in juried medical journals and in his popular paperback, *The Relaxation Response*.[16] Japanese psychiatrist Tomio Hirai correlated EEG and EMG effects of both Zen meditation and Hindu yoga with reports of spiritual and psychological states, and published his results in *Zen Meditation and Psychotherapy*.[17] Although these works represent serious research, there have been so many hyperbolic claims about the benefits of Eastern (and Western) religious systems that many professionals have reflexively dismissed such claims. Both an open mind—and an open eye—are needed to interpret data about possible mind-body benefits of religion.[18] The fact that the American Psychiatric Association hosts continuing medical education courses on yoga at annual conferences shows how accumulating scientific evidence can sway ordinarily conservative professional societies, leading to their advocating—not just tolerating—yoga.

Quite opposite of the calming meditative techniques of Eastern religions are the ecstatic dances, shaking, quaking, rocking, and rolling movement of some sects, which inspired names like "Holy Rollers, Ranters, Ravers, Quakers, Shaking Quakers, and Shakers." Such intense activity presumably relieves stress through the same mechanisms that jogging and exercise help secular devotees. The Dionysian dances of Classical Greece, described in Euripides' play about the Bacchae, were but one of many recurring manifestations of frenzied religious dancing that serves related functions.[19] Repetitive religious rituals in general are said to relieve stress as well.[20]

Side Effects of Stress Relief

Religion's efficacy at relieving stress does not come without side effects. There are times when religion relieves stress so well that its practitioners disregard demands of the "real" world, and cannot defend against impending danger. Many social scientists have said that the religious devotion of African-Americans shielded them from the pain and poverty of pre-Civil Rights America, and delayed the adoption of more appropriate political tactics. Similarly, the Tibetan practice of sending one third of its youth to Buddhist monasteries left the country defenseless against the Chinese invaders who destroyed temples, massacred monks and nuns, and sent religious leaders into exile. Some secular Zionists claimed that the insulated and self-satisfied religious infrastructure of some Eastern European Jewish communities obscured their awareness of the deadly fate that awaited Jews in Hitler's death camps. An analogous American tragedy occurred when Native American warriors went into battle unarmed, believing that their Ghost Dance ritual would protect them from the guns and arrows of approaching armies. Marx and Engels blamed the mystical and occasionally bizarre religious beliefs of Russian orthodoxy and schismatics for numbing their compatriots' reactions to the material exploitation by the capitalists and the Czars, and therefore denounced religion as "the opium of the masses."

RELIGION AS A CAUSE OF STRESS

As effective as religion can be at relieving stress, it can also reproduce stress. Threats of an afterlife full of "hellfire and brimstone" or of an impending apocalypse that will destroy the world and its inhabitants are obvious stress producers. On a more subtle level are the many moralistic demands and behavioral codes made by religion, which are often difficult to live up to, and thus tend to leave some practitioners with a near-constant state of imperfection and incompleteness. Some practitioners adopt even more zealous belief and behavior, in order to avoid that stress, in an ever spiraling pattern. In *Battle for the Mind*, psychiatrist John Sargant observed that some charismatic religions exploit this tandem stress relief-stress reproduction effect to proselytize.[21] He compared this push-pull effect of religion to behavioral conditioning, more nefarious methods of "mind control" and brainwashing, and even drug addiction.

Sargant's own participation in CIA-sponsored mind control experiments suggests that he was very well versed in these techniques, enough to meet the specs of the US spy masters.

For sure, religion-induced stress is not limited to perceived threats, nor is it confined to the personal psyche. In spite of its promises of eternal peace, organized religion is responsible for some of the most realistic threats the world has witnessed. The Inquisition, the Crusades, and the Wars of Religion, are but a few testimonies to reality-based stresses posed by religion in the past. Such stresses persist to the present day, through suicide cults like the California-based Heaven's Gate, Islamic terrorist attacks on the World Trade Center, kidnappings and beheadings by ISIS, biological weapon-wielding Aum Shariko in Japan, Hindu-Muslim-Sikh conflicts in an atomic-bomb-armed India, the tinderbox of the Balkans, where religious-ethnic conflicts set the stage for World War I, and a Jerusalem beset by rightwing religious assassinations and never-ending retaliations for real and perceived injustice against Jews and Muslims, to name just a few. Although some religions aspire to the day that "the lion will lie down with the lamb," that day has not yet arrived. Religion has been, and probably always will be, as intimately associated with the stress of war and violence as with the proverbial love and peace. Rising religious fundamentalism promises to produce more political as well as personal stresses to individuals and to the world at large. At the same time, the religions of reassurance will provide personal stress relief for individuals, creating a never-ending seesaw.[22]

RELIGION AS A CORRELATE OF STRESS

The correlation between the rapid rise of new and sometimes radical religious movements and the degree of social stress is nothing less than remarkable, and is cataloged in Norman Cohn's classic *The Pursuit of the Millennium*,[23] and in his more recent *Chaos, Cosmos, and the World to Come*.[24] We must wonder if stress related to rapid changes brought on by the Internet, the Age of Information and mass electronic communication somehow paved the path for increasing social disruptions—religious terrorism included—that plagued the world increasingly in recent years.

Glossary

Apocalyptic religion Religion based on writings prophesying a cataclysmic time when evil forces are destroyed. Apocalyptic religions often includes "millenarian religions" that focus on the end of 1000-year period described in the Book of Revelation.

Cult Pejorative term used to describe a "new religious movement", especially one which is secretive, socially isolated, often faddish, with unfamiliar rules and rituals, and coercive, manipulative, or deceptive techniques are used to win or keep new converts.

Prophetic religion Religion that maintains that truth is revealed by prophecy. Sometimes referred to as "religions of revelation," prophetic religions include "Western" monotheistic religions such as Judaism, Christianity, Islam, as well as Zoroastrianism, and Bahai.

Religion A system of beliefs centered around the concept of a supernatural being or force.

Transcendent religion Religion that teaches techniques to transcend or transform the reality of the senses or that denies the existence or importance of mundane reality. Many "Eastern" religions, as well as some Western "New Age" beliefs, are "transcendent."

References

1. Heschel AJ. *Man Is Not Alone.* Chicago, IL: University of Chicago; 2007.
2. Hilton J. *Lost Horizon.* New York, NY: Macmillan; 1933.
3. Kabot-Zinn J. *Mindfulness Meditation for Everyday Life.* London, UK: Piatkus; 2001.
4. Saunders B. *Do the Gods Wear Capes? Spirituality, Fantasy, and Superheroes.* London, UK: Bloomsbury Academic; 2011.
5. Knowles C. *Our Gods Wear Spandex.* Newburyport, MA: Weiser; 2007.
6. Packer S. *Superheroes and Superegos: The Minds Behind the Masks.* Westport, CT: Praeger; 2010.
7. Koenig HG, McCulloch M, Larson D. *Handbook of Religion and Health.* New York, NY: Oxford University Press; 2001.
8. Koenig HG. *Handbook of Religion and Mental Health.* New York, NY: Academic Press; 1998.
9. Ostow M, ed. *Judaism & Psychoanalysis.* London: Karnac Books; 1982.
10. Meissner WW. *Psychoanalysis and Religious Experience.* New Haven, CT: Yale University; 1984.
11. Klass M. *Ordered Universes.* Boulder, CO: Westview Press; 1995.
12. James W. *The Varieties of Religious Experience.* New York, NY: Penguin Books; 1982.
13. Galanter M, ed. *Cults and New Religious Movements.* Arlington, VA: American Psychiatric Association; 1989.
14. Laycock J. Where do they get these ideas? Changing ideas of cults in the mirror of popular culture. *J Am Acad Relig.* 2012;81:80–106.
15. Kaplan J. *Radical Religion in America.* New York, NY: Syracuse University Press; 1997.
16. Benson H, Klipper MZ. *The Relaxation Response.* New York, NY: Harper-Collins; 1975.
17. Hirai T. *Zen Meditation & Psychotherapy.* New York, NY: Tokyo Publications; 1989.
18. Sapolsky RM. *Why Zebras Don't Get Ulcers.* 3rd ed. New York, NY: W.H. Freeman; 2004.
19. Girard R. *Violence and the Sacred.* Baltimore, MD: Johns Hopkins University; 1993.
20. Packer S. Jewish mystical movements and the European ergot epidemics. *Isr J Psychiatry.* 1998;35:227–241.
21. Sargant W. *Battle for the Mind.* New York, NY: Penguin Books; 1957.
22. Sapolsky RM. *Why Zebras Don't Get Ulcers: An Updated Guide to Stress, Stress-Related Diseases, and Coping.* 2nd revised ed. New York, NY: W.H. Freeman; 1998.
23. Cohn N. *Pursuit of the Millennium. Revolutionary Millenarians and Mystical Anarchist of the Middle Ages.* Oxford, UK: Oxford University Press; 1957.
24. Cohn N. *Cosmos, Chaos, and the World to Come.* New Haven, CT: Yale University Press; 1993.

Further Reading

Doueiri Z. *The Attack.* 2013.

Fuller A. *Psychology & Religion.* 3rd ed. Lanham, MD: Rowman; 1982.

Kakar S. *Shamans, Mystics, and Doctors.* Chicago, IL: University of Chicago Press; 1982.

Kakar S. *The Colors of Violence.* Chicago, IL: University of Chicago Press; 1982.

Kinsley D. *Health, Healing, and Religion.* Upper Saddle River, NJ: Prentice-Hall; 1996.

Knowles C. *Our Gods Wear Spandex.* Newburyport, MA: Weiser; 2007.

Capra F. *Lost Horizon.* 1937.

Snyder Z. *Man of Steel.* 2013.

Paloutzian R. *Invitation to the Psychology of Religion.* 2nd ed. Boston, MA: Allyn & Bacon; 1996.

Schumaker JF. *Religion and Mental Health.* New York, NY: Oxford University Press; 1992.

Seligman M. *Learned Optimism: How to Change Your Mind and Your Life.* New York, NY: Vintage; 2006.

CHAPTER

60

Dental Stress

T.K. Fábián[1], *P. Fejérdy*[2], *P. Hermann*[2], *G. Fábián*[2]

[1]Private Practitioner, Faaborg, Denmark

[2]Semmelweis University, Budapest, Hungary

OUTLINE

Abstract

Stress-related problems in dentistry are a collection of various psychological and pathopsychological conditions. The most common and widely known phenomenon is dental fear, which may lead to phobic reactions, or panic attack in some cases. Further, the oral region appears to be particularly predisposed for functional and somatoform psychosomatic disorders which form another large group of stress-related challenges in dentistry. Prevention of dental fear should be integrated into the dental treatment of every patient regularly, especially because traumatizing dental events may also trigger psychosomatic oral symptoms. A normative evaluation by the dentist and a subjective evaluation by the patient related to the dental treatment may be rather different. Therefore, factors unrelated to operative/technological dental skills but that contribute to the success of dental treatments, are becoming more and more important in dentistry.

INTRODUCTION

Stress-related problems in dentistry are a collection of various psychological and pathopsychological conditions.[1] The most common and widely known phenomenon is dental fear, which may lead to phobic reactions, or panic attack in some cases. Further, the oral region is an area that appears to be particularly predisposed for a large number of different functional and somatoform psychosomatic disorders[2–5]; therefore, another large group of stress-related reactions in dentistry is formed by several oral psychosomatic manifestations.[1] The reason behind may be that there is a rich psychological coupling of the mouth, teeth, and other oral structures based on their highly important and rather unique psychological and emotional functions.[1]

PSYCHOLOGICAL COUPLINGS OF THE ORAL REGION

The rich psychological couplings of mouth, teeth, and other oral structures are rooted in their widespread and psychoemotionally highly important functions.[1] They are important aspects of an individual's facial esthetics and sexual characteristics[1]; play an important role in speech and nonverbal-communication; serve as a primary zone of interaction with the environment[6]; take part in the sexual contact; and act as organs of the senses of touch, temperature, and taste.[1]

The oral region is tremendously important in the baby's life[1,6] as an organ of pleasure, an organ of contact with the mother, as well as an organ of testing, learning, understanding, and social signaling.[6] This is especially (but not exclusively) important in the oral stage of personality development in which breastfeeding plays a prominent role in communication between mother and infant.[1]

Stress: Concepts, Cognition, Emotion, and Behavior
http://dx.doi.org/10.1016/B978-0-12-800951-2.00066-2

KEY POINTS

- There is a rich psychological coupling of the mouth, teeth, and other oral structures based on their highly important and rather unique psychological and emotional functions.
- Accordingly, there is a rich representation of the oral region in the central nervous system, and also of great importance is the symbolic value of the mouth, teeth, and other oral structures.
- The most common reasons for dental fear are bad (painful or fearful) dental experiences. A large proportion of dental fear patients have been exposed to fear-provoking dental treatment in the past.
- Hallmarks of the oral psychosomatic manifestations are a history of symptoms that are inconsistent with the physical findings and a history of a precipitating life event after which the symptom first appeared.
- Major facets of proper and efficient patient-dentist communication are a good interpersonal relationship, mutual trust, the exchange of information, and making treatment-related decisions based on mutuality.

Accordingly, there is a rich representation of the oral region in the central nervous system, and also of great importance is the symbolic value of the mouth teeth and other oral structures.[1,4] It is no wonder that tooth loss (or other oral damage) may profoundly affect the psychosocial well-being of patients, even those who are apparently coping well with dentures,[7,8] and complete edentulousness can be a serious life event.[3–5,9,10]

Besides what is stated above, it should also be considered that masticatory muscles are highly sensitive to psychoemotional processes.[3,4] These muscles react earlier, stronger, and have longer lasting muscle spasms due to psychoemotional stress compared to other muscles of the body.[11] It is also likely that increased activity of the facial and masticatory muscles have a special role in the attenuation (elimination due to motor activity) of psychoemotional stress.[12] A similar role is found in the attenuation of psychoemotional stress due to autonomic activity (i.e., similar to that of shedding tears) and was also expected in relation to the psychoemotional stress-induced alterations of saliva secretion.[3,4] Taking all above data together, it is no wonder that the orofacial region is affected by psychosomatic manifestations much more frequently compared to most other parts of the body.[1,4]

DENTAL FEAR, PHOBIC REACTIONS, AND PANIC DISORDER

Problems relating to dental fear primarily occur in childhood, adolescence, or early adulthood (before the age of 20)[13] and most patients suffering from dental fear experience at least partial remission at times.[13] The most common reasons for dental fear are bad (painful or fearful) dental experiences, lack of control over the social situation in the dental chair, lack of control over personal emotional reactions, feeling of powerlessness during treatment, and social learning processes with a negative image of dentists.[14] A large proportion of these patients have been exposed to fear-provoking dental treatment in the past.[15–18] In some cases, an unconscious psychological trauma may also be a cause of dental fear reactions.

Patients with dental fear, including patients having phobic reactions (odontophobia, dental phobia), are usually agreeable to using medications and/or several psychological techniques to reduce fear during dental treatment.[1] Patients with panic disorder and patients fearing injections (needle phobia) are exceptions, however, because they usually recognize their indisposition (usually simple collapse or panic attack) as a result of a supposed life-threatening allergic reaction to injected anesthetics, which develops as a rigid, uncompromising behavior.[1] The differential diagnosis of a real allergy to local anesthetics, panic disorder, or needle phobia is extremely important. True allergy can cause life-threatening anaphylactic shock. In such patients, the use of local anesthetics should be strictly avoided.[1] In the case of panic disorder, some data in the literature suggest that the immune reaction regulated by IgE can be increased, which means a possible higher risk of anaphylactic shock.[19] Because of this, great care should be exercised if using local anesthetics with these patients. For needle-phobic patients, the contraindication for the use of local anesthetics (if any) is only psychological.

PSYCHOSOMATIC MANIFESTATIONS IN DENTISTRY

Because of the multiple psychological couplings and importance of the mouth and teeth, many of the psychopathological mechanisms related to sexuality, aggressivity, autoaggressivity, or death anxiety can lead to orofacial manifestations, especially if these important psychological and symbolic functions are damaged by tooth or mouth disorders.[1] Acute psychological stress conditions (e.g., existential trauma, workplace problems, relationship problems with the sexual partner) or chronic conditions (e.g., depression, neuroses, chronic anxiety, death anxiety, schizophrenic, or paranoid reactions) may serve as a background.[1] These manifestations more frequently appear in women, usually in the second half of life, although they can appear in children or in young adults as well.[1] The most frequent symptoms are atypical facial pain, burning mouth syndrome, myofascial pain, temporomandibular dysfunction, bruxism or other parafunctions, gagging, psychogenic denture intolerance, psychogenic taste disorders, certain recurrent oral ulcerations or inflammations, some oral allergic reactions, psychogenic occlusal problems, tic, psychogenic salivation problems, and oral discomfort, but any other symptoms mimicking somatic symptoms of jaws, mouth, and teeth may appear. The symptoms may appear singly or in combination.[1]

Since the appearing psychogenic symptoms may mimic a great variety of (most) somatic symptoms of jaws, mouth, and teeth, a clear-cut diagnosis and proper differential diagnosis could be rather difficult.[3–5] Even if a history of symptoms is inconsistent with the physical findings[20] and a history of a precipitating life event after which the symptom first appeared[20,21] are hallmarks of such psychogenic manifestations[20]; the diagnosis of psychogenic symptoms remains a presumptive one in many cases.[3–5]

The most important tool for diagnostic purposes is collecting the patient's detailed history and careful evaluation of all relevant psychosocial, medical, and dental anamnestic data in the context of a biopsychosocial model of orofacial disorders.[3–5,21,22] To assess whether a symptom is of psychogenic origin or not, the next five characteristics of psychogenic symptoms[23] may also be considered: (1) well-marked divergence between symptoms and clinical findings, (2) unsuccessful previous somatic treatments, (3) fluctuation of symptoms, (4) conspicuous emotional involvement of the patient in the dental problem, and (5) the presumable relationship between the symptoms and the psychosocial history.[23] Symptoms that meet at least four of the above five criteria are very likely to be of psychogenic origin.[3–5]

In dentistry, psychogenic symptoms appear frequently in relation with orthodontic and prosthodontic treatments (psychogenic denture intolerance), since orthodontic treatments and dentures have a great impact on the esthetic, sexual, nutritional, phonetic, and, consequently, symbolic function of the teeth.[1] In addition, orthodontic and prosthodontic treatments can be rather uncomfortable, extremely expensive, and time-consuming, which can cause the patient to develop strong emotions toward the dentist (and/or assistant). These factors may induce pressure, aggression, and complication of the dentist-patient-assistant relationship.[1]

GENERAL PRINCIPLES OF PREVENTION

Prevention of psychological stress-related problems appearing in dentistry should be based on introducing an approach to mental hygiene which is an interdisciplinary view of maintaining mental health and preventing mental illness during all kinds of human services.[1] This approach should include, among others, especially skilled communication with the patient and monitoring of the patient-dentist-assistant interpersonal relationships.[1] Besides the screening of patients at risk, proper treatment planning, prevention of treatment-induced pain, high quality preparative dental skills, and technical background are also cornerstones of prevention.[3–5] Great care should be taken to recognize the wishes of the patient regarding his or her treatment, both conscious and unconscious.[1,3,4] Conscious wishes (e.g., to achieve a nice smile, for treatment to be inexpensive and painless) are usually easily detectable.[1] Understanding unconscious wishes (e.g., to look younger, to stop the appearance of aging, to be loved by the dentist) may be more challenging, but is similarly important for preventing the manifestation of symptoms.[1] It may occur that the patient's conscious or unconscious wishes may not harmonize with the reality and possibilities of treatment.[3–5] It is a matter of considerable significance that the clinician carefully weighs the option of nontreatment in such cases.

MANAGEMENT OF DENTAL FEAR

The most important method for the management of dental fear is communication. Great care should be taken to avoid fear-producing terms or phrases, to make only those promises that can be backed,[24] and to give the perception of being in control to the patient.[24] The "tell-show-do" technique[25,26] may also be used efficiently.

Methods for distraction to refocus patient's attention away from the potentially painful/fearful stimulus or procedure may also be used advantageously.[24,26] Music expressing positive emotions[24] as well as audiovisual methods like video games,[26] two-dimensional DVD-glasses and three-dimensional virtual reality technology[27] may be used for such purposes. Mind-body therapies including hypnosis,[28] self-hypnosis, photo-acoustic stimulation,[29] relaxation, and several biofeedback methods are also suitable for distraction during the dental treatment.[4,30]

In some cases, pharmacological methods including premedication with anxiolytics, relative analgesia (inhalation of nitrous oxide-oxygen mixture), conscious sedation (intravenously administered benzodiazepines), and occasionally general anesthesia as well as psychotherapeutic approaches[30–34] may also be used to manage dental fear.

MANAGEMENT OF PSYCHOSOMATIC MANIFESTATIONS

Communication with the patient is also a key factor in the management of psychosomatic manifestations. Major facets of proper and efficient patient-dentist communication are good interpersonal relationships, mutual trust, the exchange of information, and making treatment-related decisions based on mutuality.[3–5] The dentist should avoid hiding behind the facade of professional authority and should share the problem-solving with the patient, especially because these patients usually regard his or her symptoms as a somatic or technical problem and ask for a repetition or correction of previous dental treatment.[1] An early diagnosis is also crucial, because repeated somatic dental treatment worsens the prognosis, fortifies the patient's false conception, and may lead to further loss of oral tissues.[1]

According to the above, majority of patients with oral psychosomatic manifestations first refuse to accept the psychological nature of their symptoms.[1] Therefore, a simple referral to a psychiatrist and/or psychotherapist would not solve the problem and in most cases, special techniques should be used to lead the patient to accept psychotherapy or psychiatric treatment.[1] Therefore, an initial psychosomatic therapy is needed prior to definitive therapy, which is within the scope of the dental professional's duty.[3–5,35,36] The most important aim of the initial psychosomatic therapy are avoidance of further useless, invasive dental treatments, as well as the decrease of (recovery from) symptoms and an increase in the patients' motivation to participate in definitive psychosomatic therapy, which is the highest level of care for patients with psychogenic symptoms.[3–5]

Gradual escalation of therapy and the avoidance of irreversible forms of treatment are "cornerstones" of initial psychosomatic therapy,[37] which utilizes several palliative and/or placebo methods (e.g., physiotherapies, medication, medicinal herb therapy, diet therapy, complementary/alternative therapy, etc.) combined with certain psychotherapeutic approaches and the administration of any mind-body therapies (i.e., hypnosis, self-hypnosis, photo-acoustic stimulation, relaxation, biofeedback methods, etc.) which are the "basic therapeutics" for psychosomatic disorders.[3–5,35,36] In many cases, a dentist skilled in the above methods can effect a notable improvement in the patient's symptoms or even complete recovery.[1] In other instances, a referral to definitive psychosomatic therapy may be necessary.[1]

In contrast to initial therapy, definitive psychosomatic therapy is the highest level of care for patients with psychogenic symptoms, utilizing any available dental, medical, and psychotherapeutical treatment possibilities in an evidence-based manner.[3–5] This highest level of care for oral psychosomatic manifestations should be carried out by a specialized psychosomatic team, including experienced dentists, psychiatrists, neurologists, psychologists, psychotherapists, and other related professionals.[3–5] For religious patients, a religious-based facade of treatment (e.g., pastoral therapy, pastoral counseling) may also be needed.[1] Definitive psychosomatic therapy should be offered for those patients which do not respond properly to the initial therapy (i.e., patients with frequent relapses after the initial therapy).[3–5]

CONCLUSION

Prevention (or at least a reduction) of dental fear should be integrated into the dental treatment of every patient regularly, and especially into the treatment of patients with increased dental fear values.[4,5,30] This is especially important because traumatizing dental events may also trigger psychosomatic oral symptoms.[4,20] Even though a dental treatment may have been carried out properly, there is no assurance that the patient will be satisfied with the therapy.[4,5,38,39] A normative evaluation by the dentist and a subjective evaluation by the patient related to the dental treatment may be rather different.[4,5,40] Therefore, factors unrelated to operative/technological dental skills, but that contribute to the success of dental treatments, are becoming more and more important in dentistry.[4,5]

References

1. Fábián TK, Fábián G, Fejérdy P. Dental stress. In: Fink G, ed. 2nd enlarged ed. Oxford: Academic Press; 2007:733–736. Encyclopedia of Stress; vol. 1.
2. Kreyer G. Psychosomatics of the orofacial system. *Wien Med Wochenschr.* 2000;150:213–216.
3. Fábián TK, Fejérdy P. *Psychogenic Denture Intolerance. Theoretical Background, Prevention and Treatment Possibilities.* New York, NY: Nova Science; 2010.
4. Fábián TK, Beck A, Gótai L, Hermann P, Fejérdy P. Psychogenic complications of making dentures. Theoretical background, prevention and treatment possibilities. In: Fábián TK, Fejérdy P, Hermann P, eds. *Dentures. Types, Benefits and Potential Complications.* New York, NY: Nova Science; 2012:199–241.
5. Fábián TK, Hidalgo Wulff HC. *Implant Dentistry. Theory and Praxis.* New York, NY: Nova Science; 2014. pp. 163–168, 169–178.
6. Ament P, Ament A. Body image in dentistry. *J Prosthet Dent.* 1970;24:362–366.
7. Fiske J, Davis DM, Frances C, Gelbier S. The emotional effects of tooth loss in edentulous people. *Br Dent J.* 1998;184:90–93.
8. Allen PF, McMillan AS. A review of the functional and psychosocial outcomes of edentulousness treated with complete replacement dentures. *J Can Dent Assoc.* 2003;69:662.
9. Bergendal B. The relative importance of tooth loss and denture wearing in Swedish adults. *Community Dent Health.* 1989;6:103–111.

10. Trulsson U, Engstrand P, Berggren U, Nannmark U, Brånemark PI. Edentulousness and oral rehabilitation: experiences from the patients' perspective. *Eur J Oral Sci*. 2002;110:417–424.
11. Heggendorn H, Voght HP, Graber G. Experimentelle Untersuchungen über die orale Hyperaktivität bei psychischer belastung, im besonderen bei Aggression. *Schweiz Mschr Zahnheilk*. 1979; 89:1148–1161.
12. Sato C, Sato S, Takashina H, Ishii H, Onozuka M, Sasaguri K. Bruxism affects stress responses in stressed rats. *Clin Oral Investig*. 2010;14:153–160.
13. Hällström T, Halling A. Prevalence of dentistry phobia and its relation to missing teeth, alveolar bone loss and dental care habits in an urban community sample. *Acta Psychiatr Scand*. 1984;70:438–446.
14. Moore R, Birn H. Phenomenon of dental fear. *Danish Dent J*. 1990;94:34–41.
15. Berggren U, Meynert G. Dental fear and avoidance: causes, symptoms and consequences. *J Am Dent Assoc*. 1984;109:247–251.
16. Moore R, Birn H, Kirkegaard E, Brodsgaard I, Scheutz F. Prevalence and characteristic of dental anxiety in Danish adults. *Community Dent Oral Epidemiol*. 1993;21:292–296.
17. Berggren U, Carlsson S, Hakeberg M, Hägglin K, Samsonowitz V. Assessment of patients with phobic dental anxiety. *Acta Odontol Scand*. 1997;55:217–222.
18. Skaret E, Raadal M, Berg E, Kvale G. Dental anxiety among 18-years-olds in Norway. Prevalence and related factors. *Eur J Oral Sci*. 1998;106:835–843.
19. Schmidt-Traub S, Bamler KJ. Psychoimmunologischer Zusammenhang zwischen Allergien, Panik und Agoraphobie. *Z Klin Psychol Psychiatr Psychother*. 1994;40:325–339.
20. Müller-Fahlbusch H, Sone K. Präprotetische Psychagogik. *Dtsch Zahnarztl Z*. 1982;37:703–707.
21. Dworkin SF, Sherman J. Chronic orofacial pain: biobehavioral perspectives. In: Mostofsky DI, Forgione AG, Giddon DB, eds. *Behavioral Dentistry*. Ames, IA: Blackwell Munksgaard; 2006:99–113.
22. Green CS, Laskin DM. Temporomandibular disorders: moving from a dentally based to a medically based model. *J Dent Res*. 2000;79:1736–1739.
23. Marxkors R, Müller-Fahlbusch H. Zur Diagnose psychosomatischer Störungen in der zahnarztlich-prothetischen Praxis. *Dtsch Zahnarztl Z*. 1981;36:787–790.
24. Botto RW. Chairside techniques for reducing dental fear. In: Mostofsky DI, Forgione AG, Giddon DB, eds. *Behavioral Dentistry*. Ames, IA: Blackwell Munksgaard; 2006:115–125.
25. Addelston H. Child patient training. *CDS Rev*. 1959;38:27–29.
26. Allen KD. Management of children's disruptive behavior during dental treatment. In: Mostofsky DI, Forgione AG, Giddon DB, eds. *Behavioral Dentistry*. Oxford: Blackwell; 2006: 175–184.
27. Askay SW, Patterson DR, Sharar SR. Virtual reality hypnosis. *Contemp Hypn*. 2009;26:40–47.
28. Thompson S. Hypnosis in the modification of dental anxiety. *Hypn Int Monogr*. 1997;3:33–48.
29. Fábián TK, Kovács KJ, Gótai L, Beck A, Krause WR, Fejérdy P. Photo-acoustic stimulation: theoretical background and ten years of clinical experience. *Contemp Hypn*. 2009;26:225–233.
30. Beck A, Varga G, Hermann P, Fábián G, Fejérdy P, Fábián TK. Methods for the treatment of denture induced psychogenic symptoms. In: Fábián TK, Fejérdy P, Hermann P, eds. *Dentures. Types, Benefits and Potential Complications*. New York, NY: Nova Science; 2012:165–197.
31. Carlsson SG, Linde A, Öhman A. Reduction of tension in fearful dental patients. *J Am Dent Assoc*. 1980;101:638–641.
32. Berggren U, Carlsson SG. A psychophysiological therapy for dental fear. *Behav Res Ther*. 1984;22:487–492.
33. Berggren U, Carlsson SG, Hägglin C, Hakeberg M, Samsonowitz V. Assessment of patients with direct conditioned and indirect cognitive reported origin of dental fear. *Eur J Oral Sci*. 1997; 105:213–220.
34. Berggren U, Hakeberg M, Carlsson SG. Relaxation vs. cognitively oriented therapies for dental fear. *J Dent Res*. 2000;79: 1645–1651.
35. Pomp AM. Psychotherapy for the myofascial pain-dysfunction syndrome: a study of factors coinciding with symptom remission. *J Am Dent Assoc*. 1974;89:629–632.
36. Fábián TK. *Mind-Body Connections. Pathways of Psychosomatic Coupling Under Meditation and Other Altered States of Consciousness*. New York, NY: Nova Science; 2012. pp. 131–172.
37. Laskin DM, Block S. Diagnosis and treatment of myofacial pain-dysfunction (MPD) syndrome. *J Prosthet Dent*. 1986;56:75–85.
38. Mazurat NM, Mazurat RD. Discuss before fabricating: communicating the realities of partial denture therapy. Part I: patient expectations. *J Can Dent Assoc*. 2003;69:90–94.
39. Mazurat NM, Mazurat RD. Discuss before fabricating: communicating the realities of partial denture therapy. Part II: clinical outcomes. *J Can Dent Assoc*. 2003;69:96–100.
40. Lechner SK, Roessler D. Strategies for complete denture success: beyond technical excellence. *Compend Contin Educ Dent*. 2001; 22:553–559.

Index

Note: Page numbers followed by *b* indicate boxes, *f* indicate figures, *t* indicate tables, and *ge* indicate glossary terms.

D

E